Chemical Processes for a Sustainable Future

Chemical Processes for a Sustainable Future

Edited by

Trevor M. Letcher
Emeritus Professor, School of Chemistry,
University of KwaZulu-Natal, Durban, South Africa
Email: trevor@letcher.eclipse.co.uk

Janet L. Scott
Centre for Sustainable Chemical Technologies,
University of Bath, Bath, UK
Email: j.l.scott@bath.ac.uk

Darrell A. Patterson
Department of Chemical Engineering,
University of Bath, Bath, UK
Email: d.patterson@bath.ac.uk

THE QUEEN'S AWARDS
FOR ENTERPRISE:
INTERNATIONAL TRADE
2013

Print ISBN: 978-1-84973-975-7

A catalogue record for this book is available from the British Library

© The Royal Society of Chemistry 2015

Published by The Royal Society of Chemistry,
Thomas Graham House, Science Park, Milton Road,
Cambridge CB4 0WF, UK

Registered Charity Number 207890

Visit our website at www.rsc.org/books

Printed and bound by CPI Group (UK) Ltd, Croydon, CR0 4YY

Preface

This book is a companion book to *Materials for a Sustainable Future* published by the Royal Society of Chemistry in 2012. In the *Materials* book we focused on the elements and compounds that are becoming 'endangered', the ways of dealing with increasing quantities of methane and carbon dioxide, the use of biomass as a chemical feedstock and specialised materials for use in advanced technologies such as water splitting and energy conversions. This book, *Chemical Processes for a Sustainable Future,* is about chemical processes aimed at sustainability with a focus on developments in new technologies and on processes using renewable materials as chemical feedstocks. The chapters give an up-to-date account of some of the new processes that are either currently in operation or are planned for the future. The overarching theme is to showcase new ideas in the chemical industry related to a sustainable existence.

The book is divided into five themes:

- An Introduction
- Chemical Transformations
- Biochemical Transformations and Reactors
- Separations and Purifications
- Process Integration

The topics we have chosen for our 27 chapters are not exhaustive. They cover areas that we feel are important in the quest for more

Chemical Processes for a Sustainable Future
Edited by Trevor M. Letcher, Janet L. Scott and Darrell A. Patterson
© The Royal Society of Chemistry 2015
Published by the Royal Society of Chemistry, www.rsc.org

sustainable processing and production of chemicals and materials. There are many areas that are not mentioned such as biomass engineering, further topics on biomass processes, adsorption and absorption processes, and advances in distillation. Furthermore, waste and emission treatment technologies are not covered since these would be better dealt with in a stand-alone volume. Extensive coverage of different process intensification technologies (such as spinning disc reactors) is also not given, as this is well covered in recently published volumes.

To help readers, each theme section is preceded by an overview chapter setting the scene for the chapters within the section. Moreover, each chapter begins with an introduction and finishes with a conclusion written with non-experts in mind.

Sustainability is slowly becoming more and more important in our everyday lives. We are aware of renewable forms of energy, more eco-friendly and sustainable ways of disposing of our rubbish collections, and of the chemical industry's move to more sustainable processes and 'greener' ways of doing things as espoused in their advertising campaigns and annual reports. These issues are important, but there is much more to sustainable processes, as will become clear in *Chemical Processes for a Sustainable Future.* We hope that this book will demonstrate that a close connection between sustainability and chemical processes is both possible and important.

Over the past 200 years, chemistry has been at the forefront in reducing poverty and making our lives not only longer, but easier, with improved pesticides, fertilisers, pharmaceuticals, fuels, plastics and even micro-electronics. Hand-in-hand with this, chemical engineers have developed processes to manufacture these chemicals and products quickly, cost-effectively and in useful quantities. All this has unfortunately come at a cost. Our generation is not only rapidly dispersing some vital elements and compounds thinly across the planet, but is filling the environment with pollutants, some of which are difficult to contain. We are compromising the lives of future generations. In the same way that chemistry and chemical engineering have led the way to an easier life, chemists and chemical engineers are being called upon to lead the world in organising a sustainable lifestyle for all, so that our children and their children will have the same advantages as the present generation. Our society was born out of the Industrial Revolution. It changed our way of life. The next revolution will be to develop a way of life that can be sustained from one generation to another. We hope that this book will show the contribution made by new chemical processes aimed at sustainability. This involves the cooperation of chemists and chemical engineers, working towards this common goal. *Chemical Processes for a Sustainable Future* aims to be part of this process.

One of the great advantages of this book is that the chapters have been written by scientists or engineers who are experts in their field and include up-to-date statistics, recent research and references to the latest work. The combination of science and engineering allows both perspectives and expertise to be voiced. We have chosen to write the book in a very readable format, with the hope that all readers, whether experts or not, will gain much in knowledge and appreciation of the concept of sustainable living. A second advantage of this book is that it brings together many apparently disparate topics, but all are related to sustainability, so that comparisons can be made, synergies developed and issues put into perspective. The book should encourage more and more people to investigate ways of ensuring the survival of future generations.

The audience we hope to reach includes: industrialists and investors looking at future developments with an eye on suitable investments; policymakers in local and central governments who need to become acquainted with the latest developments in the chemical industry related to sustainability; students, teachers, researchers, professors and scientists, engineers and managers working in the field of chemical processing who need information, direction and references to new developments; and last but not least, editors, journalists and the general public who need information on the vitally important concept of a sustainable future.

The International System of Quantities is reflected in the book with the use of SI units where ever possible. Furthermore, International Union of Pure and Applied Chemistry (IUPAC) recommendations on notation and spelling have been used throughout. The book has been supported by IUPAC.

In true IUPAC style the editors have attempted to give the book an international flavour with authors coming from many different countries including Australia, Austria, Brazil, Canada, China, Denmark, Finland, France, Germany, India, Malaysia, New Zealand, Spain, South Africa, the United Kingdom and the United States of America.

The success of the book ultimately rests with the 47 authors and, we the editors, would like to thank all of them for their cooperation and their valued, willing and enthusiastic contributions. We would also like to thank Professor Ron Weir who, on behalf of the IUPAC subcommittee, the Interdivisional Committee on Terminology,

Nomenclature and Symbols (ICTNS), advised us on the correct use of thermodynamic quantities, units and symbols. Finally we wish to thank the Royal Society of Chemistry, whose representatives were helpful *and* patient in producing this monograph on *Chemical Processes for a Sustainable Future.*

Trevor M. Letcher
Janet L. Scott
Darrell A. Patterson

Contents

Chemical Processes for a Sustainable Future
Edited by Trevor M. Letcher, Janet L. Scott and Darrell A. Patterson
© The Royal Society of Chemistry 2015
Published by the Royal Society of Chemistry, www.rsc.org

Part A: Chemical Transformations

Processes to Facilitate Chemical Transformations

**Chapter 10   New Chemical Processes aimed at Sustainable
Development in Brazil**   288
Telma Teixeira Franco and Ricardo Baldassin Jr

Part B: Biochemical Transformations and Reactors

Chapter 11   Overview: Biochemical Transformations and Reactors   317
David J. Leak

Extractions and Preparations

About the Editors

Trevor M. Letcher is Emeritus Professor of Chemistry at the University of KwaZulu-Natal, Durban, and a Fellow of the Royal Society of Chemistry. He is a past director of the International Association of Chemical Thermodynamics and his research involves the thermodynamics of liquid mixtures and energy from landfill. He has published over 270 publications in peer review journals and edited, written or co-edited 13 books related to his research fields. His latest books are: *Materials for a Sustainable Future* (RSC, 2012), *Unraveling Environmental Disasters* (Elsevier, 2012), *Future Energy: Improved, Sustainable and Clean Options for our Planet,* 2nd edition (Elsevier, 2013) and *Volumes Properties: Liquids, Solutions and Vapours* (RSC, 2014).

Janet L. Scott is a Training Director at the Centre for Sustainable Chemical Technologies (CSCT), University of Bath. She has previously worked in both industry and academia in three different countries: South Africa (University of Cape Town, 1992–1995; R&D Manager, Fine Chemicals Corporation, 1996–1998); Australia (Monash University 1999–2006) where she was the deputy director of the Centre for Green Chemistry; and the UK, where she held a Marie Curie Senior Transfer of Knowledge Fellowship at Unilever R&D, Port Sunlight, UK (2006–2008). She maintains an active consulting company working with industry on sustainable chemical solutions and research interests currently centre on bio-derived chemicals and materials. She works closely with chemical engineers and industrial partners at CSCT.

Chemical Processes for a Sustainable Future
Edited by Trevor M. Letcher, Janet L. Scott and Darrell A. Patterson
© The Royal Society of Chemistry 2015
Published by the Royal Society of Chemistry, www.rsc.org

Darrell A. Patterson is currently a senior lecturer in chemical engineering and a member of the Centre for Sustainable Chemical Technologies (CSCT) at the University of Bath. He leads the Bath Process Intensification Laboratory and the cross-faculty Bath Membrane research cluster Membranes@Bath, the UK's largest academic cluster of academics focusing on membrane science and technology research. He has previously worked at WS Atkins Consultants (2001–2003), Imperial College London (2003–2005) and the University of Auckland (2005–2011). His research is in three main (but related) areas, all aiming to characterise and produce process intensification to develop more sustainable technologies: membrane science and engineering; catalytic reactions and reactor engineering; and waste-water treatment technologies. He has over 70 papers in these areas (including over 40 peer reviewed journal papers).

INTRODUCTION

CHAPTER 1

General Concepts in Sustainable Chemical Processes

DARRELL ALEC PATTERSON*[a,b] AND JANET L. SCOTT*[a,c]

[a] Centre for Sustainable Chemical Technologies, University of Bath, Claverton Down, Bath, UK; [b] Bath Process Intensification Laboratory, Department of Chemical Engineering, University of Bath, Claverton Down, Bath, UK; [c] Department of Chemistry, University of Bath, Claverton Down, Bath, UK
*Email: d.patterson@bath.ac.uk; darrell.patterson@gmail.com; j.l.scott@bath.ac.uk

1.1 WHAT IS A SUSTAINABLE CHEMICAL PROCESS?

The topics contained in this book represent the nexus between chemical engineering and chemistry. Covering many sustainable process technologies available, whilst outlining the chemistry that underpins them, this book differs from others which often focus on a subset of the processes or the chemistry, or on a specialist topic (*e.g.* process intensification). Process design equations and process modelling, scale up, process and energy integration, and process control are often neglected. Therefore, we aim to provide the first comprehensive coverage of sustainable chemical processes, the one-stop-shop for the area. This book therefore focuses on the chemical technologies (unit operations) from both a chemical engineering and chemistry perspective. In this chapter we discuss the overarching principles and concepts addressed in the individual chapters.

Chemical Processes for a Sustainable Future
Edited by Trevor M. Letcher, Janet L. Scott and Darrell A. Patterson
© The Royal Society of Chemistry 2015
Published by the Royal Society of Chemistry, www.rsc.org

The subject of this book is sustainable chemical processes. But what are these? For us, they are processes (*i.e.* a set of linked unit operations[†] that take in raw materials and energy, and ultimately convert these into chemicals, biochemicals, materials or products) in a resource efficient manner that allows preparation of the desired product with minimal production of waste. Sustainable chemical processes or clean technologies are those that help us meet the goals of sustainability and sustainable development in our process systems.

The essence of sustainability and sustainable development is to ensure that our use of the planet's limited material, energy and ecological resources will be such that we sustain the current standard of living for the increasing human population so that future generations are able live life at the current, or even better, standard of living. The latter is particularly applicable to the large proportion of the global human population that do not yet have access to clean water, adequate food, life-saving medication and other essentials for a happy, healthy life. Thus, sustainability and sustainable development has been defined as:[1]

> '...development which meets the needs of current generations while not compromising the ability of future generations to meet their own needs...'.

It is always worth reminding oneself that this widely quoted definition of sustainability arose from a commission asked to formulate 'a global agenda for change' and to recall Gro Harlem Brundtland's words from her foreword to the report:[1]

> '... the 'environment' is where we all live; and 'development' is what we all do in attempting to improve our lot within that abode. The two are inseparable.'

The commission called for development and this has been reiterated in numerous reports since, for example:[2]

> '... economic development, social development and environmental stewardship are interdependent and mutually reinforcing

---

[†]A unit operation is a basic technology step that is the same in any overall process. A unit operation can be designed according to the same set of design principles independent of the overall process it is in. Unit operations include reactors, membrane separations, distillation columns, chillers and heaters. Several different unit operations are connected together to create an overall process.

components of sustainable development which is the framework for our efforts to achieve a higher quality of life for all people.'

A simple way of determining if a process or technology meets the goals of sustainability and sustainable development is to probe whether or not it meets all three components of the triple bottom line (also sometimes called the Three Pillars of Sustainability): people (social bottom line); planet (environmental bottom line); and profit (economic bottom line). There are various measures of these, which include:[3]

- **People (social bottom line)**: worker happiness, industrial safety, benefits based on payroll expense, promotion rate, 'loss time accident frequency', 'expenditure on illness and accident prevention/payroll expense', 'number of complaints per unit value added', *etc.*
- **Planet (environmental bottom line)**: life cycle assessment (see Section 1.4); environmental impacts such as acidification, global warming, human health, ozone depletion, photochemical ozone, wastes – hazardous and non-hazardous, and ecological health; and resource usage such as energy use, material use, water use and land use.
- **Profit (economic bottom line)**: capital and operating costs, wealth created, value added per unit value of sales, value added per direct employee, and R&D expenditure as a percentage of sales.

When all three components of the bottom line are met (*i.e.* at the overlap of the three lobes, Figure 1.1), we have sustainability. The social

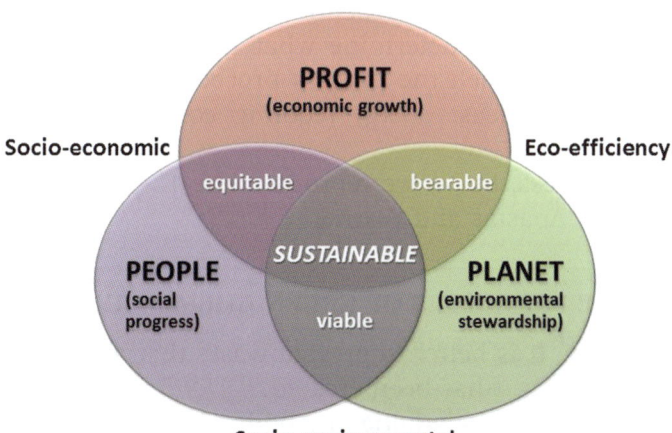

Figure 1.1 Sustainability is only achieved when all three aspects of the Triple Bottom Line are equally well satisfied.

and economic components are less directly related to the chemistry and chemical engineering emphasis of this book and therefore will not be our main focus (but they will be touched on where relevant). We will concentrate on ensuring that our sustainable chemical technologies meet the environmental bottom line. A systematic method for process choice that facilitates the selection of different technologies to meet this aim is to use the waste management hierarchy combined with the principles of green chemistry and green engineering.

1.2   THE PRINCIPLES OF GREEN CHEMISTRY AND GREEN ENGINEERING

1.2.1   The Twelve Principles of Green Chemistry

First set out by Paul Anastas and John Warner in their seminal work published in 1998,[4] it is clear that these 12 principles were defined by chemists with synthesis in mind and, as with all sets of rules or guidelines, should always be implemented in combination with careful analysis and intelligent critical thought. Nonetheless, even in cases where the process developer might decide that a particular principle is not formulated in a manner that applies to their process, these provide an excellent checklist. For example, it is certainly *not* always best to conduct a synthetic chemical process at ambient temperature and pressure, as described in principle 6. There are a plethora of examples where an exothermic reaction is best conducted at elevated temperature, as the process is rapid and the evolved heat is used to maintain the reaction temperature (usually after an initial short heating stage to initiate reaction). If put in the context of energy efficiency, it is easy to determine what the optimum temperature should be: that at which the process proceeds at a reasonable rate and consumes the least energy (cooling costs energy too), while remaining safe.

For completeness, below we reproduce the 12 principles of green chemistry from Anastas and Warner,

---

THE TWELVE PRINCIPLES OF GREEN CHEMISTRY

1. **Prevention:** It is better to prevent waste than to treat or clean up waste after it has been created.
2. **Atom Economy:** Synthetic methods should be designed to maximize the incorporation of all materials used in the process into the final product.

3. **Less Hazardous Chemical Syntheses**: Wherever practicable, synthetic methods should be designed to use and generate substances that possess little or no toxicity to human health and the environment.

4. **Designing Safer Chemicals**: Chemical products should be designed to affect their desired function while minimizing their toxicity.

5. **Safer Solvents and Auxiliaries**: The use of auxiliary substances (*e.g.*, solvents, separation agents, *etc.*) should be made unnecessary wherever possible and innocuous when used.

6. **Design for Energy Efficiency**: Energy requirements of chemical processes should be recognized for their environmental and economic impacts and should be minimized. If possible, synthetic methods should be conducted at ambient temperature and pressure.

7. **Use of Renewable Feedstocks**: A raw material or feedstock should be renewable rather than depleting whenever technically and economically practicable.

8. **Reduce Derivatives**: Unnecessary derivatization (use of blocking groups, protection/ deprotection, temporary modification of physical/chemical processes) should be minimized or avoided if possible, because such steps require additional reagents and can generate waste.

9. **Catalysis**: Catalytic reagents (as selective as possible) are superior to stoichiometric reagents.

10. **Design for Degradation**: Chemical products should be designed so that at the end of their function they break down into innocuous degradation products and do not persist in the environment.

11. **Real-time Analysis for Pollution Prevention**: Analytical methodologies need to be further developed to allow for real-time, in-process monitoring and control prior to the formation of hazardous substances.

12. **Inherently Safer Chemistry for Accident Prevention** – Substances and the form of a substance used in a chemical process should be chosen to minimize the potential for chemical accidents, including releases, explosions, and fires.

Reproduced from: P. T. Anastas and J. C. Warner, *Green Chemistry: Theory and Practice*, Oxford University Press, New York, 1998, p. 30 with kind permission from Oxford University Press.[4]

1.2.2   The Twelve Principles of Green Engineering

In some cases the 12 principles of green chemistry are difficult to apply directly in engineering applications, in particular those that do not involve chemicals or reactions. Consequently a further set of 12 principles were developed by Paul Anastas and Julie Zimmerman and published in 2003[5] as a tool to systematically apply the principles of sustainability and to achieve sustainable goals within engineering and in particular as a way of identifying and applying sustainability within engineering design (and as such these can be used as performance criteria). The principles are very general since they are intended to apply at all scales and apply to all engineering disciplines (chemical, electrical, civil, environmental, mechanical, systems, *etc.*). For engineers, these principles can be used as additional criteria that, in a complex system, can be optimised in addition to the normal parameters used in design to define and optimise a system.[6]

Since the goals of achieving sustainability are shared, the 12 principles have some commonality with the 12 principles of green chemistry. For completeness, the 12 principles are reproduced below.

THE TWELVE PRINCIPLES OF GREEN ENGINEERING

PRINCIPLE 1: Inherent Rather Than Circumstantial
   Designers need to strive to ensure that all material and energy inputs and outputs are as inherently non-hazardous as possible.

PRINCIPLE 2: Prevention Instead of Treatment
   It is better to prevent waste than to treat or clean up waste after it is formed.

PRINCIPLE 3: Design for Separation
   Separation and purification operations should be a component of the design framework and designed to minimize energy consumption and materials use.

PRINCIPLE 4: Maximize Efficiency
   Products, processes, and systems should be designed to maximize mass, energy, space, and time efficiency.

PRINCIPLE 5: Output-Pulled Versus Input-Pushed
   Products, processes, and systems should be 'output pulled' rather than 'input pushed' through the use of energy and materials.

PRINCIPLE 6: Conserve Complexity
   Embedded entropy and complexity must be viewed as an investment when making design choices on recycle, reuse, or beneficial disposition.

> **PRINCIPLE 7: Durability Rather Than Immortality**
> Targeted durability, not immortality, should be a design goal.
> **PRINCIPLE 8: Meet Need, Minimize Excess**
> Design for unnecessary capacity or capability (*e.g.* 'one size fits all') solutions should be considered a design flaw.
> **PRINCIPLE 9: Minimize Material Diversity**
> Material diversity in multicomponent products should be minimized to promote disassembly and value retention.
> **PRINCIPLE 10: Integrate Material and Energy Flows**
> Design of products, processes and systems must include integration and interconnectivity with available energy and materials flows.
> **PRINCIPLE 11: Design for Commercial 'Afterlife'**
> Products, processes, and systems should be designed for performance in a commercial 'afterlife.'
> **PRINCIPLE 12: Renewable Rather Than Depleting**
> Material and energy inputs should be renewable rather than depleting.
>
> Reprinted with permission from: J. B. Zimmerman and P. T. Anastas, *Environmental Science and Technology*, 2003, 37, 94A–101A, copyright American Chemical Society, 2003.[5]

There are also other sets of green engineering principles including the Sandestin Green Engineering Principles[7] and the Hannover Principles.[8]

For examples of how to apply the 12 principles of green engineering see 'Design Through the Twelve Principles of Green Chemistry'[5] and 'EcoWorx, Green Engineering principles in Practice'.[9]

1.3 THE WASTE MANAGEMENT HIERARCHY FOR PROCESS SELECTION

Waste minimisation is essentially embodied in the second principle of green engineering: 'It is better to prevent waste than to treat or clean up waste after it is formed'. In order to put this principle into practice, either to choose a method/technology/process for dealing with a waste stream, or to choose between alternative technologies when designing a new process or process option, so that the minimal amount of waste/environmental impact is produced, the hierarchy of waste management should be followed (Table 1.1).

Best practice is to start at the top of the hierarchy (*i.e.* eliminating the waste) and if this is not possible, to systematically test options

Table 1.1 Hierarchy of waste/emission management adapted and expanded from Crittenden and Kolaczkowski.[10]

Emission management strategy	Explanation	Possible action	Priority
Avoid/Elimination	Eliminate the source of emissions by avoiding the production in the first place	Alternative reagents Alternative processes and/or unit operations Do not undertake the activity	1 (highest)
Reduce/Source Reduction	Reduce the emission at source, generally focusing on the process or unit operation producing the emission.	Increasing process efficiency Alternative reagents Alternative processes and/or unit operations Changes in industrial practice and procedures	2
Recycling	Take unreacted or unused components and recycle them as fresh feedstocks within the same unit operation or process in order to produce the same product.	Materials separation and recovery and recycling this to the process feed.	3
Reuse	Take the unwanted components and use them elsewhere in the process or in an alternative process in a new use, *e.g.* this is commonly done with process water – hot water is used in a heat exchanger network throughout a process plant to optimise the use of process heat (known as 'pinch technology').	Materials separation and recovery and finding alternative uses with internal or external parties and processes for the unwanted components from the process operation.	4
Energy Recovery	Recover energy from the unwanted material.	Use of the waste material for combustion and/or energy generation. Anaerobic digestion for energy production.	5
Treatment	Process any potentially harmful unwanted material so that it is safe to be emitted into the environment.	The use of on-site or off-site process technologies to destroy, neutralise, detoxify any potentially harmful unwanted material.	6
Disposal	Discharge of the unwanted material into the environment (air, land, water, *etc.*) in a method compliant with local environmental and regulatory standards.	Sending the waste to a landfill site. Emission of a gaseous waste streams from a stack. Discharge of a non-harmful liquid component into a body of water.	7 (lowest)

working downwards towards the bottom (least preferable). That is, try to eliminate, reduce or recycle the emission streams from a process first by looking at the source of the emissions. Failing that, try to recover energy from the waste stream, then treat the waste stream to lessen the risk and environmental impact, and finally discharge the resulting stream to the environment. Disposal with no treatment is the least preferable option. This hierarchy has been made law in many countries and regions, including the European Union.

The nearer a chemical process and/or chemical technology is to the top of the hierarchy of waste management practice, the more sustainable it is (especially in terms of the environmental bottom line).

1.4 TAKING A LIFE CYCLE APPROACH

To determine the environmental impacts of a particular set of processes, new material or product, or to compare processes and products, one requires rigorous metrics that can be objectively applied to assist decision making. Life cycle assessment (LCA) is widely accepted as an appropriate technique to provide the data for evaluation of environmental impacts, through all stages of the lifetime of a product from raw materials extraction to final disposal or recycling (cradle to grave). LCA is defined by a set of international standards: ISO 14040: LCA – Principles and Framework and ISO 14044: LCA – Requirements and Guidelines (see ref. 11 for a review).

The briefest possible overview of LCA is, once again, provided by consideration of the principles embodied therein (now included in ISO 14040). These are reproduced below.[11]

PRINCIPLES OF LCA

- **Life cycle perspective.** LCA considers the entire life cycle of a product, from raw material extraction and acquisition, through energy and material production and manufacturing, to use and end of life treatment and final disposal. Through such a systematic overview and perspective, the shifting of a potential environmental burden between life cycle stages or individual processes can be identified and possibly avoided.
- **Environmental focus.** LCA addresses the environmental aspects and impacts of a product system. Economic and social aspects and impacts are, typically, outside the scope of the LCA. Other tools may be combined with LCA for more extensive assessments.

- **Relative approach and functional unit.** LCA is a relative approach, which is structured around a functional unit. This functional unit defines what is being studied. All subsequent analyses are then relative to that functional unit as all inputs and outputs in the LCI and consequently the LCIA profile is related to the functional unit.
- **Iterative approach.** LCA is an iterative technique. The individual phases of an LCA use results of the other phases. The iterative approach within and between the phases contributes to the comprehensiveness and consistency of the study and the reported results.
- **Transparency.** Due to the inherent complexity in LCA, transparency is an important guiding principle in executing LCAs, in order to ensure a proper interpretation of the results.
- **Comprehensiveness.** LCA considers all attributes or aspects of natural environment, human health and resources. By considering all attributes and aspects within one study in a cross-media perspective, potential tradeoffs can be identified and assessed.
- **Priority of scientific approach.** Decisions within an LCA are preferably based on natural science. If this is not possible, other scientific approaches (*e.g.* from social and economic sciences) can be used or international conventions can be referred to. If neither a scientific basis exists nor a justification based on other scientific approaches or international conventions is possible, then, as appropriate, decisions may be based on value choices.

Reproduced with kind permission from Springer Science + Business Media: International Journal of Life Cycle Assessment, 'The New International Standards for Life Cycle Assessment: ISO 14040 and ISO 14044', 2006, **11**, 80–85, Finkbeiner *et al.*[11]

LCA is a discipline requiring great expertise and there are numerous reference works and texts on the topic. Some more approachable texts for the non-expert include *Green Chemistry Metrics: Measuring and Monitoring Sustainable Processes*[12] and *Inherently Safer Chemical Processes: A Life Cycle Approach.*[13] While the process is, by necessity, very detailed and requires the collection of significant quantities of data, failure to adopt a full life cycle approach can lead to rejection of processes that are indeed less environmentally hazardous. This is a particular challenge when comparing a new (sometimes not optimised) technology, or process, with a well-established one.

For example, if a life cycle approach is not taken when choosing the optimal degree of abatement for a sustainable chemical technology used to recycle a component from a waste stream (*e.g.* a pervaporation membrane process used to remove and recycle a volatile organic compound from a wastewater stream), then there is the possibility that the environmental impact of the energy consumption and raw material use (amongst other items that have an impact beyond the process itself) will be larger than the reduction in environmental impact that removal of the pollutant has.[14] Going beyond the optimal degree of abatement by not considering the impacts from a life cycle point of view therefore does more harm than good, which is not the intention of developing or using a sustainable chemical technology.

1.5 COMMON PROCESS APPROACHES TO INCREASE SUSTAINABILITY

Two major processing approaches are commonly used to improve the sustainability of chemical processes and therefore to produce more sustainable chemical processes. These are: the use of continuous, or flow, processes; and the integration of different unit operations for process intensification. As these two approaches will feature in many of the chapters of this book, they are described here.

1.5.1 Replacing Batch with Continuous or Flow Processes

Processes can be classified as batch, continuous or semi-batch. In a batch process, the feed is introduced at the beginning of the process and products are removed at the end. The process change occurs over time. Figure 1.2 shows a typical batch stirred tank reactor. This could be used, for example, in beer fermentation where yeast and wort are added to the batch fermentation vessel, and beer (the product) and spent yeast (the waste) are removed at the end of fermentation. No flow occurs through the vessel over time—there is a loading before the start of processing and an unloading at the end of the processing.

In contrast, in a continuous, or flow, process, there are continuous inputs and outputs and the process changes over the volume and length of the vessel. Since this allows a smaller inventory of material in the process over time, the use of continuous processes typically

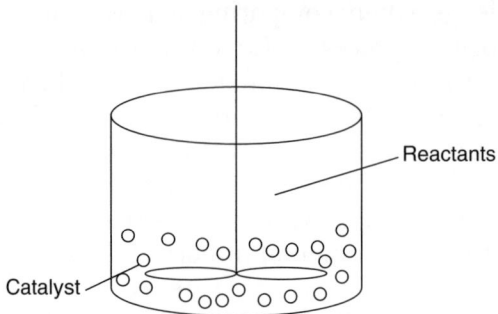

Figure 1.2 Schematic of a batch stirred tank reactor, with pellet catalyst.

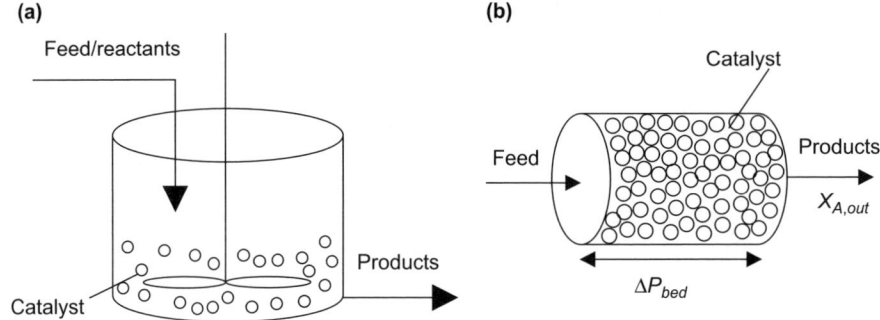

Figure 1.3 Schematics of two typical continuous stirred tank reactors using pellet catalysts: (a) continuous stirred tank reactor (CSTR); and (b) continuous packed bed reactor.

results in smaller equipment, higher throughput and better economics. However, the control of the process is usually more complicated, some operations in batch are difficult to achieve in flow (*e.g.* crystallisation) and flow equipment is not as flexible or amenable to short and changing production campaigns, as batch processes are. Examples of typical continuous reactors are shown in Figure 1.3. An example of a continuous process is the spray drying of milk powder, where milk concentrate and air are fed constantly to the spray dryer, and the products—powder and humidified air—are constantly removed. Continuous processes can be more sustainable, since the smaller reactor size can result in a reduction in energy use and the continuous throughput can help overcome limitations of equilibrium and product inhibition, making the process more efficient.

For completeness, a further type of process is the semi-batch process, which is any process that is neither batch nor continuous.

Two typical process industry examples are powder dissolution, which involves the slow addition of a powder to a solvent in a mixing vessel without removing any material, and combustion, which involves the continuous addition of air onto batch solid reactant.

1.5.2 Process Intensification

Stankiewicz and Moulin[15] defined process intensification as:

'the development of novel apparatuses and techniques that, compared to those commonly used today, are expected to bring *dramatic improvements* in manufacturing and processing, substantially decreasing equipment-size/production-capacity ratio, energy consumption, or waste production, and ultimately resulting in cheaper, sustainable technologies'.

To put this in a shorter form: process intensification is any chemical engineering development that leads to a substantially smaller, cleaner and more energy efficient technology. More commonly, a process intensification technology is one that combines several different unit operations into one (*e.g.* a membrane reactor, which combines reactions and separations), or a technology that overcomes rate limiting steps in a significant way (such as spinning disc and spinning mesh disc type reactors, where the centrifugal force and micromixing overcomes the inherent mass transfer limitations of many conventional chemical reactors). For further information on process intensification and a comprehensive treatment of the engineering of these process intensification technologies see *Process Intensification Technologies for Green Chemistry: Engineering Solutions for Sustainable Chemical Processing.*[16]

1.6 OUTLINE OF THE APPROACH TAKEN IN THE FOLLOWING CHAPTERS

The structure of the book is roughly divided into sections that follow the typical flow through a process plant starting from the raw materials and ending in products (both wanted and waste). This covers the most important overall unit operations and the new developments that have and can make these into sustainable chemical technologies and processes:

- Chemical Transformations
- Biochemical Transformations

- Separations and Purifications
- Process Integration

The concepts outlined in this chapter form part of the discussion in many of these chapters. As with any finite volume, not all unit operations and processes can be covered. However, the application of the concepts in sustainability, outlined in this chapter, to any process is a good starting point in achieving the triple bottom line benefits to these, and indeed any processes, in order to produce more sustainable processes for the future.

REFERENCES

1. World Commission on Environment and Development *Our Common Future*, Oxford University Press, Oxford, 1987.
2. United Nations, *Report of the World Summit for Social Development*, Copenhagen, Denmark, 6–12 March 1995, document A/CONF.166/9, United Nations, New York, 1995.
3. Sustainable Development Working Group, IChemE, *The Sustainability Metrics: Sustainable Development Progress Metrics Recommended for Use in the Process Industries*, Institution of Chemical Engineers, Rugby, UK, 2002.
4. P. T. Anastas and J. C. Warner, *Green Chemistry: Theory and Practice*, Oxford University Press, New York, 1998, p. 30.
5. P. T Anastas and J. B. Zimmerman, *Environ. Sci. Technol.*, 2003, **37**, 94A–101A.
6. J. B. Zimmerman and P. T. Anastas, Case studies illustrating the twelve principles of green engineering, in *Proceedings of the 1st International Conference on Sustainability Engineering and Science*, Auckland, New Zealand, 6–9 July, 2004.
7. M. A. Abraham and N. Ngyuen, *Environ. Prog.*, 2003, **22**, 233–232.
8. William McDonough & Partners, *The Hannover Principles. Design for Sustainability*, William McDonough Architects, Charlottesville, VA, 1992.
9. J. W. Segars, S. L. Bradfield, J. J. Wright and M. J. Realff, *Environ. Sci. Technol.*, 2003, **37**, 5269–5277.
10. B. Crittenden and S. T. Kolaczkowski, *Waste Minimization: A Practical Guide*. Institution of Chemical Engineers, Rugby, UK, 1995.
11. M. Finkbeiner, A. Inaba, R. B. H. Tan, K. Christiansen and H. J. Kluppel, *Int. J. Life Cycle Assess.*, 2006, **11**, 80–85.

12. A. Lapkin and D. Constable (ed.), *Green Chemistry Metrics: Measuring and Monitoring Sustainable Processes*, Wiley-Blackwell, Oxford, 2008.
13. D. E. Park, *Inherently Safer Chemical Processes: A Life Cycle Approach*, Wiley, New York, 2nd edn, 2009.
14. O. R. Hernandez, E. N. Pistikopoulos and A. G. Livingston, *Environ. Prog.*, 1998, **17**, 270–277.
15. A. J. Stankiewicz and J. A. Moulijn, *Chem. Eng. Prog.*, 2000, 22–34.
16. K. Boodhoo and A. Harvey (ed.), *Process Intensification Technologies for Green Chemistry: Engineering Solutions for Sustainable Chemical Processing*, John Wiley & Sons, Chichester, UK, 2013.

PART A
CHEMICAL TRANSFORMATIONS

PART A
CRITICAL TRANSFORMATIONS

CHAPTER 2

Overview: Chemical Transformations

JANET L. SCOTT

Centre for Sustainable Chemical Technologies and Department of
Chemistry, University of Bath, Bath, BA2 6LD, UK
Email: j.l.scott@bath.ac.uk

2.1 INTRODUCTION

The first section in this book covers sustainable processes to
facilitate chemical reactions. Frequently, in the development of a new
molecular entity, a chemical route is developed and 'optimised' in the
laboratory by research and development (R&D) chemists using typical
small batch processes conducted in laboratory glassware. This is later
passed to the process chemists for optimisation (and sometimes,
implementation of significantly greener chemical routes[1]) and, finally
to the chemical engineers for implementation at pilot and plant scale
(Figure 2.1). Often these separate groups are poorly integrated
and 'silos' of knowledge and expertise exist leading to sequential
development of, for example, chemical route and then scale up and
process development. Frequently, the more complex the chemical
transformation to be conducted, the more likely this process is to be
the case.

   This separation of expertise in the process of bringing a new
molecule to production can lead to implementation of less than
optimum final production processes just because chemistry gets

---

Chemical Processes for a Sustainable Future
Edited by Trevor M. Letcher, Janet L. Scott and Darrell A. Patterson
© The Royal Society of Chemistry 2015
Published by the Royal Society of Chemistry, www.rsc.org

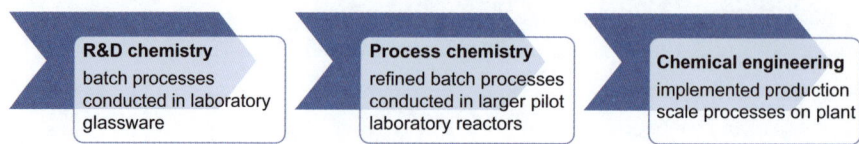

Figure 2.1   Schematic of the typical translation of a new molecular entity from the laboratory to the chemical plant.

'locked in early' and is only refined, rather than revised, during the scaling up process and transfer of knowledge to plant scale. Examples abound in the pharmaceutical industry, and the American Chemical Society (ACS), Green Chemistry Institute (GCI) Pharmaceutical Roundtable published a paper in 2010 describing the 'Key Green Engineering Research Areas for Sustainable Manufacturing', listing the following in order of priority:[2]

1. Continuous processing
2. Bioprocesses
3. Separation and reaction technologies
4. Solvent selection, recycle and optimisation
5. Process intensification.

Remarkably 'integration of chemistry and engineering' was only listed in seventh place in this publication!

The batch *versus* continuous processing argument was being discussed in the scientific literature a century ago,[3] yet, while continuous processes tend to be the norm in the bulk chemical sector, batch processes still dominate in the fine chemical and pharmaceutical sectors. Conversion of batch to continuous processes[4-6] can result in significant cost savings.[7,8] Progressing further down the line towards fully integrated end-to-end continuous manufacturing requires integration of different processes including: synthesis (often multistep); purification; drying; compounding; and final dosage form manufacture (in the example of a drug product).[9]

Because of the somewhat traditional approach adopted by many R&D chemists, particularly those skilled in the synthesis of complex organic molecules (*e.g.* monomers for polymer production, agrochemicals, flavours and fragrances, and pharmaceuticals), often the only means of speeding up a chemical reaction, such that it is viable at scale, is changing temperature. However, there are numerous other means of activation that can yield far more effective outcomes and provide cleaner, greener and more economically viable final processes.

One of the least often invoked tenets of Green Chemistry is 'Real-time analysis for Pollution Prevention—analytical methodologies need to be further developed to allow for real-time, in-process monitoring and control prior to the formation of hazardous substances'.[10] Yet a common aphorism used in management studies 'you can't control what you can't measure'[11] holds particularly true for continuous processes, so Holmes and Bourne's chapter on the 'Analysis and Optimisation of Continuous Processes' (Chapter 3) is included as the very first chapter in this section on processes for chemical transformations. Indeed, real-time, on-line (or in-line) analysis, allowing for optimisation of reaction conditions and intervention as soon as problems arise, can contribute a lot more than just 'avoiding formation of hazardous substances'. Reduction of waste, simplification of purification processes, and optimisation of space/ time yields all follow, but only if one knows what is going on in the reaction vessel. In this lead chapter a detailed discussion of various on-line and in-line analytical methods is provided. This is followed by 'Automatic Optimisation of Continuous Processes' including 'self-optimisation' and 'statistical and kinetic modelling' which, while detailed, is nonetheless accessible to the non-expert and is recommended reading ahead of any other chapters in the section.

2.2 PROCESSES TO FACILITATE CHEMICAL TRANSFORMATION

Processes to facilitate chemical transformations usually entail either catalysis (to lower the activation energy required to reach the transition state) or a means of delivering energy to the system to promote reactions (or both). The topics of homogeneous and heterogeneous catalysis are very far reaching and many books on the topic exist. However, Chapter 4 is provided in this section to give an overview of the *processes* required to effect catalytic reactions at scale. The authors consider process intensification (of catalytic reactions) and comment on the challenges of employing selective heterogeneous catalysis in the fine and pharmaceutical chemical industries. The 'innovation gap' that exists between chemistry and chemical engineering is summarised by Felicity Roberts and Klaus Hellgardt as: 'discovery, yield and mechanisms are the main drivers for chemists whereas process design, reaction engineering, heat/mass transfer and scale-up are the major interests of chemical engineers'.

Heat is often the first means of activation considered and there are many different means of adding heat energy to a reaction mixture. In some cases, it may prove desirable to provide localised high

Table 2.1 A summary of some processes that may be used to facilitate chemical transformations, or to prepare specific materials.

Means of activation	Example	Comments	References
Heat	Microwave heating	Increase in temperature of bulk reaction mixture	12, Chapter 6
	Sonication or cavitation	May cause very localised high temperature *and* pressure	13, Chapter 5
Pressure	Ultra high pressure	Pressures in the GPa range facilitate reactions that undergo changes in partial molar volumes	14,15
Heat and/or pressure	'Mechanochemistry'[a]	A catch-all term for reactions conducted using mechanical means of mixing—usually the actual mode of activation is heat or pressure	16
Ultraviolet or visible radiation	Photochemical processes	Differs from microwave chemistry in that radiation excites molecules/bonds directly	17, Chapter 7

[a]The term 'mechanochemistry' is used to denote a range of processes that promote chemical reactions based on mechanical action but, in most cases, this is highly misleading as even a cursory examination of the processes will reveal that the basis for promotion of the reaction is localised heating, melting[18] and/or simply a function of concentration! In many cases the mechanical action simply provides the means of mixing.

temperatures that far exceed bulk temperatures (but are short-lived), or to effect direct excitation of molecules or bonds. A brief summary of such methods is provided in Table 2.1 and some useful (usually recently published) resources are referred to.

Clearly it is not possible to cover all process for facilitating chemical transformations in a single, general book, but some important scalable examples are provided with particular focus on the sustainability gains in each case. Thus, in Chapter 5, Frederick Michel and Oleg Kozyuk focus on hydrodynamic cavitation as a means of efficiently delivering energy to a system, particularly to effect particle or droplet size reduction in a range of applications, including materials synthesis and control of drug particle properties. While chemical reactions per se are not their focus, the basis for promotion of reactions is clearly described and examples provided of the use in materials (including soft materials such as emulsions) and catalyst synthesis. The authors clearly describe why hydrodynamic cavitation offers more sustainable processing routes than many alternatives, including the (rather difficult to scale) acoustic cavitation (sonication) more often found in use at laboratory scale.

In Chapter 6, an industrial expert, Yvonne Wharton, demonstrates that microwave chemistry is scalable, offering opportunities in flow chemistry, which mitigates the inherent limitation of depth-of-penetration of microwaves. The long-running debate about the effects of microwave irradiation as a means of heating or the source of some manner of 'non-specific microwave effect' is easily addressed by a consideration of the wavelength of microwave irradiation (12.2 cm at 2450 MHz radiation) *versus* bond lengths, and careful and accurately measured temperatures,[19] as well as properly conducted kinetic studies.[20] In addition, scale up has been addressed by a number of other authors, including Strauss,[21,22] an early developer of this technology.

Michael Oelgemöller and co-authors (Chapter 7) focus on the most renewable of sources of electrochemical radiation accessible to humans, the Sun. Solar photochemical manufacture of fine chemicals has come a long way since Giacomo Ciamician envisaged a world of chemical manufacture based on solar radiation.[23] Modern advances in solar collectors also serve to provide the concentration of radiation required to effectively conduct chemical reactions, offering potential for distributed manufacturing of small-scale fine, or speciality, chemicals, although non-concentrating reactors may provide for more economically viable processes. Sustainability gains arise from the mitigation of carbon dioxide (CO_2) emissions inherent in replacing (usually fossil) based terrestrial energy stores with solar energy, while economic gains follow from this 'free' (once plant is installed) source of energy.

2.3 EXAMPLES OF CHEMICAL TRANSFORMATIONS AND PROCESSES

To illustrate some of the concepts discussed above, three examples of transformations and processes are provided. In Chapter 8, Robin White covers the highly topical use of CO_2 as a C1 synthon, specifically for the synthesis of methanol. Such a process would need to be realised at very large scale to be both economically viable and impactful.

Nanomaterials are rapidly becoming ubiquitous in a vast range of applications including transparent conductive coatings for use in optoelectronic devices, touch screens and solar cells,[24] drug delivery[25] and energy storage[26] to select just a few examples. While many of these applications are certainly contributors to improved sustainability of human life on Earth, the sustainable production of nanoparticles has not always been a focus. This is addressed in Chapter 9, where Sadik, Yazgan and Kariuki review the preparation of nanomaterials using renewable sources and greener processes. Finally, in Chapter 10, Telma Texeira Franco and Ricardo Baldassin Jr provide some examples of chemical processes for sustainable development in one of the world's fastest growing economies, Brazil. The use of bio-based raw materials, such as sugarcane and other crops and crop residues, is discussed, and the authors also show how a concerted approach to developing a sustainable chemical industry, driven by three factors (improved regional development, reduction of production costs, environmentally sustainable production) can lead to more sustainable production. This is underpinned by some examples of industrial production in Brazil.

Some of the more esoteric processes or technologies, such as ultra high pressure synthesis, are not covered here largely as these are not yet considered scalable, although remarkably, in the food industry, processes in 0.5–0.8 GPa range have been successfully scaled up and other biotechnological applications are suggested.[27]

REFERENCES

1. B. W. Cue and J. Zhang, *Green Chem. Let. Rev.*, 2009, **2**, 193–211.
2. C. Jiménez-González, P. Poechlauer, Q. B. Broxterman, B.-S. Yang, D. am Ende, J. Baird, C. Bertsch, R. E. Hannah, P. Dell'Orco, H. Noorman, S. Yee, R. Reintjens, A. Wells, V. Massonneau and J. Manley, *Org. Process Res. Dev.*, 2011, **15**, 900–911.
3. J. Van Nostrand Dorr, *Ind. Eng. Chem.*, 1929, **21**, 465–71.
4. (a) N. G. Anderson, *Org. Process Res. Dev.*, 2012, **16**, 852–869; (b) N. G. Anderson, *Org. Process Res. Dev.*, 2001, **6**, 613–621.
5. A. Jungbauer, *Trends Biotech.*, 2013, **31**, 479–492.
6. S. G. Newman and K. F. Jensen, *Green Chem.*, 2013, **15**, 1456–1472.
7. D. M. Roberge, B. Zimmerman, F. Rainone, M. Gottsponer, M. Eyholzer and N. Knockman, *Org. Proc. Res. Dev.*, 2008, **12**, 905–910.
8. S. D. Schaber, D. I. Gerogiorgis, R. Ramachandran, J. M. B. Evans, P. I. Barton and B. L. Trout, *Ind. Eng. Chem. Res.*, 2011, **50**, 10083–10092.
9. S. Mascia, P. L. Heider, H. Zhang, R. Lakerveld, B. Benyahia, P. I. Barton, R. D. Braatz, C. L. Cooney, J. M. B. Evans, T. F. Jamison, K. F. Jensen, A. S. Myerson and B. L. Trout, *Angew. Chem., Int. Ed.*, 2013, **52**, 12359–12363.

10. P. T. Anastas and J. C. Warner, *Green Chemistry: Theory and Practice*, Oxford University Press, New York, 1998, p. 30.
11. 'You can't manage what you can't measure' is variously ascribed to different authors, most frequently to W. E. Deming, although apparently the phrase is nowhere to be found in his writings.
12. (a) In inorganic materials chemistry: H. J. Kitchen, S. R. Vallance, J. L. Kennedy, N. Tapia-Ruiz, L. Carassiti, A. Harrison, A. G. Whittaker, T. D. Drysdale, S. W. Kingman and D. H. Gregory, *Chem. Rev.*, 2014, **114**, 1170–1206; (b) In extraction processes: A. K. Kokolakis and S. K. Golfinopoulos, *Nat. Prod. Commun.*, 2013, **8**, 1493–1504; (c) In organic chemistry: A. de la Hoz and A. Loupy (ed.), *Microwaves in Organic Synthesis*, Vol. 1 and Vol. 2, Wiley VCH, Weinheim, 2012.
13. G. Cravotto and Pedro Cintas, *Chem Soc. Rev.*, 2005, **35**, 180–196.
14. C. L. Hugelshofer and T. Magauer, *Synthesis*, 2014, **46**, 1279–1296.
15. H. Kotsuki and K. Kumamoto, *J. Synth. Org. Chem., Jpn.*, 2005, **63**, 770–779.
16. S. L. James, C. J. Adams, C. Bolm, D. Braga, P. Collier, T. Friscic, F. Grepioni, K. D. M. Harris, G. Hyett, W. Jones, A. Krebs, J. Mack, L. Maini, A. G. Orpen, I. P. Parkin, W. C. Shearouse, J. W. Steed and D. C. Waddell, *Chem. Soc. Rev.*, 2012, **41**, 413–447.
17. (a) N. Hoffmann, *ChemSusChem*, 2012, **5**, 352–371; (b) S. Protti and M. Fagnoni, *Photochem. Photobiol. Sci.*, 2009, **8**, 1499–1516.
18. G. Rothenberg, A. P. Downie, C. L. Raston and J. L. Scott, *J. Am Chem. Soc.*, 2001, **123**, 8701–8708.
19. C. O. Kappe, *Chem. Soc. Rev.*, 2013, **42**, 4977–4990.
20. C. R. Strauss and D. W. Rooney, *Green Chem.*, 2010, **12**, 1340–1344.
21. C. R. Strauss, *Org. Process Res. Dev.*, 2009, **13**, 915–923.
22. C. R. Strauss, *Aust. J. Chem.*, 2009, **62**, 3–15.
23. G. Ciamician, *Science*, 1912, **36**, 385–394.
24. M. Layani, A. Kamyshny and S. Magdassi, *Nanoscale*, 2014, **6**, 5581–5591.
25. C. Demetzos and N. Pippa, *Drug Delivery*, 2014, **21**, 250–257.
26. N. Liu, W. Li, M. Pasta and Y. Cui, *Front. Phys.*, 2014, **9**, 323–350.
27. A. Aertsen, F. Meersman, M. E. G. Hendrickx, R. F. Vogel and C. W. Michiels, *Trends Biotech.*, 2009, **27**, 434–441.

CHAPTER 3

Analysis and Optimisation of Continuous Processes

NICHOLAS HOLMES[a] AND RICHARD A. BOURNE*[b]

[a] School of Chemistry, University of Leeds, Leeds, LS2 9JT, UK;
[b] Engineering Building, University of Leeds, Leeds, LS2 9JT, UK
*Email: r.a.bourne@leeds.ac.uk

3.1 INTRODUCTION TO CONTINUOUS PROCESSING WITH ON-LINE ANALYSIS

The synthesis of compounds with batch processing techniques can be impaired by: poor reproducibility; low reaction rates; workup and isolation at each stage of the process; and the effort required for off-line analysis and reaction monitoring. Continuous processing enables facile automation, highly reproducibility and reliability, easy access to super ambient conditions (*i.e.* temperatures above the boiling points of solvents by operating at increased pressures) and hence more rapid reaction rates, greatly improved mass and heat transfer characteristics, and enhanced safety by minimisation of reactor inventory.

Analysis of reaction progress is not a new concept in chemical synthesis. For example, it is commonplace for standard batch laboratory reactions to be monitored off-line by thin layer chromatography (TLC), gas chromatography (GC) or high performance liquid chromatography (HPLC). Typically, off-line analysis requires neutralisation and possibly workup for the sample to be suitable for analysis.

Chemical Processes for a Sustainable Future
Edited by Trevor M. Letcher, Janet L. Scott and Darrell A. Patterson
© The Royal Society of Chemistry 2015
Published by the Royal Society of Chemistry, www.rsc.org

During these steps reaction intermediates could be lost and with it crucial information about the reaction.

On-line and in-line analysis can be used for a variety of applications including reaction intermediate screening and gathering kinetic data. Techniques such as infrared spectroscopy (IR),[1–8] Raman spectroscopy,[2,3,9] ultraviolet–visible (UV-Vis) spectroscopy,[2,4,10] nuclear magnetic resonance (NMR) spectroscopy,[11–14] HPLC[15–17] and GC,[18–20] as well as mass spectrometry (MS)[21–23] have been used extensively to monitor reactions in flow reactors.

The terms 'in-line' and 'on-line' describe methods of analysis without requiring manual transfer of the samples to the analytical equipment. An in-line technique indicates that the entire sample is analysed, whereas on-line analysis refers to analysis of representative aliquots. The development of micro and mesofluidic laboratory scale flow processing equipment coupled to recent development of several miniaturised analytical devices enables reliable and rapid optimisation and acquisition of process understanding.

This chapter is split into two parts. It begins with an overview of examples of in-line analysis using flow reactors which is followed by a discussion of different optimisation techniques.

3.2 SPECTROSCOPIC METHODS

3.2.1 Infrared Spectroscopy

IR offers a convenient and non-destructive method for in-line monitoring. The majority of applications of modern IR instruments for continuous monitoring utilise flow-through cells with attenuated total reflectance (ATR) technology and are capable of fast acquisition for near real-time monitoring.[24]

Fourier transform IR (FT-IR) monitoring of functional groups was used by Qian *et al.*[8] in the telescoped flow synthesis of a δ-opioid receptor antagonist (Scheme 3.1). Each step of the synthesis was carried out and optimised in individual reactor coils followed by polymeric scavenging columns to generate pure product without the need of workup or purification.[25] The optimisation studies consisted of injecting sample loops containing the substrates and reagents; an on-line FT-IR was used to measure the presence of the intermediate (2), *via* the characteristic stretching frequency of the amide, so that the addition of the Burgess reagent stream could be synchronised. It was also possible to calculate the concentration and volume of reagent required using the IR spectra.

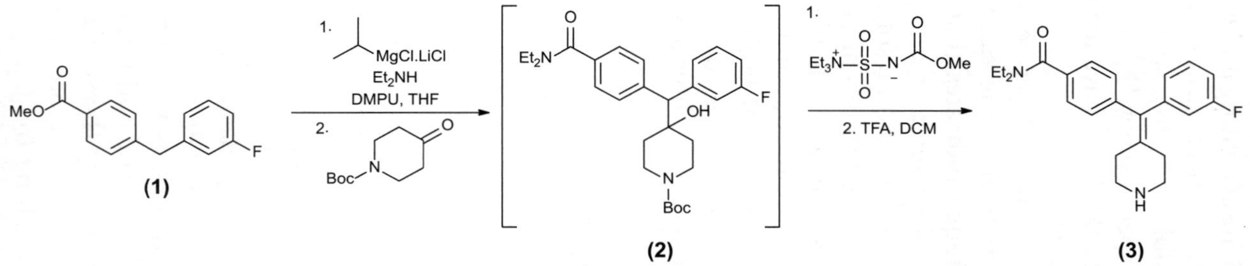

Scheme 3.1   Telescoped synthesis of δ-opioid receptor antagonist (3). On-line IR was used to observe the alcohol (2) formation and notify when to introduce the Burgess reagent for the dehydration to (3).[8]

Skilton *et al.*[26] used the fingerprint region for on-line analysis of the *O*-methylation of *n*-pentanol using dimethyl carbonate and a fixed catalyst bed containing γ-alumina. Calibration of the IR data was performed by correlation of non-convoluted peaks with calibrated GC data. The use of FT-IR significantly enhanced throughput compared with the GC experimentation. The average experiment time of 32 min (where min refers to minutes) when using GC was reduced to 3.5 min with FT-IR as the near real-time analysis could not only detect steady state performance but also quantify the outlet concentration of the desired product much more rapidly. This enabled detailed scans of the parameter space; for example, Figure 3.1 shows a contour plot of the effect of the substrate and methylating agent flow rates on the yield of the reaction.

3.2.2 Nuclear Magnetic Resonance Spectroscopy

NMR spectroscopy could be described as the 'go-to' method for compound identification by synthetic organic chemists, but its

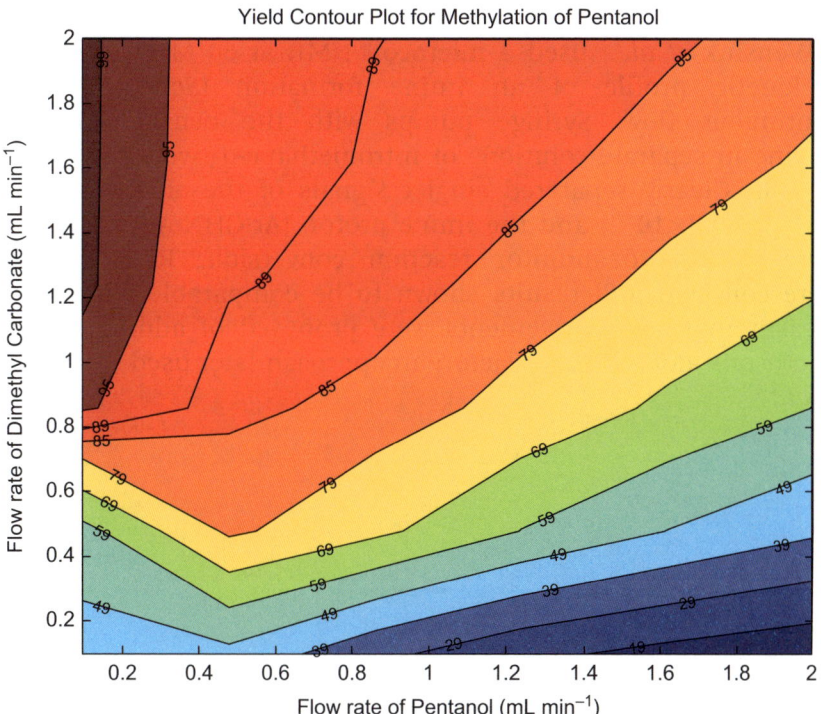

Figure 3.1 Filled contour plot showing the effect of dimethyl carbonate and n-pentanol flow rates on the yield of reaction with respect to *n*-pentanol. For full experimental details see ref. 25.

inherent insensitivity generates spectra with high signal to noise ratios, especially with samples sizes of microlitres and smaller. Recent developments have been made into using microcoils and flow capillaries to reduce the signal to noise ratio.[11,12] Olson and co-workers demonstrated that microcoils and capillaries could be used to provide high resolution NMR spectroscopy for small quantities of material.[11] Recently, miniaturised low-field NMR systems with permanent magnets have been developed for on-line analysis in the laboratory and production environments. These devices typically operate at field strengths of 2 T (where T refers to tesla) and up to 85 MHz ^{1}H frequency.[27] These low field devices typically produce complex spectra and the magnetisation of the sample is influenced by the flow velocity and further complicated by the absence of deuterated solvents in process monitoring applications. They are not, therefore, ideally suited to detailed assignment of spectral features but can be readily used for near real-time analysis of reaction mixtures. Currently, however, systems containing large numbers of compounds or materials at low concentrations may be impractical to accurately quantify.

Wensink *et al.*[12] used a microcoil NMR at 60 MHz to generate a kinetic profile of an imine formation (Scheme 3.2) in continuous flow. Syringe pumps with the benzaldehyde and aniline in separate solutions of nitromethane-d$_3$ were fed into the reactor. Clearly separated singlet signals of the aldehyde proton (CHO, δ 9.9×10^{-6}) and the imine proton (Ar-CHN-Ar, δ 8.4×10^{-6}) were chosen to monitor reaction conversion. Rate constants were collected and results shown to be comparable with off-line measurements. Improvements were proposed of adding multiple microcoils along a flow reactor; an approach later used by Ciobanu *et al.*[14]

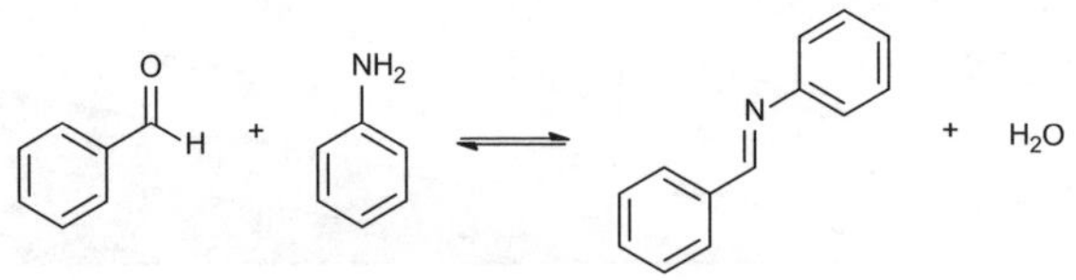

Scheme 3.2 A kinetic profile of the reversible imine formation was measured using NMR spectroscopy. Product formation and substrate degradation can be easily monitored in crude spectra due to the distinctive changes in proton environments.[12]

3.2.3   Ultraviolet–Visible Spectroscopy

UV-Vis spectroscopy is conducted in the 200–800 nm wavelength range. It is especially applicable to compounds with high conjugation and active chromophores. UV–Vis spectra, however, unlike NMR, IR and Raman spectroscopy consist of relatively broad absorption peaks due a multitude of electronic and vibrational levels, making quantification difficult in the majority of cases. Smith and co-workers showed the applicability of UV-Vis for measuring the dispersion of product collection in a react and release system for the synthesis of triazoles.[28]

3.2.4   Raman Spectroscopy

Raman spectroscopy is another form of vibrational spectroscopy, which measures the inelastic scattering of light emission. Raman spectroscopy, like ATR FT-IR spectroscopy, offers a convenient and non-destructive analytical technique for analysis of flow reactors.

Raman spectroscopy is particularly suited to systems containing water. A recent example by Chaplain and co-workers discussed the development of a Raman flow cell for the continuous Suzuki reaction shown in Scheme 3.3.[29]

Quantitative analysis was achieved by partial least squares regression to obtain a calibration model which was validated using offline GC-MS analysis. Furthermore, kinetic profiles of the reaction could be compiled to give rate constants for reaction optimisation. This achieved 61% yield of the product.

3.2.5   Near Infrared Spectroscopy

Near infrared (NIR) spectroscopy covers a spectral range between visible and mid-IR wavelengths (800–2500 nm). It measures the absorption bands resulting from overtones or combinations of the

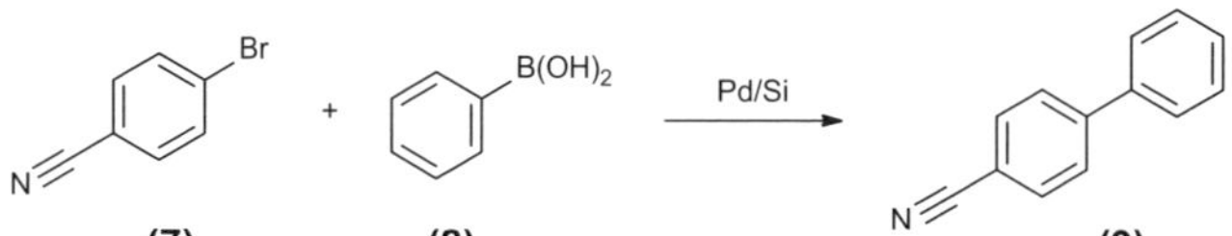

Scheme 3.3   Suzuki reaction used to develop in-line Raman spectroscopy using potassium carbonate base and palladium on a silica monolith catalyst.

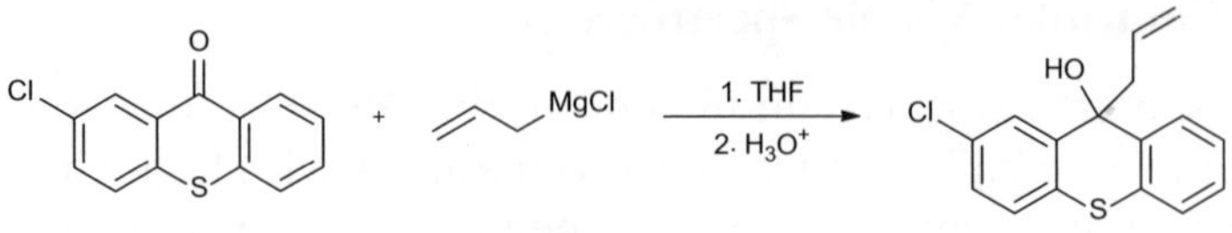

(10)   **(11)**   **(12)**

Scheme 3.4   Formation of a tertiary alcohol by Grignard reaction with a ketone. The
reaction was sensitive to reactant stoichiometry and excesses of ketone
and Grignard were monitored using on-line NIR spectroscopy.

fundamental mid-IR bands, mainly of –CH, –OH, –SH and –NH
bonds.[30] NIR spectra consist of many overlapped spectral features
which necessitate the use of chemometrics to gain quantitative data
regarding the composition of samples.[31,32]

Cervera-Padrell and co-workers used NIR spectroscopy for the on-
line monitoring of a Grignard reaction (Scheme 3.4) in a tubular reactor with a feed forward feedback loop.[31] The Grignard reaction is a
useful technique in organic synthesis for forming carbon–carbon
bonds in functionalised compounds. It can, however, be problematic
in scale up as it is incredibly fast and highly exothermic, and therefore
difficult to control. The increased heat transfer found in flow reactors
has resulted in many Grignard transformations being scaled up in
continuous reactors but heat spots are still observed at mixing sections,[31] where it is assumed the reaction occurs.[33] To counter these
potential problems, the Grignard reagent was introduced into the
reactor through multiple entries. The reaction was sensitive to
changes in the relative stoichiometry of the Grignard reagent and
ketone, with increases in the former creating impurities, and the
latter reducing the yield. NIR spectroscopy was used to monitor the
concentration of the substrates in the system, with a feedback loop
controlling the pump ratios.

3.3   CHROMATOGRAPHIC METHODS

Chromatography separates compounds based on their differential
affinity for mobile and stationary phases. It is commonly used as an
accurate measure of purity as well as product identification, if compared with known standards, or combined with other analytical
techniques (such as mass spectrometry).

The two most common forms of analytical chromatography are gas
chromatography (GC) and high-performance liquid chromatography
(HPLC). Chromatography is typically regarded as an at-line technique

as sampling must be performed discretely with analysis results typically requiring 2–30 min, compared with the continuous real-time nature of most spectroscopy techniques.

3.3.1 Gas Chromatography

GC uses a gaseous mobile phase and has been successfully used for the at-line monitoring of reactions that produce extremely volatile products or by-products that are normally lost during workup or extraction procedures. At-line GC is typically conducted using a sample loop, a switching valve between two separate flow streams, which can introduce a small volume of the sampled stream into the carrier flow of the GC.

Walsh *et al.*[18] used at-line GC analysis to create an automated continuous system to analyse the formation of highly volatile short chain ether species in the solid acid catalysed etherification of short chain alcohols (Scheme 3.5) in supercritical carbon dioxide (scCO$_2$) which could not easily be contained for off-line analysis. *O*-alkylations using short chain alcohols can be a greener alternative to their alkyl halide counterparts due to their non-toxicity.

The symmetrical etherification of alcohols with methyl to *n*-pentyl chains was investigated and the results showed that the highest yield was achieved with *n*-propanol. It was proposed that the longer carbon chain increased the nucleophilicity of the alcohol, but for the longer butyl and pentyl chains, the elimination reaction was preferred.

The implementation of at-line GC with an automated reactor resulted in a queue of experiments that could be run without any user intervention. The use of GC was fit-for-purpose for analysing the highly volatile short chain ether and alkene products. However GC is restricted to compounds that have significant volatility and has been estimated to be useful for only 20% of known organic compounds.[34]

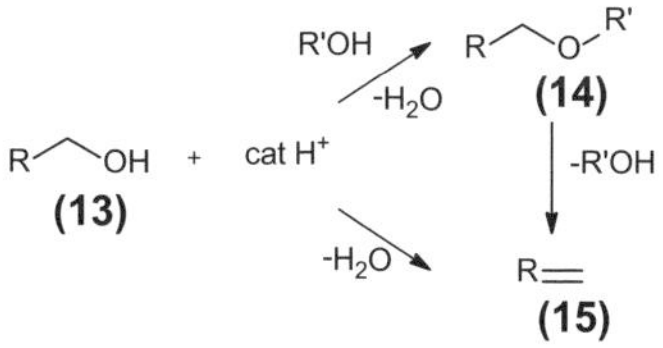

Scheme 3.5 Various reactions in the *O*-alkylation of alcohols using short chain alcohols (note the ether formed is portrayed as asymmetrical, but can also be symmetrical).[18]

3.3.2 High-Performance Liquid Chromatography

HPLC separates materials based on their differing affinity to a pressurised liquid carrier and a column filled with a solid sorbent material (typically 2–50×10^{-6} m in size). Several different detectors can be fitted including UV-Vis, photodiode array or MS.

HPLC has several advantages over GC, including:

- a greater library of compounds that can be analysed;
- a smaller diffusion coefficient, increasing the speed of analysis;
- higher viscosities of the mobile phase, reducing the diffusion coefficient;
- greater flexibility in changing the stationary and mobile phases.

At-line HPLC was used by Antes and co-workers to monitor the nitration of toluene in microreactors to assess the performance of a silicon microreactor.[17] Silicon has high heat transfer properties and high resistance to acid corrosion. At-line HPLC was chosen for the speed and reliability of analytical feedback, plus the UV detector on the machine was suitable for the aromatic substrates and products. In this research they replaced the typical mixture of aqueous sulfuric and nitric acid with fuming nitric acid, removing the need to regenerate and purify the resulting nitrating acid mixture. The effective mixing and high heat transfer that is achieved in microreactors enabled rapid reaction with effective control of the reaction exotherms. The reactor was set up with three separate syringe pumps for toluene, nitric acid and an ice water quench, with at-line HPLC sampling through a sample loop.

The nitration of toluene can generate a high number of different isomers (Scheme 3.6), but only monosubstituted products were targeted. During initial screening tests, the highest yields (89–92%) of nitrotoluene were observed at residence times of 3 s (where s refers to seconds). Because the reaction time was so short, a quench stream was introduced before sampling so that strict control over the reaction could be achieved. When the flow rate of the quench stream

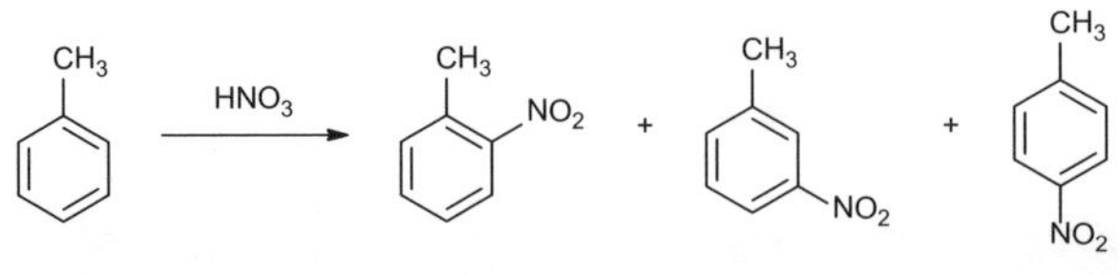

| | (16) | (17) | (18) | (19) |

Scheme 3.6 Different mono-isomers formed in the nitration of toluene.[17]

was reduced the observed yield, by on-line HPLC, fell by almost half due to the reaction progressing to di-substituted products.

3.4 MASS SPECTROMETRY

On-line MS is a standalone technique that measures the molecular mass of compounds and their constituent fragments. MS is commonly coupled to chromatographic techniques (GC-MS, LC-MS), but has been used solely for on-line analysis in reaction monitoring and has several advantages over other techniques. It can directly detect metal species, and particular elements such as chlorine and bromine can be identified by their isotope patterns. However, it does come with some disadvantages. If structural isomers don't form different fragments, then they can only be identified with ion mobility,[35] which isn't readily available on most spectrometers. Different ionisation techniques are required for different compounds, so a compromise may have to be reached in the spectra obtained in terms of how many components can be characterised.

Browne *et al.*[22] capitalised on the speed of analysis and identification opportunities to monitor short-lived and unstable intermediates formed in a tandem benzyne and Diels-Alder reaction (Scheme 3.7) using electrospray ionisation (ESI) MS. The reaction uses a nitride to convert the anthranilic acid (**20**) to a diazonium salt (**21**), which undergoes quick loss of carbon dioxide and dinitrogen to yield the benzyne (**22**). Introduction of furan initiates a Diels-Alder reaction to form the epoxy napthalene (**23**).

This follows the traditional route to benzynes, which is no longer popular in batch processes because the diazonium intermediate (**21**) is explosive if produced in large quantities or isolated to its pure crystalline form. The substrates have now been replaced with starting materials that introduce triflates and trimethylsilanes as leaving groups, but these are expensive reagents that require multistep processes, generating more waste. The use of continuous processing offers a direct route because the low inventory of the hazardous intermediate can be handled safely.

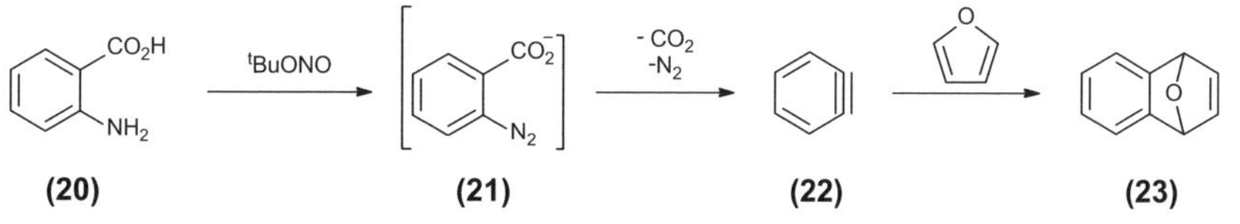

(20)   **(21)**   **(22)**   **(23)**

Scheme 3.7   Single step benzyne and Diels-Alder reaction.[22]

Preliminary studies with the mass spectrometer showed that the explosive diazonium intermediate (21) was still present in the out-stream mass spectrum when the reactor was at room temperature, so sodium thiosulfate was added to the collection vessel to reduce the intermediate and thus minimise the risk of explosion. Further optimisation of the operating conditions using the MS data enabled complete consumption of the explosive intermediate.

3.5 AUTOMATED OPTIMISATION USING CONTINUOUS SYSTEMS

3.5.1 Self-Optimisation

Self-optimisation describes the use of automated reactors combined with on-line analysis and adaptive feedback algorithms to generate optimal conditions without the need for human intervention. Flow systems are ideally suited for such experimentation as integration of analytical equipment is facile, and rapid measurement and adjustment of operating parameters such as flow rate, temperature and pressure can be accomplished quickly within a single reactor system. The concept was first used by Krishnadasan *et al.*[36] for the synthesis of CdSe quantum dot nanoparticles using a microreactor, on-line UV and SNOBFIT (stable noise optimisation by branch and fit) algorithm.[37] The reaction was optimised for a wavelength function that was termed the 'dissatisfaction coefficient' (DC), essentially converting the wavelength into a number, which the SNOBFIT algorithm could minimise. This allowed the synthesis of highly tuned CdSe quantum dot nanoparticles.

SNOBFIT is a branch and fit algorithm used in linear programming.[37] It is so-called because it will plot global (or exploratory) experimental points and then calculate polynomials based on the responses to find optimal (or local) points—essentially branching out and then finding a fit. The algorithm will continue to plot both exploratory and local points as it continues to seek optima. In addition to the work by Krishnadasan and colleagues, the algorithm has been used in optimisations of small organic compounds.[26,38]

Other optimisation algorithms used in self-optimising reactions are based on simplex algorithms, first developed by Spendley *et al.*[39] A simplex algorithm is set up by picking the number of variables, N, and then setting the initial number, $N+1$, of experimental data points to form the first simplex (Figure 3.2). The lowest response of the simplex is ignored and the midpoint between the resultant points is taken, a vertex is produced between the midpoint and lowest ranked point,

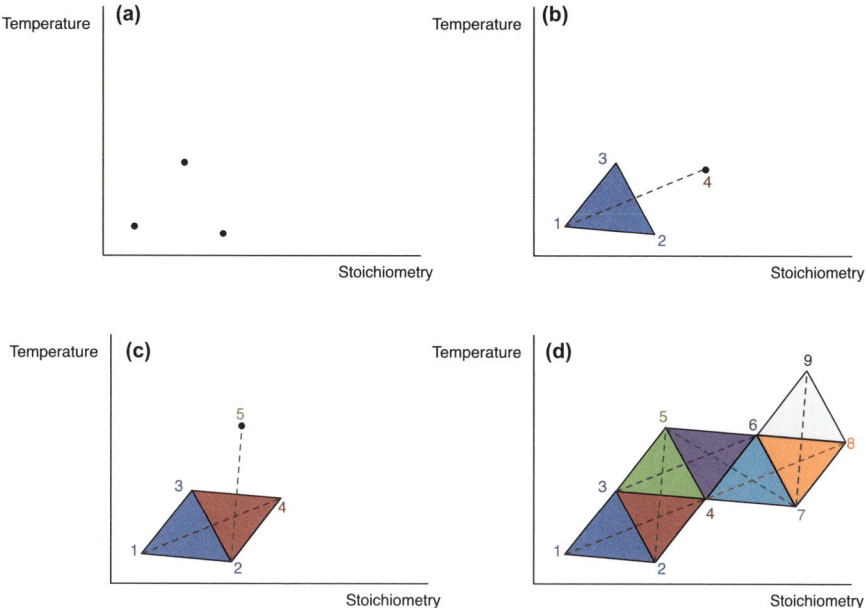

Figure 3.2 (a) The variables of temperature and stoichiometry are selected and three initial data points are plotted with three different conditions ($N=2$, $N+1=3$). (b) The three data points are ranked (1 = worst, 2 = middle, 3 = best) for how close they reach to the optimum and a vertex is plotted from the lowest ranked point through the midpoint between the top two ranked points and finishes at a new point generated by the algorithm. (c) A new simplex is formed between points 2, 3 and 4, and then these points are re-ranked. A new vertex is generated from the lowest ranked point to the newest point, 5; points 3, 4 and 5 form the new vertex. (d) The process is continued, forming new simplexes until the optimum point is reached.[39,40]

and then reflected to an equal distance in the opposite direction. The end of the vertex makes a new point, which provides the conditions for the next experiment. The new point and the resultant points make up a new simplex and then the points are re-ranked to form a new vertex and subsequent simplex. This process is repeated until the simplexes surround the optimum point.[40]

McMullen and Jensen used downhill simplex methods in their self-optimisation of a Knoevenagel condensation (Scheme 3.8) in a microreactor.[38] A two-dimensional (2D) simplex was selected using residence time and temperature as the variables, and optimised for a combined function of yield and production rate—also called space time yield (STY).

The boundary constraints for temperature and residence time were between 40 and 100 °C, and 30 and 300 s respectively. The algorithm

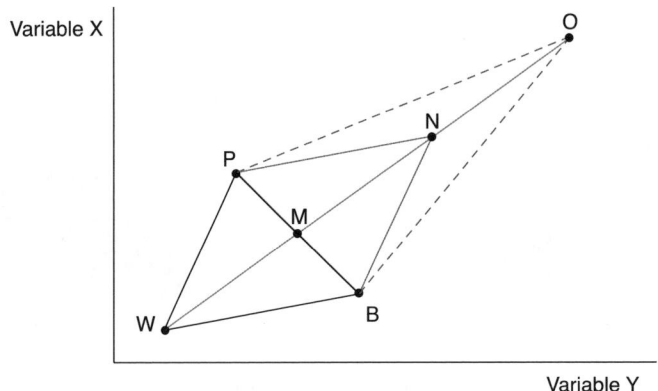

Scheme 3.8   Knoevenagel condensation used by McMullen for self-optimisation studies.[38]

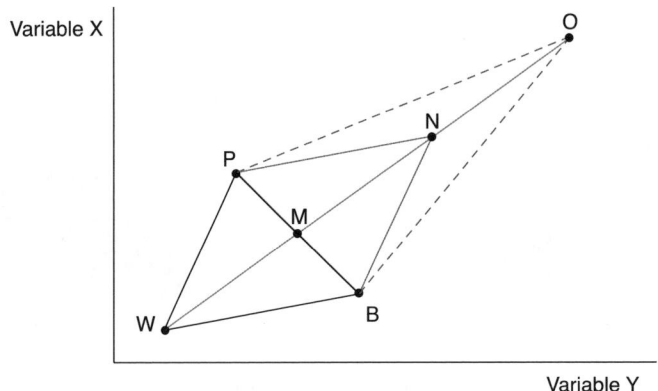

Figure 3.3   A 2D simplex reflected using the SMSIM. The initial points are ranked and the worst, *W*, is eliminated. A new simplex (blue) is reflected through the midpoint, *M*, creating a new point, *N*. The SMSIM creates a second order polynomial according to the responses of *W*, *M* and *N*. This assigns the new point, *O*, forming a simplex (red).[41]

very quickly started merging to higher temperatures and lower residence times in order to find the optimum. It reached the border range at 30 s and after trying to plot a simplex outside the defined constraints, started contracting as it moved to 100 °C. The initial rejected point outside the boundary had a temperature of 70 °C and residence time of 180 s and reached the optimum point after 30 iterations.

The super modified simplex algorithm (SMSIM) was developed by Routh *et al.*[41] and is an improved variation on the downhill simplex algorithm (see Figure 3.3). The parameters, *N*, and initial number of experimental points, $N+1$, are chosen, ranked based on their response to the optimising parameter and the lowest ranked point eliminated. A new simplex is reflected through the midpoint and a second order polynomial is generated based on the responses from the lowest ranked point; midpoint and new point. The polynomial can be concave up (parabola 'pointing' up) or concave down (parabola 'pointing' down), and its orientation determines the location of the next simplex. This crucially allows the simplex to expand or contract

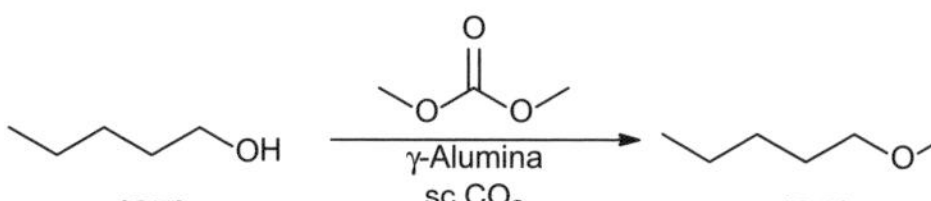

Scheme 3.9   Methylation of *n*-pentanol in supercritical CO_2 using dimethylcarbo-
nate as a methylating agent and catalysed by γ-alumina.[42,43]

enabling quicker movement across parameter space and more accurate optimum location.

Bourne and co-workers compared the approaches of a standard
simplex and SMSIM algorithms for the acid catalysed methylation of
n-pentanol in supercritical CO_2 (Scheme 3.9) using dimethyl carbonate (DMC).[42] The reaction was optimised for yield using the parameters of temperature, pressure, CO_2 flow rate and equivalence
of DMC.

A comparison of the standard simplex and SMSIM algorithms
showed that an optimum can be reached with the SMSIM in fewer
simplexes due to its ability to expand and contract as it moves towards
the optimum.[42]

3.5.2   Statistical and Kinetic Modelling

Self-optimising systems have typically been used to find the optimal
operating conditions for the particular reactor system within which
the optimisation was conducted. Direct transfer of the operating
conditions from a laboratory scale to production scale in a continuous
reactor is more applicable than with analogous batch systems. However, continuous automated reactors also enable kinetic data to be
generated quickly, providing critical information when scaling up
reactions into other reactor types which may have very different mass
and heat transfer performance characteristics. Using this rationale,
Reizman and Jensen adopted an automated model system to derive
the kinetics of a nucleophilic aromatic substitution reaction (S_NAr,
Scheme 3.10) using a D-optimal design of experiments.[44]

All four reactions were modelled as unimolecular second order, and
the reactor modelled as a perfect plug flow reactor. These assumptions could be made as it followed literature precedence for S_NAr reactions, and it was calculated that plug flow would occur at residence
times above 2 min following the dimensions of the reactor, and the
liquid–liquid diffusitivities. It was predicted that the kinetics could
still be modelled as plug flow at residence times of 30 s, but anything

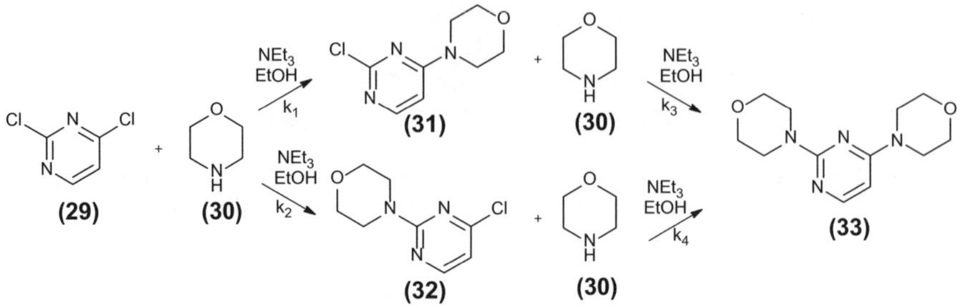

Scheme 3.10   Complex kinetic system of a nucleophilic aromatic substitution.[44]

below this would need to be de-convoluted due to dispersion effects. An experimental design was set up varying temperature, residence time and concentration of compound (**29**) in Scheme 3.10, using on-line HPLC to monitor the reaction.

Fitting of the rate constants could not be carried out accurately for all the kinetic pathways, especially the unfavourable k_4 pathway. Initial rate screenings returned infinite uncertainty in the calculated rate constants of k_3 and k_4, but these were improved when two D-optimal designs were carried out, generating 24 experimental points in total. To improve the uncertainty of the minor product pathways, the rate constants were calculated by isolating the intermediate materials and their subsequent reaction steps, with successful results. The uncertainty of the k_1 and k_2 constants decreased to within 0.1 standard deviations, and the isolation allowed full screening of the k_3 and k_4 pathways. These results could be implemented into the original D-optimal design experiments to predict conditions to optimise for each of the different products.

3.6   CONCLUSIONS AND FUTURE DIRECTIONS

The combination of continuous flow reactors, on-line analytics and algorithm-based control can enable extremely rapid and facile optimisation of reaction systems and only require minimal human effort. These automated techniques are widely applicable to a wide scope of reactions and will undoubtedly be applied to more complex reaction networks and telescoped reaction pathways. Excitingly, these automated tools should enable chemists to focus on developing more sustainable and economical processes, and enable a greater focus on the results and developing process understanding rather than investing the majority of effort into running the experiments.

REFERENCES

1. J. S. Moore and K. F. Jensen, *Org. Process Res. Dev.*, 2012, **16**, 1409.
2. W. Ferstl, S. Loebbecke, J. Antes, H. Krause, M. Haeberl, D. Schmalz, H. Muntermann, M. Grund, A. Steckenborn, A. Lohf, J. Hassel, T. Bayer, M. Kinzl and I. Leipprand, *Chem. Eng. J.*, 2004, **101**, 431.
3. W. Ferstl, T. Klahn, W. Schweikert, G. Billeb, M. Schwarzer and S. Loebbecke, *Chem. Eng. Technol.*, 2007, **30**, 370.
4. R. Herzig-Marx, K. T. Queeney, R. J. Jackman, M. A. Schmidt and K. F. Jensen, *Anal. Chem.*, 2004, **76**, 6476.
5. S. Hübner, U. Bentrup, U. Budde, K. Lovis, T. Dietrich, A. Freitag, L. Küpper and K. Jähnisch, *Org. Process Res. Dev.*, 2009, **13**, 952.
6. M. Rueping, T. Bootwicha and E. Sugiono, *Beilstein J. Org. Chem.*, 2012, **8**, 300.
7. C. F. Carter, H. Lange, S. V. Ley, I. R. Baxendale, B. Wittkamp, J. G. Goode and N. L. Gaunt, *Org. Process Res. Dev.*, 2010, **14**, 393.
8. Z. Qian, I. R. Baxendale and S. V. Ley, *Chem.–Eur. J.*, 2010, **16**, 12342.
9. S. Mozharov, A. Nordon, D. Littlejohn, C. Wiles, P. Watts, P. Dallin and J. M. Girkin, *J. Am. Chem. Soc.*, 2011, **133**, 3601.
10. A. Caceres, M. Jaimes, G. Chavez, B. Bravo, F. Ysambertt and N. Marquez, *Talanta*, 2005, **68**, 359.
11. D. L. Olson, T. L. Peck, A. G. Webb, R. L. Magin and J. V. Sweedler, *Science*, 1995, **270**, 1967.
12. H. Wensink, F. Benito-Lopez, D. C. Hermes, W. Verboom, H. J. G. E. Gardeniers, D. N. Reinhoudt and B. A. van den Berg, *Lab Chip*, 2005, **5**, 280.
13. M. V. Gomez, H. H. J. Verputten, A. Diaz-Ortiz, A. Moreno, A. de la Hoz and A. H. Velders, *Chem. Commun.*, 2010, **46**, 4514.
14. L. Ciobanu, D. A. Jayawickrama, X. Zhang, A. G. Webb and J. V. Sweedler, *Angew. Chem., Int. Ed.*, 2003, **42**, 4669.
15. A. Rehorek and A. Plum, *Anal. Bioanal. Chem.*, 2006, **384**, 1123–1128.
16. L. Zhu, R. G. Brereton, D. R. Thompson, P. L. Hopkins and R. E. A. Escott, *Anal. Chim. Acta*, 2007, **584**, 370.
17. J. Antes, D. Boskovic, H. Krause, S. Loebbecke, N. Lutz, T. Tuercke and W. Schweikert, *Chem. Eng. Res. Des.*, 2003, **81**, 760.
18. B. Walsh, J. R. Hyde, P. Licence and M. Poliakoff, *Green Chem.*, 2005, **7**, 456.
19. J. G. Stevens, R. A. Bourne and M. Poliakoff, *Green Chem.*, 2009, **11**, 409.

20. J. G. Stevens, R. A. Bourne, M. V. Twigg and M. Poliakoff, *Angew. Chem., Int. Ed.*, 2010, **49**, 8856.

21. D. Fabris, *Mass Spectrom. Rev.*, 2005, **24**, 30.

22. D. L. Browne, S. Wright, B. J. Deadman, S. Dunnage, I. R. Baxendale, R. M. Turner and S. V. Ley, *Rapid Commun. Mass Spectrom.*, 2012, **26**, 1999.

23. S. E. Hamilton, F. Mattrey, X. Bu, D. Murray, B. McCullough and C. J. Welch, *Org. Process Res. Dev.*, 2014, **18**, 103.

24. P. R. Griffiths and J. A. de Haseth, in *Fourier Transform Infrared Spectrometry*, ed. P. Griffiths and J. A. de Haseth, John Wiley & Sons, Hoboken, NJ, 2nd edn, 2006, ch. 15, pp. 321–348.

25. S. V. Ley, I. R. Baxendale, R. N. Bream, P. S. Jackson, A. G. Leach, D. A. Longbottom, M. Nesi, J. S. Scott, R. I. Storer and S. J. Taylor, *J. Chem. Soc., Perkin Trans.*, 2000, **1**, 3815.

26. R. A. Skilton, A. J. Parrott, M. W. George, M. Poliakoff and R. A. Bourne, *Appl. Spectrosc.*, 2013, **67**, 1127.

27. F. Dalitz, M. Cudaj, M. Maiwald and G. Guthausen, *Prog. Nucl. Magn. Reson. Spectrosc.*, 2012, **60**, 52.

28. C. J. Smith, N. Nikbin, S. V. Ley, H. Lange and I. R. Baxendale, *Org. Biomol. Chem.*, 2011, **9**, 1938.

29. G. Chaplain, S. J. Haswell, P. D. I. Fletcher, S. M. Kelly and A. Mansfield, *Aust. J. Chem.*, 2013, **66**, 208.

30. Y. Roggo, P. Chalus, L. Maurer, C. Lema-Martinez, A. Edmond and N. Jent, *J. Pharm. Biomed. Anal.*, 2007, **44**, 683.

31. A. E. Cervera-Padrell, J. P. Nielsen, M. Jønch Pedersen, K. Müller Christensen, A. R. Mortensen, T. Skovby, K. Dam-Johansen, S. Kiil and K. V. Gernaey, *Org. Process Res. Dev.*, 2012, **16**, 901.

32. J. Wiss, M. Länzlinger and M. Wermuth, *Org. Process Res. Dev.*, 2005, **9**, 365.

33. D. M. Roberge, L. Ducry, N. Bieler, P. Cretton and B. Zimmermann, *Chem. Eng. Technol.*, 2005, **28**, 318.

34. H. Meyer, O. Biermann, R. Faller, D. Reith and F. Muller-Plathe, *J. Chem. Phys.*, 2000, **113**, 6264.

35. A. B. Kanu, P. Dwivedi, M. Tam, L. Matz and H. H. Hill, *J. Mass. Spectrom.*, 2008, **43**, 1.

36. S. Krishnadasan, R. J. C. Brown, A. J. de Mello and J. C. de Mello, *Lab Chip*, 2007, **7**, 1434.

37. W. Huyer and A. Neumaier, *ACM Trans. Math. Software*, 2008, **35**, 1.

38. J. P. McMullen and K. F. Jensen, *Org. Process Res. Dev.*, 2010, **14**, 1169.

39. W. Spendley, G. R. Hext and F. R. Himsworth, *Technometrics*, 1962, **4**, 441.

40. P. W. Araujo and R. G. Brereton, *TrAC, Trends Anal. Chem.*, 1996, **15**, 63.
41. M. W. Routh, P. A. Swartz and M. B. Denton, *Anal. Chem.*, 1977, **49**, 1422.
42. R. A. Bourne, R. A. Skilton, A. J. Parrott, D. J. Irvine and M. Poliakoff, *Org. Process Res. Dev.*, 2011, **15**, 932.
43. A. J. Parrott, R. A. Bourne, G. R. Akien, D. J. Irvine and M. Poliakoff, *Angew. Chem., Int. Ed.*, 2011, **50**, 3788.
44. B. J. Reizman and K. F. Jensen, *Org. Process Res. Dev.*, 2012, **16**, 1770.

PROCESSES TO FACILITATE CHEMICAL TRANSFORMATIONS

Sustainable Heterogeneous Catalytic Reactions for the Fine and Pharma Industry

FELICITY ROBERTS AND KLAUS HELLGARDT*

Department of Chemical Engineering, Imperial College London, SW7 2AZ, UK
*Email: k.hellgardt@imperial.ac.uk

4.1 SUSTAINABLE PROCESSES

Sustainable, /səˈsteɪnəb(ə)l/, is defined by the Oxford English Dictionary as 'able to be maintained at a certain rate or level'. Maintained for how long? It is the consequence of the Second Law of Thermodynamics that nothing is sustainable. The expression 'Nachhaltiger Ertrag' apparently first appeared in the literature in 1713 when it was used to define sustainable yield in the context of forestry management in Germany. It is worth noting that the word 'sustainable' is here associated with biomass-based production processes—we will revisit this observation later. Some readers will remember the book, *Limits of Growth*, published by the Club of Rome in the 1972 in which, for the first time, model-based analysis was used to try to identify the limits of growth that our Earth can sustain and to identify policies that would help maintain a balance for a prosperous future. This analysis was very much triggered at the time by the first oil crisis. The Organization of the Petroleum Exporting Countries (OPEC)

Chemical Processes for a Sustainable Future
Edited by Trevor M. Letcher, Janet L. Scott and Darrell A. Patterson
Published by the Royal Society of Chemistry, www.rsc.org

subsequently reduced oil prices and all went back to normal (until the next crisis that was). The next instalment of sustainability was in the context of 'sustainable development' as demanded in the Brundtland report, *Our Common Future*, published in 1987. Here the overall goal was defined as that of environmental protection alongside economic growth and social development (the 'three pillars'). Usually only two, namely economic and environmental benefits are actively pursued in any chemical process development. Only when an industrial ecology approach is employed alongside detailed life cycle assessment (LCA) can we realistically make a statement regarding the third pillar. Nevertheless, we are still some way away from being able to find or define a common figure of merit to optimise all three objectives of sustainable development.

In 1969 the Coca-Cola Company commissioned a report to analyse resource consumption and environmental releases with respect to beverage containers. This was somewhat precipitated by incoming environmental legislation, which in many cases is the real driving force for process modification and novel process development.

With the publication of the LCA resource book in 1993,[1] a milestone was laid down in the analysis of chemical processes, leading to a quantitative life cycle impact assessment and the calculation of impact indicators. However, the problem still remains: how to compare and contrast different impact indicators, *i.e.* how to compare the associated costs and risks involved. One has yet to see an objective function that allows for an unbiased approach here.

Perhaps one day all chemical processes and synthesis routes will enjoy an independent certification, similar to the one shown in Figure 4.1. LCA and industrial ecology are already useful steps in this direction.

4.2   SUSTAINABLE REACTION ENGINEERING

Chemical engineering is the discipline that takes discoveries in chemistry to the commercial level, thereby inherently pushing processes towards being more economical and environmentally friendly. This is simply because one figure of merit can be agreed on and that is cost. Within this remit sustainable reaction engineering can be defined as:

'the conception and development of reactors, catalysts and process technology to help reduce Impact Indicators'.

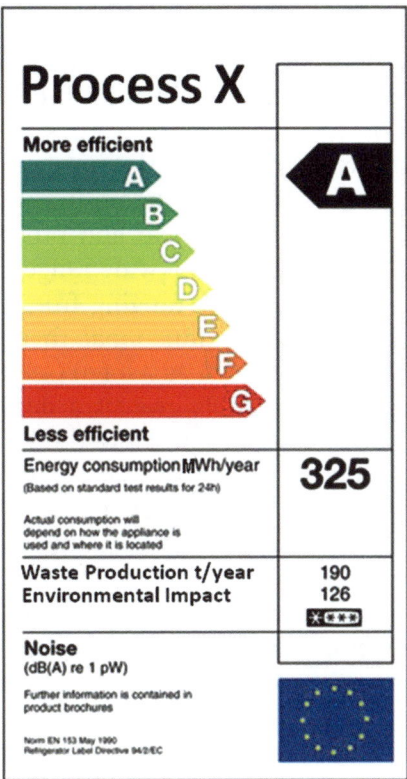

Figure 4.1 Hypothetical process certification sign.

Some alternative/disruptive technologies in this context might be:

- utilisation of solar energy for chemicals manufacture and transformations;
- processing of biomass for fuel and value added chemicals;
- development of an electron-based chemical industry employing electrochemistry for redox reactions;
- application of synthetic biology for the production of complex and personalised products.

Green and sustainable chemistry and engineering principles have been widely articulated by, for example, Anastas and co-workers[2–8] and Jimenez-Gonzalez and co-workers.[9–14] But despite the widely publicised 12 principles of green chemistry and engineering, respectively, many academic publications regarding catalytic synthesis or processes in the fine and pharma literature do not map well onto these.

4.3   INTENSIFIED CATALYTIC REACTORS

At the heart of every chemical process is a (catalytic) reactor. Catalysts are employed in approximately 85% of chemical processes and worldwide contribute some ten trillion dollars to the world's gross national product (GNP).[15] The catalytic reactor dictates the upstream (feed preparation and purification) as well as the downstream (mostly separations) operations, and thereby the theoretical yield and productivity. In order to devise successful catalytic processes, a wide range of length scales and time scales need to be understood, ranging from the molecular scale (employing, for example, density functional theory) to process scale (employing process simulators such as Aspen) (see Figure 4.2). Usually, we find that more complex models give us a deeper insight into the dynamics of a catalytic system/reactor. However, Occam's razor should be heeded to and complexity for its own sake thoroughly avoided.

The development of new, intensified reactors and new catalytic technologies directly impact on the 'sustainability' of a chemical process—in terms of economics and environmental impact. Recent key reviews in this area are by Visscher *et al.*[16] who discuss different types of rotating reactors that have the potential to help intensify chemical processes and Anxionnaz *et al.*[17] who emphasise the development of novel heat exchange reactor systems, particularly to control exothermic reactions. The control and design of internal

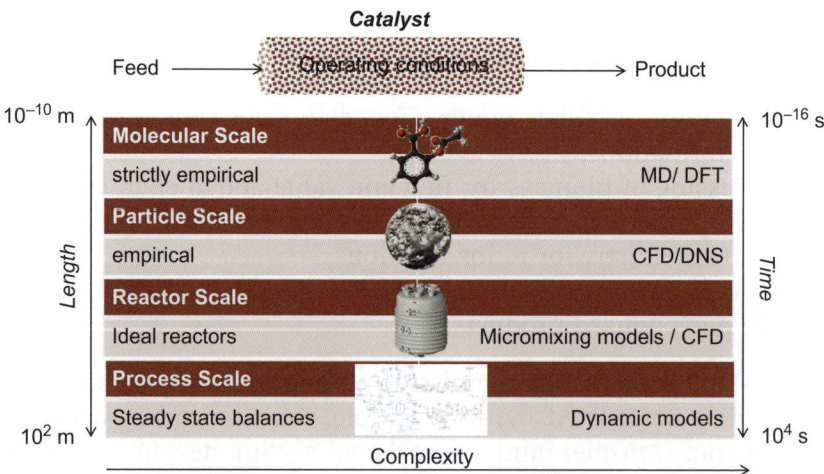

Figure 4.2   Length and timescales that need to be understood for successful catalytic reaction engineering. CFD, computational fluid dynamics; DFT, density functional theory; DNS, direct numerical simulation; MD, molecular dynamics.

structures in flow reactors to enhance mixing, heat and mass transfer is discussed and reviewed in ref. 18–24.

One major development in the past ten years has been 'flow chemistry' (covered in detail in Chapter 3). Whilst continuous flow systems are not new to chemical engineers (in fact most refinery and petrochemical processes are continuous in order to reduce cost), chemists have only recently started to adopt continuous reactors to carry out chemical synthesis in homogeneous and heterogeneous systems. Some key references in this area are provided.[20,25–51]

Flow chemistry is a useful development as its adoption improves the reproducibility and traceability of chemical synthesis. Whilst there is not much innovation in the connection of various empty tubes and then feeding these with different reactants at different temperatures, it allows the move away from ill-defined batch systems (in terms of rate of addition, thermal history, reaction time, kinetics, *etc.*). A number of challenges remain for flow systems, mainly in the area of solids (precipitates) handling and suspension conveying. Similarly, multiphase systems (gas–liquid, liquid–liquid and gas–liquid–solid) have received limited attention.

A number of manufactures now offer bespoke flow systems (*e.g.* Syrris, www.syrris.com; Uniqsys, www.uniqsis.com; Vapourtec, www.vapourtec.co.uk; Chemtrix, www.chemtrix.com; ThalesNano, thalesnano.com), albeit at significant cost, considering the actual function and complexity of these systems—consisting mainly of a high-pressure liquid chromatography (HPLC) pump, a back pressure regulator, a water bath and some stainless or perfluoroalkoxy (PFA) tubing.

Many claims are made with regard to the advantages of flow chemistry over batch processes—some are correct others are wrong. Blackmond and colleagues[52] have provided a good appraisal of what the benefits of a flow system might be compared with a batch reactor, setting straight some of the most common misconceptions, particularly with respect to mixing. The key benefits can be summarised as:

- the tight temperature control (particularly for exothermic reactions);
- the delivery of intermediates for subsequent reactions (tight control of residence time);
- the use of extended reaction space (high pressure allows liquid phase reactions at higher temperature);
- the productivity (space time yield is higher due to relatively high catalyst concentration for heterogeneous systems and the potential continuous operation of the reactor).

Flow reactors have been specifically employed in diverse C–H, C–X and C–C bond formations, for example, hydrogenations,[53] organo-fluorine chemistry,[54] nitrations,[55] heterocycles synthesis,[56] click chemistry,[57] natural products synthesis[58] and asymmetric synthesis[59] to name but a few. Flow reactors have also been combined with alternative means of energy dissipation such as microwaves,[60–65] photochemistry[66–69] and the combination of the two[70,71] to provide additional stimulus.

Continuous electrochemical reactors are being evaluated[72,73] as well as the application of unconventional solvents such as super-critical fluids[74,75] and ionic liquids.[76]

Interestingly, although being discussed in terms of intensification potential,[77,78] ultrasonic systems/reactors have only very recently entered the area of flow chemistry to enhance, for example, multi-phase reactions.[79]

A more thorough discussion of these flow systems is provided in Chapter 3.

4.4   A MORE SUSTAINABLE FINE AND PHARMACEUTICAL INDUSTRY

For many decades the fine and pharmaceutical industry did not pri-oritise sustainable and environmentally friendly processes. Profit margins were large (waste was affordable) and weak legislation did not provide sufficient incentives to address sustainability issues. Thus simple batch chemistry at large scale was pursued to synthesise the desired products. This has changed: legislation is tightening and public awareness has increased; competition from developing coun-tries has increased (production of generic alternatives); and margins are lower due to a general global turndown coupled with patent expiries not matched by the discovery of new chemical entities (NCEs). This has required the fine and pharmaceutical industry to improve on its process efficiency and impact indicators. The road map for this is difficult and a number of challenges need to be addressed and overcome.

4.4.1   Challenges Faced by the Fine and Pharmaceutical Industry

4.4.1.1   Challenge 1: Process Efficiency.   This can be exemplified by comparing E-factors, which is one metric that allows the comparison of the 'greenness' of a chemical processes for different sectors ((Equation (4.1)). One finds that the fine and pharmaceutical industry

Table 4.1   E-factors for different industrial sectors (Rothenberg, 2008).

Industry sector	Annual production/t	E-factor	Waste produced/t
Oil refining	10^6–10^8	*ca.*0.1	10^5–10^7
Bulk chemicals	10^4–10^6	<1–5	10^4 to 5×10^6
Fine chemicals	10^2–10^4	5–50	5×10^2 to 5×10^5
Pharmaceuticals	10–10^3	25–100	2.5×10^2 to 10^5

creates a considerable amount of waste compared with bulk chemical manufacture and refining operations (Table 4.1).

$$\text{E-factor} = \frac{\text{total waste(kg)}}{\text{kg product}}. \tag{4.1}$$

One has to appreciate, however, that this comparison is not particularly fair as refining does not involve many sequential catalytic reaction steps to yield a product and relies mainly on thermal separations for purification. A pharmaceutical product on the other hand might involve on average eight synthesis steps with the associated separation processes (usually non-thermal), thus naturally yielding more waste as the by-products are not necessarily saleable. Nevertheless, green chemistry has made significant advances over the past ten years to improve the mass and energy efficiency of the fine and pharmaceutical industry significantly.

4.4.1.2   Challenge 2: Reactants.   For greener, more sustainable synthesis and indeed processes to be developed we need to look at sustainable feed stocks. This brings us back to the earlier discussion on sustainable yield in forestry in Germany.

Huge steps have already been made in the use of biomass (new carbon) for fuel production to substitute for fossil fuels (old carbon) (see Figure 4.3). Much research is currently underway to identify new platform chemicals, and from these, new synthetic pathways towards bulk chemicals. The key issues are that new chemistry based on oxygenates (*e.g.* carbohydrates) needs to be developed concurrently with novel separations devised to separate (non-thermally) complex mixtures of bioderived materials.

A number of key platform chemicals have been identified to develop new bio-based chemistry such as glucose, 5-hydroxymethylfurfural (5-HMF) and glycerol.

4.4.1.3   Challenge 3: Reaction Environment.   The reaction environment challenge will require synthesis in more benign solvents

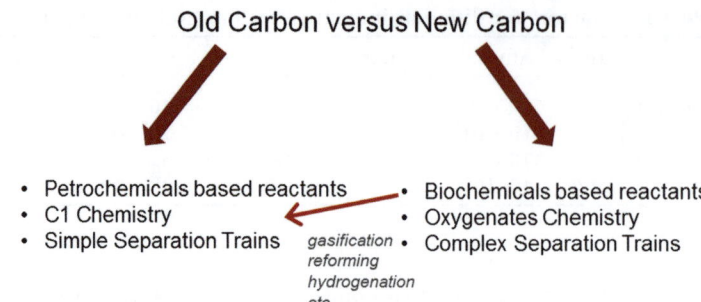

Figure 4.3   'Old carbon' *versus* 'new carbon'.

using internal recycle (so as not to impact on a mass efficiency calculation) and low inventories (environment and health and safety). Reactions should require fewer co-factors that lead to later work-up and ideally should be stoichiometric (excess requires downstream separation). Finally, from a separation and product purity point of view, it is desirable to employ heterogeneous rather than homogeneous catalysts.

The solvent issue is being addressed on a number of fronts. More sustainable solvents have been identified by the major pharma and fine chemicals manufacturers and selection guidelines have been published;[80–84] in addition numerical and process systems modelling tools are employed to aid more quantitatively in this selection process.[85–91]

The development of selective heterogeneous catalysts is still a major challenge in the fine and pharmaceutical industry. It is far more difficult to control the nature and activity of active sites of heterogeneous/inorganic systems as the versatility of ligand equivalent constructs is not available. Thus, immobilisation of homogeneous catalysts on insoluble supports (*via* different catalyst immobilisation methods such covalent tethering to different supports, adsorption of chiral modifiers, hydrogen bonding, electrostatic or ionic interaction) has been a continuing subject for many decades. In particular, the development of heterogeneous enantioselective/ chiral catalyst systems has yet to be achieved to match homogeneous/ enzyme activities. Some novel approaches are summarised in ref. 59, 92–98.

4.4.1.4   Challenge 4: Preoccupation.   For many decades it has been recognised that an innovation gap persists between the disciplines of chemistry and chemical engineering. Different national and

Table 4.2 Comparison of chemistry and chemical engineering preoccupations.

Chemist	Chemical engineer
Discovery	**Process design**
Yield	Reaction rate
Mechanism	?
Atom economy	Mass productivity
?	Heat and mass transfer
Column	Various separations
Batch reactor	Scale-up

international initiatives have sought to overcome this chasm and to integrate the work of chemists and chemical engineers. Table 4.2 identifies the main preoccupations and interests of these two disciplines. It is clear from the analysis of literature published in the area of sustainable/green chemistry/catalysis that discovery, yield and mechanisms are the main drivers for chemists whereas process design, reaction engineering, heat/mass transfer and scale-up are the major interests of chemical engineers.

John Armor recently published a short but important article[99] in which he highlights the need for self-criticism and the evaluation of the true impact and practical value of any discovery and claim relating to novel, sustainable catalytic processes. The literature is cluttered with publications making sweeping statements about the implications of the published work. This is not helpful if we really want to concentrate on the truly game changing innovations—there is too much noise!

It is clear that closer collaboration and integration of chemists and chemical engineers can lead to faster innovation and product to market. Fine and pharmaceutical companies are starting to recognise this synergy and are already establishing multidisciplinary teams to work together at the discovery stage, thereby eliminating non-viable routes early on.

4.4.1.5 Challenge 5: Synthesis Routes. In 2007 the American Chemical Society (ACS) Green Chemistry Institute (GCI) Pharmaceutical Roundtable arrived at a consensus list identifying key transformations for which all pharmaceutical companies would wish to find alternative reagents and processes (Table 4.3). It is interesting to note that, in 2014, after seven years of intense research in these areas, the challenges remain virtually unchanged and to a great extent unsolved.

Table 4.3   Key transformations for which pharma companies seek alternative reagents and processes (Constable *et al.*, 2007).

Reactions companies use now but for which they would prefer better reagents

Amide formation, avoiding poor atom economy reagents
OH activation for nucleophilic substitution
Reduction of amides without hydride reagents
Oxidation/epoxidation methods without the use of chlorinated solvents
Safer and more environmentally friendly Mitsunobu reactions
Friedel–Crafts reaction on unactivated systems
Nitrations

4.4.1.6   The Way Forward.   If we combine all the above challenges and try to devise the most sustainable approach to offer a truly green alternative for these key transformations, we set out on a quest of process-oriented discovery where:

- only preferred solvents are employed, ideally water (see solvent guides by GSK and others);
- non-petrochemical derived/sustainable feed stocks are sourced;
- flow chemistry is encouraged instead of batch operation;
- heterogeneous catalysts are used;
- co-factors (reagents that do not participate in the actual trans- formation, *e.g.* base, additives) are eliminated;
- benign reactants are selected (air, electrons);
- key kinetic data are collected early to assess scale-up;
- on-line analysis (*in situ* and *operando*) process analytical tools (PAT) are deployed routinely;
- separation strategies are devised concurrently.

With this approach in mind we briefly discuss below the four most pressing challenges (Table 4.3) and recent developments in these areas. It is interesting to note that all four areas (potentially) involve redox transformations, which are difficult to achieve conventionally.

4.4.2   Amide Formation, Avoiding Poor Atom Economy Reagents

Amides can be synthesised from alcohols and amines in two steps. Conventionally, the alcohol is first oxidised to the carboxylic acid and activated, upon which the amine is added to form the final amide. Alternatively synthesis *via* the aldehyde can also be achieved. The following is a summary of notable papers that offer a green/ sustainable approach towards amide formation.

4.4.2.1 Carboxylic Acid Amidation. Arnold and colleagues[100] have conducted a detailed kinetic study in which they show that the direct formation of amides from amines and carboxylic acids can occur *without catalyst* under reflux conditions (toluene and fluorobenzene with 3Å molecular sieve) (Scheme 4.1). Boric (1 mol%) and boronic acid based catalysts (10 mol%) were used to improve rates, particularly for less activated acids (Scheme 4.2). The authors conclude that for the purely thermal reaction the anhydride formation is the rate limiting step, whereas in the case of the catalytic reaction, carboxylate activation *via* some acyloxoboronic acid is rate limiting. Thus an electron-rich carboxylic acid will react in a more facile way *via* this intermediate. One can speculate regarding the effect that traces of water may have in this system. A flow system incorporating, for example, a pervaporation membrane might help to drive off the water at higher temperature and allow faster formation of the anhydride.

The removal of water during amide synthesis from carboxylic acids improves reaction rates and can be achieved in a number of ways from the addition of molecular sieves to azeotropic distillation. The extraction of water by azeotropic distillation depends upon the heat input into the system and was investigated by Grosjean and colleagues[101] for both an uncatalysed and boric acid catalysed amide

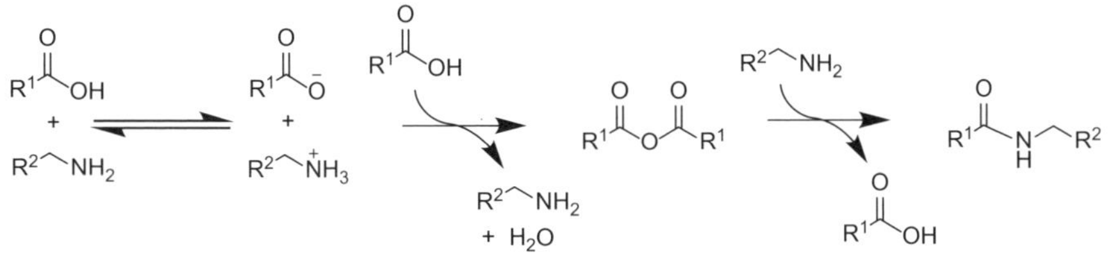

Scheme 4.1 Thermal amidation of carboxylic acids with amides.[100]

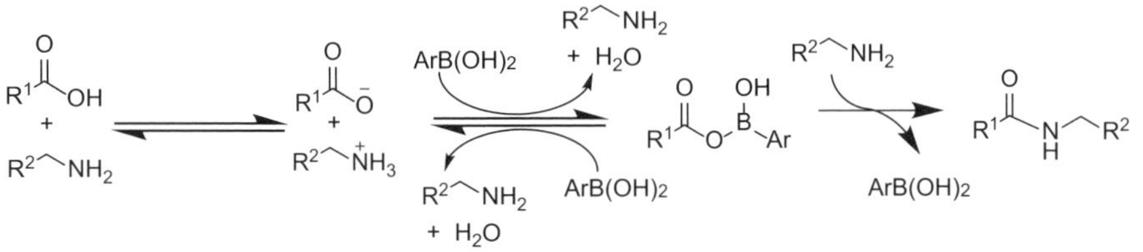

Scheme 4.2 Scheme for the amidation of carboxylic acids with amines with an arylboronic acid, simplified from ref. 100. Note that the acid can take many forms and can bind to more than one carboxylic acid.

formation reaction. One of the most striking findings was that an increased transfer of heat to the reaction vessel, and thus water removal, improved amide formation rates for both the catalysed and uncatalysed system—the effect of increased water removal on the amide yield was very large, equivalent to the addition of catalyst.

Comerford and co-workers[102] have developed a dehydrated/dehydroxylated (heat treated at 700 °C) heterogeneous silica catalyst (K60) to catalyse amide synthesis from acid and amine (in toluene). They compared their approach with other synthesis methods using either homogeneous catalyst or other promoters and found their mass intensity to be at least ten times lower. The yields are proportional to the catalyst loading and after 24 hours mostly reach over 70%. They show that due to the slow re-hydration of the catalyst, activity can be maintained over several runs. No by-products were observed and complete carbon balance was achieved. Again, this is a good candidate for flow chemistry with its higher relative catalyst loading. An integrated water trap should then improve yields even further if operated in batch recycle mode. Similar to the work by Comerford and co-workers, Ghosh *et al.*[103] used a simple oxide (activated alumina) to carry out amide synthesis from unactivated carboxylic acids and amines. Reaction scope was demonstrated with a number of reactions carried out under neat conditions. No by-products were observed and catalyst reuse was demonstrated. However, the paper fails to explain convincingly the catalytic action of the alumina. Both groups did not carry out water sorption studies to understand the sorption capacity of their catalyst and how this might affect their system.

Other metal oxides have also attracted interest including zinc oxide and nanosulfated titanium dioxide, as well as zeolites. One recurring theme is the modification of the structure to increase the pore size and so the catalyst surface area; for example, Thakuria and co-workers[104] found that macroporous materials were more active than their amorphous counterparts. Another point of interest is the use of microwaves over conventional heating methods to reduce the reaction times.[105]

Another approach has been to heterogenise a known homogeneous catalyst by attachment to a support, a method that has been successfully used for a variety of different catalysts described in a comprehensive review by Lundberg and co-workers.[105] One interesting example is Novozym 435®, a lipase used in a flow system for oleamide synthesis from oleic acid. Proteases and lipases are the most common enzymes for amide formation under mild conditions up to 60 °C and for stereoselective synthesis. One disadvantage is that the reaction times tend to be of the order of days. However Slotema and colleagues[106] demonstrated the use of Novozym 435® as a packed bed in a continuous flow reactor for amide synthesis at 60 °C, with ammonium carbamate as the nitrogen source in *tert*-amyl alcohol, a green solvent. With this method the catalyst loading could be 550 times lower than for the conventional batch process.

4.4.2.2 Ester Amidation. Although limited in scope, Pirrung and colleagues[107] demonstrated that for certain ester amidation reactions no coupling reagent was needed. The reaction of lactones with aminoalcohols at room temperature gave yields over 80% in 6 hours. Cyanomethyl esters also reacted under these conditions, but 60 hours were needed for similar yields, reduced to 1.5 hours under microwave irradiation. The hydroxyl group on the amine is important, as the amine does not attack the ester directly; instead transesterification takes place, with the new ester undergoing rearrangement to the more thermodynamically stable amide (Scheme 4.3).

Similarly, Caldwell and colleagues[108] have carried out sustainable ester amidations with amino alcohols through judicious selection of solvent and base. A number of ester derivatives were reacted with amino alcohols in the presence of a catalytic quantity (up to 30 mol%) of potassium phosphate to yield amido-alcohol derivatives. Isopropanol was used as the solvent for the substrate scope evaluation. The calculated reaction mass efficiencies (RME) were reasonably high in most cases (50–90%). The response surface, which is a three-dimensional (3D) plot of the conversion as function of catalyst loading and temperature, shows a significant effect of both. This suggests that flow chemistry could again provide a significant advantage.[108]

The formation of amides from esters in flow was demonstrated by Muñoz and colleagues.[109] The reaction range was extended beyond amino alcohols at the expense of atom efficiency by the use of stoichiometric quantities of Grignard reagents to activate the amine for nucleophilic attack of the ester, although it is thought that the magnesium salt produced may have acted as a Lewis acid to activate

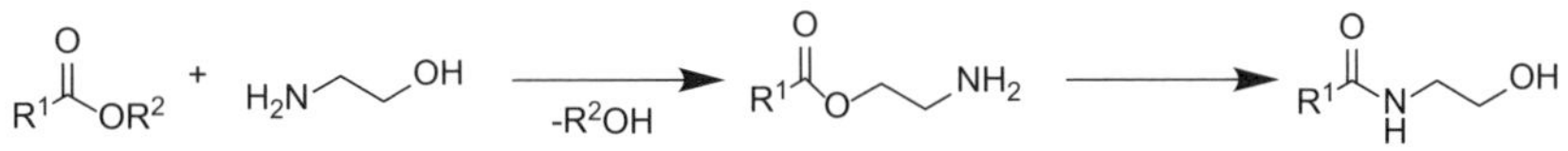

Scheme 4.3 Amidation of esters with 1,2-aminoalcohols.

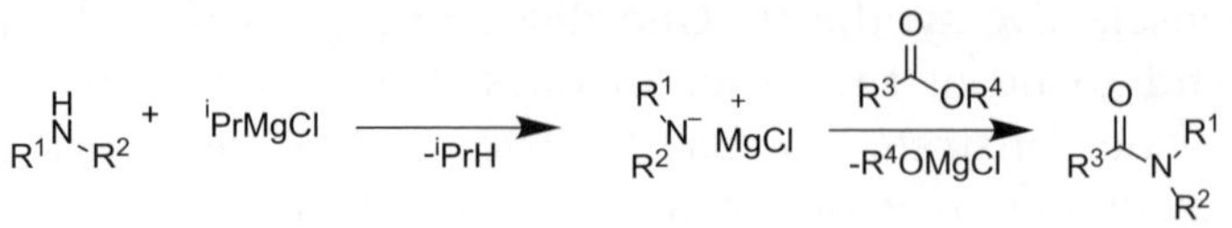

Scheme 4.4   Amidation of esters with amines and Grignard reagents.

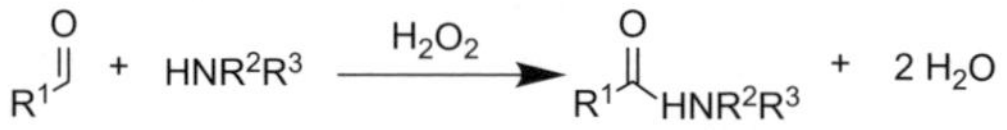

Scheme 4.5   Amide synthesis from aldehydes and amines with hydrogen peroxide.

the ester too (Scheme 4.4). The reactor had three feed streams, consisting of the coupling reagent, the amine and the ester, all in tetrahydrofuran (THF). First the Grignard reagent, isopropyl magnesium chloride and amine streams combined to form the magnesium amine, which was then mixed with the ester feed. High temperatures could not be used to raise the reaction rates as solid formation occurred above 40 °C; therefore the reaction was conducted at room temperature, with increased residence time.

4.4.2.3   Aldehyde.   The amidation of aldehydes with amines has been achieved by Tank and co-workers in a batch process with the addition of only hydrogen peroxide, a green oxidant.[110] The key step is the formation of the hemiaminal, which is an equilibrium process and therefore high concentrations of both aldehyde and amine are desirable. The reaction was run neat with 50 wt% aqueous hydrogen peroxide at temperatures between (55 and 75) °C. The reaction was adapted for a flow reactor by Liu and Jensen,[111] with an excess of amine required to compensate for the dilution by the acetonitrile solvent and a raise in temperature to 90 °C to shorten the reaction time. In both cases only the reaction of secondary amines with aromatic aldehydes was reported and it is likely that the use of primary amines results in imine formation (Scheme 4.5).

The formation of primary amides from aldehydes and aqueous ammonia was achieved by Yamaguchi and colleagues[112] over a manganese-oxide based molecular sieve, OMS-2 at 130 °C under 0.3 MPa oxygen. However, as the intermediate is a nitrile, the scope was limited to primary amides, and aliphatic aldehydes were more difficult to oxidise than aromatic aldehydes.

4.4.2.4 Alcohol. Over the last decade, green methods of amide formation from alcohols and amines with supported metal nano-particles have been developed. Despite catalyst similarities, the re-action pathways can be very different, but all involve first oxidation of the alcohol to the aldehyde (Schemes 4.6 and 4.7). For amidation with primary amines, a major competitive reaction is imine for-mation, from the elimination of water from the hemiaminal (Scheme 4.8).

Two catalysts are known for dehydrogenative amide synthesis, producing only hydrogen gas as a by-product. Shimizu and co-workers[113] developed an alumina-supported silver nanoparticle catalyst (Ag/Al_2O_3), while Zhu and colleagues[114] developed a hydro-talcite-supported catalyst of gold nanoparticles (Au/HT). Both catalysts were used under basic conditions; however, mechanistic studies indicated that the amide synthesis over the silver catalyst proceeded *via* a bound aldehyde intermediate, while the gold catalyst appeared to further oxidise the alcohol to an ester before reaction with the amine. Despite milder reaction conditions of between 70 and 90 °C rather than refluxing toluene, Au/HT appears to have higher turnover frequencies than Ag/Al_2O_3, but the reaction scope is limited

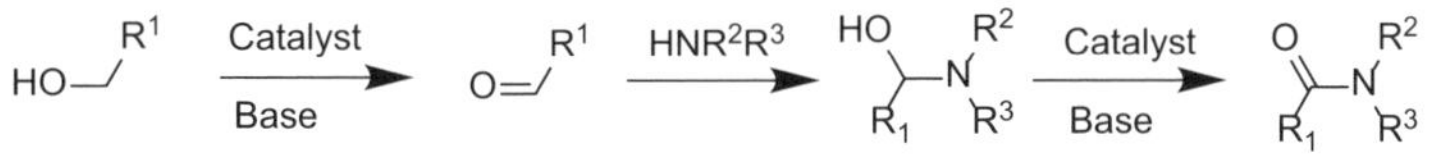

Scheme 4.6 Amide synthesis from alcohol and amines for the Ag/Al_2O_3, Au/DNA and Au-Co/PICB systems. Ag/Al_2O_3 produces hydrogen as a by-product, while reactions with oxygen produce water. PICB, polymer-incarcerated carbon black.

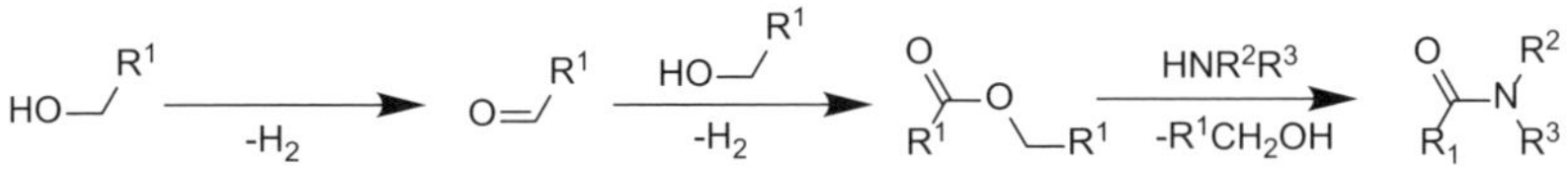

Scheme 4.7 Reaction pathway for the amidation of alcohols over Au/HT. HT, hydrotalcite.

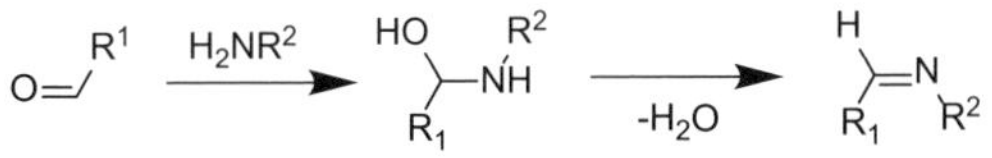

Scheme 4.8 Competitive imine formation from primary amines.

to cyclic secondary amines. One interesting observation made by Zhu and colleagues is that, under an oxygen atmosphere, the Au/HT catalyst completed the amidation of benzyl alcohol and morpholine in 3 hours at room temperature, as opposed to 1 day at 90 °C under argon. However, this great improvement was not investigated further.

Two supported gold nanoparticle catalysts for alcohol amidation under oxygen are described in the literature: gold on DNA from Wang and colleagues,[115] and a gold/cobalt alloy on polymer-incarcerated carbon black (PICB) from Soulé and co-workers.[116] Both catalysts demonstrated good yields for aromatic alcohols and moderate yields for aliphatic alcohols, under milder reaction conditions than for the dehydrogenation catalysts (40–50 °C) and at lower metal loadings. In addition the solvent for the Au/DNA catalyst was water, which is desirable as a green solvent. However, conversions were low for amines other than substituted anilines and piperidine, while both catalysts struggled with hindered amines.

A 'one-pot' two-stage synthesis route to amides was described by Kegnæs and co-workers[117] over Au/TiO_2 with oxygen and 25 mol% base. The oxidation of the alcohol was carried out in methanol, so that the first step was the synthesis of the methyl ester from the alcohol, with subsequent addition of five equivalents of amine to form the amide (Scheme 4.9). The reaction was conducted at low temperatures, unheated for the first step and 65 °C for the second. But while imine formation was avoided, the reaction required several days for moderate yields, possibly due to the gold loading of about 0.2 mol%, an order of magnitude lower than most catalysts. In addition the reaction scope was confined to primary, unhindered amines.

Another two-stage synthesis route was developed by Liu and Jensen,[118] who built a continuous flow system with two reactors (Scheme 4.10). In the first reactor, alcohol oxidation to the amide

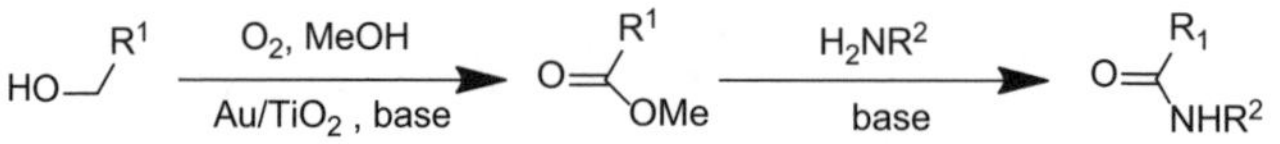

Scheme 4.9 Two-stage synthesis of amides from alcohols and amines in methanol.[117]

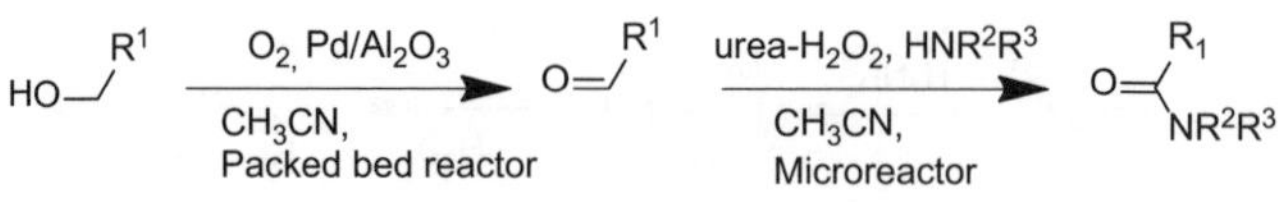

Scheme 4.10 Continuous flow synthesis route.[118]

occurred over a Pd/Al$_2$O$_3$ packed bed with molecular oxygen at 80 °C. The gas-phase oxygen was then removed with a membrane before the outlet stream was mixed with amine and urea-hydrogen peroxide, and fed into a microchannel reactor with a residence time of 22 minutes at 90–130 °C for amide formation. Despite the short residence times, the low flow rate of 5 µL min^{-1} gave a productivity for the system, but not the catalyst, equivalent to the PICB-Au/Co batch reaction above. The reaction scope was also very limited; only cyclic secondary amines such as morpholine and aromatic alcohols were tested, and the system was reported to be unsuitable for aliphatic alcohols.

4.4.3 OH Activation for Nucleophilic Substitution

The direct alkylation of an amine with an alcohol requires the activation of the OH group of the alcohol to facilitate a nucleophilic attack. One facile activation is the oxidation of the alcohol to the aldehyde and the subsequent nucleophilic substitution.

Alcohols are relatively safe and non-toxic compared with other starting materials such as alkyl halides and can be derived from natural products, leading to the possibility of sustainable sourcing. These properties make them an ideal feedstock, but while reactions of alcohols with electrophiles are well known, nucleophilic substitutions at the hydroxyl position to change the functionality tend to involve reagents that transform the hydroxyl into a good leaving group, with large quantities of by-products. In this section less wasteful methods of substitution of the alcohol group, effectively using alcohol as an alkylating reagent, will be discussed (Scheme 4.11).

One area which has been extensively studied is the formation of a carbon–carbon bond between a 1,3-diketone and the alkyl group of an alcohol (Scheme 4.12).[120] In fact for certain reagents a catalyst is not needed and coupling proceeds in high-temperature water as

$$Nu–H \ + \ R–OH \longrightarrow Nu–R \ + \ H_2O$$

Scheme 4.11 Nucleophilic substitution of hydroxyl group.

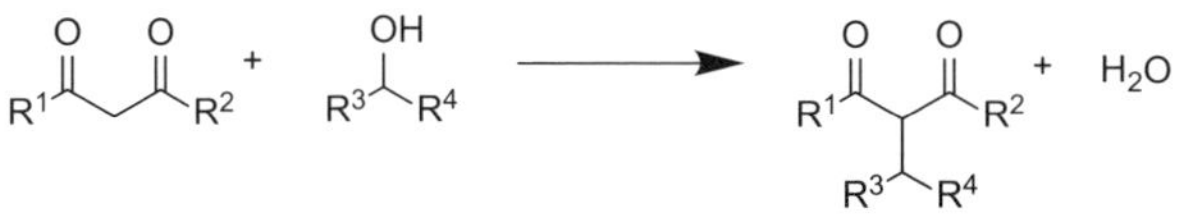

Scheme 4.12 Nucleophilic substitution of hydroxyl group.

demonstrated by Hirashita[119] and co-workers, who ran the reaction in a vessel heated to 220 °C for 6 hours. Unfortunately a mixture of products was obtained due to deacylation of the 1,3-diketone (Scheme 4.13). This process is normally base catalysed but occurs under these conditions, as at high temperatures, the ionic product if the water is greater and it can itself act as a base. The same conditions were also used for the alkylation of electron-rich aromatic rings with a range of primary, secondary and tertiary allylic or benzylic alcohols; see, for example, Scheme 4.14.

Liu and colleagues employed a catalyst, $HClO_4$ supported on silica, for the reaction of alcohols with 1,3-diketones at a lower temperature of 70 °C, either without solvent or in toluene.[121] This needed a longer reaction time of 17 hours and two equivalents of the alcohol, but no deacylation was reported. It was found, however, that the catalyst racemised a chiral alcohol added, though if the substitution mechanism is an S_N1 type, the stereochemistry would not be preserved anyway.

Amandi and co-workers[122] used a flow reactor with supercritical carbon dioxide for the Friedel-Crafts type alkylation of phenols with isopropanol (Scheme 4.15). Two different catalysts were trialled in a packed bed: Nafion® SAC-13, a commercial fluorosulfonic acid polymer on amorphous silica and γ-alumina, used at different temperatures of 175 °C and 275 °C, respectively. Despite the lower

Scheme 4.13　Deacylation of the 1,3-diketone.[119]

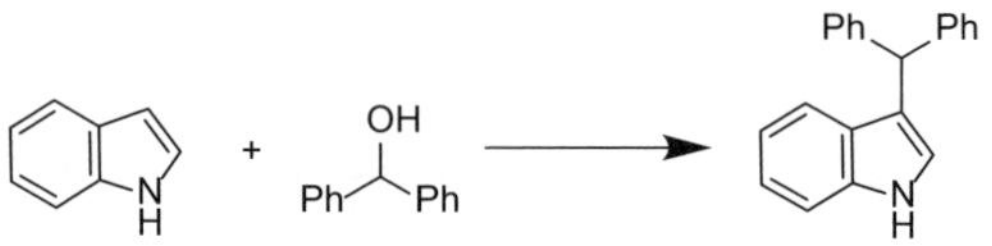

Scheme 4.14　Alkylation of indole.[119]

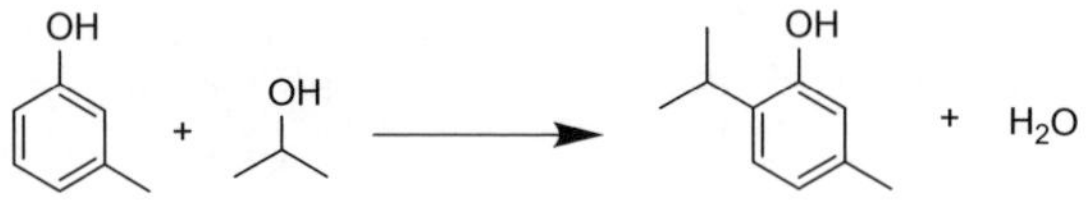

Scheme 4.15　Alkylation of m-cresol.[122]

temperature, the commercial catalyst gave higher conversions, but this was at the expense of selectivity, which was 61.2% to the main product shown in Scheme 4.15 compared with 78.9% for γ-alumina, with alkylation at other points on the ring and even dialkylations also detected. The mechanism appears to involve the coupling of the alcohol with the phenol to form an ether, which then undergoes a Fries arrangement to form the *ortho*-substituted product, and therefore the scope of the reaction is limited to phenols. Another problem is that water is produced as a by-product, but also inhibits the reaction. Amandi and colleagues prevented the formation of water by replacing the alcohol with an alkene as the alkylating agent. However, methods for extracting the water exist, such as the inclusion of molecular sieves in the catalyst bed, but these can be more difficult to implement in a flow system.

Sanz and colleagues[123] used allylic alcohols as alkylating agents for a range of different substrates over polymer-bound *p*-toluenesulfonic acid (Scheme 4.16). This included reactions with other alcohols to form ethers, primary sulphides and amines to form secondary sulphides and amines, as well as carbon–carbon bonds formed with electron-rich aromatic rings, diketones and silanes (with the loss of Me_3SiOH instead of H_2O), all formed under air at 20 °C in acetonitrile.

The alkylation of amines was also studied by Cui and co-workers[124] over a silver-molybdenum oxide at 160 °C for 12 hours with a 'borrowing hydrogen' strategy (Scheme 4.17). The direct alkylation of an amine with an alcohol requires the activation of the OH group of the alcohol to facilitate a nucleophilic attack. One facile activation is the dehydrogenation of the alcohol to the aldehyde and the subsequent addition of the amine to form the hemiaminal, elimination

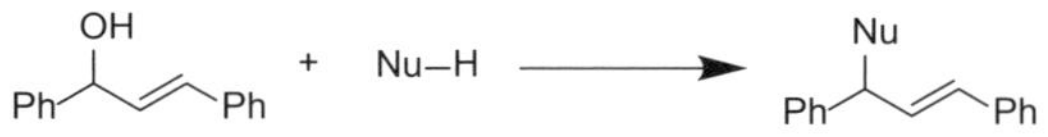

Scheme 4.16   Alkylation of *m*-cresol.[123]

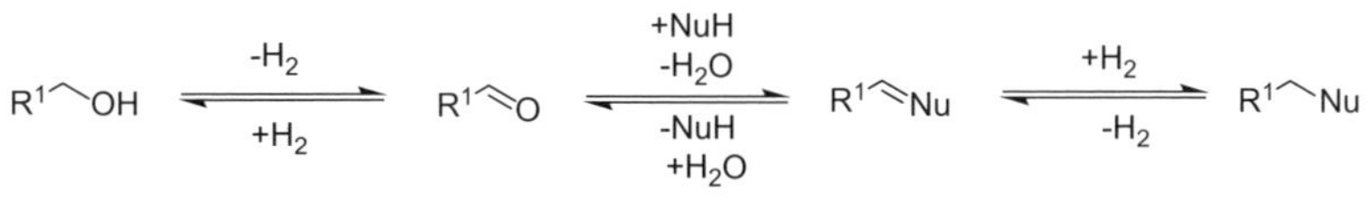

Scheme 4.17   'Borrowing hydrogen' strategy, where $Nu = NHR^2$, $NHCOR^2$, $NHSO_2R^2$ or CH_2COR^2.[124]

of water to the imine, and then the re-hydrogenation of the imine to obtain the substituted amine. Some hydrogen is lost from the catalyst before re-hydrogenation and therefore the reaction is run with a five-fold excess of the alcohol, with the aldehyde being detected as impurity. The reaction also requires 20 mol% of base and, although a range of alcohols were used, only aromatic amines were alkylated: Using *n*-octylamine as a substrate gave a selectivity of >99% to the imine. The catalyst also carried out a range of other alkylation reactions, transforming primary aromatic amides and sulfonamides into their secondary counterparts with benzylic alcohols, and lengthening the chains of methyl ketones.

A more atom-efficient method for amine alkylation was developed by Zotova and colleagues,[125] also employing the 'borrowing hydrogen' strategy, which did not require either amine or alcohol to be in excess or need any added base. The reaction was carried out with a commercial flow reactor run in batch recycle mode, with a packed bed of Au/TiO$_2$, with the reagents in toluene at 50 bar with temperatures between 180 and 200 °C. The alkylation of various secondary amines with benzyl alcohols were investigated with selectivity to the tertiary amine close to 100% in most cases.

4.4.4 Reduction of Amides to Amines

There has not been much work in this area in terms of heterogeneous catalyst development. Product selectivity is an issue as the addition of hydrogen forms a hemiaminal with the loss of either ammonia followed by further hydrogenation to an alcohol, or loss of water to form an imine or aldimine and further hydrogenation to the amine (Scheme 4.18). In addition, the imine can react with the amine product to form more substituted amines.

One promising set of catalysts are the bimetallic clusters of ruthenium or rhodium with molybdenum or rhenium described by Beamson and colleagues.[126] These are formed *in situ* from their

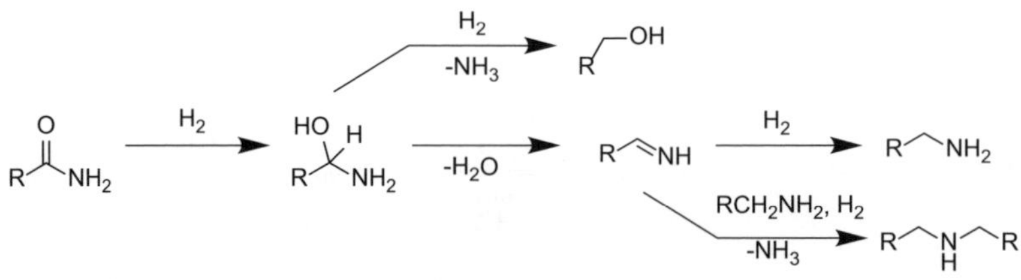

Scheme 4.18 Chemical reduction of amides.

carbonyls, although the clusters can be recovered and reused. The reaction is very atom economical as the reductant is hydrogen gas. However, hydrogen pressures of up to 10.0 MPa are needed with high temperatures of between (145 and 160) °C for good conversions and selectivity, with reaction times of 16 hours needed to convert 1.85 mmol of amide. The product selectivity to the amine is between (80 and 90)%, with alcohol as the major side product. Interestingly, primary amides were more easily reduced than tertiary amides, thought to be due to the availability of a lower energy route through a nitrile intermediate, while conversions of secondary amides were poor. Another major drawback was that unsaturated carbon–carbon bonds, including benzene rings were reduced.

A similar catalyst of PtRe/TiO$_2$ was used for the reduction of a tertiary amide, *N*-methylpyrrolidin-2-one, under slightly milder conditions of 120 °C and 2.0 MPa H$_2$.[127] In this case, the reaction was found to be hydrogen limited, suggesting mass transfer issues, a problem that could be solved by careful reactor design. A larger scope was investigated for a similar catalyst of graphite-supported platinum–rhenium, which could be used at temperatures down to 100 °C, with molecular sieves and 3.0 MPa hydrogen gas.[128] However, even at higher temperatures, 20 hours were needed to reduce 1 mmol *N*-acetylpiperidine with 12 mol% metal, and while the reduction of a range of tertiary and secondary alcohols was carried out successfully, primary amides gave only the secondary amine, and again there was complete reduction of any carbon–carbon double bonds in the substrate.

Amide reduction under milder conditions may be achieved with a silane reductant. Motoyama and co-workers[129] designed an organometallic ruthenium cluster for use with polymethylhydroxysilane. The silane polymer had two roles: Firstly as the final reductant and secondly to encapsulated the homogeneous catalyst to enable separation of ruthenium from the product. However, it was found that the used, encapsulated complex was still reactive and able to reduce 1 mmol of *N,N*-dimethyl-3-phenylpropionamide at 30 °C in 15 hours, a higher turnover frequency per mole of metal than the bimetallic catalysts.

Another example of silane for amide reduction was presented by Mikami and collaborators[130] for a catalyst of gold nanoparticles supported on hydroxyapatite with dimethylphenylsilane at 110 °C under an argon atmosphere. While less atom-efficient than the bimetallic complexes above, turnover frequencies of 416 h^{-1} were measured compared with 4.7 h^{-1} for Ru/Mo, and aromatic rings were not reduced.

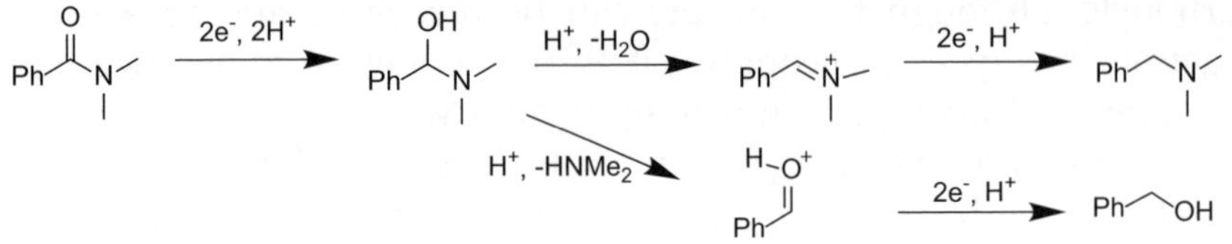

Scheme 4.19 Electrochemical reduction of amides.

Apart from the sparse chemical methods of amide reduction above, a review by Popp and Schultz[131] in 1962 notes that other groups have used electrochemical methods (Scheme 4.19), with lead cathodes in acidic solutions. More recently, Rix and co-workers have successfully carried out the reduction of amides in a recirculating electrochemical flow cell with boron doped diamond electrodes. To achieve amide reduction to amines, rather than alcohols and aldehydes, high current densities of 2000 A m^{-2} at -2.8 V were required, with an electrolyte of 1 M sulfuric acid, either aqueous or in a methanol–water mixture, pH 0. Under these conditions, large volumes of hydrogen gas were evolved, and alcohols and aldehydes were still formed. Despite this, a range of amides were reduced to amines, with primary amides being more easily reduced than secondary and tertiary, and electron donating groups giving increased conversions. Intramolecular double bonds and triple bonds were not affected.

4.4.5 Oxidation/epoxidation reactions without the use of chlorinated solvents

Of the many transformations that could have been listed it is surprising that oxidation/epoxidation reactions appear so high up the table.

In 2006, colleagues from GSK, AstraZeneca and Pfizer carried out a survey of the chemists deployed in the manufacture of 128 drug candidate molecules in order to identify synthesis/technology gaps.[132] They concluded that:

> '…there are relatively few atom efficient, chemo-selective and environmentally acceptable oxidation methods … oxidations are often designed out of syntheses. The discovery of new chemo-selective oxidations, particularly if catalytic, would greatly increase flexibility in synthetic design.'

The key issues here are the need for selective and robust catalysts, and for effective heat and mass transfer as oxidations/epoxidations tend to be rather exothermic. How would sufficient oxidant be delivered and/or removed, keeping in mind that organic solvents are flammable? What type of oxidant should be used?

Keeping with our premise of process-oriented discovery, which will lead us to the most sustainable solution for this transformation, this discussion focuses on the most desirable terminal oxidant, namely oxygen (or air) and, alternatively, for ease of delivery, hydrogen peroxide (H_2O_2).

The term 'oxidation' may refer to a large number of reactions, including dehydrogenations and the addition of oxygen to a molecule. Therefore in this section only a few heterogeneous catalysts or processes, promising for their wide scope or interesting methods, are discussed to illustrate the range of possibilities.

4.4.5.1 Supported Metals. The catalytic activity of supported metals, especially nanoparticles, is a field that has expanded rapidly over the past decade. The most commonly used metal nanoparticles for oxidation are gold, palladium, platinum and ruthenium supported on a wide range of materials from metals oxides to polymers. These are described in many reviews,[133-136] with a few examples for reactions discussed below.

One facile transformation and therefore an area of intense research for supported metal nanoparticles has been their use as catalysts for the partial oxidation of alcohols. Very high catalyst turnover frequencies are produced: it was reported that a gold supported on a gallium–alumina oxide catalyst reached turnover frequencies of 25 000 per hour for the oxidation of 1-phenylethanol to acetophenone under an O_2 atmosphere, 160 °C, solvent free[137] (Scheme 4.20).

Primary alcohols are first oxidised to the aldehyde, but can then be oxidised further. If the reaction is carried out in methanol solvent, the methyl ester is produced (Scheme 4.21). The reaction kinetics for methyl esterification of benzyl and aliphatic alcohols in methanol with 0.5 MPa of O_2 over Au/β-Ga$_2$O$_3$ were studied by Su and others,[138] who concluded that the oxidation of the alcohol was the rate determining step, with aldehyde oxidation facile in comparison. Studies of other gold catalysts, such as ethyl acetate over Au/TiO$_2$ by Jørgensen

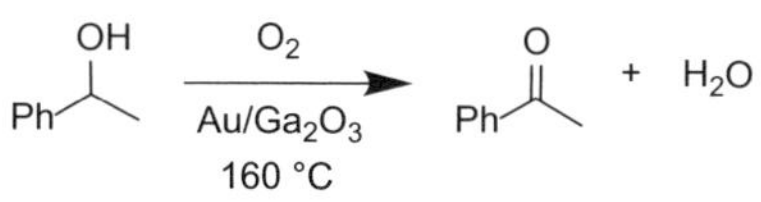

Scheme 4.20 Oxidation of 1-phenylethanol over a supported gold catalyst.[137]

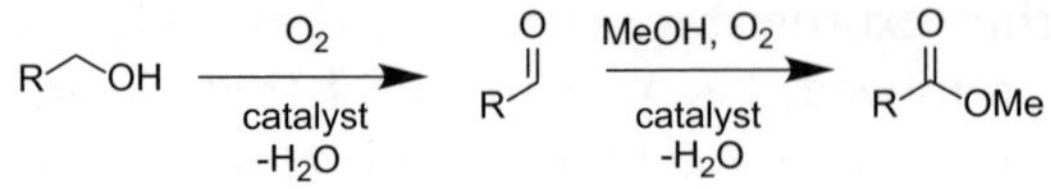

Scheme 4.21 Oxidation of primary alcohols to aldehydes and methyl esters.

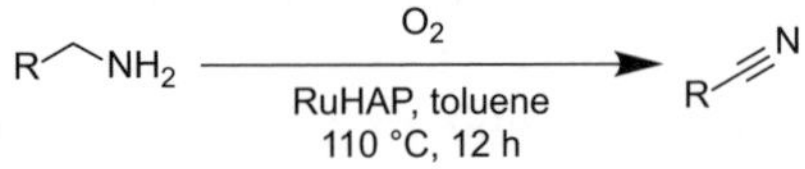

Scheme 4.22 Oxidation of primary amines to nitriles over RuHAP.

and colleagues,[139] gave the same result. To obtain the aldehyde, the reaction is normally carried out in an inert solvent, such as water. Uozumi and Yamada[140] synthesised a catalyst of palladium nano-particles supported on an amphiphilic resin and refluxed benzyl alcohol in water for 90 minutes at 1 mol% loading of palladium to produce 97% benzaldehyde. In addition to choosing green synthesis methods, Uozumi and Yamada chose a green extraction method using supercritical carbon dioxide to separate out the carboxylic acid product for a similar catalyst.

Amines can also undergo oxidation over supported metals, as shown by Mori and co-workers[141] for a catalyst of ruthenium mono-dispersed on hydroxyapatite (RuHAP), with the amine refluxed in toluene for 12 hours under an oxygen atmosphere (Scheme 4.22). A large scope was investigated, including aliphatic and benzylic amines, and it was found that primary amines were oxidised to nitriles with selectivities over 95% for most reagents, although up to 10% of an imine side product was formed for certain species. One interesting feature is that hydroxyl groups on the amine do not react, thought to be due to the greater binding strength of the amine group to the catalyst. Secondary amines were oxidised to imines, with some hydrolysis to the aldehyde seen. Interestingly, with the addition of water to the reactor, the hydration of the nitriles occurred to yield the corresponding primary amides. Based on this observation, a one-pot synthesis route from the amine to the primary amide over RuHAP was designed (Scheme 4.23).

A more difficult reaction to carry out selectively is epoxidation of alkenes, which is undertaken on an industrial scale for ethylene over silver catalysts, but fewer examples exist for green oxidations of other substrates. Hughes and co-workers[142] investigated the epoxidation of cyclohexene and cyclooctene over gold supported on carbon, under

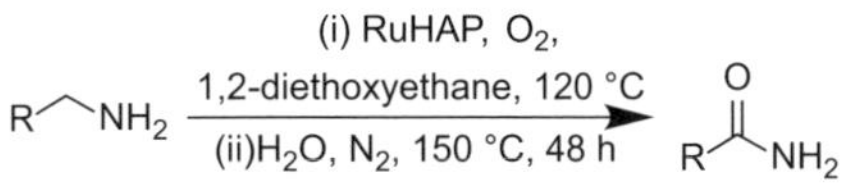

Scheme 4.23 Oxidation of primary amines to primary amides over RuHAP.

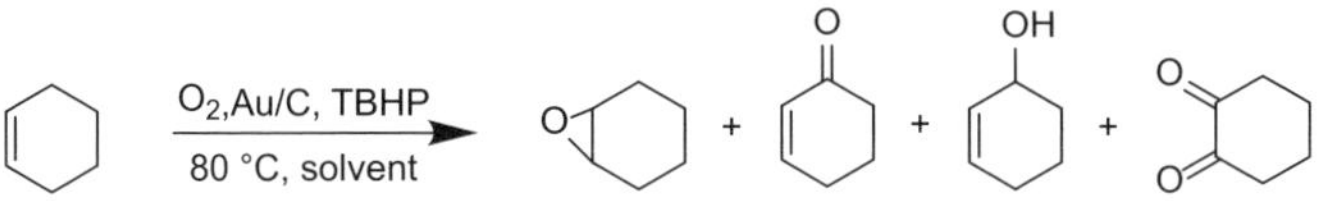

Scheme 4.24 Oxidation/epoxidation of cyclohexene over Au/C.

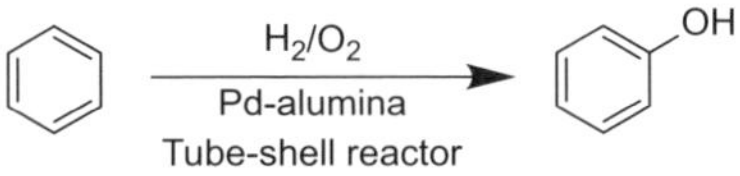

Scheme 4.25 Hydroxylation of benzene in a tube-and-shell reactor.

oxygen but requiring small amounts of *tert*-butyl hydroperoxide (TBHP) as a radical initiator. Scheme 4.24 shows the many potential products of cyclohexene oxidation, including the epoxide, allyl alcohol, enone and 1,2-diketone. The solvent choice had a great effect on the reaction selectivity, although no kinetic isotope effect was observed. The use of polar solvents such as water leads to the break-up of the ring, while in substituted benzenes, a mixture of the products in Scheme 4.25 was observed. A reaction without a solvent attempted with cyclooctene produced a good selectivity of about 80% to the epoxide, with evidence that small amounts of bismuth added to the catalyst may further enhance selectivity. A turnover frequency of 35 moles of product produced per mole of gold per hour was calculated, which compared favourably with other systems, with Hughes and co-workers stating that further improvements could be made through changes to the reactor system.

An interesting tube-and-shell reactor was built by Niwa and colleagues[143] to control the oxidation of benzene to phenol (Scheme 4.25) and toluene to cresols. The main feature was a porous alumina membrane coated with palladium to safely produce radical oxidants from oxygen and hydrogen gas. The bulk of the two gases were kept separate by the membrane, but hydrogen could adsorb onto the membrane and split into its atoms. These adsorbed atoms then diffused through to the other side and combined with the oxygen to produce peroxide and hydroxyl radicals, which then reacted with the aromatic substrate. This process enabled selectivities to phenol of over 80% at conversions of 10–15% at 150 °C, with a possible throughput of 1.5 kg of phenol per kilogram of catalyst per hour.

4.4.5.2 Carbon Catalysts. Scaling up precious metal catalysts can be very expensive and in addition leaching can occur, leading to contamination of the product by the metal, requiring extra process steps for purification. As a result, there has been increasing interest in developing metal-free catalysts, such as carbon-based networks.

Alkane oxidation is a difficult reaction, partly because alkanes do not adsorb well onto many catalysts.[145] Dhakshinamoorthy and colleagues[146] investigated the oxidation of tetralin over graphene oxide and boron and nitrogen-doped graphene under oxygen at 120 °C, without solvent (Scheme 4.26). As expected for alkane oxidation, yields were low with conversions of about 36% in 24 hours for the doped graphenes, but the selectivity to the alcohol and ketone was close to 90%. Other substrates such as cyclooctane and diphenylmethane gave similar conversions. Therefore graphenes are promising catalysts for the metal-free oxidation of alkanes.

The use of graphene oxide for the oxidative homocoupling of primary amines to form imines (Scheme 4.27) was investigated by Huang and colleagues[147] who chose to use a pressure of 0.5 MPa of molecular oxygen at 100 °C for a range of amines, with conversions of about 5 mmol oxidised over 2–8 hours, although the catalyst loading was high (50 wt%). The reaction was successful for many aromatic amines, even sterically hindered species, but aliphatic amines did not couple with each other. Interestingly, cross-coupling of benzylamine with other amines was achieved with 80–100% selectivity by adding the less reactive amine in a four-fold excess.

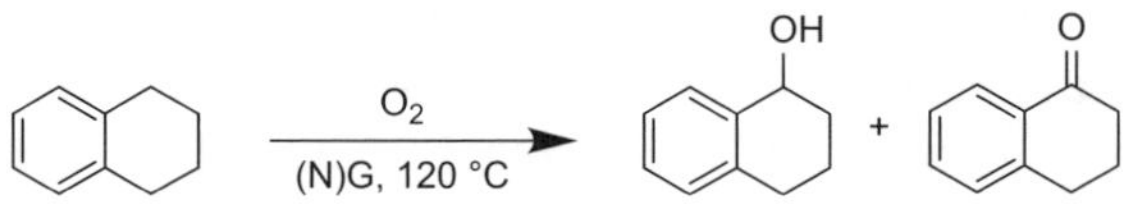

Scheme 4.26 Oxidation of tetralin over N-doped graphene.

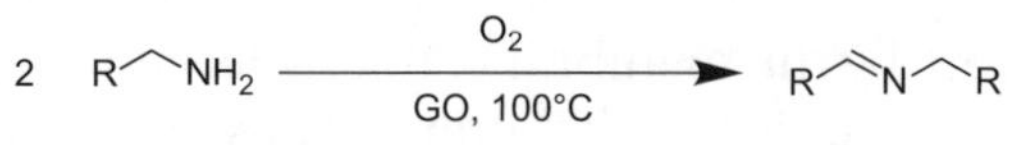

Scheme 4.27 Oxidation of amines to coupled imines over graphene oxide.

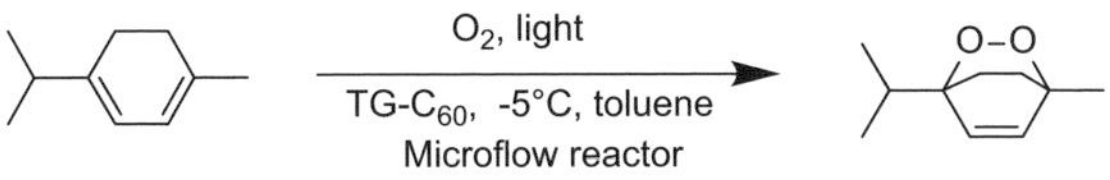

Scheme 4.28 Oxidation of α-terpinene to ascaridole.[144]

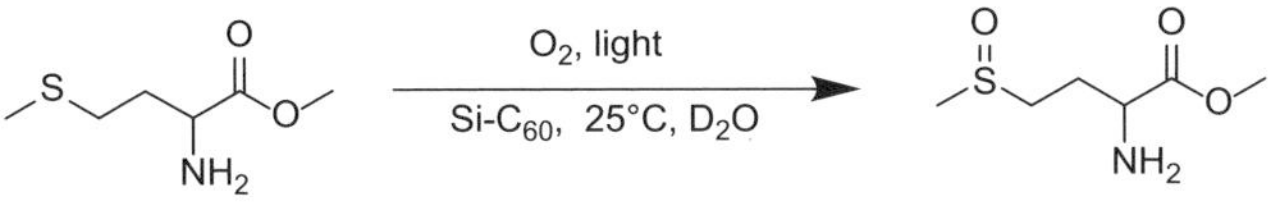

Scheme 4.29 Oxidation of L-methionine to methionine sulfoxide.[144]

Another form of carbon, supported C_{60} fullerene, was employed for two different photo-oxidation reactions in a microflow system. Carofiglio and colleagues[144] built a flow reactor with two inlets: one for oxygen gas, the other for the substrate, with the reaction channels cut into glass plates and illuminated with a 300 W tungsten lamp. The first reaction studied was the photo-oxidation of α-terpinene to ascaridole (Scheme 4.28), over Tentagel®-supported fullerene. Maximum flow rates of 1 mL h^{-1} of the substrate in toluene and 2 mL h^{-1} of molecular oxygen were used, with residence times between 27 s and 109 s under the light source. The temperature was kept low at -5 °C to raise the solubility of oxygen in toluene, but while almost complete conversion of the α-terpinene was achieved, the yield of ascaridole was at most only 53%, thought to be due to its photodegradation, suggesting that further optimisation of the residence time needs to take place.

The second reaction studied was the oxidation of L-methionine to methionine sulfoxide, over C_{60} fullerene supported on silicon dioxide, with deuterated water as the solvent (for ^{1}H-NMR analysis) (Scheme 4.29). Shorter residence times were used of 30–40 s, and only a small drop in conversion was seen on raising the temperature to 25 °C compared with the larger change for the previous reaction. One explanation is that the singlet oxygen produced by the light has a shorter lifetime in toluene than in water; hence the first reaction is oxidant limited.

4.4.5.3 Metal Oxides. Mixed metal oxide catalysts are used extensively in the chemicals industry[148] due to their low cost and tolerance to harsh process conditions such as high temperatures.

Although often used as a support for metal nanoparticles, titanium oxide itself can be activated for alcohol oxidation. When TiO_2 is

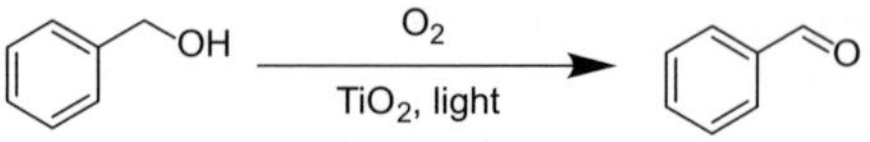

Scheme 4.30   Photooxidation of benzyl alcohol over TiO_2.

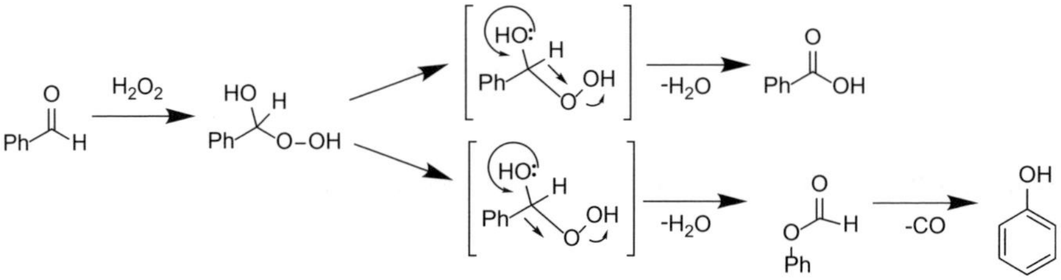

Scheme 4.31   Baeyer–Villiger oxidation of aromatic aldehydes.

irradiated with ultraviolet light corresponding to its band gap, electrons can be promoted from the valence to the conduction band, and then transferred to oxygen molecules, which are then activated for reaction. Higashimoto and colleagues[149] found that substituted benzyl alcohols in acetonitrile could be oxidised over TiO_2-anatase under oxygen to the aldehydes with almost perfect selectivity, except for the 4-hydroxy-substituted alcohol, which bound strongly to the catalyst and so was subjected to further irradiation leading to degradation (Scheme 4.30).

The Baeyer–Villiger oxidation involves the insertion of oxygen into a carbon–carbonyl bond, transforming ketones into esters or lactones. This process conventionally uses peracids, but these are unstable and dangerous to store. Therefore Renz and colleagues[150] investigated a system for generating the peracid in situ from hydrogen peroxide by employing a reusable catalyst of tin atoms in a zeolite framework. This method was successfully employed for the oxidation of a range of ketones and aldehydes, was well as lactone formation from cyclic ketones, with a turnover number of 88 at 80 °C in dioxane over 7 hours being obtained for cyclohexanone. This method gave excellent selectivity to the lactone of over 98%, but had one major drawback compared with the peracid systems in that a mixture of the two region-isomers of the lactone is obtained. This can be seen more clearly in the Baeyer–Villiger oxidation of aromatic aldehydes (Scheme 4.31). After the nucleophilic attack of the aldehyde to form the tetrahedral intermediate, the carbonyl bond re-forms with one of two groups migrating to the inserted oxygen. Either migration of the hydride group occurs to form the carboxylic acid, or migration of the phenyl group occurs to form the ester, with subsequent hydrolysis to the phenol. The selectivity of the group migration depends upon the substituents on the phenyl ring, but the phenol to ester ratio can be altered by changing the reaction time or solvent.

4.5 CONCLUSIONS

Within the vast arena of catalytic processes and reactors for sustainable processes, it has been necessary to make certain choices. We have tried to put sustainable processes in context and concentrated on recent developments of interesting heterogeneous catalytic reaction concepts. As many novel reactors and contactors are clearly of a flow type, this subsequently led us to discuss and make some comments on a number of aspects of modern flow chemistry. For many reasons, but mainly to do with the need for catalyst separation and utilisation of catalysts in continuous fine and pharmaceutical flow processes, we have focused on the recent development of heterogeneous catalytic systems. The main applications/key transformation challenges were selected from the list produced by the ACS GCI Pharmaceutical Roundtable, which today is as topical as it was when it was first published. We have discussed recent discoveries and potential approaches for the four top transformations as a current snapshot of the state-of-the-art in these areas.

Many exciting new discoveries, concepts and solutions have emerged and are continuing to emerge. With the advent of flow chemistry and the slow breaking down of discipline boundaries between chemistry and chemical engineering, we can envisage a step change in the speed of translation from discovery to final process.

As to prove the above assertion, the Massachusetts Institute of Technology (MIT) recently spun out the company, Continuus Pharmaceuticals, which is the result of a ten year, 65×10^6 collaboration between MIT and Novartis on the development of integrated continuous manufacturing technology for pharmaceutical products. This is the first ever demonstration of a continuous processing plant that combines active pharmaceutical ingredient (API) synthesis all the way to the final tablet manufacture.[151]

Whilst delivery of this process was a tour de force (if you believe those involved), we are only at the beginning of a steep learning curve—and with learning comes reduced cost and increased speed to market.

REFERENCES

1. J. Elkington, A. MacGillivray, L. Varcoe, J. Hailes, A. Dimmock, *The LCA Sourcebook. A European Business Guide to Life-Cycle Assessment*, Sustainability, SPOLD, and BiE, UK, 1993.
2. P. T. Anastas and J. C. Warner, *Green Chemistry: Theory and Practice*, Oxford University Press, New York, 1998.
3. P. Anastas and J. Zimmerman, *Environ. Sci. Technol.*, 2003, **37**, 94A.
4. P. Anastas, *Chemsuschem*, 2009, **2**, 391.
5. E. Beach, Z. Cui and P. Anastas, *Energy Environ. Sci.*, 2009, **2**, 1038.
6. I. Horvath and P. Anastas, *Chem. Rev.*, 2007, **107**, 2167.
7. C. Li and P. Anastas, *Chem. Soc. Rev.*, 2012, **41**, 1413.
8. M. Mulvihill, E. Beach, J. Zimmerman, P. Anastas, A. Gadgil and D. Liverman, *Annual Rev. Environ. Resour.*, 2011, **36**, 271.
9. C. Jimenez-Gonzalez, A. Curzons, D. Constable and V. Cunningham, *Int. J. of Life Cycle Assess.*, 2004, **9**, 114.
10. C. Jimenez-Gonzalez, P. Poechlauer, Q. Broxterman, B. Yang, D. Ende, J. Baird, C. Bertsch, R. Hannah, P. Dell'Orco, H. Noorrnan, S. Yee, R. Reintjens, A. Wells, V. Massonneau and J. Manley, *Org. Process Res. Dev.*, 2011, **15**, 900.
11. C. Jimenez-Gonzalez, C. Ponder, Q. Broxterman and J. Manley, *Org. Process Res. Dev.*, 2011, **15**, 912.
12. C. Jimenez-Gonzalez, D. Constable and C. Ponder, *Chem. Soc. Rev.*, 2012, **41**, 1485.
13. C. Jimenez-Gonzalez and P. Dell'Orco, *Chim. Oggi*, 2013, **31**, 62.
14. C. Jimenez-Gonzalez, C. Ollech, W. Pyrz, D. Hughes, Q. Broxterman and N. Bhathela, *Org. Process Res. Dev.*, 2013, **17**, 239.
15. C. H. Bartholomew and R. J. Farrauto, *Fundamentals of Industrial Catalytic Processes*, 2nd edn., John Wiley & Son, New York and Chichester, 2005.
16. F. Visscher, J. van der Schaaf, T. Nijhuis and J. Schouten, *Chem. Eng. Res. Des.*, 2013, **91**, 1923.
17. Z. Anxionnaz, M. Cabassud, C. Gourdon and R. Tochon, *Chem. Eng. Process.*, 2008, **47**, 2029.
18. P. Chin, W. Barney and B. Pindzola, *Curr. Opin. Drug Discovery Dev*, 2009, **12**, 848.
19. A. Ghanem, T. Lemenand, D. Della Valle and H. Peerhossaini, *Chem. Eng. Res. Des.*, 2014, **92**, 205.
20. T. Illg, P. Lob and V. Hessel, *Bioorg. Med.Chem.*, 2010, **18**, 3707.
21. M. Kashid and L. Kiwi-Minsker, *Ind. Eng. Chem. Res.*, 2009, **48**, 6465.

22. J. Moulijn, M. Kreutzer, T. Nijhuis, F. Kapteijn, B. Gates and H. Knozinger, *Adv. Catal.*, 2011, **54**, 249.

23. A. Sachse, A. Galarneau, B. Coq and F. Fajula, *New J. Chem.*, 2011, **35**, 259.

24. A. Rodrigues, C. Pereira and J. Santos, *Chem. Eng. Technol.*, 2012, **35**, 1171.

25. I. Baxendale, *J. Chem. Technol. Biotechnol.*, 2013, **88**, 519.

26. J. Brazier, *Platinum Metals Rev.*, 2012, **56**, 99.

27. C. Cullen, R. Wootton and A. de Mello, *Curr. Opin. Drug Discovery Dev.*, 2004, **7**, 798.

28. P. Diniz, L. de Almeida, D. Harding and M. de Araujo, *TRaC-Trends Anal.Chem.*, 2012, **35**, 39.

29. A. Gunther and K. Jensen, *Lab Chip*, 2006, **6**, 1487.

30. R. Hartman and K. Jensen, *Lab Chip*, 2009, **9**, 2495.

31. S. Haswell and V. Skelton, *TRaC-Trends Anal. Chem.*, 2000, **19**, 389.

32. V. Hessel, H. Lowe and F. Schonfeld, *Chem. Eng. Sci.*, 2005, **60**, 2479.

33. V. Hessel and Q. Wang, *Chim. Oggi*, 2011, **29**, 81.

34. V. Hessel and Q. Wang, *Chim. Oggi*, 2011, **29**, 70.

35. V. Hessel, D. Kralisch, N. Kockmann, T. Noel and Q. Wang, *Chemsuschem*, 2013, **6**, 746.

36. K. Kanno and M. Fujii, *J. Synth. Org. Chem. Jpn*, 2002, **60**, 701.

37. W. Lin, Y. Wang, S. Wang and H. Tseng, *Nano Today*, 2009, **4**, 470.

38. W. Melchert, B. Reis and F. Rocha, *Anal. Chim. Acta*, 2012, **714**, 8.

39. S. Newman and K. Jensen, *Green Chem.*, 2013, **15**, 1456.

40. T. Razzaq and C. Kappe, *Chem. – Asian J.*, 2010, **5**, 1274.

41. J. Ruzicka, *Collect. Czech. Chem. Commun.*, 2005, **70**, 1737.

42. P. Watts and S. Haswell, *Chem. Soc. Rev.*, 2005, **34**, 235.

43. P. Watts and S. Haswell, *Chem. Eng. Technol.*, 2005, **28**, 290.

44. P. Watts, *QSAR Comb. Sci.*, 2005, **24**, 701.

45. P. Watts and C. Wiles, *Org. Biomol. Chem.*, 2007, **5**, 727.

46. P. Watts and C. Wiles, *J. Chem. Res.*, 2012, 181.

47. D. Webb and T. Jamison, *Chem. Sci.*, 2010, **1**, 675.

48. J. Wegner, S. Ceylan and A. Kirschning, *Adv. Synth. Catal.*, 2012, **354**, 17.

49. G. Wild, C. Wiles and P. Watts, *Lett. Org. Chem.*, 2006, **3**, 419.

50. C. Wiles and P. Watts, *Future Med. Chem.*, 2009, **1**, 1593.

51. C. Wiles and P. Watts, *Chem. Commun.*, 2011, **47**, 6512.

52. F. Valera, M. Quaranta, A. Moran, J. Blacker, A. Armstrong, J. Cabral and D. Blackmond, *Angew. Chem., Int. Ed.*, 2010, **49**, 2478.

53. M. Irfan, T. Glasnov and C. Kappe, *Chemsuschem*, 2011, **4**, 300.

54. H. Amii, A. Nagaki and J. Yoshida, *Beilstein J. Org. Chem.*, 2013, **9**, 2993.
55. A. Kulkarni, *Beilstein J. Org. Chem.*, 2014, **10**, 405.
56. T. Glasnov and C. Kappe, *J. Heterocyclic Chem.*, 2011, **48**, 11.
57. C. Kappe and E. Van der Eycken, *Chem. Soc. Rev.*, 2010, **39**, 1280.
58. J. Pastre, D. Browne and S. Ley, *Chem. Soc. Rev.*, 2013, **42**, 8849.
59. D. Zhao and K. Ding, *ACS Catal*, 2013, **3**, 928.
60. I. Baxendale and M. Pitts, *Chim. Oggi.*, 2006, **24**, 41.
61. I. Baxendale, J. Hayward and S. Ley, *Comb. Chem. High Throughput Screening*, 2007, **10**, 802.
62. T. Glasnov and C. Kappe, *Macromol. Rapid Commun.*, 2007, **28**, 395.
63. E. Gjuraj, R. Kongoli and G. Shore, *Chem. BioChem. Eng. Q.*, 2012, **26**, 285.
64. S. Sadler, A. Moeller and G. Jones, *Expert Opin. Drug Discovery*, 2012, **7**, 1107.
65. S. Wolkenberg, W. Shipe, C. Lindsley, J. Guare and J. Pawluczyk, *Curr. Opin. Drug Discovery Dev.*, 2005, **8**, 701.
66. J. Knowles, L. Elliott and K. Booker-Milburn, *Beilstein J. Org. Chem.*, 2012, **8**, 2025.
67. M. Oelgemoller, N. Hoffmann and O. Shvydkiv, *Aust. J. Chem.*, 2014, **67**, 337.
68. M. Ralph, *Curr. Org. Chem.*, 2011, **15**, 2658.
69. G. Scholes, G. Fleming, A. Olaya-Castro and R. van Grondelle, *Nat. Chem.*, 2011, **3**, 763.
70. V. Cirkva and S. Relich, *Curr. Org. Chem.*, 2011, **15**, 248.
71. V. Cirkva and S. Relich, *Mini-Rev. Org. Chem.*, 2011, **8**, 282.
72. W. Zimmerman, *Chem. Eng. Sci.*, 2011, **66**, 1412.
73. K. Watts, A. Baker and T. Wirth, *J. Flow Chem.*, 2014, **4**, 2.
74. J. Hyde, P. Licence, D. Carter and M. Poliakoff, *Appl. Catal. A Gen.*, 2001, **222**, 119.
75. S. Marre, Y. Roig and C. Aymonier, *J. Supercrit. Fluids*, 2012, **66**, 251.
76. P. Lozano, E. Garcia-Verdugo, S. Luis, M. Pucheault and M. Vaultier, *Curr. Org. Synth.*, 2011, **8**, 810.
77. F. Keil and K. Swamy, *Rev. Chem. Eng.*, 1999, **15**, 85.
78. C. Leonelli and T. Mason, *Chem. Eng. Process.*, 2010, **49**, 885.
79. F. Castro, S. Kuhn, K. Jensen, A. Ferreira, F. Rocha, A. Vicente and J. Teixeira, *Chem. Eng. J.*, 2013, **215**, 979.
80. R. Gani, C. Jimenez-Gonzalez and D. Constable, *Comput. Chem. Eng.*, 2005, **29**, 1661.
81. M. Folic, R. Gani, C. Jimenez-Gonzalez and D. Constable, *Chin. J. Chem. Eng.*, 2008, **16**, 376.

82. R. Gani, P. Gomez, M. Folic, C. Jimenez-Gonzalez and D. Constable, *Comput. Chem. Eng.*, 2008, **32**, 2420.
83. B. Subramaniam, *Coord. Chem. Rev.*, 2010, **254**, 1843.
84. R. Henderson, C. Jimenez-Gonzalez, D. Constable, S. Alston, G. Inglis, G. Fisher, J. Sherwood, S. Binks and A. Curzons, *Green Chem.*, 2011, **13**, 854.
85. A. Cervera-Padrell, T. Skovby, S. Kiil, R. Gani and K. Gernaey, *Eur. J. Pharmaceut. Biopharmaceut.*, 2012, **82**, 437.
86. S. Elgue, L. Prat, M. Cabassud and J. Cezerac, *Chem. Eng. J.*, 2006, **117**, 169.
87. M. Harini, J. Adhikari and K. Rani, *Ind. Eng. Chem. Res.*, 2013, **52**, 6869.
88. D. Hsieh, A. Marchut, C. Wei, B. Zheng, S. Wang and S. Kiang, *Org. Process Res. Dev.*, 2009, **13**, 690.
89. H. Modarresi, E. Conte, J. Abildskov, R. Gani and P. Crafts, *Ind. Eng. Chem. Res.*, 2008, **47**, 5234.
90. C. Schaefer, D. Clicq, C. Lecomte, A. Merschaert, E. Norrant and F. Fotiad, *Talanta*, 2014, **120**, 114.
91. T. Sheldon, M. Folic and C. Adjiman, *Ind. Eng. Chem. Res.*, 2006, **45**, 1128.
92. A. Corma and H. Garcia, *Top. Catal.*, 2008, **48**, 8.
93. B. Dioos, I. Vankelecom and P. Jacobs, *Adv. Synth. Catal.*, 2006, **348**, 1413.
94. D. Han, Q. Yang and C. Li, *Chin. J. Catal.*, 2008, **29**, 789.
95. G. Hutchings, *Annu. Rev. Mat. Res.*, 2005, **35**, 143.
96. X. Li, S. Lu and A. Chan, *Prog. Chem.*, 2001, **13**, 33.
97. P. McMorn and G. Hutchings, *Chem. Soc. Rev.*, 2004, **33**, 108.
98. D. Murzin, P. Maki-Arvela, E. Toukoniitty and T. Salmi, *Catal. Rev.-Sci. Eng.*, 2005, **47**, 175.
99. J. Armor, *Catal. Today*, 2011, **178**, 8.
100. K. Arnold, B. Davies, R. Giles, C. Grosjean, G. Smith and A. Whiting, *Adv. Synth. Catal.*, 2006, **348**, 813.
101. C. Grosjean, J. Parker, C. Thirsk and A. Wright, *Org. Process Res. Dev.*, 2012, **16**, 781.
102. J. Comerford, J. Clark, D. Macquarrie and S. Breeden, *Chem. Commun.*, 2009, 2562.
103. S. Ghosh, A. Bhaumik, J. Mondal, A. Mallik, S. Sengupta and C. Mukhopadhyay, *Green Chem.*, 2012, **14**, 3220.
104. H. Thakuria, B. M. Borah and G. Das, *J. Mol. Catal. A: Chem.*, 2007, **274**, 1.
105. H. Lundberg, F. Tinnis, N. Selander and H. Adolfsson, *Chem. Soc. Rev.*, 2014, **43**, 2714.

106. W. F. Slotema, G. Sandoval, D. Guieysse, A. J. J. Straathof and A. Marty, *Biotechnol. Bioeng.*, 2003, **82**, 664.
107. M. Pirrung, F. Zhang, S. Ambadi and T. Ibarra-Rivera, *Eur. J. Org. Chem.*, 2012, 4283.
108. N. Caldwell, C. Jamieson, I. Simpson and A. Watson, *ACS Sustainable Chem. Eng.*, 2013, **1**, 1339.
109. J. Munoz, J. Alcazar, A. de la Hoz, A. Diaz-Ortiz and S. de Diego, *Green Chem.*, 2012, **14**, 1335.
110. R. Tank, U. Pathak, M. Vimal, S. Bhattacharyya and L. K. Pandey, *Green Chem.*, 2011, **13**, 3350.
111. X. Y. Liu and K. F. Jensen, *Green Chem.*, 2012, **14**, 1471.
112. X. Jin, K. Yamaguchi and N. Mizuno, *Angew. Chem., Int. Ed.*, 2014, **53**, 455.
113. K. Shimizu, K. Ohshima and A. Satsuma, *Chem. –Eur. J.*, 2009, **15**, 9977.
114. J. Zhu, Y. Zhang, F. Shi and Y. Deng, *Tetrahedron Lett.*, 2012, **53**, 3178.
115. Y. Wang, D. P. Zhu, L. Tang, S. J. Wang and Z. Y. Wang, *Angew. Chem., Int. Ed.*, 2011, **50**, 8917.
116. J.-F. Soule, H. Miyamura and S. Kobayashi, *J. Am. Chem. Soc.*, 2011, **133**, 18550.
117. S. Kegnæs, J. Mielby, U. V. Mentzel, T. Jensen, P. Fristrup and A. Riisager, *Chem. Commun.*, 2012, **48**, 2427.
118. X. Y. Liu and K. F. Jensen, *Green Chem.*, 2013, **15**, 1538.
119. T. Hirashita, S. Kuwahara, S. Okochi, M. Tsuji and S. Araki, *Tetrahedron Lett*, 2010, **51**, 1847.
120. R. Kumar and E. V. Van der Eycken, *Chem. Soc. Rev.*, 2013, **42**, 1121.
121. P. N. Liu, L. Dang, Q. W. Wang, S. L. Zhao, F. Xia, Y. J. Ren, X. Q. Gong and J. Q. Chen, *J. Org. Chem.*, 2010, **75**, 5017.
122. R. Amandi, J. R. Hyde, S. K. Ross, T. J. Lotz and M. Poliakoff, *Green Chem.*, 2005, 7, 288.
123. R. Sanz, A. Martínez, D. Miguel, J. M. Álvarez-Gutiérrez and F. Rodríguez, *Adv. Synth. Catal.*, 2006, **348**, 1841.
124. X. Cui, Y. Zhang, F. Shi and Y. Deng, *Chem. –Eur. J.*, 2011, **17**, 1021.
125. N. Zotova, F. J. Roberts, G. H. Kelsall, A. S. Jessiman, K. Hellgardt and K. K. Hii, *Green Chem.*, 2012, **14**, 226.
126. G. Beamson, A. J. Papworth, C. Philipps, A. M. Smith and R. Whyman, *Adv. Synth. Catal.*, 2010, **352**, 869.
127. R. Burch, C. Paun, X. M. Cao, P. Crawford, P. Goodrich, C. Hardacre, P. Hu, L. McLaughlin, J. Sa and J. M. Thompson, *J. Catal.*, 2011, **283**, 89.
128. M. Stein and B. Breit, *Angew. Chem., Int. Ed.*, 2013, **52**, 2231.

129. Y. Motoyama, K. Mitsui, T. Ishida and H. Nagashima, *J. Am. Chem. Soc.*, 2005, **127**, 13150.
130. Y. Mikami, A. Noujima, T. Mitsudome, T. Mizugaki, K. Jitsukawa and K. Kaneda, *Chem. –Eur. J.*, 2011, **17**, 1768.
131. F. D. Popp and H. P. Schultz, *Chem. Rev.*, 1962, **62**, 19.
132. J. Carey, D. Laffan, C. Thomson and M. Williams, *Org. Biomol. Chem.*, 2006, **4**, 2337.
133. R. J. White, R. Luque, V. L. Budarin, J. H. Clark and D. J. Macquarrie, *Chem. Soc. Rev.*, 2009, **38**, 481.
134. M. Králik and A. Biffis, *J. Mol. Catal. A: Chem.*, 2001, **177**, 113–138.
135. S. E. Davis, M. S. Ide and R. J. Davis, *Green Chem.*, 2013, **15**, 17.
136. B. R. Cuenya, *Thin Solid Films*, 2010, **518**, 3127.
137. M. B. Gawande, R. K. Pandey and R. V. Jayaram, *Catal. Sci. Technol.*, 2012, **2**, 1113.
138. F.-Z. Su, J. Ni, H. Sun, Y. Cao, H.-Y. He and K.-N. Fan, *Chem. – Eur. J.*, 2008, **14**, 7131.
139. B. Jørgensen, S. Egholm Christiansen, M. L. Dahl Thomsen and C. H. Christensen, *J. Catal.*, 2007, **251**, 332.
140. Y. Uozumi and Y. M. A. Yamada, *Chem. Record*, 2009, **9**, 51.
141. K. Mori, K. Yamaguchi, T. Mizugaki, K. Ebitani and K. Kaneda, *Chem. Commun.*, 2001, 461.
142. M. D. Hughes, Y. J. Xu, P. Jenkins, P. McMorn, P. Landon, D. I. Enache, A. F. Carley, G. A. Attard, G. J. Hutchings, F. King, E. H. Stitt, P. Johnston, K. Griffin and C. J. Kiely, *Nature*, 2005, **437**, 1132.
143. S.-i. Niwa, M. Eswaramoorthy, J. Nair, A. Raj, N. Itoh, H. Shoji, T. Namba and F. Mizukami, *Science*, 2002, **295**, 105.
144. T. Carofiglio, P. Donnola, M. Maggini, M. Rossetto and E. Rossi, *Adv. Synth. Catal.*, 2008, **350**, 2815.
145. A. Corma and H. Garcia, *Chem. Soc. Rev.*, 2008, **37**, 2096.
146. A. Dhakshinamoorthy, A. Primo, P. Concepcion, M. Alvaro and H. Garcia, *Chem. –Eur. J.*, 2013, **19**, 7547.
147. H. Huang, J. Huang, Y.-M. Liu, H.-Y. He, Y. Cao and K.-N. Fan, *Green Chem.*, 2012, **14**, 930.
148. F. Cavani and J. H. Teles, *Chemsuschem*, 2009, **2**, 508.
149. S. Higashimoto, N. Kitao, N. Yoshida, T. Sakura, M. Azuma, H. Ohue and Y. Sakata, *J. Catal.*, 2009, **266**, 279.
150. M. Renz, T. Blasco, A. Corma, V. Fornés, R. Jensen and L. Nemeth, *Chem. –Eur. J.*, 2002, **8**, 4708.
151. S. Mascia, P. Heider, H. Zhang, R. Lakerveld, B. Benyahia, P. Barton, R. Braatz, C. Cooney, J. Evans, T. Jamison, K. Jensen, A. Myerson and B. Trout, *Angew. Chem., Int. Ed.*, 2013, **52**, 12359.

CHAPTER 5

Hydrodynamic Cavitation Processing

FREDERICK C. MICHEL JR.*[a] AND OLEG KOZYUK[b]

[a] Department of Food, Agricultural and Biological Engineering,
The Ohio State University, Wooster, OH, USA; [b] Arisdyne Systems Inc.,
Cleveland, OH, USA
*Email: michel.36@osu.edu

5.1 INTRODUCTION

Hydrodynamic cavitation is a physical phenomenon that results from
a drop in pressure in a liquid that creates cavities or bubbles, followed
by an increase in pressure that leads to their violent collapse generating shock waves. The end result is tremendous 'localized' forces
which impart destructive effects on solids in dispersions, emulsions
and slurries affecting particle size and distribution. Hydrodynamic
cavitation also can increase the surface area between insoluble liquid
reactants, accelerate dissolution, disaggregate particles, and clean the
surface of a solid reactant or catalyst.

Particle size is an intrinsic property of solids that influences many
processing aspects of materials. Uniform particle size distributions
make thermal transitions and chemical reactions more predictable.
Small particle sizes reduce diffusion limitations, increase surface
area, and increase rates of reaction and thermal transfer. A promising
new area for further development of bioprocessing technologies
relates to systems for the generation of emulsions, slurries, catalysts

Chemical Processes for a Sustainable Future
Edited by Trevor M. Letcher, Janet L. Scott and Darrell A. Patterson
© The Royal Society of Chemistry 2015
Published by the Royal Society of Chemistry, www.rsc.org

and dispersions having the smallest possible particle sizes with maximum size uniformity.[1-3]

The destructive effect of cavitation has been used for technological processes for many years. The most widely used form is acoustic cavitation.[4] Acoustic cavitation is produced by sound waves in a liquid that create pressure variations. Although acoustic cavitation (or sonication) is very effective and widely used at small laboratory scales, directing sound waves into fluids is not an energy efficient process and the high frequency sounds emitted by sonication devices can damage hearing. Acoustic cavitation is also difficult to scale up.

Hydrodynamic cavitation, on the other hand, is much easier to scale which makes it better suited to the processing of emulsions, dispersions and slurries at commercial scales. From an energy transfer standpoint, hydrodynamic cavitation is also as much as 40 times more efficient compared to acoustic cavitation. It is not only safe and environmentally friendly in its application but also efficient and economical.[2,5-7] Hydrodynamic cavitation technologies have been applied to the preparation of homogeneous solutions, emulsions,[1] suspensions and dispersions of various products.[7] It has also been used to activate and accelerate chemical reactions, for water purification and sludge dispersion in waste water treatment systems,[5,8] and for the processing of corn grain for corn ethanol production[2] and for biodiesel production.

Controlled flow cavitation (CFC) refers to systems consisting of a pump and cavitator assembly that generate hydrodynamic cavitation with a minimal amount of energy input and degradation of system components by cavitation forces. It can be applied to existing processes to reduce or eliminate the need for chemicals and/or thermal processing in a variety of industrial processes. The applications for which hydrodynamic cavitation can be used range from existing processes that are enhanced by the retrofitting of controlled flow cavitation devices to new processes that up until now were not economically feasible.[9]

In this chapter, the fundamentals of hydrodynamic cavitation, as well as a wide range of applications for which it is used in different industries, are described. The use of hydrodynamic cavitation, particularly for bioprocessing, is likely to grow in importance as new processes for the sustainable use of renewable organic materials are developed. In Sections 5.1 to 5.3, the principles of hydrodynamic cavitation are described including an analysis of the specific amount of energy used for the process. The mechanisms of comminution and chemical reaction in the cavitation bubble zone are developed in

Section 5.4. Section 5.5 discusses controlled flow cavitation applied to homogenization and emulsification processes. An important effect of controlled flow cavitation; its ability to reduce particle sizes and ag-glomerations, is presented in Section 5.6. In Sections 5.7 and 5.8, applications of controlled flow cavitation to the production of nano-materials, catalysts, pharmaceuticals, corn ethanol, biodiesel and biogas are described.

5.2   PRINCIPLES OF HYDRODYNAMIC CAVITATION PROCESSES

The process of hydrodynamic generation of cavitation bubbles is shown schematically in Figure 5.1. The cavitation process begins with a flowing stream of material. A high pressure, P_1 is generated in chamber 1 by a localized flow constriction at point 2. In this localized flow constriction, the velocity of the flow increases. This causes the hydrostatic pressure to drop to a pressure below the saturated vapour pressure of the liquid phase flow. This phenomenon creates the physical conditions necessary for the development of cavitation bubbles in the stream. A high velocity jet (point 3) is formed by the localized flow constriction that becomes saturated with cavitation bubbles at point 4. This cavitating jet penetrates into the second chamber 5, which is under a hydrostatic pressure of P_2. This hydro-static pressure initiates a violent collapse of the cavitation bubbles, which results in the energy accumulated by the bubbles during the growth process being transferred to the flowing material. It is im-portant to note that the higher the velocity of the cavitating jet at point 3, the higher is the hydrostatic pressure P_2 in region 5. With respect to higher values of P_2, maintaining a low pressure in the cavitating jet is required for the development of cavitation bubbles, and this is only possible by increasing the velocity of this jet in the localized flow constriction at point 2.

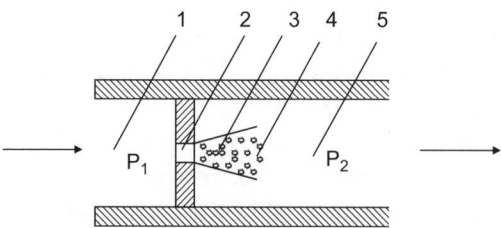

Figure 5.1   Schematic illustrating the generation of cavitation in a fluid flowing stream using a small diameter orifice from the high-pressure zone at P_1 to low pressure at P_2.

The idea of using the energy from collapsing cavitation bubbles to intensify material processing led to the creation of controlled flow cavitation devices. Hydrodynamic cavitation can be purposefully created in these devices through a variety of different designs and configurations. The most effective designs are easy to use in the laboratory as well as industrial applications and are known as static hydrodynamic cavitation mixers or reactors.[10,11] Longitudinal views of various hydrodynamic cavitation devices, each containing a bluff body, a flow-through transit channel and a local constriction for the liquid flow are shown in Figure 5.2. In order to control and specify the required structure of the cavitation bubble field, the bluff body may have various shapes. This results in the flow-through transit channel having various shapes which produce a local constriction of the flow around the baffle. The flow-through transit channel may also have a cross-section that has one linear section and a circular or irregularly shaped cross section, such as a semi-circle (Figure 5.2).

Fluids are fed into a hydrodynamic cavitation device with the aid of a pump. The type of pump selected is determined by the physicochemical properties of the fluid medium and the hydrodynamic parameters necessary to accomplish cavitation.

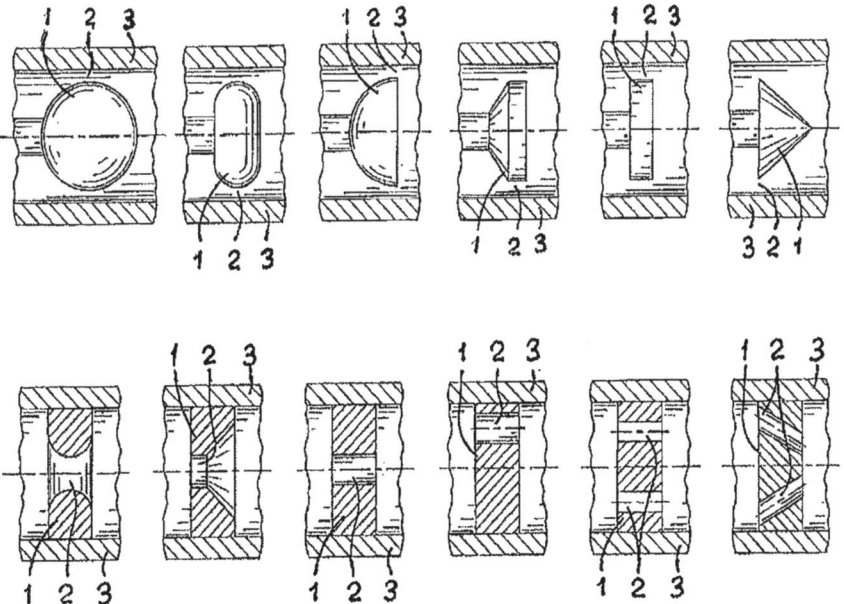

Figure 5.2 Views of the longitudinal sections of the local flow constriction in various hydrodynamic cavitation devices.

The *cavitation number* is the most useful parameter for defining cavitation flow in hydrodynamic cavitation devices. It is a scaling parameter used to design cavitation devices with similar cavitation intensities. For a given configuration, conservation of the cavitation number between two different operating conditions helps assure the development of similar degrees of cavitation. The cavitation number can be derived from Bernoulli's equation (for negligible friction loss) as shown in Equations (5.1) and (5.2), where the subscript, 0, is used to designate the pressure and a flow velocity within a local constriction, P_1 is the upstream pressure and P_2 is the downstream pressure.

$$P_1 + \frac{1}{2}\rho v^2 = P_0 + \frac{1}{2}\rho v_0^2 \tag{5.1}$$

$$P_1 - P_0 = \frac{1}{2}\rho v_0^2 - \frac{1}{2}\rho v^2 \tag{5.2}$$

$$\sigma = \frac{P_1 - P_0}{\frac{1}{2}\rho v_0^2} \tag{5.3}$$

In Equation (5.3), the pressure outside of the orifice can be defined as either the upstream or the downstream pressure. Typically, P_0 is replaced with the vapour pressure of the fluid, P_v. There are several other forms of the cavitation number σ that are also commonly used to characterize the cavitation in hydrodynamic cavitation devices (Equations (5.4) to (5.6)). Valve applications usually use Equation (5.3) while other industries use Equations (5.4) and (5.5). The units of cavitation numbers are dimensionless.

$$\sigma_1 = \frac{P_2 - P_0}{\frac{1}{2}\rho v_0^2} \tag{5.4}$$

$$\sigma_2 = P_2/(P_1 - P_2) \tag{5.5}$$

$$\sigma_3 = P_2/P_1 \tag{5.6}$$

5.3 SPECIFIC ENERGY USE DURING CAVITATION

Since cavitation forces are usually used to fracture dispersed materials, a general understanding of the magnitude of the energy emitted to the bulk liquid by the collapse of a cavitation bubble is important. Cavitation energy is transferred through the system in the form of mechanical work *via* shock waves or microjet forces.

The energy emitted during the collapse of a cavitation bubble can be analysed in terms of the adiabatic collapse cavity.[12] As a simplification, only the action of a single cavitation bubble from the moment of the beginning of its collapse is examined (Figure 5.3).

At the moment of collapse, the bubble has an initial radius, R_{max}, the hydrostatic pressure surrounding the bubble is P, and the liquid surrounding the bubble has a volume V with radius R_∞. The final bubble radius after collapse will be R_0. During complete collapse of the cavitation bubble, an instantaneous stoppage takes place of all liquid mass filling into the bubble volume creating a very large pressure impulse distributed throughout the surrounding bubble liquid. Assuming that the bubble remains spherical, the work generated upon collapse is given by Equation (5.7):

$$W = \frac{4}{3}\pi P(R_{max}^3 - R_0^3) \tag{5.7}$$

In the limit of a total collapse, the final radius, R, is zero, and thus the total work transferred to the liquid surrounding the bubble is calculated using Equation (5.8):

$$W = \frac{4}{3}\pi P(R_{max}^3) \tag{5.8}$$

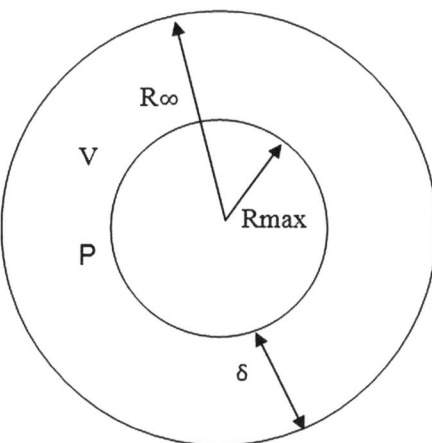

Figure 5.3 Schematic illustrating the dynamics of the collapse of a bubble generated by hydrodynamic cavitation.

To determine the power output, N, from the bubble collapse, the work is divided by the time, t, that it takes for the bubble to collapse (Equation (5.9)).

$$N = \frac{4}{3}\pi R_{max}^3 \frac{P}{t} \qquad (5.9)$$

The time for the complete collapse of the bubble was shown by Lord Rayleigh[13] to be:

$$t = 0.91 R_{max}\sqrt{\frac{\rho}{P}} \qquad (5.10)$$

where ρ is the density of the liquid around the bubble. Substituting this into Equation (5.9) yields Equation (5.11):

$$N = 4.60 R_{max}^2 \sqrt{\frac{P^3}{\rho}} \qquad (5.11)$$

The liquid mass which is affected by the energy generated in the collapse is given by Equation (5.12):

$$m = \frac{4}{3}\pi \rho R_\infty^3 \qquad (5.12)$$

As a result, the energy dissipation per mass of liquid, ε is equal to Equation 5.13.

$$\varepsilon = \frac{N}{m} = \frac{R_{max}^2}{R_\infty^3}\sqrt{\frac{P^3}{\rho^3}} \qquad (5.13)$$

This final expression indicates that the amount of energy dissipated depends on the size of the bubble, the hydrostatic pressure in the surrounding bubble liquid, the distance from the point of collapse and the liquid density.

Assuming that at the moment of collapse bubbles are attached to a slurry particle or droplet $R_\infty \rightarrow 0$, the energy dissipated can reach very high levels, sufficient to destroy the particles and emulsion droplets. By varying the hydrostatic pressure in the cavitation zone and the size of the bubble, the energy produced by cavitation can be controlled.

The magnitude of energy dissipation as a function of the thickness of the liquid layer R_{max}, subjected to processing and hydrostatic pressure P, is shown in Figure 5.4. According to Figure 5.4, the magnitude of the energy dissipated may be quite large. At such

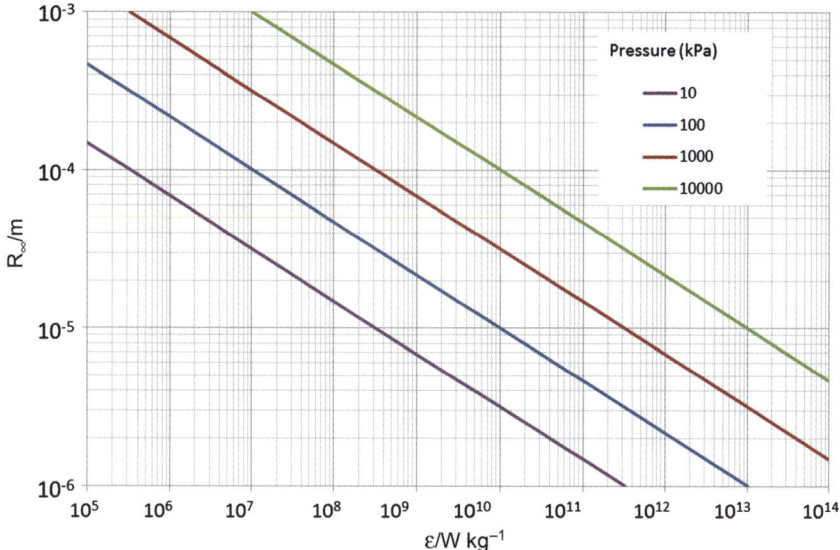

Figure 5.4 Dependence of energy dissipation on radius of liquid layer (R_∞) at different pressure.

levels of energy dissipation, extreme physical conditions are created sufficient to enhance chemical reactions and to impose mechanical effects which are difficult or impossible to achieve by other means.

5.4 MECHANISMS OF COMMINUTION AND EFFECTS ON CHEMICAL REACTION IN THE CAVITATION BUBBLE ZONE

Cavitation systems can produce billions of cavitation bubbles in a flow creating extraordinary shear, impact and shock wave forces and yielding efficient transfer of the input energy. Figure 5.5 illustrates how collapsing cavitation bubbles create these forces. When the bubble forms in the low pressure zone it will grow and move with the liquid flow towards the high pressure zone (Stage 1). The growing cavitation bubbles have the tendency to pulsate. This pulsation due to Bjerknes force creates flotation effects which draw nearby particles to the bubble surface (Figure 5.5). Particles suspended in the fluid concentrate around the surface of a pulsating bubble due to surface tension effects. There exists a limited distance beyond which capture of the particles by the bubble does not occur. This limited

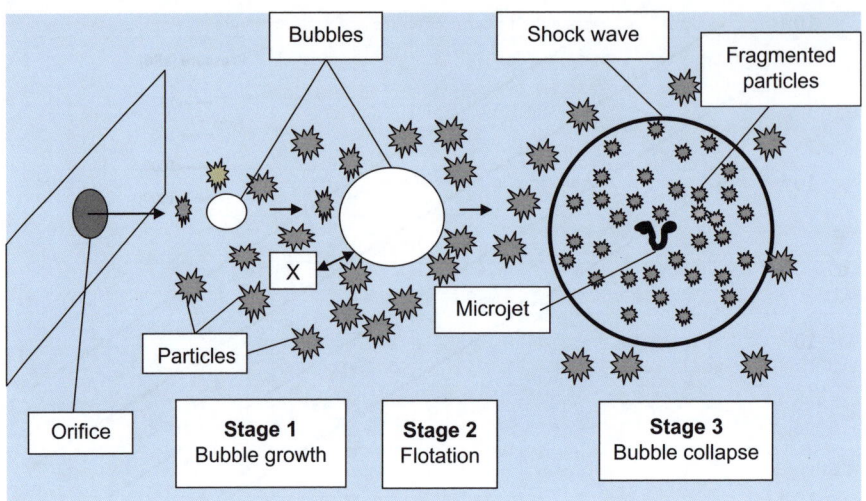

Figure 5.5 Principle of interaction of cavitation bubbles with particles in a flow.

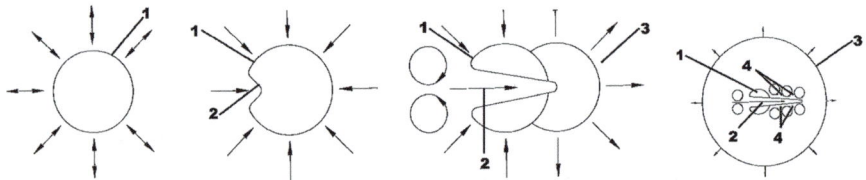

Figure 5.6 Stages of bubble pulsation, microjet formation and collapse. 1: Bubble. 2: Liquid jet. 3: Shock wave formation. 4: Ultra high shear layer.

distance, called the 'capture distance', can be calculated using Equation (5.14):[4]

$$X = \sqrt{rR^3 f \rho / \mu} \qquad (5.14)$$

where X is the capture distance, ρ is the liquid density, r is the particle radius, R is the bubble radius, f is the bubble frequency and μ is the liquid viscosity (Stage 2).

A localized increase in the surrounding pressure induces bubble collapse, usually in the form of a microjet in one part of the bubble (Figures 5.6 and 5.7). The liquid microjet impacts the opposite side of the bubble with a velocity of 1000–1500 m s^{-1}. A shock wave is thereby created (Figure 5.6). The microjet continues through the opposite side of the bubble wall and penetrates deeply into the surrounding media creating an ultra-high shear layer between the jet and the adjacent liquid. The result is extreme turbulence with shear rates as high as

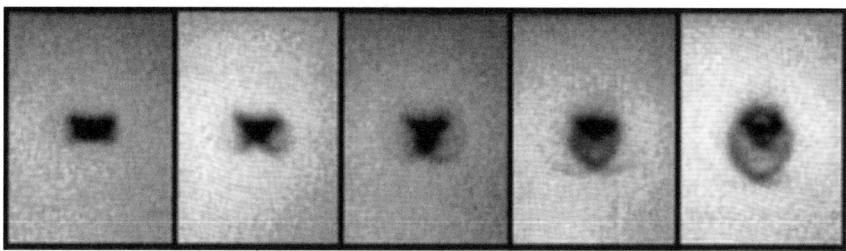

Figure 5.7   High-speed photograph of an asymmetric bubble collapse shock wave
emitted by a laser-generated bubble.[14]
(Permission Granted By Author Claus Dieter Ohl.)

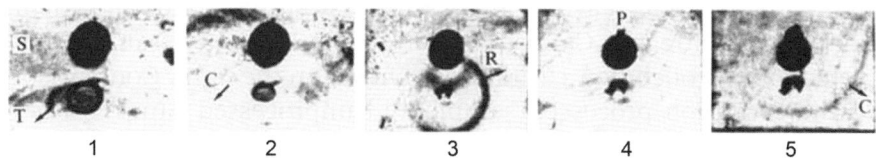

Figure 5.8   A 3 mm bubble upstream of a 3 mm lead particle inclusion on an axis
perpendicular to the shock. P vortex shed downstream of the particle.
The slider enters the sequence from below in frames 3–5. Interframe
time is 5 μs.[15]
(Permission Granted By Author Claus Dieter Ohl.)

100×10^6 s^{-1} and higher, which generate extreme shear forces on
particles and droplets (Stage 3, Figure 5.6). The shock wave develops
enormous pressure. Pressure impulses created by the shock wave can
reach up to 1000 MPa. If the shock wave encounters an obstacle, it
breaks the surface of the obstacle. Shock waves first destroy ir-
regularities on the surface of particles. A microjet can break a particle
apart if the jet impinges the particle surface. These powerful forces
are created by each bubble in a cavitation zone. Each cavitation zone
in turn may contain millions of bubbles. The bubbles act upon each
other to create additional forces.

Borne and Field recorded the collapse of bubbles in proximity to
solid particles and surfaces using high-speed photography.[15] The
images show the interactions between a collapsing bubble and a
particle (Figure 5.8). To create this phenomenon, a 3 mm bubble was
placed upstream of a 3 mm lead particle and collapsed by an external
pressure impulse. Collapse occurred by frame 3 and the resulting
shock, *R*, can be seen propagating out from the collapse site, which
appears as a dark area in Figure 5.8. The secondary compressive
shock wave C is reflected from the upstream surface of the particle

and travels back through the shocked material. This shock reflection acts on the collapsing bubble and increases the speed of its collapse. In the final stage of collapse, the rate of temperature increase inside the bubble can reach 10^{10} to 10^{11} K s^{-1}.[16] The gas within the collapsing bubble can be heated to temperatures of approximately (2000–5000) K as a result of adiabatic compression.[17]

Evidence for shock wave heating has been shown by subjecting crystals with known melting temperatures to hydrodynamic cavitation.[17] In one study, a suspension of calcium fluoride (CaF_2) crystals, with a melting point of 1633 K, was passed through a hydrodynamic cavitation processor and recirculated for 30 minutes at room temperature. Scanning electron micrographs (SEM) of untreated CaF_2 and cavitation processed samples are shown in Figure 5.9, untreated CaF_2 is shown on the left and a sample subjected to controlled flow cavitation is shown on the right. Comparison of the cavitation processed sample and unprocessed sample shows that the processed sample contained CaF_2 crystals that were rounded off and many particles were attached to one another by melted necks. One can expect that CaF_2 would soften at the Tamman temperature of about 50% its normal melting point. Thus, a shock wave heat up temperature of *ca.* 873 K is consistent with these data.

Under the physical conditions inside the cavitation bubble, various chemical reactions take place between the gases found within the bubble. In the final stage of bubble collapse, heating also occurs adjacent to the bubble in the liquid sphere having a layer thickness of

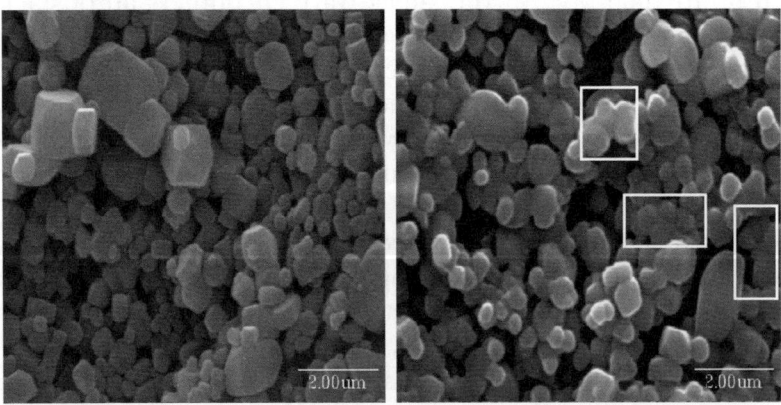

Figure 5.9 Effects of cavitation on solid crystals. Left: unprocessed CaF_2. Right: cavitation processed CaF_2.
Reprinted from ref. 18, Copyright (2001), with permission from Academic Press.

approximately (20–100) nm. The temperature to which this liquid layer is heated is on the order of (30–40)% of the temperature of the steam-gas phase inside of the bubble. The pressure at the boundary of the bubble is equal to the pressure inside the bubble. The conditions at the boundary of the cavitation bubble in the liquid phase (pressure and temperature) are sufficient to enhance chemical reactions in this zone. Each cavitation bubble behaves as an 'autonomous system'. Because of this high energy level, cavitation has been studied not only for its ability to disperse particles but to also aid in chemical reactions. An additional effect that may enhance chemical reaction is the disruption of the mass transfer limiting boundary layer that exists at the surface of particles.

Dispersion effects of the cavitation result from a substantial plurality of force effects due to the collapse of cavitation bubbles. The collapse of cavitation bubbles near the boundary of 'liquid–solid particle' phases results in dispersion of these particles in the fluid and in the formation of a suspension. In 'liquid–liquid' systems, one fluid is atomized in the other fluid resulting in the formation of an emulsion. In both cases, the mass transfer boundary layer is destroyed, chemical reactions are enhanced, solids are eroded, and a dispersive medium and a dispersed phase are formed.

5.5 HOMOGENIZATION AND EMULSIFICATION PROCESSES

Hydrodynamic cavitation can also be applied to mix effectively, to homogenize liquids and to form emulsions. Relatively low cavitation energy input can lead to the formation of stable emulsions with very small, uniform sized droplets. This application has been commercialized in the chemical, petrochemical, polymer, chemical, cosmetics, food and pharmaceutical industries and may be a useful unit operation for sustainable chemical processing.[19,20]

The preparation of emulsions in static hydrodynamic cavitation mixers can be described using Kolmogorov's theory of locally isotropic turbulence.[21] This theory predicts that the diameter of the most stable droplets in an emulsion are a function of the amount of energy dissipated based on an equation derived from Kolmogorov's 'law of two-thirds':

$$d_{max} \approx 3.5 \left(\frac{\sigma}{\rho} \right)^{0.6} (\varepsilon)^{-0.4} \tag{5.15}$$

where σ is the interfacial surface tension, ρ is the density of the continuous medium, ε is the local value of the dissipation of energy per unit of mass of the liquid medium and d_{max} is the maximum size stable droplets of the emulsion.

For an emulsion prepared in a cavitation mixer, the disperse phase is broken up mainly by the energy liberated during the collapse of cavitation bubbles. The dissipation of the energy of the liquid flow ε occurs as a result of the collapse of cavitation bubbles:[19]

$$\varepsilon = \varepsilon_c \tag{5.16}$$

where ε_c is the value of the dissipation of energy per unit of mass of the liquid medium in case of collapse of n cavitation bubbles within it. Therefore, for a case of emulsification in a hydrodynamic cavitation mixer, Equation (5.15) with considerations of the ratio in Equation (5.16) becomes Equation (5.17):

$$d_{\max} \approx 3.5 \left(\frac{\sigma}{\rho} \right)^{0.6} (\varepsilon_c)^{-0.4} \tag{5.17}$$

Under real conditions, the field of cavitation bubbles consists of a significant number of bubbles of random dimensions. To simplify such conditions, it can be assumed that all bubbles have an average size of R_{av} and are distributed evenly in the volume (V) of the liquid medium inside the chamber of the cavitation mixer unit. With this assumption and using Equations (5.11) to (5.13), ε_c is defined by Equation (5.18):

$$\varepsilon_c = 4.6 \, \Theta \, R_{av}^2 \sqrt{\frac{P^3}{\rho^3}} \tag{5.18}$$

where $\Theta = \dfrac{n}{V}$, the average concentration cavitation bubbles in a hydrodynamic cavitation mixer. Substituting Equation (5.18) into Equation (5.17) and simplifying, the following equation is obtained:

$$d_{max} \approx 1.9 \left(\frac{\sigma}{P} \right)^{0.6} (\Theta)^{0.4} R_{av}^{-0.8} \tag{5.19}$$

This equation shows the influence of the basic parameters of the field of collapsing bubbles: their average size, the concentrations and the static pressure in the surrounding bubbles in the liquid on the maximum sized droplets of the emulsion.

The influence of hydrodynamic cavitation on liquid–liquid emulsification has been shown using an experimental cavitation loop mixer apparatus[22] such as that shown in Figure 5.10. The apparatus consists of a series connection in a circulating loop tank (1), with a

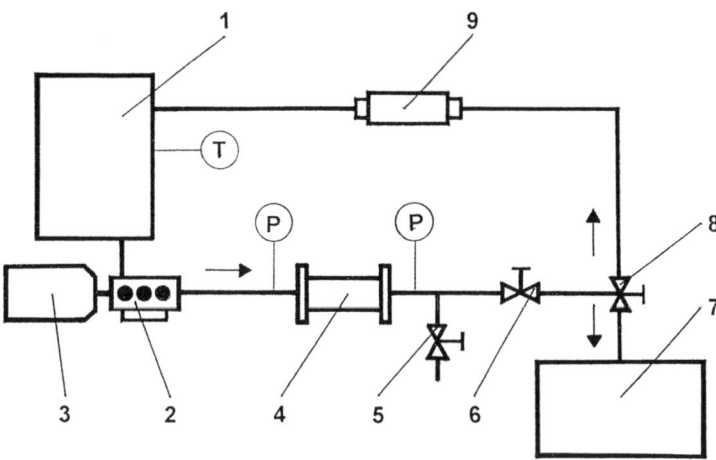

Figure 5.10 Schematic of experimental cavitation emulsification apparatus. 1: Tank A. 2: Plunger pump. 3: Electric motor. 4: Cavitation mixer. 5: Sampler. 6: Valve A. 7: Tank B. 8: Valve B. 9: Flowmeter.

plunger pump (2), cavitation mixer (4), adjustable valves and a flow meter (9). The plunger pump has a maximum pressure capacity of 10 MPa and flow rate of 1.1×10^{-3} m^3 s^{-1}. The cavitation mixer (4) has changeable flow-through chambers, made from transparent acrylic material and changeable cone-shaped cavitators of various dimensions. The use of transparent flow-through chambers allows visual monitoring of the development of cavitation. In the course of the experiments, three flow-through chambers were used having channel diameters of (6, 10 and 22) mm. Static pressures at the inlet and at the outlet from the mixer were measured with two pressure gauges (P). The degree of development of cavitation was control by adjusting the backpressure using valve A (6). Valve B served the purpose of shifting the flow regime from recirculation to single pass into tank B. The flow rate was measured by a flow meter (9). The measured parameters allow calculation of the cavitation index, $æ = P_2/P_{2c}$, where P_2 is the static pressure downstream from the flow-through chamber and P_{2c} is the static pressure downstream from the flow-through chamber when cavitation occurs.

The process of emulsification was investigated during a single pass and multi passes of the emulsion through the cavitation mixer. Oil was pre-mixed with distilled water in tank A for 1 minute with the aid of a portable turbine mixer. After that, the pump was switched on and the emulsion was pumped in a single pass through the cavitation mixer into tank B. During this, a sample of the emulsion was drawn from the sampler for subsequent dispersion analysis. The droplet size analysis was conducted using a laser diffraction particle size analyser 'Mastersizer Micro' and the Sauter diameter d_{32} was chosen for characterization of the results. The time between sample collection and measurement was less than 1 minute. During the course of the experiments, vegetable and silicone oils (Dow Corning 200 fluids) of various viscosities were used. The properties of the oils at a temperature of 297.65 K are shown in Table 5.1. A very low dispersed phase concentration of 2% was chosen for all experiments.

The development of cavitation in the mixer affects the emulsion droplet size. As shown in Figure 5.15, as the flow transitions from the turbulent regime to the cavitation regime (cavitation index $æ \leq 1.0$), the Sauter mean diameter of oil drops, d_{32} decreases. The smallest droplet sizes were achieved in the cavitation index range from 0.6 to 1.0 when cavitation exists in the form of transient microcavities and bubbles. Further reduction in the cavitation index due to a reduction in backpressure (Figure 5.11) changes the structure of the cavitation field in the cavitation mixer. First the concentration of individual bubbles decreases and a series transient cavity is formed in the flow. This regime exists over the cavitation index range from 0.4 to 0.6. Below a cavitation index value of 0.4, cavitation evolves into a super-cavitation flow regime in which a single cavity, called a super-cavity, is formed that is attached to the cone-shaped cavitator.

Super-cavity generated cavitation bubbles form from a pulsated tail. These bubbles have a large size but their concentration is low. A super-cavitation flow regime leads to a rise in droplet sizes. However, even under super-cavitation flow, emulsion droplet sizes are less than droplets produced under the turbulent regime at a cavitation index >1. The best results are achieved when cavitation mixers operate with a cavitation index from 0.6 to 1.0. In this regime, the field of cavitation bubbles have the highest pressure and bubble

Table 5.1 Properties of oils used in cavitation emulsification trials.

Dispersed phase	Viscosity/mPa s^{-1}	Density/kg m^{-3}
Vegetable oil	52	920
Silicone oil	20	954
	50	963
	100	968
	375	972
	545	973

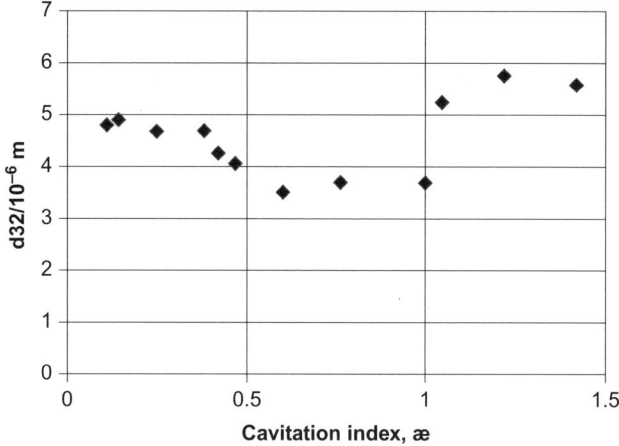

Figure 5.11 Effect of cavitation index on the mean Sauter diameter of an oil emulsion. Cone-shaped 30° cavitator. Channel diameter 6 mm.

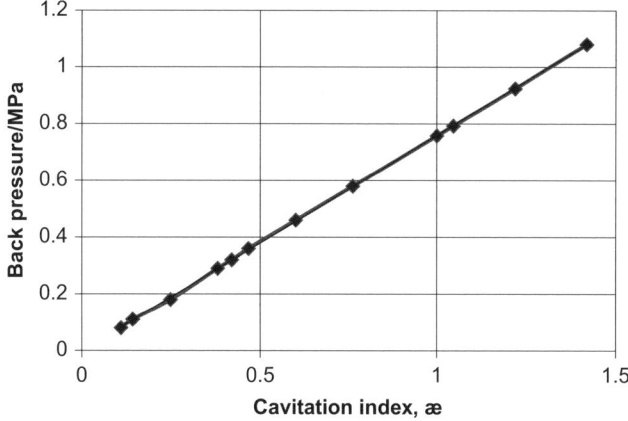

Figure 5.12 The effect of the backpressure on the cavitation index in the cavitation mixer.

concentration and medium bubble sizes compared with regimes with a lower cavitation index. This observation is supported by Equation (5.19).

Freuding *et al.* have studied the influence of cavitation on droplet breakup in high pressure homogenizers.[1] By applying and varying backpressure behind an orifice type homogenizing valve, various cavitation index values could be obtained (Figure 5.12). Using this apparatus, the influence of cavitation on droplet disruption was determined. The system employed a two-stage orifice valve. The first stage had a constant borehole diameter $D_2 = 0.4$ mm and $D_2 = 2$ mm for generation of different backpressures. By applying a varying back pressure behind an orifice type homogenizing valve, the influence of cavitation on droplet disruption can be determined. During this research a two-stage orifice valve was employed. The first stage had a constant borehole diameter (D_1) of 0.5 mm whereas the valve borehole diameter (D_2) varied from 0.4 mm to 2 mm to generate different back pressures. A 1% vegetable oil water emulsion was produced using the non-ionic surfactant Tween 80. Oil concentrations were 1%.

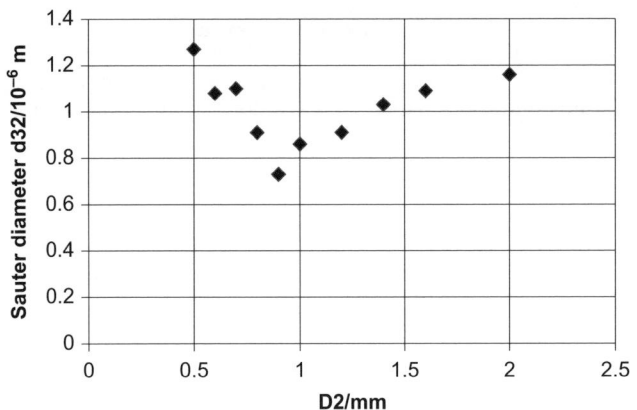

Figure 5.13   Influence of backpressure on droplet distribution.
(Reprinted from ref. 1 with permission from Wiley.)

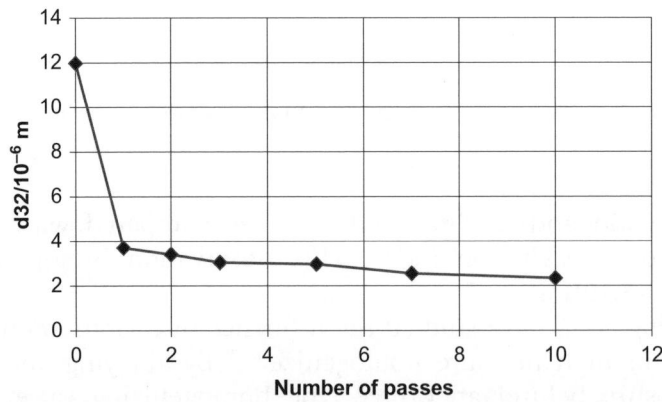

Figure 5.14   Effect of number of passes on Sauter mean diameter for vegetable oil.
Cavitation index, æ = 0.74.

Figure 5.13 shows the results of these experiments. This result correlates well with data shown in Figure 5.11.

Figure 5.14 shows the dependence of emulsion droplet size on the number of passes through a cavitation mixer. Only three passes are sufficient through the cavitation mixer for process completion. Further increases in the number of passes leads to an insignificant rise in the degree of homogenization.

With an increase in the concentration of the dispersed phase, the effect of cavitation dispersions is reduced. Figure 5.15 shows the dependence of the emulsion drop size from a viscosity dispersed phase. This dependence is linear; higher viscosities lead to larger oil droplets.

Viscosity of the continuous phase affects the process of drop breakage. However, the viscosity of the continuous phase has a more significant impact on turbulence emulsification compared with cavitational emulsification. Table 5.2 shows the results of studies on the effect of the viscosity of the continuous phase on the process of droplet breakage in the high shear turbulent and cavitational flows. The research used 1% silicon oil in water emulsions with lower and higher viscosity of the aqueous phase. The viscosity of the continuous phase was varied by adding a thickening agent (Carpool 941) to the aqueous phase. These emulsions were not stabilized by surfactants. Emulsification was performed using an IKA rotor–stator system and a low pressure cavitation mixer.

It is seen from these data that an increase in the viscosity of the continuous phase leads to inefficient emulsification in the high shear rotor–stator system. The flow declines by approximately 45% and the

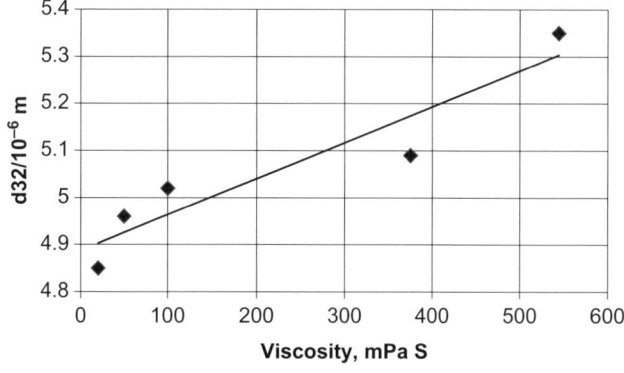

Figure 5.15 Effect of dispersed phase viscosity on Sauter mean diameter for silicone oil. Cavitation index, æ = 0.6.

Table 5.2   Effect of the viscosity of the continuous phase on droplet breakage in high shear turbulent and cavitational flows.

Process condition	Premix		Rotor–stator homogenizer, UTL 25 Ultra-Turrax®, one pass, 24 000 rpm		Cavitation mixer, single orifice stage, one pass, operating pressure 2.8 MPa	
Energy density $(J\ m^{-3})^a$			4.20×10^7		2.76×10^6	
Viscosity of the continuous phase (MPa)	10	350	10	350	10	350
Capacity $(m^3\ s^{-1})$	–	–	8.3×10^{-6}	4.6×10^{-6}	1.4×10^{-4}	1.4×10^{-4}
Mean droplet size d_{43} (μm)	114.8	175.9	14.9	36.9	9.4	10.2
Maximum droplet size d_{90} (μm)	182.1	292.8	27.8	97.4	16.4	21.3

[a]Energy densities were calculated using equations from Schubert.[23]

mean drop size is 145% larger. At the same time, an increase in viscosity of the continuous phase does not significantly affect processes of drop breakage in the cavitation mixer. The mean drop size decreases by just 8.5%.

Most likely, effective emulsification with a high viscosity continuous phase in a cavitation mixer is due to a shock wave mechanism of drop breakage compared with turbulence mechanisms. Different mechanisms of emulsification impact the energy requirement for droplet break up. In contrast to cavitation emulsification, in a rotor–stator homogenization system the energy requirements are higher because only a small fraction of the energy input is used for droplet break up and about 99.9% of the energy is lost as heat.

5.6   PARTICLE SIZE REDUCTION/DE-AGGLOMERATION

A variety of different types of technologies are employed for dispersing solid particles into liquids. These include colloid and disk mills, jet mills, rotor–stator mixers and high pressure homogenizers. However hydrodynamic cavitation devices offer an alternative to these technologies that is especially useful for dispersing particles in the nanometre particle size range.

The fragmentation of particles and agglomerates in cavitation devices is the result of dynamic fracturing of solid material in response to mechanical stresses. The primary mechanism for the generation of stress on solid particles is cavitation bubble collapse.[24] Surface damage results from the shock waves and microjet generated during such collapses. The semi-empirical relationship for the stresses generated during bubble implosion can be estimated from the following equation:

$$\tau_c = \alpha \rho c u_i \tag{5.20}$$

where α is an empirical constant, c is the velocity of compressional wave in the liquid and u_i is the microjet velocity given by Equation (5.21):

$$u_i = \frac{(P - P_v)^{1/2}}{0.915 \rho^{1/2}} \tag{5.21}$$

Wolfrum *et al.*[25] investigated bubble dynamics after laser induced shock wave exposure in the vicinity of salt crystals suspended in water. They discovered that bubble dynamics depends strongly on the position of neighbouring bubbles and on the number of boundaries given by the surrounding salt grains. Bubbles have a tendency to be drawn to the closest particles in their vicinity. External shock waves lead to jet formation against the rigid boundaries. The bubbles often tend to form in or migrate into cracks on the crystal surfaces and sometimes lead to the breakage of particles due to rapid bubble dynamics.[25]

Zeiger and Suslick[26] investigated the possible mechanisms for the breakage of molecular crystals under high-intensity ultrasound using acetylsalicylic acid (aspirin) crystals as a model compound.[26] The collapsing bubble emits a shock wave that, in water, has pressures up to 6000 MPa and velocities on the order of 4000 m s^{-1}. It was shown that a major contributor to particle breakage is particle-shock wave interactions. Using high speed recordings with a duration of 1 s containing up to 300 000 frames per second, Wagterveld *et al.* visualized the effects of cluster and streamer cavitation bubbles on calcite crystals.[27] Cavitation clusters collapsed, caused attrition, disrupted aggregates and de-agglomerated, whereas streamer cavitation which occurs at lower cavitation numbers, caused only de-agglomeration. SEM analysis of irradiated calcite seeds showed deep circular indentations. It was suggested that these indentations might be caused by shock waves induced by jet impingement.[27]

In commercial processing, most particles processed using grinding equipment have flaws or micro cracks randomly distributed at the interface or at the grain boundary. During cavitation, the mechanical stresses imposed by the shock waves and the microjet cause these pre-existing micro cracks to exceed a threshold value which eventually leads to the fragmentation of particles or particle agglomerates. Microjet-induced damage is primarily surface erosion and does not penetrate much into the bulk of the solid material. Despite this, such erosion seems to weaken the surface structure of the particles and makes them much more sensitive to the impact of shock waves.

One of the advantages of hydrodynamic cavitation technology compared with existing technologies is the equal and consistent treatment of all particles within a slurry. In hydrodynamic cavitation, particle size reduction occurs in a continuous flow mode. In this flow all particles have to follow a predetermined path and thus are treated with the same intensity of cavitation in the cavitation zone.

Tokumitsu[28] prepared TiO_2 and ZrO_2 particles using a nozzle cavitation apparatus. It was shown that the cavitation method was able reduce particles to several tens of nanometres (Figure 5.16). Cavitation fracturing of powder differed from pulverization caused by ball milling.[28] Hydrodynamic cavitation has been successfully used for particle size reduction of metal oxides and metals by Five Star Technologies, Cleveland, OH, USA. Most of the work performed by Five Star Technologies in the area of catalysts, advanced materials, formulations and dispersions involves cavitation. An SEM photograph of a commercially available ZnO with average crystallite size of 67 nm, agglomerated in the 2–50 μm range is shown in Figure 5.16. Particles of various sizes are agglomerated.

When prepared as slurry in water and processed through a cavitation device, the mean particle size of zinc oxide agglomerates was reduced from 16.73 to 0.27 μm. The particle size range (D_{90} to D_{10}) is also drastically reduced and the particle distribution curves tighten (Table 5.3).

Figure 5.16   Commercially available ZnO showing agglomerated particles.

Table 5.3 Effect of the number of passes through a cavitation device on particle size range D90 to D10.

Number of passes	Mean $D_{43}/\mu m$	$D_{90}/\mu m$	$D_{50}/\mu m$	$D_{10}/\mu m$
Control	16.73	51.93	5.37	2.33
1	0.50	0.91	0.45	0.19
5	0.31	0.61	0.26	0.10
10	0.27	0.51	0.24	0.10

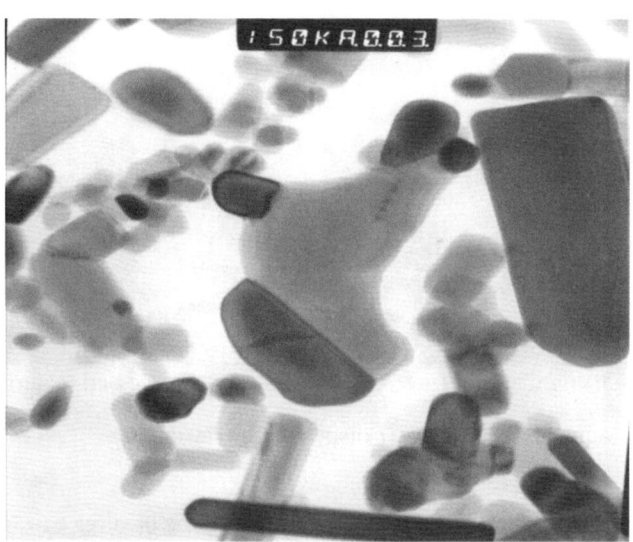

Figure 5.17 TEM of deagglomerated ZnO. Note separation of small particles from agglomerates.

The TEM photograph shown in Figure 5.17 is from a sample taken from the final slurry. This image shows that the large agglomerates are broken down to smaller aggregates and individual particles. More space is apparent in Figure 5.17.

In another example, 5% bentonite clay slurry in water was processed through a cavitation device operating at different working pressures. The starting material had a mean particle size diameter of 567 nm, which was reduced by 51% after one pass at a working pressure of 103 MPa (Table 5.4).

Another example is silver powder processing. The same processing techniques have been successfully used for silver powder dispersing and de-agglomeration in different media. Processed silver powders in aqueous dispersions showed a reduction in D_{50} from 25 μm to

Table 5.4 Bentonite clay slurry processed through a cavitation device at different pressures.

Working pressure/MPa	Mean D_{43}/nm	Reduction one pass/%
Control	567.4	–
34.47	431.9	23.9
68.95	343.7	39.4
103.42	275.8	51.4

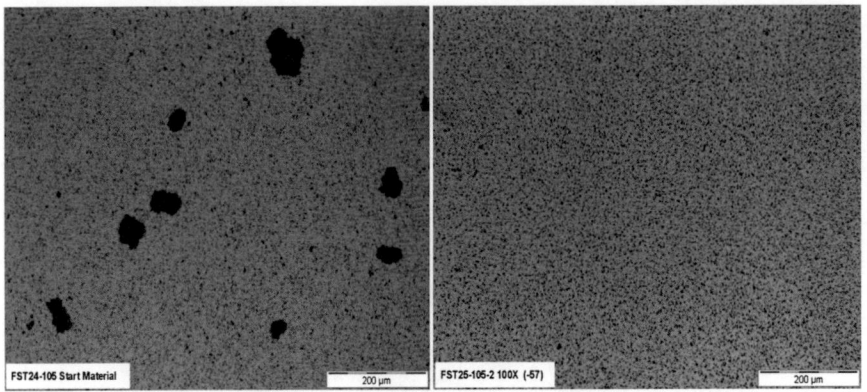

Starting material Processed material

Figure 5.18 Aqueous silver powder dispersion by cavitation.

Table 5.5 Hydrodynamic cavitation processing of silver powder.

	D_{10}/μm	D_{50}/μm	D_{90}/μm
Control	1.06	14.32	42.27
Processed	0.10	0.19	0.36

0.73 μm. Surface area increases by more than 1000 times. Microphotographs of the pre-processed and processed materials are shown in Figure 5.18. After cavitation, nearly an order of magnitude reduction in surface impurities such as sodium was observed.

Data in Table 5.5 illustrate the ability of hydrodynamic cavitation to convert conventionally produced silver powder into μm material. As shown, the starting material has particle sizes in the tens of nanometres; it is so aggregated that its mean particle size is 14 μm, with many aggregates larger than 42 μm. Processed powder in the epoxy resin medium has a D_{50} of 190 nm. The largest size was controlled to a D_{90} of 360 nm, enabling the use of the material in applications where extremely fine particulate sizes are required.

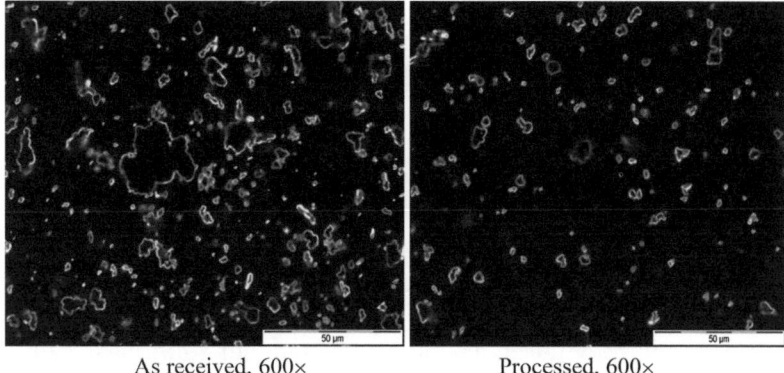

As received, 600×   Processed, 600×

Figure 5.19   Silver flake in a polymer matrix.

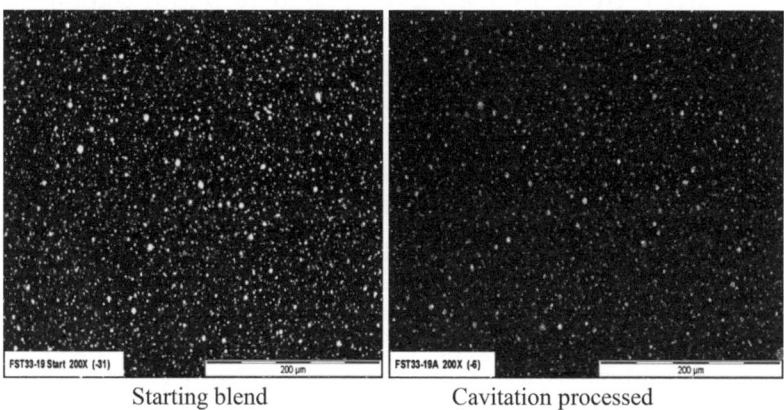

Starting blend   Cavitation processed

Figure 5.20   Silica in resin processed by cavitation.

Hydrodynamic cavitation is able to disperse silver flake in a wide range of solvents (Figure 5.19). Flake sizes are reduced and size distributions are tightened in this application.

Similar effects of cavitation on silica particles are observed (Figure 5.20). Additionally viscosity was lower compared with conventional mixing.

5.7   CAVITATION SYNTHESIS OF NANOSTRUCTURED MATERIALS AND CATALYSTS

A wide variety of chemical processes are used for the synthesis of advanced ceramic and catalytic materials. Recently there has been increased interest in novel methods for the synthesis of high purity nanostructured materials. However, many of the existing synthesis techniques for these materials require extensive chemical processing to achieve nanostructured grains of high phase purity while providing little to no capability to systematically control grain size and particle size distribution.

Research by Moser and co-workers has shown the capability of chemical synthesis within a cavitation field to provide exceptional mixing and *in situ* shock wave calcination during the co-precipitation of metal salts by a precipitating agent.[3,29–36] The experimental investigation showed that in most cases the primary grain sizes of catalysts prepared by hydrodynamic cavitation are smaller than classically prepared samples. Furthermore, in many catalyst preparations the technology enables the preparation of grains of catalyst in the important 1–20 nm crystallite size range. This is important due to the fact that catalytic rates increase from a factor of 10 to 1000 as the grain size of the active catalyst is reduced from 10 nm to 2–3 nm. Other studies have shown the capability of this technology to introduce a high degree of crystallographic strain in catalyst materials.[18]

Both supported metal catalysts and pure metal oxide catalysts have been synthesized as nanostructured grains in high dispersions and at high phase purities. Some of the catalysts prepared in this way are titania, alumina, cobalt molybdate supported on alumina and silica or unsupported nanostructured crystallites, copper-modified zinc oxide, noble metals on titania, perovskites, bismuth molybdates, platinum and rhodium on stabilized zirconia, La–Ce–Ni catalysts for SO_2 removal, nanostructured silver on α-alumina, dispersions of nano-grains of silver or lead in aqueous or hydrocarbon media and nano-grains of simple metal oxides such as iron, cobalt, molybdenum, cerium, nickel and zirconium oxides.

The new synthesis technology uses two cavitational devices manufactured by Five Star Technologies of Cleveland: CaviPro and CaviMax Reactor. The CaviPro 300 is a high pressure processor (a double orifice processor having variable pressure adjustment capabilities up to 206 MPa) and is shown in Figure 5.21. The CaviMax processor is a single orifice processor, operating at a constant flow rate and variable pressures over a range from 0.1 to 6.9 MPa and a nominal flow rate of 8.3×10^{-3} m^3 min^{-1} (Figure 5.22). Processing with the CaviPro 300 affords synthesis under Reynolds numbers between 10 000 and 80 000 and cavitation numbers $\sigma = 1$–8. The CaviMax Reactors use Reynolds numbers of 50 000–200 000 and cavitation numbers $\sigma = 1$–2.

Processing strategy for the synthesis of nanostructured finished metal oxide grains of complex catalytic materials involves

Figure 5.21 CaviPro 300 processor.

Figure 5.22 CaviMax processor.

precipitating all the metal salts from solution immediately before the hydrostatic stream pressure is elevated to the set operating head pressure of the cavitational processors in the range of 2 MPa to 206 MPa depending on the processor used. This soft gel is then carried a short distance as a slurry where it passes at high velocities through a set of orifices in the processor. A high fluid velocity through an orifice generates a wake of cavitational bubbles according to Bernoulli's principle. This results in exceptionally high shear, high Reynolds numbers, and shock wave heating. The result of this processing is to mix all the metallic components, break the soft gel into nanostructured grains, and *in situ* calcine them with salt decomposition to their homogeneous metal oxides. The result after solids separation is fine, nanostructured grains of agglomerated crystallites in the (1–20) nm size range which are homogeneous solid solutions and are well mixed crystallographically in the form of their metal oxides.

An important aspect of the formation of nanostructured catalyst grains by cavitational processing is the *in situ* calcination that the processors afford during synthesis. This enables the precipitated metal nitrate, chloride, hydroxide, *etc.* salts to be decomposed during synthesis directly to the finished metal oxide. This freezes the nanostructure toward grain growth and eliminates the need for a post synthesis calcination step to decompose metal salts.

The evidence that shock wave *in situ* calcination indeed occurs using these processors is three fold. First, many metal oxides taken directly from the product slurries show selected area electron diffraction (SAED) patterns of the pure metal oxide. Second is a comparison of the decomposition of ammonium molybdate at different temperatures in a standard oven for 4 hours to dried products simply dried from three cavitation experiments in an insoluble media where the heating residence times are extremely short (Figure 5.23). A detailed analysis of the X-ray diffraction (XRD) data shows that the bulk heating of the particles during cavitation synthesis was over $T = 473$ K. Under the different calcination temperatures, peak intensity and locations shift (Figure 5.23). From the kinetic data and activation energies extracted from the oven heating data, it can be estimated that in the CaviMax processor heating resulted in temperatures exceeding 900 K for a brief period due to shock wave heating.

To illustrate the capabilities of the process to produce high purity complex metal oxide nanocrystals while classical syntheses do not, the XRD analysis resulting from the synthesis of a perovskites, $La_{0.6}Sr_{0.4}FeO_3$, is illustrated in Figure 5.24. XRD shows that the use of the CaviMax processor with the 1.85 mm orifice and the CaviPro

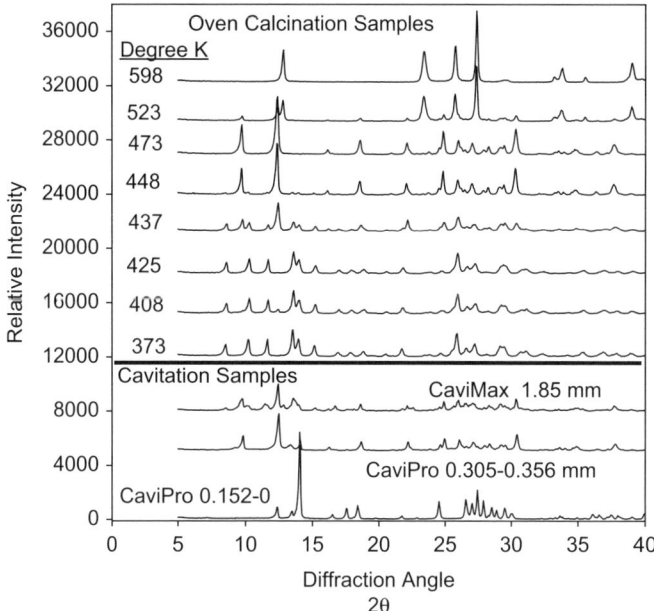

Figure 5.23 XRD of ammonium molybdate calcined in oven (top section) and decomposition in cavitation processors at different settings.
Reprinted from ref. 18, Copyright (2001), with permission from Academic Press.

processor resulted in the desired complex metal oxide having no impurities. The classical synthesis resulted in substantial phase impurities (top intensity plot in Figure 5.24) and all experiments in the CaviPro 300 processor using a variety of orifice settings resulted in pure phase materials.

Another example is the synthesis of a series of methanol synthesis catalysts of the composition $Cu_{0.22}Zn_{0.68}Al_{0.10}O_x$. The synthesis was performed using a wide range of cavitation processor settings. After calcining in air at 350 °C, XRD analysis showed that the grain size of the CuO component could be systematically changed in the important (1–10) nm range by a simple adjustment of processor parameters. In addition, the crystallographic strain in the crystals increased as the grain size decreased (Figure 5.25). These data demonstrate a phenomenon never reported before by any type of synthesis, that certain conditions cause CuO components to expand to match the ZnO lattice.

Crystallographic strain is thought to lead to more active catalyst particles and other experiments have shown that classically prepared materials exhibit very little strain. Thus the introduction of strain seems to be unique to processing under these high shear conditions.

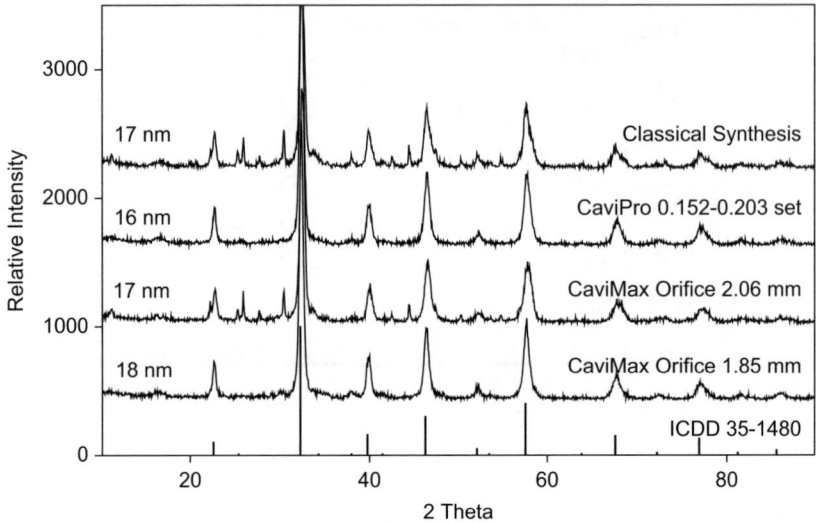

Figure 5.24   Cavitational synthesis of the phase pure perovskite, $La_{0.6}Sr_{0.4}FeO_3$.
Reprinted from   ref. 18, Copyright (2001), with permission from
Academic Press.

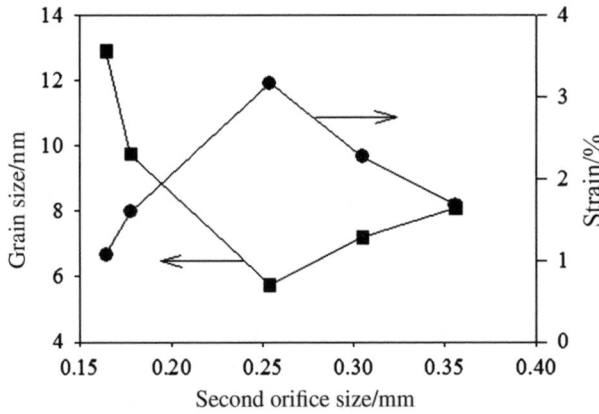

Figure 5.25   Relationship between Copper oxide grain size and strain in
$Cu_{0.22}Zn_{0.68}Al_{0.10}O_x$.

The small CuO grain sizes are important since it has been shown that
methanol synthesis activity greatly increases for (4–6) nm size grains.[37]
   A major advantage of cavitational processing for the synthesis of
supported metals is the capability to synthesize metal grain sizes in
the important (1–10) nm grain sizes. By carrying out the synthesis of
fine gold particles in a CaviPro 300 processor followed by deposition
onto titania, Moser *et al.*[34] were able to synthesize several Au grain

sizes on titania in the 1–10 nm region (Figure 5.26). The exhibited XRD analysis was performed on 400 °C air calcined materials.

Following the cavitation synthesis of nanostructured grains of silver, the slurry of particles was deposited onto α-alumina under cavitational processing. XRD analysis of the materials after air drying at 100 °C showed the grain sizes were between 16 and 21 nm, and could be generated in four sizes by altering the process parameters. Interestingly, the grains grew to only 40 nm when calcined at 450 °C.

By carrying out the reduction of a metal salt within a cavitating zone, reduced metals could be prepared at variable grain sizes. Figure 5.27 shows the results for the preparation of palladium black

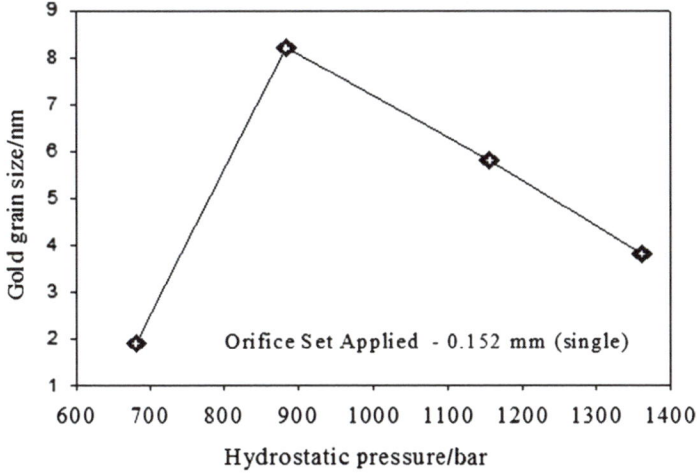

Figure 5.26 A 2% Au on titania catalyst prepared in varying grain sizes.[34]

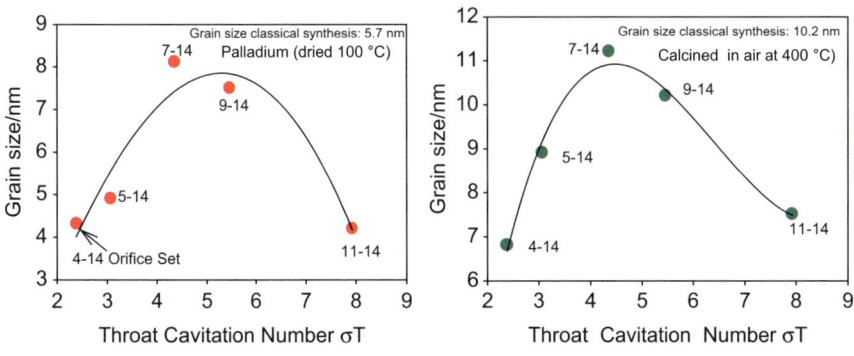

Figure 5.27 Effect of cavitation number on grain size of palladium black (left) and palladium oxide (right). Orifice settings 4-14 to 11-14 correspond to 100–355 micron openings.[34]

in water after drying to 100 °C. The figure shows that the grain sizes varied as a function of the throat cavitation number derived from mechanically changing the processor parameters on the CaviPro 300. Calcining these samples to 400 °C in air resulted in the formation of nanostructured palladium oxide with the same profile as in Figure 5.27 where the grains grew only (2–3) nm larger.

A wide range of cavitation synthesis methods for the preparation of nanostructured metal oxides on both inert and reactive supports have been developed.[3,18,29–33] These studies show that, by altering the processor parameters and the concentrations of reactants, the grain sizes of active supported material can be altered to a modest degree. In all cases the processed materials exhibit different colours compared with classically prepared materials, though in some cases, their XRD shows little difference compared with classical preparations. In all cases where the degree of strain was evaluated, cavitation prepared materials exhibit significantly higher crystallographic strain.

For example, a SO_2 reduction catalyst having the composition of 5 at% Ni, 8 at% La and 87 at% Ce was synthesized using a variety of cavitational processor conditions. The samples were calcined to progressively higher temperatures and their grain sizes and crystallographic strains were measured by XRD. XRD results show that the strain introduced into samples which have been calcined to 225 °C ($T = 498$ K) demonstrate a large amount of strain which decreases upon calcination at 650 °C ($T = 923$ K) (Figure 5.28). The calcining

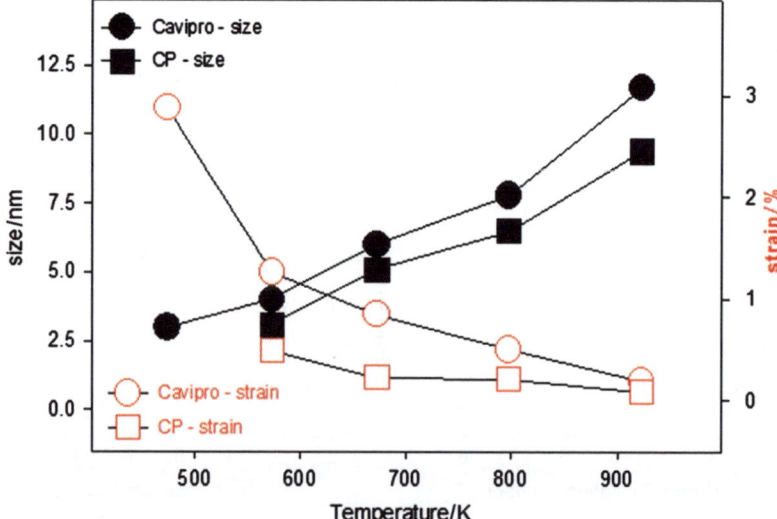

Figure 5.28   Introduction of strain and variable grain sizes in the cavitational and classical synthesis of $Ni_{0.05}La_{0.08}Ce_{0.87}O_x$.[34]

technique resulted in variable catalyst grain sizes systematically varying between 3 and 11 nm.

Cobalt molybdate on alumina was synthesized by cavitation on commercial γ-alumina using techniques which were designed to afford nanocrystalline $CoMoO_4$ rather than a monolayer of the active components on the support. These syntheses showed that processing by cavitation resulted in a much higher amount of pure phase $CoMoO_4$ deposited on the support in contrast to the classical method of synthesis. The XRD of the catalysts prepared using different orifice sets in the CaviPro 300 and calcined to 350 °C are shown in Figure 5.29. The classical synthesis afforded significant amounts of separate phase MoO_3. The preparation of the same catalyst by classical incipient wetness resulted in no formation of discrete $CoMoO_4$ crystallites. The grain size of the cavitation prepared materials was also smaller than the classically prepared samples. Thus the cavitation prepared materials are much different from similar preparations of $CoMoO_4$ carried out on Cabosil at active metal oxide concentration levels between (10 and 60)%w/w Co–Mo (Figure 5.30). The cavitational preparations resulted, in all cases, in higher purities of the $CoMoO_4$ phase than the classical preparation (by precipitation using good mechanical stirring), and the use of the 6–10 orifice sets (150–255) μm virtually eliminated separate phase MoO_3.

The synthesis of unsupported cobalt molybdate was examined using a wide variety of processors and process parameters. One of the

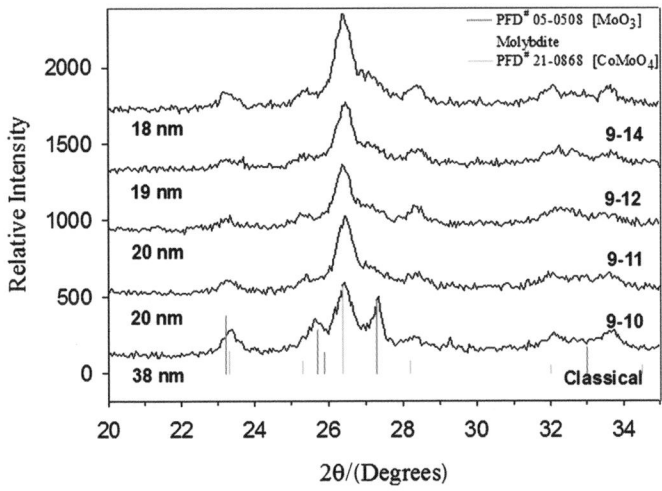

Figure 5.29 Cavitation and classical synthesis of cobalt molybdate supported on γ-alumina. 9-14 orifice sets (225–355 micron).[34]

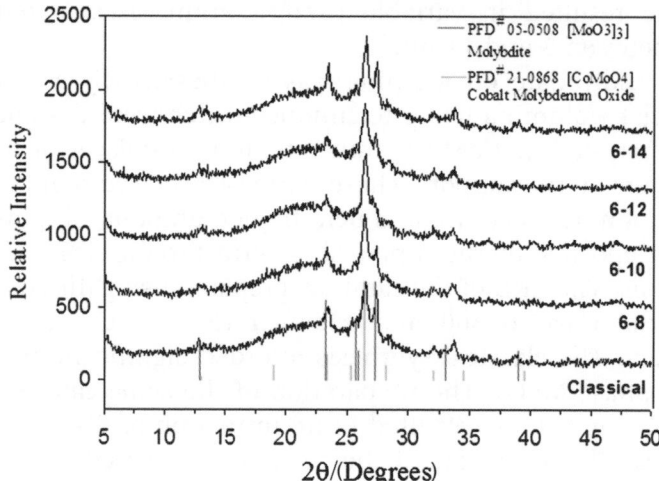

Figure 5.30   Cavitational and classical synthesis of CoMoO₄ on Cabosil after 350 °C calcination.[34]

most interesting results was obtained using the CaviMax equipped with a back-pressure regulator. The syntheses were carried out using a Mo to Co ratio of 2.42 to approximate commercial catalysts. The XRD diffraction patterns of the CaviMax prepared catalyst were determined using the 1.85 mm orifice and four different back pressures (0, 0.62, 1.24 and 1.72) MPa. Following calcination at 325 °C, the XRD phase analysis indicated that the classical preparation resulted in significant amounts of separate phase MoO_3. The most striking result was the 1.24 MPa back-pressure experiments which resulted in a high fraction of $CoMoO_4$. Secondly, the MoO_3 in the other cavitation experiments showed that the excess MoO_3 was formed in a preferred crystal orientation.

The synthesis of titania was investigated to determine whether different cavitation processor parameters would introduce differing amounts of crystallographic strain into titania and whether grain sizes could be systematically regulated. One set of experiments used the CaviMax processor equipped with different orifice sizes. Figure 5.31 shows the relationship between the different size orifices used in the CaviMax processor, and the introduction of crystallographic strain after the materials had been calcined to 400 °C. The surprising observation is that the strain increases smoothly to very high values as the size of the orifice increases. The strains measured for classically prepared titania were in the league of (0.22–0.25)%. Crystal strain is expected to increase the activity of

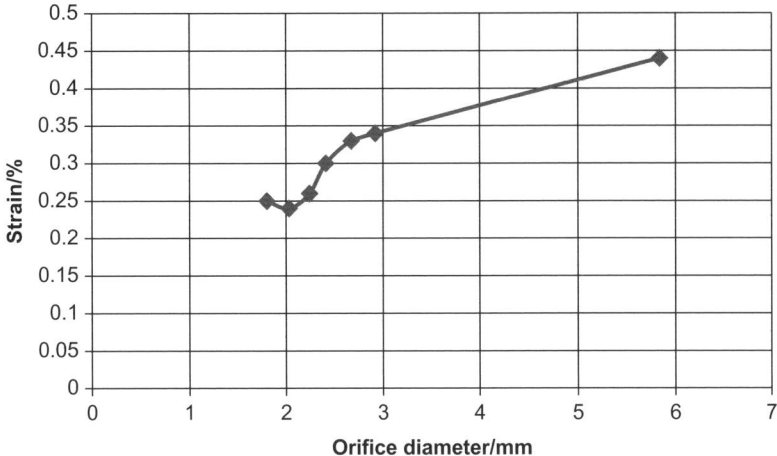

Figure 5.31 Introduction of crystallographic strain into titania by CaviMax processing.[34]

titania surfaces for catalytic reactions such as photo decomposition of organics.

A wide range of studies in the single (CaviMax) and double orifice processors (CaviPro 300) on the piezoelectric (PZT) system in the family $PbZr_xTi_{1-x}O_3$, showed that the processing parameters could greatly enhance the amount of micro-strain to values as large as 0.93%. Figure 5.32 illustrates the range of micro-strain that resulted from cavitational processing in both the single orifice and double orifice processors. The micro-strain was measured on samples calcined in air at $T = 873$ K. Figure 5.32 shows that the high flow rate, single orifice processor resulted in a much higher degree of induced strain than the low flow, high pressure single orifice processor. The minimum to maximum pressure for the double orifice, variable flow, maximum pressure experiments was 20.6 MPa to 169 MPa. The flow rate in the constant flow, variable pressure experiments used constant flow rates of 400 mL min^{-1} by adjusting the operating pressure of the processor. Other experiments showed that all of the cavitational synthesized PZT compositions were essentially 100% pure phase while the classically prepared materials all contained modest amounts of secondary phases. The grain sizes of the cavitational samples at temperatures above 873 K air calcination were around 25 nm. Although it has not been reported experimentally in the literature, it would be expected that these high levels of micro-strain would affect both the piezoelectric properties and rates of grain growth.

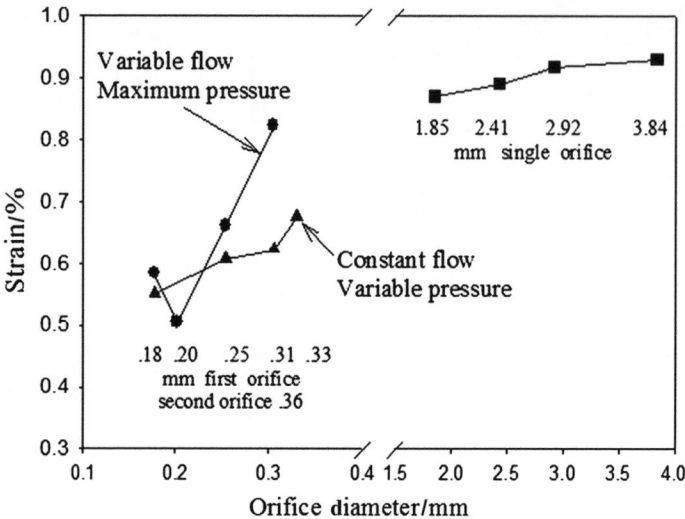

Figure 5.32   Strain introduced into PZT crystallites of Pb $Zr_{0.5}$ $Ti_{0.5}O_3$ by cavitational processing.[34]
Reprinted from ref. 18, Copyright (2001), with permission from Academic Press.

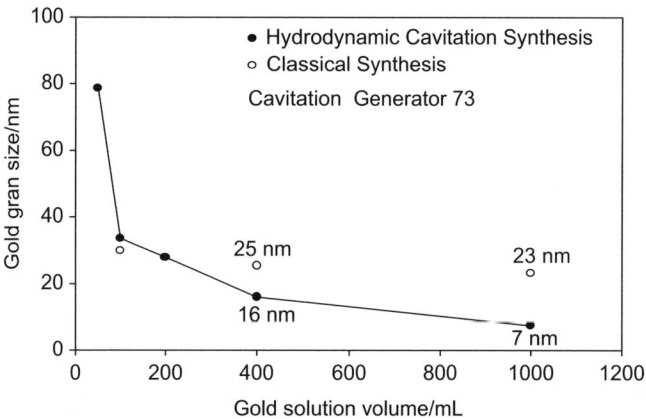

Figure 5.33   Variation in grain size as a function of gold salt solution volume. The system consisted of 2% gold on titania CO oxidation catalysts.[18]

A further application of cavitation processing is in gold catalysts (Figure 5.33). Particles of systematically varying grain sizes result from a combination of cavitation processing and applied gold concentration during the synthesis. The graph shows that the parallel classical synthesis all resulted in about 25 nm Au particles.

These catalysts are potentially important in the selective oxidation of CO in the presence of hydrogen.

Nanocalcite can also be synthesized using a hydrodynamic cavitation reactor.[38] The orifice diameter and its geometry have significant effects on the carbonation process. The average grain size of the calcite synthesized without cavitation was found to be approximately 100 nm while, with cavitation processing at the optimal reactor geometry, particle sizes of approximately 37 nm were achieved.[38]

In summary, the application of hydrodynamic cavitation to the synthesis of advanced catalysts was found to be successful for the production of nanostructured catalysts grains in variable grain sizes and in exceptional purity. The intervention of *in situ* shock wave heating during synthesis normally results in products that require little or no post synthesis calcination. This type of processing was found to introduce unusually high degrees of crystallographic strain which should enhance the catalytic performance of cavitation prepared catalysts. The data illustrate the capability of the hydrodynamic cavitation to prepare industrially important catalysts. It is expected that this type of processing will become important in the reproducible synthesis of commercial catalysts, both old and new, due to the high degree of mixing involved in the syntheses.

5.8 HYDRODYNAMIC CAVITATION AS A TOOL TO CONTROL PROPERTIES OF ACTIVE PHARMACEUTICAL INGREDIENTS

Cavitation processing has applications in the creation of new proprietary dosage forms in the pharmaceutical industry due to its ability to precisely control the particle size (or droplet size), distribution and morphology of pharmaceutical preparations at the nano, sub-micron and micron scale, and to coat or encapsulate active pharmaceutical ingredients (APIs) to enhance their targeting and controlled-release functionality.

Hydrodynamic cavitation is a powerful tool for creating new dosage forms, as well as improving the bioavailability of low solubility compounds. Cavitational processes can integrate well with conventional pre- and post-processing and material handling steps, and often reduce the number of these steps in pharmaceutical production. Cavitational processes also share certain limitations with, and reliance on, conventional approaches. These include the need for effective surface chemistries to stabilize small particles and the current state-of-the-art for drying and separating sub-micron and nano-sized particles.

The energy transfer from cavitation bubble collapse is the primary mechanism for breaking up particles and droplets. The use of shock waves represents a unique means of imparting energy to particle processing. Even if this level of energy impulse could be applied to materials through traditional means (*e.g.* media mills, high-speed blades, gap or jet homogenizers), the results would typically be degradation of the processed particles because their primary means of imparting energy is shear (*i.e.* tearing) and impact (*i.e.* hammering). However, cavitation applies these energy levels through the mechanism of a focused, short-lived, high-energy shock wave created by bubble implosion, resulting from a homogeneous, cavitation field. As a result, cavitation technology is both higher energy and 'gentler' as a 'top–down' particle processing method and may avoid the negative effects of traditional shear and impact processes methods, especially with more sensitive biological materials.

Generally, the preparation of drug particles using hydrodynamic cavitation can be classified into 'top–down' and 'bottom–up' strategies. Top–down strategies refer to mechanical attrition of coarse drug powders. Bottom–up strategies involve fabrication of nano/micro scale crystals from molecular scale materials.

The use of hydrodynamic cavitation to reduce the particle size of APIs has been tested at Merck & Co. Inc. by Felmet *et al.*[39] at laboratory and pilot scales. Particle size reduction was tested using two orifice type cavitation mills implemented using the experimental setup shown in Figure 5.34. The lab scale cavitation mills were double

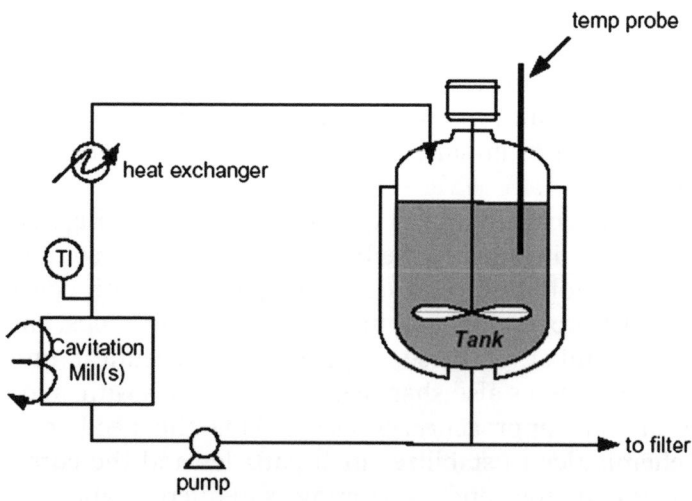

Figure 5.34 Cavitation mill for pharmaceuticals processing, experimental setup.[39]

orifice processors having variable pressure adjustment capabilities up to 206 MPa and orifice diameters ranging from 0.2 mm to 0.4 mm. The pilot plant scale cavitational mills were single orifice processors having variable pressure adjustment capabilities up to 34.5 MPa and orifice diameters ranging from 1.17 mm to 2.35 mm.

A proprietary anti-convulsant pharmaceutical ingredient was processed under two sets of operating conditions (the primary variable being operating pressure): at a low working pressure of approximately 0.7 MPa and at a high pressure of approximately 10.3 MPa. By increasing the operating pressure, a more uniform suspension of the pharmaceutical and smaller-sized crystals was obtained (Figure 5.35).

Using relatively modest working pressure (less than 10.3 MPa) allowed the production of APIs with average particle sizes in the range of (2 to 10) µm. The use of surfactants, stabilizers or encapsulating agents enabled the creation and stabilization of sub-micron and nanostructured particles from the cavitation process.

A second important parameter that impacts the results of cavitation processing is the number of times the material is passed through a cavitation chamber (*i.e.* recirculated through a single chamber or multiple chambers in series). Data for dispersing a material with a single, and then with multiple passes through the cavitation zone while holding other processing parameters constant are shown in Figure 5.36. At a constant pressure, an increase in the number of cycles results in smaller particle sizes and narrower particle size

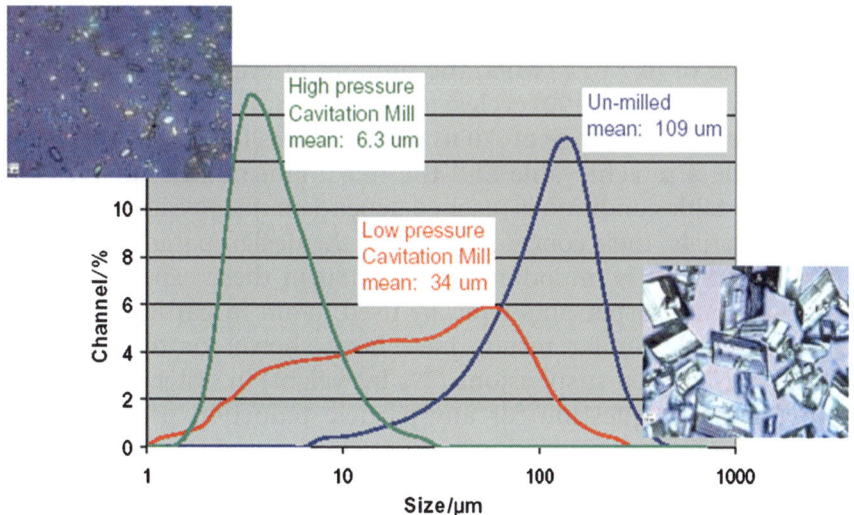

Figure 5.35 Pharmaceutical particle size reduction *via* hydrodynamic cavitation.[39]

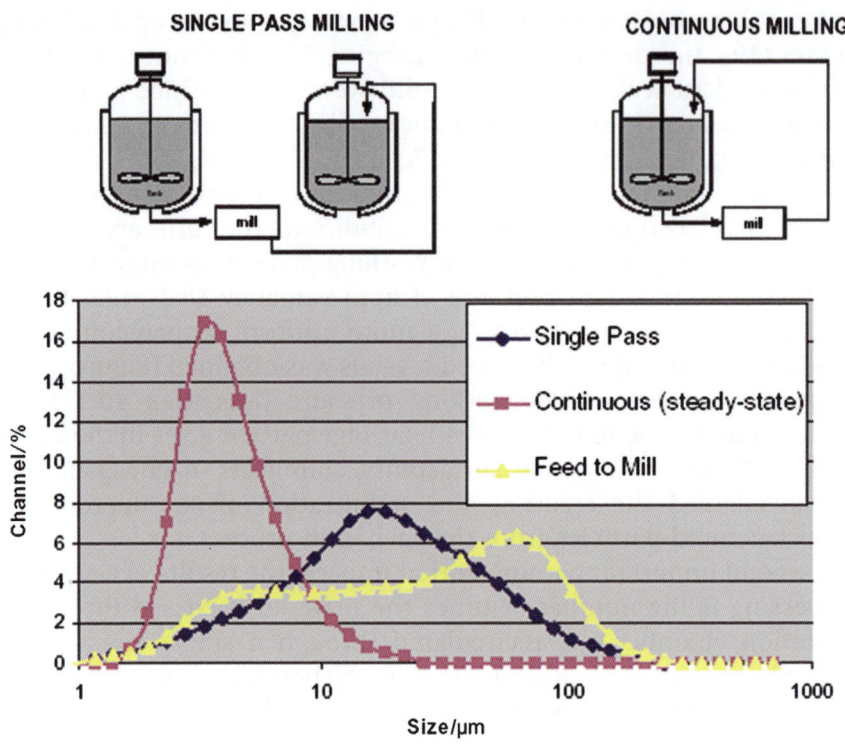

Figure 5.36   Single pass *vs.* continuous milling.[39]

distributions. As would be expected, for every working pressure (and a given material and formulation) there exists some number of cycles at which little or no additional particle reduction occurs. For pressures in the range of (0.7 to 7) MPa, the number of cycles to arrive at this stable condition is (60–90) cycles; at 34.4 MPa it is (15–30) cycles, and for pressures in the range of (70 to 100) MPa it is from (2 to 10) cycles. The particle size achievable and the maximum number of cycles at which the stable condition is reached depend on the physical properties of the materials, their concentration and the design parameters of the cavitation mill. It is important to note that in these experiments the surfactants and dispersants typically used to aid in particle reduction were not used. Similar results have been shown for the cavitation milling of Naproxen suspensions (7% by weight) in water (Table 5.6).

An important question for any particle reduction process is the impact of solids concentration on particle size and distribution. For any given dispersion or milling process, there is a limit to the solids concentration that can be effectively and economically processed. For example, in a high-pressure homogenizer, it is generally

Table 5.6 Naproxen milling processes (Naproxen, HPC-SL, SLS, water system).

Processing pressure/MPa	Cumulative number of passes	D_{43}/μm	D_{10}/μm	D_{50}/μm	D_{90}/μm
Control	–	10.43	3.07	8.27	20.28
17.2	50.7	5.66	2.16	4.73	10.37
62.0	102.3	2.76	1.18	2.41	4.83
62.0	204.7	2.31	1.10	2.10	3.85
172.4	35.0	2.83	1.19	2.41	5.03

HPC-SL, hydroxypropyl cellulose (super low viscosity); SLS, sodium lauryl sulphate.

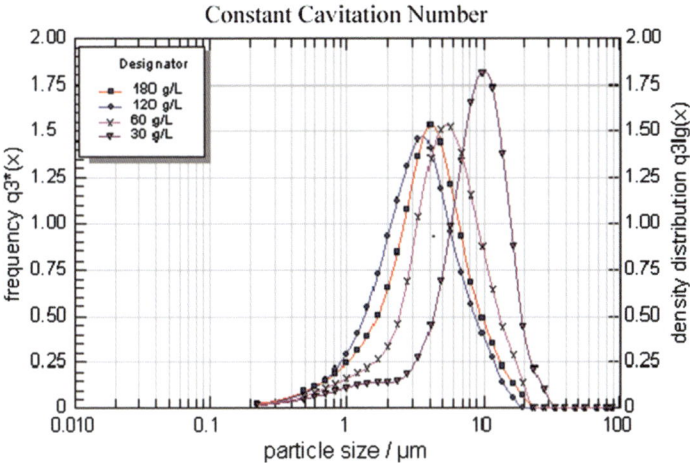

Figure 5.37 Influence of the content of crystals on the crystal size.[39]

recommended that the solids concentration does not exceed 10% and the preferred level is under 5%. This is due to the increase in viscosity that occurs as the solid concentration increases, which in turn reduces turbulence during mixing, and thus results in poor dispersion of materials.

In contrast to the homogenizer, during cavitation processing, dispersion results improve concentration increases (Figure 5.37). Of even greater interest is that the best results can be achieved at solids concentrations of 12–18%, significantly higher than the maximum practical or recommended for a high-pressure homogenizer. In other words, cavitation technology allows the processing of more concentrated suspensions, which can significantly impact process economics

For the cavitation process to be effective in this type of production, it must provide consistent results from batch to batch, and as it is scaled from the bench to pilot and production scales. Reproducibility from one batch to another is even more critical in the pharmaceutical industry because of the high cost of non-compliance and of the cost of the pharmaceutical materials. Loss of any batch because of a fluctuation in particle size can bring direct economic losses and increase the costs of production.

Particle size distribution curves for three different batches (identical size and formulation) of an API processed in a pilot scale cavitation mill (11.3 L min^{-1}) are shown in Figure 5.38.

Traditionally, particle size reduction of pharmaceutical ingredients is carried out wet using a rotor stator mill, or dry in a jet mill. Comparison of these technologies with cavitation technology (Figure 5.39) shows that cavitation technology can achieve smaller particle sizes and narrower size distributions. All tests were run with a crystal having a pre-processing mean particle size of 109 μm. The mean particle size after dry milling was approximately 28 μm, and after rotor stator processing was approximately 17 μm. However, the mean crystal size of the API after cavitation processing (at pressures of 10 MPa) was 6.3 μm. The particle sizes achieved with wet and dry

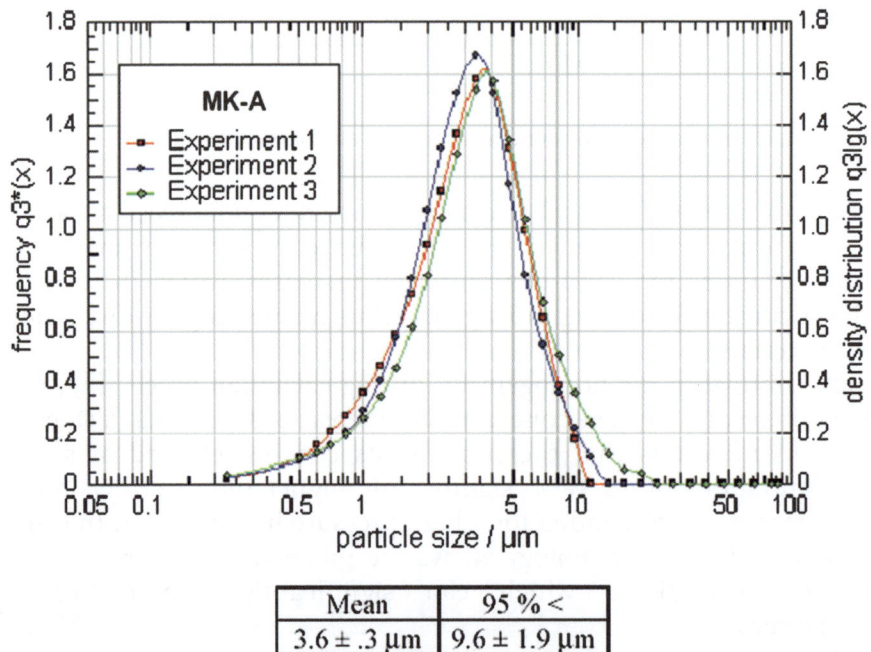

Mean	95 % <
3.6 ± .3 μm	9.6 ± 1.9 μm

Figure 5.38   Reproducibility between batches.[39]

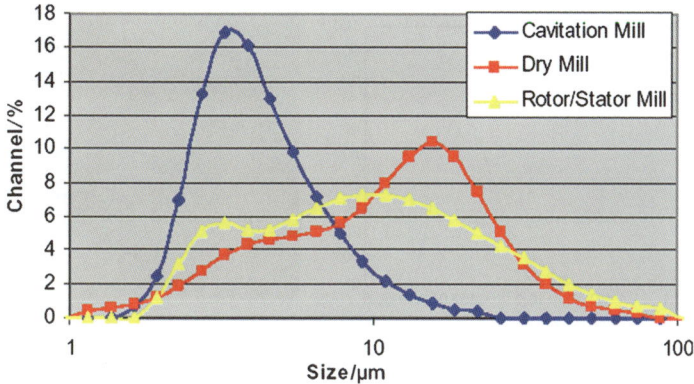

Figure 5.39 Comparison of the effects of milling technology on the particle size distribution of pharmaceutical ingredients.[39]

milling can be matched with results from cavitation milling at working pressures of less than 0.7 MPa. An example is the processing of the antibiotic florfenicol (Figure 5.44). This drug is available commercially with a mean particle size around 100 μm relatively broad particle distribution. The drug is currently prescribed in tablet form.

Figure 5.40 shows micrographs and particle size distribution curves for florfenicol before and after cavitation processing. A cavitation system with a working pressure of 70 MPa was employed. During processing, the crystal concentration of florfenicol was 12% by weight. As a result of cavitation processing, the particles of initial material were readily reduced from a mean of 100 μm to 10 μm, and had a narrower particle size distribution. Initial toxicology work indicated that the smaller particle florfenicol produced less side effects (*e.g.* site wound in animals) than the formulations currently on the market. With this particle size, there is the potential to develop new dosage forms of this drug (*e.g.* topical serum drops and patch-delivered drug).

A second example is the processing of the drug propofol. Figure 5.41 shows micrographs and the particle size distribution of propofol produced by an impingement jet homogenizer with an operating pressure of 140 MPa. As with the high-pressure gap homogenizer system, there still exist a significant number of encapsulated droplets of the API with a particle size greater than 1 μm. In contrast, the micrograph and particle size distribution of the propofol suspension produced with a cavitation device with a working pressure of 103 MPa show that cavitation creates smaller size droplets with narrower size distribution compared with the homogenizers.

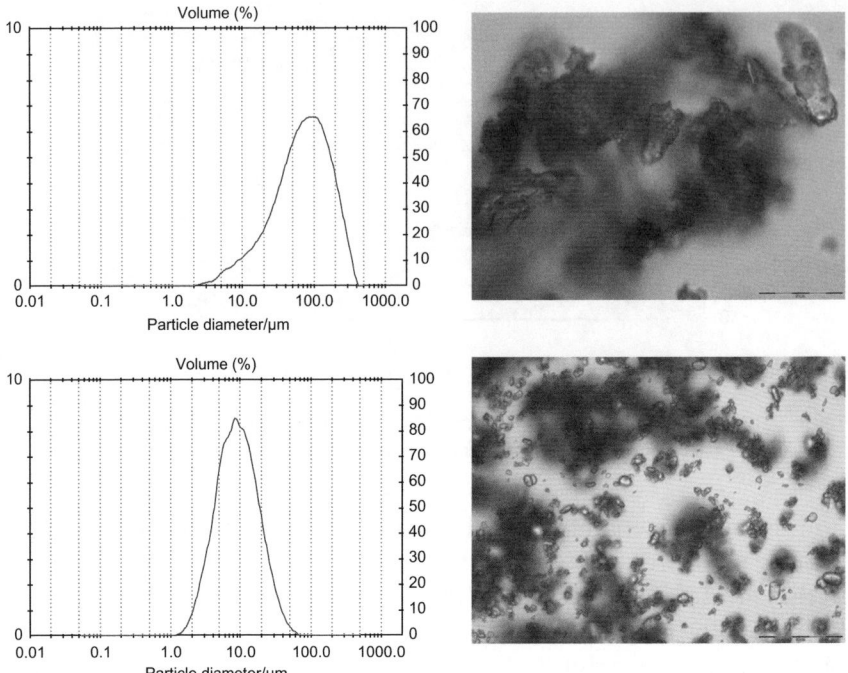

Figure 5.40 Florfenicol particle size distribution and photo before (top) and after (bottom) cavitation processing.

It is important to note that in a homogenizer, under certain process conditions, hydrodynamic cavitation may occur as result of the velocity of the fluid within and as it exits the homogenizer channels and gaps.[40] This cavitation has the effect of improving process results. However, the uncontrolled and variant nature of homogenizer cavitation can result in uncontrolled variations in process conditions and cavitation erosion in the equipment. This variability can result in inconsistencies in the process results and contamination from eroded parts, which is especially intolerable for pharmaceutical materials.

The 'top–down' approach of milling is a common technique for reducing the particle size of APIs, but may not always produce the optimum or desired results. In addition, since traditional milling generates considerable heat, it may not always be appropriate, for example, where the API has a low melting point or may otherwise degrade the materials. Prolonged milling in a high-energy media mill results in crystal lattice defects, which may be manifested, for example, in lower stability or hydroscopicity.

An alternative technique to milling is to precipitate the solid (in crystalline or amorphous forms) from solution. Standard

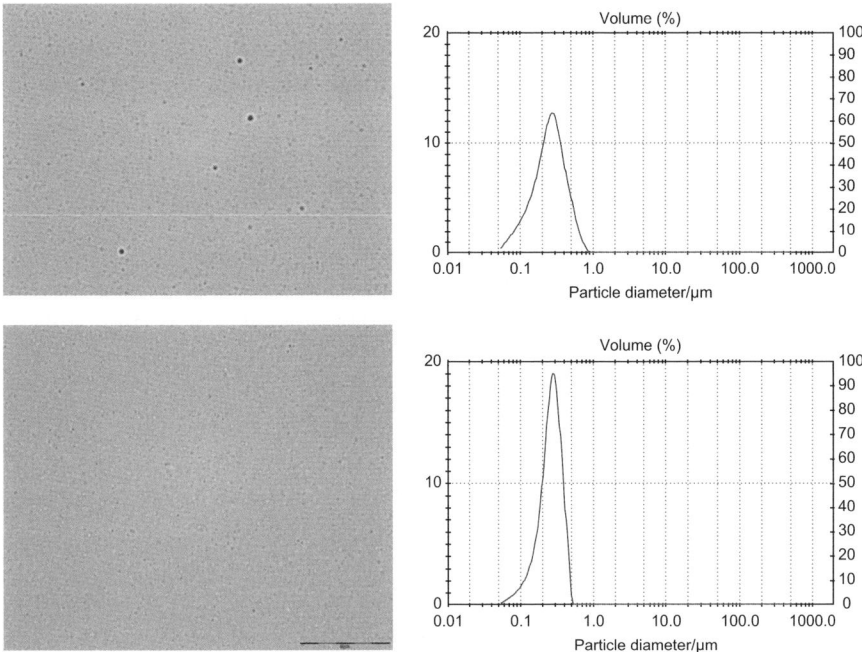

Figure 5.41 Micrograph and particle size distribution of propofol produced by an 'impingement jet' homogenizer (top) and cavitation processing (bottom).

crystallization procedures involve contacting a supersaturated solution of the compound to be crystallized with an appropriate 'anti-solvent' in a stirred vessel. Within the stirred vessel, the anti-solvent initiates primary nucleation, which leads to crystal formation and crystal growth during an aging step. Mixing within the vessel can be achieved with a variety of agitators. Current crystallization processes produce particles in a relatively large particle size distribution of (50 to 2000) µm, usually requiring a subsequent milling step, and the problems and issues described above. In addition, the uncontrolled nature of conventional crystallization processes can often result in variations in crystal morphologies within each batch and between batches. A strategy of conventional crystallization followed by milling does not allow the desired level of control of morphology, and the subsequent surface and stability treatment of the crystals. Since process control is essential, and different crystal morphologies of a single compound can impact pharmacokinetics and may represent differentiable pharmaceutical products, these crystal variations may create serious negative consequences for developers and manufacturers.

The anti-solvent precipitation process steps include dissolution, nucleation, solution diffusion and particle growth. The potential for hydrodynamic cavitation to improve mass transfer, enhance nucleation and suppress crystals growth or agglomeration can improve this process.[41,42] Results from classical and cavitational crystallization of a nano dispersion of Naproxen composition with a copolymer of PVP and Pluronic® in water is shown in Table 5.7. The crystallization process was carried out in a cavitation device with a single orifice and a diameter of 2 mm. Initially, deionized water was added to the hopper of the cavitation device. The cavitation device was then started to permit the water to flow through the flow-through channel of the cavitation device, during which PVP and Pluronic was added to the hopper and dissolved in the water to form a water phase mixture (anti-solvent, fluid stream). The cavitation device was then turned off temporarily. Next naproxen was dissolved in ethanol at 7% to prepare a feed solution, which was kept at room temperature.

The cavitation device was then restarted with the water phase mixture already in it and the water phase mixture was supplied to the cavitation device at a pressure of 0.7 MPa and flow rate of 12.0 L min^{-1}. Next, the naproxen solution was placed in the dosing hopper. Crystallization was achieved by adjusting the dosing nozzle valve manually to allow the entire amount of solution to be introduced in the cavitation zone. The naproxen : PVP : Pluronic ratio in suspension was 1 : 3.43 : 1.6. For comparison, a classical crystallization process was carried out in a laboratory mixer with 3000 revolutions per minute (rpm) mixer rotation. Cavitation prepared naproxen crystals of sizes ranging from 0.12 to 1.87 μm were produced. The median particle size of the naproxen crystals was 0.35 μm. Classically prepared naproxen crystals ranged in size from 0.31 to 18.84 μm and were highly agglomerated. The median particle size of the naproxen crystals was 2.91 μm (Table 5.7).

The application of hydrodynamic cavitation to crystallization systems offers significant potential for modifying and improving both the process and products. The most important mechanism by which cavitation can influence crystallization is the ability of cavitation to generate nuclei in a highly uniform and reproducible way. Nucleation is a formation of the nucleus in a previously crystal-free solution. To initiate nucleation, conventional processes require a large super-saturation driving force. This can result in low yields of the desired pharmaceutical form.

Once nuclei have been formed, crystal growth begins. The ability of cavitation bubbles to generate primary nuclei is based on the unusual physical effects which occur inside, on the surface, and adjacent to collapsing cavitation bubbles. At the point of collapse, extreme

Table 5.7 Comparison between classical and cavitational methods of crystallization.

Method	Processing conditions Recirculation time after dosing completion 1.0 min						Naproxen particle size after crystallization measured 3–4 hours after processing					
	Processing pressure		Processing temperature		Solvent dosing rate/g min^{-1}							
	P$_1$/MPa	P$_2$/MPa	Aqueous phase/°C	Organic phase/°C		D$_{43}$/μm	D$_{10}$/μm	D$_{50}$/μm	D$_{80}$/μm	D$_{90}$/μm
Cavitational	0.7	0.14	1.3	24.0	152.3	0.72	0.12	0.35	1.13	1.87
Classical	0	0	1.5	21.6	10.0	6.73	0.31	2.91	10.19	18.84

gradients of temperature, pressure and concentration create local points of supersaturation. The highly localized heat arising from cavitation close to crystal surfaces may enhance crystal purity by causing temporary local supersaturation. This results in the dissolution of impurities from the crystal surface. Secondarily, when the mixture of solvents and anti-solvents passes through the cavitation zone, which has a local lower pressure, instant vaporization of part of the solvents occurs inside the cavitation cavern. The partial vaporization of the solvents throughout the cavitation zone can create super-saturation conditions, which helps grow nuclei to crystals.

Figure 5.42 shows various designs for introducing solvent, anti-solvent and other modifiers into a cavitation system where the

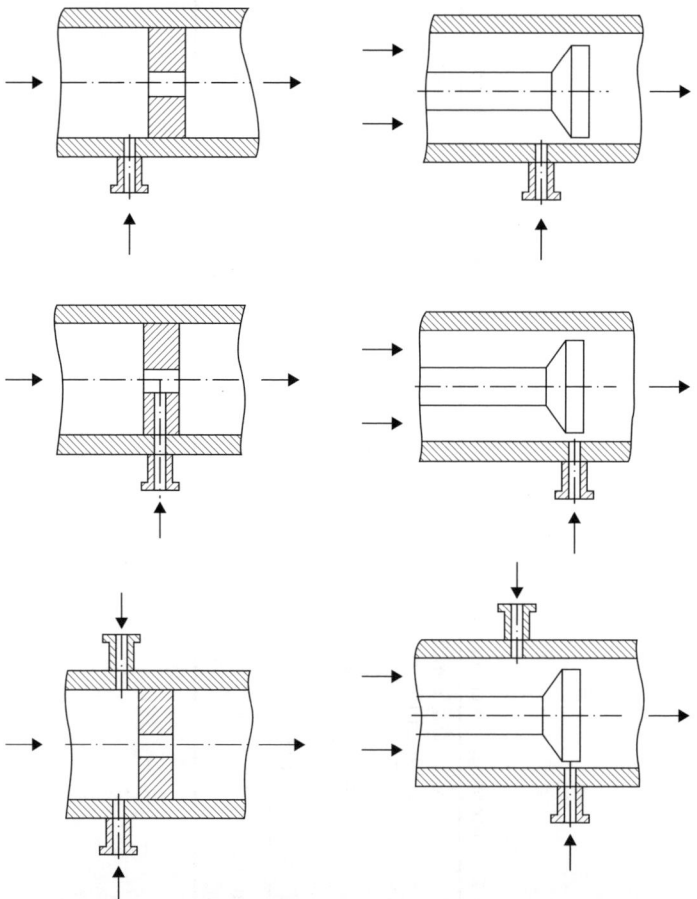

Figure 5.42 Diagrams illustrating the basic principle of hydrodynamic cavitational crystallization processing.

nucleation and crystal growth begins. In the traditional approach, the contacting of solvent and anti-solvent, nucleation and crystal growth occurs in a stirred vessel. Arrows show points of introduction of solvents and anti-solvents. Precise levels of solvents and anti-solvents can be introduced immediately before the cavitation generator or directly inside the chamber where cavitation is developed. Cavitation creates highly intimate contacting of the solvent and anti-solvent resulting in fine and highly uniform nucleation. A cavitation crystallizer can have several cavitation zones positioned sequentially in the direction of the flow. This mixture then enters subsequent cavitation zones (by recirculation or by passing into additional cavitation chambers in sequence). During recirculation, additional nuclei and crystals develop throughout the cavitation zone. Additionally, cavitation induces crystal breakage, thus increasing the overall rate of nucleation *via* secondary nucleation as the crystal breakage creates fragments that act as additional nuclei.

5.9 APPLICATIONS OF HYDRODYNAMIC CAVITATION IN BIOPROCESSING

Applications where hydrodynamic cavitation has been used as part of processes for renewable energy production include the processing of corn slurries to improve ethanol production, activated sludge to increase biogas yield from recalcitrant secondary solids, and for the continuous production of biodiesel.

5.9.1 Use of Cavitation to Improve Corn Ethanol Production

Corn ethanol production supplies nearly 10% of the gasoline transportation fuel in the USA. Production of corn ethanol at dry milling ethanol plants begins by mixing milled corn, water and enzymes to create high solids slurries (30–35)%. The particle size distribution of corn slurries are dependent on the type of mill used, the screen size and the corn kernel hardness (hard or soft). Most plants use hammer mills containing screens with relatively small openings to mill corn grain and allow more complete and rapid starch release. However, the energy cost of reducing particle size using hammer milling much beyond 1 mm increases exponentially and becomes prohibitive to the point that it cannot be justified on an energy use basis. Controlled flow cavitation has been applied to hammer milled corn grain to further reduce the particle size distribution and to liberate recalcitrant starch.

Both acoustic and hydrodynamic cavitation have been applied to corn slurries to improve starch release.[2,43–45] Acoustic cavitation can increase particle surface area and degrade crystalline starch making it more susceptible to amylase hydrolysis.[43–45] This approach has been tested at ethanol plants, but found to be energy inefficient and unwieldy at large scale. In contrast, the scale-up of hydrodynamic cavitation equipment is much simpler, making it well suited to industrial-scale processing.[46]

In a recent study, the effects of commercial scale hydrodynamic cavitation on corn particle size distribution and starch release were determined by comparing the particle size distribution in cavitated corn slurry samples with that in uncavitated samples.[2] Whole kernel no. 2 yellow-dent corn was hammer-milled at a commercial-scale plant and passed through a 2.78 mm screen. The milled corn was transported to a mixer, and hot water and alpha-amylase were added to create a slurry. The slurry was then cooked and then passed through a controlled flow cavitation unit (Figure 5.43).

The particle size distribution for the control treatment particle showed two peaks with apexes centered at 11 μm and 900 μm. These sizes are commonly obtained after corn hammer milling. Cavitated slurry showed two similar peaks (Figure 5.44). However, there was a significant increase in the smallest particle size group indicating that the large particles had been converted to individual starch granules in the slurry. The amount of solids was reduced after cavitation

Figure 5.43   A controlled flow cavitation system installation at a 100 million gallon per year corn ethanol plant. Cavitation occurs within the stainless steel column section of the unit.

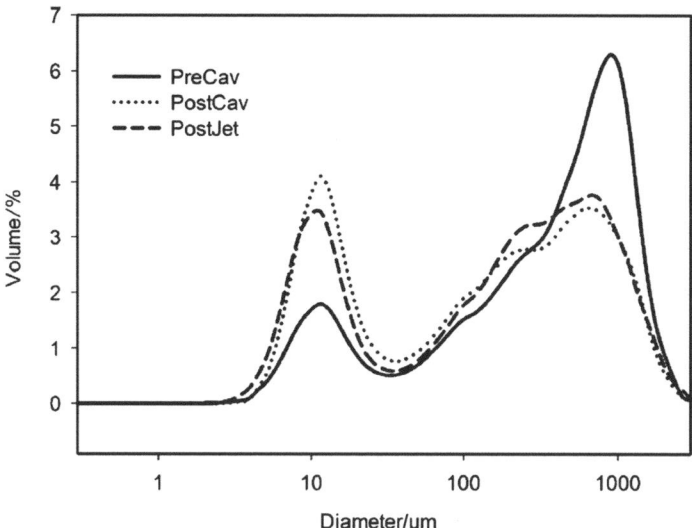

Figure 5.44 Effects of cavitation on corn slurry particle size distribution. PreCav, corn slurry immediately prior to cavitation; PostCav, corn slurry after cavitation treatment; PostJet, corn slurry after cavitation and jet cooking.[2]

indicating that starch granule clusters and granule pieces were solubilized due to cavitation treatment. These effects contributed to increased starch release and hydrolysis as well as higher levels of DP4 +, maltotriose and maltose in liquefied cavitated slurries.[2]

After fermentation, ethanol production was significantly greater in cavitated treatments compared with uncavitated treatments.[2] The increase in ethanol production typically ranges from 1 to 4% depending on the plant type and cavitation conditions. Additional benefits of greater starch conversion and smaller more uniform particle size distributions include more nutrient dense distiller grain by-products and improved ability to pump the slurry.

An important consideration in the use of new technologies like hydrodynamic cavitation to increase biofuel yield is the energy return on investment (EROI). The electricity expended for cavitation has been shown to be much less than the additional ethanol energy generated by the process.[2] Since electrical energy is generally cheaper than energy in the form of liquid fuel, the electricity cost for running a CFC unit also is a small fraction of the value of the additional ethanol fuel generated by cavitation.

In summary, cavitation applied to corn slurries can alter the particle size distribution of corn leading to qualitative changes in cell structure, increases in total sugars after liquefaction, reductions in total solids after liquefaction, and significant increases in ethanol production and solids conversion.

5.9.2 Use of Cavitation to Reduce Particle Size Distribution during Anaerobic Digestion of Wastewater Treatment Plant Primary and Secondary Sludge

Anaerobic digestion (AD) is one of the primary methods used for sludge stabilization and organics recycling in modern wastewater treatment plants. It has the potential to reduce solids by transforming part of the organic material into biogas, a mixture of methane (CH_4), carbon dioxide (CO_2) and trace gases. Methane is the principal component of natural gas and it is currently used for electricity production, pipeline quality natural gas, and as a transportation fuel.

Biogas production during anaerobic digestion is influenced by reactor design, operational parameters and feedstock characteristics. In general, wastewater treatment plants use mixtures of sludge from primary and secondary treatment as feedstock for anaerobic digestion. The feedstock ratio of primary to secondary sludge varies from plant to plant and is affected by the design and efficiency of the overall wastewater treatment process, sludge disposal fates and the design of the anaerobic digestion system.

The physical and chemical properties of primary and secondary sludge are very different in characteristics and behaviour. Raw primary sludge consists of undigested settled solids with a high potential for biogas production including food waste, faecal matter, fibre, and floatable solids such as grease. Raw primary sludge degrades very well in anaerobic digestion systems and its conversion to biogas per kg of sludge is greater than for any other municipal sewage sludge combination. In contrast, secondary sludge, the by-product of the biological conversion of suspended and soluble organic materials present in the liquid effluent from primary treatment, usually under aerobic conditions, is more recalcitrant. Secondary sludge is an energy-depleted material composed of microbial flocs and a mixture of biopolymers, polysaccharides, uronic acids and humic substances. The recalcitrant characteristics of secondary sludge limit its hydrolysis during anaerobic digestion.[47] Therefore, much research is directed toward pre-treatment that increases the solubilisation and disintegration of the organic materials in secondary sludge.[48] Research in this regard has been conducted on biological,[49] thermal hydrolysis,[50] chemical oxidation[51] and alkali treatments,[52] and using mechanical[53] and ultrasound treatments.[6] At present there is a great need to develop sewage sludge pre-treatment methods that improve biogas production and volatile solids reduction from secondary sludge. There is also a need to improve the ability to remove water and thickening during wastewater treatment. However, most of the pre-treatment to date are not energy efficient, requires long retention times and/or the use chemicals that could represent an added impact to the environment for the overall wastewater treatment process.

Ultrasound pre-treatment is one of the most promising pre-treatment methods for secondary sludge disintegration. Both laboratory and full-scale results have shown that this technology efficiently improves disintegration rates,[54] biogas production and dewaterability,[55] and reduces volatile solids[56] compared with other chemical and thermal methods.[57] It also is environmentally benign since chemicals are not added to the sludge. However, the advantages of ultrasound pre-treatment at full-scale are diminished by its high energy demand[55] and difficulties adapting equipment to variable sewage sludge characteristics.[6]

Hydrodynamic cavitation is a type of pre-treatment that is more energy efficient, simpler and less costly than ultrasound pre-treatment.[58,59]

Cavitation causes physical and chemical changes in the sludge exposing light organic materials to further anaerobic digestion.[8,60] In this sense, hydrodynamic cavitation offers new possibilities to wastewater treatment facilities to improve their operations and the cost-effectiveness of the overall process.

In recent studies,[5] hydrodynamic cavitation has been evaluated as a pre-treatment of both primary and secondary sludge. Initially, primary and secondary sludge are characterized by a particle size distribution centred at 77 or 150 µm, respectively (Figure 5.44). Cavitation at a range of pressure drops and passes resulted in a particle size distribution centred at less than 22 µm. The particle size distribution was both reduced and narrowed as a result of cavitation.[5] Different pressures and passes through the cavitation device further reduced and narrow the particle size distribution (Figure 5.45).

Analysis of SEM images of sludge before and after cavitation show that agglomerates of material have been reduced in size substantially (Figure 5.46). However, during anaerobic digestion, only secondary sludge exhibits higher rates and extents of biogas production as a result of cavitation. This is likely due to the limiting factor controlling biogas production rate being hydrolysis for secondary sludge but other factors for primary sludge. In essence the conversion of primary solids to biogas is already occurring at an optimal rate. Cavitation

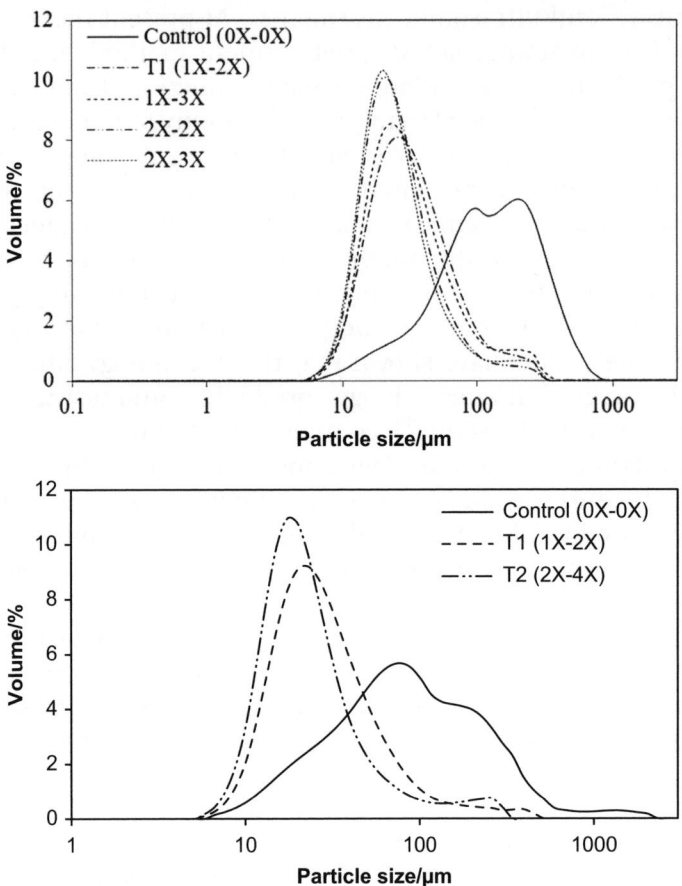

Figure 5.45 Particle size analysis of (a) primary and (b) secondary sludge samples from a municipal waste water treatment plant before and after controlled flow cavitation treatment.[5]
(Reprinted with permission from Eddie Gomez.)

processing thus primarily improves the processing of the more recalcitrant secondary but not primary sludge.[5]

5.9.3 Continuous Biodiesel Production using Controlled Flow Cavitation

Biodiesel is a transportation fuel made by the transesterification of vegetable oils. Its production can be broken into four stages: reaction, separation, alcohol recovery and washing. In the reaction stage a fatty acid triglyceride (oil or fat) is reacted with a primary alcohol (usually methanol, sometimes ethanol) in the presence of a catalyst to produce

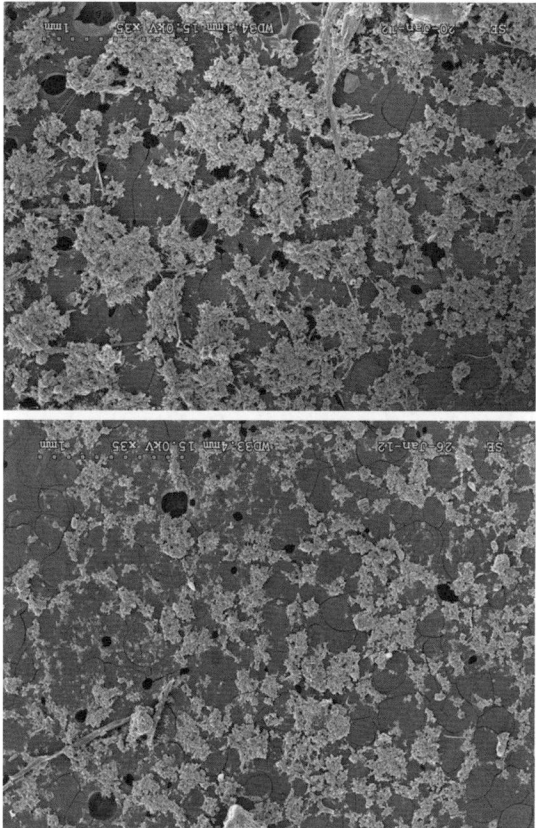

Figure 5.46 SEM image of activated sludge feedstock collected before (upper panel) and after (lower panel) cavitation treatment.

fatty acid methyl ester (FAME or biodiesel) and the by-product glycerine.

The transesterification process is affected by the alcohol type, molar ratio of alcohol to oil, type and amount of catalyst, temperature and purity of the reactants. The transesterification of vegetable oils with an alcohol is a heterogeneous reaction and blending the reagents is of crucial importance. For this reason one of the most important parameters affecting biodiesel transesterification is mixing intensity and efficiency.[61]

The vast majority of biodiesel processes use a base catalyst for the transesterification reaction. This has the advantage of being many times faster than an acid catalyst, but has the disadvantage of being intolerant to free fatty acids (FFAs) in the feedstock. FFAs are produced by various mechanisms. The FFA content is dependent on the feedstock and generally higher FFA content feedstock is lower in value than low FFA feedstock. FFAs inhibit base catalysed transesterification reactions by reacting with the base to form soap and water. This removes the catalyst from the reaction and also encourages the formation of more FFA exacerbating the problem.

FFAs in feedstock can be reduced if is pre-treated before the base catalysed reaction by esterification of the free fatty acids through an acid-catalysed esterification. By reducing the FFA content of the feedstock to less than (0.5–1)%, the amount of soap created in the base catalysed reaction can be minimized. Acid catalysed esterification reactions are usually carried out homogeneously in batch processes using Bronsted acids such as sulfuric acid or Lewis acids. Acid catalysed esterification with alcohol has water as a by-product. Water limits the efficiency of the reaction. Due to the lengthy conversion times and low rates of conversion, resulting in poor economics, acid catalysed esterification pre-treatment is rarely used.

The use of hydrodynamic cavitation in biodiesel production process can both reduce the transesterification reaction time for FFAs and increase the conversion rate of fatty acids to fatty acid esters. This improvement can allow the acid catalysed pre-treatment batch process to be made continuous. By allowing acid catalysed pre-treatment, hydrodynamic cavitation further enables continuous base catalysed transesterification. Systems for sequential acid and base catalysed continuous hydrodynamic cavitation processing for biodiesel production have been developed commercially (Figure 5.47).

Figure 5.47   Commercial-scale continuous cavitation reactor for biodiesel production.

The apparatus used for continuous biodiesel production is capable of generating controlled flow cavitation during the reaction phases of biodiesel production, imparting a high level of energy into the reaction mixture and increasing the transesterification rate of fatty acids to fatty acid alkyl esters. The energy generated from CFC acts to mix effectively and to introduce alcohol to the reaction sites on the fatty acid chains at a more rapid rate than conventional methods, allowing the acid catalysed esterification reaction to be used. Controlled flow cavitation can have additional benefits, such as increasing the mixing efficiency during base catalysed transesterification, decreasing reaction time, increasing conversion rate, and reducing production costs and time.[61,62]

5.10 HYDRODYNAMIC CAVITATION, A PROCESS TOOL FOR A SUSTAINABLE FUTURE

Society's transition away from fossil sources of energy and materials to renewable plant derived sources has brought with it a need to develop advanced methods for processing and transforming these feedstocks. Hydrodynamic cavitation is one such process that can serve as a useful tool in the development of efficient processes for the utilization of renewable feedstock. It can improve starch release, the particle size distribution of organic products, and the mixing and reaction rate of immiscible two phase liquid systems as in biodiesel production from vegetable oils. In addition it has useful applications in the production of catalysts and nanomaterials that will play additional roles in the processing of renewable feedstock.

The properties that make hydrodynamic cavitation a useful unit operation include its ability to create destructive shear forces that homogenize, emulsify and transform recalcitrant materials and its ability to be efficiently applied at commercial scales.

REFERENCES

1. B. Freudig, S. Tesch and H. Schubert, *Eng. Life Sci.*, 2003, **3**, 266.
2. D. A. Ramirez-Cadavid, O. V. Kozyuk and F. C. Michel Jr., *Biomass Conv. Bioref.*, 2014, **4**, 211.
3. W. R. Moser, J. Find, S. C. Emerson, I. Krauz and O. V. Kozyuk, Controlled flow hydrodynamic cavitation for the synthesis of advanced catalyst, in *Proceedings of the 2000 ACS National Meeting*, San Francisco, CA, 29 March 2000.
4. B. G. Novitsky, *Applications of acoustic oscillation in chemical technological processes*, M.: Chemistry, 1983, 192 pp. [in Russian].

5. E. Gomez, *Biodegradation of bio-based plastics and anaerobic digestion of cavitated municipal sewage sludge*, PhD thesis, The Ohio State University, Columbus, OH, USA, 2013.
6. S. Pilli, P. Bhunia, S. Yan, R. J. LeBlanc, R. D. Tyagi and R. Y. Surampalli, *Ultrason. Sonochem.*, 2011, **18**, 1.
7. P. R. Gogate and A. B. Pandit, *Ultrason. Sonochem.*, 2005, **12**, 21.
8. T. I. Onyeche, O. Schläfer, H. Bormann, C. Schröder and M. Sievers, *Ultrasonics*, 2002, **40**, 31.
9. P. R. Gogate, R. K. Taayal and A. B. Pandit, *Curr. Sci.*, 2006, **91**, 35.
10. (a) O. V. Kozyuk, US Pat., 5,810,052, 1998; (b) US Pat., 5,971,601, 1999; (c) US Pat., 5,937,906, 1999.
11. E. B. Brennen, in *Hydrodynamics of Pumps*, ch. 5, 1994 (updated 2000), web resource, http://authors.library.caltech.edu/25019/1/chap5.htm.
12. W. R. Moser, T. Giang, S. Nguyen and O. V. Kozyuk, A new route to cavitational chemistry and chemical processing by controlled flow cavitation, in *Process Intensification for the Chemical Industry*, BHR Group, Antwerp, 1999, pp. 173–187.
13. Lord Rayleigh, *Philos. Mag.*,1917, **34**, 94.
14. C. D. Ohl, T. Kurz, R. Geisler, O. Lindau and W. Lauterborn, *Philos. Trans. R. Soc. London*, 1999, **A357**, 269.
15. N. K. Borne and J. E. Field, *J. Fluid Mech.*, 1994, **259**, 149.
16. E. B. Flint and K. S. Suslick, *Science*, 1991, **253**, 1397.
17. K. S. Suslick, *Science*, 1990, **247**, 1439.
18. W. R. Moser, S. C. Emerson, I. Krausz and J. Find, Engineered synthesis of nanostructured materials and catalysts, in *Nanostructured Materials, Volume 27, Advances in Chemical Engineering*, ed. J. Ying, Academic Press, San Diego, CA, 2000, pp. 1–48.
19. O. V. Kozyuk and I. M. Fedotkin, *Chem. Appur. Eng.*, 1989, **49**, 37.
20. A. S. Machinskii, O. V. Kozyuk and D. N. Shishlov, *Cavitation Mixers,* Central Scientific-Research Information Institute and Technical- Economic Research of Petroleum Refining and Petrochemical Industries, Moscow, 1990.
21. A. N. Kolmogorov, *Proc. USSR Acad. Sci.*, 1941, **32**, 16.
22. O. V. Kozyuk. Breakup of drops in hydrodynamic cavitation flow, in *Proceedings of the North American Mixing Forum, AIChE Annual Conference*, 26 June 1997.
23. H. Schubert, Mechanical emulsification – new development and trends, in *Proceedings of the 2000 AIChE Annual Meeting*, Los Angeles, 12–17 November 2000.

24. J. Baldyga and K. Malik, *Pol. J. Chem. Technol.*, 2009, **11**, 5.
25. B. Wolfrum, T. Kurz, R. Mettin and W. Lauterborn, *Phys. Fluids*, 2003, **15**, 2916.
26. B. W. Zeiger and K. S. Suslick, *J. Am. Chem. Soc.*, 2011, **133**, 14530.
27. R. M. Wagterveld, L. Boels, M. J. Mayer and G. J. Witkamp, *Ultrason. Sonochem.*, 2011, **18**, 216.
28. K. Tokumitsu, *J. Metastable Nanocryst. Maert.*, 2000, **8**, 568.
29. W. R. Moser, O. V. Kozyuk, J. Find, S. C. Emerson and I. Krausz, *US Pat.*, 6,365,555, 2002.
30. W. R. Moser, O. V. Kozyuk, J. Find, S. C. Emerson and I. Krausz, *US Pat.*, 6,589,501, 2003.
31. W. R. Moser, O. V. Kozyuk, J. Find, S. C. Emerson and I. Krausz, *US Pat.*, 6,869,586, 2005.
32. W. R. Moser, O. V. Kozyuk, J. Find, S. C. Emerson and I. Krausz, *US Pat. Appl.*, 2007/0066480, 2007..
33. W. R. Moser, T. Giang, S. Nguyen and O. V. Kozyuk, in *Process Intensification for the Chemical Industry, 3rd International Conference*, ed. A. Green, BHR Group, London, 1999.
34. W. R. Moser, S .C. Emerson, I. M. Krauz, J. Find and O. V. Kozyuk, Synthesis, Characterization and Catalytic Properties of Nanostructured Catalysts by Controlled Flow Cavitational Synthesis, Presented at the 2000 United Engineering Foundation Conference on Processing and Catalytic/Chemical Properties of Nanostructured Materials, Lahaina, Hawaii, 17 January 2000.
35. J. Find, S. C. Emerson, I. M. Krausz and W. R. Moser, *J. Mater. Res.*, 2001, **16**, 3503.
36. J. Find and W. R. Moser, *J. Mater. Sci.*, 2003, **38**, 1917.
37. E. B. M. Doesburg, R. H. Hoppener, B. de Koning, X. Xiaoding and J .J. F. Scholten, in *Preparation of Catalysts IV*, Elsevier, Amsterdam, 1987, *31*, pp. 767–783.
38. S. H. Sonawane, S. P. Gumfekar, K. H. Kate, S. P. Meshram, K. J. Kunte, R. Laxminarayan, C. M. Mahajan, M. G. Parande and A. Muthupandian, *Int. J. Chem. Eng.*, 2010, **2010**, ID 242963, 8ID 242963, 8.
39. K. Felmet, M. Olsofsky, S. Robertson, C. Starbuck, J. Tom, H. H. Tung and J. Wang, An evaluation of cavitation milling to achieve particle size reduction of active pharmaceutical ingredients, presented at AIChE Annual Meeting 2001, Reno, NV.
40. R. H. Muller, R. Becker, B. Kruss and K. Peters, *US Pat.*, 5,858,410, 1999.
41. O. V. Kozyuk, *US Pat.*, 7,041,144, 2006.

42. O. V. Kozyuk, A. S. Myerson and R. Weinberg, *US Pat.,* 7,314,516, 2008.
43. M. Montalbo-Lomboy, *Ultrasonic pre-treatment for enhanced saccharification and fermentation of ethanol production from corn,* PhD thesis, Iowa State University, 2010.
44. M. Montalbo-Lomboy, L. Johnson, S. K. Khanal, J. van Leeuwen and D. Grewell, *Bioresour. Technol.,* 2010, **101**, 351.
45. M. Montalbo-Lomboy, S. K. Khanal, J. van Leeuwen, D. Raj Raman, L. Dunn, Jr and D. Grewell, *Ultrason. Sonochem.,* 2010, **17**, 939.
46. P. R. Gogate, *Chem. Eng. Process: Process Intensif.,* 2008, **47**, 515.
47. A. Carvajal, M. Peña and S. Pérez-Elvira, *Biochem. Eng. J.,* 2013, **75**, 21.
48. H. Carrère, C. Dumas, A. Battimelli, D. J. Batstone, J. P. Delgenès, J. P. Steyer and I. Ferrer, *J. Hazard. Mater.,* 2010, **183**, 1.
49. H. Ge, P. D. Jensen and D. J. Batstone, *Water Res.,* 2010, **44**, 123.
50. S. Tanaka, T. Kobayashi, K.-I. Kamiyama and M. L. N. Signey Bildan, *Water Sci. Technol.,* 1997, **35**, 209.
51. I. T. Yeom, K. R. Lee, Y. H. Lee, K. H. Ahn and S. H. Lee, *Water Sci. Technol.,* 2002, **46**, 421.
52. J. Kim, C. Park, T.-H. Kim, M. Lee, S. Kim, S.-W. Kim and J. Lee, *J. Biosci. Bioeng.,* 2003, **95**, 271.
53. C. D. Muller, M. Abu-Orf and J. T. Novak, *Proc. Water Environ. Fed.,* 2003, **12**, 558.
54. M. P. J. Weemaes and W. H. Verstraete, *J. Chem. Technol. Biotechnol.,* 1998, **73**, 83.
55. X. Yin, P. Han, X. Lu and Y. Wang, *Ultrason. Sonochem.,* 2004, **11**, 337.
56. C. M. Braguglia, G. Mininni, M. C. Tomei and E. Rolle, *Water Sci. Technol.,* 2006, **54**, 77.
57. H. Carrère, C. Dumas, A. Battimelli, D. J. Batstone, J. P. Delgenès, J. P. Steyer and I. Ferrer, *J. Hazard. Mater.,* 2010, **183**, 1.
58. S. S. Save, A. B. Pandit and J. B. Joshi, *Food Bioprod. Process.,* 1997, **75**, 41.
59. P. Senthil Kumar, M. Siva Kumar and A. B. Pandit, *Chem. Eng. Sci.,* 2000, **55**, 1633.
60. P. R. Gogate and A. M. Kabadi, *Biochem. Eng. J.,* 2009, **44**, 60.
61. L. C. Meher, D. Vidya Sagar and S. N. Naik, *Renew. Sustain. Energy Rev.,* 2004, **10**, 248.
62. O. V. Kozyuk, *US Pat.,* 7,754,905 B2, 2010.

CHAPTER 6

Microwave Chemistry

YVONNE WHARTON

C-Tech Innovation Ltd, Capenhurst Technology Park, Chester, CH1 6EH, UK
Email: yvonne.wharton@ctechinnovation.com

6.1 INTRODUCTION

Using microwave energy to drive chemical reactions has been known since 1948 and has been routinely used at laboratory scale since the 1980s. It is now often the method of choice especially when synthesising small quantities of large numbers of compounds, for example, in drug discovery and pharmaceutical chemistry. There are now thousands[1–3] of literature examples describing the advantages that using microwave energy as the heat source for a reaction can give, such as increased yields, reduction in reaction times and the ability to use small amounts of catalysts or solvents in reactions.[4]

Microwave chemistry refers to the use of microwave radiation in the heating of chemical reactions. Microwave heating uses a frequency of either 900 MHz or 2.45 GHz and uses a magnetron as the source (Figure 6.1). Using microwave radiation as a heat source has been widely used in industry for many years, for example, in the food industry, for polymer processing, for drying processes and for pyrolysis.[5,6] The principle advantage of microwave heating is based on the fact the heat is generated within the reacting substance itself. The microwave field is an oscillating field and molecules with a dipole align with the magnetic field and continuously rotate to align with the

Chemical Processes for a Sustainable Future
Edited by Trevor M. Letcher, Janet L. Scott and Darrell A. Patterson
© The Royal Society of Chemistry 2015
Published by the Royal Society of Chemistry, www.rsc.org

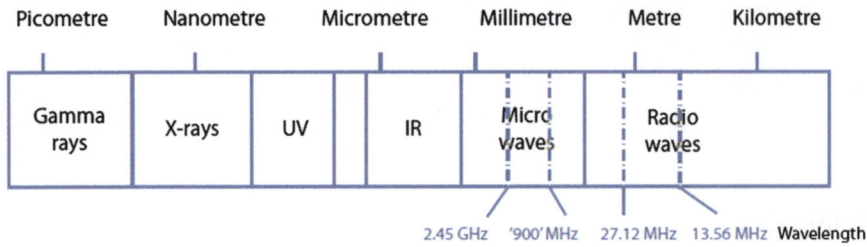

Figure 6.1   Electromagnetic spectrum.

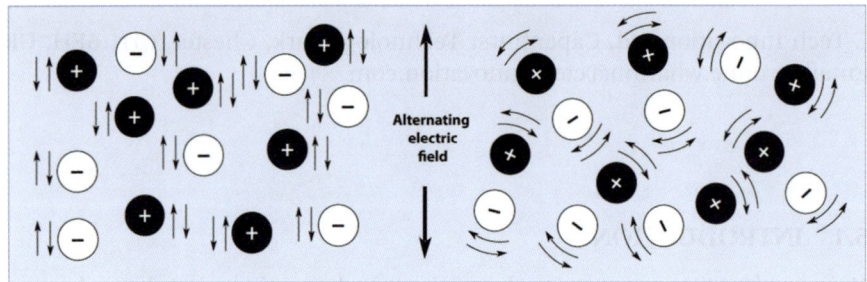

Figure 6.2   Dielectric heating.

field. The rotating molecules push, pull and collide with other molecules and it is these collisions that result in the observed heating (see Figure 6.2).[7]

Solvents heated inside a sealed vessel in a microwave reactor can quickly superheat above their boiling point, which is why reactions that typically take hours or days when heated conventionally can be completed in seconds or minutes in a microwave reactor. This effect has been documented in numerous academic publications since microwave assisted chemistry was first introduced in 1986; it has been postulated that any reaction requiring heat can probably be accelerated by microwave heating.[8,9] It is this convenience that has led to the widespread use of microwave heating by chemists requiring to generate large numbers of compound libraries (*e.g.* in medicinal chemistry).[10] These fast reactions allow chemists to run numerous reaction conditions, do quick analysis and get valuable feedback that would otherwise take weeks to obtain.[11]

When heating materials by microwave it is important to consider the penetration depth of the material by the microwave energy. The penetration, and therefore the heating, stops when all the microwave energy has been absorbed, and so the greater the absorbing power of the material, the shorter will be the penetration depth. Most organic solvents only have a penetration depth of a few centimetres from the outer surface, and as a result this is an important consideration when thinking about the design of a microwave reactor; if the heating only takes place at the surface you will get the same non-uniform heating that is observed when using conventional heating methods.

The ability of molecules to couple with microwave radiation and heat is related to their molecular polarisability; only polar molecules interact with microwave energy. Compounds that have high dielectric constants (*e.g.* water, acetonitrile, ethylene glycol, *N*,*N*-dimethylformamide and ethanol) heat rapidly in a microwave field whereas compounds with low dielectric constants (*e.g.* aromatic or aliphatic hydrocarbons) do not heat well.[12]

6.2 MICROWAVE ASSISTED ORGANIC REACTIONS

The uptake in the use of microwave chemistry for organic reactions was relatively slow to begin with, probably due to the lack of control and difficulty in reproducing results in the modified equipment that was being used.[6] Once specialised microwave reactors were produced and the safety increased, chemists could reliably reproduce results and increase the use and understanding of microwave chemistry. The driving factors for uptake of this technology were the reduced reaction times and the improved yields. This technique was ideally suited to the increasing requirements in the pharmaceutical industry to produce large libraries of novel compounds within short time frames. Microwave chemistry allowed rapid heating of reaction mixtures with uniform heating profiles; as a result the reactants and solvents are heated directly and there is no temperature gradient throughout the reaction mixture which can often lead to by-product formation.[13–16] Many chemists have also utilised the ability to heat solvents above their boiling points in pressurised microwave reactors which allows access to products which would be difficult to obtain by traditional processes.[17]

In order to rationalise the short reaction times possible and different reaction product profiles observed when carrying out reactions under microwave heating, a number of theories of 'non thermal microwave effects' have been put forward. These theories suggest that the microwave energy induces a non-thermal effect which changes the properties of the reaction profile. Several comparisons between microwave and conventional heating have been made under identical conditions. Results show that the differences are due to heating alone and it is the rapid heating rates and taking solvents above their boiling points that is responsible for the success of microwave heating.[18] Not all microwave reactions can be classed as sustainable by any means but microwave chemistry can go some way to help achieving the goals of developing cleaner and 'greener' routes to compounds.[19]

The main advantages of microwave chemistry are as follows.[20,21]

- Reaction times are shorter.
- It has higher yields, leading to cleaner reaction profiles.
- Reactions are carried out in sealed vessels and so can be taken above their boiling points.
- Microwave heating is a direct heating technique which leads to an even temperature profile in the reaction mixture. This eliminates hot spots and gives more reproducible conditions.
- Energy efficiency is increased due to direct molecular heating.
- High temperatures can easily be achieved due to the ability to take materials above their boiling points.
- Microwave chemistry can be readily adapted to parallel synthesis for generating libraries or to flow chemistry for scale up.

6.3 LARGE-SCALE MICROWAVE SYNTHESIS AS A TOOL FOR SUSTAINABLE CHEMISTRY

6.3.1 Scale Up of Microwave Synthesis

Although microwave chemistry is widely used in academia and industry on a milligram scale, there has until recently been little work on the scale up of microwave processes from laboratory to production. Until very recently there has been no large-scale microwave option available so while medicinal chemists synthesised compounds using microwave assisted chemistry in the laboratory, the process and scale-up team had to then re-design the reaction scheme to avoid using microwave heating.[10] The options for process chemists were to either run multiple microwave assisted reactions until they had synthesised the required quantities or to modify the conditions to avoid microwave heating. This could mean lower temperatures or different reagents, or in the worst case, having to redesign the entire synthetic route to avoid microwave heated steps.[22]

Sustainable process chemistry is becoming increasingly important to the chemical industry.[23] To achieve this there needs to be a fundamental change in the way that chemical processes are carried out; this might include new reactor concepts. One area that is being explored is carrying out reactions more efficiently by rapidly heating reactions, as this leads to shorter reaction time and fewer side reactions. Microwave chemistry is an ideal way of achieving this on a laboratory scale but moving to production scale is complex as the heating is far less effective on the large scale due to the penetration depth of the microwaves into the reaction mixture. Batch microwave units have been used to scale up reactions to kilogram laboratory scale, but this is reaching the limits of their capabilities and does not represent a viable way of scaling up the processes.[24] One way of circumventing this is to use flow chemistry as this has the benefits of being able to replicate the rapid heating rates that are achievable on small scale and has the additional benefit of being much safer to operate at high temperature and pressures on a large scale, as there is a much smaller inventory of hazardous materials being heated at any time. The short reaction times that can be attained using microwave chemistry mean that short residence times are required which can lead to much higher throughputs.

6.3.2 Scale Up Using Flow Chemistry

If the heating profile of the reaction is crucial to the outcome of the reaction then this must be replicated in the scale up. Continuous flow is an ideal way of managing to reproduce this heating profile, even on large scale, as only a small amount of the reaction is heated at any time.

Flow chemistry has the benefits of reaction monitoring and process versatility, and avoids the problems associated with large batch vessels.[25] Microwave assisted flow chemistry may be able to offer advantages over conventional flow processing, for example, when heating viscous solutions that do not have good heat transfer properties but which would heat evenly through the pipe using microwave chemistry. This could, for example, have applications in polymerisation reactions.[10] Flow reactions are scalable and reproducible because the reactants all see identical conditions unlike in round bottomed flasks where inefficient mixing is likely, which generates temperature and concentration gradients.[11,26]

The safety benefits of flow chemistry are also a major advantage; in a flow reactor only a small amount of material is being heated at any time so the consequences of an accident are not as acute as in a batch reactor. This makes flow chemistry ideally suited to reactions which involve hazardous reactants or products.[11] The small amount of material inside the microwave reactor at any one time also has the advantage of avoiding a loss of an entire batch of material if anything goes wrong.

6.4   MICROWAVE ASSISTED SYNTHESIS UNDER CONTINUOUS FLOW CONDITIONS

A major shortcoming of microwave chemistry has been the lack of scale-up equipment available to scale from lab to production scale. Until now most microwave chemistry has been carried out in scales of <1 g.[27] Scale up of microwave chemistry is limited by the limited penetration depth of the microwave energy into the reaction media which is generally only a few centimetres (depending on the reaction media and its dielectric properties). This phenomenon means that direct batch scale-up is not a suitable method. If microwave heating were used for batch processes, the reacting material would heat at the surface and the centre of the reaction would be heated by convection (as with conventional heating), resulting in temperature gradients within the reaction. Avoidance of temperature gradients is one of the main benefits of using microwave chemistry and has led to the development of continuous flow microwave chemistry.

Sustainable chemistry is becoming of prime importance to the chemical industry. Most countries have programmes in place to achieve energy use and energy efficiency. Energy savings that can be achieved easily with no extra cost have been estimated to be as high as 30% of the total energy use.[28,29] This has been done without using microwave heating. Once this low hanging fruit has been taken, companies have to look towards new innovations and look to reorganise their production processes including looking at new reactor configurations and/or novel process conditions. The industry has begun to investigate process intensification technologies which look for changes in the traditional processing techniques that allow chemists to access harsh conditions. Until recently, there has been very little choice of alternative methods or equipment. Continuous flow microwave chemistry overcomes many of these problems and goes a long way to meeting the demands that companies set for new processes.

Recently published work has demonstrated the successful scale up of reactions to kilogram scale using microwave flow chemistry.[30–32] These reactions show that it is possible to scale up reactions from lab scale up to kilogram lab scale while retaining high yields and introducing short reactions times. The energy efficiency of many reactions was also evaluated and was found to be very efficient. Scheme 6.1 gives examples of transformations that were performed by Morschhäuser and co-workers.[30] The times given in the reaction scheme indicate the time the reaction mixture spent in the microwave heated zone of the reactor. These reactions show the extreme temperatures that can be achieved in very short time frames using microwave heating and highlights the benefits that microwave processing can have in organic transformations requiring very high temperatures. Based on this several companies are now looking into the further scale up of this technology to full production capability of >1000 tonnes per annum (t a^{-1}).

Microwave heating becomes more efficient as the scale increases in part due to using multimode microwave reactors which are more efficient at transferring energy into large masses, although energy efficiency has to be assessed on a case by case basis. Microwave heating is already used fairly routinely for industrial applications such as drying and food processing, demonstrating that energy efficient processes can be achieved when applied to microwave chemistry at production scale.[28]

In the past few years, there has been an increase in the research into the scale up of microwave chemistry which has led to several companies developing equipment for this specific need. Several microwave flow reactors have been marketed to facilitate the scale up; Upscale Microwave,[22] Milestone, Pueschner,[30,33] Sairem and C-Tech Innovation[34] have all developed reactors that can be used to scale up to pilot plant scale and beyond. All of these reactors use a continuous flow design and generally reach temperatures of up to 250 °C and pressures of up to 2.5 MPa. These can therefore be used to carry out very high temperature reactions that are beyond the capabilities of traditional batch reactors.

6.5 CASE STUDIES

Many of the routes to organic compounds targets were discovered many years ago and, while these routes are well tested, they may not be particularly high yielding or efficient. There is a need to find more sustainable ways of making compounds in order to streamline processes; one way of achieving this is by microwave synthesis. What follows are some examples of how microwave chemistry has been used to increase the efficiency of chemical transformations.

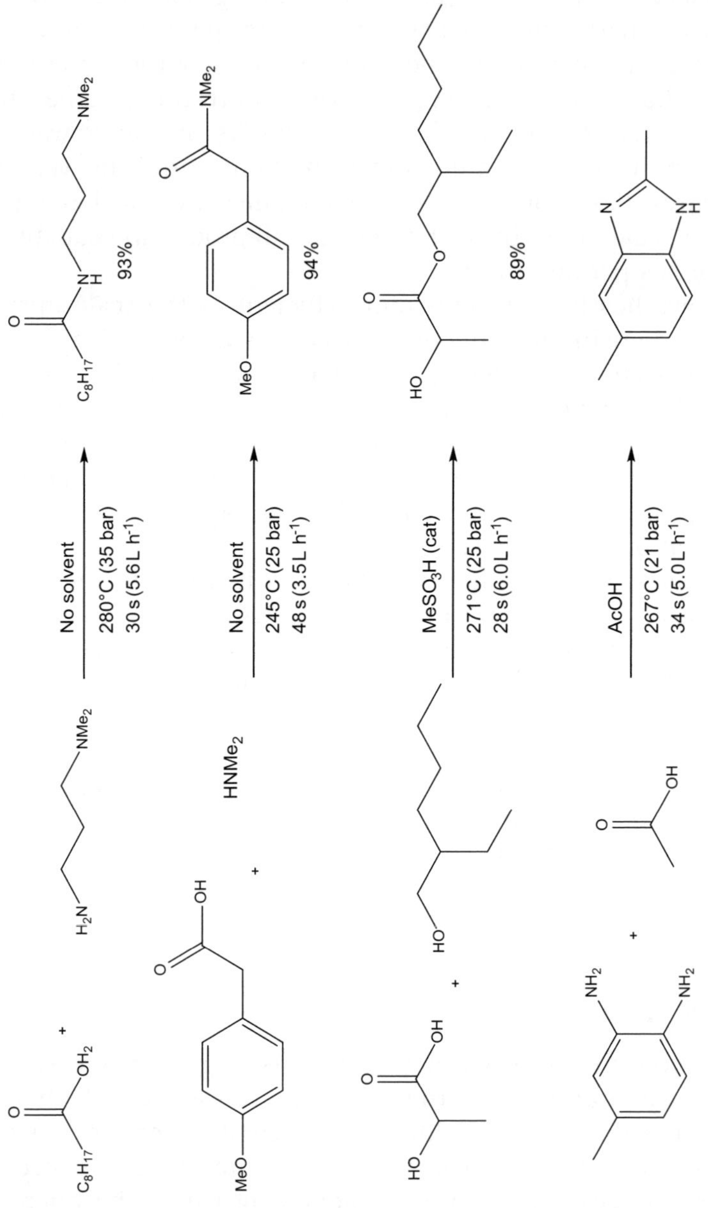

Scheme 6.1   Demonstration of large scale synthesis using microwave heating.

6.5.1 Parallel Synthesis

The following example involves a synthesis which was carried out by two researchers: one researcher had access to a microwave reactor whereas the other did not and both devised the same three-step route (see Scheme 6.2) based on a literature precedent. Using conventional heating, all three steps were slow and gave only moderate yields even after optimisation. Overall 37 days were spent trying to optimise the non-microwave route with only a 4% overall yield being ultimately achieved. Due to this low yield this researcher concluded that this would not be a viable route for parallel synthesis. However, using the microwave heated synthesis the reactions were fast and high yielding. The target compound was produced in 88% yield in 2 days, which represents an 18-fold increase in productivity.[20] Furthermore due to the faster reaction times achieved in the microwave synthesis, far more optimisation could be carried out in future trials.

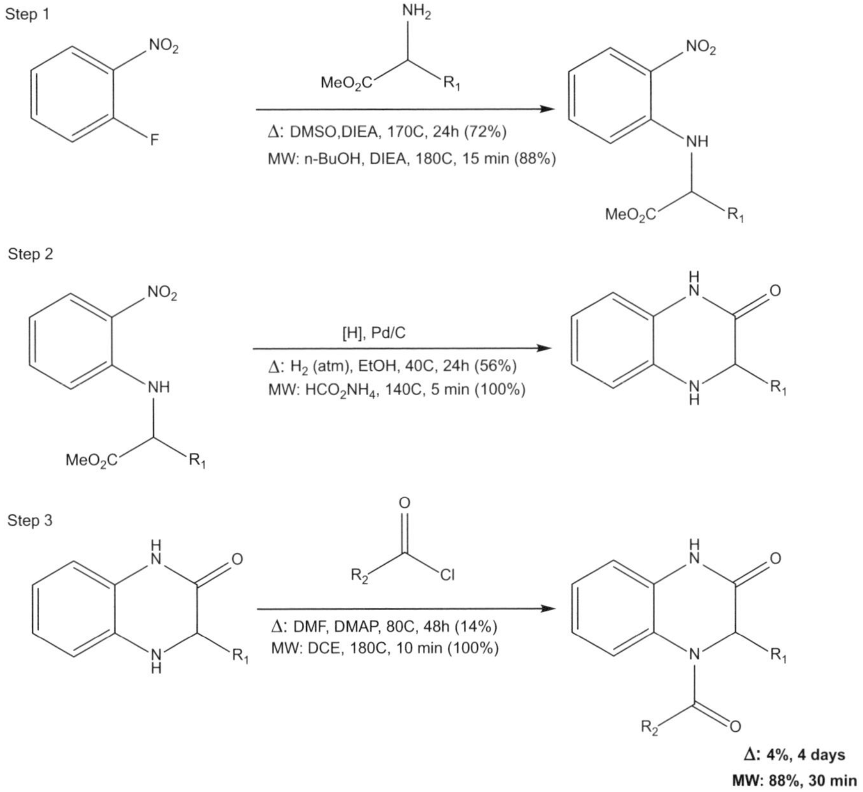

Scheme 6.2 Direct comparison of the microwave (MW) and non-microwave (marked Δ) synthesis of 4-acyl-1,2,3,4-tetrahydroquinoxalin-2-one.

6.5.2 Transition Metal Free Suzuki and Sonogashira Coupling Reactions

Transition metal coupling reactions are widely used in organic synthesis to form carbon–carbon bonds. These couplings usually involve a transition metal catalyst such as palladium which can be expensive and difficult to recover from the process. Recently it has been demonstrated using microwave heating that these couplings can be carried out without the need for metal catalysts.[35] The Suzuki reaction between an activated aryl bromide and an aryl boronic acid was carried out in water using tetrabutylammonium bromide (TBAB) as a phase transfer catalyst (see Scheme 6.3).

The Sonogashira reaction between an aryl bromide (or iodide) and phenylacetylene uses polyethylene glycol as the phase transfer catalyst (see Scheme 6.4). Both these reactions gave reasonable yields without the need for expensive catalysts.

6.5.3 Multi-component Reactions

Multi-component reactions are of value in organic synthesis due to the possibility of increased speed and efficiency compared with conventional linear synthesis. This example, the Biginelli dihydropyrimidine synthesis, is increasingly employed in medicinal chemistry as

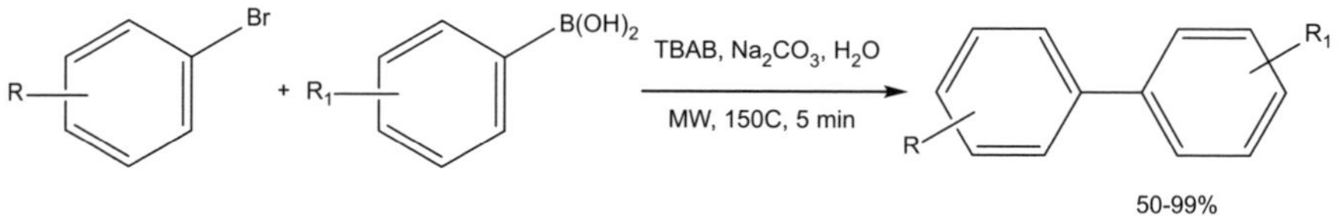

R= H, 2-Me, 4-Me, 2-OMe, 4-OMe, 4-Ac, 4-CHO, 4-NO$_2$, 4-NH$_2$, 4-CO$_2$H, 4-CO$_2$Me, 2,6- Me$_2$, 2-Pyr

R$_1$ = H, 4-Me, 4-Ac

Scheme 6.3   Transition metal free Suzuki coupling reaction.

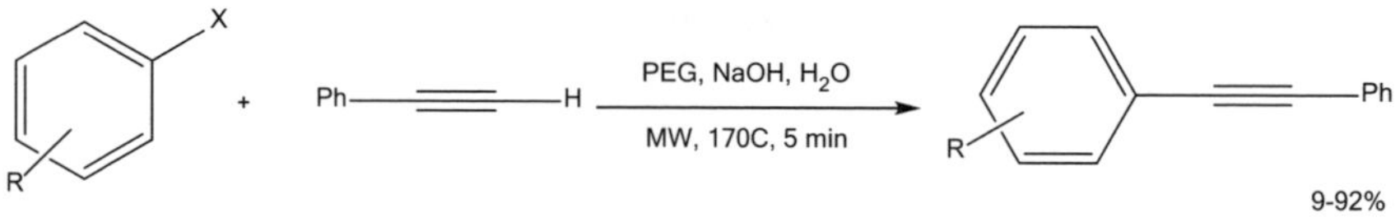

X = Br, I
R = H, 4-Me, 4-Ac, 4-OMe

Scheme 6.4   Transition metal free Sonogashira coupling.

the resulting dihydropyrimidine scaffold has displayed biological properties and its core can be further converted into derivatives to produce new compounds for medicinal chemistry. The traditional procedure for the Biginelli synthesis is a one-pot condensation of the three components in a solvent with a strong acid catalyst. The reaction typically involves high temperatures, long reaction times and usually only gives moderate yields, especially when using more complex building blocks. Carrying out this reaction in a microwave (see Scheme 6.5), the reaction was complete in 10 min, rather than the 3–4 hours when using conventional heating, and gave the dihydropyrimidine in 88% yield and 98% purity. This reaction was carried out on a multi-gram scale.[24]

6.5.4 Comparison of Energy Consumption in a Benzamide Hydrolysis

A model reaction was used to study the comparative energy consumption of a conventional and a microwave heating reaction; benzamide hydrolysis using sulfuric acid at reflux was chosen for the study (see Scheme 6.6). The conventionally heated process takes 12 h

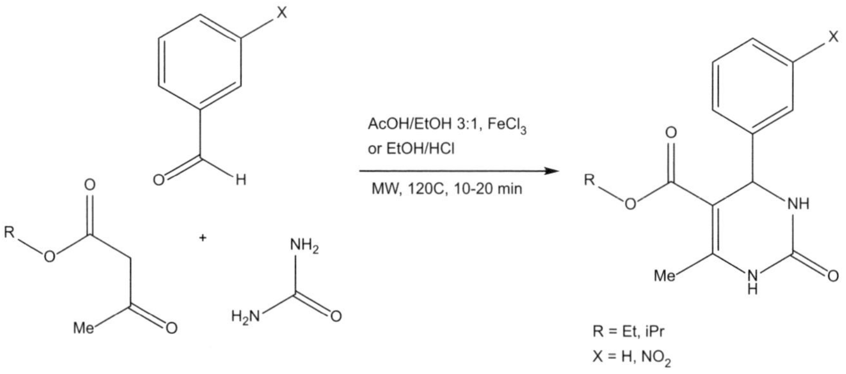

Scheme 6.5 Biginelli dihydropyrimidine synthesis.

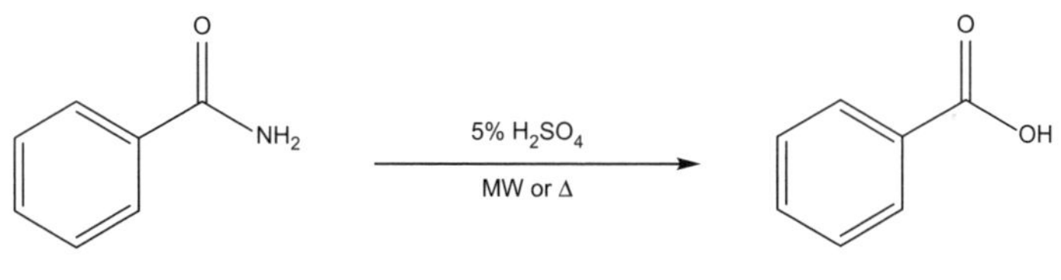

Scheme 6.6 Benzamide hydrolysis.

at 100 °C to reach completion whereas the microwave reaction takes 7 min at 180 °C.[23]

The energy consumed by the reaction was measured with a domestic electricity meter to measure the consumed energy in kW h and kW h mol^{-1} and the results are presented in Table 6.1.

It can be seen that, when microwave reactors were used to heat reactions in sealed vessels to temperatures above the boiling point of the reactants, this led to significantly faster reaction times which in turn led to more energy efficient reactions compared with the conventional reflux processes performed in oil baths. This was due to the shortened reaction times which were only made possible when heating reactions in sealed vessels. This result was made clear by carrying out a reaction in a microwave oven in an open vessel at the reflux temperature and comparing it with the same reaction done in a conventional way in an oil bath. It was found that the two reactions had comparable reaction times. Furthermore the oil bath and open vessel microwave reactions were analysed by high-pressure liquid chromatography (HPLC) after 2 h and found to have comparable conversions; the microwave reactions were stopped at this point (to avoid running the microwave for 12 h) and the results calculated by extrapolation. The energy consumption of the microwave reaction for equivalent reaction times was found to be higher than that of the oil bath reaction; this was due to the fact that the electrical energy conversion in most magnetrons was only about 50–65% efficient.

From such experiments, it can be seen that the increased energy efficiency found in using microwave heating can be attributed to

Table 6.1 Comparison of the energy consumed in the hydrolysis of benzamide by microwave and conventional synthesis.[23]

Heating method[a]	Scale/g	Volume of solvent/ mL	t/°C	t_R/min[b]	Yield/%	E/kW h	t_{total}/ min[c]	E/kW h mol^{-1}
Oil bath	1	10	100	720	91	0.605	735	80.1
Oil bath	5	5	100	720	94	0.709	750	18.26
Sealed MW-A	0.4	4	180	7	89	0.004	8.4	1.36
Sealed MW-B	0.4	4	180	7	94	0.026	8.5	8.38
Sealed MW-C	1	10	180	7	90	0.009	8.6	1.2
Open MW-A	5	50	100	120	76	1.793	720	48.23
Open MW-B	5	50	100	120	75	6.001	720	161.44

[a]MW-A is an Initiator Eight EXP 2.0, single mode reactor, 400 W maximum output. MW-B is a Discover Labmate, single mode reactor, 300 W maximum output. MW-C is a MicroSYNTH, multimode reactor, 1000 W maximum output.
[b]t_R is the fixed hold time of the reaction, *i.e.* the time spent at the required reaction temperature.
[c]t_{total} is the total heating time of the reaction.

decreased reaction times. These are related to the possibility of working at high pressures and hence at temperatures which are higher than the boiling points of some of the solvents.

6.6 CONCLUSIONS

Microwave chemistry is a well-used technique in laboratory-scale chemistry and has proved invaluable in generating small quantities of large numbers of compounds. Most chemists understand the use of microwaves to drive chemical reactions and are becoming increasingly familiar with the use of flow chemistry.

Most of the significant reductions in energy consumption have already been achieved by the large chemical manufacturers without using microwave technologies. These companies can afford to invest in energy-saving equipment and to employ specialists in energy management. However, even these companies are finding that they cannot make further energy savings without radically changing the way they produce chemicals. Chemical manufacturers will have to adopt new technologies if they are going to continue to reduce their energy consumption and environmental impact in an ever growing market. Microwave flow chemistry offers one way of contributing towards this and presents a convenient way to scale up laboratory-scale reactions to plant scale with relatively minor process development. With companies and governments ever more mindful of the safety implications of chemical production, microwave flow chemistry also offers a safer way of heating large volumes of flammable materials as there is a much smaller inventory being heated at any one time compared with batch chemistry.

As equipment becomes available for carrying out microwave chemistry on a large scale, this should lead to further research and may open up routes to previously inaccessible compounds and chemistries. As companies try to develop new products and compounds, the ability to carry out high temperature or hazardous chemistry with relative ease will be invaluable.

REFERENCES

1. M. Taylor, *Developments in Microwave Chemistry*, Royal Society of Chemistry, London, 2005.
2. B. Wathey, J. Tierney, P. Lidström and J. Westman, *Drug Discov. Today*, 2002, 7, 373.
3. S. Caddick, *Tetrahedron*, 1995, **51**, 10403.

4. S. K. Ritter, *Chem. Eng. News*, 2012, **90**, 32.
5. C. R. Buffler, *Microwave Cooking and Processing: Engineering Fundamentals for the Food Scientist*, Van Nostrand Reinhold, New York, 1993.
6. N. Leadbetter, *Microwave Heating as a Tool for Sustainable Chemistry*, CRC Press, Boca Raton, FL, 2010.
7. C. Gabriel, S. Gabriel, E. H. Grant, B. S. J. Halstead and D. Mingos, *Chem. Soc. Rev.*, 1998, **27**, 213.
8. P. Lidström, J. Tierney, B. Wathey and J. Westman, *Tetrahedron*, 2001, **57**, 9225.
9. H. M. Kingston and S. J. Haswell, *Microwave Enhanced Chemistry*, American Chemical Society, Washington DC, 1997.
10. R. Van Noorden, *Chem. World*, October 2008, 40.
11. A. M. Thayer, *Chem. Eng. News*, 2012, **90**, 12.
12. M. A. Surati, S. Jauhari and K. R. Desai, *Arch. Appl. Sci. Res.*, 2012, **4**, 645.
13. S. A. Galema, *Chem. Soc. Rev.*, 1997, **26**, 233.
14. M. Gaba and N. Dhingra, *Ind. J. Pharm. Edu. Res.*, 2011, **45**, 175.
15. C. O. Kappe and D. Dallinger, *Mol. Divers.*, 2009, **13**, 71.
16. C. O. Kappe, *Angew. Chem., Int. Ed.*, 2004, **43**, 6250.
17. C. O. Kappe, *Chem. Soc. Rev.*, 2008, **37**, 1127.
18. A. de la Hoz, A. Díaz-Ortiz and A. Moreno, *J. Microw. Electromagn.c Energy*, 2007, 41.
19. M. A. Herrero, Jennifer M. Kremsner and C. Oliver Kappe, *J. Org. Chem.*, 2008, **73**, 36.
20. C. O. Kappe and D. Dallinger, *Nat. Rev. Drug Discov.*, 2006, **5**, 51.
21. J. Jacob, *Int. J. Chem.*, 2012, **4**, 29.
22. J. R. Schmink, C. M. Kormos, W. G. Devine and N. E. Leadbeater, *Org. Process Res. Dev.*, 2010, **14**, 205.
23. T. Razzaq and C. O. Kappe, *ChemSusChem*, 2008, **1**, 123.
24. A. Stadler, B. H. Yousefi, D. Dallinger, P. Walla, E. Van dcr Eycken, N. Kaval and C. O. Kappe, *Org. Process Res. Dev.*, 2003, 7, 707.
25. A. De La Hoz, J. Alcázar, J. Carrillo, M. A. Herrero, J. De M. Muñoz, P. Prieto, A. De Cózar and A. Diaz-Ortiz, in *Microwave Heating*, ed. U. Chandra, InTech, Rijeka, Croatia, and Shanghai, 2011, ch. 7, pp. 137–162.
26. M. Damm, T. N. Glasnov and O. C. Kappe, *Org. Process Res. Dev.*, 2010, **14**, 215.
27. T. N. Glasnov and C. O. Kappe, *Macromol. Rapid Commun.*, 2007, **28**, 395.
28. Organisation for Economic Co-operation and Development, *Environmental Outlook for the Chemicals Industry*, OECD, Paris, 2011.

29. S. Milmo, *Chem. World*, October 2007, 62.
30. R. Morschhäuser, M. Krull, C. Kayser, C. Boberski, R. Bierbau, P. A. Püschner, T. N. Glasnov and C. O. Kappe, *Green Process Synth.*, 2012, **1**, 281.
31. E. Gjuraj, R. Kongoli and G. Shorec, *Chem. Biochem. Eng. Q.*, 2012, **26**, 285.
32. P. T. Baraldi and V. Hessel, *Green Process Synth.*, 2012, **1**, 149.
33. R. W. Wagner and J. H. Brownell, *Spec. Chem. Mag.*, November 2012, 24.
34. Y. Wharton, *Spec. Chem. Mag.*, November 2012, 27.
35. B. L. Hayes, *Aldrichim. Acta*, 2004, **37**, 66.

CHAPTER 7

Solar Photochemical Manufacturing of Fine Chemicals: Historical Background, Modern Solar Technologies, Recent Applications and Future Challenges

SAIRA MUMTAZ,[a] CHRISTIAN SATTLER[b] AND
MICHAEL OELGEMÖLLER*[a]

[a] James Cook University, College of Science, Technology and Engineering, Townsville, Queensland 4811, Australia; [b] German Aerospace Centre (DLR), Institute of Technical Thermodynamics, Solar Research, D-51170 Cologne, Germany
*Email: michael.oelgemoeller@jcu.edu.au

7.1 INTRODUCTION TO SYNTHETIC ORGANIC PHOTOCHEMISTRY

Photochemistry is the branch of chemistry concerned with the chemical effects of ultraviolet (UV), visible or infrared radiation.[1] It thus covers a broad range of disciplines and finds applications in analytical, inorganic, material, medicinal, organic and physical chemistry. For photochemical synthesis, light is used to activate molecules either directly or *via* the use of a photosensitizer or photocatalyst (Scheme 7.1). This activation is typically achieved

Chemical Processes for a Sustainable Future
Edited by Trevor M. Letcher, Janet L. Scott and Darrell A. Patterson
© The Royal Society of Chemistry 2015
Published by the Royal Society of Chemistry, www.rsc.org

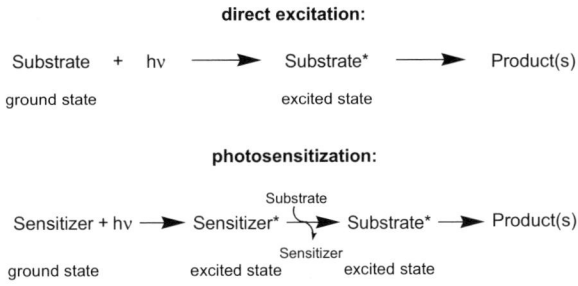

Scheme 7.1 Direct and sensitized photoreaction modes.

within a part of the substrate or sensitizer, *i.e.* the chromophore. If different chromophoric groups are present within separate parts of the molecule, activation can be accomplished individually ('chromatic orthogonality concept' by Bochet[2]). The amount of energy is simply controlled with the wavelength as expressed in the Planck relation:[3]

$$E = h \times v = h \times c/\lambda = h \times c \times \tilde{v} \qquad (7.1)$$

where E is the energy of light, h is the Planck constant, v is the frequency, c is the velocity of light, λ is the wavelength and $\tilde{v}$ is the wavenumber.

As a result of this activation, the substrate reaches an excited state from which numerous physical and chemical pathways are possible, which are commonly depicted in a Jablonski diagram.[4] Physical processes such as fluorescence, phosphorescence or internal conversion typically result in deactivation. While these pathways do not yield any 'chemical products', they are nevertheless important in spectroscopy, analysis, sensing and photomedicine. Alternatively, the extra energy can be used for chemical transformations that are thermally not feasible. These include, among others, cleavages, rearrangements, isomerizations, additions and redox reactions. Over the past decades, a range of highly selective photochemical transformations has been developed.[5] Photochemical reactions are also commonly used in the synthesis of natural products.[6,7]

The efficiency of a photochemical process at a given wavelength is described by its quantum yield (Φ_λ) (Equation (7.2)).[8] Its value is unity ($\Phi_\lambda = 1$) when each photon absorbed results in the formation of a product molecule. Most photochemical reactions have much smaller quantum yields due to competing deactivation processes ($\Phi_\lambda < 1$), although the chemical yield of the product may still be high.

In contrast, photoinduced chain reactions can show very large quantum yields $(\Phi_\lambda \gg 1)$ since light is only required for the initiation step.

$$\Phi_\lambda = \frac{\text{amount of product formed}}{\text{amount of photons absorbed}} \qquad (7.2)$$

The penetration of light through a photochemical reaction mixture depends on the photophysical properties of the chromophore, as described by the Beer–Lambert law (Equation (7.3)).[9] To enable complete light penetration, photochemical reactions are thus typically conducted in high dilutions or in narrow reaction vessels.

$$A = -\log T = \log (I/I_0) = \varepsilon \times l \times c \qquad (7.3)$$

where A is the absorbance of a solution at a given wavelength, T is the transmittance, I is the intensity of light exiting a medium, I_0 is the intensity of light entering a medium, ε is the molar absorption coefficient, l is the thickness of solution traversed by light (path length) and c is the molar concentration of absorbing species.

Due to their operational features and protocols, photochemical processes are often regarded as 'green'.[10,11] Some of the main reasons to support this statement are:

- light (or a photon) is a 'clean and traceless reagent' that does not produce any waste by itself;
- reactions are typically controlled with a 'flick of a switch', thus allowing instant termination of reactions and increasing safety;
- activation is selectively achieved within the chromophoric group and controlled by the wavelength, thus avoiding the need of protecting groups and consequently additional synthesis steps;
- activation and often reaction outcomes can be controlled by direct or sensitized (catalysed) process options.

Despite these sustainable features, photochemistry suffers from a variety of 'non-green' drawbacks that are often linked to light itself, while others originate from the photophysical properties of the substrate or reaction. Some of the most significant limitations are as follows.

- The generation of light comes with significant electrical losses due to the generation of heat. Energy- and water-demanding cooling is thus required.
- Light, especially UV light, is harmful to humans and therefore strict safety protocols must be developed for its usage. Reactor

materials, pipes and gaskets are furthermore exposed to light stresses, thus necessitating their frequent replacement.

- Most light sources show broad emissions of wavelengths of which some are activating while others are destructive or ineffective. Optical filters must thus be applied, which further reduces the energy efficiency of the photochemical process.
- Many photoreactions suffer from lower quantum yields (Φ_λ) and therefore require constant irradiation in order to achieve high conversions.
- Photochemical transformations are often performed in solvents such as benzene, acetonitrile, carbon tetrachloride or methanol. While chemically inert and transparent to light, these solvents are hazardous.
- Light penetration is typically limited to a narrow layer within the reaction mixture, thus requiring high dilutions and producing large amounts of solvent waste.

Various of these drawbacks can be overcome in the laboratory by applying near-monochromatic (such as excimer lamps[12]) or low-energy light sources (such as light-emitting diodes, LEDs[13]), the introduction of suitable leaving groups (such as gaseous CO_2[14,15]), the implementation of alternative media (such as water,[16] micelles[17] or supercritical CO_2[18]) and/or the utilization of continuous flow processes in microreactors,[19-21] respectively.

In terms of industrial applications, however, the limiting key factor is energy,[22] in particular to power the artificial light sources. Industrial bulk photochemistry is thus only economical if inexpensive electricity is readily available. With a few exceptions, most industrial photoreactions are therefore limited to low-volume fine chemicals such as fragrances, flavours or vitamins.[23,24]

The need for cheap electricity and the currently small volumes of many photochemical products make natural sunlight an attractive, alternative energy and light source.[25-28] In fact, the energy from sunlight reaching the surface of the earth within one hour (4.3×10^{20} J) would be sufficient to cover the energy consumed globally within one year (for 2001: 4.1×10^{20} J).[29] To allow for the efficient implementation of sunlight into preparative photochemical processes, suitable solar reactors should be utilized. In addition, the chromophoric substrate or photosensitizer/photocatalyst must emit in the useable solar spectrum, *i.e.* between 300 and 700 nm. Due to its discontinuous availability, however, sunlight has been widely neglected as a sustainable energy input. Notable exceptions are applications for the production of electrical power (photo-voltaics[30,31]) and wastewater treatment.[32] Recently, solar synthesis on laboratory to demonstration scales has received growing attention[33] and has seen a further boost with the appearance of visible light photoredox catalysis.[34]

7.2 EARLY HISTORY OF SOLAR PHOTOCHEMISTRY

Scientific interest in photochemical applications commenced in the eighteenth century when the fundamental laws of photochemistry—the Grotthaus–Draper law (1842) and the Stark–Einstein law (1913)—were established.[35] Due to the lack of suitable artificial light sources, early photochemical experiments were conducted by exposing reaction mixtures in simple glass bottles to natural sunlight. Early pioneers in solar photochemistry include Stanislao Cannizzaro (1826–1910),[36] Emanuele Paternó (1847–1935),[37] Giacomo Ciamician (1857–1922)[38] and Alexander Schönberg (1892–1985).[39] Despite the technical limitations of the time, photochemistry was already recognized as an environmentally friendly methodology. This is especially notable from the spectacular vision entitled 'The Photochemistry of the Future' by Ciamician, presented at the International Congress of Applied Chemistry in New York in 1912. In his speech, he propagated the use of natural sunlight as a replacement of coal, the dominant source of energy at that time:[40]

'On the arid lands there will spring up industrial colonies without smoke and without smokestacks; forests of glass tubes will extend over the plains, and glass buildings will rise everywhere; inside of these will take place the photochemical processes that hitherto have been the guarded secret of the plants, but that will have been mastered by human industry which will know how to make them even more abundant fruit than nature, for nature is not in a hurry and mankind is'.

In his outdoor laboratory at the University of Bologna in Italy and together with his long-term collaborator Paolo Silber, Ciamician studied and discovered a range of photochemical transformations, among them photoreductions, photopinacolizations, photo-cycloadditions and photocleavages.[41] The solar exposure of carvone (1) in aqueous ethanol over a period over several months, for example, furnished a solid isomeric product in a yield of approximately 6%, for which they proposed the tricyclic structure 2 with a new cyclobutane

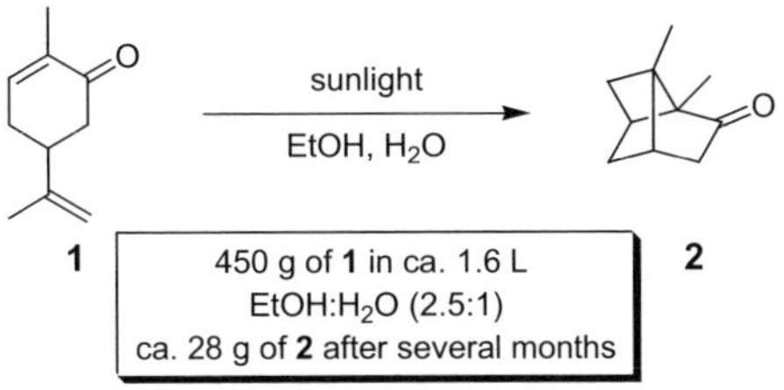

Scheme 7.2 Intramolecular [2 + 2]-cycloaddition of carvone (**1**) in aqueous ethanol.

ring (Scheme 7.2).[42] Using modern analytical techniques, the structure of **2** was later confirmed as correct.

With the development of reliable artificial light sources, preparative photochemistry moved 'indoors' and became an independent research area.[43] Recently, solar photochemistry has seen a remarkable renaissance within the growing field of green chemistry, as evident from a number of summarizing review papers.[33,44–48]

7.3 SOLAR TECHNOLOGY

Early solar photochemical reactions followed a simple 'flask in the sun' approach.[35,41] Reaction mixtures were typically sealed in thick-walled glass tubes and exposed to direct sunlight without cooling. Losses of reaction vessels due to severe weather conditions or explosions were not uncommon.[49] Despite these somewhat primitive and potentially hazardous conditions, a wealth of solar chemical reactions has been discovered. Solar reactions are still commonly conducted by placing a flask in the Sun, although cooling and pressure release is nowadays provided.

Schenck's pilot plant for the solar synthesis of ascaridole (**4**), erected after the Second World War in Germany, represents the first milestone in solar technology. Using an array of 200 exposed 10 L glass bottles, α-terpinene (**3**) in air-saturated ethanol solution was converted into the at that time important anthelmintic drug ascaridole (Scheme 7.3). The reaction mixtures were periodically shaken to maintain oxygen saturation and the whole plant was sprayed with water for cooling. Singlet oxygen was generated with the help of chlorophyll, which was sourced from non-eatable plant material such as stinging nettles.[50] The entire plant was capable of producing 2 kg of ascaridole (**4**) in just two sunny summer days.[24] Although ascaridole was later banned due to its side effects and its production subsequently ceased, it has recently found renewed interest due to its mild antimalarial properties.[51]

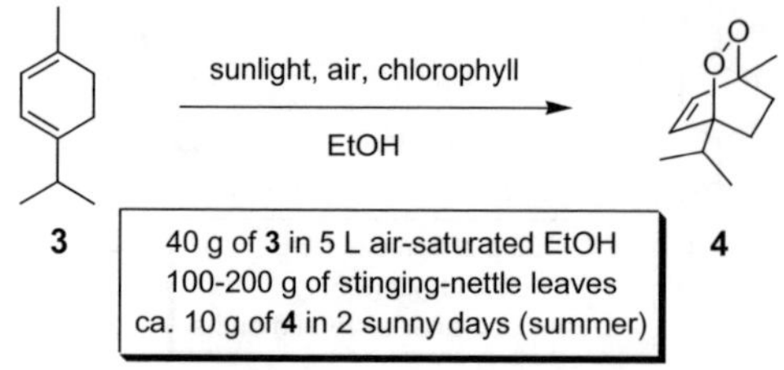

Scheme 7.3 [4 + 2]-Photo-oxygenation of α-terpinene (**3**) to ascaridole (**4**).

With the development of modern solar reactors, originally for the generation of electrical power,[52–54] various non-concentrating and concentrating reactors have become available for solar chemical applications.[55–57] Dedicated solar research facilities furthermore exist, for example, at the Plataforma Solar de Almería (PSA) in Spain[58] or the German Aerospace Centre (DLR) in Germany.[59] It is hoped that the growing implementation of sustainable energy resources by chemical companies will lead to the application of these environmental friendly technologies in the foreseeable future.[60–62] However, the economic advantages of solar processes must be evaluated for each industrial application.

7.3.1 Non-Concentrating Reactors

Solar reactors must be designed in a way that the available radiation, direct and diffuse, is most efficiently utilized. Non-concentrating reactors like flatbeds or glass tubes are thus a cost efficient way to couple this radiation into reactions (Figure 7.1). These simple reactor types are characterized by their large surface areas and shallow depths, and are typically tilted to the local latitude to maximize the harvest of sunlight. More advanced reactors may have cooling water coils running through the reactor body. In an extension of this concept, Liu and co-workers have designed floating solar reactors, *i.e.* 'labs on surfboards', for small scale syntheses in sunlight.[63] These devices use natural water reservoirs as heat sinks, thus completely avoiding the need of artificial coolants.

7.3.2 Compound Parabolic Collectors (CPC)

To improve light utilization, a static mirror system that reflects all available sunlight on the surface of the receiver tube, a so-called compound parabolic collector (CPC), has been developed (Figure 7.2).[64] CPC systems are normally constructed from aluminium sheets, which

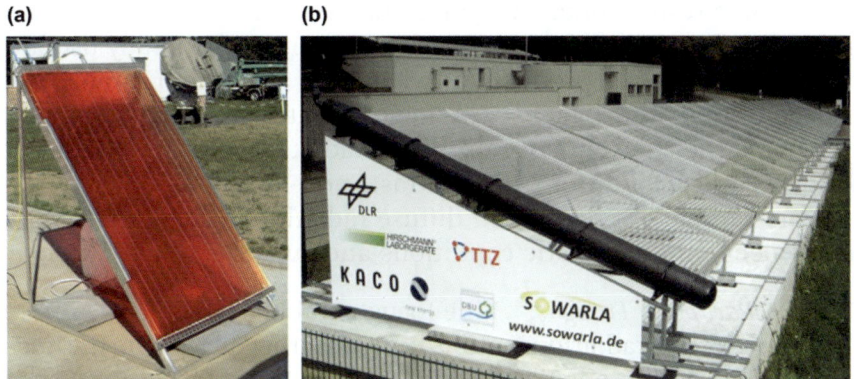

Figure 7.1 (a) Flatbed reactors (DLR, Germany) and (b) 240 m² SOWARLA-receiver (Lampoldshausen, Germany).

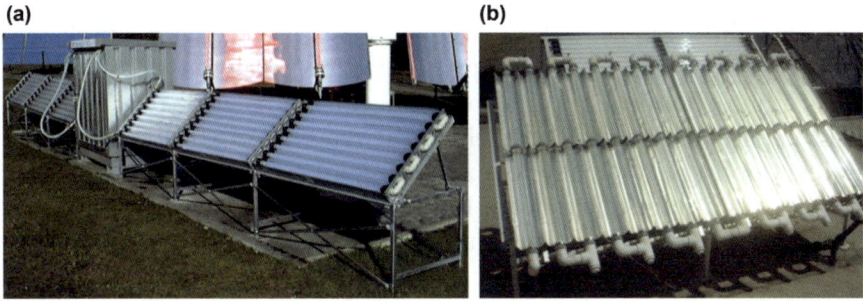

Figure 7.2 (a) Horizontal CPC-reactors (DLR, Germany). The feeding and cooling system is in the central metal container. (b) Vertical CPC-reactor (PSA, Spain).

capture almost all of the UV radiation arriving at the CPC aperture. The involute 'round W' geometry of the reflective surface distributes the reflected light around the back of the tubular reactor, thus enabling illumination of most of the tubes' circumference. The ratio of CPC aperture and tube diameter corresponds to a concentration factor of one sun, similar to simple flatbed reactors. CPC reactors are typically operated in circulation mode and cooling is achieved in a heat exchanger unit after the reaction mixture leaves the exposed tubes. Their simple construction and operation make these devices an excellent choice for solar operations.

7.3.3 Concentrating Reactors

Solar concentrating systems use mirrors to collect the direct solar irradiation over a large area and concentrate it onto a small receiver surface. As a result, concentration factors of several hundred suns and higher can be easily achieved. These collectors are generally used for solar thermal power production, where the irradiation is absorbed and converted into heat.[52-54] A major limitation of concentrating systems is their dependence on direct radiation and thus clear and sunny weather conditions. In contrast, non- and low-concentrating reactors can harvest direct and diffuse radiation, which makes them less dependent on climatic conditions and subsequently locations.

7.3.3.1 Parabolic Troughs. Line-focus parabolic trough reactors reflect the direct irradiation onto a tube absorber placed in the focal line.[65] Their design enables a concentration factor of (20–150) suns, although optical losses can reduce these values significantly. Sun tracking is typically realized throughout the day, either by rotation about the longitudinal axis (single-axis) or by three-dimensional tracking about both axes (two-axes). Large installations of these concentrators, up to 150 m long and 5 m wide, with single-axis tracking are operating for energy production purposes.

For photochemical applications, the central receiver tube is made of transparent glass, through which the reaction mixture is passed in cycles. Cooling is typically realized using an external heat exchanger. This configuration has been followed for simple laboratory-scale parabolic trough reactors (Figure 7.3(a)) as well as advanced production-scale prototypes. The PROPHIS-loop (PaRabolic trough-facility for Organic PHotochemical synthesIS) at DLR in Germany (Figure 7.3(b)), a modification of the SOLARIS-reactor (SOLAR photochemical synthesIS of fine chemicals) at PSA in Spain, consists

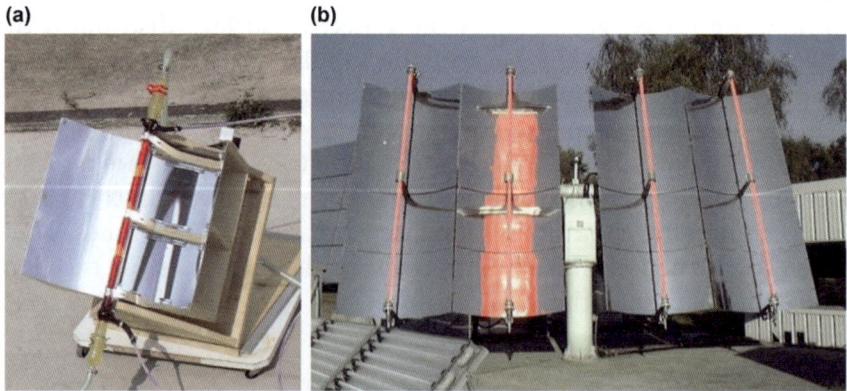

(a)   **(b)**

Figure 7.3   (a) Laboratory-scale reactor utilizing a Liebig condenser (DLR, Germany).
(b) PROPHIS loop (DLR, Germany).

Figure 7.4 Parabolic dish test facility (PSA, Spain).

of four line-focusing parabolic troughs of 8 m^2 each, which enable a concentration factor of (30–32) suns.[66] The individual 1 m^2 mirror elements are manufactured from silver-coated glass and the whole concentrator follows the Sun with a two-axis sun tracking system. The loop allows operations on (35–120) L scales but requires direct sunlight. The storage tank, pump, heat exchanger and gas feeding equipment are stored in a nearby shed.

7.3.3.2 Dish Reactors. Solar dish reactors use parabolic mirrors to concentrate sunlight onto a receiver in their focal points (Figure 7.4).[55–57] The dish concentrator requires a two-axis tracking system that follows the Sun throughout the day. Subsequently, high concentration factors of up to 5000 suns can be achieved. This high degree of concentration results in extreme temperatures, which limits their application to thermochemical processes.[67] The point focus and the need to track the Sun furthermore represent significant engineering challenges as the receiver must remain small and the connecting tubing flexible. Due to these limitations, solar dish reactors for preparative photochemistry are limited to small-scale applications. While larger scale chemical production may be achieved by operating several dishes in parallel or series, this numbering-up strategy causes significant investment and operation costs.

7.3.3.3 Highly Concentrating Reactors. Concentration factors of (5000–20 000) suns can be realized using large solar furnaces. The solar furnace facility at DLR in Cologne, for example, utilizes a flat 52 m^2 heliostat that reflects sunlight onto a concentrator consisting

Figure 7.5   Solar furnace (DLR, Germany).

Figure 7.6   Photooxygenation reactor in the solar furnace (DLR, Germany).

of 147 spherical mirrors (Figure 7.5).[68] Its multi-facet design generates a highly concentrated beam that is focused onto the corresponding experimental setup inside the laboratory building. The furnace can achieve concentration factors of 4800 suns with a total power of 20 kW and is mainly used for component development in material science, but also for solar chemistry. Highly concentrated sunlight enables significantly enhanced space-time yields and is thus attractive for reactions with low quantum yields. Major disadvantages are the extreme temperature stresses on reactor materials, the high installation costs and the typically complex operation procedures.

The high concentration factors demand the construction of specialized reactor systems with efficient cooling. Figure 7.6 shows a reactor block that was designed and utilized for photooxygenation reactions in the solar furnace at DLR. The reaction cell inside the reactor is positioned in the focal point of the incoming beam.

Similar solar furnaces have been constructed at PSA in Spain (69 kW and 40 kW) and the National Renewable Energy Laboratory (NREL) in Golden, CO, USA (10 kW). The largest solar furnace in the world of approximately 1000 kW has been erected in Odeillo, France. The facility consists of a field of 10 000 mirrors, mounted on the surrounding hillside, which reflect sunlight onto a large concave mirror. This focuses a highly concentrated beam onto an experimental setup in a separate building. Both the solar furnace in Golden and at Odeillo have been successfully used for the solar thermal synthesis of fullerenes.[69–72]

7.4 SOLAR PHOTOCHEMICAL PRODUCTION OF FINE CHEMICALS

The appearance of modern solar reactors and 'green chemistry' has caused a renaissance in solar photochemistry. While advanced water treatment and energy production processes are already established in practice, solar photochemical syntheses of chemicals have been much less researched.[73] The 'chromatic limitations' caused by the solar spectrum (necessary absorption between 300 and 700 nm) and the discontinuous nature of sunlight have thus far prevented an adaptation of this technology for large-scale continuous productions. The fine-chemical industry, however, offers attractive opportunities for small-scale and decentralized manufacturing.[74,75] Fine chemicals are produced in small volumes (<1000 tonnes per year) and at relatively high prices (US$ >10 per kg). As a result, most of the current lamp-driven industrial processes are in the flavour, fragrance and supplementary medicine (vitamins) industries.[23,24]

This chapter focuses on photochemical syntheses conducted on small- to demonstration-scales in specifically designed solar reactors. Key experimental details and results are summarized within the various schemes.

7.4.1 Solar Reactions in Non- to Low-concentrating Reactors

Photoinduced cyclodehydrogenations of 1,2-diheteroarylethylenes (**5a–c**) to yield thiohelicenes (**6a–c**) have been studied by Caronna *et al.* (Scheme 7.4).[76] The authors used the SOLFIN-loop (SOLar synthesis of FINe chemicals), a CPC-reactor with 2 m^2 aperture, a concentration factor of 1 sun and an illuminated volume of up to 10 L (total volume 10–20 L) at PSA in Spain. Compared with a conventional chamber reactor equipped with 16×8 W fluorescent tubes, the much

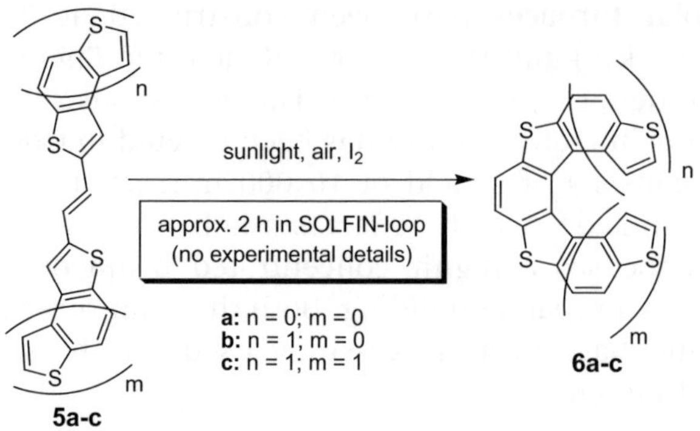

Scheme 7.4 Synthesis of thiohelicenes (**6a–c**) *via* photoinduced cyclodehydrogenation of **5a–c**.

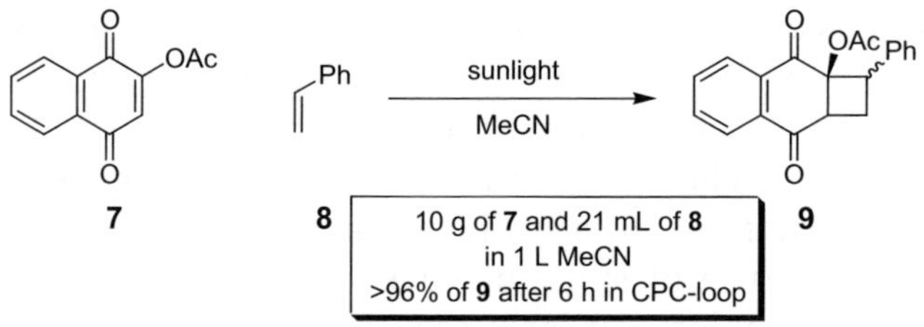

Scheme 7.5 [2 + 2]-Cycloaddition of styrene (**8**) to 2-acetoxy-1,4-naphthoquinone (**7**).

larger SOLFIN reactor significantly reduced irradiation times to 2 h. The amounts of energy (in W h) required for the cyclodehydrogenations were determined for both systems and the solar reactor gave values approximately 100-lower than that of the chamber reactor. The latter finding was primarily explained by the circulating flow conditions within the solar loop. No experimental details regarding the illumination procedure were provided, however, the original laboratory irradiation procedure with either UVB or visible light, employed benzene as solvent and catalytic amounts of iodine as oxidant.[77]

The solar [2 + 2]-photocycloadditions of 2-acetoxy-1,4-naphthoquinone (**7**) with styrene (**8**) in acetonitrile was investigated by Covell and colleagues in a small-scale CPC reactor at PSA in Spain (Scheme 7.5).[78] The reactor loop had a concentration factor of 2 suns and a total volume of 1 L. When operated in circulation mode with process cooling, the reaction mixture containing multigram quantities of naphthoquinone **7** was converted to the cyclobutane **9** after just 6 h of solar exposure. The product was obtained in high purity and in a yield of 96% after simple evaporation of the solvent. When conducted in the laboratory using an immersion-well batch reactor equipped with a 125 W medium-pressure mercury lamp, the reaction proceeded more sluggishly and larger amounts of by-products were obtained. Subsequently, solvent-demanding column chromatography had to be performed, which reduced the yield of **9** to approximately 50%. The solar process thus gave the desired cyclobutane **9** in a multigram quantity, excellent yield and purity with a minimum work-up procedure.

Using a modified SOLFIN facility at PSA in Spain with a concentration factor of approximately 4 suns and a total volume of 1.2 L, Dondi and co-workers investigated a series of solar-sensitized additions to electrophilic alkenes and alkynes.[79] Disodium benzophenondisulfonate (BPSS) was chosen as a water-soluble photocatalyst that could be simply removed by extraction during workup. As an example of an efficient multigram synthesis, solar exposure of maleic acid (**10**) in aqueous isopropanol (**11**) for 14 h furnished terebic acid (**12**) in an isolated yield of 75% (Scheme 7.6). High-pressure liquid chromatography (HPLC) analysis revealed a conversion of 90% for **10** at the end of the experiment. Reaction monitoring during the three-day illumination period confirmed that the poor weather conditions during the second day did not contribute significantly to the overall conversion. The solar illumination protocol was expanded to other unsaturated acids and aldehydes, and yielded similar addition products in yields of (31–55)%. In a previous project, the same group investigated a TiO_2 catalysed version of this addition reaction.[80]

Various solar photoisomerizations of dienes and trienes of the vitamin A series were examined at the University of Hawaii, USA, by Zhao and co-workers in floating 'kick-board' or 'boogie-board' reactors.[63] Several test tubes filled with reactions mixtures were placed inside a sample frame, which was fitted into a hole that was cut into the body of the 'kick-board'. Depending on the absorption of the reagents, an exchangeable filter plate (Plexiglass or Pyrex) was attached

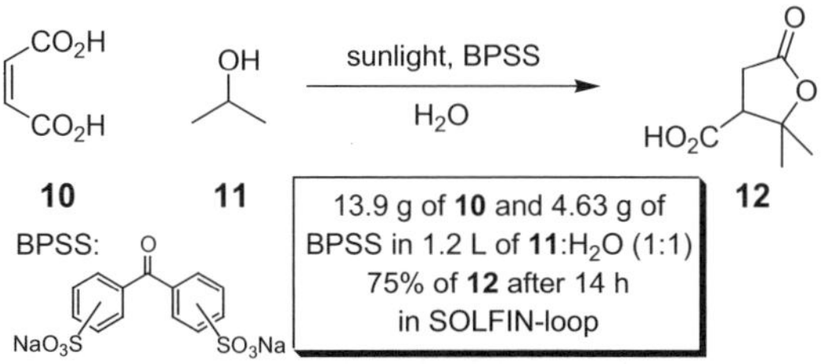

Scheme 7.6 BPSS-catalysed addition of isopropanol (**11**) to maleic acid (**10**).

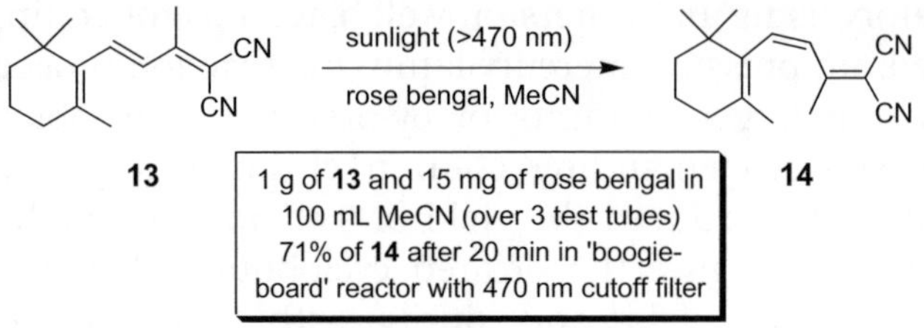

Scheme 7.7   Rose bengal-sensitised isomerization of **13**.

as a cover. The cover and frame ensured that the sample tubes were about half immersed when the board was placed on water. The larger 'boogie-board' reactor did not have a holding frame and the test tubes were held in place with wire instead. The reactor was used either in an artificial or natural water reservoir. Although simple in its design, the reactor system remained completely independent from any external power or cooling supply.

Using a 470 nm cut-off filter plate, the rose bengal sensitized isomerization of the 7-*trans* dinitrile **13** in the 'boogie-board' reached a stable conversion of 82% after just 20 min (where 'min' refers to minute) of solar exposure (Scheme 7.7). The corresponding 7-*cis* isomer **14** was subsequently isolated by column chromatography in a yield of 71%. Analogue results were obtained with other trienes and β-ionol. The 'boogie-board' reactor was also successfully adopted for other solar isomerizations.[81]

7.4.2   Solar Reactions in Moderately-Concentrating Reactors

Esser and co-workers conducted a variety of photochemical reactions using the advanced SOLARIS loop at PSA in Spain. The reactor enables a concentration factor of 20 suns and has an operating volume of (35–70) L, thus allowing multi-kilogram scale experiments. Of the four available troughs, only one with a reflective area of approximately 8 m^2 was used for the experimental series. The three transformations chosen covered the entire usable spectrum of the Sun.[46,47]

The acetone-sensitized [2 + 2]-photocycloaddition between 5-ethoxy-[5*H*]-furan-2-ones **15** and ethene (**16**), for example, was investigated as a 'solar borderline' reaction since acetone only absorbs light <330 nm (Scheme 7.8).[46] The experiment was performed over a period of two days, of which only the second day showed favourable radiation conditions. Using a solution of **15** in acetone, a moderate

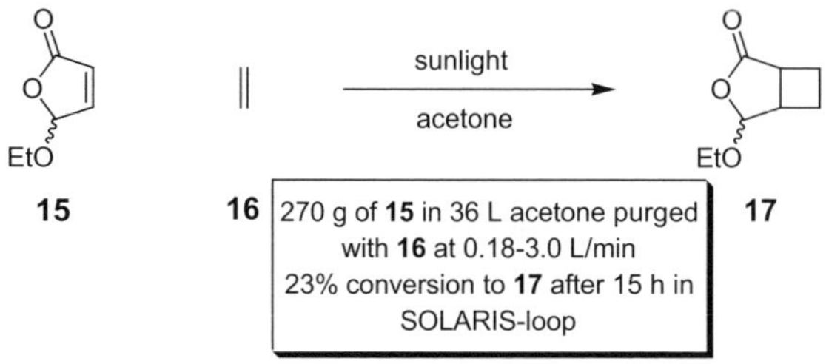

Scheme 7.8 Acetone-sensitized [2 + 2]-cycloaddition of 5-ethoxy-[*5H*]-furan-2-ones (**15**) and ethene (**16**).

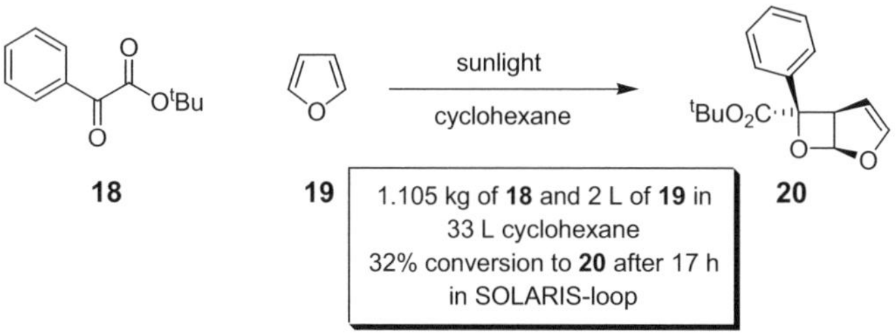

Scheme 7.9 Paternò–Büchi reaction of (*tert*-butyl)phenylglyoxylate (**18**) with furan (**19**).

conversion to the cyclobutane **17** (as a 2.6 : 1 mixture of diastereo-isomers) of 23% was achieved after a total exposure time of 15 h. Due to the volatility of acetone, the reaction mixture was cooled to 10 °C inside the feeding facility. Depending on the solar conditions and thus the conversion rate, ethene gas was blown into the reaction mixture at variable flow rates of (0.18–3.0) L min^{-1} before entering the concentrator. Using a flow rate for the reaction medium of 20 L min^{-1}, a maximum heat-up of 6 °C was recorded between the inlet and outlet of the reaction tube.

Likewise, the Paternò–Büchi reaction of (*tert*-butyl)phenylglyoxylate (**18**) with furan (**19**) was realized in the SOLARIS reactor (Scheme 7.9).[46] Due to the absorption of **18**, the transformation can utilize sunlight up to <400 nm, which makes it more suitable for solar photochemistry. A mixture of **18** and **19** in cyclohexane was illuminated for two consecutive days with good solar conditions. The temperature of the reaction mixture was kept at 20 °C. Using a flow rate of 23 L min^{-1}, the heat up along the receiver tube was minimal at just 3 °C. After an expose of 17 h, a conversion to the corresponding oxetane **20** of 32% was determined.

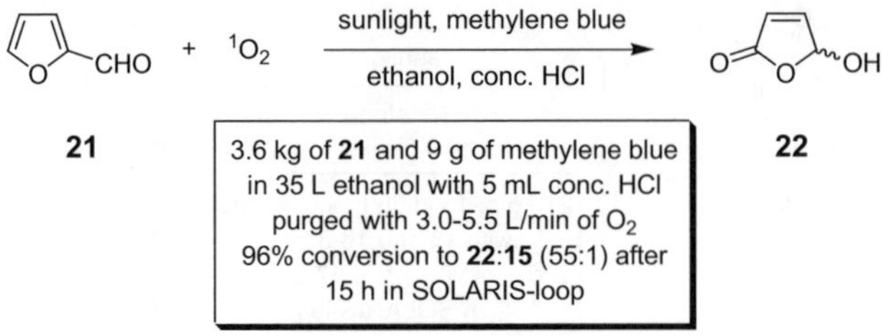

Scheme 7.10   Photo-oxygenation of furfural (**21**) to 5-hydroxyfuranone (**22**).

Photooxygenation reactions are particularly interesting for solar applications as these transformations are conducted industrially,[24] many from renewable materials as starting reagents.[82] Singlet oxygen is produced using organic dyes as sensitizers that absorb strongly within the solar spectrum.[83] For easy recovery and recycling, sensitizers can be immobilized on a solid support.[84]

Consequently, the photooxygenation of furfural (**21**) to 5-hydroxy-furanone (**22**) was studied systematically using the SOLARIS loop (Scheme 7.10).[46] The reaction conditions were varied to investigate the influence of sensitizer (rose bengal *vs.* methylene blue), fluid flow rate, starting material (**21** *vs.* its diethylacetal) and additives. Methylene blue was determined as a more suitable sensitizer due to its increased stability. Likewise, the addition of catalytic amounts of concentrated hydrochloric acid proofed beneficial. Under ideal conditions on a multi-kilogram scale, **21** were almost completely consumed within a single day and 15 h of solar exposure. Depending on the degree of solvent evaporation, oxygen was injected with variable flow rates of $(3.0–5.5)$ L min^{-1}, while the reaction mixture was circulated at 45 L min^{-1}. Although the reaction was kept at 20 °C and heat-up within the reactor tube was minimal with 3 °C, **22** partially converted to its corresponding ethoxy derivative **15** due to the acidic reaction conditions.

The SOLARIS reactor was later relocated to DLR in Germany and re-commissioned as the improved PROPHIS loop.[66] This modified reactor has a concentration factor of $(30–32)$ suns and an operating volume of $(35–120)$ L. It was subsequently used by various researcher groups for demonstration-scale solar syntheses.

The photoacylation of 1,4-naphthoquinone (**23**) with butyraldehyde (**24**) yields the important building block chemical **25** (Scheme 7.11) and was thus investigated by Schiel and co-workers in the PROPHIS plant using all four troughs (aperture of 32 m^2).[85,86] The original laboratory protocol required benzene as solvent,[87] which was

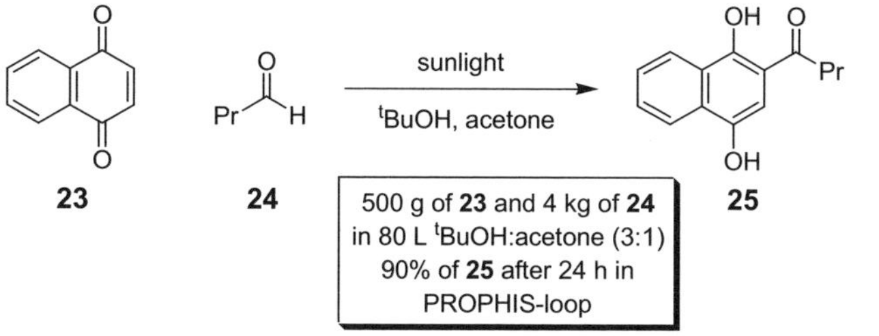

Scheme 7.11 Photoacylation of 1,4-naphthoquinone (**23**) with butyraldehyde (**24**).

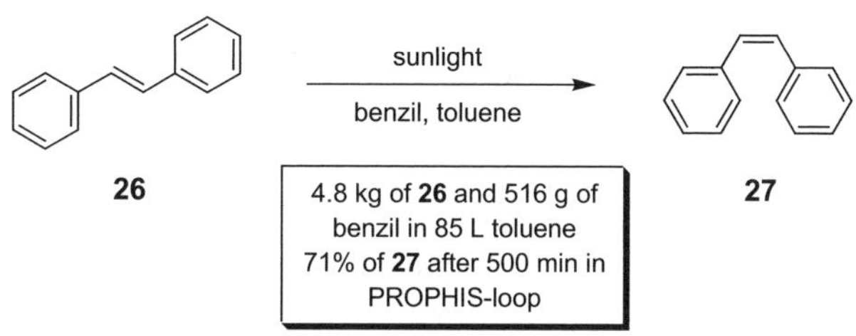

Scheme 7.12 Benzil-sensitized photoisomerization of trans-stilbene (**26**).

subsequently replaced by *tert*-butanol or its mixture with acetone. Due to the strong absorption of the acylated naphthohydroquinone **25**, the transformation demands long irradiation times. A mixture of **23** and **24** in a 3 : 1 *tert*-butanol–acetone mixture was exposed to sunlight for 24 h over a period of 3 days. Only the first day offered good illumination conditions, whereas the remaining two days were characterized by cloudy conditions. At the end of the experiment, a yield of 90% was determined for compound **25**.

The photoisomerization of trans-stilbene (**26**) was likewise studied in the PROPHIS loop on demonstration scale (Scheme 7.12).[66] Benzil was used as a suitable sensitizer that absorbs within the solar spectrum. Solar exposure of **26** in toluene for just under 8.5 h (over two separate days) and using four troughs furnished a maximum conversion of 83%, of which 77% was determined as the desired *cis*-stilbene (**27**). The transformation thus showed a high selectivity of 93% with very few by-products. Isolation subsequently produced **27** in a yield of 71%.

Heller and co-workers realized a series of large scale Co(I)-catalysed [2 + 2 + 2]-cycloadditions in the PROPHIS reactor.[88–93] The reaction of benzonitrile (**28**) with acetylene (**29**) to 2-phenylpyridine (**30**) served as a model reaction (Scheme 7.13). Cyclopentadienyl-1,5-cyclooctadiene-cobalt(I) (CpCoCOD) was employed as a readily available catalyst and

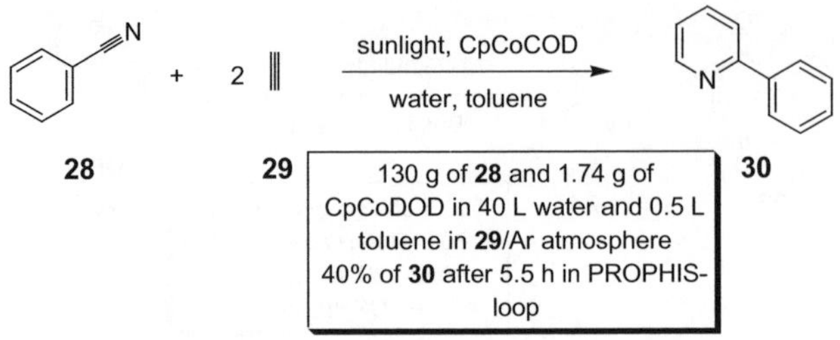

28   29

130 g of **28** and 1.74 g of CpCoDOD in 40 L water and 0.5 L toluene in **29**/Ar atmosphere 40% of **30** after 5.5 h in PROPHIS-loop

30

Scheme 7.13 Co(I)-catalysed [2 + 2 + 2]-cycloaddition of benzonitrile (**28**) with acetylene (**29**) to 2-phenylpyridine (**30**).

water as an environmentally benign solvent, although small amounts of toluene were found to be advantageous. Due to the sensitivity of the reaction towards oxygen, the reaction was performed in an argon–acetylene atmosphere and at 20 °C. From the best run obtained with one trough module (8 m^2), a conversion of 41% was achieved after just 5.5 h. At the end of the experiment, CpCoCOD remained active and besides 1.3% of benzene, no by-products could be detected by gas chromatography (GC) analysis. The desired pyridine derivative **30** was subsequently isolated in a yield of 40%. Further cycloaddition experiments were conducted at PSA with different alkynes and nitriles and using an early SOLFIN facility. The corresponding pyridine derivatives were obtained in acceptable to high yields of 12–91% and with excellent selectivities.[90,91]

The Scheck-ene-type photooxygenation of citronellol (**31**) was likewise investigated by Oelgemöller and colleagues in the PROPHIS plant.[94] During operation, oxygen gas was injected into the reaction mixture at a flow rate of 200 L h^{-1}. Initially, the reaction produces two regioisomeric hydroperoxides, which are subsequently reduced to their corresponding diols **32a** and **b**. The transformation is the key step in the synthesis of the important fragrance rose oxide (**33**), which is obtained from the regioisomer **32a** *via* acid mediated cyclization (Scheme 7.14).[95] Two technical-scale experiments were conducted using either one or all four troughs of the PROPHIS reactor and both reached complete conversion within 3 h of exposure to sunlight. From these, rose oxide (**33**) was subsequently obtained in an excellent quality and overall yield of 55% (from **31**).

The same authors studied the photooxygenation of 1,5-dihydroxynaphthalene (**34**) to the important natural product and building block juglone (**35**) in a small parabolic trough reactor (Scheme 7.15).[94,96] The reactor achieved a concentration factor of 15 suns and was suitable for operations on (0.2–1.0) L scales. Oxygen was

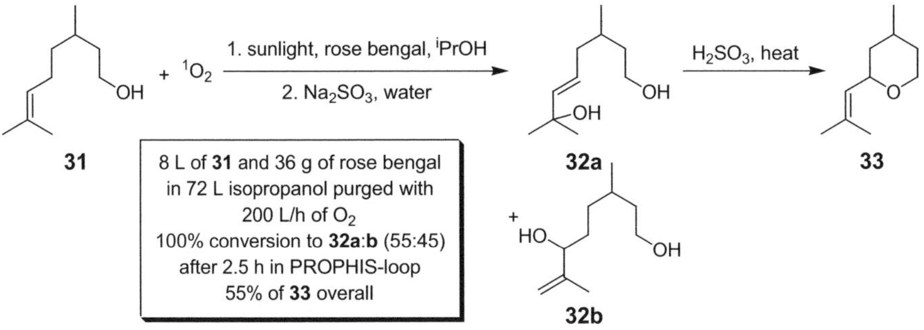

Scheme 7.14 Rose bengal-sensitized photo-oxygenation of citronellol (**31**) and subsequent conversion to rose oxide (**33**).

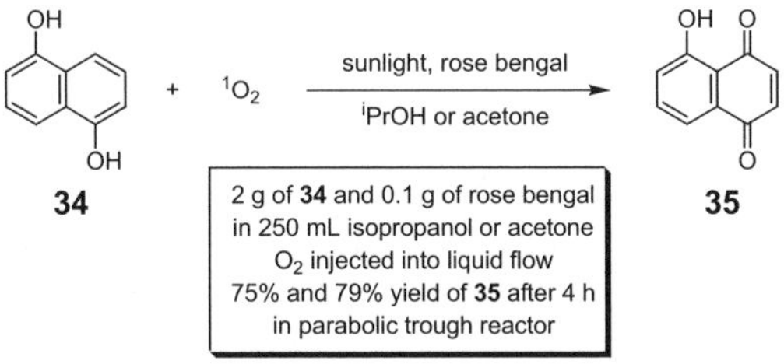

Scheme 7.15 Rose bengal-sensitized photooxygenation of 1,5-dihydroxynaphthalene (**34**) to juglone (**35**).

injected into the reaction mixture *via* a simple Y-connector, thus creating a liquid–gas slug flow. In a first set of solar experiments, the reactor was equipped with holographic mirror elements that showed a reflectivity range of (550 ± 140) nm optimal for the application of rose bengal. Holographic mirrors are designed to reduce extreme heat-up in the receiver tube and thus cooling water costs.[97,98] Their limited life time and expensive construction, however, make them currently not economical. Using isopropanol as a solvent, (1–2) gram-scale transformations of **34** were achieved within 12 h. Conversions were high with >83% and juglone (**35**) was isolated in yields of (54–79)%.[94] In a second series of experiments, the mirror elements were made out of highly reflective eloxated aluminium sheets. Using a fixed illumination time of 4 h, various sensitizers (soluble and im-mobilized) and different solvents were tested. The best results were achieved with rose bengal as sensitizer in either isopropanol or acetone and juglone (**35**) was isolated in 75% and 79% yield, respectively.[96]

Likewise, this transformation was studied in a simple parabolic trough reactor constructed from a Liebig condenser (Figure 7.3(a)).[99]

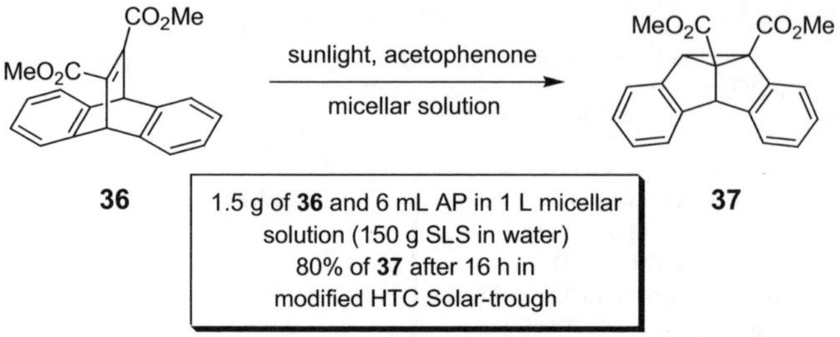

Scheme 7.16   Di-π-methane rearrangement of **36** to the polycyclic compound **37** in micellar solution.

All three illumination reactions conducted for (2.5–4.5) h furnished high conversions of (86 to >95)% and gave juglone (**35**) in isolated yields of (46–71)%.

Ritter and co-workers conducted a series of rearrangement reactions in an line-focussing reactor (designed by HTC Solar, Lörrach) at the Max-Planck Institute (MPI) in Mühlheim, Germany.[100] The device used an aluminium coated Hostaflon film, which was stretched over a chamber as a reflector. Focusing of sunlight was achieved by applying a vacuum and concentration factors of 60 suns could be reached in practice.[101] The feature also allowed a controlled defocusing and hence gave a superior temperature control. The entire loop incorporated a cooling tower and had a maximum operating volume of 2 L. Several di-π-methane and oxadi-π-methane rearrangements in aqueous micellar solutions, made from sodium lauryl sulphate (SLS) in water, were studied under direct and sensitized conditions. With acetophenone (AP) as a sensitizer, reaction times could be significantly shortened. As an example of an efficient transformation, the sensitized di-π-methane rearrangement of **36** furnished the polycyclic compound **37** in 80% after 16 h of solar exposure (Scheme 7.16). Under direct illumination conditions, the reaction gave **37** in a reduced yield of 75% and after a much prolonged reaction time of 104 h.

7.4.3   Solar Reactions in Highly-Concentrating Reactors

Preparative solar photoreactions in concentrated sunlight are rather rare due to the extreme temperatures generated, which demand sophisticated cooling facilities. As a result, these reactor systems are mainly used for solar thermal applications including hydrogen production.[102,103]

İçli's group successfully applied a solar dish concentrator developed by HTC Solar to a series of dehydrogenation reactions.[104,105]

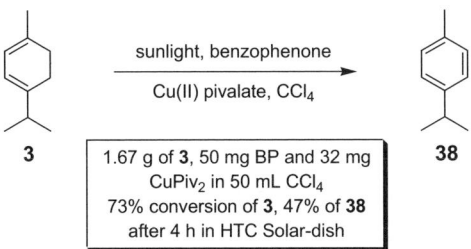

Scheme 7.17 Solar catalytic dehydrogenation of α-terpinene (**3**) to *p*-cymene (**38**).

Similar to the line-focusing reactor by Ritter and co-workers, the concentrator was made from a reflective, aluminium-coated polymer film that enabled concentration factors of (40–150) suns at its focal point. Using a combination of a copper(II) pivalate (CuPiv$_2$) catalyst and benzophenone (BP) as photosensitizer, abietic acid and α-terpinene were converted into their corresponding dehydrogenated products dehydroabietic acid and *p*-cymene. The reaction involving acenaphthene required a high concentration factor of 110 suns and prolonged illumination times to achieve measurable conversions to acenaphthylene of up to 2.2%. The conversion of α-terpinene (**3**) to *p*-cymene (**38**) is exemplarily shown in Scheme 7.17. After 4 h of exposure with a concentration factor of 40 suns, a conversion of 73% was achieved and the reaction mixture contained 47% of the de-hydrogenated product **38**.[104]

Photoreactions in highly concentrated sunlight were studied by Pohlmann and co-workers using the solar furnace at DLR in Germany.[47] The high concentration factor of >1000 suns required the construction of a special reactor unit. The reactor consisted of a cylindrical metal body, which was divided into an infrared (IR) filter cell (front) and a reaction cell (back) by three quartz glass panels. The reaction mixture was passed through the reaction cell with a powerful pump and was cooled by a plate heat exchanger. Reaction monitoring was performed by on-line UV spectroscopy.

The reactor was subsequently used to study the diastereoselective Paternò–Büchi reaction of (−)-menthyl 2-thienylglyoxylate (**39**) with furan (**19**).[47] The reaction mixture was pumped through the reaction cell at a flow rate of 20 L min^{-1}, enabling a reaction temperature below 30 °C. Using two different compositions for the reaction mixture, high conversions were rapidly achieved within (1.5–2) h (Scheme 7.18). The reactions proceeded remarkably clean and no by-products could be detected. Likewise, diastereoselectivities (d.e.) for

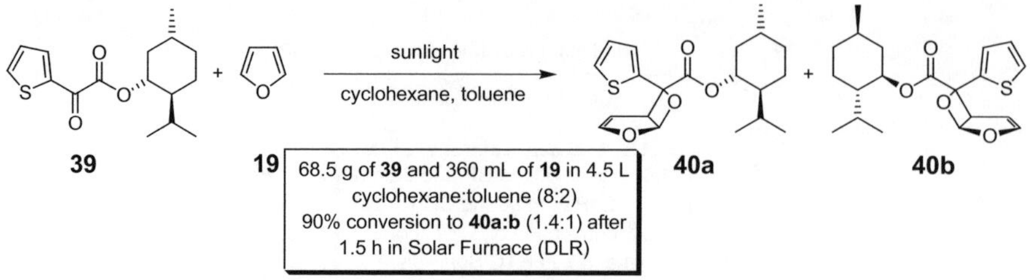

Scheme 7.18   Diastereoselective Paternò–Büchi reaction of (−)-menthyl 2-thienyl-glyoxylate (**39**) with furan (**19**).

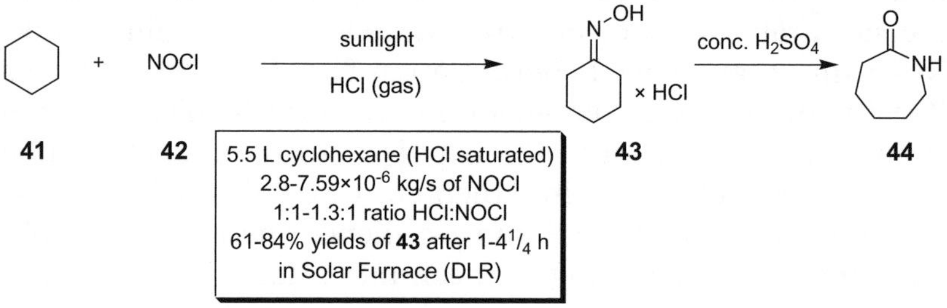

Scheme 7.19   Photo-oximation of cyclohexane (**41**) to cyclohexanone oxime hydro-chloride (**43**) and subsequent conversion to ε-caprolactam (**44**).

the oxetanes **40a** and **b** were comparable with laboratory reactions with approximately 58% (in favour of **40a**). The overall solar photon yields for the utilization of solar radiation up to 400 nm (the absorption limit of **39**) remained low at (3–4)%, which was explained by DLR's northern location and the time of the experiment (October). Similar results were obtained for the reaction of (*tert*-butyl)phenyl-glyoxylate (**18**) with **19**.[47]

The photooximation of cyclohexane (**41**) with nitrosyl chloride (**42**) to cyclohexanone oxime hydrochloride (**43**) was conducted by Funken and co-workers in the solar furnace of DLR (Scheme 7.19).[106] This transformation is currently performed industrially using artificial light on very large scales by Toray in Japan (1991: 160 000 tonnes per year).[23] Oxime **43** is the precursor for ε-caprolactam (**44**) and consequently nylon-6. Nitrosyl chloride (**42**) absorbs light of wavelengths shorter than 635 nm, which makes it feasible for solar applications.

A specialized reactor vessel with a volume of 6.5 L was constructed from titanium and fitted with glass windows for the incoming and exiting solar beam. The whole reactor was immersed in a glass cylinder, which was cooled by circulating water. The surrounding water mantle additionally functioned as an effective IR filter. During illuminations, the focal area of the incoming solar beam was positioned in the centre of the reactor. Dry and oxygen free cyclohexane (**41**) was placed inside the reaction vessel and saturated with hydrochloric acid. A fine gas stream of **42**, freshly produced from sodium nitrite, and hydrochloric acid entered the reactor from the bottom of the reaction vessel. Cyclohexanone oxime hydrochloride (**43**) formed during solar exposure and precipitated as an oily substance. At the end of the experimental run, **43** was converted into the free oxime by neutralization. Using an approximately equimolar ratio of NOCl and HCl, good to excellent yields of (61–84)% for the desired **43** were achieved after (1–4.25) h of solar irradiation. The efficiency of the reaction dropped remarkably when the amount of hydrochloric acid was increased. All solar reactions showed excellent selectivity and only small amounts of chlorocyclohexane could be detected as the main by-product. A similar photooximation using simulated sunlight has been described by Talukdar *et al.*[107]

7.4.4 Solar Reactor Comparison Studies

Isolated examples of solar reactor comparison studies have been reported over the years. These studies typically compare concentrating *vs.* non-concentrating reactor types to establish the best reactor model for a specific location. In particular, non-concentrating reactors can utilize the diffuse fraction of the sunlight, which makes them advantageous for operations in less sun-blessed countries. In contrast, concentrating solar reactors generally show superior performances in countries with high levels of annual sunshine hours.[59]

At MPI in Mühlheim in Germany, Heinemann and co-workers compared an advanced line-focusing trough reactor (HTC Solar, Lörrach) with a simple flatbed reactor.[108] The improved HTC trough reactor enabled a concentration factor of (5–60) suns, incorporated a two-axis sun tracking system and allowed flow rates of *ca.* 100 L h^{-1}. In contrast, the flatbed reactor was positioned flat on the ground, had a simple glass or transparent plastic cover with a safety valve and a reflective bottom plate. The reactor rested on a simple heat exchanger for process cooling. The photoinduced electron-transfer reaction of the 1,1-dicarbonitrile **45** to the corresponding cyclization product **46** served as a model reaction (Scheme 7.20), although no experimental details on reagent scales were reported. The transformation was conducted separately on a partially sunny and on a completely cloudy day in both reactors. Under both illumination conditions, the simple

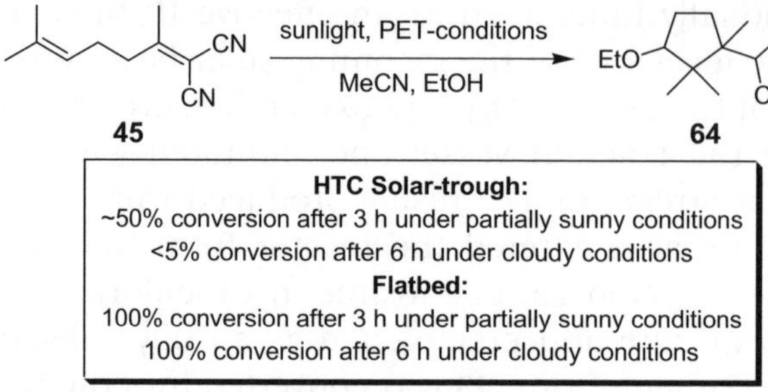

Scheme 7.20   Photoinduced electron-transfer cyclization of 1,1-dicarbonitrile **45** to the corresponding cyclization product **46**.

flatbed reactor showed complete conversions of **45** after 3 h and 6 h, respectively. In contrast, the concentrating trough reactor showed a lower conversion of approximately 50% after 3 h under partially sunny conditions. In the absence of direct sunlight, the conversion stayed below <5% after 6 h of exposure. The superior performance of the flatbed reactor was caused by its ability to utilize diffuse radiation, which accounts for approximately 40% of the incident sunlight at the location of the MPI.

The photoacylation of 1,4-benzoquinone (**47**) with 2,4-dimethoxy-benzaldehyde (**48**) was used by Schiel and colleagues to evaluate the performances of three different solar reactors at the DLR in Germany (Scheme 7.21).[86,109] Using artificial light, the corresponding acylation product **49** was obtained in 66% yield after 18 h of irradiation in *tert*-butanol.[85] For the solar study, the PROPHIS plant, the CPC reactor and a simple double sheet flatbed reactor were chosen. All three reactors were loaded with a standard solution and the aperture of each reactor system was fixed at 3 m^2 to allow for a direct comparison. Illumination conditions during the solar campaign varied and were not optimal to reach high conversions. In addition, all reactors proceeded less selectively than the corresponding laboratory experiment and several unknown by-products were detected in the crude reaction mixtures. After a total exposure period of 19 h, the PROPHIS loop furnished a low conversion of **47** of *ca.* 40%, of which only 7% were assigned to the photoproduct **49**. The CPC reactor gave the best result with a conversion of **47** of 53% and a yield for **49** of 17% after 17 h of illumination. The tilted flatbed reactors produced the lowest yield for **49** of 3.6% and conversion of **47** of 30%, respectively. The poor selectivity of the reaction was explained by the presence of oxygen during the solar reactions, as the reaction mixtures had not been degassed with an inert gas prior to solar exposure. Likewise, the

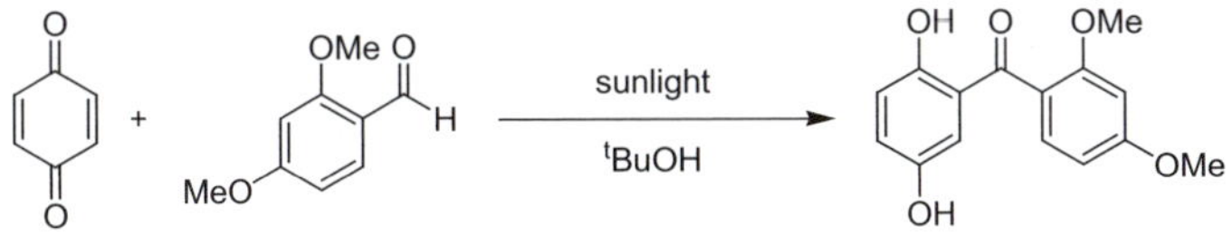

Standard solution:
175 g of **47** and 566 g of
48 in 58 L ^{t}BuOH
3 m^2 aperture per reactor

PROPHIS-loop (CF: 16 suns):
40% conversion after 19 h; 7% of **49**
CPC-reactor (CF: 1 sun):
53% conversion after 17 h; 17% of **49**
Flatbed (CF: 1 sun):
30% conversion after 30 h; 3.6% of **49**

Scheme 7.21 Photoacylation of 1,4-benzoquinone (**47**) with 2,4-dimethoxy-benzaldehyde (**48**).

Figure 7.7 Photooxygenation campaign in different solar reactors (DLR, Germany).[44]

reactor material and the static operation mode of the flatbed reduced its effectiveness significantly.[109] The superior performance of the CPC reactor was justified by its ability to collect and consequently utilize diffuse radiation.

Due to its industrial relevance, the photooxygenation of citronellol (**31**) has been investigated by Oelgemöller and co-workers at DLR in Germany in five different solar reactors and under different radiation conditions (Figure 7.7, Scheme 7.22).[44] Each reactor was loaded with a 10 vol% solution of **31** in isopropanol and rose bengal (0.5 g L^{-1} isopropanol). Under sunny conditions, the advanced PROPHIS loop gave a complete conversion after just 2.5 h. The CPC reactor required

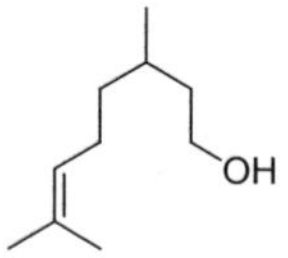

31	**PROPHIS-loop (CF: 32 suns):** **sunny:** 100% conversion after 2.5 h; **cloudy:** not operated **CPC-reactor (CF: 1 sun):** **sunny:** 100% conversion after 7 h; **cloudy:** 100% conversion after 30 h **Flatbed (CF: 1 sun):** **sunny:** 100% conversion after 15 h; **cloudy:** 100% conversion after 30 h **horizontal Tube-reactor (CF: 1 sun):** **sunny:** 100% conversion after 12 h; **cloudy:** 65% conversion after 33 h **vertical Tube-reactor (CF: 1 sun):** **sunny:** 100% conversion after 13 h; **cloudy:** 65% conversion after 33 h
Standard solution: 10 vol-% of **31** and 0.5 g (RB)/L in isopropanol	

Scheme 7.22   Solar reactor comparison study using the photooxygenation of citro-nellol (31) as a model.

7 h to reach complete consumption of **31**, whereas the flatbed, horizontal tube and vertical tube systems needed 15, 12 and 13 h, respectively. During cloudy conditions, the PROPHIS plant was not operated due to its dependency on direct radiation. The CPC system and the flatbed performed best and both reached full conversion after 30 h. The advantageous reflector design of the CPC reflector and the large aperture of the flatbed enabled an effective operation under diffuse light conditions. At the end of the campaign after 33 h, both tube reactors showed incomplete consumptions of **31** of 65%. The study confirmed that non-concentrating reactors can operate under a variety of light conditions, whereas concentrating reactors rely on direct sunlight. Due to their robust performances, CPC reactors are now becoming widespread for solar applications.

7.5   SUMMARY, CONCLUSION AND OUTLOOK

The examples presented in this chapter clearly show that sunlight can be efficiently harvested to drive preparative photochemical reactions. In many cases, products are produced in higher yields and better qualities than under laboratory conditions with artificial light. This is mainly due to the low UV content of natural sunlight. To achieve faster reactions, *i.e.* space-time-yields, solar concentrators have be utilized. Reactor comparison studies have nevertheless demonstrated that non- or low-concentrating systems operate more reliably as they do not depend on direct radiation and thus favourable weather conditions.[86,108] These static devices are furthermore more economical in their installation and operation, which makes them advantageous for on-demand productions of low-volume chemicals. Naturally, the location of any solar manufacturing plant is crucial. Solar photochemistry should thus be adapted in countries that offer good solar conditions, but at the same time have the necessary infrastructure to support a production plant. The possible implementation of solar-powered electrical components makes this technology attractive for developing countries, in particular when access to cheap starting martials from local biomass is given.[110]

Cost evaluations performed for the solar *vs.* lamp-driven synthesis of ε-caprolactam (**44**)[111,112] and rose oxide (**33**)[113] have clearly shown that the solar syntheses of these important chemicals are indeed economical in the long term. At present, the high development and installation costs, and the insecurity regarding the availability of sunlight, may be the biggest hurdles that prevent a realization on technical scales. The installation costs for solar manufacturing with limiting 'retrofitting' options are naturally high, but are commonly compensated by the annual savings on energy, cooling water and replacement lamps. Smart and cost-effective solar collectors or tandem solar-lamp driven processes may furthermore offer solutions to these challenges. The possible implementation of cheap and energy efficient light sources such as LEDs[13] or organic light-emitting diodes (OLEDs)[114] during periods of poor insolation or at night appears additionally appealing in terms of 'delivery security'. Likewise, continuous flow technologies represent an attractive opportunity for solar applications.[19–21] The small inner dimensions of these devices give high photonic efficiencies, which would enable them to operate effectively even under low solar power conditions. As the same time, their continuous flow mode avoids accumulation of potentially hazardous reagents or products, thus increasing process safety. The development of new photocatalysts or photosensitizers that absorb in the visible range may furthermore allow for a transfer of previously UV-driven protocols to solar conditions.[34] Immobilization will additionally permit the usage of smaller ('localized') amounts and will enable reuse of these materials, which in turn may result in additional cost savings.[84]

The environmental benefits of solar manufacturing are much more obvious. Solar operation reduces carbon dioxide emissions and subsequently the environmental burden of the process significantly. This has been recently confirmed by Ravelli and co-workers for the synthesis of rose oxide (**33**) using environmental assessment calculations.[115] As notable from the examples presented, many solar chemical experiments have been conducted by simply transferring the existing laboratory protocols outdoors.[116] Volatile or hazardous solvents, reagents gases or photocatalysts are thus commonly applied.

In terms of the overall greenness and operational safety of solar photochemistry, it is highly desirable to develop protocols that address all aspects of the process, *i.e.* its entire life cycle.[117–119] Visible light photoredox catalysis, which has seen significant interest in past years, is a good example of this problem.[34] While these transformations can be conducted with natural sunlight, they are typically operated in high dilutions using toxic solvents and often require the presence of sacrificial reagents, which are added in access amounts. Thus far, most photoredox reactions were furthermore performed with artificial light. A recent environmental assessment by Ravelli and co-workers consequently questioned the general sustainability of this popular methodology.[120] Large-scale examples of solar-driven photoredox reactions have also not been realized yet.

Notable from the literature, preparative solar chemistry with modern reactors peaked between 1990 and 2005 and has since seen a decline.[121] The main reason is the decreasing interest in this synthesis methodology and a shift to solar powered technologies for energy productions at the large solar sites in German and Spain which in the past attracted numerous collaborations with external researchers. It is hoped that this trend will be reversed, also due to new developments in solar technologies and photochemical protocols. After over 100 years, Ciamician's prophetic warning 'for nature is not in a hurry and mankind is' still stands today. In order to realize his vision of a clean and green 'photochemistry of the future',[40] the authors urge current and future generations of organic photochemists to bring this exciting methodology 'back to the roofs!'.

ACKNOWLEDGEMENTS

The authors thank Dr Peter Esser, Dr Achim Hülsdünker, Dr Bettina Pohlmann, Dr Jürgen Ortner and Dr Christian Jung for valuable discussions, information and resources. The corresponding author (M.O.) additionally thanks his diploma mentor Professor Jochen Mattay (University of Bielefeld), his collaborators and all students who have contributed to the success of his solar research projects in the past. Financial support for projects from the authors came from the Arbeitsgemeinschaft Solar des Landes Nordrhein-Westfalen (Themenfeld Solare Chemie und solare Materialforschung), the Irish Research Council for Science, Engineering and Technology (IRCSET; PhD Scholarship 2006), the Irish Environmental Protection Agency (STRIVE; PhD Scholarship 2007-PhD-ET-7), James Cook University (JCU Faculty Grants Schemes 2009 and 2011), the Australian Institute of Tropical Health and Medicine (AITHM; Development Grant 2012) and the Queensland Department of Employment, Economic Development & Innovation (DEEDI; Proof of Concept Grant 2012).

REFERENCES

1. S. E. Braslavsky, *Pure Appl. Chem.*, 2007, **79**, 293.
2. C. G. Bochet, *Synlett*, 2004, 2268.
3. A. M. Braun, M. Maurette and E. Oliveros, *Photochemical Technology*, Wiley, Chichester, UK, 1991.
4. D. Frackowiak, *J. Photochem. Photobiol. B: Biol.*, 1988, **2**, 399.
5. N. Hoffmann, *Chem. Rev.*, 2008, **108**, 1052.
6. T. Bach and J. P. Hehn, *Angew. Chem., Int. Ed.*, 2011, **50**, 1000.
7. J. Iriondo-Alberdi and M. F. Greaney, *Eur. J. Org. Chem.*, 2007, **29**, 4801.
8. M. B. Rubin and S. E. Braslavsky, *Photochem. Photobiol. Sci.*, 2010, **9**, 670.
9. J. M. Parnis and K. B. Oldham, *J. Photochem. Photobiol. A: Chem.*, 2013, **267**, 6.
10. A. Albini, M. Fagnoni and M. Mella, *Pure Appl. Chem.*, 2000, **72**, 1321.
11. C. L. Ciana and C. G. Bochet, *Chimia*, 2007, **61**, 650.
12. E. A. Sosnin, T. Oppenländer and V. F. Tarasenko, *J. Photochem. Photobiol. C: Photochem. Rev.*, 2006, 7, 145.
13. S. Landgraf, *Spectrochim. Acta Part A*, 2001, **57**, 2029.
14. A. G. Griesbeck, W. Kramer and M. Oelgemöller, *Synlett*, 1999, 1169.
15. D. Budac and P. Wan, *J. Photochem. Photobiol. A: Chem.*, 1992, **67**, 135.
16. M. Martín-Flesia and Al Postigo, *Curr. Org. Chem.*, 2012, **16**, 2379.
17. N. J. Turro, M. Grätzel and A. M. Braun, *Angew Chem., Int. Ed. Engl.*, 1980, **19**, 675.
18. J. M. Tanko, J. F. Blackert and M. Sadeghipour, in *Benign by Design: Alternative Synthetic Design for Pollution Prevention*, ed. P. T. Anastas and C. A. Farris, ACS Symposium Series 577, American Chemical Society, Washington DC, 1994, pp. 98–113.
19. M. Oelgemöller, *Chem. Eng. Technol.*, 2012, **35**, 1144.
20. M. Oelgemöller and O. Shvydkiv, *Molecules*, 2011, **16**, 7522.
21. E. E. Coyle and M. Oelgemöller, *Photochem. Photobiol. Sci.*, 2008, 7, 1313.

22. R. Schlögl, *ChemSusChem*, 2010, **3**, 209.
23. M. Fischer, *Angew. Chem., Int. Ed.*, 1978, **17**, 16.
24. K. Gollnick, *Chim. Ind.*, 1982, **64**, 156.
25. F. W. Wilkins and D. M. Blake, *Chem. Eng. Prog.*, 1994, **90**, 41.
26. V. K. Mathur and E. H. Wong, *Energy*, 1987, **12**, 311.
27. A. Stankiewicz, *Chem. Eng. Res. Des.*, 2006, **84**, 511.
28. J. R. Bolton and D. O. Hall, *Ann. Rev. Energy*, 1979, **4**, 353.
29. R. M. Navarro, M. C. Sànchez-Sànchez, F. de Valle and J. L. G. Fierro, *Energy Environ. Sci.*, 2009, **2**, 35.
30. G. K. Singh, *Energy*, 2013, **53**, 1.
31. B. Parida, S. Iniyan and R. Goic, *Renew. Sustain. Energy Rev.*, 2011, **15**, 1625.
32. D. Bahnemann, *Sol. Energy*, 2004, **77**, 445.
33. S. Protti and M. Fagnoni, *Photochem. Photobiol. Sci.*, 2009, **8**, 1499.
34. D. M. Schultz and T. P. Yoon, *Science*, 2014, **343**, 985.
35. H. D. Roth, *Angew. Chem., Int. Ed. Engl.*, 1989, **28**, 1193.
36. H. D. Roth, *Photochem. Photobiol. Sci.*, 2011, **10**, 1849.
37. M. B. Rubin, *Eur. Photochem. Assoc. Newslett.*, 1982, **15**, 4.
38. N. D. Heindel and M. A. Pfau, *J. Chem. Educ.*, 1965, **42**, 383.
39. A. A. Nada, *J. Chem. Educ.*, 1983, **60**, 451.
40. G. Ciamician, *Science*, 1912, **36**, 385.
41. A. Albini and V. Dichiarante, *Photochem. Photobiol. Sci.*, 2009, **8**, 248.
42. G. Ciamician and P. Silver, *Ber. Dtsche. Chem. Ges.*, 1908, **41**, 1928.
43. V. Balzani and F. Scandola, *Quím. Nova*, 1996, **19**, 542.
44. M. Oelgemöller, C. Jung and J. Mattay, *Pure Appl. Chem.*, 2007, **79**, 1939.
45. M. Oelgemöller, C. Jung, J. Ortner, J. Mattay, C. Schiel and E. Zimmermann, *The Spectrum*, 2005, **18**, 28.
46. P. Esser, B. Pohlmann and H.-D. Scharf, *Angew. Chem., Int. Ed. Engl.*, 1994, **33**, 2009.
47. B. Pohlmann, H.-D. Scharf, U. Jarolimek and P. Mauermann, *Sol. Energy*, 1997, **61**, 159.
48. K.-H. Funken, *Nachr. Chem. Techn. Lab.*, 1992, **40**, 793.
49. M. Oelgemöller and J. Mattay, in *CRC Handbook of Organic Photochemistry and Photobiology*, ed. W. M. Hoorspool and F. Lenci, CRC Press, Boca Raton, 2nd edn, 2004, vol. 2, ch. 88, pp. 1–45.
50. G. O. Schenck, K. G. Kinkel and H.-J. Mertens, *Justus Liebigs Annal. Chem.*, 1953, **584**, 125.

51. O. Dechy-Cabaret, F. Benoit-Vical, C. Loup, A. Robert, H. Gornitzka, A. Bonhoure, H. Vial, J.-F. Magnaval, J.-P. Séguéla and B. Meunier, *Chem. Eur. J.*, 2004, **10**, 1625.
52. S. A. Kalogirou, *Prog. Energy Combust. Sci.*, 2004, **30**, 231.
53. M. Roeb, M. Neises, N. Monniere, C. Sattler and R. Pitz-Paal, *Energy Environ. Sci.*, 2011, **4**, 2503.
54. D. Mills, *Sol. Energy*, 2004, **76**, 19.
55. S. Malato, J. Blanco, A. Vidal and C. Richter, *Appl. Catal. B*, 2002, **37**, 1.
56. K.-H. Funken and J. Ortner, *Z. Phys. Chem.*, 1999, **213**, 99.
57. S. Malato, J. Blanco, C. Richter, D. Curcó and J. Giménez, *Water Sci. Technol.*, 1997, **35**, 157.
58. C. Richter and D. Martinez, *DLR-Magazin*, 2005, **109**, 22.
59. K.-H. Funken and M. Becker, *Renew. Energy*, 2001, **24**, 469.
60. R. B. N. Baig and R. S. Varma, *Chem. Soc. Rev.*, 2012, **41**, 1559.
61. E. S. Beach, Z. Cui and P. T. Anastas, *Energy Environ. Sci.*, 2009, **2**, 1038.
62. W. J. W. Watson, *Green Chem.*, 2012, **14**, 251.
63. Y. P. Zhao, R. O. Campbell and R. S. H. Liu, *Green Chem.*, 2008, **10**, 1038.
64. J. I. Ajona and A. Vidal, *Sol. Energy*, 2000, **68**, 109.
65. A. Fernández-García, E. Zarza, L. Valenzuela and M. Pérez, *Renew. Sustain. Energy Rev.*, 2010, **14**, 1695.
66. C. Jung, K. H. Funken and J. Ortner, *Photochem. Photobiol. Sci.*, 2005, **4**, 409.
67. R. Dunn, K. Lovegrove and G. Burgess, *Proc. IEEE*, 2012, **100**, 391.
68. A. Neumann and U. Groer, *Sol. Energy*, 1996, **58**, 181.
69. L. P. F. Chibante, A. Thess, J. M. Alford, M. D. Diener and R. E. Smalley, *J. Phys. Chem.*, 1993, **97**, 8696.
70. C. L. Fields, J. R. Pitts, M. J. Hale, C. Bingham, A. Lewandowski and D. E. King, *J. Phys. Chem.*, 1993, **97**, 8701.
71. G. Flamant, D. Luxembourg, J. F. Robert and D. Laplaze, *Sol. Energy*, 2004, **77**, 73.
72. T. Guillard, G. Flamant, J.-F. Robert, B. Rivoire, J. Giral and D. Laplaze, *J. Sol. Energy Eng.*, 2002, **124**, 22.
73. K.-H. Funken, *Sol. Energy Mat.*, 1991, **24**, 370.
74. W. J. W. Watson, *Green Chem.*, 2012, **14**, 251.
75. A. Bruggink, *Chim. Oggi*, 1998, **16**, 44.
76. T. Caronna, M. Catellani, S. Luzzati, L. Malpezzi, S. V. Meille, A. Mele, C. Richter and R. Sinisi, *Chem. Mater.*, 2001, **13**, 3906.
77. R. M. Kellogg, M. B. Groen and H. Wynberg, *J. Org. Chem.*, 1967, **32**, 3093.

78. C. Covell, A. Gilbert and C. Richter, *J. Chem. Res. (S)*, 1998, 316.
79. D. Dondi, S. Protti, A. Albini, S. M. Carpio and M. Fagnoni, *Green Chem.*, 2009, **11**, 1653.
80. L. Cermenati, C. Richter and A. Albini, *Chem. Commun.*, 1998, 805.
81. Y.-P. Zhao, L.-Y. Yang and R. S. H. Liu, *Green Chem.*, 2009, **11**, 837.
82. M. Bolte, K. Klaeden, A. Beqiraj and M. Oelgemöller, *Eur. Photochem. Assoc. Newslett.*, 2013, **83–84**, 79.
83. C. Schweitzer and R. Schmidt, *Chem. Rev.*, 2003, **103**, 1685.
84. J. Wahlen, D. E. De Vos, P. A. Jacobs and P. L. Alsters, *Adv. Synth. Catal.*, 2004, **346**, 152.
85. C. Schiel, M. Oelgemöller and J. Mattay, *Synthesis*, 2001, 1275.
86. C. Schiel, M. Oelgemöller, J. Ortner and J. Mattay, *Green Chem.*, 2001, **3**, 224.
87. M. Oelgemöller, C. Schiel, R. Fröhlich and J. Mattay, *Eur. J. Org. Chem.*, 2002, 2465.
88. B. Heller, D. Heller, H. Klein, C. Richter, C. Fischer and G. Oehme, *J. Inf. Rec.*, 2000, **25**, 15.
89. B. Heller, *Nachr. Chem. Tech. Lab.*, 1999, **47**, 9.
90. B. Heller, D. Heller and G. Oehme, *Eur. Photochem. Assoc. Newslett.*, 1998, **62**, 37.
91. G. Oehme, B. Heller and P. Wagler, *Energy*, 1997, **22**, 327.
92. B. Heller, P. Wagner and G. Oehme, *J. Inf. Rec.*, 1996, **22**, 443.
93. P. Wagler, B. Heller, J. Ortner, K.-H. Funken and G. Oehme, *Chem. Ing. Tech.*, 1996, **68**, 823.
94. M. Oelgemöller, C. Jung, J. Ortner, J. Mattay and E. Zimmermann, *Green Chem.*, 2005, 7, 35.
95. W. Rojahn and H.-U. Warnecke, *Dragoco-Report*, 1980, **27**, 159.
96. M. Oelgemöller, N. Healy, L. de Oliveira, C. Jung and J. Mattay, *Green Chem.*, 2006, **8**, 831.
97. J. Ortner, D. Faust, K.-H. Funken, T. Lindner, J. Schulat, C. G. Stojanoff and P. Fröning, *J. Phys. IV France*, 1999, **9**, Pr3-379.
98. J. Hull, J. Lauer and D. Broadbent, *Energy*, 1987, **12**, 209.
99. O. Suchard, R. Kane, B. J. Roe, E. Zimmerman, C. Jung, P. A. Waske, J. Mattay and M. Oelgemöller, *Tetrahedron*, 2006, **62**, 1467.
100. A. Ritter, A. Hülsdünker, P. Ritterkamp, F. Heimann, W. Binder, S. Klinge and M. Demuth, in *Solare Chemie und solare Materialforschung*, ed. M. Becker and K.-H. Funken, Verlag C. F. Müller, Heidelberg, 1997, pp. 204–216.
101. A. Hülsdünker, A. Ritter and M. Demuth, *Eur. Photochem. Assoc. Newslett.*, 1992, **45**, 23.

102. E. A. Fletcher, *J. Sol. Energy Eng.*, 2001, **123**, 63.
103. A. Steinfeld, *Sol. Energy*, 2005, **78**, 603.
104. N. Avcibaşi, S. Içli and A. Gilbert, *Turk. J. Chem.*, 2003, **27**, 1.
105. S. Içli, A. Bulut and Y. Gül, *Turk. J. Chem.*, 1992, **16**, 289.
106. K.-H. Funken, G. Luedtke, J. Ortner and K.-J. Riffelmann, *J. Inf. Rec.*, 2000, **25**, 3.
107. J. Talukdar, E. H. S. Wong and V. K. Mathur, *Sol. Energy*, 1991, **47**, 165.
108. C. Heinemann, X. Xing, K. D. Warzecha, P. Ritterskamp, H. Görner and M. Demuth, *Pure Appl. Chem.*, 1998, **70**, 2167.
109. M. Oelgemöller, J. Mattay, C. Jung, C. Schiel, C. Ortner and E. Zimmermann, in *Proceedings of International Solar Energy Conference*, Portland, OR, 2004, pp. 523–531.
110. P. Gallezot, *Chem. Soc. Rev.*, 2012, **41**, 1538.
111. C. Sattler, F.-J. Müller, K.-J. Riffelmann, J. Ortner and K.-H. Funken, *J. Phys. IV*, 1999, **9**, Pr3-723.
112. K.-H. Funken, F.-J. Müller, J. Ortner, K.-J. Riffelmann and C. Sattler, *Energy*, 1999, **24**, 681.
113. N. Monnerie and J. Ortner, *J. Sol. Energy Eng.*, 2001, **123**, 171.
114. N. T. Kalyani and S. J. Dhoble, *Renew. Sustain. Energy Rev.*, 2012, **16**, 2696.
115. D. Ravelli, S. Protti, P. Neri, M. Fagnoni and A. Albini, *Green Chem.*, 2011, **13**, 1876.
116. J. Mattay, *Chem. unserer Zeit*, 2002, **36**, 98.
117. P. Anastas and N. Eghbali, *Chem. Soc. Rev.*, 2010, **39**, 301.
118. C. Jiménez-González, D. J. C. Constable and C. S. Ponder, *Chem. Soc. Rev.*, 2012, **41**, 1485.
119. S. Böschen, D. Lenoir and M. Scheringer, *Naturwissenschaften*, 2003, **90**, 93.
120. D. Ravelli, S. Protti, M. Fagnoni and A. Albini, *Curr. Org. Chem.*, 2013, **17**, 2366.
121. K.-H. Funken, J. Ortner, K.-J. Riffelmann and C. Sattler, *J. Inf. Rec.*, 1998, **24**, 61.

EXAMPLES OF CHEMICAL
TRANSFORMATIONS AND PROCESSES

CHAPTER 8

The Sustainable Synthesis of Methanol – Renewable Energy, Carbon Dioxide and an Anthropogenic Carbon Cycle

ROBIN J. WHITE

Earth, Energy and Environment – E^3 Cluster, Institute for Advanced Sustainability Studies e. V., Berliner Str. 130, D-14467 Potsdam, Germany
Email: robin.white666@googlemail.com

8.1 INTRODUCTION

The 'anthropocene' describes the current interval of global environmental change, induced as a consequence of human activity predominantly the result of fossil fuel combustion and the emission of the greenhouse gas (GHG), carbon dioxide (CO_2).[1] Unprecedented technological advancement over the past two centuries has in general resulted in rapid population growth and improvements in the quality of life for the majority of the Earth's population. This advancement has been powered by the exploitation of natural resources and increasingly over the last century, the exploitation of non-renewable fossil fuel reserves (*i.e.* oil, natural gas, coal). The use, and in particular combustion of these high-energy hydrocarbons, has been at the expense of environmental, climatic and arguably political stability. The maintenance of a stable climatic system is fundamental for

Chemical Processes for a Sustainable Future
Edited by Trevor M. Letcher, Janet L. Scott and Darrell A. Patterson
Published by the Royal Society of Chemistry, www.rsc.org

our civilisation to prosper, whilst the supply of a consistent and affordable energy supply is also increasingly important. For example, the global primary energy consumption in 2012 was 538.68 EJ, where E (exa) refers to 10^{18} (510.6 quadrillion Btu), with the European Union (EU) accounting for 14.7% of this total.[2] Therefore, one of the major challenges facing modern society is the search for increasingly efficient energy alternatives to the use of diminishing fossil fuel resources, which are accessible for all. The ultimate intention will be to limit or reverse the environmental and climatic consequences resulting from the combustion of these fuels and the associated emissions such as carbon oxides (CO_X), nitrogen oxides (NO_X) and sulfur oxides (SO_X). In this context, a coupled scientific, political and socio-economic revolution is required if the tightly linked problems of energy security and climate change are to be comprehensively addressed.

The importance of hydrocarbon feedstocks and fuels in the world today[3] is evident from the structuration of our industries (*e.g.* petrochemical), financial markets and transport infrastructure. Hydrocarbon fuels are near-perfect energy carriers (*e.g.* regarding volumetric energy density and physical state), but due to their finite reserves (*i.e.* from fossil sources), global distribution and associated emissions, they should only really be considered primary energy sources in the short term (*i.e.* for the next 50–100 years). Hydrocarbon fossil fuels are essentially stored sunlight (photonic energy) and sequestered carbon produced *via* photochemical, biochemical and geochemical processes over millennia. Crude oil, coal and natural gas are the end result of processes that began with a biological organism capturing and converting sunlight into chemical energy *via* the catalytic conversion of CO_2 and H_2O to (oxy)hydrocarbons such as glucose and oxygen (*i.e.* photosynthesis). With the added products of metabolism (*e.g.* by herbivores), the resulting biomass is reduced to hydrocarbons under conditions of high temperature and pressure. The reserves of some of these compounds are still relatively plentiful, though maybe not (as yet) economically viable. In this regard, the more abundant fossil fuels such as methane (CH_4) can potentially provide a bridge in the transition from a fossil to renewable energy based society (provided the net CO_2 emission approaches zero). There is also an ongoing debate about whether these reserves should be extracted or simply 'left in the ground'.[4]

One thing is clear; with an ever increasing global population and energy demand, their availability will continue to dwindle and will not be replenished in time scales relevant to humanity. Thus to maintain our present standard of living and to allow the developing world to attain similar levels, it is imperative that attempts are made to establish socio-economic and environmentally appropriate energy provision platforms that facilitate a sustainable future for all. In this chapter, sustainability will follow the definition of the United Nations Brundtland Commission of 1987:[5]

'Sustainable development is development that meets the needs of the present, without compromising the ability of future generations to meet their own needs'.

One potential approach to solve this energy/chemical dilemma involves the implementation of carbon-neutral energy and production systems that do not increase atmospheric (*i.e.* carbon-neutral) or may even reduce CO_2 concentration. Furthermore, if new sustainable energy systems incorporate CO_2 (*e.g.* as a vector for energy transfer), then provided it is efficiently trapped from waste flue gases (and in the future, the atmosphere), it may be possible to reverse the current increase in atmospheric CO_2 concentration—a potentially carbon-negative scenario. This approach, generally referred to as 'carbon capture and utilisation' (CCU), can provide industry and society with a number of environmental benefits.[6] *Via* the use of CO_2 as a C1 feedstock (*e.g.* in fuel or chemical synthesis), CCU can in principle bring reductions in GHG emissions.[7] In terms of global GHG emissions, in 2012, CO_2 was emitted at a rate of 34.5 Gt a^{-1} (an increase from 32 Gt a^{-1} in 2009) and is predicted to increase to >45 Gt a^{-1} by 2050 (based on 'business as normal' models).[8] The increasing use of industrial, large-scale, multi-tonne CO_2 scrubbing equipment to minimise these emissions also means that relatively pure, concentrated streams of CO_2 will become increasingly available. The utilisation of these streams in synthesis provides an advantageous (*e.g.* environmental and economic) alternative to carbon capture and sequestration (CCS) schemes.[9] Therefore, the CCU concept has the potential to transform CO_2 from a commercial and environmental liability to an industrial and economic asset, such that CO_2 can be viewed ultimately as a non-depleting, C1 renewable feedstock for the chemical and energy industries.[10] In this regard it has the potential to become the energy carrier of the future.

In terms of chemical and particularly fuel production, the CO_2 molecule represents a significant challenge. At this point, it is worth chemically distinguishing the different opportunities afforded by the CCU platform. Transforming CO_2 into a new molecule can be achieved *via* its entire incorporation [*e.g.* a poly(carbonate)],[11] a comparatively low energy process), or where one or more of the C–O bonds is broken [*e.g.* in the synthesis of CO, methanol (CH_3OH) and hydrocarbons], requiring chemical reduction (*e.g.* with hydrogen), a potentially more energy consumptive process. In terms of the overall energy and carbon balance, processes employed in the capture and activation of CO_2 should require as little (non-renewable) energy as possible, thus minimising indirect GHG emissions. The energy for these processes should be supplied by solar, (geo)thermal, nuclear fission or other 'renewable' non-emitting forms of heat energy, which when coupled with appropriate catalysis, may open up CO_2 based chemistry in a sustainable manner. This 'chemical mitigation' of carbon emissions may result (as is discussed in this review) in the synthesis of the simplest of the alcohols, CH_3OH.

8.2   THE HYDROGEN ECONOMY: SOURCES AND LIMITATIONS

To establish a clean, GHG emission free and sustainable energy provision for all, the development of a global chemo-economic system based on hydrogen (H_2), has been proposed and is still receiving significant interest.[12] On a first principle basis, this seems like a very reasonable proposition. From an availability point of view, hydrogen is one of the most plentiful and abundant elements on Earth. Nature has exploited this and the reactivity of hydrogen to combine it with other elements to form the most abundant molecules including H_2O, fossil hydrocarbons and of course biomass (plants, vegetation, *etc.*). Therefore hydrogen is, in principle, available in enormous quantities but unfortunately is typically incorporated into more stable 'transport' molecules such as carbohydrates ($C_xH_{2x}O_x$). These molecules must therefore be converted if this stored energy is to be successfully exploited. Nature is the expert of this procedure and utilises complex bio-catalytic systems to transfer hydrogen and convert energy from one form to another, *i.e.* photosynthesis, cell metabolism, nicotinamide adenine dinucleotide; many research efforts now aim to 'mimic' these principles in synthetic chemical systems.[13]

The hydrogen economy has received extensive attention primarily as a consequence of its clean combustion (*i.e.* the only product is H_2O).[12,13] If hydrogen is produced *via* an appropriate (*e.g.* CO_2-free) pathway, it can be viewed as an environmentally friendly energy source. However, its scarcity naturally (*e.g.* as free H_2), represents a serious limitation if energy provision is to be based on this bimolecular compound. It has been argued to be the best fuel/energy source based on resources and emissions, whilst the use of proton exchange membrane fuel cells (PEMFC) can potentially provide high efficiency electricity production. Given the increasing demand for the mobile 'on-the-go' electrical power (cell phones, mp3 players, mobile computing, *etc.*), PEMFCs are considered a more efficient power delivery platform than existing (and future) lithium ion battery technology.[14] However, the high operating temperatures of current PEMFCs and the direct 'on-board' storage of the hydrogen appear (at the time of writing) impractical.

The production of hydrogen is not a new science and the use of electrolysis (*e.g.* of H_2O) can be traced back as far as the 1800s.[15] During the same era, the reversal of electrolysis (*i.e.* the fuel cell) was found to generate electricity when H_2 and O_2 combine to form H_2O.[16] Today, current industrial H_2 production predominantly requires fossil fuels using methods including steam methane reforming (SMR), partial oxidation of hydrocarbons and coal gasification.[17] The fossil fuel used in hydrogen production will depend on geo-economic conditions (*e.g.* natural gas in Saudi Arabia[18] and coal in China[19]). Currently, >80% of current hydrogen demand is met with fossil fuels whilst electrolysis accounts for ~4% of total output. Most commonly the production of hydrogen is conducted *via* SMR (Figure 8.1 equation 1) and the water gas shift (WGS) reaction (Figure 8.1, equation 4), which are both energy consuming (*e.g.* endothermic) and

1) Steam reforming

$$CH_4 + H_2O \overset{\text{Ni Cat.}}{\rightleftharpoons} CO + 3H_2 \qquad \Delta H_{298K} = 205.4 \text{ kJ mol}^{-1} \quad S_r = 3$$

2) Partial oxidation

$$CH_4 + 1/2\,O_2 \rightleftharpoons CO + 2H_2 \qquad \Delta H_{298K} = -38.0 \text{ kJ mol}^{-1} \quad S_r = 2$$

$$CO + 1/2\,O_2 \rightleftharpoons CO_2 \qquad \Delta H_{298K} = -282.8 \text{ kJ mol}^{-1}$$

$$H_2 + 1/2\,O_2 \rightleftharpoons H_2O \qquad \Delta H_{298K} = -241.4 \text{ kJ mol}^{-1}$$

3) Dry reforming

$$CO_2 + CH_4 \overset{\text{Ni Cat.}}{\rightleftharpoons} 2CO + 2H_2 \qquad \Delta H_{298K} = 247.3 \text{ kJ mol}^{-1} \quad S_r = 1$$

4) (Reverse) Water Gas Shift Reaction

$$CO + H_2O \rightleftharpoons CO_2 + H_2 \qquad \Delta H_{298K} = -49.8 \text{ kJ mol}^{-1} \quad S_r = \infty$$

NB: High temperatures (> 800 °C)

Figure 8.1 Production of syn-gas mixtures as a precursor to Hydrogen production based on the conversion of methane ($S_r = \text{mol}(H_2)/\text{mol}(CO)$).

performed typically at $T > 800$ °C. The molar ratio of H_2/CO (S_r) (as is discussed later) is an important factor determining hydrogen production efficiency and also significantly syn-gas (a mixture of CO, and H_2 with varying amounts of CO_2 depending on the carbon source) composition, which is important for specific chemistries such as Fischer–Tropsch (FT) hydrocarbon synthesis.

Over the past few decades, alternative (*i.e.* non-fossil based) routes to the production of hydrogen have received increasing attention, with a significant effort focused on the splitting of the H_2O into its constituents. A variety of approaches have been proposed and a number of detailed reviews document the development of chemical technologies relating to this process.[13,20] Briefly, the most discussed approaches include thermal scission using solar energy, photo-chemical or photo(electro)chemical splitting, other thermochemical cycles and (*e.g.* high temperature) electrolysis.[21] From a process point of view and generally in terms of overall efficiency and technological status, the use of electrolysis is particularly attractive as it can under certain circumstances represent the most efficient transfer of electrical to chemical energy (*e.g.* solid oxide cells).

Denmark, for example, views electrolysis as a critical technology in its ambition to reach complete renewable energy provision by 2035.[22] This is particularly relevant when the storage/use of intermittent, renewable energy (*e.g.* wind and solar power) is considered. It is important to note that hydrogen can also be produced from biomass through pyrolysis, liquefaction or gasification processes.[23] However, using biomass to produce hydrogen (at least on the scales needed) creates a competition with land use, food and materials production. There is also the corresponding energy demand for hydrogen separation from the gaseous product mixtures (CO, CO_2, CH_4, *etc.*), although it is important to note that such side stream gases may be advantageous with regard to the production of liquid fuels[24] in certain geographical locations (*e.g.* Sweden, ChemRec/BioDME project).[25] The reader is referred to references 23–25 and references therein for further details regarding biomass conversion.

Irrespective of production method (*e.g.* renewable *vs.* non-renewable), the efficient storage, transfer and distribution of hydrogen will be problematic, for example, in the large scales needed to support society. Under ambient conditions, gaseous H_2 is a poor energy carrier as a result of a low volumetric/weight energy relation-ship. Although still offering significant benefits compared with standard lithium ion battery technology, at 15 °C and at a pressure of 0.1 MPa, gaseous H_2 has a density of 0.09 kg m^{-3} or in terms of energy

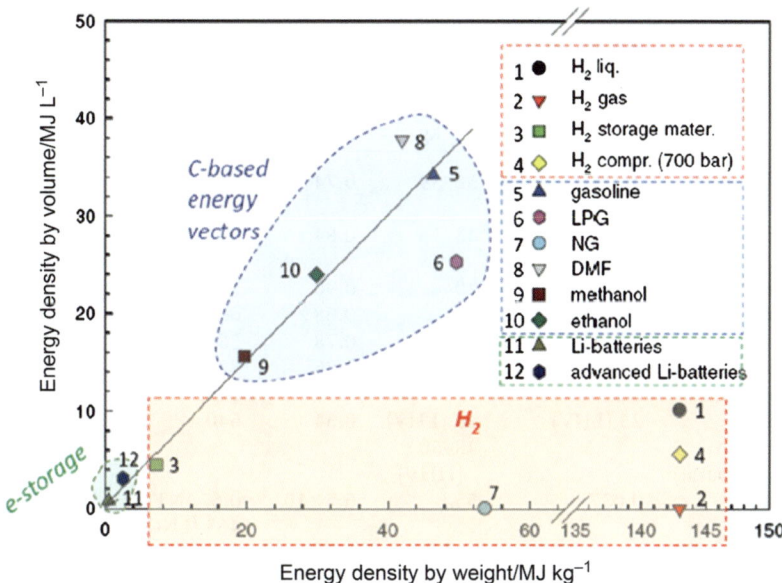

Figure 8.2   Energy density per unit weight (MJ kg^{-1}) *versus* per unit volume (MJ L^{-1}) for a range potential storage media.
Reproduced and adapted from reference 26.

density $= 143$ MJ kg^{-1} (Figure 8.2). These values can be dramatically increased if H$_2$ gas is compressed [*e.g.* 5.6 MJ L^{-1} (at 70.0 MPa and 15 °C) or 4.5 kg m^{-3} at 69.0 MPa] or cooled (*e.g. t* < -200 °C), but is still not a match for hydrocarbon fuels (based on an energy volume/weight relation; Figure 8.2).[26] However, the associated energy and infrastructure investment needed to cope with the transport of liquefied hydrogen must also be considered in the overall description of a new fuel cycle. To overcome these limitations, materials are being developed [*e.g.* metal organic frameworks (MOF) and complex hydrides][27] that are capable of storing large(r) quantities of hydrogen, thus dramatically improving energy volume/weight relationship.[28] The application, and indeed synthesis, of these materials requires further energy/synthesis steps, enlarging the overall fuel/chemical life cycle. The use of specialised on-board vehicle equipment, as well as the necessary adaptation of existing fuel transport infrastructure to handle hydrogen, increases complexity and adds investment costs which are potentially prohibitive. Independent of these apparent limitations, it is important to note the importance of hydrogen in all past and future energy schemes. Likewise, it is important to contemplate how this highly energetic gas could be produced in a sustainable manner.

Table 8.1 Properties of a range of different energy carriers/fuels.[a]

Energy carrier/ fuel	Energy density (volume)/ MJ L^{-1}	Specific energy (weight)/ MJ kg^{-1}	Energy (volume/ weight) ratio	Density/ kg m^{-3}	Standard enthalpy of combustion/ kJ mol^{-1}
Gasoline fuel	33	52–39	0.74	630–830	− 47.3 (kJ g^{-1})
Diesel	36	43	0.84	840	− 44.8 (kJ g^{-1})
Methanol	17.9	20	0.90	791	− 726
Dimethyl ether	19	28	0.68	666	− 1460
Ethanol	21	27	0.78	789	− 1371
Jet fuel (kerosene)	37.6	43.8	0.86	820	− 4620.0
Liquid petroleum gas (propane)	25 (LHV)	46 (LHV) 48–50 (HHV)	0.54	540	− 2202.0
Methane	0.037	55.6	6.7×10^{-4}	0.67 (NTP) 465 (LNG)	− 890.3
Hydrogen	0.01 (NTP) 4.5 (liquefied) (at 69.0 MPa)	123	8.1×10^{-5} (0.037)	0.09 (NTP)	− 286
Ammonia	14.1	32.8	0.43	0.682 (at $T = 240$ K and 0.1 MPa) 0.603 (at 298 K and 1.0 MPa)	− 382.0

[a]HHV = higher heating value; LHV = lower heating value; LNG = liquefied natural gas; NTP = normal temperature and pressure.

In terms of mobility fuels, traditional (*e.g.* fossil-based liquid) transport fuels have significantly higher energy densities than hydrogen (Figure 8.2, Table 8.1) [*e.g.* gasoline: 33 MJ L^{-1}; ethanol: 21 MJ L^{-1}; methanol: 17.9 MJ L^{-1}; hydrogen (g): 0.01 MJ L^{-1}].[29] Liquid fuels are also easier to handle with the current fuel distribution network and, in many cases, the infrastructure for one liquid fuel is suitable for the distribution of another (or at least with minor modifications). There are also significant safety issues regarding the use and transportation of pure hydrogen. Hydrogen is capable of diffusing through most materials used in the construction of conventional storage tanks and as a result there is significant risk of containment failure (*i.e.* long-term storage) and the continuous risk of ignition upon contact with air. Industrially, it would also be more convenient and safer to store hydrogen as an accessible and multi-use platform molecule which could cover a range of chemical and energy needs. Furthermore, if this platform molecule could be derived from industrial waste new potentially carbon-neutral energy/synthesis streams could be implemented. These challenges facing the elaboration of the hydrogen economy concept and the general support infrastructure currently in place for liquid energy provision suggest that a more practical, realistic and feasible solution, ideally without incurring excessive investment costs, is required to provide sustainable energy (*e.g.* for mobility). In this context and for completeness, ammonia (NH_3) has recently received interest, primarily based as the ease with which it is liquefied and also the extensive existing infrastructure (*e.g.* fertiliser production) currently available for its production.[30]

8.3 RENEWABLE ELECTRICITY PROVISION: THE CHALLENGE OF INTERMITTENCY AND NORTHERN EUROPEAN AMBITIONS

Particularly within northern European countries, the contribution of renewable electrical power (wind, photovoltaics, geothermal, *etc.*) to the national energy mix is increasing year by year. Power generation technologies that utilise solar and wind resources will in turn reduce the contribution of the respective electrical power networks to GHG emissions. Taking Denmark and Germany as examples, these two northern European countries have high ambitions regarding renewable electricity provision *via* wind turbine or photovoltaic units.

In Germany, a political movement, termed 'Der Energiewende' or Energy Transition (proposed initially in the 1980s by the German Öko-Institut),[31] has transformed the country's future energy perspective. Initially, a very divisive and controversial policy, the Öko-Institut's approach has gathered support and is now common knowledge in the German political and scientific landscape. To demonstrate its significance, the policy has resulted in removal of nuclear power from future German energy planning (Figure 8.3). Current policy indicates a desire to install wind power capacity capable of supplying ~250 TWh per year by 2050.[32] The Energiewende policy in its current form demands the reorientation of energy provision from the classical demand to one of supply—with an associated transition from traditional centralised generation—to distributed, decentralised systems ultimately resulting in a democratisation of energy supply. Such an approach should in principle overcome issues relating to overproduction and avoidable energy consumption with energy-saving measures and increased efficiency. In this context, Germany intends to reduce its CO_2 emissions by 95% in 2050,[33] whilst providing >80% of electricity demand from renewable sources (notably excluding nuclear options)[34] (Figure 8.3).

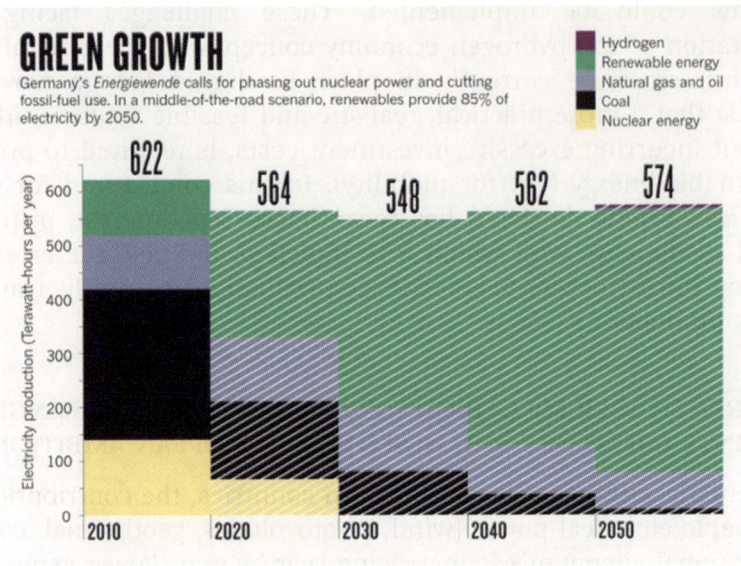

Figure 8.3   Expectations for renewable energy provision in the German energy mix
up to 2050.
Reproduced from reference 34.

Denmark has equally high ambitions and intends to have fossil-free
electricity provision by 2035, primarily based on wind power generation.[35] The intention is to install wind electricity generation such
that in 2020 (∼6 GW) and 2035 (>9 GW), 50% and 100% of supply
will be generated from the wind. Denmark has made a very brave
political decision and does not intend at present to build any more
fossil fuel electricity plants and will have discontinued existing plants
by 2035. In the context of the following chapter, the Danish plan
envisages that electrolysis will play a significant role in achieving a
fossil fuel-free state.[23]

It is important to note that renewable electricity sources (*e.g.*
wind or solar) suffer from the same major problem. The wind does
not always blow and the Sun does not always shine, such that renewable electricity provision is an extremely difficult challenge given
the large variation in supply and its relationship with consumption
(Figure 8.4).[36] Whilst the total renewable electricity production (*e.g.*
wind) may meet national annual demand, there will be timeframes
(*e.g.* over a week) where there is limited supply and conversely periods
where a surplus of electricity exists (*e.g.* >30% of gross consumption)
(Figure 8.4). The energy density of current battery technology is not
sufficient to address such a (large-scale) storage challenge, whilst

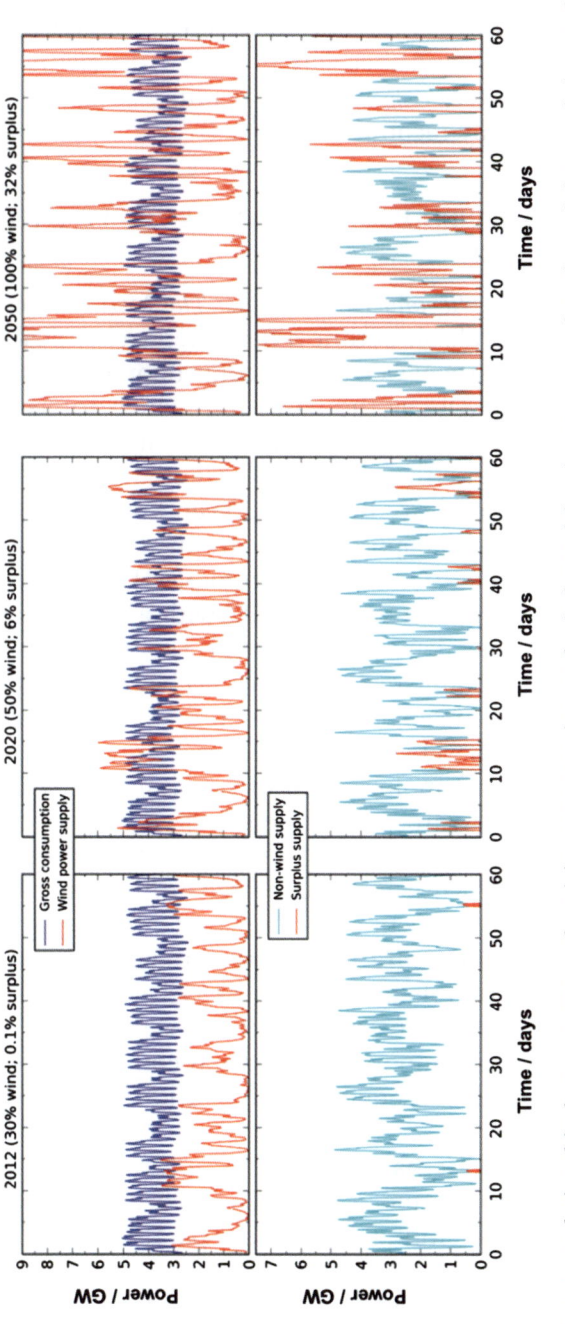

Figure 8.4 Relationship between gross electricity consumption and wind provision in 2012 as a function of time and the predicted relationship in 2020 and 2050 in Denmark. Data sourced from ref. 22, image provided by Dr Chris Graves, Danish Technical University.

storage as liquid fuels (if efficiently converted) perhaps seems more appropriate (Figure 8.2). The question therefore remains of how to effectively and efficiently convert surplus electrical energy to chemical energy (and vice versa), and therefore overcome problems of supply, demand and intermittency. For further reading, the work of Graves and colleagues is highly recommended.[37]

8.4   A METHANOL-BASED ECONOMY AND A VIABLE HYDROGEN ENERGY CARRIER?

In their seminal book, *Beyond Oil and Gas: The Methanol Economy*®, Olah, Goeppert and Prakash outline a new energy/chemical economy based around the capture of CO_2 and its conversion to CH_3OH.[38] This concept advocates CH_3OH as the base from which to access future chemical and fuel needs. Under standard, ambient conditions CH_3OH is a liquid and therefore can potentially be viewed as a practical and safer alternative to H_2 (and even liquefied natural gas. CH_3OH can be stored, transported and 'dropped in' relatively easily using current fuel infrastructure. Furthermore, CH_3OH is an excellent fuel for internal combustion engines (ICE) with a high octane rating [Octane No. (ON) $\geq$ 100; conventional petrol = 95–100]. In additional, relatively simple and established dehydration chemistry (*e.g.* based on zeolite catalysis) can be used to synthesise dimethyl ether (CH_3OCH_3; DME), a high cetane diesel substitute (Cetane No. $\geq$ 55).[39]

The chemistry and technology used to synthesise CH_3OH can also be used to produce hydrocarbon fuels [*e.g.* methanol-to-gasoline (Mobil's MtG process) or synthesis gas (syn-gas) to hydrocarbons *via* Fischer–Tropsch (FT) synthesis]. Such transformations will be required to meet the demands of other transport sectors (*e.g.* aviation). The chemical possibilities are also more attractive (*i.e.* than from LNG), such that this simple alcohol can be viewed as a C1 platform chemical.[40] Established (industrial) chemistry can be used to convert CH_3OH into a variety of important industrial compounds (*e.g.* ethylene, propylene)[41] and subsequently a significant majority of our consumer plastics.

Elaboration of the Methanol Economy® concept requires new chemical technologies if atmospheric CO_2 levels are ultimately to be reduced (*e.g.* efficient membrane flue gas separations or air capture). New means by which CO_2 can also be utilised as a platform chemical are also under significant consideration and ultimately may represent a significant mechanism for longer 'CO_2' storage, potentially providing the chemical industry with socio-economical revenue from conventional waste streams.[42] In this regard, if the supply of CO_2-free electricity (wind, tidal, solar, nuclear, *etc.*) is integrated into combined schemes for CO_2 conversion/CH_3OH production (and associated fuels/chemicals), then resulting products (*e.g.* liquid fuel) will represent a store of carbon, hydrogen and intermittent/surplus renewable electricity. It is the integration of these topics that provides a potential opportunity for the chemical and energy industries to 'green' their public image, whilst providing a degree of 'future proofing' when fossil reserves are no longer available or indeed accessible. New chemical technologies [*e.g.* photo(electro)chemical water splitting and CO_2 capture materials] can then be further developed until techno-economic/market conditions are appropriate for their introduction to enhance the sustainable credentials of this emerging platform.

If CH_3OH is produced using renewable electricity, then electrical energy stored in this liquid can be converted to kinetic energy (*e.g.* the internal combustion engine) or viewed as a chemical flow battery whereby electrical energy is recovered using emerging technologies such as direct alcohol fuel cells such as direct methanol fuel cells (DMFCs). The Methanol Economy® scheme would then represent a closed (*i.e.* non-carbon emitting) energy provision loop provided the resulting CO_2 is recycled. Therefore, if the energy used to capture and convert CO_2 is provided *via* renewables, the resulting platform will represent highly flexible, renewable liquid feedstock that can take advantage of our existing transportation fuel and industrial infrastructure.

The implementation of the Methanol Economy® therefore attempts to reverse the indulgent and wasteful 'carbon' consumption of the last century *via* an anthropogenic carbon cycle, analogous to the natural carbon cycle.[43] The beauty of an anthropogenic CH_3OH based carbon cycle is that this simple alcohol can be synthesised from a variety of routes including fossil and non-fossil sources, offering the opportunity to integrate intermittent renewable energy and CO_2 capture (Figure 8.5). This means that the capture and recycling of CO_2 will provide the basis for the synthesis of future CH_3OH supplies, while fossil-based CH_3OH is phased out. This is the challenge and opportunity for chemists and chemical engineers—the efficient integration of renewable energies with efficient, selective and cost-effective chemistry to provide the basis for a sustainable economy and society.

The industrial production of CH_3OH exists today as does the technology required for its subsequent conversion to DME, synthetic

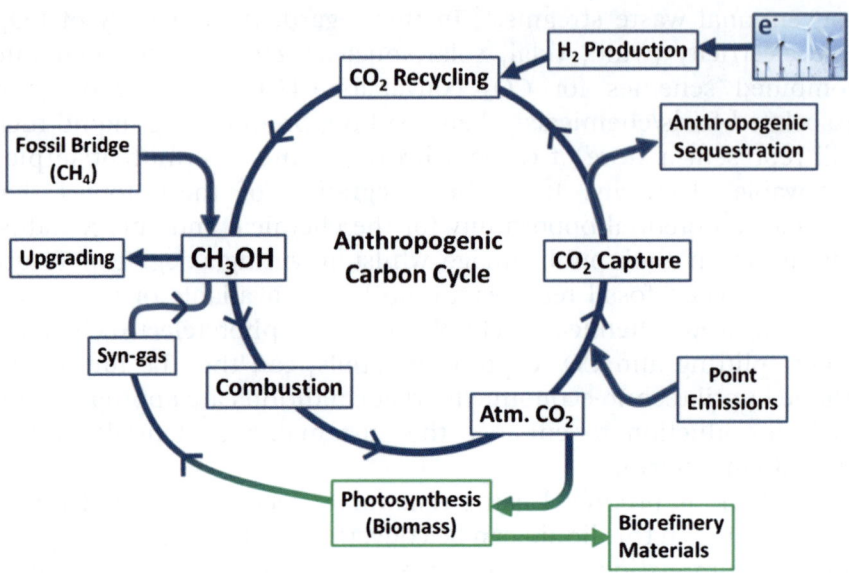

Figure 8.5 A simplified depiction of the anthropogenic carbon cycle based on CH_3OH cycling, as proposed by Olah, Prakash and Geppart, indicating a (green) link to the Biorefinery and materials production.

FT hydrocarbons and products (olefins, formaldehyde, acetic acid, *etc.*). As is discussed later in this chapter, CH_3OH can also be considered as a liquid transport vector for hydrogen supply (*e.g.* to a fuel cell) and a convenient feedstock for the synthesis of light olefins (*e.g.* ethylene and propylene) and subsequently hydrocarbons (*e.g.* MtG). Furthermore, if hydrogen can be extracted efficiently from CH_3OH or other chemical vectors (*e.g.* a non-syngas direct steam reforming of CH_3OH to H_2 and CO_2), CO_2 can be viewed as an 'energy gas' to supply fuel cells (*e.g.* PEMFCs in mobile applications) without invoking any CO shift or additional clean-up stages.[44] Alcohols (*e.g.* methanol and ethanol) can also be employed as fuels directly in direct alcohol fuel cells (*e.g.* DMFCs), transforming the stored chemical energy into electricity at potentially higher efficiencies.[45] DMFC powered automobiles have been reported as being 2.6–3.5 times more efficient than their ICE equivalent.[46] Thus, if renewable energy is integrated into CH_3OH production, the aforementioned problems of intermittency (*e.g.* storage of surplus electricity) can potentially be mitigated. The Methanol Economy® concept therefore has the ability to render fuel and chemical production potentially carbon neutral (or even negative). With regard to CO_2 mitigation and its recycling to CH_3OH (higher fuels and associated chemical products), inputs of hydrogen and/or heat/electrical energy are required. It is therefore necessary to ask which technologies and chemistries are best suited at the present time (and in the future) to achieve this goal in the most efficient manner possible.

Currently, the CH_3OH industry is very dynamic especially in developing countries where oil resources are scarce. However, the main feedstock for production globally is still fossil fuels (*i.e.* natural gas or coals) with associated emissions of CO_2. In 2010, the annual global demand for CH_3OH and derivatives (such as DME) was *ca.* 45.6×10^6 t (where 't' refers to metric tonnes) and is expected to exceed 70×10^6 t in 2016 (excluding ethylene and propylene production),[47] driven by the increased demand for cleaner chemistry (*e.g.* synthetic fuels) and increased regulatory devices designed to encourage the adoption of renewable fuels and chemicals (*e.g.* the Renewable Energy Directive 2009/28/EC). However, it must be noted that if the concept of the Methanol Economy® is to be fully implemented then global production must be dramatically increased.

As the synthesis of CH_3OH (as introduced in Section 8.5) is possible industrially from syn-gas or *via* the direct hydrogenation of CO_2, it seems logical that future CH_3OH demand should be provided *via* the conversion of captured CO_2 (*e.g.* from flue gas exhaust or the air), such that GHG emissions can be mitigated and potentially reversed. Briefly, in such a scheme there would be five general process steps constituting a general synthesis of sustainable CH_3OH, fuels and chemicals based on the recycling of waste CO_2 and renewable electricity utilisation:

1. CO_2 capture/concentration/purification
2. H_2 and/or syn-gas production
3. synthesis, *e.g.* methanol, MtG, or FT-to-hydrocarbons
4. raw product purification (if required)
5. (a) conversion/upgrading of methanol (*e.g.* to DME, olefins and other fuels/chemicals) and/or (b) recovery of 'stored' electrical energy (*e.g.* reforming, DMFC or dehydrogenation to H_2) for mobility/power generation.

A wide variety of scenarios have been proposed and predominantly focus on optimising H_2 or syn-gas production for a desired end product (*e.g.* CH_3OH *vs.* hydrocarbon fuels), with electrical or heat energy being supplied ideally from (*e.g.* renewable) CO_2-free sources.[48] This concept can be generalised as a multicomponent flow sheet(s) depicting the key processes and also the potential to integrate associated recycling of energy (*e.g.* exothermic CH_3OH synthesis) and emissions (Figure 8.6).

The following sections aims to provide a general overview of 'sustainable CH_3OH' and the chemical technologies required to elaborate this concept. Discussion focuses on the latest reports regarding high efficiency electrolysis (*e.g.* for H_2 and syn-gas

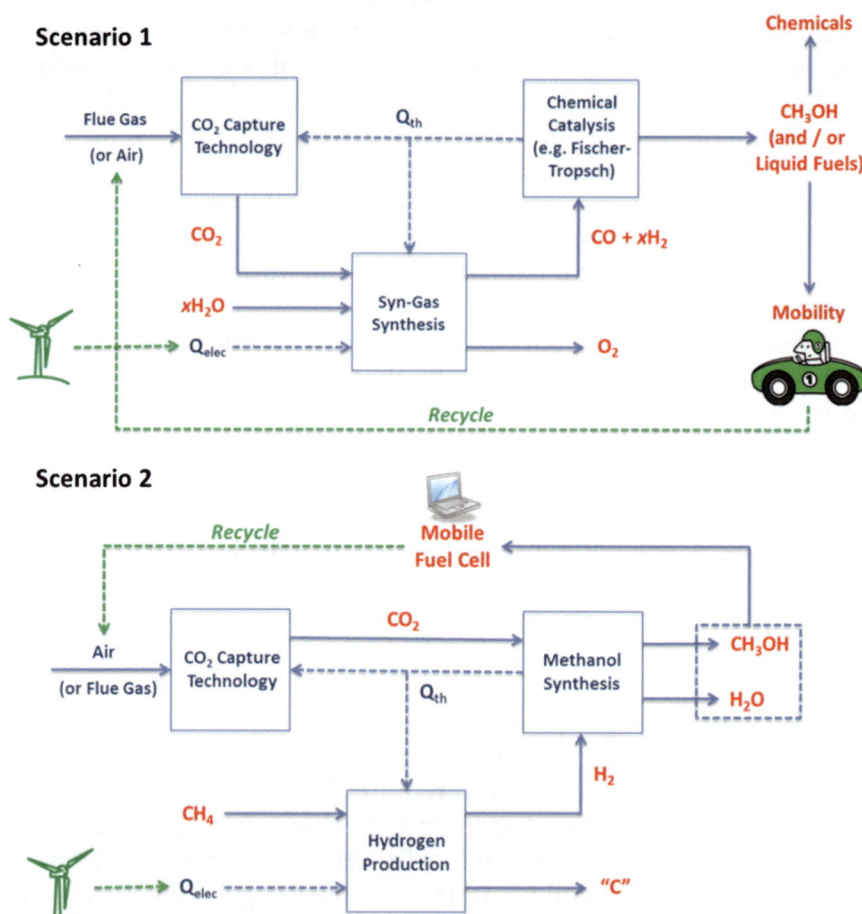

Figure 8.6 Example scenario for the sustainable synthesis of CH_3OH (and fuels) based on (1) capture and co-conversion (*e.g.* electrolysis) of CO_2 with H_2O to provide a syn-gas platform for the synthesis of methanol and higher alkane liquid fuels; and (2) capture and direct hydrogenation of CO_2 using a sustainable source of H_2 (*e.g.* hydrocarbon decarbonisation). Inspired and adapted from reference 48.

production), methanol synthesis (based on CO_2 and gas-shift related works) and dehydrogenation catalysts. The purification of methanol (and hydrocarbon fuels) is a well-established, commercialised process and in some scenarios (*e.g.* Figure 8.6, scenario 2) may not even be required. Whilst reference will be made to associated processes, the reader is referred to a number of authoritative works regarding the development of materials for CO_2 capture,[49] and hydrogen production *via*, for example, photo(electro)chemical) water splitting.[13,20,21] The utilisation of CH_3OH as a hydrogen vector for fuel cell applications and the development of DMFC technology is also highlighted. Finally, a number of commercial endeavours that aim to bring sustainable fuels from renewables to the market place are introduced.

8.5 METHANOL SYNTHESIS

Fortunately, the production of CH_3OH is already a serious industrial endeavour and there are a number of existing, large-scale industrial plants located around the world. The largest of these typically operates to produce (2500–3000) t d^{-1} (where 'd' refers to day). New technology and plant design is also facilitating the construction of new installations capable of producing >5000 t d^{-1}, with the Methanex plant in Trinidad & Tobago capable of producing 2.5 Mt a^{-1} (where 'a' refers to year). More ambitious plants incorporating combined reforming (*i.e.* steam + autothermal reforming) are being constructed, with the intention to produce ≥10 000 t d^{-1}.

Whilst the majority of the global CH_3OH demand is currently met *via* SMR to provide syn-gas, there is an ever increasing contribution from coal-to-liquid processes, especially in Asia (*e.g.* China). As indicated in Section 8.2, syn-gas composition is dependent on the carbon source. As high temperature SMR produces a $S_r = 3$, additional processing is therefore required to adjust this ratio to the optimum (*e.g.* $S_r = 2$) for CH_3OH synthesis (requiring an additional energy input). CH_3OH synthesis from a syn-gas (*e.g.* SMR) or CO_2 base is defined by two closely related exothermic reactions (excluding any contribution from the WGS) (Figure 8.7, equations 5 and 6). As mentioned, in the future it would be desirable if CH_3OH could be synthesised where CO_2 can be reduced to the alcohol either *via* a syn-gas or direct hydrogenation pathway.

Both reactions are favoured thermodynamically at low temperatures and this has consequences in terms of process design as it will reduce the total one-pass conversion. Industrially, CH_3OH is

$$\Delta H_{298K} \qquad \Delta G_{298K}$$

5) $CO + 2H_2 \rightleftharpoons CH_3OH$ $\qquad$ -90.8 kJ mol^{-1} $\quad$ -25.3 kJ mol^{-1} $\quad$ $S_r = 2$

6) $CO_2 + 3H_2 \rightleftharpoons CH_3OH + H_2O$ $\qquad$ -49.5 kJ mol^{-1} $\quad$ +3.3 kJ mol^{-1}

Figure 8.7 The synthesis of methanol based on the hydrogenation of carbon monoxide (5) or carbon dioxide (6) and associated thermodynamic parameters.

traditionally produced based on equation (5), *i.e.* syngas-to-CH$_3$OH, developed over half a century ago by ICI. Industrially, syn-gas is fed to the reactor at <250 °C and ∼6.0 MPa. Unconverted syn-gas is separated from CH$_3$OH *via* condensation and recycled back to the reactor. This 'low pressure synthesis' is typically operated at one-pass conversions of (15–20)%, negating extra plant engineering to manage exotherms at higher conversion rates. As mentioned above, the syn-gas feed is produced *via* the SMR reaction (Figure 8.1, equation 1; $S_r = 3$). It is important to note that the syn-gas feed may contain varying quantities of CO$_2$ and H$_2$O (depending on the carbon source) and hence the hydrogenation of CO$_2$ may also occur producing additional H$_2$O (Figure 8.7, equation 6). This means the WGS reaction (Figure 8.1, equation 4) also occurs, consuming H$_2$O to shift the equilibrium in favour of CH$_3$OH synthesis.[50]

As a consequence of emission penalties, and given the anticipated access to high purity streams (*e.g.* from carbon capture), CO$_2$ is increasingly viewed as a beneficial, alternative feedstock (*i.e.* to aid plant integration and mitigate overall GHG site emissions). The catalysis to convert syn-gas to CH$_3$OH is well documented and commercially relies on, for example, Al$_2$O$_3$ (5–10 mol%) supported Cu nanoparticles (NPs) with ZnO acting as a 'chemical promoter'.[51] Significantly, the same catalysts can be used in CH$_3$OH synthesis *via* the direct hydrogenation of CO$_2$.[52] Various promoters have been investigated to increase productivity and catalyst stability (*e.g.* ZrO$_2$, Ga$_2$O$_3$, SiO$_2$),[53] though the systematic investigation of new catalyst compositions is made complicated by the intricate interdependency of the aforementioned synthesis reactions. Contrary to traditional syn-gas based routes, the direct hydrogenation of CO$_2$ to CH$_3$OH is actually inhibited by the formation of H$_2$O. Therefore, continuous H$_2$O removal by distillation or (*e.g.* zeolite-based) membranes is required to shift reaction equilibrium in favour of higher CH$_3$OH production (*e.g.* the CAMERE process).[54]

Recent research efforts have focused on the design and characterisation (*e.g.* of the active site) of CH$_3$OH synthesis catalysts that are more active and not inhibited by H_2O. Based on this work, new reaction mechanisms have also been proposed whereby the formation of syn-gas (*i.e. via* the reverse WGS reaction) does not occur. Characterisation of the active site (*e.g.* industrially used Cu/ZnO/Al_2O_3) suggest that a combination of crystal lattice strain (*e.g.* in Cu nanoparticles at the interface with ZnO) and a high number of crystal defects (*e.g.* edge sites) leads to increased activity in CH_3OH synthesis (*i.e. via* WGS pathways).[55] A variety of new materials have been proposed, with systems based on carbon nanotubes (*e.g.* Pd/ZnO/carbon nanotubes),[56] of particular interest due to their capability to reversibly adsorb a greater quantity of hydrogen. A number of recent publications highlight the latest developments in catalyst design for CH_3OH synthesis.[57–63]

In the context of catalyst characterisation, the work of Behrens and colleagues is of particular note.[58] Following on from the work of Klier, Chinchen and Hansen and colleagues,[59] Behrens and colleagues have extensively investigated different chemical promoter/support chemistry (*e.g.* MgO, Al_2O_3,)[60] and preparation procedures (*e.g.* nanostructuring, co-precipitation, ageing)[61] on the activity of predominantly binary Cu/ZnO based model systems (Figure 8.8). The complex interplay between

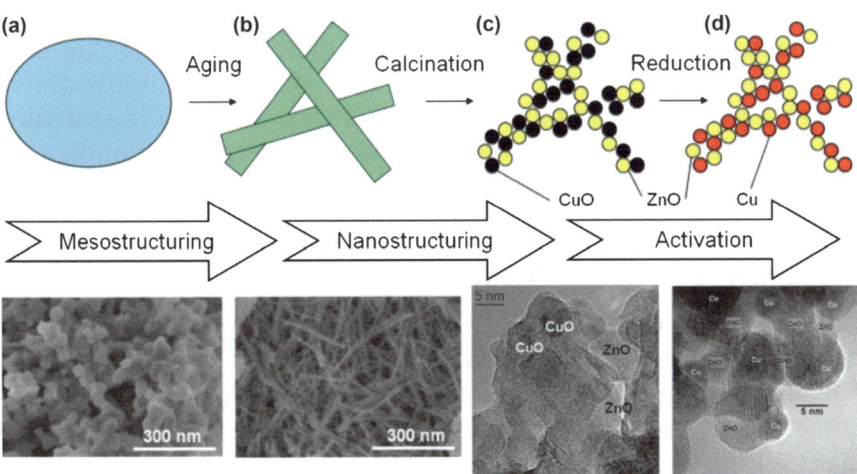

Figure 8.8   Simplified preparation scheme for the Cu/ZnO catalyst and electron microscopy images of the different stages of preparation based on (a) ageing/co-precipitation procedures of the parent solution, to produce (b) a zincian malachite phase, followed by (c) calcination/thermal decomposition to yield an intimate CuO/ZnO mixture. (d) CuO is reduced in H_2 to yield highly porous, microstructured nanoparticulate Cu/ZnO with high Cu dispersion.
Reproduced with permission from reference 60.

these parameters and their relationship with catalytic activity in the synthesis of CH_3OH is often referred to as the 'chemical memory'.[62]

Characterisation of Cu nanoparticles in the binary Cu/ZnO catalyst for CH_3OH synthesis has revealed, in agreement with density functional theory (DFT) studies, that steps at the Cu surface are necessary to obtain high activity in the hydrogenation reaction.[63] These steps can be stabilised by bulk crystal defects (*e.g.* stacking faults or twin boundaries) terminating at the nanoparticle surface. The observed activity enhancement is proposed to be the result of stronger binding of the synthetic intermediates at 'stepped sites' and the associated lower energy barrier between them. For the highest activity, it was indicated that $Zn^{\delta+}$ is required to interface with such Cu steps to induce a dynamic strong metal–support interaction (SMSI) and partial coverage of the Cu nanoparticle by ZnO. Substitution of Cu for Zn at the step sites is proposed to further enhance the intermediate states and therefore catalytic activity. This critical insight into catalyst structure opens opportunities for fundamental manipulation and direction of active site chemistry, importantly on highly abundant metals (*e.g.* Cu, Zn). The work of Behrens and colleagues highlights the intricate relationship between a high Cu surface area, defects/step/edge site chemistry and a dynamic SMSI to generate catalysts with optimal activity. It is important to note that such catalytic technology is already available at the commercial level (even for pure CO_2 utilisation), negating the requirement for significant investment in new catalyst development, although investment in infrastructure/hardware installation will be required to meet, for example, the volume production) demands of the Methanol Economy®.

8.5.1   Synthesis of Dimethyl Ether

DME, considered as a diesel substitute, is the dehydration product of two moles of CH_3OH. If it is derived (as CH_3OH) from purified streams of syn-gas or captured CO_2, then there will be reduced emissions (*e.g.* SO_X), when fuelling a diesel engine.[64] Interestingly, it is already known to be synthesised directly from CO_2 in a single step industrial process, involving a bi-functional catalytic system combining a traditional CH_3OH synthesis catalyst (*e.g.* Cu/ZnO) with an acid, dehydration catalyst (*e.g.* γ-Al_2O_3, H-ZSM-5, NaH-ZSM-5).[65] DME may also be synthesised in a two-step process, where CH_3OH is produced individually before being dehydrated in a separate reactor containing the dehydration catalyst.[66] Whilst the second option may provide ultimately a flexible plant design (*e.g.* in the context of a large

Table 8.2   Reported bi-functional catalysts for the one step synthesis of DME from CO_2.

Bi-functional catalyst	Reaction temperature $t/°C$	Yield (DME)/%	Conversion (CO_2)/%	Ref.
(Pd/Cu/Zn/Al/Zr)/H-ZSM-5	200	13.7	18.6	68
(Cu/Mo)/H-ZSM-5	240	9.5	12.3	69
(Cu/Zn/Al/Zr)/H-ZSM-5	250	21.2	30.9	70
(Cu/Ti/Zr)/H-ZSM-5	250	7.4	15.6	71
(Cu/Zn/Al)/H-ZSM-5	250	12.5	–	72
(Cu/Zn/Al)/NaH-ZSM-5	275	26.0	35.0	73

demand of CH_3OH *vs.* DME), the one-step process has a lower thermodynamic limitation and therefore may be energetically beneficial. The necessity to produce DME from pure CH_3OH (*i.e.* due to the negative effects of H_2O on the equilibrium), has been circumvented recently by AG Lurgi's MegaDME® plant.[67] This is a significant advancement in the context of the Methanol Economy®, as this will reduce the process step number in the direct hydrogenation of CO_2, (*i.e.* the energy intensive separation of CH_3OH and H_2O is not required). Reported catalysts used in the one-step process are summarised in Table 8.2.

Whilst not as acidic as some other materials, traditionally (MFI) H-ZSM-5 based catalysts have been employed as the dehydration catalyst in DME synthesis, as its activity is not inhibited by substantial quantities of H_2O, unlike, for example, γ-Al_2O_3.[74] This is advantageous, particularly regarding olefins or higher hydrocarbon fuel synthesis (*e.g.* kerosene for the aviation industry), as H-ZSM-5 can provide a flexible catalyst support chemistry, capable of converting CH_3OH/DME feeds that contain H_2O (*e.g.* from the hydrogenation of CO_2), even at high Si/Al ratios.[75] Likewise for CH_3OH synthesis, additional chemical promoters can be added to the catalyst to improve DME production (*e.g.* Ga_2O_3),[76] whilst mixed metal systems have shown adequate activity at lower reaction temperatures (*e.g.* Pd, 200 °C).[68] However, careful control of acidity or the introduction of larger transport pores (*i.e.* mesopores) into the zeolite structure will help reduce the possibility of carbon deposition and catalyst deactivation.

8.5.2   Synthesis of Associated Hydrocarbons: Accessing Higher Fuels from Methanol

The synthesis platforms for CH_3OH and DME can also be utilised for the synthesis of higher hydrocarbons. Furthermore, the worldwide

7) $nCH_3OH \rightarrow C_nH_{2n} + nH_2O$ $\Delta H_{298K} = -85 - 90 \text{ kJ mol}^{-1}$

8) $6C_3H_6 \rightarrow C_9H_{12} + 3C_3H_8$

Figure 8.9   Simplified reaction equations for the synthesis of olefins and associated products from a methanol base.

CH_3OH capacity is projected to increase by $\sim 63\%$ to $\sim 125\,000$ kt a^{-1} by 2015.[77] This means that any CH_3OH surplus could in principle be converted to liquid hydrocarbons. This chemistry will be required to meet demands from certain fuel sectors (*e.g.* kerosene for aviation).[78] Again, with regard to the elaboration of the Methanol Economy®, such plant and conversion chemistry is well known having formed the basis for Mobil's MtG process, operated in New Zealand during the oil crisis of the late 1970s (Figure 8.9).[79] The conversion of syn-gas streams to hydrocarbons is also a well-known process based on classical FT chemistry, catalysed by various solid acid supported non-noble metals (*e.g.* γ-Al$_2$O$_3$ supported Fe/Co).[80] Briefly, after the initial hydrogenation of CO, C–O bond scission occurs followed by the formation of a C–C bond. Multiple dehydration and carbon bond forming reactions lead to an elongation and growth of the hydrocarbon chain. As this approach has been well discussed by previous authors, the reader is referred to a number of authoritative reports on the topic.[80]

As the MtG process goes through a number of individual steps; other high value products can also be obtained. The first step of the MtG process involves the dehydration of the alcohol to generate a spectrum of olefins. These products, including useful polymer precursors such as propylene are converted typically over solid acid catalysts (*e.g.* H-ZSM-5, SAPO-17, SAPO-34), with the product profile having a degree of similarity (in terms of molecular weight) to the naphtha isomer fraction obtained *via* high temperature FT chemistry (*e.g.* based on syn-gas). Mechanistically, the dehydration of nCH_3OH proceeds *via* DME, followed by the dehydration of DME to yield olefins.[81] The olefins can then undergo aromatisation to yield a complex product mixture of paraffins and aromatics (Figure 8.9).

In this simplified process, CH_3OH or DME (or mixtures thereof) can be feed to the reactor with conditions optimised to favour a strong equilibrium shift to successfully remove H_2O. The composition of the resulting product mixture/distribution (*i.e.* C_n) is determined by a limited equilibrium, such that the formation of heavy products is restricted. Under these conditions, the formation of kerosene is the highest C_n product formed. In this approach, zeolite catalyst selectivity leads to highly efficient refining and direction to a specific C_n

product distribution. In this regard, the MtG process was developed to produce on-specification, good quality, petrol fuel from CH_3OH with a high iso-pentane content (ON > 90), which is sufficient for ICE operation. For specific operational details (*e.g.* plant design) and catalyst design, the reader is referred to references 75, 82 and 83. The flexibility of this approach is worth noting as syn-gas, CH_3OH and/or DME can all be feed to the reactor. By varying conditions on one catalyst (*e.g.* H-ZSM-5), it is possible to direct production to a specific composition.

The MtG process can also provide an alternative route to local high octane petrol production (*e.g.* as opposed to importation and associated CO_2 costs). As in the late 1970s, this process will become increasingly more viable with ever decreasing and viable supplies of oil in the future. It is also particularly attractive in remote communities and as will be mentioned (see Section 8.10), a number of commercial ventures are exploring 'power-to-liquid' schemes which can utilise this or similar FT chemistry combined with renewable electricity. As mentioned earlier, the MtG process can be also be adapted to produce olefins (*e.g.* ethylene, propylene) in the 'methanol-to-olefins (MtO)' process[84] for the production of CH_3OH-derived plastics; a longer term form of CO_2 storage and contribution to the anthropogenic carbon cycle. Industrially, a number of MtO processes are currently in operation including those from UOP/Norsk Hydro (*e.g.* ethylene/propylene production in Nigeria) and AG Lurgi (*e.g.* propylene production in Trinidad & Tobago).[85]

8.6   CARBON DIOXIDE CAPTURE, CONCENTRATION AND PURIFICATION

Whilst in the long term the aim will be to remove CO_2 from the atmosphere, this remains a significant technological challenge, though recent reports indicate its potential efficacy.[86] Indeed, air capture could decouple GHG utilisation from its source, potentially leading to net atmospheric CO_2 reduction. Recent reports indicate that the cost of such approach results in a price of €(55–95) per tonne of CO_2,[87] which is still considered not sufficiently competitive.[86] Energetically, such processes are not prohibitively consumptive, with air capture requiring $\sim$(3–5)% of the energy carried by any subsequent CO_2 derived fuels.[88]

Over the past decade or so there has been interest in the development of new technologies and materials chemistry specifically focused on increasing the efficiency of CO_2 capture. This is a rapidly developing area and the reader is encouraged to read references 86 and 89. The development of metal organic frameworks (MOFs) is of particular note, as these extremely high surface materials show the potential for high CO_2 capacities (*e.g.* >1.5 g CO_2 per g MOF) as well as being particularly selective, for example, *vs.* N_2.[90] These new technologies will require maturation to compete at a comparable techno-economic level to existing processes. To be ultimately sustainable, they must also perform better whilst demonstrating (ideally) a reduced energy consumption and CO_2 footprint.

The technology currently employed to remove/concentrate CO_2 at the industrial/power plant level is relatively well established,[91] and with legislation encouraging CO_2 removal from industrial waste flues, this source will at least in the short term mean there is an abundance of relatively high purity CO_2. As industrial emissions account for 3400 ± 100 Mt a^{-1}, streams of CO_2 are already available at existing commercial points of potential new process integration (*e.g.* process integration; reducing costs associated with air captured CO_2). Processes including NH_3 and alcohol (*e.g.* fermentation) production can also provide access to high purity CO_2 and typically are devoid of problematic contaminants (*e.g.* H_2S). In fact, CO_2 purification to suitable levels of purification (*e.g.* $H_2S < 10 \times 10^{-6}$), aside from the inherent chemical challenges, is a serious consideration in terms of overall process economics. Whilst the utilisation of industrial emissions is capable of representing $\sim 10\%$ reduction in CO_2 entering the atmosphere, the conversion to a liquid fuel would generate a recycling loop, resulting in an overall higher reduction and additional revenue streams.

There are three main methods by which CO_2 can be concentrated, particularly from fossil fuel power plants, but such approaches can be extended to other industrial sectors:

- post-combustion capture (PCC);
- pre-combustion capture (PrCC);
- oxygen (oxy-) combustion capture (OC).

PCC, perhaps the most established industrial technology, removes CO_2 from the waste flue gases based on chemisorption using a chemical solvent—most commonly an organic amine such as monoethanol amine (MEA) (aq). In this process, the exhaust gas flows to a low-pressure gas–liquid contactor. A partition is established whereby CO_2 enters the solvent forming a carbamate salt with two molecules of MEA. The CO_2 is then recovered by heating and salt dissociation. However, there are a number of energy demanding steps, namely amine regeneration and compression of CO_2 from ambient pressure ≥ 10.0 MPa. Questions regarding the high volume use of amine-based solvents and their potential interactions with the atmosphere are still to be resolved.

PrCC separates CO_2 by reacting a fossil fuel with steam and air (or oxygen) before it is combusted, essentially the preparation of a syngas, and is often referred to as 'Integrated Gasification Combined Cycle' (IGCC). WGS chemistry increases the yield of H_2 and CO_2. Physical separation techniques are then applied to purify H_2 for power generation and a purified CO_2 flow. This is still a relatively young technology with less than 10 plants globally and requires extensive process integration to produce an overall high efficiency.

As discussed in Section 8.7, if the electrolysis of H_2O (and/or CO_2) powered by renewable electricity is extensively adopted to produce H_2 (and/or CO) for synthesis, then there will be an associated volume of pure O_2 available. Whilst this is valuable to the electronics industry, it could also be used to increase the efficiency of combustion processes. Here, OC employs pure O_2 instead of air to combust fossil fuels to produce exhaust flues composed of only H_2O, CO_2 and $<10\%$ O_2/N_2 (and depending on the fossil fuel, small amounts of SO_X and NO_X). One of the drawbacks of OC is the energy needed to produce pure O_2 but this can be solved (*i.e.* electrolysis), improving the long-term outlook. A recent review highlights various scenarios that are envisaged for the increased introduction of such carbon capture technology.[92]

In the context of the Methanol Economy®, a PCC process currently under operation industrially is worth noting. The AG Lurgi Rectisol® process is a physisorption method for of waste gas purification using pure CH_3OH as solvent and (electrically powered) refrigeration to control partition coefficients.[93] The electrical energy required to control the refrigeration temperature can potentially be supplied from renewable sources (*e.g.* wind). The technique has the potential to produce very high purity individual gas streams (*e.g.* separation to individual streams of CO_2, H_2S, CO, CH_4, *etc.*) and is particularly suited to cope with the complex product mixtures of (*e.g.* coal or biomass) gasification. The Rectisol® process has the capability of reducing the sulfur content of a desired gas to $<0.1\times10^{-6}$, an important consideration with regard to catalyst lifetime/activity and product purity if the gas is used synthetically (*e.g.* CH_3OH synthesis). Sulfur-containing side streams can also generate revenue (*e.g. via* the Klaus process/H_2SO_4 production). If large quantities of CH_3OH are to be produced in the future to meet demands of a new anthropogenic carbon cycle, the use of the Rectisol® process to purify CO_2 (or syn-gas) seems logical particularly if renewable electricity powers the process.

8.7 SUSTAINABLE HYDROGEN AND SYN-GAS PRODUCTION

As introduced in Section 8.3, there is a desire (*e.g.* in Germany) to increase the proportion of electricity generated by perennial renewable energy sources (*e.g.* wind, solar, geothermal).[31,32,35] As mentioned above, it is often very difficult to match such intermittent sources with general demand. There may be periods of surplus electricity generated (*e.g.* from wind) and therefore the question remains how to store or convert this energy (*i.e.* for periods of limited electricity supply). In this context, the relative merits (*i.e.* in terms of a techno-economic assessment) of each pathway to CH_3OH (*e.g.* syn-gas *vs.* direct CO_2 hydrogenation; Figure 8.6) must be considered. For example the production of syn-gas or H_2 should be conducted in as efficient manner as possible. If this production uses renewable electricity, then conversion of electrical to chemical energy would help address the challenge of intermittent renewable electricity. Whilst in the long term, the photocatalytic splitting of H_2O (and/or CO_2) to produce H_2 (and/or CO) may be the way forward; this technology is far from commercialisation. The reader is referred to a number of reviews currently available on the subject.[94]

As mentioned earlier, a sustainable liquid fuel/chemical production (*e.g.* CH_3OH) requires GHG-free production of H_2 (or syn-gas). Based on the direct hydrogenation pathway (Figure 8.7, equation 6), 1.0 kg of H_2 is required to convert 7.3 kg of CO_2 to produce 5.3 kg of CH_3OH (and 2.65 kg of H_2O). The cost of sustainable CH_3OH will therefore predominantly depend on costs of the carbon precursor and the production of hydrogen (*e.g.* based on values in 2007).[95] Of course, the cost of fossil-derived CH_3OH will depend on the same factors (*e.g.* SMR/WGS methanol production in 2013 was €0.08 per kg).[96] Therefore to reduce the price of CO_2 derived CH_3OH, it may be beneficial to produce H_2 (or syn-gas) using renewable electricity, the price of which will continue to decrease with increasing installed capacity (*e.g.* wind < €0.04 per kWh in 2013). Taking this one step further it may be even more beneficial to combine H_2 production and CO_2 reduction in one step to produce syn-gas, which can then be converted into CH_3OH or liquid hydrocarbons (and associated chemicals) using existing the industrial catalysis described above.

8.7.1 High Temperature Electrolysis

The use of electrolysis for hydrogen production dates back to the 1800s.[97] The electroreduction of CO_2 is also now receiving increasing attention, providing a mechanism to store intermittent electrical energy and cycle large volumes of CO_2 (*e.g. via* reduction to CO).[6,98] However, the conversion of renewable electricity to chemical energy (*e.g.* CH_3OH or liquid fuels) faces a problem of energy conservation.

High temperature electrolysis of steam and/or CO_2 offers a potentially efficient route to convert, low emission, and intermittent, renewable electricity to chemical energy (*i.e.* not limited by Carnot efficiency).[99] Conducting H_2O electrolysis at high temperatures (>400 °C) is also thermodynamically and kinetically more efficient compared with the same processes performed under standard conditions (*e.g.* in the liquid phase). The product can be H_2 or syn-gas—both suitable precursors for CH_3OH (and/or higher fuels) synthesis. The electrolytic splitting of H_2O (and/or CO_2) also produces high purity O_2 with economic value and potential use in associated chemistry (*e.g.* to improve gasification efficiency). Therefore the use of renewable electricity *via* high temperature electrolysis in synthesis means the products (*e.g.* CH_3OH) can be thought of as high density energy stores to counteract intermittency. A number of studies highlight this possibility,[100] with process models demonstrating high overall heat-to-syn-gas efficiency. Fu and colleagues have demonstrated an energy conversion efficiency using co-electrolysis of $\sim 90\%$ (based on HHV syn-gas), 10–15% higher than the same process using low temperature H_2O electrolysis.[101]

High temperature electrolysis involves the use of a solid oxide electrolysis cell (SOEC; Figure 8.10). Briefly, with the application of a direct current, the co-(electro) reduction of CO_2 (slower kinetics) and H_2O (kinetics favoured)—a $2e^-$ process for each reduction—occurs at the cathode, catalysed by supported [*e.g.* yttrium-stabilised zirconia (YSZ)] Ni nanoparticles to yield H_2 and CO (Figure 8.10, equations 9 and 10). The O^{2-} diffuses through the solid electrolyte from the cathode to the anode as it is formed, oxidising on arrival to form pure O_2 and leaves the cell (*e.g. via* an optional carrier gas; Figure 8.10, equation 11). In addition to the electrochemical production of H_2, CO and O_2 respectively, the reverse water gas shift (RWGS) reaction (Figure 8.1, equation 4) may also occur in the presence of the Ni catalyst and the cell operating temperature. It is important to note that co-electrolysis can provide any desired S_r ratio, ideal for a flexible fuel production platform. For an $S_r = 2$ (Figure 8.7,

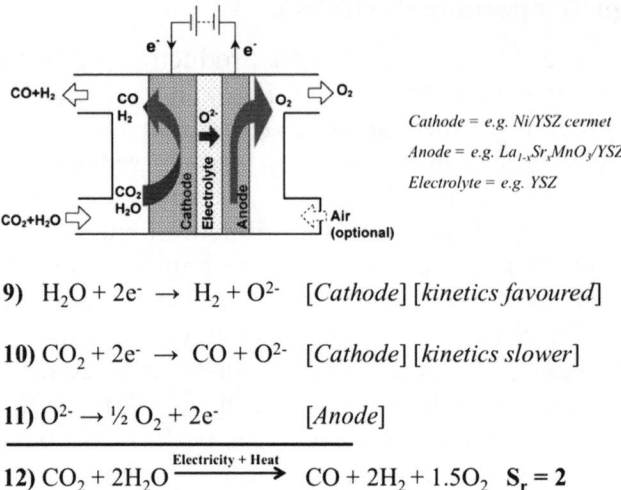

9) $H_2O + 2e^- \rightarrow H_2 + O^{2-}$ [*Cathode*] [*kinetics favoured*]

10) $CO_2 + 2e^- \rightarrow CO + O^{2-}$ [*Cathode*] [*kinetics slower*]

11) $O^{2-} \rightarrow \frac{1}{2} O_2 + 2e^-$ [*Anode*]

12) $CO_2 + 2H_2O \xrightarrow{\text{Electricity + Heat}} CO + 2H_2 + 1.5O_2$ $S_r = 2$

Figure 8.10 Simplified construction principle of a solid oxide electrolysis cell (*e.g.* for co-electrolysis of steam and CO_2) and the corresponding half-cell equations and overall reaction for the production of syn-gas. YSZ = yttrium-stabilised zirconia. Adapted from reference 101.

equation 5), a feed ratio of CO_2 to $H_2O = 0.5$ produces a high purity (*e.g.* SO_x-free) syn-gas which can then be fed to a (*e.g.* CH_3OH) synthesis reactor. Furthermore, heat can be recovered from the out-gas to heat the cathode feed streams (*e.g.* H_2O/CO_2). It is important to note that the Boudard reaction may lead to the formation of carbon at the Ni cathode under the SOEC operating conditions reducing cell lifetime, although the addition of a small amount of H_2 to the feed gas can counteract this reaction.

The total enthalpy (ΔH) required to reduce H_2O or H_2O/CO_2 in a SOEC is supplied *via* an electrical (ΔG) and thermal ($Q = T\Delta S$) component; the relationship between these two sources being inversely proportional (see Figure 8.11 for syn-gas production); the applied cell voltage is linked to the energy component *via* Figure 8.11 (equation 13). Under standard conditions, the total energy to produce a syn-gas output ($S_r = 2$) is equal to an applied cell voltage of 1.48 V; the thermo-neutral voltage (V_{tn}; *i.e.* no heat energy is applied to the system) (Figure 8.11). The enthalpy of H_2O evaporation is equivalent to 0.14 V. Following the respective curves, the potential advantage of the SOEC system becomes self-evident; increasing temperature results in a reduction in the overall electricity demand, while the total energy demands remains the same. This behaviour is in contrast with

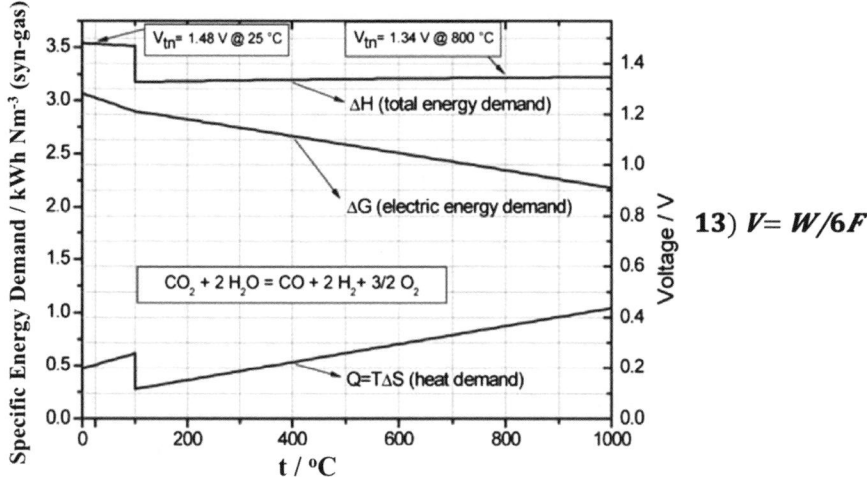

Figure 8.11   The thermodynamic relationship between electrical, thermal and total energy demand for the co-electrolysis of carbon dioxide and water for syn-gas production and the corresponding cell voltage demand, as defined by equation 13 where V refers to cell voltage (volts), W to energy ($J\ mol^{-1}$) and F to Faraday's constant ($96\ 485\ C\ mol^{-1}$)).
Adapted from ref. 101.

low temperature electrolysis (*e.g.* alkaline or PEM based cells). The SOEC can therefore reduce the electrical energy consumed per unit product (*e.g.* syn-gas), potentially reducing the overall production prices (*e.g.* of CH_3OH) as thermal energy can be provided more economically than electrical energy. Such cells could therefore potentially be integrated with other production lines (*e.g.* exothermic CH_3OH synthesis) or other electricity generation schemes (*e.g.* waste heat from existing nuclear reactors).[102]

The energy conversion efficiencies of SOEC have been shown to be very high (*i.e.* >90% (based on $HHV_{syn-gas}$), dependent on the operating conditions (*e.g.* temperature, applied current, *etc.*).[103] Whilst scale-up and pilot plant systems are currently under development (see Section 8.10), there are as yet no commercially operated plants using SOEC. However, the platform has enormous potential to produce efficient, cost-competitive high volumetric outputs of H_2 or syn-gas, with notably the use of generally inexpensive material components (*e.g.* no noble metal catalysis) being particularly attractive in terms of scale-up.[104] Promisingly, the same cell can be operated in H_2O electrolysis or H_2O/CO_2 co-electrolysis mode without any change in materials demand or chemistry offering a flexible, efficient solution to renewable energy storage. Furthermore, the SOEC can be operate at either above or below the V_{tn} (*i.e.* in exothermal or endothermal mode), opening up the scope for reducing the electrical demand further or using excess electrical supply to generate heat for a connected process. SOEC technology also has the base provided by previous research and development on solid oxide fuel cells (SOFCs) to overcome issues pertaining to long-term stability and degradation (*e.g.* anode delamination). For further details and references, the work of Laguna-Bercero, Lackner, Graves and Mogensen is particularly recommended (*e.g.* references 37, 99 and 104).

8.7.2 Fossil Fuel Decarbonisation

It may still be possible to utilise fossil fuels in a CO_2 free manner. This approach to decarbonise H_2 production relies on the dissociation of hydrocarbons to yield H_2 and a carbon solid by-product, thus in principle avoiding direct GHG emissions. The simplest form of this direct decarbonisation (which is receiving increasing attention in part due to the increasing availability of natural gas) is termed 'methane cracking' or 'methane decarburation'.[105] This involves the decomposition of one mole of CH_4 to produce two moles of H_2 and a carbon solid in a completely air/water free environment (Figure 8.12). Energetically, the simplified reaction equation is mildly endothermic, whilst the overall energy requirement $= 37.8$ kJ mol_{H2}^{-1}. This is notably less than the currently industrialised process of steam reforming (63.3 kJ mol_{H2}^{-1}).

This reaction is normally performed at high temperatures (*e.g.* >1200 °C) due to strong C–H bonds (the C–H bond in H_3C–H is 436 kJ mol^{-1}) and a lack of molecular polarisability, requiring a large energy input. Whilst the technology is relatively simple and an (semi-continuous) industrial process is known,[106] the energy demand per cubic metre of H_2 was estimated as 1.9 kW $h_{electricity}$.[107] Catalysts can lower the activation energy/temperature of the process. Catalysts have predominantly been metal-based (*e.g.* Ni, Fe or Co) and carbon-based materials.[108] However, the use of metal catalysts promotes the formation of carbon deposits at the active site resulting in deactivation, shutdown time, potential H_2 contamination, and ultimately a

Methane Decarburation or "Cracking"

$$14) \ CH_4 \rightarrow C + 2H_2 \qquad \Delta H_{298K} = 74.85 \ \text{kJ mol}^{-1}$$

Figure 8.12   Methane Decarburation or "Cracking" reaction as a source of CO_2-free H_2.

reactivation process involving combustion of the carbon residue (and in turn CO_2). The carbon side product has value and can be utilised in traditional applications (*e.g.* tyre fillers, catalysts supports, battery electrodes, filters, *etc.*). One significant advantage is that the solid carbon can be easily stored (as compared with CO_2), with the associated chemical energy not being lost. This advantage increases in significance when renewable energies are integrated into the process (*e.g.* to provide the heat). If the hydrocarbon (*e.g.* CH_4) supplied is also generated from captured CO_2 (*i.e.* methanation), the overall significance for hydrogen production/storage increases dramatically (*e.g.* a closed loop cycle). Comparison of energy efficiencies between methane decarburation and SMR (+CCS), indicate they are practically identical.[109] It is worth noting the potential of integrating this approach with existing energy infrastructure (nuclear power stations, geothermal vents, *etc.*) or new solar installations to provide inexpensive heat to drive the reaction.[110] Likewise, new reactor concepts are necessary to separate the forming solid (ideally continuously). Previous reports include the development of fluid wall aerosol reactors[111] and more recently liquid metal systems.[112] Interest at the industrial level has been revived by BASF®.[113] From a number of feasibility studies regarding production from fossil fuel decarbonisation, H_2 must cost <\$4 (€3) per kg_{H2} to be economically competitive, for example, with SMR.[114]

This is a constantly evolving topic with a number of international projects looking at the feasibility of this approach as a sustainable CO_2 free source of H_2 and the reader is referred to the listed references for more details.

8.8 METHANE AS A FOSSIL BRIDGE IN THE METHANOL ECONOMY

It is perhaps a fault that the concept of sustainability and the use of fossil fuels instead of natural renewable sources of energy, for example, as wind, solar energy and biomass, are mutually exclusive. It is simply the ecological footprint associated with a given process and GHG emission which creates the observed climatic problems (excluding the impacts of industrial accidents, spillage, *etc.*). There is the continuing economic viability (*e.g.* as a consequence of legislation— local or global) of fossil fuel reserves with time, which will make accessibility and use ultimately inhibitory. Therefore, whilst reserves will eventually reach levels which are no longer sustainable, provided they can be extracted and used in a non-GHG emitting manner, then clean energy extraction from fossil fuels can be used to aid the transition to a CO_2 derived Methanol Economy®. Of the remaining fossil fuels, given the current interest in shale reserves (*e.g.* in the USA and the UK), clean burning properties, higher energy yield (as compared with coal) and existing transport infrastructure, CH_4 appears as a very useful energy gas to bridge the fossil and methanol eras. As mentioned in Section 8.2, CH_4 is the predominant industrial feedstock for syn-gas production (*e.g.* for CH_3OH synthesis), However, SMR is energy intensive and represents a relatively inefficient use of CH_4. Therefore, it would seem more appropriate to valorise CH_4 as a platform other than the traditional syn-gas based chemistry. Additionally, in certain geographical locations (*e.g.* in the USA), use of CH_4 as a C1 feedstock may be more economically favourable in the short term. The chemical inertness of CH_4 as a consequence of the H_3C–H bond dissociation energy (*i.e.* 435 kJ mol^{-1}) and the very low bond polarity derived from the similar electronegativity of carbon and hydrogen (w_C : 2.55; w_H : 2.20) makes its utilisation in this context a significant challenge. There is also the associated problem of selectivity/conversion as the product(s) is often more reactive than CH_4 itself.[115] Therefore, the development of catalysts with sufficient activity and selectivity to address this challenge is a significant area of research.

8.8.1   Homogeneous Catalysis

Since the 1990s, the activation of the H_3C–H bond by homogeneous transition metal complexes has proven to be a fruitful area of research.[116] The most active homogeneous catalysts are those based on dichlorobipyrimidyl platinum(II) complex [*i.e.* $Pt(bpym)Cl_2$] developed Periana and co-workers.[117] In this system, the catalytic oxidation is performed in fuming or concentrated sulfuric acid, achieving high yields of CH_3OH with selectivities $\geq 90\%$. In this process the solvent is used simultaneously as both a protecting and an oxidising agent and ultimately can be recycled (Figure 8.13). Similar systems based on Pd-, Au- and Hg-based complexes with superior stability have also been reported but suffer from impractical turnover frequencies (TOFs) <1 h^{-1}.

In this mechanistic cycle, H_3C–H is activated *via* an electrophilic pathway to give a methyl–platinum intermediate complex, followed by the transformation of Pt(II) (the activating oxidation state) to Pt(IV) (the functionalising oxidation state) *via* oxidation in the presence of oleum (SO_3/H_2SO_4). This produces SO_2 as a side product which is recycled to regenerate H_2SO_4. A reductive elimination step releases methyl bisulfate and the regeneration of the initial active Pt(II) species.

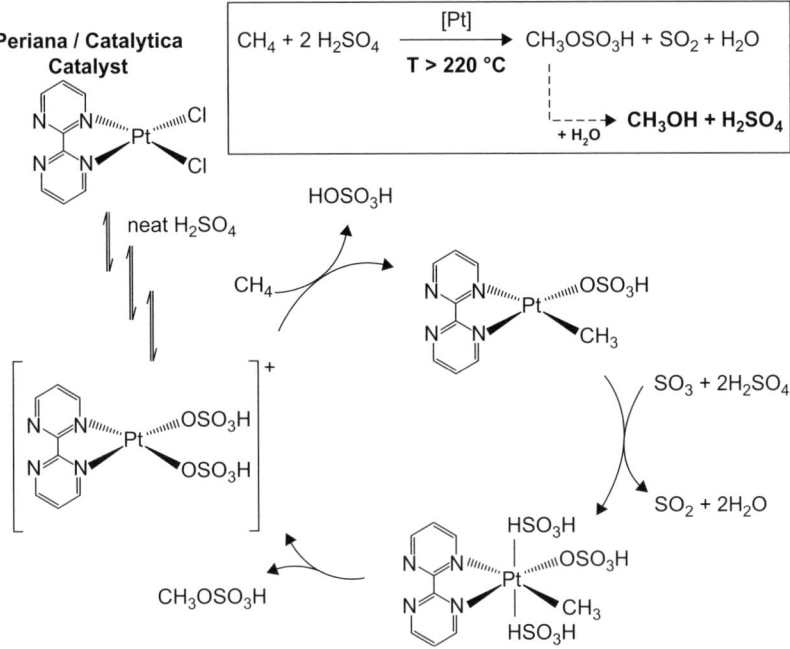

Figure 8.13 Methane electrophilic functionalisation using a homogeneous platinum catalyst as proposed by the Periana and Catalytica systems. Adapted from reference 117(b).

Hydrolysis of methylbisulfate yields CH_3OH with further H_2SO_4 recycling possible. A similar strategy has also been reported using Au^0-based catalysts.[118]

This protecting strategy is interesting as it inhibits over oxidation of CH_4 to form CO_2. In terms of process integration, H_2SO_4 could potentially be produced on-site at a traditional methane-to-methanol plant using the sulphur waste streams (*e.g.* desulphurisation, sweet gas removal, Klaus process). However, the limited catalytic activity as well as difficult product and catalyst separation pose challenges regarding industrial use of these approaches.[119]

8.8.2 Heterogeneous Catalysis

Attempts to develop heterogeneous catalysts (*e.g.* to avoid product–catalyst separation difficulties) have in the majority of cases used reaction temperatures ≥ 250 °C, utilising a variety of different materials including basic oxides,[120] transition metal oxides[121] and zeolite-supported Fe complexes.[122] Unfortunately these early heterogeneous systems presented very poor selectivity to the desired product (typically the result of over-oxidation of CH_4 to CO_2), with maximum CH_3OH yields of <10% being reported.

More recently, an interesting approach to the heterogenisation of the Periana catalyst has been demonstrated by Schüth *et al.* (Figure 8.14).[123] This approach, based on the trimerisation of dicyano compounds such as 2,6-dicyanopyridine (DCP) to produce 'Covalent Organic Frameworks', for example, Covalent Triazine Frameworks (CTFs),[124] is performed at high temperatures (>400 °C) in molten $ZnCl_2$ to shift the equilibrium in favour of a porous polymeric network formation. The materials have high surface areas and pore volumes (S_{BET}>1000 m^2 g^{-1}/V_{pore}>0.8 cm^3 g^{-1}), and when formed from monomers such as DCP, the CTF network present a high number of heteroatoms (*e.g.* Schiff base analogues), ideal for supported catalyst preparation (*i.e.* with high loadings). These CTF materials have been used to support Pt (*e.g.* Pt@CTF), with the final catalyst resembling a polymerised version of the Periana platinum bipyridinium catalyst.

More recently, White and colleagues have prepared Pt-based catalysts using nitrogen-doped carbonaceous materials (*e.g.* derived from lobster shell waste) as supports (Figure 8.15).[125] The low value chitin/calcium carbonate shell composite is converted to a nitrogen-doped carbon material *via* the hydrothermal carbonisation of the organic component, temperature annealing at the desired temperature and removal of $CaCO_3$ (*e.g.* with acetic acid washing), to produce hierarchically porous carbons (S_{BET}>400 m^2 g^{-1}) with tuneable surface chemistry (polarisability, hydrophobicity, nitrogen and carbon condensation, *etc.*). Carbonisation of the lobster shell derived material at 900 °C followed by Pt^{2+} coordination yielded extremely active materials. Remarkably, the catalytic performance was found to be substantially better (under the same experimental conditions) than the Pt@CTF and Periana catalysts (Table 8.3). Deactivation, however, was more pronounced for the biomass-derived support (but not necessarily as a result of Pt leaching), with extremely high TOFs (relative to the Periana and Pt@CTF materials) being recorded (*i.e.* >1000 h^{-1}) after five recycles. It is worth noting that producing catalyst supports (with indeed very high promise) can add value to low value, waste biomass whilst hydrothermal carbonisation converts less stable biomass into more stable carbonised forms—a mechanism of carbon sequestration. These approaches open further investigations and optimisation of support–metal interaction for the efficient conversion of CH_4 to CH_3OH.

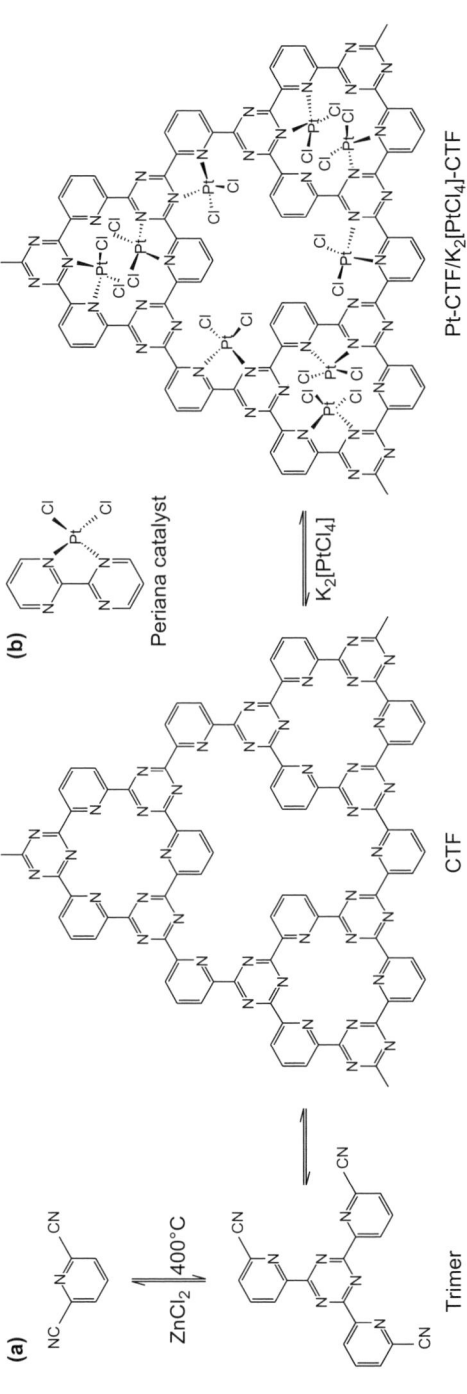

Figure 8.14 (a) Trimerization of 2,6-dicyanopyridine (DCP) in molten ZnCl$_2$, conversion to a covalent triazine-based framework (CTF), and subsequent platinum coordination (Pt-CTF); (b) Periana's platinum bipyrimidine complex. Reproduced with permission from reference 123.

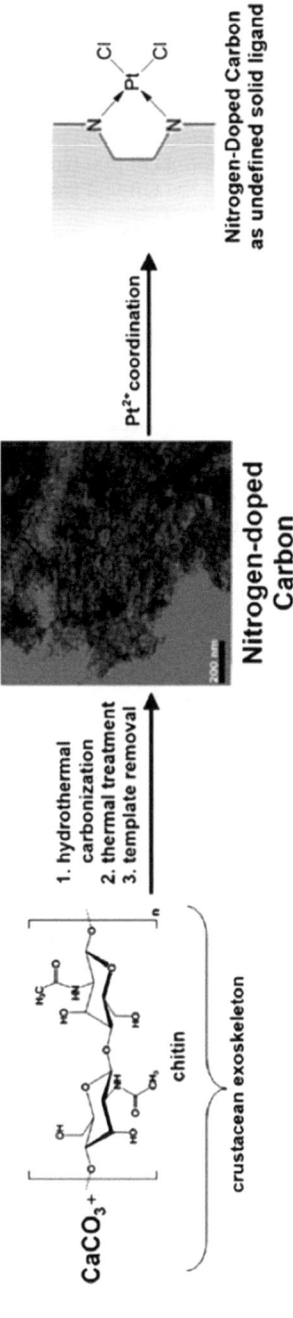

Figure 8.15 Preparation of coordinatively modified Pt@NDC materials derived from crustacean exoskeleton of lobsters (ExLOB). Reproduced with permission from ref. 125.

Table 8.3 Conversion (X), yield (Y), selectivity to CH_3OH (*via* methylbisulfate) (S) and catalytic activity (*TOF*) achieved under the given reaction conditions. Reproduced from reference 125.

Entry	Catalyst	X/%	Y/%	S/%	TOF^a/h^{-1}	TOF^b/h^{-1}
1	Pt@CTF	7.0	6.0	85.4	174	233
2	Pt(bpym)Cl$_2$	17.9	17.2	96.3	912	779
3	Pt@*Ex*LOB-900 (1st run)	33.8	31.7	94.0	2074	1227
4	Pt@*Ex*LOB-900 (run)	18.5	17.6	95.2	1938	1516
5	Pt@*Ex*LOB-900 (run)	6.0	5.6	91.6	1826	1802

^{a}Determined from a pressure drop from 6.9 to 6.75 MPa (ESI).
^{b}Determined from the amount of methanol produced within 30 minutes.

8.8.3 (Oxidative) Bi-reforming Approaches

As discussed in Section 8.5, the CH_3OH synthesis from syn-gas requires $S_r = 2$. Steam reforming, carbon dioxide reforming and the partial oxidation of CH_4 in this context produce $S_r = 3$, 1 and 2, respectively. Therefore, each of these routes requires either additional H_2 or CO to modify the syn-gas composition or there are safety considerations, for example, as a consequence of partial oxidation energetics. This in turn results in a loss of overall efficiency and hydrogen utilisation (*e.g.* high number of process steps, including separation, hydrogen adjustment), increasing process/plant and final product costs. One potential route to circumvent these issues has been put forward by Olah and colleagues who proposed a coupling of SMR and partial oxidation commonly practised in industry to produce a final $S_r = 2$. This process has been termed (oxidative) bi-reforming (Figure 8.16).[126]

Based on earlier reports,[127] the combination of dry (Figure 8.1, equation 3) and SMR (Figure 8.1, equation 1) proposed by Olah and colleagues was investigated using a pressure of 0.7 MPa (7 atm) and a reaction temperature of 830 °C. The gas mixture produced is termed 'met-gas' (Figure 8.16, equation 16), identifying the perfect S_r for CH_3OH synthesis formed in this approach (*i.e.* 2). A mixture of CH_4, H_2O and CO_2 (ratio: 3 to 2.4 to 1.2) is converted to 'met-gas' using a hydrogen-activated Ni catalyst supported on MgO. Under the reported conditions, single pass conversions of $\sim 70\%$ were reported, which could be improved by adjusting the feed gas partial pressures. Operation at higher pressures would also be beneficial to remove the need to compress the resulting met-gas for CH_3OH synthesis.

This approach is attractive as it converts all hydrogen into the final product, excluding the hydrogen required for catalyst (re)activation. Any natural gas source can also be used and is particularly suitable for

15) Methane combustion

$$CH_4 + 2O_2 \longrightarrow CO_2 + 2H_2O \quad \Delta H_{298K} = -802.9 \text{ kJ mol}^{-1}$$

16) After admixing fresh methane, bi-reforming is carried out

$$3CH_4 + 2H_2O + CO_2 \longrightarrow \underbrace{4CO + 8H_2}_{\text{metgas}} \quad \Delta H_{298K} = 659.0 \text{ kJ mol}^{-1}$$

17) Methanol synthesis

$$4CO + 8H_2 \longrightarrow 4CH_3OH \quad \Delta H_{298K} = -363.2 \text{ kJ mol}^{-1}$$

18) Overall reaction

$$4CH_4 + 2O_2 \longrightarrow 4CH_3OH \quad \Delta H_{298K} = -507.1 \text{ kJ mol}^{-1}$$

Figure 8.16   The (oxidative) bi-reforming of methane to methanol as represented by the combination of equations 15 to 17 to give the overall synthetic equation 18.
Adapted from reference 126.

sources containing CO_2 (*e.g.* bio-gas). The heat needed to drive the endothermic met-gas production can be provided from the exo-thermic combustion of CH_4, with the resulting CO_2 and H_2O cycled to the synthesis reactor. This approach will become increasingly at-tractive as shale gas reserves are increasingly exploited (*e.g.* in the USA). (Oxidative) bi-reforming potentially represents a neat, efficient coupling of a 'bridging' fossil fuel and captured CO_2 with a strong potential to enable short-term CH_3OH production, based on past and future energy gases.

8.9   METHANOL AS A HYDROGEN AND ENERGY VECTOR

8.9.1   Catalysts for Dehydrogenation

As highlighted in this chapter, hydrogen is a potential fuel for the provision of clean energy, (see Section 8.2). However there are a number of problems with hydrogen storage, infrastructure costs and physical properties (flammability, low energy density gas, *etc.*). In an attempt to circumvent such problems, liquid fuels (*i.e.* CH_3OH) can be viewed as vectors whereby chemical catalysis is used to recover 'stored' hydrogen (*e.g.* for fuel cell applications). For CH_3OH, the low 'C' number means that efficient H_2 recovery (*e.g.* >80%) can be achieved *via* reforming, importantly (given the purity of the described synthesis above) without any S residues, which can inhibit the fuel cell catalysts. Therefore, CH_3OH is a 'liquid organic hydrogen carrier' (LOHC),[128] providing a route (if achieved efficiently) to 'on demand' hydrogen, given its high hydrogen content (12.6% weight) and its state at room temperature. *Via* the use of well-established technology such as PEMFCs, hydrogen can be efficiently converted (>60%) to electrical energy.[129] Therefore, the conversion of CO_2 to CH_3OH using H_2 (*i.e.* produced using renewable electricity) and appropriate catalysis, may be a particularly efficient use of waste GHGs to transport intermittent electricity.

The state-of-the-art recovery of H_2 from CH_3OH is performed in the gas phase at high temperatures and pressures (>200 °C; >2.5 MPa), using a supported (noble) metal catalyst (*e.g.* Pt/Al_2O_3).[130] However, these conditions are not especially practical particularly regarding mobile energy provision (*e.g.* handheld devices). Furthermore, these systems struggle to produce gas feeds for the PEMFC with [CO] $<10\times10^{-6}$. Therefore, there has been a renaissance in research describing CH_3OH dehydrogenation at more convenient conditions. Early reports detail the production of H_2 from CH_3CH_2OH and secondary alcohols at rates corresponding to TONs ~ 100.[131] These reports were followed by more detailed studies extending homogeneous Rh and Ru complexes to the dehydrogenation of a range of alcohols including CH_3OH with $[RuH_2(N_2)(PPh_3)_3]$ proving particularly active (@ 150 °C).[132]

The efficient production of hydrogen from isopropanol under mild conditions has been recently reported with TOF and TON of 14 145 h^{-1} and 40 000, respectively.[133] The group of Beller and Junge have recently extended the use of molecularly defined Ru catalysis for hydrogen production.[134] Their work describes an efficient low temperature aqueous phase CH_3OH dehydrogenation using homogeneous (Milstein-like) pincer ligand-based Ru complexes. The production of hydrogen was found to occur within a practical range of (65–95) °C (under ambient pressure). A dehydrogenation mechanism was described involving initial abstraction of 1 molecule H_2 from CH_3OH to give formaldehyde (Figure 8.17). The H_2O promoted dehydrogenation of formaldehyde to formic acid then occurs generating another mole of H_2. Finally formic acid is dehydrogenated to H_2 and CO_2. This mechanism is significant as it represents the production of three moles of H_2 per mole of CH_3OH (resembling classical high temperature steam reforming), whereas previous molecular-defined organometallic catalysts have only been capable of generating one mole of H_2 per mole of CH_3OH.[135] Furthermore, the work of Beller and colleagues demonstrates product streams featuring

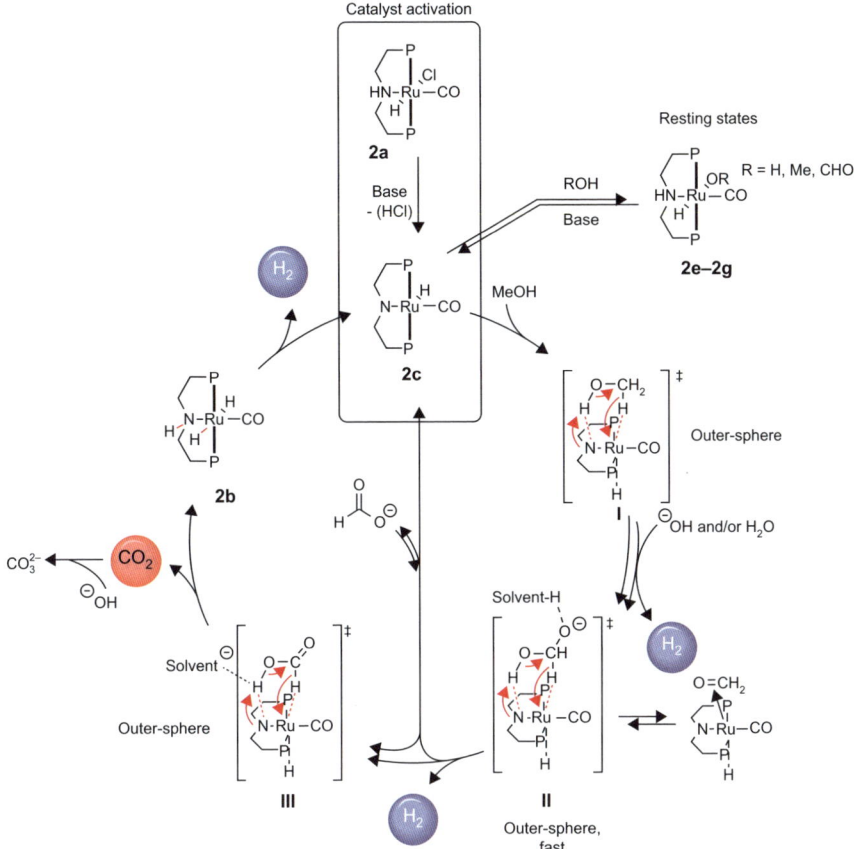

Figure 8.17 Proposed catalytic cycle for the aqueous phase dehydrogenation of methanol to give 3 moles of H_2 (and 1 mole of CO_2) per mole of CH_3OH. Reproduced with permission from ref. 134.

contaminants (*i.e.* CO and CH_4) at levels $<10\times10^{-6}$, important with regard to PEMFC catalyst activity/lifetime. The optimised catalyst was the molecular-defined $[RuHCl(CO)(HN(C_2H_4PiPr_2)_2)]$ where $iPr = CH(CH_3)_2$, exhibiting high activities (TOF > 4700 h^{-1} for pure CH_3OH) and productivities (TON > 350 000). The presented system was found also to be stable for up to three weeks. A similar system was also reported by Trincado and Gruetzmacher and colleagues, who described the use of a Ru complex with a chelating bis(olefin) dia-zadiene ligand.[136]

A drawback in both these cases, however, is the necessity to include (albeit an abundant) base to maintain the system activity. Positively, mixtures of CH_3OH and H_2O can be used with either component in excess. This may be highly beneficial from an overall process design and actually may correspond to a reduced energy demand (*i.e.* separation of H_2O/CH_3OH mixtures is not necessary). Such homogeneous systems are practically workable (particularly given their slightly endothermic nature) as the products themselves are gaseous and will simply leave the system upon production (*i.e.* $t > 70$ °C).[137] It is also worth noting in the context of using CO_2 as a vector, the chemistry employed in the dehydrogenation of CH_3OH, it is also applicable to the dehydrogenation of formic acid [$H_2C(O)OH$],[138] although the overall H_2 yield is lower but this may be compensated for in the overall production energy demand. The emitted CO_2 can then recycle back to the original synthesis, creating a closed loop.

8.9.2 Direct Methanol Fuel Cells: Liquid Electrical Energy Carriers

In the context of mobility, the direct conversion of CH_3OH ($+ O_2$ / air) to CO_2 and H_2O, without the need for an initial dehydrogenation/ reforming step (or other routes to hydrogen for fuel cell applications) would be highly advantageous. Work to realise this concept, termed the 'direct methanol fuel cell (DMFC)', was initiated by the Olah group and the Jet Propulsion Laboratory in the early 1990s.[139] The basic cell is composed of two electrodes partitioned by a polymer electrolyte membrane such as poly(styrene)-sulfonic acid/poly(vinylidenefluoride) composite or Nafion-H®.[140] During operation at $t = 20$–130 °C, CH_3OH/H_2O mixtures are fed to a carbon-supported metal (or alloy) (*e.g.* Pt-Ru/C) anode, with the conformation/composition of the electrocatalyst considered performance determining.[141] At the anode, oxidation occurs to release protons, which migrate through the PEM, with electrons travelling to the cathode *via* an external circuit. At the cathode, a noble metal catalyst (*e.g.* Pt) on a conductive support such as poly(vinylpyrrolidone)-modified graphite or carbon nanotubes[142] is exposed to flowing air or pure O_2, which can be supplied from electrolysis.

CH_3OH can provide theoretical power densities of 5 kW h L^{-1} in the DMFC, *i.e.* twice the theoretical output for pure hydrogen. As the intention is to supply such cells with a sustainable fuel (*e.g.* CH_3OH produced *via* co-electrolysis), the process of electricity generation does not result in the production of pollutants; H_2O and CO_2 are the only (importantly recyclable) by-products. Furthermore, the direct reaction of CH_3OH eliminates the reforming step to produce hydrogen, reducing fuel cell stack and overall system complexity, whilst the overall operation is near silent and highly efficient (*e.g.* there are no moving parts); factors important when considering mobile applications. The use of an easily stored and transported, flowing liquid feed to the DMFC is also beneficial for heat removal/thermal management as well as preventing membrane dry-out. Significantly, the DMFC can operate using CH_3OH and H_2O mixtures, removing the need to separate these two components if synthesis is based on the direct hydrogenation of captured CO_2, aiding overall process efficiency.

In the context of new electrode design, the combined necessity for conductivity, electrochemical stability and suitable porosity (*i.e.* to facilitate fast diffusion coefficients essential for fast kinetics) to maximise the triple phase boundary at the liquid–anode interface and the gas–cathode interface should be borne in mind. At the stage of writing the use of Pt and Pt-based alloys are the most commonly used catalysts in DMFCs. Pt has been shown to provide the highest catalytic activity, chemical stability, exchange current density and work function.[143] However, the cost and global scarcity of Pt represents a challenge to the large-scale adoption of DMFC (and indeed PEMFC) technology. The development of metal-free systems based on heteroatom-doped carbon nanostructures (*e.g.* graphene; Figure 8.18),[144] may be a promising avenue of research (*e.g.* analogous to oxygen reduction reaction catalysis).[145]

The work of Xu and colleagues demonstrates the potential of such materials.[144] Reduced graphene oxide (RGO)-supported PtAuRu NP displayed high electrocatalytic activity towards methanol electro-oxidation (in alkaline medium), as evidenced by the low onset oxidation potential and the higher peak current density (Figure 8.18). Au is introduced into the classical PtRu alloy to modify the electronic structure of the catalyst, to reduce the propensity of Ru to leach and to enhance electrocatalytic activity. The activity of the reported RGO supported PtAuRu NP electrocatalyst was found to be 190% that of the first scan after 500 cycles, indicating good long-term stability for methanol oxidation.

Future technology developments in this area will need to address problematic CH_3OH crossover and electrode design/stability—important with regard to the long-term performance (*e.g.* of the cathode catalyst), fuel use and an overall reduction in the short- to intermediate-term DMFC performance. An excellent review by Sharma and Pollet is recommended for further reading.[146] The conductive, nanostructured carbon materials needed for the preparation of next generation electrodes for the DMFC could in principle be synthesised from carbon produced from other processes in the Methanol Economy® scheme, using the carbon by-product from fossil fuel decarburation.

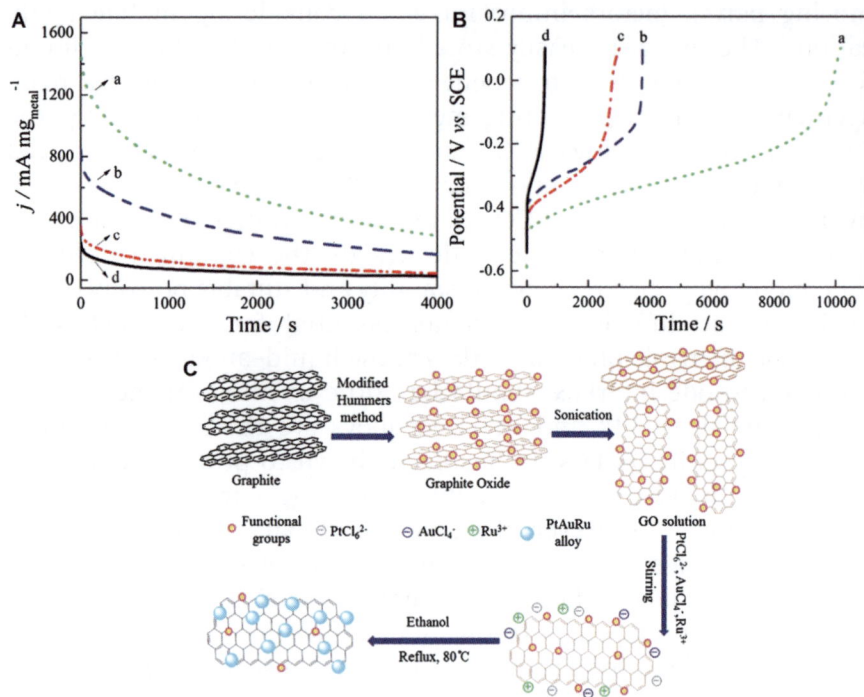

Figure 8.18 (A) Chronoamperometric curves of 1.0 M CH$_3$OH oxidation on the PtAuRu/reduced graphene oxide (RGO)/glassy carbon electrode (GC) (a), PtAu/RGO/GC (b), PtRu/RGO/GC (c) and Pt/RGO/GC (d) electrodes in a 1.0 M KOH solution at −0.3 V. (B) Chronopotentiometric curves of 1.0 M methanol oxidation on the PtAuRu/RGO/GC (a), PtAu/RGO/GC (b), PtRu/RGO/GC (c) and Pt/RGO/GC (d) electrodes in a 1.0 M KOH solution at 0.4 mA cm^{-2} (metal to total metal loading of Pt, Au and Ru). (C) Procedure for the synthesis of the PtAuRu/RGO catalyst. Adapted from ref. 144.

8.10   COMMERCIAL EXAMPLES

Due to the ever increasing abundance of CO$_2$ emissions and our desire to recycle this waste GHG, new CH$_3$OH markets are unlikely to suffer from traditional market factors such as feedstock availability. Furthermore, the overall cost of CO$_2$ will presumably reduce in the future as the cost of capture and other additional measures (*e.g.* legislation) encourage recycling and act to make CH$_3$OH (or other CO$_2$ derived fuels and chemicals) increasingly viable and economically attractive. Based on the potential of sustainable CH$_3$OH (or fuel), a number of small-to-medium sized enterprises (SMEs) are breaking ground and are entering the market place using technology either based on gasification (*e.g.* of biomass) or the direct hydrogenation of CO_2 to CH_3OH (or higher FT products) and chemicals.

Carbon Recycling International® (CRI), founded near Reykjavik in Iceland in 2006, is perhaps the most well-known exponent of the Methanol Economy® concept.[147] Perhaps the most advanced commercial venture for the synthesis of CH_3OH from CO_2, CRI exploits geographical conditions in Iceland to combine geothermal (and in principle industrial) CO_2 with H_2 produced using (low temperature) PEM electrolysis powered by renewable electricity. The first fully commercial plant at the CRI site was completed in 2011; the George Olah Renewable Methanol Plant is currently operating at a 5×10^6 L a^{-1} scale (Figure 8.19). CRI claims that its process has the ability to reclaim 4.5×10^3 t of CO_2 per year from the atmosphere. Recent investment from Methanex® of US$5 million demonstrates the recognised potential of sustainable CH_3OH in future markets, with the intention to continue to grow production capability with a 50×10^6 L a^{-1} facility currently under construction. CRI® has designed its production system in modules which, in principle, can be exported to any desired location. The company has recently started exporting to the European market, where the sustainable CH_3OH is intended to be blended with conventional petrol.

Newer commercial ventures such as SunFire GmbH (Germany)[148] and Air Fuel Synthesis (UK)[149] in the European market are looking to convert renewable electricity to liquid hydrocarbons ('power-to-liquids') and CH_3OH, respectively. In the United States, Ceramatec Ltd[150] is developing solid oxide electrolysers to produce syn-gas from the co-electrolysis of H_2O and CO_2 powered by the abundant wind electricity in the state of Wyoming. Both SunFire GmbH and Ceramatec Ltd are the first commercial entities to develop and introduce SOEC/SOFC technology to the market for sustainable fuel synthesis from renewable electricity.

With regard to alternative sources of syn-gas for CH_3OH or DME production, BioMCM and the SuperMethanol/GtM concept project are utilising waste glycerol from bio-diesel production as a precursor. The glycerol is gasified to syn-gas for the synthesis of CH_3OH.[151] The current plant in the Netherlands is currently capable of producing 200×10^6 t a^{-1} of CH_3OH with the intention to expand capacity in the near future.

Waste lignosulfates from the paper industry can also be gasified to syn-gas, as exploited by ChemRec® (Sweden) and partners as part of the 'BioDME' project (Figure 8.20).[152] The black liquor paper mill

Figure 8.19   The George Olah Renewable Methanol plant at the HS Orka Svartsengi installation near Rejkavik, Iceland. Methanol synthesis at Carbon Recycling International® is based on the combination of geothermal or industrial CO_2 hydrogenation using H_2 produced from H_2O (PEM) electrolysis.
Adapted from reference 147.

waste is gasified and the syn-gas product is converted using a one-step reactor to DME. The diesel substitute produced in this manner is claimed to be one of the most efficient of all bio-fuels, as a consequence of extensive plant integration. The use of lignosulfates as a precursor means it is necessary to clean the syn-gas to remove sulfur impurities. This fuel is being extensively tested in a fleet of Volvo® trucks across Sweden.[153]

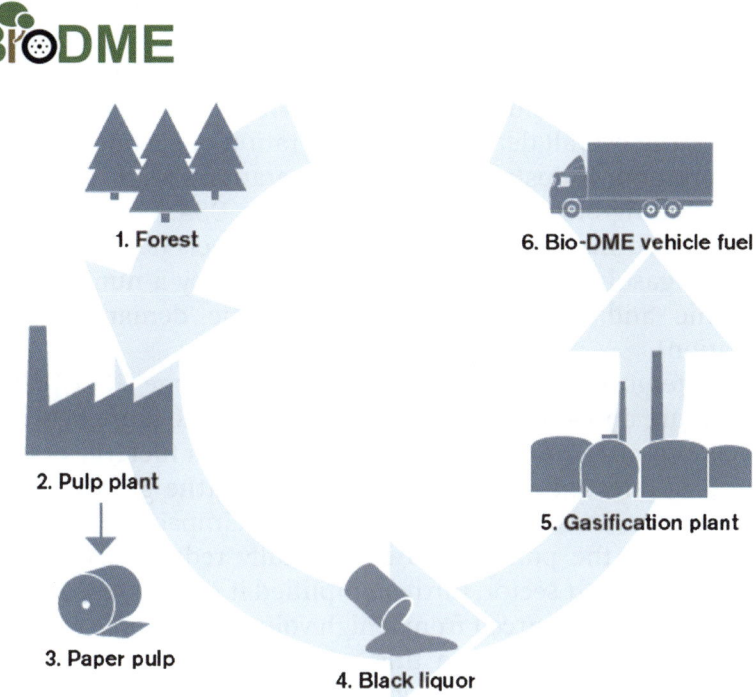

BIODME

1. Forest

6. Bio-DME vehicle fuel

2. Pulp plant

5. Gasification plant

3. Paper pulp

4. Black liquor

Figure 8.20 The BioDME 'Wood-to-Wheel' cycle.
Reproduce and adapted from reference 153.

8.11 SUMMARY AND OUTLOOK

There is no one 'solve all' solution to the question of energy/fuel/chemical demand and there are numerous questions that must be addressed if a 'methanol economy' is to be successfully implemented. Specific geographic and socio-economic conditions may encourage specific routes to synthetic CH_3OH, fuel and chemicals (*e.g.* abundance of wind energy in Denmark, shale gas in the USA and limited fossil resources in Germany). The demonstration of 'sustainable' energy provision by leading nations should also encourage developing nations, where adaptation of existing infrastructure or the deployment of new technology systems may be less prohibitive than in older, more entrenched fossil-based economies. However, there is a general underlying theme. If fuel and chemicals are to be produced synthetically, it must be done with as minimal a carbon footprint as possible (ideally neutral or negative), with the incorporation of renewable electrical energy and recycled CO_2 highly desirable.

Compared with other fuels and chemical precursors, a number of developing 'sustainable' (with a particular accent on the flexible base compound CH_3OH) fuel enterprises are highlighted in this chapter, each having their individual merits. Whilst markets are currently opening, the overall demand and penetration of sustainable fuels for transportation, industry and power generation is still relatively limited, excluding bio-ethanol in Brazil and the USA. With this in mind, the increased market uptake of sustainable fuels (*e.g.* CH_3OH, DME, synthetic gasoline) will largely be determined by a number of socio-economic and political factors (*e.g.* public demand, price and legislation).

The Intergovernmental Panel on Climate Change (IPCC) report on CO_2 capture and storage[154] suggests that the time lapse between CO_2 conversion into a product such as a polymer or a fuel (the cradle) and when it is released again to the atmosphere (the grave) is a crucial consideration. In this regard, one of the high impacts of the Methanol Economy® is the potential to dramatically reduce GHG emissions from the transport sector, further amplified if the CO_2 used in CH_3OH or FT synthesis is sourced from a high volume emitter (*e.g.* fossil fuel-burning power stations). In this context, CH_3OH or DME has the potential to reduce carbon-based emissions from €(65–95) on a wheel-to-tank basis—extremely promising values compared with other alternative fuels.[156] An anthropogenic carbon cycle for fuels will be completed over shorter time scales than the natural equivalent and it will therefore be very useful to sequester large volumes of CO_2 in longer life polymers and plastics (*e.g.* polycarbonates). For a structural polymer, 'the cradle-to-grave' cycle will be much longer than for a fuel. The conversion of CO_2 to polymers has been reported as having the capability to remove $(250–350) \times 10^6$ t a^{-1} or 10% of the total global reduction necessary to meet the CO_2 reduction targets outlined by the IPCC.[155] The advent of commercially viable and practical air capture systems will also be critical in closing this CH_3OH life cycle, and has the potential to decouple capture from source and reverse the consequence of a high atmospheric CO_2 loading. In this regard, the significant benefit of developing a methanol-based economy is the ability to produce both a fuel and solid store from a single precursor.

The production of sustainable CH_3OH, and associated synthetic fuels and chemicals, has the potential to integrate different energy sectors and to recycle low value, problematic CO_2 utilising this GHG as a C1 carbon source and energy carrier gas. It also creates a potentially closed multi-fuel anthropogenic carbon cycle. With the integration of renewable energy (wind, solar, *etc.*), it can represent a solution to intermittent renewable electricity supply and demand. Given current political desires, it can be anticipated that the exploitation of shale gas, and consequently the methane hydrates, will mean that the price and/or availability of CH_4 remains an economically attractive crude oil alternative at least in the short to medium term. The conversion of these CH_4 reserves into CH_3OH (*e.g.* directly *via* bi-reforming or indirectly *via* H_2 production) will therefore remain of interest, potentially providing a bridge to a CO_2 based CH_3OH economy as the necessary technologies become increasingly viable. This will ultimately be based on the market dynamics (*e.g.* shale gas cost), but recent reports demonstrate the techno-economic feasibility of such an approach.[156] The question of whether these reserves are used is of course a delicate one. It seems, however, that the conversion of CO_2 and H_2O to CH_3OH or higher hydrocarbons, and the subsequent establishment of a closed 'energy' and 'chemical' carbon recycling loop should be the final target, particularly when CO_2 air capture becomes practical.

CH_3OH can be produced *via* the catalysed hydrogenation of CO_2 utilising similar conditions and catalysts to those used for the production of CH_3OH from syn-gas in a well-known and commercial process. In this case, the use of hydrocarbons can be avoided by using hydrogen from renewable sources (*i.e.* H_2O splitting *via* electrolysis). However, equilibrium yields based on the direct hydrogenation of CO_2 are much lower than CH_3OH production from syn-gas. In addition, a third of the H_2 is converted to H_2O; this process is inefficient to produce pure CH_3OH, particularly when considering that H_2 production *via* electrolysis can be energy intensive. However, in DMFC and dehydrogenation catalysis, the presence of H_2O is workable and indeed necessary in the case of dehydrogenation with high TOFs. Therefore, the situation may occur where syn-gas-derived CH_3OH will provide high purity volumes for ICE or turbine combustion, whilst CH_3OH and H_2O mixtures will be suitable for fuel cell powered devices. Electrolysis technology has the opportunity in these synthetic schemes to perform the role of an 'enabling technology' particularly with regard to co-electrolysis, allowing the substitution of fossil fuels with fuel alternatives generated from renewable electricity to provide a flexible renewable syn-gas platform for fuel and chemical production. However, the long-term application and life time of high temperature SOEC remains a partial limitation (delamination, carbon formation, *etc.*). The reversibility of the SOEC (*e.g.* to SOFC) is a very attractive option and something that should be considered strongly as an option to address problems of renewable electricity intermittency.

There is also the question of how best to utilise CH_3OH, DME or hydrocarbons. The production of hydrocarbons for certain sectors (*e.g.* aviation) and the development of deliverable mobility options based on fuel cells is surely desirable, particularly in an urban context where these clean fuels would be highly beneficial (*e.g.* SO_X free). A number of research programmes now focus on the optimisation of PEMFC performance at low temperatures (<150 °C) as well as the development of new micro fabrication techniques on the premise that hydrogen can be supplied safely and affordably.[157] Reports on dehydrogenation catalysts have been briefly discussed in this chapter and appear to have promise to supply hydrogen on demand, although it may be more practical (*e.g.* for mobile electronics) to use DMFC technology to provide electrical power. In this context, there is still the need to move away from noble metal based catalysts, and the development of non-rare metal substitutes and/or new electrode chemistry to provide enhance CH_3OH oxidation activity is required. Likewise, the problematic issue of long-term use and CH_3OH crossover limitations need to be addressed *via* the development of new membrane design and chemistry.

In the context of new fuel markets, it is important to note that fossil fuels dominate the global energy mix, supported by US\$$523 \times 10^9$ (€384×10^9 or £319×10^9) subsidies in 2011, up ~30% on 2010 and six times larger than subsidies for renewable fuels.[158] Therefore, support from regulatory bodies such as the European Union to establish new legal frameworks to define and promote fuels derived from CO_2 is needed. The provision of subsidies, perhaps akin to feed-in tariffs established for renewable electricity adoption, will be required to overcome fossil dominance and to provide incentives for technology implementation and market creation. Therefore, the production of sustainable fuels must be demonstrated at increasingly large volumes *via* commercially viable production pathways, which have an economic advantage compared with conventional ones. This can in part be achieved by well-designed, highly efficient, integrated processes, assisted by the introduction of new chemical technologies, but in the short term, further statutory instruments (*e.g.* directives, regulations) or financial incentives (*e.g.* tax breaks, subsidies, feed-in-tariff equivalents) are needed to support recycling CO_2 and sustainable fuels. Such measures will ultimately be enhanced if associated markets (*e.g.* the chemical industry) are also in turn encouraged to adopt sustainable feedstock based processes.

Whilst the future of synthetic CH_3OH and associated fuels from renewable energies looks optimistic, a gap exists between the strength of scientific argument and technology development for such sustainable alternative on one hand and a lack of political awareness and action on the other hand. Therefore, not only must scientific research and innovation continue in the area, but there must also be a simultaneous increase in advocacy directed at policymakers and indeed the general public to enhance awareness if there is to be a true, long-term adoption of a sustainable methanol economy concept in Germany, the European Union and beyond. There is, however, one critical premise, *i.e.* that the electrical energy used to power any of these processes is a non-emitting GHG source (*e.g.* wind, tidal, nuclear, solar). Each energy source has its pros and cons, but presumably we must be forward thinking and consider the consequences for future generations (*e.g.* in terms of waste disposal). Northern European countries such as Germany and Denmark are leading the way *via* innovative nationwide programmes such as Energi2050 and Energiewende to introduce renewable electricity to the energy mix and to provide solutions to intermittency. Furthermore, if SMEs such as CRI and Sunfire can demonstrate their respective synthetic platforms at larger scale facilities capable of meeting their production goals, whilst demonstrating market penetration, this will establish the benchmark and hopefully encourage other similar geo-economic areas where renewable energy resources and large CO_2 emissions are co-located such as in the UK.

REFERENCES

1. (a) J. Zalasiewicz, M. Williams, A. Smith, T. L. Barry, A. L. Coe, P. R. Brown, P. Brenchley, D. Cantrill, A. Gale, P. Gibbard, F. J. Gregory, M. W. Hounslow, A. C. Kerr, P. Pearson, R. Knox, J. Powell, C. Waters, J. Marshall, M. Oates, P. Rawson and P. Stone, *GSA Today*, 2008, **18**, 4–8; (b) J. Zalasiewicz, M. Williams, W. Steffen and P. Crutzen, *Environ. Sci. Technol.*, 2010, **44**(7), 2228–2231.
2. US Energy Information Administration, International Energy Statistics, Total Energy, http://www.eia.gov/.
3. (a) P. Sadorsky, *Energy Econ.*, 1999, **21**(5), 449–469; (b) M. E. Dry, *Catal. Today*, 2002, **71**, 227–241; (c) T. H. Fleisch, R. A. Sills and M. D. Briscoe, *J. Nat. Gas Chem.*, 2002, **11**, 1–14; (d) N. Armaroli and V. Balzani, *Angew. Chem., Int. Ed.*, 2007, **46**, 52–66; (e) H. Khatib, *Energy Policy*, 2014, **64**, 71–77.

4. (a) M. A. Adelman, J. C. Houghton, G. Kaufman and M. B. Zimmerman, in *Energy Resources in an Uncertain Future*, Ballinger Publishing, Cambridge, MA, 1983; (b) M. I. Hoffert, K. Caldeira, G. Benford, D. R. Criswell, C. Green, H. Herzog, A. K. Jain, H. S. Kheshgi, K. S. Lackner, J. S. Lewis, H. D. Lightfoot, W. Manheimer, J. C. Mankins, M. E. Mauel, L. J. Perkins, M. E. Schlesinger, T. Volk and T. M. L. Wigley, *Science*, 2002, **298**, 981–987; (c) C. Hall, P. Tharakan, J. Hallock, C. Cleveland and M. Jefferson, *Nature*, 2003, **426**, 318–322; (d) A. Hern, It's not safe to leave fossil fuels in the ground, *New Statesman*, 2012, http://www.newstatesman.com/sci-tech/2012/10/its-not-safe-leave-fossil-fuels-ground.

5. (a) United Nations Report of the World Commission on Environment and Development, General Assembly Resolution 42/187, 11 December 1987; (b) World Commission on Environment and Development, *Our Common Future*, 1987. Published as Annex to General Assembly document A/42/427, Development and International Co-operation: Environment, 2 August 1987; (c) G. C. Andrews, *Canadian Professional Engineering and Geoscience: Practice and Ethics*, Nelson Education, Toronto, 4th edn, 2009.

6. (a) W. Leitner, *Coord. Chem. Rev.*, 1996, **153**, 257–284; (b) M. Mazzotti, in *IPCC Special Report on Carbon Dioxide Capture and Storage*, ed. B. Metz, O. Davidson, H. de Coninck, M. Loos and L. Meyer, Cambridge University Press, Cambridge, UK, 2005, ch. 7, p. 319–338; (c) T. Sakakura, J.-C. Choi and H. Yasuda, *Chem. Rev.*, 2007, **107**, 2365–2387; (d) M. Aresta, *Carbon Dioxide as Chemical Feedstock*, Wiley-VCH, Weinheim, Germany, 2010; (e) N. MacDowell, N. Florin, A. Buchard, J. Hallett, A. Galindo, G. Jackson, C. S. Adjiman, C. K. Williams, N. Shah and P. Fennell, *Energy Environ. Sci.*, 2010, **3**, 1645–1669; (f) M. Cokoja, C. Bruckmeier, B. Rieger, W. A. Herrmann and F. E. Khn, *Angew. Chem., Int. Ed.*, 2011, **50**, 8510–8537; (g) E. A. Quadrelli, G. Centi, J.-L. Duplan and S. Perathoner, *ChemSusChem*, 2011, **4**, 1194–1215; (h) M. Peters, B. Koehler, W. Kuckshinrichs, W. Leitner, P. Markewitz and T. E. Müller, *ChemSusChem*, 2011, **4**, 1216–1240; (i) D. J. Darensbourg and S. J. Wilson, *Green Chem.*, 2012, **14**, 2665–2671; (j) P. Markewitz, W. Kuckshinrichs, W. Leitner, J. Linssen, P. Zapp, R. Bongartz, A. Schreiber and T. E. Müller, *Energy Environ. Sci.*, 2012, **5**, 7281–7305; (k) G. Centi, E. A. Quadrelli and S. Perathoner, *Energy Environ. Sci.*, 2013, **6**, 1711–1731.

7. M. Romero and A. Steinfeld, *Energy Environ. Sci.*, 2012, **5**, 9234–9245.

8. (a) J. G. J. Olivier, G. Janssens-Maenhout, M. Muntean and J. A. H. Peters, *Trends in Global CO_2 Emissions – 2013 Report*, PBL Netherlands Environmental Assessment Agency, The Hague, and European Commission Joint Research Centre, Ispra, Italy, 2013; (b) T. Boden, G. Marland and R. Andres, *Global, Regional, and National Fossil Fuel CO2 Emissions*, Carbon Dioxide Information Analysis Centre, Oak Ridge National Laboratory, US Department of Energy, Oak Ridge, TN, USA, 2012; (c) V. Marchal, R. Dellink, D. van Vuuren, C. Clapp, J. Château, E. Lanzi, B. Magné and J. van Vliet, in *OECD Environmental Outlook to 2050: The Consequences of Inaction*, OECD Environment Directorate and PBL Netherlands Environmental Assessment Agency, 2011, ch. 3.

9. (a) H. J. Herzog, *Environ Sci. Tech.*, 2001, **35**, 148A–153A; (b) M. Z. Jacobson, *Energy Environ. Sci.*, 2009, **2**, 148–173; (c) A. A. Olajire, *Energy*, 2010, **35**, 2610–2628; (d) M. E. Boot-Handford, J. C. Abanades, E. J. Anthony, M. J. Blunt, S. Brandani, N. MacDowell, J. R. Fernández, M. C. Ferrari, R. Gross, J. P. Hallett, R. S. Haszeldine, P. Heptonstall, A. Lyngfelt, Z. Makuch, E. Mangano, R. T. J. Porter, M. Pourkashanian, G. T. Rochelle, N. Shah, J. G. Yao and P. S. Fennell, *Energy Environ. Sci.*, 2014, **7**, 130–189.

10. (a) M. Mikkelsen, M. Jørgensen and F. C. Krebs, *Energy Environ. Sci.*, 2010, **3**, 43–81; (b) Y. Patil, P. Tambade, S. Jagtap and B. Bhanage, *Front. Chem. Eng. China*, 2010, **4**, 213–235.

11. (a) D. J. Darensbourg, *Chem. Rev.*, 2007, **107**, 2388–2410; (b) D. J. Darensbourg and S. J. Wilson, *Green Chem.*, 2012, **14**, 2665–2671.

12. (a) G. W. Crabtree, M. S. Dresselhaus and M. V. Buchanan, *Physics Today*, 2004, **57**, 39–44; (b) A. Züttel, A. Remhof, A. Borgschulte and O. Friedrichs, *Phil. Trans. R. Soc. A*, 2010, **368**, 3329–3342.

13. (a) A. J. Bard and M. A. Fox, *Acc. Chem. Res.*, 1995, **28**, 141–145; (b) P. V. Kamat, *J. Phys. Chem. C*, 2007, **111**, 2834–2860; (c) D. G. Nocera, *Energy Environ. Sci.*, 2010, **3**, 993–995; (d) T. A. Faunce, W. Lubitz, A. W. Rutherford, D. MacFarlane, G. F. Moore, P. Yang, D. G. Nocera, T. A. Moore, D. H. Gregory, S. Fukuzumi, K. B. Yoon, F. A. Armstrong, M. R. Wasielewski and S. Styring, *Energy Environ. Sci.*, 2013, **6**, 695–698.

14. T. S. Zhao (ed.), *Micro Fuel Cells: Principles and Applications*, Elsevier, Amsterdam, 2009.

15. (a) J. A. Turner, *Science*, 2004, **305**, 972–974; (b) J. D. Holladay, J. Hu, D. L. King and Y. Wang, *Catal. Today*, 2009, **139**, 244–260.

16. (a) W. R. Grove, On a new voltaic combination', The London and Edinburgh Philosophical Magazine and Journal of Science, 1838; (b) C. R. Alder Wright and C. Thompson, Note on the development of voltaic electricity by atmospheric oxidation of combustible gases and other substances, *Proc. R. Soc. London*, 1889, **46**, 372–376; (c) B. C. H. Steele and A. Heinzel, *Nature*, 2001, **414**, 345–352; (d) D. Stolten and B. Emonts (ed.), *Fuel Cell Science and Engineering: Materials, Processes, Systems and Technology*, Wiley-VCH, Weinheim, Germany, 2012.

17. (a) R. M. Navarro, M. A. Peña and J. L. G. Fierro, *Chem. Rev.*, 2007, **107**, 3952–3991; (b) K. Liu, C. Song and V. Subramani (ed.), *Hydrogen and Syngas Production and Purification Technologies*, Wiley-VCH, Weinheim, Germany, 2010.

18. L. A. Pellegrini, G. Soave, S. Gamba and S. Lange, *Appl. Energy*, 2011, **88**, 4891–4897.

19. X. Hao, G. Dong, Y. Yang, Y. Xu and Y. Li, *Chem. Eng. Tech.*, 2007, **30**, 1157–1165.

20. (a) J. Barber, *Chem. Soc. Rev.*, 2009, **38**, 185–196; (b) C. Herrero, A. Quaranta, W. Leibl, A. W. Rutherford and A. Aukauloo, *Energy Environ. Sci.*, 2011, **4**, 2353–2365; (c) X. Chen, C. Li, M. Grätzel, R. Kostecki and S. S. Mao, *Chem. Soc. Rev.*, 2012, **41**, 7909–7937.

21. (a) M. Ni, M. K. H. Leung, D. Y. C. Leung and K. Sumathy, *Renew. Sustain Energy Rev.*, 2007, **11**, 401–425; (b) A. Kudo and Y. Miseki, *Chem. Soc. Rev.*, 2009, **38**, 253–278; (c) P. G. Loutzenhiser, A. Meier, D. Gstoehl and A. Steinfeld, *Advances in CO_2 Conversion and Utilization*, ACS Symposium Series, Volume 1056, American Chemical Society, Washington DC, 2010, ch. 3, pp. 25–30; (d) M. Romero and A. Steinfeld, *Energy Environ. Sci.*, 2012, **5**, 9234–9245; (c) J. R. McKone, N. S. Lewis and H. B. Gray, *Chem. Mater.*, 2014, **26**, 407–414; (f) T. Hisatomi, J. Kubota and K. Domen, *Chem. Soc. Rev.*, 2014, DOI: 10.1039/C3CS60378D.

22. (a) Energinet.dk, 'Download of market data', http://www. energinet.dk/EN/El/Engrosmarked/Udtraek-af-markedsdata/ Sider/default.aspx; (b) Danish Energy Agency, *Energy Strategy 2050 – From Coal, Oil and Gas to Green Energy*, Danish Ministry of Climate and Energy, Copenhagen, 2011.

23. (a) A. J. Turner, *Science*, 2004, **305**(5686), 972–974; (b) B. E. Logan, *Environ. Sci. Tech.*, 2004, **38**, 160A–167A; (c) J. Turner, G. Sverdrup, M. K. Mann, P. C. Maness, B. Kroposki, M. Ghiradi, R. J. Evans and D. Blake, *Int. J. Energy Res.*, 2008, **32**,

379–407; (d) R. M. Navarro, M. C. Sanchez-Sanchez, M. C. Alvarez-Galvan, F. del Valle and J. L. G. Fierro, *Energy Environ. Sci.*, 2009, **2**, 35–54.

24. (a) G. W. Huber, S. Iborra and A. Corma, *Chem. Rev.*, 2006, **106**, 4044–4098; (b) H. de Lasa, E. Salaices, J. Mazumder and R. Lucky, *Chem. Rev.*, 2011, **111**, 5404–5433; (c) M. Besson, P. Gallezot and C. Pinel, *Chem. Rev.*, 2014, **114**, 1827–1870.

25. (a) M. Naqvi, J. Yan and E. Dalhquist, *Bioresour. Technol.*, 2010, **101**, 8001–8015; (b) T. Hammar, *Forest Chem. Rev.*, 2012, **122**, 12–13.

26. G. Centi and S. Perathoner, *Greenhouse Gas Sci. Technol*, 2011, **1**, 21–35.

27. (a) N. L. Rosi, J. Eckert, M. Eddaoudi, D. T. Vodak, J. Kim, M. O'Keeffe and O. M. Yaghi, *Science*, 2003, **300**, 1127–1129; (b) U. Mueller, M. Schubert, F. Teich, H. Puetter, K. Schierle-Arndt and J. Pastré, *J. Mat. Chem.*, 2006, **16**, 626–636.

28. (a) M. P. Suh, H. J. Park, T. K. Prasad and D. W. Lim, *Chem. Rev.*, 2012, **112**, 782–835; (b) R. B. Getman, Y. S. Bae, C. E. Wilmer and R. Q. Snurr, *Chem. Rev.*, 2012, **112**, 703–723; (c) J. Sculley, D. Yuan and H.-C. Zhou, *Energy. Env. Sci.*, 2011, **4**, 2721–2735; (d) J. R. Li, J. Sculley and H. C. Zhou, *Chem. Rev.*, 2012, **112**, 869–932; (e) J. L. C. Rowsell and O. M. Yaghi, *Microporous Mesoporus Mater.*, 2004, **73**, 3–14; (f) A. F. Dalebrook, W. Gan, M. Grasemann, S. Moret and G. Laurenczy, *Chem. Commun.*, 2013, **49**, 8735–8751.

29. (a) R. J. Pearson, J. W. G. Turner, and A. J. Peck, presented at Low Carbon Vehicles 2009, Institution of Mechanical Engineers, London, 20–21 May 2009, IMechE Conference C678; (b) Z. Jiang, T. Xiao, V. L. Kuznetsove and P. P. Edwards, *Phil. Trans. R. Soc. A*, 2010, **368**, 3343–3364.

30. (a) A. Klerke, C. H. Christensen, J. K Nørskov and T. Vegge, *J. Mater. Chem.*, 2008, **18**, 2304–2310; (b) F. Schüth, R. Palkovits, R. Schögl and D. S. Su, *Energy Environ. Sci.*, 2012, **5**, 6278–6289.

31. F. Krause, H. Bossel and K.-F. Müller-Reißmann, *Energiewende – Wachstum und Wohlstand ohne Erdöl und Uran* [Energy Transition – Growth and Prospertiy without Petroleum and Uranium], S. Fischer Verlag, Frankfurt am Main, Germany, 1980.

32. (a) Deutsches Zentrum für Luft- und Raumfahrt, Fraunhofer Institut für Windenergie und Energiesystemtechnik, Ingenieurbüro für neue Energien: Langfristszenarien und Strategien für den Ausbau der erneuerbaren Energien in Deutschland bei Berücksichtigung der Entwicklung in Europa und global, im Auftrag des BMU, Schlussbericht, 29. März 2012; (b) *Renewable*

Energy Sources in Figures – National and International Development, ed. D. Böhme, W. Dürrschmidt and M. van Mark, Federal Ministry for the Environment, Nature Conservation and Nuclear Safety (BMU), Berlin, Germany, 2012.

33. R. Benndorf, M. Bernicke, A. Bertram, W. Butz, F. Dettling, J. Drotleff, C. Elsner, E. Fee, C. Gabler, C. Galander, Y. Hargita, R. Herbener, T. Hermann, F. Jäger, J. Kanthak, H. Kessler, Y. Koch, D. Kuntze, M. Lambrecht, C. Lehmann, H. Lehmann, S. Leuthold, I. Lütkehus, K. Martens, F. Müller, K. Müschen, D. Nissler, S. Plickert, K. Purr, A. Reichart, J. Reichel, H. Salecker, J. Schuberth, D. Schulz, U. Strenge, M. Sieck, B. Westermann, K. Werner, C. Winde, D. Wunderlich and B. Zietlow, *Germany 2050 – A Greenhouse Gas-Neutral Country*, ed. K. Werner, D. Nissler and K. Purr, Federal Environment Agency, Dessau-Roßlau, Germany, 2013.

34. Q. Schiermeier, *Nature*, 2013, **496**, 156–158.

35. Danish Energy Agency, *Energy Strategy 2050 – From Coal, Oil and Gas to Green Energy*, Danish Ministry of Climate and Energy, Copenhagen, 2011.

36. D. W. Keith, J. F. DeCarolis, D. C. Denkenberger, D. H. Lenschow, S. L. Malyshev, S. Pacala and P. J. Rasch, *Proc. Natl. Acad. U. S. A.*, 2004, **101**, 16115–16120.

37. C. Graves, S. D. Ebbesen, M. Mogensen and K. S. Lackner, *Renew. Sustain. Energy Rev.*, 2011, **15**, 1–23.

38. (a) G. A. Olah, *Angew. Chem., Int. Ed.*, 2005, **44**, 2636–2639; (b) G. A. Olah, A. Goeppert and G. K. S. Prakash, *Beyond Oil and Gas: The Methanol Economy*, Wiley-VCH, Weinheim, Germany, 2nd edn, 2009.

39. C. Arcoumanis, C. Bae, R. Crookes and E. Kinoshita, *Fuel*, 2008, **87**, 1014–1030.

40. (a) D. L. King and J. H. Grate, *ChemTech*, 1985, 244–251; (b) W. Keim, *Pure App. Chem.*, 1986, **58**, 825–832; (c) M. Bertau, F. X. Effenberger, W. Keim, G. Menges and H. Offermanns, *Chemie-Ingenieur-Technik*, 2010, **82**, 2055–2058.

41. (a) D. Chen, K. Moljord and A. Holmen, *Microporous Mesoporous Mater.*, 2012, **164**, 239–250; (b) U. Olsbye, S. Svelle, M. Bjrgen, P. Beato, T. V. W. Janssens, F. Joensen, S. Bordiga and K. P. Lillerud, *Angew. Chem., Int. Ed.*, 2012, **51**, 5810–5831.

42. (a) M. Aresta and A. Dibenedetto, *Dalton Trans.*, 2007, **28**, 2975–2992; (b) M. Aresta, *Carbon Dioxide as Chemical Feedstock*, Wiley-VCH, Weinheim, Germany, 2010; (c) M. Peters, B. Köhler, W. Kuckshinrichs, W. Leitner, P. Markewitz and T. E. Müller, *ChemSusChem*, 2011, **4**, 1216–1240; (d) P. Markewitz,

W. Kuckshinrichs, W. Leitner, J. Linssen, P. Zapp, R. Bongartz, A. Schreiber and T. E. Müller, *Energy Environ. Sci.*, 2012, **5**, 7281–7305.

43. (a) G. A. Olah, G. K. S. Prakash and A. Goeppert, *J. Am. Chem. Soc.*, 2011, **133**, 12881–12898; (b) G. A. Olah, *Angew. Chem., Int. Ed.*, 2013, **52**, 104–107.

44. K. M. K. Yu, W. Tong, A. West, K. Cheung, T. Li, G. Smith, Y. Guo and S. C. E. Tang, *Nat. Commun.*, 2012, **3**, 1230.

45. S. Wasmus and A. Kuver, *J. Electroanal. Chem.*, 1999, **461**, 14–31.

46. K. Scott, W. M. Taama and P. Argyropoulos, *J. Membr. Sci.*, 2000, **171**, 119–130.

47. (a) IHS Chemical, Methanol Report, August 2013; (b) Methanol Investor Presenation, Methanex®, August 2013, http://www. methanex.com/investor/documents/MXPresentation-Aug2013_ 000.pdf.

48. I. Ridjan, B. Vad Mathiesen, D. Connolly and N. Duic, *Energy*, 2013, **57**, 76–84.

49. (a) S. Choi, J. H. Drese and C. W. Jones, *ChemSusChem*, 2009, **2**, 796–854; (b) S. Ma and H. C. Zhou, *Chem. Commun.*, 2010, **46**, 44–53; (c) N. Hedin, L. Chen and A. Laaksonen, *Nanoscale*, 2010, **2**, 1819–1841; (d) A. J. Hunt, E. H. K. Sin, R. Marriott and J. H. Clark, *ChemSusChem*, 2010, **3**, 306–322; (e) D. M. D'Alessandro, B. Smit and J. R. Long, *Angew. Chem., Int. Ed.*, 2010, **49**, 6058–6082.

50. K. L. Ng, D. Chadwick and B. A. Toseland, *Chem. Eng. Sci.*, 1999, **54**, 3897–3592.

51. S. A. French, A. A. Sokol, S. T. Bromley, C. R. A. Catlow and P. Sherwood, *Top. Catal.*, 2003, **24**, 161–172.

52. (a) M. Saito and K. Murata, *Catal. Surv. Asia*, 2004, **8**, 285–294; (b) J. Strunk, K. Kaehler, X. Xia and M. Muhler, *Surf. Sci.*, 2009, **603**, 1776–1783.

53. (a) X.-M. Liu, G. Q. Lu, Z.-F. Yan and J. Beltramini, *Ind. Eng. Chem. Res.*, 2003, **42**, 6518–6530; (b) R. Raudaskoski, E. Turpeinen, R. Lenkkeri, E. Pongracz and R. L. Keiski, *Catal. Today*, 2009, **144**, 318–323.

54. (a) O.-S. Joo, K.-D. Jun and J. Yonsoo, *Stud. Surf. Sci. Catal.*, 2004, **153**, 67–72; (b) M.-J. Choi and D.-H. Cho, *Clean: Soil, Air, Water*, 2008, **36**, 426–432; (c) M. Kazemimoghadam and T. Mohammadi, Desalination, 2010, **55**, 196–200.

55. (a) I. Kasatkin, P. Kurr, B. Kniep, A. Trunschke and R. Schlögl, *Angew. Chem., Int. Ed.*, 2007, **46**, 7324–7327; (b) P. Kurr, I. Kasatkin, F. Girgsdies, A. Trunschke, R. Schlögl and T. Ressler, *Appl. Catal., A*, 2008, **348**, 153–164.

56. (a) X.-L. Liang, X. Dong, G.-D. Lin and H.-B. Zhang, *Appl. Catal., B*, 2009, **88**, 315–322; (b) Q. Zhang, Y.-Z. Zuo, M.-H. Han, J.-F. Wang, Y. Jin and F. Wei, *Catal. Today*, 2010, **150**, 55–60.

57. (a) X. M. Lui, G. Q. Lu, Z. F. Yan and J. Beltramini, *Ind. Eng. Chem. Res.*, 2003, **42**, 6518–6530; (b) K. C. Waugh, *Catal. Letters*, 2012, **142**, 1153–1166.

58. M. Behrens, *J. Catal.*, 2009, **267**, 24–29.

59. (a) K. Klier, *Adv. Catal.*, 1982, **31**, 243; (b) G. C. Clinchen, P. J. Denny, D. G. Parker, G. D. Short, M. S. Spencer, K. C. Waugh and D. A. Wan, *Prepr. Am. Chem. Soc. Div. Fuel Chem.*, 1984, **29**, 178; (c) P. L. Hansen, J. B. Wagner, S. Helveg, J. R. Rostrup-Nielsen, B. S. Clausen and H. Topsøe, *Science*, 2002, **295**, 2053–2055.

60. (a) S. Zander, E. L. Kunkes, M. E. Schuster, J. Schumann, G. Weinberg, D. Teschner, N. Jacobsen, R. Schlögl and M. Behrens, *Angew. Chem., Int. Ed.*, 2013, **52**, 6536–6540; (b) M. Behrens, S. Zander, P. Kurr, N. Jacobsen, J. Senker, G. Koch, T. Ressler, R. W. Fischer and R. Schlögl, *J. Am. Chem. Soc.*, 2013, **135**, 6061–6068.

61. M. Behrens, S. Kißner, F. Girsgdies, I. Kasatkin, F. Hermerschmidt, K. Mette, H. Ruland, M. Muhler and R. Schlögl, *Chem. Commun.*, 2011, **47**, 1701–1703.

62. B. Berns, M. Schur, A. Dassenoy, H. Junkes, D. Herein and R. Schlögl, *Chem. Eur. J.*, 2003, **9**, 2039–2052.

63. M. Behrens, F. Studt, I. Kasatkin, S. Kühl, M. Hävecker, F. Abild-Pedersen, S. Zander, F. Girgsdies, P. Kurr, B. L. Kniep, M. Tovar, R. W. Fischer, J. K Nørskov and R. Schlögl, *Science*, 2012, **336**, 893–897.

64. R. A. Koppel, C. Stocker and A. Baiker, *J. Catal.*, 1998, **179**, 515–527.

65. (a) F. Yaripour, F. Baghaei, I. Schmidt and J. Perregaard, *Catal. Commun.*, 2005, **6**, 542–549; (b) A. T. Aguayo, J. Erena, D. Mier, J. M. Arandes, M. Olazar and J. Bilbao, *Ind. Eng. Chem. Res.*, 2007, **46**, 5522–5530.

66. (a) T. Wang, J. Wang and Y. Jin, *Ind. Eng. Chem. Res.*, 2007, **46**, 5824–5847; (b) T. A. Semelsberger, K. C. Ott, R. L. Borup and H. L. Greene, *Appl. Catal., A*, 2006, **309**, 210–223; (c) M. Nilsson, L. J. Pettersson and B. Lindstrom, *Energy Fuels*, 2006, **20**, 2164–2169.

67. F. Pontzen, W. Liebner, V. Gronemann, M. Rothaemel and B. Ahlers, *Catal. Today*, 2011, **171**, 242.

68. K. P. Sun, W. W. Lu, M. Wang and X. L. Xu, *Catal. Commun.*, 2004, **5**, 367–370.

69. G. X. Qi, J. H. Fei, X. M. Zheng and Z. Y. Hou, *Catal. Lett.*, 2001, **72**, 121–124.

70. X. An, Y. Z. Zuo, Q. Zhang, D. Z. Wang and J. F. Wang, *Ind. Eng. Chem. Res.*, 2008, **47**, 6547–6554.

71. S. Wang, D. S. Mao, X. M. Guo, G. S. Wu and G. Z. Lu, *Catal. Commun.*, 2009, **10**, 1367–1370.

72. J. L. Tao, K. W. Jun and K. W. Lee, *Appl. Organomet. Chem.*, 2001, **15**, 105–108.

73. J. Erena, R. Garona, J. M. Arandes, A. T. Aguayo and J. Bilbao, *Catal. Today*, 2005, **107–108**, 467–473.

74. T. Takeguchi, K. Yanagisawa, T. Inui and M. Inoue, *Appl. Catal., A*, 2000, **192**, 201–209.

75. M. Stöcker, *Microporous Mesoporous Mater.*, 1999, **29**, 3–48.

76. K. W. Jun, K. S. R. Rao, M. H. Jung and K. W. Lee, *Bull. Korean Chem. Soc.*, 1998, **19**, 466–470.

77. Stanford Research Institute, *Chemical Economics Handbook*, SRI International, Menlo Park, CA, 2010.

78. (a) C. D. Chang, *Catal. Rev. Sci. Eng.*, 1983, **25**, 1–118; (b) F. J Keil, *Microporous Mesoporous Mater.*, 1999, **29**, 49–66.

79. A. Y. Kam, M. Schreiner and S. Yurchak, in *Handbook of Synfuels Technology*, ed. R. A Meyers, McGraw-Hill, New York, 1984, pp. 2.75–2.111.

80. (a) H. H. Storch, N. Golumbic and R. B. Anderson, *The Fischer-Tropsch and Related Syntheses*, Wiley, New York, 1951; (b) P. H. Emmett(ed.), *Catalysis Vol. IV. Hydrocarbon synthesis, Hydrogenation and Cyclization*, Reinhold, New York, 1956; (c) R. B. Anderson, *The Fischer-Tropsch Synthesis*, Academic Press, Orlando, FL, 1984; (d) A. P. Steynberg and M. E. Dry (ed.), Fischer-Tropsch Technology (*Stud. Surf. Sci. Catal.* 152), Elsevier, Amsterdam, 2004; (e) E. B. Fox, Z. W. Liu and Z. T. Liu, *Energy Fuels*, 2013, **27**, 6335–6338; (f) A. Taheri Najafabadi, *Int. J. Energy Res.*, 2013, **37**, 485–499; (g) C. Higman and S. Tam, *Chem. Rev.*, 2014, **114**, 1673–1708.

81. J. F. Haw, W. Song, D. M. Marcus and J. B. Nicholas, *Acc. Chem. Res.*, 2003, **36**, 317.

82. (a) D. Liederman, S. M. Jacob, S. E. Voltz and J. J. Wise, *Ind. Eng. Chem. Process Des. Dev.*, 1978, **17**, 340; (b) S. Yurchak, S. E. Voltz and J. P. Warner, *Ind. Eng. Chem. Process Des. Dev.*, 1979, **18**, 527; (c) F. J. Kiel, *Microporous Mesoporous Mater.*, 1999, **29**, 49–66.

83. (a) G. Seo, J. H. Kim and H. G. Jang, *Catalysis Surveys from Asia*, 2013, **17**, 103–118; (b) K. Hemelsoet, J. Van Der Mynsbrugge, K. De Wispelaere, M. Waroquier and V. Van Speybroeck, *ChemPhysChem*, 2013, **14**, 1526–1545.

84. (a) J. Q. Chen, A. Bozzano, B. Glover, T. Fugleerud and S. Kvisle, *Catal. Today*, 2005, **106**, 103–107; (b) S. Illias and A. Bhan, *ACS Catalysis*, 2013, **3**, 18–31.

85. (a) M. Stocker, J. Weitkamp (special issue editors), Microporous Mesoporous Mater., 1999, **29**, 1–2; (b) UOP, Expanding routes for olefins production, http://www.uop.com/processing-solutions/petrochemicals/olefins/; (c) Air Liquide, Methanol and derivatives: Technology got methanol and derivatives, http://www.engineering-solutions.airliquide.com/en/about-us-global-engineering-and-construction-solutions/technologies-air-liquide-lurgi-cryo-zimmer/methanol-and-derivatives.html.

86. (a) A. Goeppert, M. Czaun, G. K. S. Prakash and G. A. Olah, *Energy Environ. Sci.*, 2012, **5**, 7833–7853; (b) G. H. Rau and K. S. Lackner, *Science*, 2013, **340**, 1522–1523.

87. G. Holmes and D. W. Keith, *Phil. Trans. R. Soc., A*, 2012, **370**, 4380–4403.

88. Energy Information Administration, *Annual Energy Outlook 2005 with Projections to 2025*, Energy Information Administration, Washington, DC, 2005.

89. (a) H. J. Herzog, *Environ. Sci. Tech.*, 2011, **35**, 148A–153A; (b) M. M. Hossain and H. I. de Lasa, *Chem. Eng. Sci.*, 2008, **63**, 4433–4451; (c) M. Z. Jacobson, *Energy Environ. Sci.*, 2009, **2**, 148–173; (d) Y. S. Bae and R. Q. Snurr, *Angew. Chem., Int. Ed.*, 2011, **50**, 11586–11596; (e) K. S. Lackner, S. Brennan, J. M. Matter, A. H. A. Park, A. Wright and B. Van Der Zwaan, *Proc. Natl. Acad. U. S. A.*, 2012, **109**, 13156–13162.

90. (a) J. L. C. Roswell and O. M. Yaghi, *Microporous Mesoporous Mater.*, 2004, **73**, 3–14; (b) Y. Liu, Z. U. Wang and H. C. Zhou, *Greenhouse Gas Sci Technol.*, 2012, **2**, 239–259; (c) H. Furukawa, K. E. Cordova, M. O'Keeffe and O. M. Yaghi, *Science*, 2013, **341**, 1230444.

91. (a) R. Steeneveldt, B. Berger and T. A. Torp, *Chem. Eng. Res. Design*, 2006, **84**, 739–763; (b) T. C. Merkel, H. Lin, X. Wei and R. Baker, *J. Membr. Sci.*, 2010, **359**, 126–139.

92. N. MacDowell, N. Florin, A. Buchard, J. Hallett, A. Galindo, G. Jackson, C. S. Adjiman, C. K. Williams, N. Shah and P. Fennell, *Energy Environ. Sci.*, 2010, **3**, 1645–1669.

93. Lurgi, *The Rectisol® Process: Lurgi's Leading Technology for Purification and Conditioning of Synthesis Gas*, Lurgi GmbH, Frankfurt am Main, Germany, 2008.

94. (a) M. Ni, M. K. H. Leung, D. Y. C. Leung and K. Sumathy, *Renew. Sustain Energy Rev.*, 2007, **11**, 401–425; (b) S. Linic,

P. Christopher and D. B. Ingram, *Nat. Mater.*, 2011, **10**, 911–921; (c) K. Maeda, *J. Photochem. Photobiol. C: Photochem. Rev.*, 2011, **12**, 237–268; (d) A. Kubacka, M. Fernández-García and G. Colón, *Chem. Rev.*, 2012, **112**, 1555–1614; (e) P. Du and R. Eisenberg, *Energy Environ. Sci.*, 2012, **5**, 6012–6021; (f) P. D. Tran, L. H. Wong, J. Barber and J. S. C. Loo, *Energy Environ. Sci.*, 2013, **5**, 5902–5918.

95. (a) V. Roan, D. Betts, A. Twining, K. Dinh, P. Wassink and T. Simmons, *An Investigation of the Feasibility of Coal-based Methanol for Application in Transportation Fuel Cell Systems*, Report, University of Florida, Gainesvill, FL, 2004; (b) R. H. Williams, E. D. Larson, R. E. Katofsky and J. Chen, *Energy Sustainable Dev.*, 1995, **1**, 18–34; (c) National Energy Technology Laboratory, Commercial-scale Demonstration of the Liquid Phase Methanol (LPMEOH) Process, Report No. DOE/NETL-2004/1199, Office of Fossil Energy, US Department of Energy, Washington DC, 2003; (d) M. Specht and A. Bandi, *The Methanol Cycle – Sustainable Supply of Liquid Fuels*, Centre for Solar Energy and Hydrogen Research (ZSW), Stuttgart, Germany, 1999; (e) M. Specht, A. Bandi, F. Baumgart, C. N. Murray and J. Gretz, in *Greenhouse Gas Control Technologies: Proceedings of the 4th International Conference*, ed. B. Eliasson, P. W. F. Riemer and A. Wokan, Pergamon, Amsterdam, 1998, pp. 723–728; (f) M. Specht, A. Bandi, M. Elser and F. Staiss, in *Advances in Chemical Conversions for Mitigating Carbon Dioxide*, ed. T. Inui, M. Anpo, K. Izui, S. Yanagida and T. Yamaguchi, Elsevier, Amsterdam, 1998, pp. 363–367.

96. Methanex, Methanol price, Methanex, Vancouver, Canada, http://www.methanex.com/products/methanolprice.html.

97. W. Kreuter and H. Hofmann, *Int. J. Hydrogen Energy*, 1998, **23**, 661–666.

98. (a) G. Centi, S. Perathoner, G. Wine and M. Gangeri, *Green Chem.*, 2007, **9**, 671–678; (b) Y. Chen, C. W. Li and M. W. Kanan, *J. Am. Chem. Soc.*, 2012, **134**, 19969–19972; (c) J. L. Di Meglio and J. Rosenthal, *J. Am. Chem. Soc.*, 2013, **135**, 8798–8801; (d) J. Qiao, Y. Liu, F. Hong and J. Zhang, *Chem. Soc. Rev.*, 2014, **43**, 631–675.

99. M. A. Laguna-Bercero, *J. Power Sources*, 2012, **203**, 4–16.

100. (a) M. Mogensen, S. H. Jensen, A. Hauch, I. Chorkendorff and T. Jacobsen, Performance of reversible solid oxide cells: a review, presented at 7th European SOFC Forum, Lucerne, Switzerland, 2006, paper P0301; (b) S. H. Jensen, P. H. Larsen and M. Mogensen, *Int. J. Hydrogen Energy*, 2007, **32**, 3253–3257; (c) C. M.

Stoots, J. E. O'Brien, J. S. Herring and J. J. Hartvigsen, *J. Fuel Cell Sci. Technol.*, 2009, **6**, 011014–011017;(d) C. Stoots, J. O'Brien and J. Hartvigsen, *Int. J. Hydrogen Energy*, 2009, **34**, 4208–4215; (e) Z. Zhan, W. Kobsiriphat, J. R. Wilson, M. Pillai, I. Kim and S. A. Barnett, *Energy Fuels*, 2009, **23**, 3089–3096.

101. Q. Fu, C. Mabilat, M. Zahid, A. Brisse and L. Gautier, *Energy Environ. Sci.*, 2010, **3**, 1382–1397.

102. J. E. O'Brien, M. G. McKellar, E. A. Harvego and C. M. Stoots, *Int. J. Hydrogen Energy*, 2010, **35**, 4808–4819.

103. (a) C. Graves, S. Ebbesen, M. Mogensen and K. S. Lackner, *Renew. Sustain. Energy Rev.*, 2011, **15**, 1–23; (b) Coherent Energy and Environmental System Analysis, *CEESA 100% Renewable Energy Scenarios towards 2050*. Aalborg University, Aalborg, Denmark, 2011.

104. A. Hauch, S. D. Ebbesen, S. Højgaard Jensen and M. Mogensen, *J. Mater. Chem.*, 2008, **18**, 2331–2340.

105. N. Z. Muradova and T. N. Veziroglu, *Int. J. Hydrogen Energy*, 2008, **33**, 6804–6839.

106. S. Lynum, R. Hildrum, K. Hox and J. Huglahl, Kværner-based technologies for environmentally friendly energy and hydrogen production, presented at the 12th World Hydrogen Energy Conference, Buenos Aires, 1998.

107. L. Fulcheri and Y. Schwob, *Int. J. Hydrogen Energy*, 1995, **20**, 197–202.

108. (a) M. Callahan, in *Proceedings of the 26th Power Sources Symposium*, PSC Publishers, Red Bank, NJ, 1974, p. 181; (b) M. Poirier and C. Sapundzhiev, *Int. J. Hydrogen Energy*, 1998, **22**, 429–433.

109. M. Steinberg, *Int. J. Hydrogen Energy*, 1999, **24**, 771–777.

110. S. Rodat, S. Abanades and G. Flamant, *Sol. Energy*, 2011, **85**, 645–652.

111. J. K. Dahl, K. J. Buechler, A. W. Weimer, A. Lewandowski and C. Bingham, *Int. J. Hydrogen Energy*, 2004, **29**, 725–736.

112. A. Abánades, C. Rubbia and D. Salmieri, *Int. J. Hydrogen Energy*, 2013, **38**, 8491–8496.

113. (a) Statuskonferenz zur BMBF-Fördermaßnahme 'Technologien für Nachhaltigkeit und Klimaschutz - Chemische Prozesse und stoffliche Nutzung von CO2, 9–10 April 2013, Berlin, Germany, http://www.chemieundco2.de/de/317.php; (b) F. Sietz, *Rohstroffwandel bei BASF. Stoffliche Nutzung von CO_2*, 2013, http://www.chemieundco2.de/_media/Seitz_Herausforderung_Rohstoffwandel_bei_BASF.pdf.

114. (a) *Proceedings of the 2000 DOE Hydrogen Programme Review*, NREL, 9–11 May, 2000, US Department of Energy, San Ramon, CA; (b) T. Pregger, D. Graf, W. Krewitt, C. Sattler, M. Roeb and S. Moller, *Int. J. Hydrogen Energy*, 2009, **34**, 4256–4267.

115. H. D. Gesser, N. R. Hunter and C. B. Prakash, *Chem. Rev.*, 1985, **85**(4), 235–244.

116. (a) A. E. Shilov and G. B. Shul'pin, *Chem. Rev.*, 1997, **97**, 2879–2932; (b) A. E. Shilov and G. B. Shul'pin, *Activation and Catalytic Reactions of Saturated Hydrocarbons in the Presence of Metal Complexes*, Kluwer, Dordrecht, 2000; (c) J. A. Labinger and J. E. Bercaw, *Nature*, 2002, **417**, 507–514; (d) A. Caballero and P. J. Perez, *Chem. Soc. Rev.*, 2013, **42**, 8809–8800.

117. (a) A. Sen, M. A. Benvenuto, M. Lin, A. C. Hutson and N. Basickes, *J. Am. Chem. Soc.*, 1994, **116**, 998–1003; (b) R. A. Periana, D. J. Taube, S. Gamble, H. Taube, T. Satoh and H. Fuji, *Science*, 1998, **280**, 560–564; (c) D. Wolf, *Angew. Chem., Int. Ed.*, 1998, **37**, 3351–3353; (d) B. L. Conley, W. J. Tenn, K. J. H. Young, S. Ganesh, S. Meier, V. Ziatdinov, O. Mironov, J. Oxgaard, J. Gonzales, W. A. Goddard and R. A. Periana, *Methane Functionalization*, 2006, Wiley-VCH, Weinheim, Germany; (e) I. H. Hristov and T. Ziegler, *Organometallics*, 2003, **22**, 1668–1674.

118. C. J. Jones, D. Taube and R. A. Periana, *Angew. Chem., Int. Ed.*, 2004, **43**, 4626–4629.

119. H. Arakawa, M. Aresta, J. N. Armor, M. A. Barteau, E. J. Beckman, A. T. Bell, J. E. Bercaw, C. Creutz, E. Dinjus, D. A. Dixon, K. Domen, D. L. DuBois, J. Eckert, E. Fujita, D. H. Gibson, W. A. Goddard, D. W. Goodman, J. Keller, G. J. Kubas, H. H. Kung, J. E. Lyons, L. E. Manzer, T. J. Marks, K. Morokuma, K. M. Nicholas, R. Periana, L. Que, J. Rostrup-Nielson, W. M. H. Sachtler, L. D. Schmidt, A. Sen, G. A. Somorjai, P. C. Stair, B. R. Stults and W. Tumas, *Chem. Rev.*, 2001, **101**, 953–996.

120. (a) M. Baerns and J. R. H. Ross, in *Perspectives in Catalysis*, 1992, ed. J. A. Thomas and K. I. Zamaraev, Blackwell, Oxford, UK, 1992; (b) O. Forlani and S. Rossini, *Mater. Chem. Phys.*, 1992, **3**, 155–158.

121. R. A. Periana, D. J. Taube, S. Gamble, H. Taube, T. Satoh and H. Fujii, *Science*, 1993, **259**, 340–343.

122. R. Raja and P. Ratnasamy, *Appl. Catal. A*, 1997, **158**, L7–L15.

123. R. Palkovits, M. Antonietti, P. Kuhn, A. Thomas and F. Schüth, *Angew. Chem. Int. Ed.*, 2009, **48**, 6909–6912.

124. P. Kuhn, M. Antonietti and A. Thomas, *Angew. Chem., Int. Ed.*, 2008, **47**, 3450–3453.

125. M. Soorholtz, R. J. White, T. Zimmermann, M. M. Titirici, M. Antonietti, R. Palkovits and F. Schüth, *Chem. Commun.*, 2013, **49**, 240–242.

126. (a) G. A. Olah, A. Goeppert and G. K. S. Prakash, *J. Org. Chem.*, 2009, **74**, 487–498; (b) G. A. Olah, G. K. S. Prakash and A. Goeppert, *J. Am. Chem. Soc.*, 2011, **133**, 12881–12298; (c) G. A. Olah, A. Goeppert, M. Czaun and G. K. S. Prakash, *J. Am. Chem. Soc.*, 2013, **135**, 648–650; (d) G. A. Olah, G. K. S. Prakash, A. Goeppert, M. Czaun and T. Mathew, *J. Am. Chem. Soc.*, 2013, **135**, 10030–10031.

127. (a) M. C. J. Bradford and M. A. Vannice, *Catal. Rev. Sci. Eng.*, 1999, **41**, 1–42; (b) M. S. Fan, A. Z. Abdullah and S. Bhatia, *ChemCatChem*, 2009, **1**, 192–208.

128. D. Teichmann, W. Arlt, P. Wasserscheid and R. Freymann, *Energy Environ. Sci.*, 2011, **4**, 2767–2773.

129. (a) D. Stolten (ed.), *Hydrogen and Fuel Cells*, Wiley-VCH, Weinheim, Germany, 2010; (b) N. Z. Muradov and T. N. Veziroglu, *Int. J. Hydrogen Energy*, 2008, **33**, 6804–6839.

130. (a) D. R. Palo, R. A. Dagle and J. D. Holladay, *Chem. Rev.*, 2007, **107**, 3992–4021; (b) R. D. Cortright, R. R. Davada and J. A. Dumesic, *Nature*, 2002, **418**, 964–967; (c) J. W. Shabaker, R. R. Davada, G. W. Huber, R. D. Cortright and J. A. Dumesic, *J. Catal.*, 2003, **215**, 344–352; (d) B. A. Peppley, J. C. Amphlett, L. M. Kearns and R. F. Mann, *Appl. Catal. A: Gen.*, 1999, **179**, 31–49.

131. (a) A. Dobson and S. D. Robinson, *Inorg. Chem.*, 1977, **16**, 137–142; (b) C. W. Jung and P. E. Garrou, *Organometallics*, 1982, **1**, 658–666; (c) D. Morton and D. J. Cole-Hamilton, *Chem. Commun.*, 1987, 248–249; (d) D. Morton, D. J. Cole-Hamilton, J. A. Schofield and R. J. Pryce, *Polyhedron*, 1987, **6**, 2187–2189.

132. D. Morton and D. J. Cole-Hamilton, *Chem. Commun.*, 1988, 1154–1156.

133. M. Neilsen, A. Kammer, D. Cozzula, H. Junge, S. Gladiali and M. Beller, *Angew. Chem., Int. Ed.*, 2011, **50**, 9593–9597.

134. M. Nielsen, E. Alberico, W. Baumann, H. J. Drexler, H. Junge, S. Gladiali and M. Beller, *Nature*, 2013, **495**, 85–89.

135. (a) J. Zhang, G. Leitus, Y. Ben-David and D. Milstein, *J. Am. Chem. Soc.*, 2005, **127**, 10840–10841; (b) D. Spasyuk, S. Smith and D. G. Gusev, *Angew. Chem., Int. Ed.*, 2012, **51**, 2772–2775.

136. R. E. Rodrigguez-Lugo, M. Trincado, M. Vogt, F. Tewes, G. Santiso-Quinones and H. Gruetzmacher, *Nat.Chem.*, 2013, **5**, 342–347.

137. (a) M. Nielsen, H. Junge, A. Kammer and M. Beller, *Angew. Chem., Int. Ed.*, 2012, **51**, 5711–5713; (b) D. Spasyuk and D. G. Gusev, *Organometallics*, 2012, **3**, 5239–5242.

138. (a) M. Grasemann and G. Laurenczy, *Energy Environ. Sci.*, 2012, **5**, 8171–8181; (b) Y. Himeda, *Green Chem.*, 2009, **11**, 2018–2022.

139. (a) S. Surampudi, S. R. Narayanan, E. Vamos, H. Frank, G. Halpert, A. LaConti, J. Kosek, G. K. Surya Prakash and G. A. Olah, *J. Power Sources*, 1994, **47**, 377–385; (b) S. Surampudi, S. R. Narayanan, E. Vamos, H. Frank, G. Halpert, G. K. Surya Prakash and G. A. Olah, *US Pat.*, 5,599,638, 1997; (c) S. Surampudi, S. R. Narayanan, E. Vamos, H. Frank, G. Halpert, G. K. Surya Prakash and G. A. Olah, *Eur. Pat.*, 0755 576 B1, 2008.

140. (a) N. W. Deluca and Y. A. Elabd, *J. Polym. Sci. B: Polym. Phys.*, 2006, **44**, 2201–2225; (b) V. Neburchilov, J. Martin, H. Wang and J. Zhang, *J. Power Sources*, 2007, **169**, 221–238.

141. (a) J. H. Choi, K. W. Park, B. K. Kwon and Y. E. Sung, *J. Electrochem. Soc.*, 2003, **150**, A973–A978; (b) E. S. Steigerwalt, G. A. Deluga, D. E. Cliffel and C. J. Lukehart, *J. Phys. Chem. B*, 2001, **10**, 8097–8101; (c) J. J. Wang, Y. T. Liu, I-L. Chen, Y. W. Yang, T. K. Yeh, C. H. Lee, C. C. Hu, T. C. Wen, T. Y. Chen and T. L. Lin, *J. Phys. Chem. C*, 2014, **118**(5), 2253–2262.

142. Y. L. Hsin, K. C. Hwang and C. T. Yeh, *J. Am. Chem. Soc.*, 2007, **129**(32), 9999–10010.

143. (a) J. Kua and W. A. Goddard III, *J. Am. Chem. Soc.*, 1999, **121**, 10928–10941; (b) Y. Lin, X. Cui, C. Yen and C. M. Wai, *J. Phys. Chem. B*, 2005, **109**, 14410–14415; (c) J. Chen, M. Wang, B. Liu, Z. Fan, K. Cui and Y. Kuang, *J. Phys. Chem. B*, 2006, **110**, 11775–11779; (d) A. Halder, S. Sharma, M. S. Hegde and N. Ravishankar, *J. Phys. Chem. C*, 2009, **113**, 1466–1473.

144. F. Ren, C. Wang, C. Zhai, F. Jiang, R. Yue, Y. Du, P. Yanga and J. Xu, *J. Mater. Chem. A*, 2013, **1**, 7255–7261.

145. S. A. Wohlgemuth, R. J. White, M. G. Willinger, M. M. Titirici and M. Antonietti, *Green Chem.*, 2012, **14**, 1515–1523.

146. S. Sharma and B. G. Pollet, *J. Power Sources*, 2012, **208**, 96–119.

147. Carbon Recycling International, http://www.carbonrecycling.is

148. Sunfire, http://www.sunfire.de

149. Air Fuel Synthesis, http://www.airfuelsynthesis.com/lrm

150. Ceramatec, http://www.ceramatec.com/

151. BioMCN, http://www.biomcn.eu/

152. (a) Chemrec, http://www.chemrec.se/; (b) BioDME, http://www.biodme.eu/

153. Volvo BioDME, *Unique Field Test in Commercial Operations, 2010–2012*, Volvo Truck Corporation, Gothenburg, Sweden, 2009, http://www.volvotrucks.com/SiteCollectionDocuments/VTC/Corporate/News%20and%20Media/publications/Volvo%20BioDME.pdf

154. B. Metz, O. Davidson, H. de Coninck, M. Loos and L. Meyer, *Carbon Dioxide Capture and Storage*, IPCC Special Reports, Cambridge University Press, Cambridge, 2006.

155. E. A. Quadrelli, G. Centi, J. L. Duplan and S. Perathoner, *ChemSusChem*, 2011, **4**, 1194–1215.

156. V. M. Ehlinger, K. J. Gabriel, M. M. B. Noureldin and M. El-Halwagi, *ACS Sustainable Chem. Eng.*, 2014, **2**(1), 30–37.

157. (a) R. Borup, J. Meyers, B. Pivovar, Y. S. Kim, R. Mukundan, N. Garland, D. Myers, M. Wilson, F. Garzon, D. Wood, P. Zelenay, K. More, K. Stroh, T. Zawodzinski, J. Boncella, J. E. McGrath, M. Inaba, K. Miyatake, M. Hori, K. Ota, Z. Ogumi, S. Miyata, A. Nishikata, Z. Siroma, Y. Uchimoto, K. Yasuda, K. I. Kimijima and N. Iwashita, *Chem. Rev.*, 2007, **107**, 3904–3951; (b) Y. Shao, G. Yin, Z. Wang and Y. Gao, *J. Power Sources*, 2007, **167**, 235–242; (c) W. Schmittinger and A. Vahidi, *J. Power Sources*, 2008, **180**, 1–14.

158. International Energy Agency, *IEA World Energy Outlook 2012 – Executive Summary*, IEA, Paris, 2012.

CHAPTER 9

Sustainable Nanotechnology: Preparing Nanomaterials from Benign and Naturally Occurring Reagents

O. A. SADIK,* I. YAZGAN AND V. KARIUKI

Department of Chemistry, Center for Advanced Sensors & Environmental Systems (CASE), State University of New York – Binghamton, PO Box 6000, Binghamton NY 13902-6000, USA
*Email: osadik@binghamton.edu

9.1 INTRODUCTION

Nanotechnology deals with the creation of a range of useful nano-materials for consumer products, healthcare, automotive, energy, agriculture, electronics, and environmental remediation, including the entertainment industry.[1] However, the current challenge is that the synthesis of many of the existing nanomaterials is not sustainable on account of their dependence on high temperature and pressures, the use of large quantities of hazardous reagents and/or solvents coupled with the insufficient understanding of their environmental health and safety.[2,3] In addition to being excessively intensive with respect to resources and energy, some existing nanomanufacturing processes utilize non-renewable materials.[3,4] As a result, a complete

Chemical Processes for a Sustainable Future
Edited by Trevor M. Letcher, Janet L. Scott and Darrell A. Patterson
© The Royal Society of Chemistry 2015
Published by the Royal Society of Chemistry, www.rsc.org

set of synthetic protocols are needed to support the development of sustainable nanomaterials.

Numerous approaches have been reported for the development of sustainable nanomaterials from both economic and environmental viewpoints. A commonly used approach involves the concept of 'safer-by-design'.[3–9] Murphy's group reported a one-step synthetic method for gold nanoparticles at room temperature using water as the solvent.[4,5] They utilized mild reducing agents with cheap cationic surfactants to facilitate crystal growth into rods. Another approach involves the use of biologically inspired experimental processes for the synthesis of nanoparticles. One such example is the synthesis of gold nanoparticles using *Cinnamomum zeylanicum* leaf broth as the reducing agent.[10] Other greener approaches for the development of nanomaterials include the use of ambient temperatures, exclusion of hazardous reagents, microwave irradiation and sonication, as well as life cycle, and risks assessments, whenever possible.[10–20]

The development of sustainable nanomaterials has also been proposed that focuses on the replacement, substitution and/or functionalization of various types of nanomaterials with biological or naturally occurring macromolecules such as carbohydrates, proteins or polypeptides.[21] In our laboratories, we have performed pioneering work on the development of novel functional materials.[22–29] We have reported the development of functional polymeric nanostructures using electrochemical synthesis, phase inversion and photopolymerization.[22,23]

We are currently exploring the concept of 'sustainable by design' (S^bD) in which the synergistic interactions of nanomaterials are combined with those of biomacromolecules, small linkers and/or biodegradable materials (*e.g.* glutaraldehyde, thiols, triphosphates, oligosaccharides) to create non-toxic and sustainable nanocomposites.[27] The rationale is that the S^bD approach should result in materials that are sustainable or biodegradable and therefore will not be a new source of contamination after usage. This chapter provides a review of the preparation of sustainable nanomaterials using benign and naturally occurring reagents from both economic and environmental viewpoints. In addition, we demonstrate that S^bD based nanomaterials such as the poly(amic) acid (PAA) system and fullerene–cyclodextrin are low cost, non-cytotoxic and biodegradable.

9.2 SUSTAINABILITY AND NANOMATERIALS

Sustainable nanotechnology is the research and development of nanomaterials that have economic and societal benefits with little or no negative environmental impacts.[28] Elements of sustainable nanotechnology includes green synthesis, green energy, sustainable manufacturing, nanomedicine, nano life assessment, environmental applications, toxicology of nanomaterials, fate/transport of nanomaterials, sensors and tools for achieving nanomaterials, legal aspects, and societal/policy considerations.[29] Although it is broad in scope, the focus of this chapter is to examine one aspect of sustainable nanotechnology, *i.e.* green synthesis or the preparation of sustainable nanomaterials that uses benign and naturally-occurring reagents.

Green chemistry is the design of chemical products and processes that reduce or eliminate the generation and use of hazardous substances.[20] Green chemistry seeks to redesign processes and materials in ways that are benign to humans and the environment, and possess intrinsic sustainability. Green chemistry has evolved over the past 20 years with the 'design' emerging as the most important concept. Sustainability is a concept which, in the context of material synthesis, coincides with many principles of green chemistry.[6]

The components of green chemistry that are related to sustainability include: using less toxic precursors to prepare nanomaterials; using water as a solvent where possible; using minimal number of reagents and few synthetic steps as possible; reducing the amount of by-products and waste; and using a reaction temperature close to room temperature.[6,8] We have reviewed some of the significant milestones that have been achieved towards sustainable development of nanomaterials. A case study is presented for PAA using the S^bD concept. S^bD of nanomaterials can be designed by integrating biodegradable polymers and controlling the chemistry, porosity and the method of desolvation. These include the incorporation of macromolecules—cyclodextrins, proteins, polymeric amphiphiles and chitosan (Figure 9.1).

9.2.1 Minimally Toxic Quantum Dots

Quantum dots (QDs) are semiconductor nanoparticles with diameter in the range of 1 to 10 nm.[9] Compared with organic fluorophores, QDs are highly photoluminescent and resistant to photobleaching, and the synthetic process is amenable to tunable particle diameter. The most popular quantum dots are the II–VI materials such as CdS, CdSe and CdTe. The first significant synthetic method for preparing CdSe quantum dots was reported by Murray, Norris and Bawendi in 1993 as shown in Equation (9.1).[30]

The synthetic process begins with the rapid introduction of organometallic reagents into a hot coordinating solvent to produce a

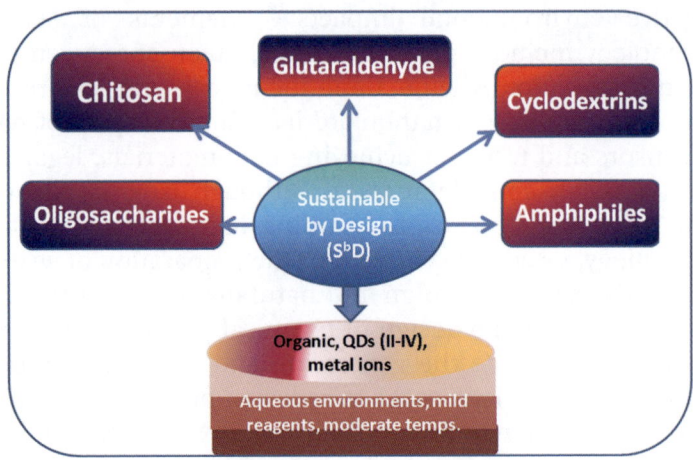

Figure 9.1   Sustainable by design (S^{b}D) combines the synergistic properties of nanomaterials with those of biomacromolecules, small linkers and/or biodegradable materials (*e.g.* glutaraldehyde, thiols, triphosphates, oligosaccharides) to create non-toxic and sustainable nanocomposites.

temporally discrete, homogeneous nucleation. Slow growth and annealing in the coordinating solvent result in a uniform surface of derivatization and regularity in the core structure. A size-selective precipitation results in powders of nearly monodispersed nanocrystallites, which can be dispersed in a variety of solvents. The average size of the crystallite is tunable from $(-1.2$ to $-11.5)$ nm and is defined by its major axis.[31]

$$\text{Me}_2\text{Cd} + \begin{array}{c} (\text{octyl})_3\text{P}=\text{E} \\ \text{or} \\ [(\text{alkyl})_3\text{Si}])_2\text{E} \end{array} \xrightarrow{\text{TOPO, 300}^\circ\text{C}} \text{CdE} \qquad (9.1)$$

where E = S, Se, or Te, (octyl)$_3$, P = trioctlylphosphine and Me$_2$Cd = dimethyl cadmium.[18]

In this reaction, trioctlylphosphine oxide (TOPO) is both a solvent and a capping agent; it binds to the nanoparticle surface so that the particles do not grow too large. Me$_2$Cd is chosen as the Cd source and (TMS)$_2$E (where E = S, Se, Te) or TOPSe and TOPTe are selected as chalcogen sources with TOPSe and TOPTe preferred due to their stability and ease of preparation.[31]

From green chemistry point of view, there are abundant shortcomings with the above reaction; the precursor, especially dimethyl cadmium is very toxic. If inhaled, dimethyl cadmium is actually more toxic than the other non-organic, cadmium compounds being a poisonous, polytropic substance with particular damaging effects on the kidney, the liver, the central nervous system and the respiratory organs.[32] The solvent is also obnoxious; the reaction requires a high temperature, and hence the energy cost is high. Moreover, leaching of the final product, CdE could result in some biological effects.[13] Other workers have developed greener synthetic methods for making CdSe quantum dots. For example, in 2007, Pradhan and co-workers reported that the synthesis could be achieved without the use of cadmium and they prepared Mn(II)-doped ZnSe quantum dots.[14,15]

9.2.2 'Green Gold'

Gold nanoparticles have been used for major applications in biochemical sensing, biological imaging and biological therapies.[16,19] The original preparation of gold nanoparticles uses benzene as a solvent and gaseous diborane as a reducing agent. Researchers such as Hutchison and co-workers have developed another greener method to make ultra-small (1.5 nm) gold nanoparticles. These authors replaced diborane with sodium borohydride and benzene with toluene. They considered these reagents 'greener' benzene *vs.* toluene. The purification step, which involves liters of organic solvent, can now be achieved using filtration membranes.[11] This has led to a drastic reduction in the cost of preparing the gold nanoparticles from $300 000 per gram of product to $500 per gram.[12]

Biologically inspired experimental processes for the synthesis of nanoparticles have been developed. One such example is the synthesis of gold nanoparticles using *Cinnamomumzeylanicum* leaf broth as the reducing agent, which is considered a highly 'green' process.[33] The leaves were cut and boiled in de-ionized water before being added to 99.9% hydrogen tetracholoroaureate (III) hydrate (HAuCl$_4$.3H$_2$O). After about an hour, the reduction was evident by the formation of stable purple, colloidal gold nanoparticles.[10] Green synthesis of gold nanoparticles using *Hibiscus rosasinensis* has also been achieved.[31]

Murphy's group reported a one-step synthetic method for gold nanoparticles at room temperatures using water as the solvent. The authors utilized mild reducing agents with cheap cationic surfactants to facilitate the crystal growth into rods.[5] This was achieved in two main steps. The first step was the reduction of HAuCl$_4$ in water by sodium borohydride in presence of sodium citrate as a capping agent to make 3.5 nm seed nanoparticles. This step was followed by an aliquot of this seed solution to grow more metal with a weak reducing agent such as vitamin C in the presence of more sodium citrate. This process can be further improved since the HAuCl$_4$ is toxic according to the material safety data sheet (MSDS).

9.2.3 'Green Silver'

Approaches to green synthesis have been reported for nanosilver having advantages over conventional methods. Methods for green synthesis include mixed-valence polyoxometallates, polysaccharide, Tollens', irradiation and biological reagents. The green synthesis of silver nanoparticles involves three main steps, which must be evaluated based on green chemistry perspectives. These include: (i) the selection of solvent medium; (ii) the selection of environmentally benign reducing agent; and (iii) the selection of non-toxic substances for the stability of silver nanoparticles.[34]

9.2.3.1 Polysaccharide Method. In this method, silver nano-particles were prepared using water as an environmentally benign solvent and polysaccharides as a capping agent, or in some cases polysaccharides served as both a reducing and a capping agent. For instance, the synthesis of starch–silver nanoparticles was carried out using starch as a capping agent and β-D-glucose as a reducing agent in a gently heated system.[35] The starch in the solution mixture avoids the use of relatively toxic organic solvents.[36]

9.2.3.2 Tollens' Method. Basically, Tollens' reaction involves the reduction of Ag(NH$_3$)$_2$$^+$(aq), a Tollens' reagent and an aldehyde as shown in Equation (9.2):

$$Ag(NH_3)_2{}^+(aq) + RCHO\ (aq) \rightarrow Ag(s) + RCOOH(aq) \qquad (9.2)$$

In the modified Tollens' procedure, Ag$^+$ ions are reduced by saccharides in the presence of ammonia, resulting in silver nanoparticle films with particle sizes from (50 to 200) nm, Ag hydrosols with particles in the order of (20–50) nm, and silver nanoparticles of different shapes.[37] The replacement of chemical reducing agents by saccharides gives this reaction the 'green' aspect.

Silver nanoparticles have also been successfully synthesized from AgNO$_3$ through a simple green route using the latex of *Jatropha curcas* as the reducing/capping agent. An extensive literature survey has revealed that the major constituent of the latex of *J. curcas* are curcain (an enzyme), curcacycline A (a cyclic octapeptide) and curcacycline B (a cyclic nonapeptide).[38] Briefly, crude latex was added to an aqueous

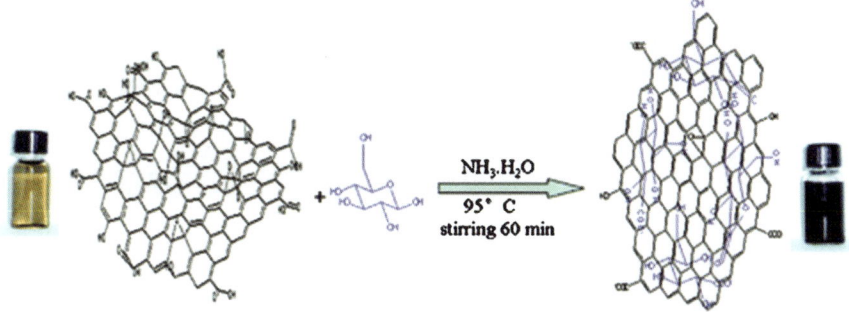

Figure 9.2   Synthesis of green nanosheets using sugars as reducing agents.[39]

silver nitrate solution, and the mixture was heated to 85 °C with constant stirring for 4 hours in an oil bath and silver nanoparticles were gradually obtained.

9.2.4   Green Graphene Nanosheets

Zhu and co-workers have developed a green approach to the synthesis of chemically converted graphene nanosheets (GNS) based on reducing sugars, such as glucose, fructose and sucrose using exfoliated graphite oxide (GO) as precursor (Figure 9.2).[39]

The advantage of this method is that both the reducing agents (glucose) and the oxidized products are environmentally friendly.[39] Previously Zu and co-workers[39] had reported that graphene oxide sheets could be readily reduced under a mild condition using L-ascorbic acid. This results in an environmentally friendly approach for large-scale production of water-soluble graphene. However, the length of time to conduct this experiment (about 48 h) and the limited reducing agent (L-ascorbic acid) are a major constraint when using it for practical applications.

9.2.5   Biomass Extracts as Precursor for the Synthesis of Nanomaterials

In the pursuit of sustainability, attention is shifting to the use of biomass as precursors (Figure 9.3). Plant and algal extracts are currently being used to reduce metal ions to metal nanoparticles in water. Examples of biomass-based reducing agents include starch,[31] extracts from lemon-grass,[40,41] seaweed,[42] algae,[43] and mushroom.[44] Nanoparticles from gold, silver, cobalt, nickel and cobalt–nickel alloy have been prepared with biomass-based extracts.[5] This brings about

Figure 9.3   Use of biomass extracts as precursors for the synthesis of nanomaterials; from top left to bottom right: Starch,[a] lemongrass,[b] seaweed,[c] algae,[d] mushroom[e] and *Cinnamon camphora* leaf.[f]

[a]http://academic.brooklyn.cuny.edu/biology/bio4fv/page/starch.html
[b]http://aidanbrooksspices.blogspot.com/2007/10/lemon-grass.html
[c]http://www.acadianseaplants.com/edible-seaweed-nutritional-supplements-ingredients/botanical-ingredients/major-seaweed-classes
[d]http://www.microscopy-uk.org.uk/mag/indexmag.html?http://www.microscopy-uk.org.uk/mag/wimsmall/green.html
[e]http://www.telegraph.co.uk/foodanddrink/foodanddrinknews/9644065/Britons-go-mushroom-mad.html
[f]http://www.eattheweeds.com/camphor-tree-cinnamon%E2%80%99s-smelly-cousin/

an advantage in the recycling of food wastes for chemical production from a sustainability point of view.

9.2.6   Safer-by-Design (S[b]D) Concept

The safer by design (S[b]D) concept is also used to ensure sustainability. Conceptually it is relatively easy to eliminate known hazards from a synthesis or a process, but it is much more difficult to ensure that the alternative is benign. The future of green chemistry requires that chemists take a more pro-active role in the design of products, at the same time taking into account the need for reduced hazards by avoiding the kind of cycle that leads to accidental environmental damage, toxicity surprises and daunting cleanup problems.

Demokritou and co-workers have developed a promising approach towards the engineering of safer engineered nanomaterials (ENMs).[45] In their approach, potentially toxic nanomaterials are coated with a biologically inert layer of amorphous SiO_2. Armophous SiO_2 is used since it is considered a biologically and environmentally inert

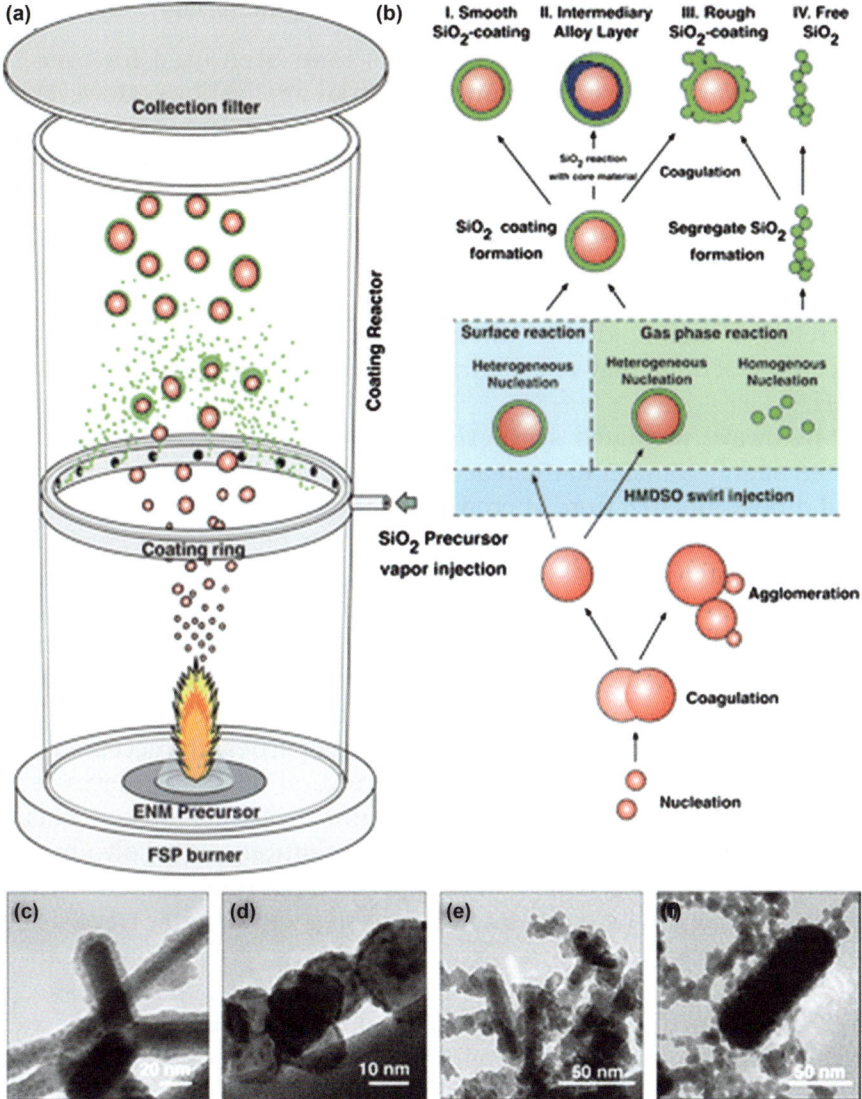

Figure 9.4 Synthesis of engineered nanomaterials using SiO_2 coating.
Reprinted with permission from ref. 45, copyright (2013) American
Chemical Society.

material.[45] These researchers have formulated a concept for flame
generated ENMs based on a one-step, inflight SiO_2 encapsulation
process, as illustrated in Figure 9.4. This process is versatile and
they have utilized similar procedure to coat four engineered nanomaterials: CeO_2, ZnO, Fe_2O_3 and Ag.[45]

9.2.7 Microwave Mediated Synthesis of Nanomaterials

Microwave assisted chemical synthesis have been used for over 30 years in a variety of research and industrial applications. However, it has not been used in pilot scale production.[46] Microwave-mediated synthesis was first utilized in kitchen-based microwave ovens.[47] Today, dedicated microwave ovens are widely preferred because specific reactions can require specific temperatures, pressures, stirring and highly controlled irradiation. Microwave irradiation is one of the most important parameters in the fate of synthesis, but commercially available kitchen microwave provides only 2.45 GHz while specific reactions may require irradiation in the range of (0.3 to 300) GHz.[48]

Microwave mediated synthesis is based on dielectric heating assisted energy transfer to the reactant or reactants *via* ionic conduction and dipolar polarization. Microwave irradiation provides an advanced acceleration in the reaction rate, which can be accepted as electromagnetic catalyst. Its catalytic effect is selective even in heterogeneous media due to the fact that different substances have different resonances. Specifically, microwave can be combined with every possible liquid–phase synthesis technique used in nanomaterial production. The combination will not adversely affect the size of the crystal and the shape of the nanomaterials whose properties are accepted key factors in nanomaterial production. Several methods have been combined with microwave irradiation and these include sonochemistry, biomimetic approaches, co-precipitation and solvo/hydrothermal. Gold, silver, palladium, copper and platinum as well as their combinations can be synthesized in one step with microwave assisted methodologies.[48]

The synthesis of silver nanostructures has been reported using a microwave mediated process, which provide environmentally and eco-friendly production by bringing selected solvent on the green side using alcohol instead of toxic organic solvents, as well as providing uniform heating and less material usage. For example, instead of organic solvents or environmentally problematic reducing agents, simple sugars such as mono- and disaccharides can be preferred as reducing agents in microwave mediated noble metal synthesis as silver.[49] Besides noble metallic and bimetallic nanostructures, complex nanostructures such as $ZnFe_2O_4$ reduced graphene oxide can be synthesized using microwave assisted synthesis.[50] Currently, microwave irradiation is combined with supersonic process in the synthesis of nanomaterials.[51]

9.2.8   Life Cycle Assessment

Product life cycle is nowadays accepted as an indispensable parameter when it comes to having a realistic perspective about materials.[52,53] Life cycle assessment (LCA) defines the potential impact of the materials and their decomposition products in the environment at different life cycle stages. Figure 9.5 shows the overall risk assessment approaches currently used by the US Environmental Protection Agency (USEPA) for conventional chemicals. These approaches are thought to be generally applicable to engineered nanomaterials.

Currently, a variety of nanomaterials have been used in commercially available consumer products. For instance, titanium-based nanoparticles are now an indispensable part of sunscreen and paints, while biomolecules such as protein-based nanostructures are widely used in cosmetics and detergents. While advantages of nanomaterials are widely appreciated, they may have strong negative impacts on human health and the environment. Nanomaterials can enter into human the body *via* inhalation, ingestion and dermal transition.

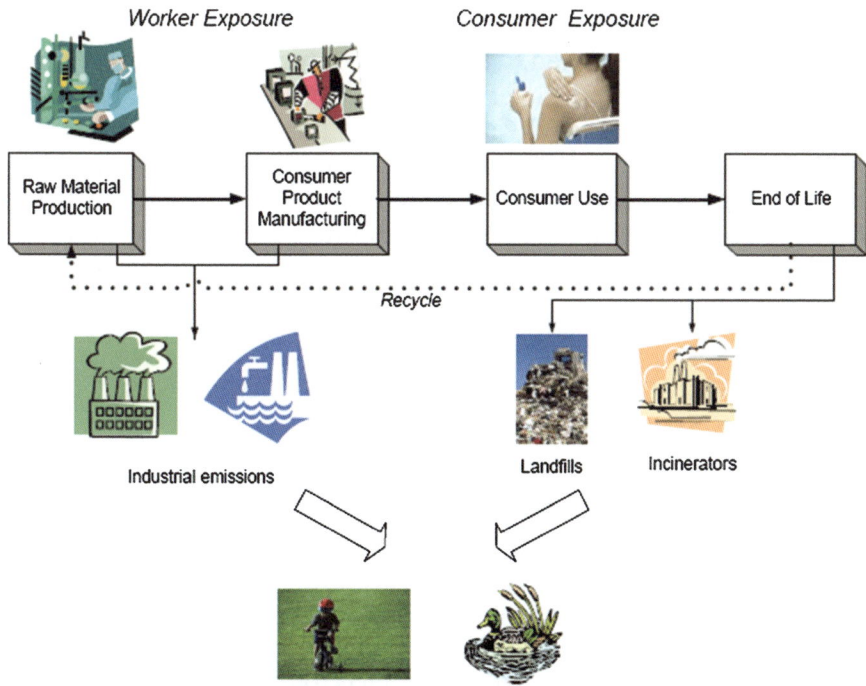

Figure 9.5   Life cycle perspective to risk assessment.[54]

Nanomaterial intake is also possible by consuming foods which are already exposed to nanomaterial contamination through packaging or other means. However, there is no defined critical value at the moment for nanomaterials in terms of environmental and human health.[55] It is clear from Figure 9.5 that manufactured nanoparticles found in consumer products will eventually enter into the environment. Nanoparticles released into the air, dissolved into the water, or absorbed by the soil will be inhaled by animal, digested by fish and absorbed by plants, and in turn they will affect the whole food chain.

However, there are three major problems in defining the LCAs of nanomaterials. First, it is not easy to totally characterize how much and what type of impurities the nanomaterials possess. It is therefore challenging to determine the accurate concentration of metal catalysts released into the environment. In common LCA approaches, materials that are toxic to humans and/or the environment are evaluated in a couple or sometime little more midpoints. The fate and mobilization of nanomaterials in the environment and their potential toxicities to biological systems including humans have not been adequately characterized.

In order to overcome the failures of classical LCA approaches in defining LCAs of nanomaterials, more extensive criteria and approaches have been offered.[52] European countries have been funding a collaborative study for determining the LCAs of nanomaterials by emphasizing such research questions as analyzing the interaction between nanomaterials, and living and non-living systems, mechanisms of nanomaterial intake, mobilization and toxicity in living systems, and acute and chronic environmental impact of nanomaterials, as well as improving public and governmental understanding of possible outcomes.[45]

9.3 NANOSTRUCTURED POLY(AMIC) ACID MEMBRANES: A CASE STUDY

Nanoscale materials can be used to achieve wastewater purification, desalination, and inactivation of pathogenic bacteria and viruses, as well as for the adsorption or degradation of contaminants.[55-60] We have discovered that a new class of amphiphilic conducting polymers such as poly(amic) acid (PAA) could be used to fabricate nanoporous membranes with potential applications in water treatment.[61-63]

Conducting PAA is a conducting electroactive polymer and its properties can be tuned by manipulating the delocalized π electron system for chemical and electrocatalytic applications. Its molecular

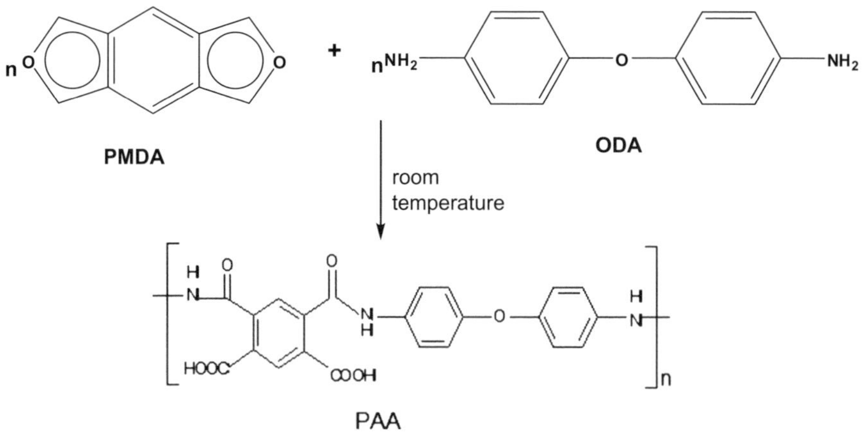

Scheme 9.1   Synthesis and molecular structure of poly(amic) acid (PAA).

structure is given in Scheme 9.1. PAA provides a means of generating nanocomposites containing monodispersed metal particles while retaining its physical and chemical properties. It is versatile both in organic and inorganic solvents due to its chemical resistance. The carbonyl and amide functionalities in PAA act as anchors that have been previously employed in the fabrication of flexible nanostructured PAA-silica (PSG) films.[61]

Depending on the solvent used, PAA can be prepared as a viscous liquid or as powder and the resulting polymers were soluble either in aqueous or organic solvent medium.[23] For instance, when tetrohydrofuran (THF) was used as solvent the resulting PAA was a viscous liquid. When acetonitrile (ACN) was used as solvent, another PAA derivative was obtained in a powdered form using the same 4,4'-oxydianiline (ODA) and pyromellitic dianhydride (PMDA). This homogeneous powder was found to be soluble in dimethyl acetamide (DMAc), DMF and dimethyl sulfoxide (DMSO).

Unlike other polymers (such as self-assembled monolayers on mesoporous supports, dendrimers, carbon nanotubes and metalloporphyrinogens), PAA has the following advantages: it is easy to synthesize; soluble in both organic and inorganic solvents; possesses both hydrophobic and hydrophilic groups (amphiphilic); and is able to retain the orientation of its functional groups. PAA also displays both reduction and chelation properties due to the presence of both carboxylic and amide groups that can reduce and chelate metal ions.[64,65] We have shown that PAA can be used to generate nanocomposites containing monodispersed metal particles, *i.e.* in this case it is acting as a stabilizer.[64,65]

9.3.1   PAA for Membrane Filtration

PAA membranes are promising candidates for membrane filtration because of their transport characteristics.[63,68] When prepared using a phase-inverted process, PAA membranes have similar morphology as the typical phase-inverted membranes with a thin, smooth dense (Figure 9.6, left column) supported by porous sub-layer (Figure 9.6, left column, row 2). The smooth hydrophobic surfaces (Figure 9.6, left column, rows 3 and 4) of the membranes lead to frictionless flow of water across the surface, enabling the transport that is an order of magnitude higher than the transport in conventional pores. Figure 9.6 (mid column, from top to bottom) also shows the scanning electron micrograph (SEM) data obtained for PAA membranes using different concentrations of PAA solution. This synthetic approach affords well controlled membranes, leading to various desired sizes from <5 nm to >100 nm.

9.3.2   Experimental Control of Pore Size

Our approach to experimentally controlled pore sizes using PAA membranes utilizes a wet phase inversion process. The mechanism is somewhat complex for this process which often leads to a porous structure on the surface and inside of the membrane. A thin film of a homogeneous polymer solution is contacted with a second liquid which is a nonsolvent for the polymer but miscible in all proportions with the polymer solvent. The exchange of solvent and nonsolvent across the interface introduces phase separation in the polymer film, which can lead to a variety of characteristic asymmetry or symmetric structures.[66]

The main route for the phase-inversion process involves two different types of phase transition as illustrated in the ternary phase diagram shown in Figure 9.7.[68] These are: liquid–liquid phase separation, in which the completely miscible solution crosses the binodal boundary to enter the two-phase region (from I to II); and solidification (from I to III or from II to III). Since the viscosity of the polymer solution increases to a certain assumed value, the motion of polymer chains will be limited and the system can be regarded as a solid. Young *et al.* explained the formation mechanism of pores in detail.[67] A two-step mechanism model depicts that the membrane consists of two layers: a top layer and a sublayer. The top layer is formed first at the casting solution–nonsolvent interface. The ratio of nonsolvent inflow to solvent outflow is of the utmost importance for the top layer

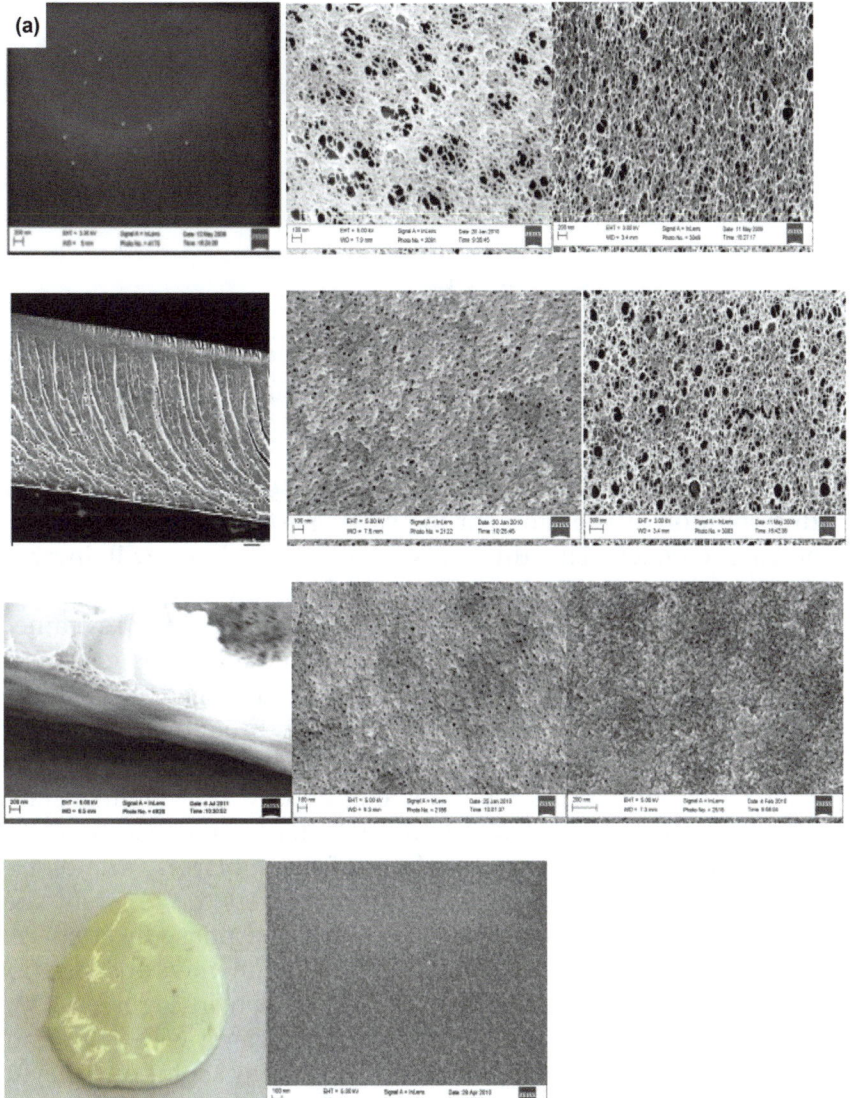

Figure 9.6   SEM and pictograph images of phase-inverted PAA membranes at varying pore sizes (magnification 100 000×).

structure. The structure of the sublayer depends on the local composition at the instant of phase transition, but this is directly influenced by the degree of aggregation of the polymer molecules at the top layer.

In the two-step mechanism model, a dense skin is formed because the solvent in the casting solution diffuses rapidly into the

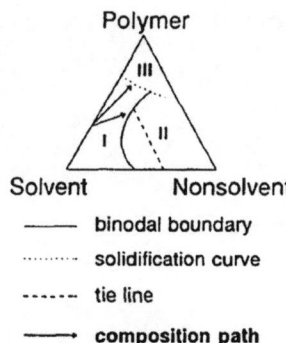

Polymer

Solvent   Nonsolvent

———— binodal boundary

·········· solidification curve

------ tie line

———→ **composition path**

Figure 9.7   Ternary phase diagram. I: Homogeneous solution region; II: Two-phase region; III: Solidification region.[68]

coagulation bath, and the relatively little nonsolvent diffuses into the casting solution at the moment the casting solution and nonsolvent come into contact. The thickness of the skin layer will increase gradually until the diffusion of solvent from the sublayer solution through the dense skin layer into the coagulation bath is restrained. As a result, the composition of the sublayer does not change as rapidly, and the solution is within the initial homogeneous solution region. In the meantime nonsolvent diffuses through the dense top layer. When a certain amount of nonsolvent has caused the composition variation to cross the binodal boundary to become unstable somewhere, the first nucleus of the polymer-poor phase occurs to form the nascent pore in the first layer of the sublayer.

More pores occur when more nonsolvent enters the different sites of the first layer.[67] The surface pore size of the phase-inverted PAA membrane is found to be greatly dependent on the concentration of the corresponding casting solution. As shown in Figure 9.6, the surface pore size of the PAA membrane decreases with increasing concentration. Membranes made from casting solutions having concentrations between 0.2 M and 0.47 M were imaged. A casting solution having a concentration <0.2 M was too dilute to form a standalone membrane, but with increasing concentration, the casting solution became too viscous to be dispersed on the substrates. The casting solution with a concentration >0.47 M was not considered. Between 0.2 M and 0.23 M, the pore size decreased dramatically; the pore size changed slowly, leveling off with increasing concentration after 0.23 M.

Generally, the PAA coating layer on filter paper has a smoother surface than the PAA standalone membrane (Figure 9.6, column 1, bottom row). At the lowest concentrations (0.2–0.22 M), the PAA coating layer may have bigger pore size than the PAA of the stand-alone membrane. This is attributed to the absorption of the filter paper and hence the presence of less PAA casting solution on its surface. For the concentration range of 0.23 to 0.30 M, the PAA coating layer generally had a smaller pore size range and more uniform pores than the standalone membranes. At a concentration of 0.31–0.47 M, the pore sizes were quite similar for both the PAA coating layer and the PAA of the standalone membrane. The surface pores of the high concentration membranes were more uniform and in a smaller size range than the pores at low concentrations. Using statistical analysis software (SigmaPlot 12.0 by Systat Software Inc.), we found that a single component exponential decay fitting is the best curve to interpret the relationship between concentration and pore size according to:[68]

$$Y = a + b * \exp\{-cX\} \tag{9.3}$$

where X is the concentration of the PAA in mols, Y is the corresponding pore size in nm, and a, b and c are the coefficients. For the PAA of the standalone membranes, these constants were determined to be 11.77, 2.37×10^5 and 38.77, respectively; an R value of 0.9748 and R^2 value of 0.9502 were recorded. For the PAA of the coated filter paper, these constants were 16.06, 9.01×10^5 and 45.97, with an R value of 0.9271 and R^2 value of 0.8594. Finally, the structures of the PAA permitted targeted and specific modifications of the pore without destroying the unique properties of the membranes.[68] The combination of these factors could enable a new generation of membranes whose transport efficiency, rejection properties and lifetimes drastically exceed those of the current membranes.

9.3.3   Sustainable by Design using Chitosan–PAA System

We have prepared bio-functionalized PAA membranes with biological decontamination. We have described the synthesis of biodegradable polymeric membranes of PAA, glutaraldehyde-derivatized PAA (PAA-GA) and chitosan-modified poly(amic) acid (PAA-CS) using phase-inversion procedures.[27] PAA is being used to prepare membranes for removing bacteria, viruses and toxic molecules. Although the filtration mechanically traps and removes the microbes, it does not kill them nor can it remove some of the toxins they produce. The bio-functionalized PAA membranes have achieved antibacterial functions by impacting antibacterial moieties such as glutaraldehyde and chitosan.[27] The following section describes the cytotoxicity characterization and biodegradation properties of the resulting membranes.

9.3.3.1 Cytotoxicity Studies.   The relative cytotoxicity of the membranes was performed using conventional colorimetric MTT assays.[69] A 1 mg mL^{-1} PAA membrane was used in the experimental concentration to assess the membrane toxicities. The four cell lines studied included A549 lung, HT29-MTX, Caco-2 colon [Dulbecco's modified Eagle's medium (DMEM) with 10% fetal bovine serum (FBS)], and non-cancerous IEC-6 cells. Four membrane types were examined: PAA (alone), chitosan-functionalized PAA (PAA-CS), glutaraldehyde-functionalized PAA (PAA-GA) and glutaraldhyde-linked chitosan PAA (PAA-CS-GA). The control studies were performed in cell free Eagle's minimal essential medium (EMEM) and DMEM media for all the membranes. DMSO and/or cells were prepared simultaneously following the same experimental procedure. PAA solution (at the same experimental concentration) was added to each well containing cells which had been grown for one day to give about (70–80)% confluence. The plates were thereafter incubated for 24 hours after which cell viability was tested using MTT assays. Pure solvents were used as control and the results were correlated with the number of cells, absorbance of the medium, and the concentration of PAA in each well.

Both the parent PAA and the PAA-CS membranes did not reform upon exposure to the cell culture medium whereas PAA-GA and PAA-CS-GA were reformed after 2 hours of exposure. Although the reformed membranes lost their stability in 2 hours, light microscopy showed the presence of dispersed fragments of PAA-GA and PAA-CS-Ga membranes at up to 6 hours in the media. In addition, 50 mg mL^{-1} and 40 mg mL^{-1} PAA-GA and PAA-CS-GA concentrations, respectively, were prepared in DMSO stocks. When 10 µL from the stocks were transferred to 190 µL cell-free media, this resulted in membrane reformation. However, no membrane reformation was observed between 25 µL and 50 mg mL^{-1} stock solutions of dissolved PAA solution. Overall, the parent PAA membranes showed little or no toxic effects (5–20%) when tested in all of the four cell lines within a 24-hour incubation period. However, the PAA-CS-GA exhibited the highest toxic effects for all tested cell lines (Figure 9.8). The PAA-CS-GA treated cells showed suppressed or slower growth than the PAA-GA treated membranes. Interestingly, A549 lung cancer cell lines were observed to be more sensitive while HT29-MTX and IEC-6 cells did not

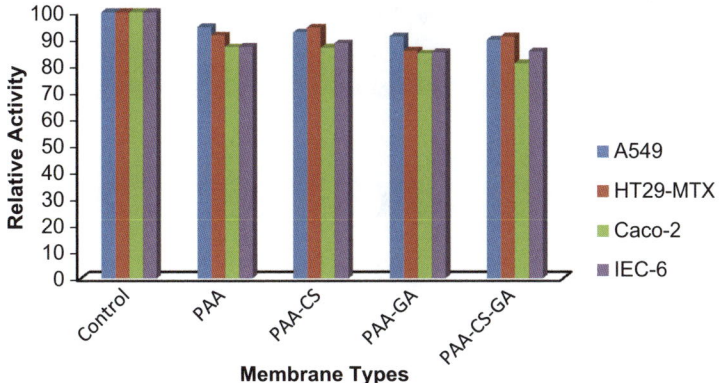

Figure 9.8 Relative cytotoxicity of PAA membranes assessed using MTT analysis to monitor cell death. MTT measurements were taken with a Biotek Fluorometer (Winooski, VT, USA) at an experimental concentration of 1 mg mL^{-1} PAA membranes.

show significant sensitivity. This is attributed to mucin producing ability of HT29-MTX cell lines,[70] which probably protected them against possible toxic effects of the PAA. Hence the toxicity study confirmed that the PAA membranes appeared benign to cell lines, and the observed low MTT metabolic activity may simply be related to the temporary three-dimensional (3D) structure of the PAA membranes.

9.3.3.2 Biodegradation Studies. PAA-GA and PAA-CS-GA that were synthesized as reported in ref. 27 were used in the biodegradation study with a pure water mediated membrane preparation. It was observed that the PAA and the PAA-CS dissolved easily in the medium and so it was not possible to monitor the disintegration of these membranes in parallel to biomass accumulation. Biodegradation of the PAA membranes was studied using mixed microbial cultures obtained from a collection of rotten sticks taken from the Binghamton University garden (Figure 9.9(a)). The rotten sticks were identified as branches of the elm tree (*Ulmus Americana*). The sticks were selected because they naturally contain a variety of fungi species such as *Trametes versicolor*.[71] The optimized protocol employed for the biodegradation of PAA membranes is summarized in Figure 9.9(b).

Trametes defined media (TDM) were selected to obtain starter culture from the sticks. Nutrient Broth, Mueller Hinton broth and a

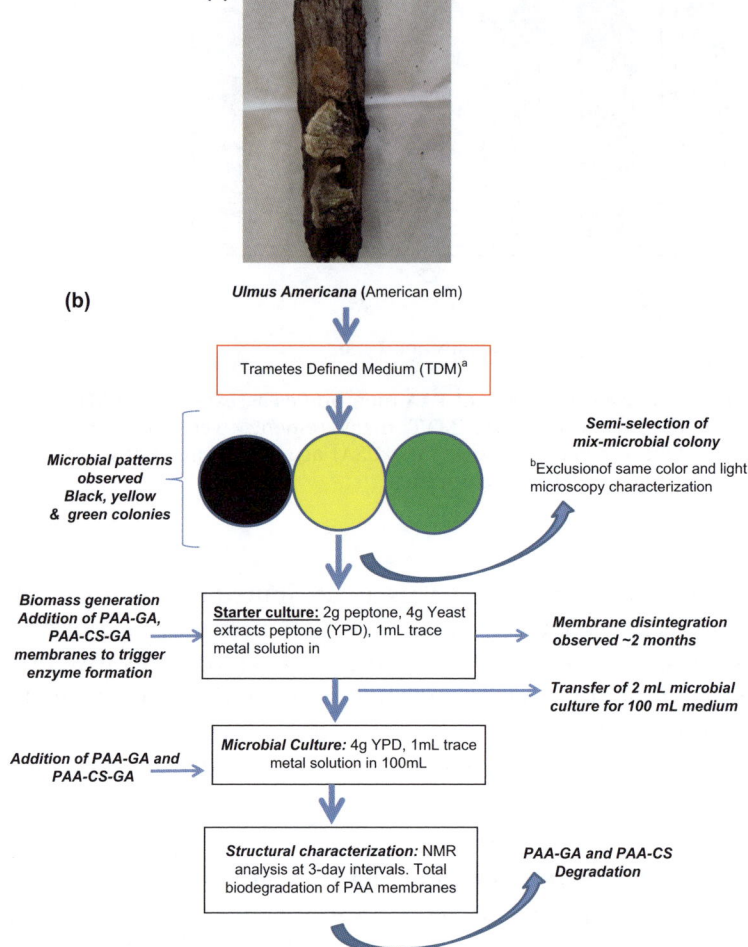

Figure 9.9 (a) Digital image of *Ulmus Americana* and the fungi employed for bio-degradation of PAA membrane (*Ulmus Americana* was obtained from Binghamton University Gardens). (b) Procedures employed for bio-degradation of PAA membranes.

[a]Three different medium initially selected for biodegradation studies. These are Trametes Defined Medium, Candid albicans medium, and Mueller Hinton Broth (MHB) [Sigma-Aldrich, MA]. MHB contains beef infusion solids, casein hydrolysate and starch at pH 7.4 and it provides eligible conditions for a variety of bacteria.[75] [b]Bacteria were eliminated to decrease competition between bacteria and fungi at possible maximum level degradation level for PAA membranes. Candida albicans was selected to eliminate possible pathogenic fungi[76] usage in PAA membrane degradation. Any similar microbial patterns from Trametes Defined Medium were observed in MHB and Candida albicans medium was subsequently excluded from the study. Trametes Defined Medium was prepared according to ref. 77. YPD medium, Yeast Extracts Peptone Dextrose Agar [Sigma-Aldrich, MA] is a non-selective nutrient-rich medium for fungi.[75]

medium for *Candida albicans* were tested. As expected, the mixed culture of white-rot fungi did not grow in the *Candida albican* medium. In the Nutrient Broth, the odor emanating from the culture was similar to that of a rotten egg, and the resulting microbes appeared to be very different from the microbes grown in TDM. This is normal because the pH in the TDM was 5.7, while the pH in the Nutrient Broth was 7.4. TDM is a semi-selective medium for fungi because fungi grow at low pH while bacteria do not grow well in acidic conditions.[72]

There were three different microbial patterns observed in the TDM medium, which showed black, yellow and green colors (Figure 9.9(b)). However, all these colonies developed hype, which is a characteristic of fungi development.[73,74] Then the colonies were used as starter culture. In the first step, peptone with peptone yeast medium containing trace metal solution was used to generate the microbial biomass. PAA membranes were then added to this medium to trigger enzyme production that can breakdown the membranes. The microbes are expected to synthesize the enzyme to utilize the PAA membranes as carbon source after consuming the easily accessible carbon source peptone yeast. Following this procedure, we observed that membrane disintegration required over 2 months for this step.

A 2 mL microbial culture was subsequently transferred into a new flask which contained enough amount of peptone yeast medium. This step requires significantly less amount of peptone yeast medium. It was observed that 4 g per 100 mL was sufficient to generate a significant amount of biomass, resulting in the complete breakdown of the membrane in less than 2 months. At this stage, 2 g per 100 mL of glucose was used instead of 4 g per 100 mL of peptone yeast medium. We observed that the glucose did not provide enough biomass and so its use was discontinued. The transferred microbes from the first step were expected to have already induced the enzyme formation that will degrade the PAA membranes. The PAA-GA and PAA-CS-GA membranes were totally degraded in 35 days while total physical disintegration was observed at 20 days. Nuclear magnetic resonance (NMR) was performed at 3 days intervals after the 20th day. We observed that the blank from the peptone yeast medium was very similar to that of the NMR spectrum at the 20th day. On this day, the characteristic aromatic proton shifts were clearly seen, while on the 35th day, the aromatic peaks totally disappeared (results not shown). This is not surprising because the PAA has a very simple structure compared with the complex organic molecules such as lignin which were degraded by the white-rot fungi.

Physical disintegration of the PAA membrane was observed by its presence in the medium. A confirmation that the membranes have been completely degraded was determined by monitoring its chemical structures using a 600 MHz Bruker NMR spectroscope with TopSpin 3.0 Software (Bruker, MA, USA). The PAA polymers were characterized by shifts in the $(6-8\times)10^{-6}$ range.[27,61] On the 20th day, there was no visible membrane in the medium, which is an indication that total a disintegration of the PAA membranes had occurred. NMR data showed that PAA polymers were in the medium; hence we continued the NMR monitoring. On the 35th day, a 2 mL sample from the medium was taken, which was subjected to freezing and a lyophilization process to eliminate the effect of the solvent. A high amount of the solvent blocking the peaks was noticed on the 20th day. Also, the NMR data showed no peak between the $(6-8)\times10^{-6}$ regions, confirming that full degradation of PAA polymers had occurred.

9.3.3.3 Sustainable by Design using Fullerene-Cyclodextrin System.

Fullerenes were discovered in 1985 and since then there has been a tremendous growth in the production and applications of ENMs and products.[78-82] Fullerenes are also referred to as a 'buckminster'. Fullerenes are hollow carbon molecules joined together by single and double bonds to form a closed cage like or cylinder like molecule. Fullerene C_{60} in particular exhibits a wide range of novel chemical properties in that it can readily accept or donate electrons; hence it is widely used in batteries and other electronic applications.[80,81] Fullerene C_{60} can also readily undergo both hydrogen and halogen addition reactions, thereby opening up possible routes for derivatization of the fullerene parent molecule.[80,81] Given the unique properties of these ENMs, much emphasis has been given to their application with little or no regard to the potential impacts of the nanomaterials on human health and the environment.

Fullerenes are typically prepared *via* the thermal reactions of appropriate carbon sources under various conditions and always exist as a mixture of C_{60}, C_{70} and higher homologues. Their separation is a challenge for chemists. Several methods have been applied to separate C_{60} from fullerene mixtures including sublimation, liquid chromatography, column chromatography, capillary electrophoresis and recrystallization. Despite this progress, these methods usually require a large amount of organic solvents and are not environment friendly. Recent studies have reported simple approaches involving inclusion of unmodified fullerenes into water-soluble polymers and synthetic receptors such as poly(*N*-vinylpyrrolidone)[82,83] cyclodextrin (CD),[84,85] or calixarene[86] to form guest–host complexes or supramolecular assemblies. Based on this concept, Liu *et al.*[86] prepared a thio[2-(benzoylamino)ethylamino]- 6-CD fragment modified gold nanoparticle, and investigated its selective association and dissociation behavior with C_{60}. These authors successfully reported the capture and release of C_{60} fullerenes by 6-cyclodextrin (6-CD).

6-cyclodextrin is a cyclic oligosaccharide with seven D-glucose units linked by α-1,4-glucose bonds. It can only form a 2 : 1 inclusion complex with C_{60} but does not interact with C_{70} and higher fullerene analogues.[86] Using a similar approach we modified gold quartz crystal microbalance (QCM) with 6-CD and measured the concentration levels of C_{60} in water using simple QCM mass changes.[86] We can also explored the possibility of using γ-CD, which has a larger cavity, to capture C_{70} or higher fullerene analogues in water.

9.3.3.3.1 Cyclodextrin Derivatives for Fullerenes C_{60}.

We have used these synthesized derivatives to develop the first mass-sensitive nanosensor for the isolation and quantitative analyses of engineered fullerene (C_{60}) nanoparticles while excluding mixtures of structurally similar fullerenes. Amino-modified β-cyclodextrin (β-CD-NH$_2$) was synthesized and confirmed using ^{1}H NMR as the host molecule to isolate the desired fullerene C_{60}. This was subsequently assembled onto the surfaces of gold-coated QCM electrodes using *N*-dicyclohexylcarbodiimide/*N*-hydroxysuccinimide (DCC/NHS) surface immobilization chemistry to create a selective molecular configuration described as (Au)-S–$(CH_2)_2$-CONH-β-CD sensor. The change in mass of the sensor configuration on the QCM was monitored for selective quantitative analysis of fullerene C_{60} from a C_{60}–C_{70} mixture and soil samples. About $\sim 10^{14}$ to 10^{16} C_{60} particles per cm^2 were successfully quantified by QCM measurements. A continuous spike of 200 μl of 0.14 mg C_{60} per mL produced changes in frequency $(-\Delta f)$ that varied exponentially with concentration. Field emission scanning electron microscopy (FESEM) and time-of-flight secondary ion mass spectrometry (ToF-SIMS) confirmed the validity of the sensor surface chemistry before and after exposure to fullerene C_{60}. The utility of this sensor for real-world spiked soil samples has been demonstrated. Comparable sensitivity was obtained using both the soil and purified toluene samples. This work demonstrates that the sensor has potential application in complex environmental matrices (Scheme 9.2).[87,88]

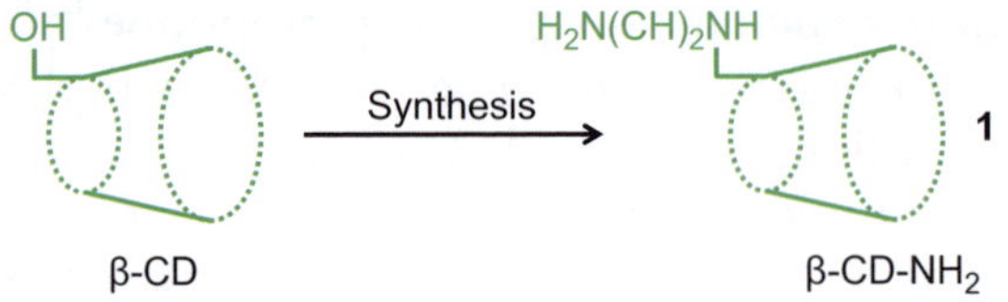

OH   H$_2$N(CH)$_2$NH

Synthesis ⟶   1

β-CD   β-CD-NH$_2$

Scheme 9.2 Synthesis of β-CD.[87,88]

9.4 CONCLUSIONS

We have shown that the development of nanomaterials could be sustainable if the economic and societal benefits are considered. By utilizing safe-by design concepts, researchers are exploiting novel ways to prepare safe and biodegradable nanomaterials with the ultimate goal of minimizing its negative environmental impacts. Case studies have been described for nanostructured poly(amic) acid membranes. When combined with other molecules such as chitosan or glutaraldehyde, the resulting membranes have been shown to be nontoxic and biodegradable. Biodegradation of PAA membranes were studied using mixed microbial cultures from elm trees (*Ulmus Americana*) because these naturally contained a variety of *Trametes* species. Trametes, fungi-degrading microbes were selected from the culture media. NMR results showed that PAA-GA and PAA-CS-GA membranes were totally degraded in 35 days while total physical disintegration was observed 20 days. This is not surprising because PAA has very simple structure compared with the complex organic molecules such as lignin degraded by white-rot fungi.

ACKNOWLEDGEMENTS

The authors acknowledge the National Science Foundation CBET 1230189 for funding. I.Y. also acknowledges the support of a Turkish government fellowship. The Regional NMR facility (600 MHz instrument) at SUNY-Binghamton is supported by NSF (CHE-0922815). All members of the Sadik group who have worked on poly(amic) acid are hereby acknowledged, including Dr Nian Du, Dr Veronica Okello, Dr Marcells Omole, Dr Daniel Andreescu, Dr Ailing Zhou and Ms Ola Haboub. Abdoulaziz Tour is also acknowledged for his assistance with the degradation studies.

REFERENCES

1. M. Diallo and C. J. Brinker (with contributions from A. Nel, M. Shannon, N. Savage, N. Scott and J. Murday), in *Nanotechnology Research Directions for Societal Needs in 2012, Science Policy Reports*, 2011, **1**, 221–259.
2. R. Dhingra, S. Naidu, G. Upreti and R. Sawhney, *Sustainability*, 2010, **2**, 3323–3338.
3. R. Luques and R. S. Varma (ed.), *Sustainable Preparation of Nanoparticles: Methods and Applications*, RSC Green Chemistry No. 19, Royal Society of Chemistry, Cambridge, UK, 2013.
4. C. J. Murphy, *J. Mater. Chem.*, 2008, **18**, 2173–2176.
5. C. J. Murphy, T. K. Sau, A. M. Gole, C. J. Orendorff, J. Gao, L. Gau and T. Li, *J. Phys. Chem. B*, 2005, **109**, 13857–13870.
6. P. T Anastas and M. M. Kirchhoff, *Acc. Chem. Res.*, 2002, **35**, 686–694.
7. P. T. Anastas and J. Charles, *Green Chemistry*, Oxford University Press, Oxford, 1996.
8. J. A. Dahl, B .L. Maddux and J. E. Hutchison, *Chem. Rev.*, 2007, **107**, 2228–2269.
9. M. L. Steigerwald and L. E. Brus, *Acc. Chem. Res.*, 1990, **23**, 183–188.
10. S. D. Smitha, D. Philip and K. Gopchandran, *Spectrochim. Acta A: Mol. Biomol. Spect.*, 2009, **74**, 735–739.
11. S. F Sweeney, G. H. Woehrle and J. E. Hutchison, *J. Am. Chem. Soc.*, 2006, **128**, 3190–3197.
12. J. Weng, C. Wang, H. Li and Y. Wang, *Green Chem.*, 2006, **8**, 96–99.
13. A. M. Derfus, W. C. Chan and S. N. Bhatia, *Nano Lett.*, 2004, **4**, 11–18.
14. N. Pradhan, D. M. Battaglia, Y. Liu and X. Peng, *Nano Lett.*, 2007, **7**, 312–317.
15. N. Pradhan and X. Peng, *J. Am. Chem. Soc.*, 2007, **129**, 3339–3347.
16. M.-C. Daniel and D. Astruc, *Chem. Rev.*, 2004, **104**, 293–346.
17. (a) N. L. Rosi, and C. A. Mirkin, *Chem. Rev.*, 2005, **105**, 1547–1562; (b) C. N. Murthy and K. E. Geckeler, *Chem. Commun.*, 2001, **13**, 1194–1195; (c) Y. Liu, Y. W. Yang and Y. Chen, *Chem. Commun.*, 2005, **33**, 4208–4210.
18. C. Loo, A. Lowery, N. Halas, J. West and R. Drezek, *Nano Lett.*, 2005, **5**, 709–711.
19. J. Chen, D. Wang, J. Xi, L. Au, A. Siekkinen, A. Warsen and X. Li, *Nano Lett.*, 2007, **7**, 1318–1322.

20. P. T. Anastas, L. G. Heine and T. C. Williamson (ed.), ACS Symposium Series 767, ACS Publications, 2000.
21. A. P. Akhlaghi, B. Peng, Z. Yao and K. C. Tam, *Soft Matter*, 2013, **9**, 7905–7918.
22. M. Breimer, E. Yevgheny and O. A. Sadik, *Nano Letters*, 2001, **1**, 305–308.
23. D. Andreescu, A. Wanekaya, O. A. Sadik and J. Wang, *Langmuir*, 2005, **21**, 6891–6899.
24. O. Marcells, V. Okello, V. Lee, L. Zhou and O. Sadik, *ACS Catal.*, 2011, **1**, 139–146.
25. V. A. Okello, A. Zhou, J. Chong, M. T. Knipfing, D. Doetschman and O. A. Sadik, *Environ. Sci. Technol.*, 2012, **46**, 10743–10751.
26. S. K. Kikandi, Q. Wong, V. Okello, O. Sadik, K. Varner and S. Burns, *Environ. Sci. Technol.*, 2011, **45**, 5294–5300.
27. I. Yazgan, N. Du, R. Congdon and O. A. Sadik, *J. Memb. Sci.*, 2014, **472**, 261–271.
28. B. Karn and O. Sadik, *ACS Sustainable Chem. Eng.*, 2013, **1**, 685.
29. Sustainable Nanotechnology Organization, http://www.susnano. org [accessed 20 January 2014].
30. C. D. Murray, D. Norris and M. G. Bawendi, *J. Am. Chem. Soc.*, 1993, **115**, 8706–8715.
31. P. J. Raveendran, H. Fu and S. L. Wallen, *J. Am. Chem. Soc.*, 2003, **125**, 13940–13941.
32. S. Jin, Y. Yang, J. E. Medvedeva, J. R. Ireland, A. W. Metz, J. Ni and T. J. Marks, *J. Am. Chem. Soc.*, 2004, **126**(42), 13787–13793.
33. D. Philip, *Phys. E*, 2010, **42**, 1417–1424.
34. V. K. Sharma, R. A. Yngard and Y. Lin, *Adv. Colloid Interface Sci.*, 2009, **145**, 83–96.
35. K. M. M. Abou El-Nour, A. Eftaiha, A. Al-Warthan and R. A. A. Ammar, *Arabian J. Chem.*, 2010, **3**, 135–140.
36. Y. Saito, J. J. Wang, D. N. Batchelder and D. A. Smith, *Langmuir*, 2003, **19**(17), 6857–6861.
37. A. van den Berg, *FEBS Lett.*, 1995, **358**, 215–218.
38. C. Auvin, C. Baraguey, A. Blond, F. Lezenven, J. L. Pousset and B. Bodo, *Tetrahedron Lett.*, 1997, **38**(16), 2845–2848.
39. C. Zhu, S. Guo, Y. Fang and S. Dong, *ACS Nano*, 2010, **4**(4), 2429–2437.
40. S. S. Shankar, A. Rai, B. Ankamwar, A. Singh, A. Ahmad and M. Sastry, *Nat. Mater.*, 2004, **3**(7), 482–488.
41. S. S. Shankar, A. Rai, A. Ahmad and M. Sastry, *Chem. Mater.*, 2005, **17**(3), 566–572.

42. B. Liu, J. Xie, J. Y. Lee, Y. P. Ting and J. P. Chen, *J. Phys. Chem. B*, 2005, **109**(32), 15256–15263.
43. R. Brayner, M.-J. Vaulay, F. Fiévet and T. Coradin, *Chem. Mater.*, 2007, **19**, 1190–1198.
44. N. Vigneshwaran, A. A. Kathe, P. V. Varadarajan, R. P. Nachane and R. H. Balasubramanya, *Langmuir*, 2007, **23**(13), 7113–7117.
45. S. Gass, J. M. Cohen, G. Pyrgiotakis, G. A. Sotiriou, S. E. Pratsinis and P. Demokritou, *ACS Sustainable Chem. Eng.*, 2013, **1**(7), 843–857.
46. M. N. Nadagouda, T. F. Speth and R. S. Varma, *Acc. Chem. Res.*, 2011, **44**, 469–478.
47. I. Bilecka and M. Niederberger, *Nanoscale*, 2010, **2**, 1358–1374.
48. I. Bilecka, P. Elser and M. Niederberger, *ACS Nano*, 2009, **3**, 467–477.
49. M. C. Moulton, L. K. Braydich-Stolle, M. N. Nadagouda, S. Kunzelman, S. M. Hussain and R. S. Varma, *Nanoscale*, 2010, **2**(5), 763–770.
50. F. A. Jumeri, H. N. Lim, S. N. Ariffin, N. M. Huang, P. S. Teo, S. O. Fatin and I. Harrison, *Ceram. Int.*, 2013, **40**(5), 7057–7065.
51. J. Lin, J. B. Mueller, Q. Wang, G. Yuan, N. Antoniou, X. C. Yuan and F. Capasso, *Science*, 2013, **340**(6130), 331–334.
52. T. P. Eager and I. Linkov, *J. Ind. Ecol.*, 2008, **12**, 282–285.
53. M. R. Wiesner, G. V. Lowry, P. Alvarez, D. Dionysiou and P. Biswas, *Environ. Sci. Technol.*, 2006, **40**(4), 4336–4345.
54. Nanotechnology Working Group, Science Policy Council, US Environmental Protection Agency, *Nanotechnology White Paper*, EPA 100/B-07/001, US Environmental Protection Agency, Washington DC, 2007.
55. M. Shannon, P. Bohn, M. Elimelech, J. Georgiadis, B. Marinas and A. Mayes, *Nature*, 2008, **452**, 301–310.
56. T. E. Cloete, M. de Kwaadsteniet, M. Botes and J. M. López-Romero (ed.), *Nanotechnology in Water Treatment Applications*, Caister Academic Press, Wymondham, Norfolk, UK, 2010.
57. B. Marelize Botes and T. Cloete, *Crit. Rev. Microbiol.*, 2010, **36**, 68–81.
58. T. Masciangioli and W. Zhang, *Environ. Sci. Technol.*, 2003, **37**, 73A–112A.
59. W. X. Zhang, *J. Nanopart Res.*, 2003, **5**, 323–332.
60. M. R. Wiesner, G. V. Lowry, P. Alvarez, D. Dionysios and P. Biswas, *Environ. Sci. Tech.*, 2006, **40**, 4336–4345.
61. N. Du, C. Wong, M. Feurstein, O. Sadik, C. Umbach and B. Sammakia, *Langmuir*, 2010, **26**, 14194–14202.

62. D. S. Andreescu, S. Andreescu and O. A. Sadik, in *Comprehensive Analytical Chemistry*, ed. L. Gorton, Elsevier, Amsterdam, 2005, vol. 44, pp. 285–327.

63. M. A. Omole, I. K'Owino and O. A. Sadik, in *Nanotechnology Applications: Solutions for Improving Water Quality*, ed. A. Street, J. Duncan, M. Diallo, N. Savage and N. Sustich, William Andrew Inc., Norwich, New York, 2009, ch. 17.

64. M. A. Omole, I. K'Owino and O. A. Sadik, *Appl. Catal. B Environ.*, 2007, **76**, 158–176.

65. M. Omole, V. Okello, V. Lee, L. Zhou and O. Sadik, *ACS Catal.*, 2011, **1**, 139–146.

66. C. A. Smolders, A. J. Reuvers, R. M. Boom and I. M. Wienk, *J. Membr. Sci.*, 1992, **73**, 259–275.

67. T. H. Young and L. W. Chen, *Desalination*, 1995, **103**, 233–247.

68. N. Du, PhD dissertation, Binghamton University, 2012.

69. E. Osibote, N. Noah, O. Sadik, D. McGee and M. Ogunlesi, *Reprod. Biol. Endocrinol.*, 2011, **9**, 65.

70. J. M. Laparra, R. P. Glahn and D. D. Miller, *Cell Biology International*, 2009, **33**, 971–977.

71. A. K. Gautam, *J. New Biol. Rep.*, 2013, **2**, 67–70.

72. W. Weyman-Kaczmarkowa and Z. Pedzivilk, *Microbiol. Res.*, 2000, **155**, 107–112.

73. C. T. Voisey, *Fungal Biol. Rev.*, 2010, **24**, 123–131.

74. A. Brand and N. Gow, *Microbiology*, 2009, **12**, 350–357.

75. Sigma-Aldrich, http://www.sigmaaldrich.com/united-states.html.

76. R. Dumitru, J. M. Hornby and K. W. Nickerson, *Antimicrob. Agents Chemother.*, 2004, **48**, 2350–2354.

77. B. P. Roy and F. Archibald, *Appl. Environ. Microbiol.*, 1993, **59**, 1855–1863.

78. A. Avent, G. Birkett, P. R. Kroto, W. Harold Taylor, Roger and D. R. M. Walton, *Chem. Commun.*, 1998, 2153–2154.

79. T. M. Scown, E. M. Santos, B. D. Johnston, B. Gaiser, M. Baalousha, S. Mitov, J. R. Lead, V. Stone, T. F. Fernandes, M. Jepson, R. van Aerle and C. R. Tyler, *Toxicol. Sci.*, 2010, **115**, 521–534.

80. H. Yamawaki and N. Iwai, *Am. J. Physiol.*, 2006, **290**, C1495–C1502.

81. X. L. Yang, C. H. Fan and H. S. Zhu, *Toxicol. in Vitro*, 2002, **16**, 41–46.

82. C. W. Isaacson, M. Kleber and J. A. Field, *Environ. Sci. Technol.*, 2009, **43**, 6463–6474.

83. Y. N. Yamakoshi, T. Yagami, K. Fukuhara, S. Sueyoshi and N. Miyata, *J. Chem. Soc., Chem. Commun.*, 1994, 517–518.

84. (a) F. Diederich and M. Gomez-Lopez, *Chem. Soc. Rev.*, 1999, 28, 263;   (b) Y. Liu, H. Wang, P. Liang and H.-Y. Zhang, *Angew. Chem.*, 2004, **116**, 2744-2748; *Angew.Chem., Int. Ed.*, 2004, **43**, 2690–2694.

85. A. Ikeda, T. Hatano, M. Kawaguchi, H. Suenaga and S. Shinkai, *Chem. Commun.*, 1999, 1403–1404.

86. L. Yu, Y. Ying-Wei and Y. Chen, *Chem. Commun.*, 2005, 4208–4210.

87. S. N. Kikandi, V. Okello, Q. Wang, O. A. Sadik, K. E. Varner and S. A. Burns, *Environ. Sci. Technol.*, 2011, **45**, 5294–5300.

88. V. Okello, PhD dissertation, Binghamton University, 2013.

CHAPTER 10

New Chemical Processes aimed at Sustainable Development in Brazil

TELMA TEIXEIRA FRANCO*[a,b] AND RICARDO BALDASSIN JR[c]

[a] Interdisciplinary Centre for Energy Planning, State University of Campinas, Campinas, São Paulo, Brazil; [b] School of Chemical Engineering, State University of Campinas, Campinas, São Paulo, Brazil; [c] School of Agricultural Engineering, State University of Campinas, Campinas, São Paulo, Brazil
*Email: franco@feq.unicamp.br

10.1 INTRODUCTION

Brazil has a broad ranging climate and a varied geomorphology which is responsible for several important biomes and ecosystems. It is fortunate in having abundant and diverse agricultural and forestry resources, occupying a total area of 8.5×10^6 km^2 between 5°16'N and 33°44'S. The Neotropical region that stretches from the southern part of North America through to the southern tip of South America, thus encompassing most of the Latin American countries, is one of the most diverse biogeographic regions on earth.[1,2]

Over the past 50 years, Brazilian agriculture has developing into modern, efficient and competitive sector. The Brazilian climate, the reasonable rain distribution, the solar energy, the abundance of water resources and the size of the available land are all factors that make 390×10^6 ha suitable for agriculture. However, not all this land is used

Chemical Processes for a Sustainable Future
Edited by Trevor M. Letcher, Janet L. Scott and Darrell A. Patterson
© The Royal Society of Chemistry 2015
Published by the Royal Society of Chemistry, www.rsc.org

for agricultural purposes. In 1973 the Ministry of Agriculture created the Brazilian Agricultural Research Corporation (EMBRAPA) with the aim of improving Brazilian agriculture by speeding up of the development of agricultural technologies and encouraging innovation. It helped Brazil to increase food production, while at the same time, conserving natural resources and the environment. It also helped to reduce the dependence on technological know-how from outside the country and the importation of basic products and genetic material.[3]

Another important and direct effort was targeted at the study of Brazilian biodiversity. The BIOTA programme was set up in 1999 in Sao Paulo state and is supported by the State of São Paulo Research Foundation (FAPESP). FAPESP is a non-political, taxpayer-funded foundation, one of the main funding agencies for scientific and technological research in Brazil. The community making up this programme (approximately 500 PhDs, plus 400 graduate students and more than 100 collaborators from other Brazilian states and approximately 80 collaborators from abroad) have described more than 1800 new species, acquired and archived more than 12 000 species, and made data available more than 35 major collections online. One of the most striking influences of the programme is the recommendation of a joint resolution by the secretaries of state of environment and agriculture to establish an agro-ecological zoning ordinance that prohibits sugarcane expansion to areas that are priorities for biodiversity conservation and restoration.[2]

Other Brazilian states have long-term programmes based on the BIOTA-FAPESP guidelines. The Brazilian National Research Council (CNPq) is planning a similar initiative and, likewise, the US National Science Foundation recently launched the programme, Dimensions of Biodiversity.[2]

In response to the increasing need for research and development (R&D) in the area of biofuels, in 2008 FAPESP created the programme for research on bioenergy, BIOEN, which aims to link public and private R&D, using academic research institutions and industrial laboratories to advance and apply knowledge in fields related to ethanol production. The BIOEN program has five main divisions: (1) biomass, (2) biofuel technologies, (3) biorefinery, (4) engines and (5) impacts. The main objectives of the biomass division are to identify new paths to genetically manipulate the energy metabolism of crops creating new biofuel alternatives and the main objectives of the biofuels technology division are to increase productivity (amount of ethanol per tonne of sugarcane), improve energy and water saving, and minimize environmental impacts. The objectives of the biorefinery division are to find substitutes for fossil-derived compounds and the main objective of the engines division are to develop new engine configurations that efficiently use renewable sources of energy focusing on combustion and fuel cells. Finally, the main objective of the impact division is to study the impact on areas such as social, economic, environmental and land. To consolidate biofuel production, one needs to understand the local and global market, patterns of bioenergy diffusion, institutional incentives, rules and industrial public policies.[4]

A number of projects on plant genomics have been developed in Brazil over the past 10 years with the support of FAPESP and CNPq. These projects include research related to oranges, sugarcane, coffee, eucalyptus and bananas. Thus the development and the maturity of the Brazilian sugarcane industry should be considered a result of a concentrated effort by many organized initiatives, national and local administrations, associations of producers and scientists working at research centres.[2]

A team of Brazilian scientists sequenced the sugarcane genome in 2003.[5] Their findings, published in *Genome Research*, suggest that 2000 of the more than 33 000 genes in the sugarcane genome are associated with sugar production. Transgenic sugarcane plants have been studied by Brazilian laboratories in cooperation with local industries. Several other publications identified genes of sugarcane and correlated to their functional characterization. Arruda[6] has reported that sugarcane breeding has progressed over the past 30 years but attempts to further increase crop yield have been limited due to the complexity of the sugarcane genome. According to this work, an alternative to boost crop yields is the introduction of genes encoding desirable traits in the elite sugarcane cultivars. Genetically modified sugarcane with increased yield and pest and disease resistance has proved its value not only by its increased sugar content but also by the improvement in crop performance. However, transgene stability is still a challenge since transgene silencing seems to occur in a large proportion of genetically modified sugarcane plants. In addition, regulatory issues associated with the crop propagation model will be a challenge for the commercial approval of genetically modified sugarcane.[6]

Brazil is one of the few leaders in the development of modern technology in tropical agriculture and in the introduction of renewable energy on a large scale. A process of complete integration of the sugar milling process has taken place. Specific legislation, negotiation between the main sectors (ethanol producers, car manufacturers, governmental regulatory sectors and the oil industry) were all part of the process.[7]

One of our major challenges today is to convert biomass into compounds for the chemical industry and partially replace petrochemicals.[8] Several processes which today would be considered as examples of biorefinery have been known since the 19th century. These include the saccharification of wood, the solubilization of cellulose, sugar refinery, starch and maltodextrin production for industrial purposes, vegetable oil and chlorophyll extraction, furfural production from bran distillation and the isolation of vanillin from lignin (developed in 1874). More recently, the production of lactic acid by fermentation can be added to this list.[9] The main value-adding technologies associated with plant biorefineries are complex, offering great potential given the huge variety and quantity of renewable biomass and derived products. In addition, the biorefinery provides the possibility to develop non-polluting and sustainable industries with a low environmental impact.

10.2   A SUSTAINABLE CHEMICAL PROCESS

Sustainable chemical processes are not just linked to those processes that use biomass as a feedstock, though this is a vital part of the sustainable chemistry platform in Brazil. Green chemistry refers to the practice of chemistry with the goal of protect the environment and health of the biosphere, and may contribute to building a sustainable society and future.[7–10] Green chemistry comprises two main components. First, it addresses the efficient use of resources and the concomitant minimization of waste and secondly it deals with the ecological, health and safety issues associated with the manufacture, use and disposal or re-use of chemical products.[11] An environmentally sustainable production target is a vital issue for industry which should aim to produce chemicals, including feedstocks, in compliance with environmental and social requirements. This will increasingly become a requirement for accessing markets and as the certification processes become more complex, they will require adaptations by the supply chain.[12,13]

From 2011 to 2013, an initiative by Boeing, Embraer and FAPESP[14] to assess the technological, economic and sustainability challenges and opportunities associated with the development and commercialization of sustainable aviation biofuels in Brazil lead to a roadmap for Brazilian aviation biofuels This project was coordinated by the State University of Campinas (UNICAMP) and developed in workshops with participation of 30 different stakeholders from feedstock producers, agriculture, supply chain, refining technology companies, government, non-governmental organizations (NGOs) and academia. The project lays out the pathways to establish a new biofuels industry to replace petroleum jet fuels, including recommendations for: filling research and development gaps in the production of sustainable feed stocks; more incentives to overcome conversion technologies barriers including scaling up issues; greater involvement and interaction between private and government stakeholders; and the creation of a national strategy to make Brazil a leading country in the development of aviation biofuels.[14]

A similar roadmap for sustainable chemicals is being developed in Brazil, taking into account the strong sustainability criteria used in the roadmap for biofuels for aviation. In summary, the sustainable production of chemicals and biochemicals in Brazil (in this chapter these two are referred to as chemicals) should be guided by three important targets: improved regional development; reduction of production costs; and environmentally sustainable production (Figure 10.1).

In 2004 a joint study between the Pacific Northwest National Laboratory of the United States (PNNL) and the National Renewable Energy Laboratory of Brazil (NREL) was published which listed and described a selection a of potential chemicals that could be obtained from biomass and that would be essential for the gradual replacement of the petrochemical platform.[15] It was recognized that those compounds could lead to a partial replacement of the known

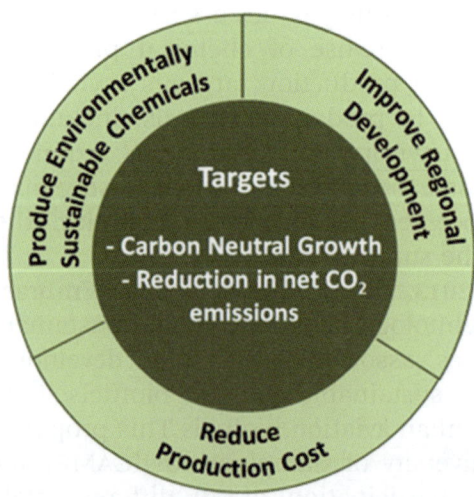

Figure 10.1 Targets for a sustainable chemical platform in Brazil.

chemical platform and were assigned the name 'building blocks'. From these compounds with multiple functional groups (bio-commodities platform), other compounds could then be obtained. The preliminary study suggested 300 chemicals (building blocks), later reduced to 30 and then finally to 14 molecules, to be used to produce high-value chemicals: 1,4-diacids (succinic, fumaric and malic); 2,5-furan dicarboxylic acid; 3-hydroxy propionic acid; aspartic acid; glucaric acid; glutamic acid; itaconic acid; levulinic acid; 3-hydroxybutyrolactone; glycerol; sorbitol; and xylitol/arabinitol.

10.3   SUGARCANE AND OTHER FEEDSTOCKS IN BRAZIL

Brazil has a unique combination of: suitable land already cleared and available for agriculture; a dynamic agriculture sector showing strong productivity growth; large tracts of legally protected native vegetation; strong conservation laws; and human health and safety regulations already in place for rural activities. This places Brazil in a good position from the point of view of feedstock supply and, if policies are carried out, to develop a programme to produce chemicals in compliance with responsible principles and sustainable requirements.[14]

Sugarcane only occupies 1% of the Brazilian territory and represents one of the most important Brazilian agriculture products, while pasture (mainly cattle breeding) uses approximately one quarter of Brazilian territory, with a cattle density of around of one head per ha. The Amazon forest, Pantanal, and many other areas, represent 65.1% of Brazil, all of which are protected by strong conservation laws (Figure 10.2).[16]

Sugarcane was first introduced into Brazil in 1532 by the Portuguese crown. However, even though sugarcane was a major crop in Brazil, it was only after 1970 that production and productivity increased considerably (Figure 10.3).[17,18] In 2009–2010 the harvest registered was a record with the production of 690×10^6 t of sugarcane and a productivity of 80.2 t ha^{-1}. In addition to sugar and bioethanol, electricity is generated in Rankine cycles (cogeneration system) using sugarcane bagasse as the fuel. In 2009–2010, the surplus of bagasse was approximately 10% (estimated to be 14.7×10^6 t). Brazil has become the largest producer and exporter of sugar and the second largest producer of bioethanol in the world.[18,19]

The Brazilian energy matrix is made up of largely renewable sources. In 2012, 42.4% of the total energy supply and 84.6% of the electricity supply came from renewable sources. Most of the electricity supply in Brazil comes from hydro (70–85)%, while biomass (mainly

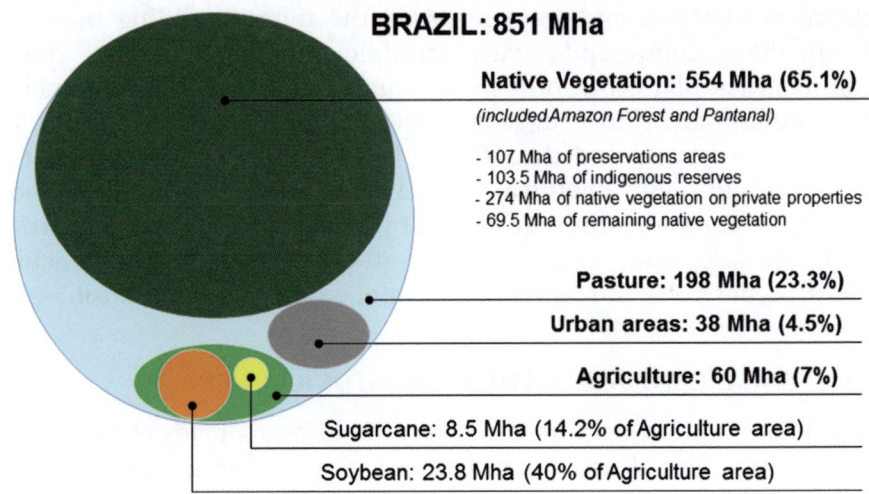

Figure 10.2   Land use in Brazil.

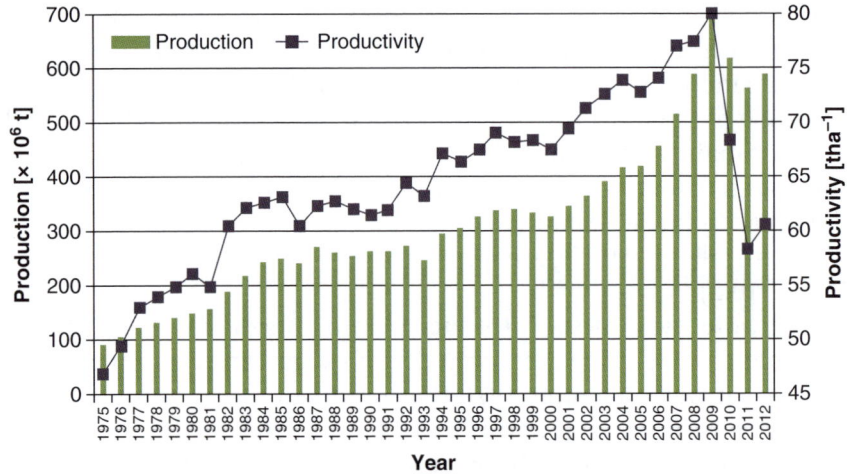

Figure 10.3   Evolution of sugarcane production and productivity in Brazil, 1975–2012.

sugarcane) contributes 6.6% (figures from 2012).[20] In the same year, bioethanol was responsible for 12.5% of the total energy consumption in transportation and approximately 27% of energy consumption for light vehicles. Biodiesel is a renewable fuel introduced in Brazil (made mainly by soybean), and in 2012 total production was 2.7×10^9 L (2.7 billion litres).[20] Besides bioethanol and biodiesel, the production and use of charcoal is also known in Brazil as bioenergy. Most of the Brazilian charcoal is used in the production of steel and iron based products.[21] In 2012, firewood and charcoal represented 9.1% of the total energy supply.[20]

Biomass feedstock is a critical element in the production of chemicals; biomass can be rich in sugar, starch, fat, oil, fibre and can be obtained directly or indirectly from cultivated crops, algae or residues. Using this definition, a large number of potential feedstocks could be used in the production of chemicals, each with their own unique process and technology to be converted into chemicals.[3]

Brazil is in a good position to supply large amounts of feedstocks for chemical production. Abandoned pasture land is being partially recovered to produce large amounts of grains and food, and could potentially be used plant species to produce feedstock for chemicals rather than to produce sugar. Table 10.1 shows the potential feedstocks actually produced in Brazil and their main characteristics.

The use of lignocellulose materials (lignin, hemicellulose and cellulose) for second-generation production of chemicals has not yet become an industrial reality due to the lack of efficiency and low-cost technologies. Even the massive use of the first generation chemicals (from sugars and vegetable oils) from biomass has not yet reached its full potential in Brazil, due to a series of difficulties, mainly related to sustainability, food *versus* fuel competition and land use.[25]

One of the biggest by-products of the ethanol industry is sugarcane bagasse. It is a cellulosic residue obtained after extracting the juice from the sugarcane. Currently, bagasse is used as a primary source of energy in most Brazilian mills and as raw material in paper production in some countries. Bagasse is a heterogeneous material that varies in its composition as well as in their morphological structure depending on the procedures used in cutting and processing. Typically, it contains cellulose (32–48)%, hemicellulose (27–32)%, lignin (19–24)%, silica (0.7–3.5)% and ash (1–5.5)%.[26]

Mechanical (fractionation, crushing, extraction), chemical (acid, alkaline, organosolv), physicochemical (steam explosion, hydrothermolysis) and biological (enzymes, microorganisms) or a combination of pretreatments that prepare large volumes of biomass and condense them to lower volumes can be found in small units and many chemicals can be derived from this biomass feedstock.

Agricultural by-products in Brazil are abundant, *e.g.* crop wastes, forest and wood residues, and from wastes from the biodiesel industry (*i.e.* press cake from seeds, fruit bodies, empty fruit bunches as well as from the leaves of the plants without any commercial

Table 10.1 Main feedstocks in Brazil.

Feedstock		Production/10^6 t	Area/10^6 ha	Productivity/ t ha^{-1} a^{-1}	Main states where produced	Status and potential
Fats and oils	Tallow	0.947 (estimated)	198[16] (total pasture area)	23[16] (kg head^{-1})	Mato Grosso, Mato Grosso do Sul, São Paulo e Goiás	Tallow is an important feedstock for energy production in Brazil, and responsible for 14% of amount of biodiesel produced in 2013. Considering that Brazil is the second most important world beef producer, the use of tallow to produce chemical and fuels will increase in future years.
	Soybean[16]	81.7	27.9	2.93	Mato Grosso, Paraná, Rio Grande de Sul e Goiás	Soybean is the basis of the Brazilian Biodiesel Programme, responsible for 80% of the amount of biodiesel produced in 2013. The disadvantages of this biomass to produce energy/fuels are the low productivity and the high competition with food supply.
	Cotton[16]	3.5 (seed)	0.94	3.7	Mato Grosso, Bahia e Goiás	Cotton (fibres) has an important role in Brazilian textile industry and the seeds are used to produce vegetal oil.

Palm[16]	1.2	0.1	22	Pará, Bahia e Amapá	Palm is considered one of the main options for Brazil to improve the production of biofuels and chemicals due to the potential high productivity, low costs (production and transformation) and the high quality of the oil. However, large-scale production of palm is not yet a reality in Brazil.
Wastes — Municipal solid waste (MSW)	34,432[22] (collected waste)	851[16] (Brazilian territory)	94,335[22] (t d^{-1} for organic waste)	São Paulo, Minas Gerais, Rio de Janeiro e Bahia	Highs amounts of MSW are available in all countries around the world. For the use of these feedstock to be feasible for the production of energy/fuel (or chemicals), however, it is necessary to pretreat the material (separation, drying/grinding, *etc.*), and the high costs of this process is the main barrier to be overcome.
Sugar — Sugarcane[16]	588.5	8.81	74.77	São Paulo, Goiás, Minas Gerais e Paraná	Sugarcane is the basis of the Brazilian Biofuel Programme and is considered the most important feedstock for energy production in Brazil. In 2013, the production was converted into 23.2×10^9 L of bioethanol and 38.2×10^6 t of sugar.[18] The bioethanol

Table 10.1 (*Continued*)

Feedstock		Production/10^6 t	Area/10^6 ha	Productivity/ t ha^{-1} a^{-1}	Main states where produced	Status and potential
						produced was responsible for 27% of the transport energy consumption (light vehicles only).[20]
Starches	Sweet Sorghum	NA	NA	30	NA	Sweet sorghum is not yet a reality in Brazil. However, it is considered a potential feedstock for energy/fuel production due the high productivity when planted in the land not used for sugarcane.
	Rice[16]	11.76	2.38	NA	Rio Grande do Sul, Santa Catarina e Maranhão	Rice plays an important role in food supply in Brazil. However, no identified commercial use was identified for its waste (straw). Logistic barriers need to be overcome (high costs).
	Corn[16]	80.4	15.67	5.0	Paraná, Mato Grosso e Minas Gerais	Corn has an important role in food supply in Brazil and the wastes (straw and cob) have been used for animal feed or steam production. However, some places simply leave them over the soil as vegetal cover.

Cassava[16]	21.24	2.09	13.9	Pará, Paraná, Bahia e São Paulo	Cassava has an important role in food supply in Brazil. However, its use for energy/chemicals production has not been studied. The main characteristic is the high productivity and the disadvantage is the competition with food supply.
Algae	Waste and Sucrose	NA	NA	NA	NA	Algae are not yet a reality in Brazil. However, it is considered an important feedstock to be considered in future years due to the high productivity and low area demand.
Lignocellulose	Sugarcane trash and bagasse	126.86 and 123.34, respectively *(estimated)*	-	14.4 and 14, respectively *(dry base)*[23]	São Paulo, Goiás, Minas Gerais e Paraná	Today, two feedstocks are considered waste/residues from the sugarcane industry. Bagasse is used in cogeneration systems (old technologies with low efficiency) to produce electricity and steam to be used in the production processes. Sugarcane trash (dry and green leaves) is left on the fields as natural cover. The high costs of harvesting and transportation of trash are the main barriers to be overcome for the use of trash for electricity, fuels and chemicals production in the mills.

Table 10.1 (*Continued*)

Feedstock	Production/10^6 t	Area/10^6 ha	Productivity/ t ha^{-1} a^{-1}	Main states where produced	Status and potential
Eucalyptus[24]	207.77×10^6 m^3	5.1	40.7 (m^3 ha^{-1} a^{-1})	Minas Gerais, São Paulo, Bahia e Mato Grosso do Sul	Brazil is one of the most important producer of wood, mainly for pulp and paper, and lumber and processed wood production. The main characteristics are the high productivity and the low production cost.
Pinus[24]	62.75×10^6 m^3	1.6	40.1 (m^3 ha^{-1} a^{-1})	Paraná, Santa Catarina, Rio Grande do Sul e São Paulo	Similar to eucalyptus, pinus is an important culture in Brazil, mainly for lumber and processed wood production. The main characteristics are the high productivity.

value.[3–21] Lignocellulosic biomass is the most abundant raw material from agricultural residues and is usually left in the fields after harvesting. Its utilization is an important option to be considered in industrial biotechnology and biorefining, as it could result in a reduction in costs, avoid competition with the food and feed industries, and minimize environmental problems resulting from excessive volume of disposed waste.[27]

It is estimated that 350 million tonne of lignocellulose waste is available annually in Brazil.[28]

10.4 BIOREFINERIES

A biorefinery is similar in nature to a petroleum refinery in that it takes in raw materials, in this case, consisting primarily of renewable polysaccharides and lignin, and then by fractionation and other chemical processes converts the feedstock into transportation fuels, co-products and direct energy. This should be supported by economies of large scale and by efficient use of all bioresources. Biorefineries at present operate on a limited product range of simple material (bioethanol, electricity, sugar). Second-generation biorefineries need to be built to explore the full concept of sustainable production of chemicals by bioprocessing, thermochemical conversion, including pyrolysis, Fischer–Tropsch and other catalytic processes. Also high value products should be extracted directly from the biomass; these include fragrances, flavouring agents and natural pigments.[9] Figure 10.4 shows the relevant Brazilian products already derived from sugarcane after being processed by the operations described in Sections 10.4.1 to 10.4.6 below.

10.4.1 Fast Pyrolysis and Solvent Liquefaction

Fast pyrolysis converts lignocellulose biomass into a liquid product (pyrolysis oil or bio-oil), which can be upgraded to produce liquid chemicals. Fast pyrolysis of biomass produces bio-oil, which is easily transported and stored, and can be upgraded by hydroprocessing or fractionation (distillation).[29] The products can then be commercialized according to the properties of each of the fractions obtained. Bio-oil could also be transported to a syngas refinery to be further processed by Fischer–Tropsch technologies.[30] Several decentralized pyrolysis plants have been envisaged as the best logistics model to pretreat and transport the bio-oil to a centralized large scale gasification and Fisher–Tropsch site, allowing the production of electricity, fuels and chemicals at competitive costs.[21–31]

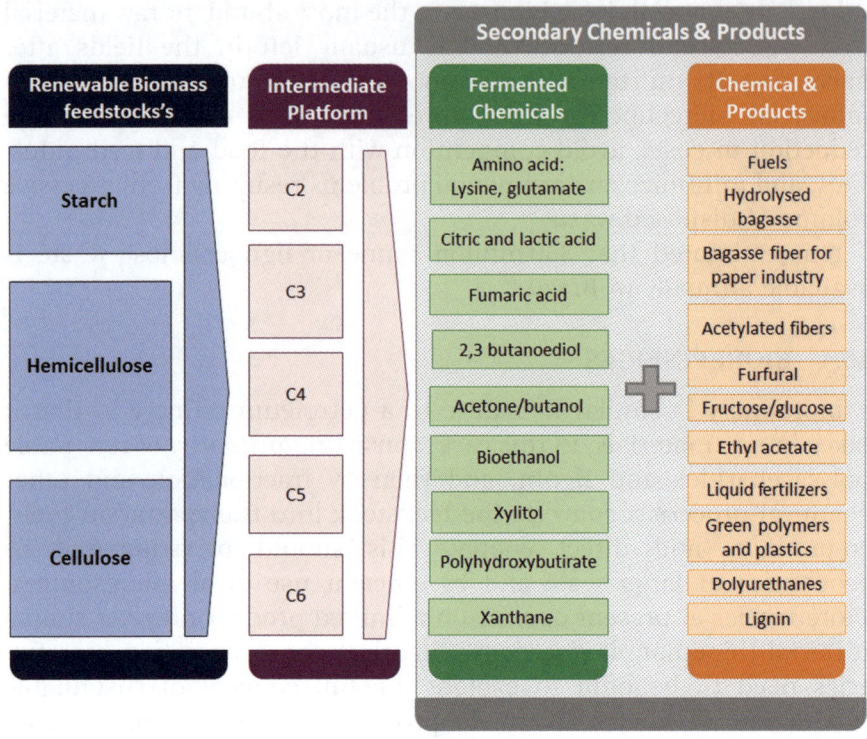

Figure 10.4 Main products potentially produced in a Brazilian biorefinery from lignocellulose and starch.

Biomass conversion to liquid (BTL) fuels (bio-oil) *via* solvent liquefaction, pyrolytic processes and catalytic hydroprocessing achieves good yields of liquid hydrocarbons. Solvent liquefaction seems to be a promising technique as no hydrogen or catalysts are needed and the conversion conditions are rather moderate. Patents have shown that the level of upgrading of bio-oil depends on the final use of the product, which in turn depends on the properties of the desired liquid fuel.[29] BTL combines elements of pyrolysis, gasification, catalytic conversion and the entire platform can be implemented in a biorefinery, according to some authors.[32,33] BTL typical bio-oil is an emulsion in which an aqueous solution of low molecular oxygenated compounds (cetic acid, methanol, acetone) is stabilized by an organic phase.

10.4.2 Hydrolysis

Hydrolysis with a low concentration of acid and steam explosion processing is able to disrupt the hemicellulose fraction and, to some

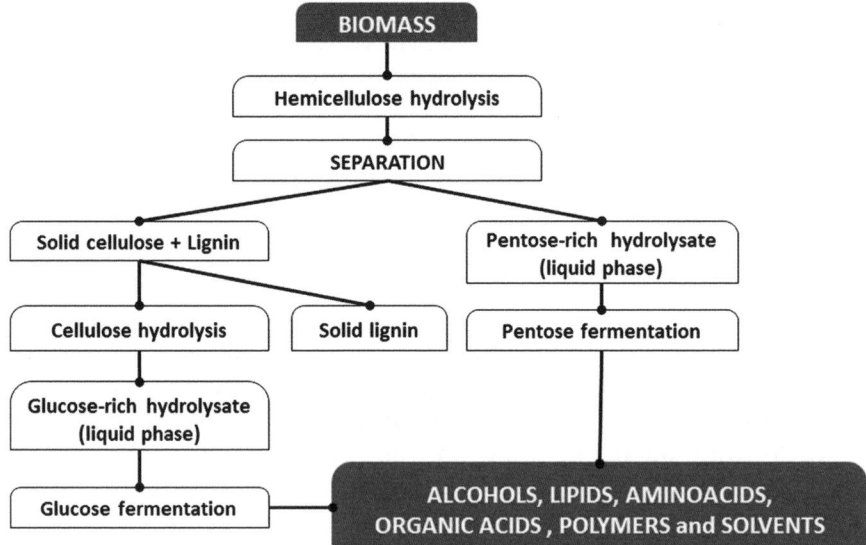

Figure 10.5   Flow chart for the production of two hydrolysates: hemicellulose pentose-rich hydrolysates and cellulose glucose-rich hydrolysate. Products are generated from both hydrolysates *via* fermentation or chemical pathways.

extent, release lignin and cellulose. The two main products obtained are the lignin combined with a cellulose fraction and the hemicellulose-rich hydrolysate (Figure 10.5).[34] Acid hydrolysis is an important process for pre-hydrolysis. The pre-hydrolysed cellulose can then be hydrolysed enzymatically. Different technological concepts aim to increase the amount of solids in the reactors.[35]

By-products such as furfural, hydroxymethylfurfural, organic acids and others which inhibit fermentation can be formed during hydrolysis. The concentration of inhibitory by-products, as well of the desired sugars (pentose sugars in monomeric or oligomeric forms) in the hydrolysis streams, obtained after the steam explosion pretreatment can be adjusted by setting the process conditions. In this way the hemicelluloses sugars can be released into the liquid phase and the solid cellulose can be further treated enzymatically. Despite the differences in acid pretreatments (inorganic acids, organic acids, impregnation) and steam treatments, it is possible to find a set of pretreatment conditions that allows the biomass to be converted with satisfactory yields, resulting in solid cellulose ready for enzymatic treatment.

In order to reduce the amount of microbial growth inhibitors, enzyme hydrolysis technology is being optimized owing to the high price of the enzymes. Various pilot and demonstration plants around the world are using different technological concepts aimed at increasing the amount of solids in the reactors to produce these compounds in-house and to recycle enzymes and to perform simultaneous saccharification and fermentation (SSF).[36–38]

The first cellulosic bioethanol plant in Brazil and in the southern hemisphere will use non-food sources such as sugarcane bagasse and straw, and is expected to begin operation in 2014. Operated by GranBio, the plant will have an annual production capacity of 82×10^6 L of ethanol, turning non-food vegetable material into fuel.

GranBio, a Brazilian biotechnology company, and Rhodia, part of Belgian chemical group Solvay, have signed an agreement to create a partnership to produce bio *n*-butanol in Brazil. The bio *n*-butanol will be made from sugarcane straw and bagasse, the same raw material that is used to manufacture ethanol and other biochemicals in Brazil. Under the partnership, the companies plan to build the world's first biomass-based *n*-butanol plant in Brazil, which will enter into operation in 2015.[36]

10.4.3   Fermentation to Alcohols, Lipids and Organic Acids

The Brazilian experience with fermentation to biofuels started in 1931 with the mandatory blending (5%) of ethanol from all gasoline commercialized in gas stations. In 1975, the Brazilian Ethanol Programme induced a large expansion of production by the progressive improvement of agro-industrial productivity, with the use of 25% ethanol gasoline blends and the introduction of pure ethanol cars. Production stagnated until the introduction of flex-fuel cars in 2003. Today, these cars represent about 93% of sales of new cars. In 2013, straight ethanol could be used by 20×10^6 Brazilian vehicles (mostly cars with flex-fuel engines) and by around 50% of the national light vehicle fleet.[39] In the 2012–2013 harvest season, 8.9×10^6 ha of sugarcane fields (approximately 1% of the Brazilian national area) produced 588.5×10^6 t of feedstock for sugar, ethanol and electricity. About 50% of the available sugar was used to produce 23.2×10^9 L of bioethanol.[18]

Although the main feedstocks for producing ethanol are sugars and starches, other alternatives including lignocellulose, industrial waste gases and municipal solid wastes (MSW) are feasible. Starches and sugars may be converted to alcohols through direct fermentation and industrial waste gases, rich in CO, can potentially be converted to alcohols by gas fermentation.[40]

The feasibility of efficiently converting waste material and dedicated vegetable material into biofuels, valuable chemicals or with a higher added value or into building blocks for industrial needs depends on the capability of the microorganisms to feed on pentoses as well as hexoses and to withstand the most common impurities found in the starting materials.

Organic compounds produced by fermentation can be considered as alternatives or substitutes for fossil-derived chemicals, since the substrates from nature (*i.e.* carbohydrates) can be converted into building blocks such as organic acids, alcohols, *etc.*, which can then be used to produce polymers, resins, fuels, *etc.*[41,42]

Amyris (Paraíso Bioenergia mill, São Paulo state, Figure 10.6) typically employs a strain of a modified yeast to produce oils from sucrose. Amyris[43] has launched a set of six large-scale 200 000 bioreactors to produce biofene (trans-ß-farescene) from sugarcane sucrose using genetically engineered yeast, *Saccharomyces*. Biofene, a terpenoid, is the chemical building block which will be used to make diesel and jet fuel, perfumes, cosmetics and detergents.[44]

Solazyme has recently announced two plants to produce oils from sugar through algae fermentation, one in the USA and one in Brazil. In the USA, Solazyme intends to produce 20×10^3 t a^{-1} of oil from dextrose in a joint project with Archer-Daniels-Midland (ADM), primarily for the industry and food markets.[45] In Brazil, at UsinaMoema, in São Paulo state, Solazyme has launched a project for 100×10^3 t a^{-1} production of oils from sugarcane for the Brazilian industry and fuel market, starting operations in 2014, in a joint venture with Bunge Limited.[46]

Corbion (formely Purac) produces bio-based materials, including a variety of biochemical building blocks such as lactic acid and lactates,

Figure 10.6 Amyris industrial plant for bio products (Paraíso Bioenergia mill, São Paulo state).

replacing materials based on fossil carbohydrates. The plant is located in Campos dos Goytacazes, Rio de Janeiro state, in plants located next to sugar mills. Sucrose is used as main substrate in a relatively simple fermentation route and the resultant lactic acid is then recovered by ion exchange.[47]

Citric acid is also produced in a microbiological process by submersed fermentation using *Aspergillus niger* by the company Tate & Lyle in Ribeirão Preto (São Paulo state)[48] and in another plant operated by Cargill located in Uberlandia, Minas Gerais state.[49] Currently there is no chemical process for citric acid production in Brazil.

10.4.4 Alcohol Chemistry and Polymers from Bioethanol

Different reactions of alcohol chemistry potentially lead to a variety of chemicals and materials (Figure 10.7).[50]

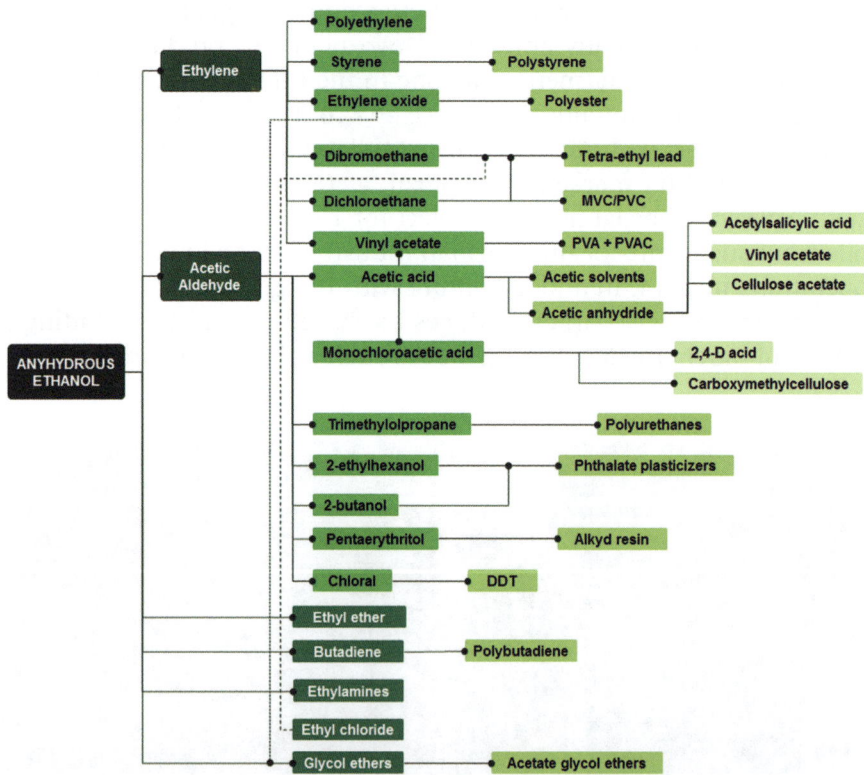

Figure 10.7 Products potentially produced in a Brazilian biorefinery from bioethanol.

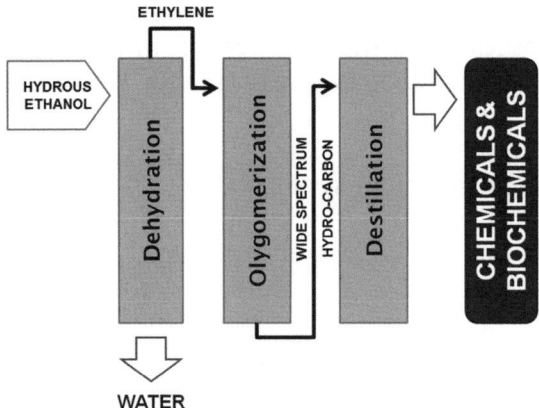

Figure 10.8   Oligomerization and purification of 'ethanol to chemicals' process.

The production of polyethylenes *via* alcohol dehydration/ oligomerization involves linking short-chain alcohol molecules (*e.g.* methanol, ethanol and others) together to form a broad range of hydrocarbons (Figure 10.8). Several chemical reactions can be employed to oligomerize alcohols. However, the economic rationale of this process still needs to be assessed.[51] Polyethylene and among other polypropylenes produced from sugarcane ethanol has also been announced by the Brazilian company Braskem.[52]

Different research groups have developed specific catalysts for the direct and high-yield conversion of bioethanol to fuel additives, rubbers and solvents. With increased availability and reduced cost of bioethanol, conversion of this particular bio-based feedstock to high-value fuels and chemicals has been a particularly important research goal. Currently, research on bioethanol conversion to value-added chemicals focuses mainly on ethanol dehydration to ethylene, or ethanol dehydrogenation to acetaldehyde and then to acetone *via* Aldol condensation pathways.[53]

When cellulosic sugar is available, the commercial risks are reduced. The alcohol intermediates may be methanol, ethanol, butanol, isopropanol, other alcohols or a mixture of alcohols.[54] A significant cost advantage would be achieved with the reduction of operational steps for this process.

10.4.5   Production of Biodegradable Plastic in Brazil

Polyhydroxybutyrate (PHB) is an environmentally degradable material belonging to the polyhydroxyalkanoate (PHA) family. Its manufacture involves a complete cycle starting from sugarcane and bacterial fermentative synthesis. In the absence of microorganisms, the hydrolysis of PHB in aqueous environments is slow because of its hydrophobicity. In principle, the lifetime of a stored PHB product is unlimited; however, after disposal, PHB becomes biodegradable in domestic effluent treatment systems.[55]

It has been shown that the composition of the culture media influences the weight average molar mass, M_w, and the number average molar mass, M_n, of pure PHB, with a molecular mass of 180–270 kDa.[56] Composites of this biopolymer associated with other material have been extensively described.[57] It can be processed as a conventional thermoplastic in most industrial transformation processes, including extrusion, injection and thermopressing, and can be transformed into rigid shapes (*e.g.* pipes) and films for packaging. PHB can also be modified by extrusion by incorporation of additives (stabilizers, plasticizers and pigments), immiscible additives (*e.g.* wood and starch powder), or by mixing with other plastics.

PHB Industrial S.A. is a Brazilian company created by the Biagi and Balbo groups, two large producers of the sugar and bioethanol, located in Serrana, São Paulo state (Figure 10.9) The joint land area for sugarcane production is approximately 100×10^3 ha. The company currently produces 20×10^6 packs of sugar (50 kg packs) and 500×10^6 L of bioethanol per year. In 2014, the company announced an investment of US\$ 50 million to scale up the production of 2×10^3 t a^{-1} of polyhydroxybutyrate, which would allow the preparation of 4×10^3 t a^{-1} of composites.[58]

Figure 10.9   PHB Industrial S.A. plant in Serrana, São Paulo state.

Currently there are approximately 10 companies able to produce PHA in the world, but only one in Brazil.[59]

10.4.6 Production of Amino Acids from Sugar

Amino acids are the most often produced chemicals in Brazil by fermentation of sugar (Table 10.2).[59] Evonik Industries has inaugurated the production of amino acids and ingredients for the production of cosmetics in two new plants in Brazil which should come on stream

Table 10.2 Relevant chemicals and products produced from sugarcane in Brazil.[60]

Product	Process	Company/Location
Organic acids		
Acetic	Ethanol oxydation	– Santa Maria sugar mill, São Paulo state
		– Rhodia Poliamida, Paulínia, São Paulo state
		– Cloroetil, Mogi Mirim, São Paulo state
		– Butilamil, Piracicaba, São Paulo state
Citric	Direct fermentation of sucrose	– Tate & Lyle, São Paulo, São Paulo state
Citric acid	Direct fermentation of sucrose	– Cargill, Uberlandia, Minas Gerais state
Lactic and lactates	Direct fermentation of sucrose	– Corbion-PuracSinteses, Campos, Rio de Janeiro state
Amino acids		
Glutamate	Fermentation of sucrose	– Ajinomoto, Laranjal Paulista, São Paulo state
Lysine	Fermentation of sucrose	– Ajinomoto, Valparaiso, São Paulo state
		– Evonik, Castro, Paraná state
		– CJ Cheil Jedang, São Paulo state
Material for polymers		
Poly(ethylene) and poly(propylene)	Chemical reaction from ethanol	– Braskem, Triunfo, Rio Grande do Sul state
Solvents		
Ethyl acetate	Esterification of acetic acid	– Oxiteno, Mauá, São Paulo state.
		– Braskem-Rhodia Poliamida, Paulínia, São Paulo state
		– Cloroetil, Mogi Mirim, São Paulo state
		– Butilamil, Piracicaba, São Paulo state

Figure 10.10   Evonik's plant for amino acids at the Cargill biorefinery complex in
                    Castro, Paraná state.

in 2014. The cosmetic ingredients (50×10^3 t a^{-1}) will be produced
in Americana, São Paulo state.[61] Evonik's production of lysine, in
partnership with Cargill, for animal feed is located in Castro at
the industrial complex and has a total capacity of 85×10^3 t a^{-1}
(Figure 10.10).

CJ CheilJedang (CJ Group) inaugurated a large plant in Sao Paulo
state in 2007 to produce lysine. The plant produces 50×10^3 t a^{-1} from
100×10^3 brown sugar. This company originally came from South
Korea.[62]

Ajinomoto has a few plants in Brazil to produce large supplies of
amino acids for food additives and for feed nutrition. This company
produces glutamate in Laranjal Paulista and lysine and threonine in
Valparaiso. Exclusive production of animal ingredients is located in
Pederneiras. All of its industrial plants are located close to sugar mills
in at Sao Paulo state, the largest Brazilian supplier of sugarcane.[63]

10.5   CONCLUSIONS AND COMMENTS

The separation (fractionation) of lignocellulose biomass into hemicellulose, cellulose and lignin for chemical processing is still a major
obstacle to the establishment of refineries; their improvement, optimization of operation and minimization of costs need to be encouraged.

Sugarcane bagasse, efficiently processed for pretreatment, hydrolysed (acid, enzyme or in combination) is the most abundant

lignocellulose in the estates of the south-east and centre west of Brazil. The processes to obtain good quality of hydrolysates (depolymerized) bagasse, rich into hexoses and pentoses, still need to be improved; there are problems related to the efficiency of fermentation and the chemical reactions related to enzymatic catalysis. Other feedstocks are abundant in different Brazilian geographical regions.

Microbiological routes from glucose, fructose and xylose should also be encouraged (because of their abundance) involving new microorganisms. Other types of catalysts have been described for a variety of new chemical routes.

Tools of genetic engineering and analysis of metabolic fluxes should be strongly encouraged to obtain a viable and economical route for obtaining chemicals from renewable sources.

Routes to obtain biopolymers from renewable and sustainable building blocks (existing or brand new) need to be developed, given the huge deficit of polymers from renewable resources in industrial production.

Simulation tools for metabolic processes and industrial processes, optimization techniques, modelling decision-making and exploration of 'process systems engineering' should also be encouraged to reduce the development time to obtain the necessary knowledge.

REFERENCES

1. T. T. Franco, in *Industrial Perspectives for Bioethanol*, Instituto Uniemp, São Paulo, Brazil, 2006, pp. 107–120.
2. C. A. Joly, R. R. Rodrigues, J. P. Metzger, C. F. B. Haddad, L. M. Verdade, M. C. Oliveira and V. S. Bolzani, *Science*, 2010, **328**, 1358.
3. S. Vaz Jr (ed.), *Biomassa para Química Verde*, Embrapa Agroenergia, Brasília, Brazil, 2013.
4. BIOEN-FAPESP, *Brazilian Research on Bioenergy*, São Paulo Research Foundation (FAPESP), São Paulo, Brazil, 2010.
5. A. L. Vettore, F. R. da Silva, E. L. Kemper, G. M. Souza, A. M. da Silva, M. I. T. Ferro, F. H. Silva, É. A. Giglioti, M. V. F. Lemos, L. L. Coutinho, M. P. Nobrega, H. Carrer, S. C. França, M. Bacci Junior, M. H. S. Goldman, S. L. Gomes, L. R. Nunes, L. E. A. Camargo, W. J. Siqueira, M. A. Van Sluys, O. H. Thiemann, E. E. Kuramae, R. V. Santelli, C. L. Marino, M. L. P. N. Targon, J. A. Ferro, H. C. S. Silveira, D. C. Marini, E. G. M. Lemos, C. B. Monteiro-Vitorello, J. H. M. Tambor, D. M. Carraro, P. G. Roberto, V. G. Martins, G. H. Goldman, R. C. de Oliveira, D.

Truffi, C. A. Colombo, M. Rossi, P. G. de Araujo, S. A. Sculaccio, A. Angella, M. M. A. Lima, V. E. de Rosa Junior, F. Siviero, V. E. Coscrato, M. A. Machado, L. Grivet, S. M. Z. Di Mauro, F. G. Nobrega, C. F. M. Menck, M. D. V. Braga, G. P. Telles, F. A. A. Cara, G. Pedrosa, J. Meidanis and P. Arruda, *Genome Res.*, 2003, **13**, 2725.

6. P. Arruda, *Curr. Opin. Biotechnol.*, 2012, **23**, 315.
7. I. C. Macedo, *Estud. Av.*, 2007, **21**, 157.
8. V. F. Ferreira and F. Carvalho da Silva, *Quim. Nova*, 2013, **36**, 1514.
9. M. H. Thomsen, M. Andersen and P. Kiel, in *Biorefineries – Industrial Processes and Products. Status Quo and Future Directions, Vol. 1*, ed. B. Kamm, P. R. Gruber and M. Kamm, Wiley-VCH, Weinheim, 2006, pp. 295–314.
10. R. A. Bourne and M. Poliakoff, *Mendeleev Commun.*, 2011, **21**, 235.
11. J. Clark, R. Sheldon, C. Raston, M. Poliakoff and W. Leitner, *Green Chem.*, 2014, **16**, 18.
12. J. Gárcia-Serna, L. Pérez-Barrigóna and M. J. Cocero, *Chem. Eng. J*, 2007, **133**, 7.
13. J. C. Warner, A. S. Cannon and K. M. Dye, *Environ. Impact Assesss.*, 2004, **24**, 775.
14. Boeing, Embraer and FAPESP, *Flightpath to Aviation Biofuels in Brazil: Action Plan*, São Paulo Research Foundation (FAPESP), São Paulo, Brazil, 2013.
15. T. Werp and G. Petersen, *Top Value Added Chemicals From Biomass – Volume I: Results of Screening for Potential Candidates from Sugars and Synthesis Gas*, National Renewable Energy Laboratory, US Department of Energy, Golden, CO, 2004.
16. Brazilian Institute of Geography and Statistics (IBGE), *Sidra – Sistema IBGE de Recuperação Automática*, IBGE, Rio de Janerio, Brazil, 2014.
17. Ministry of Agriculture, Livestock and Supply (MAPA), *Anuário Estatístico da Agroenergia 2010*, MAPA, Brasília, Brazil.
18. Brazilian Sugarcane Industry Association (UNICA), *UnicaData*, UNICA, São Paulo, Brazil, 2014.
19. UN Food and Agriculture Organization (FAO), *Statistical Yearbook 2013: World Food and Agriculture*, FAO, Rome, 2013.
20. Ministry of Mines and Energy (EPE), *Brazilian Energy Balance – Year 2012*, EPE, Brasília, Brazil, 2013.
21. E. Virmond, J. D. Rocha, R. F. P. M. Moreira and J. H. Jose, *Braz. J. Chem. Eng.*, 2013, **30**, 197.
22. Ministry of the Environment (MMA), *Plano Nacional de Resíduos Sólidos*, MMA, Brasília, Brazil, 2012.

23. L. A. B. Cortez (ed.), *Sugarcane Bioethanol: R&D Productivity and Sustainability*, Editora Edgard Blucher, São Paulo, Brazil, 2010.
24. Associação Brasileira de Produtores de Florestas Plantadas (ABRAF), *Anuário Estatístico 2013 – Ano base 2012*, Brasília, Brazil, 2013, p. 148.
25. M. O. Dias, M. P. Cunha, C. D. F. Jesus, G. J. M. Rocha, J. G. C. Pradella, C. E. V. Rossell and A. Bonomi, *Bioresour. Technol.*, 2011, **102**, 8964.
26. W. A. Bizzo, P. C. Lenço, D. J. Carvalho and J. P. S. Veiga, *Renew. Sustainable Energy Rev.*, 2014, **29**, 589.
27. V. Gopinath, A. Murali, K. S. Dhar and K. M. Nampoothiri, *J. Am. Chem. Soc.*, 2011, **133**, 11096.
28. C. A. M. Santana and F. O. M. Durães, in *Biomassa para Química Verde*, ed. S. Vaz, Jr, Embrapa Agroenergia, Brasília, Brazil, 2013, pp. 17–46.
29. D. Mohan, C. U. Pittman and P. H. Steele, *Energy Fuels*, 2006, **10**, 848.
30. J. Zhang, Z. Lou, Q. D. J. Wang and W. Chen, *Energy Fuels*, 2012, **26**, 2990.
31. L. A. B. Cortez, R. A. Jordan, J. M. M. Pérez and J. D. Rocha, in *Sugarcane Bioethanol: R&D Productivity and Sustainability*, ed. L. A. B. Cortez, Editora Edgard Blucher, São Paulo, Brazil, 2010, pp. 919–936.
32. A. V. Bridgwater and G. V. C. Peacocke, *Renew. Sust. Energ. Rev.*, 2000, **4**, 1.
33. A. V. Bridgwater, *Biomass Bioenergy*, 2012, **38**, 68.
34. N. Mosier, C. Wyman, B. Dale, R. Elander, Y. Y. Lee, M. Holtzapple and M. Ladisch, *Bioresource Technol.*, 2005, **96**, 673.
35. M. O. S. Dias, T. L. Junqueira, C. E. V. Rossell, R. Maciel Filho and A. Bonomi, *Fuel Process. Technol.*, 2013, **109**, 84.
36. GranBio, http://www.granbio.com.br.
37. Centro de Tecnologia Canavieira (CTC), http://www.ctcanavieira.com.br/etanol2g.html.
38. Petróleo Brasileiro S.A-Petrobras, http://www.petrobras.com.br.
39. Brazilian Automotive Industry Association (ANFAVEA), *Brazilian Automotive Industry Yearbook – 2012*, ANFAVEA, São Paulo, Brazil, 2012.
40. LanzaTech, http://www.lanzatech.com.
41. J. Venus, *Res. J. Biotechnol.*, 2009, **4**, 15.
42. T. Ghaffar, M. Irshada, Z. Anwar, T. Aqil, Z. Zulifqar, A. Tariqa, M. Kamrana, N. Ehsana and S. Mehmooda, *J. Rad. Res. Appl. Sci.*, 2014, **7**, 222.

43. Amyris, http://www.amyris.com.
44. Amyris, Amyris meets major milestone at its Farnesece production facility, press release, 10 October 2013, http://www.amyris.com/News/337/Amyris-Meets-Major-Milestone-at-Its-Farnesene-Production-Facility.
45. Solazyme, http://solazyme.com.
46. L. A. B. Cortez, R. Baldassin Jr., E. Almeida, F. Santos, in *Bioenergia & Biorrefinaria: Cana-de-Açúcar & Espécies Florestais*, ed. F. Santos, J. Colodette and H. de Queiroz, Viçosa, Brazil, 2013, pp. 59–104.
47. Carbion Purac, http://www.purac.com.
48. Tate & Lyle, http://www.tateandlyle.com.
49. Cargill, http://www.cargill.com.br.
50. Center for Strategic Studies and Management (CGEE), *Química Verde no Brasil 2010-2030*, CGEE, Brasília, Brazil, 2010.
51. J. Street, F. Yu, J. Wooten, E. Columbus, M. G. White and J. Warnock, *Fuel*, 2012, **96**, 239.
52. Plástico Braskem, http://www.braskem.com.br/site.aspx/plasticoverde.
53. J. Sun, K. Zhu, F. Gao, C. Wang, J. Liu, C. H. F. Peden and Y. Wang, *J. Am. Chem. Soc.*, 2011, **133**, 11096.
54. K. D. Parghi, J. R. Satam and R. V. Jayaram, *Green Chem. Lett. Rev.*, 2011, **4**, 143.
55. M. T. Gutierrez-Wing, B. E. Stevens, C. S. Theegala, I. I. Negulescu and K. A. Rusch, *J. Environ. Eng-ASCE*, 2010, **136**, 709.
56. C. Zhu, C. T. Nomura, J. A. Perrotta, A. J. Stipanovic and J. P. Nakas, *Biotechnol. Prog.*, 2010, **26**, 424.
57. N. Gogotov, V. A. Gerasin, Y. V. Knyazev, E. M. Antipov and S. K. Barazo, *Appl. Biochem. Micro +*, 2010, **46**, 607.
58. PHB Industries, http://www.biocycle.com.br/site.htm.
59. PRWEB, Polyhydroxyalkanoatc market provides ample opportunities for investors – a study by Allied Market Research, 21 May 2014, Polhttp://www.prweb.com/releases/polyhydroxyalkanoate/market/prweb11873098.htm.
60. Associação Brasileira da Indústria Química (ABIQUIM) http://canais.abiquim.org.br/braz_new/.
61. Evonik Industries, *Sustainability Report 2013*, Evonik Industries, Essen, Germany, 2014.
62. CJ CheilJedang, http://www.cjbio.net/html/product/amino.asp.
63. Ajinomoto, http://www.ajinomoto.com.br/unidades.

PART B
BIOCHEMICAL TRANSFORMATIONS
AND REACTORS

CHAPTER 11

Overview: Biochemical Transformations and Reactors

DAVID J. LEAK

Department of Biology & Biochemistry, University of Bath, Claverton Down, Bath, BA2 7AY, UK
Email: d.j.leak@bath.ac.uk

In the quest for improved sustainability, the exploitation of biological systems as a source of renewable energy, chemicals and catalysts seems obvious. While there is clearly huge potential and there are already some classic success stories, this is not always as straightforward as it might seem. The concept of exploiting the natural ability of algae, which use sunlight and CO_2 as their sources of energy and carbon, to produce oils which can be converted to biodiesel, is appealing. But, as Chuck *et al.* explain in Chapter 16, problems of productivity and recovery of the natural oils mean that this is still not an economic proposition, despite 30 years of research, and the field appears to be turning away from exploiting the natural oil to looking at thermochemical conversion of the entire biomass to bio-oils. Ironically this could mean the best algal species to use are those that grow much faster and do not naturally produce oils!

In a similar vein, Schroder (Chapter 13) provides a perspective on the use of microbial bio-electrochemistry to both clean up wastewaters and provide electrical energy directly in the form of a microbial

Chemical Processes for a Sustainable Future
Edited by Trevor M. Letcher, Janet L. Scott and Darrell A. Patterson
© The Royal Society of Chemistry 2015
Published by the Royal Society of Chemistry, www.rsc.org

fuel cell, or provide a potential difference to drive other processes such as desalination. Compared with the body of research on algal technology this is a fledgling area, with a massive increase in publications in the past 10 years following the seminal discovery of direct electron transfer from bacteria to electrodes *via* bio-nanowires,[1] and also the discovery of naturally produced electron shuttling mediators, both of which allow higher cell densities to be used at the electrode surfaces. Coupled with innovations in electrode design this had led to dramatic increases in the reported current density of microbial anodes to 40 mA cm^{-2}. The field clearly has some way before it reaches commercial reality at scale, but is clearly developing fast.

Leak *et al.* (Chapter 12), Domínguez de María (Chapter 14) and Malhotra *et al.* (Chapter 15) all deal with different aspects of biocatalysis. The first covers fundamental principles underlying methods for exploiting enzymes in synthesis while the second reviews some recent successful applications, particularly in industrial fine chemistry. Malhotra *et al.* then cover biocatalysis in the context of sustainability and green chemistry, and ask some searching questions as to why biocatalysis has not had a greater impact on academic synthetic chemistry. The application of enzymes for biotransformation can be traced back to food manufacturing processes in ancient China and Japan, but with no understanding of the underlying process, this must have been a mysterious art. The enzymes involved would have been extracellular amylases, proteases and lipases, which are now produced commercially on a large scale for a diverse range of applications, but still including traditional food use.

The development of large-scale commercial enzyme production by the likes of Novozymes has benefitted (and continues to benefit) applied biocatalysis hugely. First, it provided confidence that enzymes can be produced at scale and so estimation of process costs is straightforward. Second, enzyme companies are usually prepared to supply academic laboratories with samples of these enzymes for evaluation in the confidence that these are robust and fairly simple to handle. As a result, much of the early work on applied biocatalysis revolved around lipases, esterases and proteases. However, developments in molecular biology, DNA sequencing and a reduction in the cost of DNA synthesis have changed this landscape dramatically. With a basic level of molecular biology skill it is possible to produce and purify a protein of interest from an easily grown heterologous host (*e.g. E. coli*, yeast). Furthermore, it is relatively simple to make variants of this protein, enabling creation of much better biocatalysts for the reaction under study.

Domínguez de María highlights a number of commercial applications, predominantly in the pharmaceutical area, where this approach has been remarkably successful. Obviously, the primary driver for this has been to exploit the potential regio- and stereo-selectivity of enzymes, rather than sustainability *per se*, and often this work is done in collaboration with specialist enzyme companies such as Maxygen. However, evidence that the larger pharmaceutical companies are looking seriously at issues of sustainability (Malhotra *et al.* point out that currently they have one of the poorest E-factors) and renewing their commitment to in-house biocatalysis teams is promising. The review by Malhotra *et al.* is a timely reminder that, despite the potential sustainability benefit of using enzymes, issues such as choice of solvent, use of water (in enzyme production), energy costs of transport and sourcing of precursors can also have a major impact on life cycle and sustainability assessments.

In the final chapter of this section, Kerton (Chapter 17) reminds us that nature provides a wide range of non-oil based chemical precursors and functional polymers, and that marine sources of these materials are relatively underexploited. In certain geographical regions the development of marine product and co-product biorefineries would make sense. While there already large markets for polysaccharides such as carrageenans and alginates from macroalgae, and chitosans from the shells of crustaceans, co-products from these operations are still going to landfill and could be further exploited.

REFERENCE

1. G. Reguera, K. D. McCarthy, T. Mehta, J. S. Nicoll, M. T. Tuominen and D. R. Lovley, *Nature*, 2005, **435**, 1098.

CHAPTER 12

Enzyme Biotransformations and Reactors

DAVID J. LEAK,[*][a] XUDONG FENG[c] AND EMMA A. C. EMANUELSSON[b]

[a] Department of Biology & Biochemistry, University of Bath, Claverton Down, Bath, BA2 7AY, UK; [b] Department of Chemical Engineering, University of Bath, Claverton Down, Bath, BA2 7AY, UK; [c] Department of Chemical and Materials Engineering, University of Auckland, Private Bag 92019, Auckland Mail Centre, Auckland 1142, New Zealand
*Email: d.j.leak@bath.ac.uk

12.1 INTRODUCTION

Enzymes are remarkable catalysts which, in some instances, have been shown to provide rate enhancements that far exceed chemical catalysis. Indeed, with some enzymes, catalysis is limited by the rate of substrate diffusion. Furthermore, some enzymes can exhibit exquisite reaction-, regio- and stereo-selectivity. However, these properties are not universal and despite the availability of some remarkably robust enzymes, particularly in immobilized form, many others seem fragile and prone to inactivation. In the first part of this chapter we provide an introduction to the fundamental principles which underpin these properties, followed by a survey of their immobilization and application in different reactor formats.

Chemical Processes for a Sustainable Future
Edited by Trevor M. Letcher, Janet L. Scott and Darrell A. Patterson
© The Royal Society of Chemistry 2015
Published by the Royal Society of Chemistry, www.rsc.org

Enzymes are comprised of single or multiple polypeptides, each composed of a linear chain of amino acids joined *via* peptide bonds. Through the processes of transcription (production of messenger RNA) and translation (production of a polypeptides based on the sequence of the mRNA), the DNA sequence of the gene which encodes a polypeptide directly determines this linear assembly of amino acids, *i.e.* the primary structure. The primary sequence, in turn, determines the localized formation of regions of secondary structure in the polypeptide which are organized, based on polar, non-polar and charge–charge interactions together with occasional disulfide cross-linking into characteristic three-dimensional tertiary structures. Thus, the catalytic site of an enzyme, which typically forms inside this structure, can be created by the juxtaposition of different parts of the polypeptide, often quite distant on the primary sequence. Some enzymes require interaction between polypeptides (quaternary structure) in order to exhibit activity. This can range from simple homo-dimers to complex assemblages of different polypeptides. Homo- and hetero-multimerisation can provide an enzyme with regulatory as well as catalytic properties or enable multiple steps to be conducted within the enclosed cage of the active site(s), providing pathway channelling and protection of sensitive intermediates from the surrounding aqueous environment.

A fundamental feature of enzyme catalysis, which results from these tertiary and quaternary structures and contributes to the large rate enhancements observed, is flexibility during the catalytic cycle. Enzyme catalysis can involve quite dramatic changes in three-dimensional structure during catalysis and some hetero-multimers are effectively molecular production lines, able to pass intermediates from one polypeptide to another. As a consequence, enzyme catalytic rates are very sensitive to temperature and there is typically a fairly narrow range between poor activity because of the lack of flexibility and loss of activity due to excess flexibility and disruption of the tertiary structure (for further explanation of these processes see ref. 1 or similar texts).

Genetic engineering allows the movement of genes between organisms, enabling the over-production of enzymes in commonly used host organisms such as the bacterium *Escherichia coli* or the yeast *Saccharomyces cerevisiae*. Facile alteration of DNA sequences enables the modification of the wild-type sequence to either add additional amino acids to the start (*N* terminal end) or end (*C*-terminal end) of the protein, or to make single or multiple changes in the primary amino acid sequence. Addition of a short 'signal sequence' to the

N-terminal end can target the enzyme to be secreted out of the cell, while addition of 'tags' at either end of the protein sequence can be used to facilitate affinity purification (*e.g.* immobilized metal affinity chromatography[2]). Site-directed mutagenesis can be used to change specific amino acids by design and is often used in mechanistic studies or when trying to change the catalytic properties of an enzyme. However, given that the primary sequence influences both protein folding and ultimate three-dimensional structure, it is often difficult to accurately predict the outcome of such experiments and attempts to improve the properties of enzymes as biocatalysts frequently involve the generation of large numbers of mutants for subsequent screening and selection.

The relative ease and speed of genome sequencing and access to excellent online databases (*e.g.* http://www.ncbi.nlm.nih.gov) and software for comparative analysis means that, starting with a single gene sequence, it is relatively straightforward to identify similar sequences present in other organisms, generate a suite of similar biocatalysts and even mix-and-match properties. Additionally, the cost of DNA synthesis is such that it is often cheaper to get the genes synthesized *de novo* rather than grow the parent organism and isolate the relevant gene. While it is still not possible to design a functional biocatalyst *ab initio* and our understanding of features such as instability or catalytic inactivation needs to be improved, it is evident that access to different types of enzyme is not the limiting factor in the application of biocatalysis in synthesis.

12.2   APPLIED BIOCATALYSIS AND BIOTRANSFORMATION

Although a useful enzyme may derive from a range of sources, developments in genetic manipulation mean that for ease of production and scale up, a single enzyme for use in purified or partially purified form will normally be produced in recombinant form in *E. coli*, yeast or some other organism that can be grown at scale. Although there are many reports in the literature of bacteria, yeasts, filamentous fungi and even higher organisms having interesting enzyme activities, the use of a natural host often limits the possibility of manipulating over-expression and risks the possibility of side reactions of either the substrate or product. However, natural hosts may be useful where multi-step biotransformations are involved[3] or where the host offers specific advantages (*e.g.* substrate or product uptake, generation of reducing power).

Despite the possibility of targeting secretion or adding purification tags, unless the enzyme is going to be used in a whole cell biotransformation, there will be a cost associated with recovery and purification. Together with the potentially limited lifetime of the catalyst this may restrict the potential niches where biocatalysts can be economically applied. Lifetimes and reusability may be extended through immobilisation (see below), where there is a general rule of thumb which suggests that if an immobilized enzyme can be used for 10 cycles this typically justifies the cost of immobilization. This scenario is producing four patterns of application: (1) high volume, relatively low added value biotransformations typically using intact or permeabilized whole cells (*e.g.* acrylamide,[4] high fructose corn syrup[5]); (2) low volume niche/complex biotransformations typically using whole cells (*e.g.* steroid biotransformations[6]); (3) free/immobilized enzyme biotransformations based on easily produced (typically secreted) enzymes with enhanced lifetimes; and (4) high added value biotransformations which exploit unique characteristic of the enzyme, typically regio- and stereo-selectivity of reaction, where a single enzyme step may replace multiple chemical steps (*e.g.* protection and deprotection reactions).

12.3 FEATURES OF ENZYMES USEFUL FOR APPLIED BIOCATALYSIS

Enzymes are classified into six broad categories by the Enzyme Commission: (1) oxido-reductases, which catalyse the transfer of hydrogen and oxygen atoms or electrons from one substrate to another; (2) transferases, which catalyse the transfer of a functional group (*e.g.* a phosphate group) from one compound to another; (3) hydrolases, which catalyse hydrolytic cleavage of C–O, C–N, C–C and some other bonds; (4) lyases, which cleave similar substrates to hydrolases but by elimination; (5) isomerases, which catalyse geometric or structural rearrangement in a molecule; and (6) ligases, which catalyse the joining together of two molecules, typically coupled to a source of energy generation, such as ATP hydrolysis. Lyases and isomerases may involve single substrate reactions, but in all other enzyme types, two substrates are involved, typically the substrate of interest and a second (and potentially a third) substrate which is used in the reaction. Where that substrate is inexpensive it can be supplied in stoichiometric amounts (in the case of hydrolases it is clearly present in excess). However, where the additional substrates are relatively expensive (*e.g.* sources of reducing equivalents such as

NADH/NADPH or reactions requiring ATP or other tri-nucleotides as a source of energy or in phosphate transfer), consideration has to be given to issues of *in situ* substrate regeneration, which may influence decisions between using whole cells or a free enzyme format.

12.3.1 Enzyme Kinetics

In a simple enzyme reaction an enzyme (E) binds a substrate (S) reversibly to form an intermediate (ES) complex (in reality there may be many interconverting forms of the ES complex). This is then converted to product which is released, regenerating the free enzyme (Equation (12.1)). Typically the rate of substrate binding (rate constant k_1) and dissociation (rate constant k_{-1}) is faster than the rate of conversion to product (rate constant k_2), and it is the latter which determines the initial rate of reaction. One feature which is evident from this description is that there must be maximum rate of a reaction (V_{max}) at the point at which all the enzyme is bound as ES (*i.e.* enzymes can become saturated with substrate). The maximum conversion rate observed is then a fundamental property of the enzyme turnover rate and the amount of enzyme in the reaction. Given that enzymes have widely different molar masses, representing the maximum rate in terms of the weight or volume of protein present makes comparison difficult. Where enzymes are impure or present in whole cells, however, this may be the only option. However, with a pure protein of known molar mass it is possible to calculate k_2 as a pure turnover rate [mole product (mole enzyme)$^{-1}$ s^{-1}], referred to as k_{cat} (s^{-1}). This can vary widely between enzymes, with some turning over 10^6 times a second while others can be quite sluggish. A useful economic parameter, the total turnover number (TTN), may then be generated by multiplying k_{cat} by the average lifetime of the enzyme:

$$E + S \underset{k_{-1}}{\overset{k_1}{\rightleftharpoons}} ES \overset{k_2}{\longrightarrow} E + P \tag{12.1}$$

While all enzymes exhibit saturation, the relationship between reaction rate and sub-saturating enzyme concentration depends on the affinity between the substrate and enzyme and the kinetic mechanism. Many enzymes display Michaelis–Menten kinetics, which derive from considering the reaction scheme shown in Equation (12.1) at steady state. However, some oligomeric enzymes exhibit co-operativity, where binding of the substrate to one sub-unit affects (positively or negatively) the affinity of binding of a second substrate. The relationship between reaction rate and substrate concentration then becomes more complex than the Michaelis–Menten equation[7] and is not considered further here. Equation (12.1) considers a single substrate–enzyme interaction, whereas it is clear than many enzymes react with more than one substrate. Assuming the relevant assays are done at saturating concentrations of all the other substrates, the treatment below can be considered for each substrate individually giving a K_m value for each one.

The Michaelis–Menten consideration of the effect of substrate concentration S on the rate of an enzyme reaction assumes that, during the assay, the concentrations of ES, S and P remain constant, *i.e.* the system rapidly reaches a steady state. It also assumes that the concentration of substrate is considerably higher than that of ES, allowing the actual value of ES to be ignored. Because most enzyme catalysed reactions are reversible, this typically means that enzyme kinetic parameters are calculated by performing an initial time course assay, starting with $P=0$ and assuming that the rate of the reverse reaction is negligible. These assumptions then simplify the analysis to give the Michaelis–Menten relationship:

$$v = \frac{v_{max}S}{K_m + S} \tag{12.2}$$

where K_m, the Michaelis constant is given by Equation (12.3) and is defined as the concentration of substrate that gives half the maximum rate assuming saturating concentrations of all other substrates and a product concentration of zero:

$$K_m = (k_{-1} + k_2)/k_1 \tag{12.3}$$

Where k_2 is small compared with k_1 and k_{-1} then K_m will have a value close the substrate–enzyme dissociation constant $K_d = k_{-1}/k_1$ and the K_m is often loosely referred to as reflecting the (inverse of) the affinity of an enzyme for its substrate. From the perspective of applied biocatalysis, given the relationship between the value of K_m and V_{max}, it is important to work with enzymes with low K_m so that V_{max} may be achieved at relatively low substrate concentrations. This is particularly important for batch processes as, with an enzyme that has a high K_m, a significant amount of substrate may remain unconverted at the end of the process.

12.3.2 Enzyme Specificity and Competing Substrates

From an alternative working of the Michaelis–Menten equation the rate of reaction can be represented in terms of the concentration of free enzyme (Equation (12.4)). In a situation where an enzyme has two on more competing substrates in solution, clearly [E] will be identical for all substrates so the observed rate of reaction will depend on [S] and (k_{cat}/K_m), and at identical substrate concentrations on k_{cat}/K_m only. This term is therefore a measure of the specificity for a particular substrate and is known as the 'specificity constant'. Starting with an equimolar mixture of two substrates, the substrate that has the highest k_{cat}/K_m will react preferentially, but as the concentration of this substrate declines the second substrate may start to be metabolized depending how large the differences are in their respective k_{cat}/K_m values. This is a particularly useful concept when we come to consider discrimination between enantiomers (see below):

$$V = (k_{cat}/K_m)[E][S] \qquad\qquad (12.4)$$

12.3.3   Reversibility and Equilibrium

Technically, all enzyme-catalysed reactions are reversible, although in some cases the equilibrium constant is so high that these enzymes are considered to be irreversible. However, many enzymes which catalyse biocatalytically valuable reactions are genuinely reversible; hence, reactions will proceed to equilibrium and rates will decrease as this is approached. This means that either a separation or enrichment step is required or a strategy is necessary to drive the reaction beyond equilibrium. In the example of making high-fructose corn syrup, the glucose isomerase equilibrium mixture does not have the optimum sweetness, so it is necessary to chromatographically enrich a fraction of the fructose to add to this mixture to bring it up to 55% of the carbohydrate.

In nature, metabolism does not reach equilibrium because the product of one reaction is the substrate for the next and thus is removed from the reaction. (Although in times of metabolic imbalance, biology does use mass action due to product accumulation to slow down metabolic reactions.) Therefore, in order to achieve good conversions in reactions with small equilibrium constants it is necessary to employ strategies which pull the reaction in one direction, typically linking a near equilibrium reversible process to one which is effectively irreversible. In that way, a product from the first reaction can be removed from the process. Typical examples are spontaneous or enzyme catalysed decarboxylation,[8] with release of a gaseous product or employing transesterification with isopropenyl/vinyl esters which exploits the effectively irreversible chemical keto-enol tautomerism of vinyl alcohol (Figure 12.1).

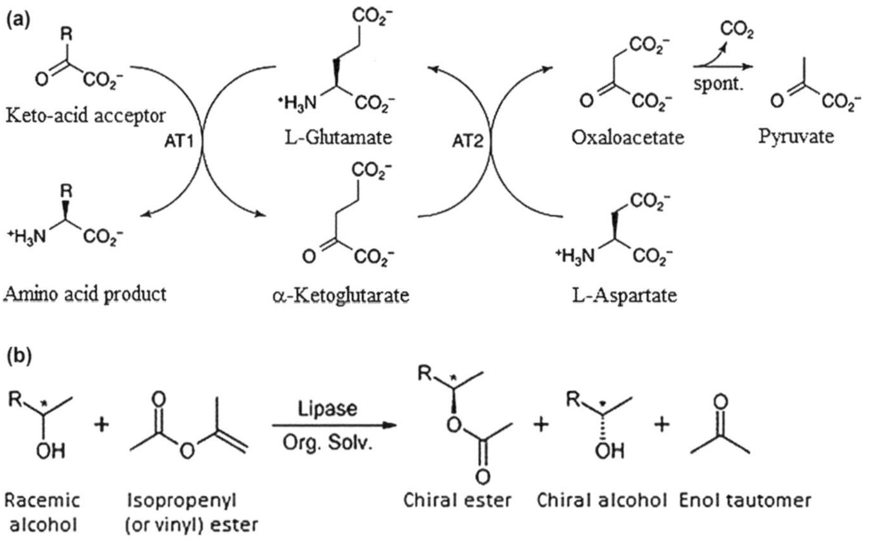

Figure 12.1 Examples of biocatalytic strategies to pull reversible reactions in the direction of the desired product: (a) exploiting spontaneous decarboxylation of oxaloacetate (adapted from ref. 8); and (b) exploiting the conversion of a product to its unreactive tautomer.

12.4 APPLICATIONS OF BIOCATALYSIS

12.4.1 Whole Cell Biocatalysis

Native or recombinant whole cell biotransformations come to the fore when multiple consecutive steps are involved, co-factor recycling is necessary, the product is high volume but low unit value or when complex, particularly membrane bound enzymes are involved.

One of the earliest commercial success stories is in the transformation of steroids for use in the pharmaceutical industry. The ability of various bacteria such as *Mycobacterium* spp., *Nocardia* spp. and *Rhodococcus* spp. (all of which have waxy cell walls) to degrade the side chains of readily available plant steroids such as diosgenin, stigmasterol and sitosterol provided an alternative route to steroid precursors than slaughterhouse derived cholesterol or deoxycholic acid.[6,9] While this may now be done chemically, the selective 11α or 11β hydroxylation of the steroid nucleus by fungi such as *Rhizopus* spp., *Aspergillus* spp. and *Curvularia lunata* provided access to the key 11β hydroxyl functionality in hydrocortisone and related corticosteroids, while 1-dehydrogenation catalysed by *Arthrobacter simplex* opened the route to the anti-inflammatories and contraceptive hormones.

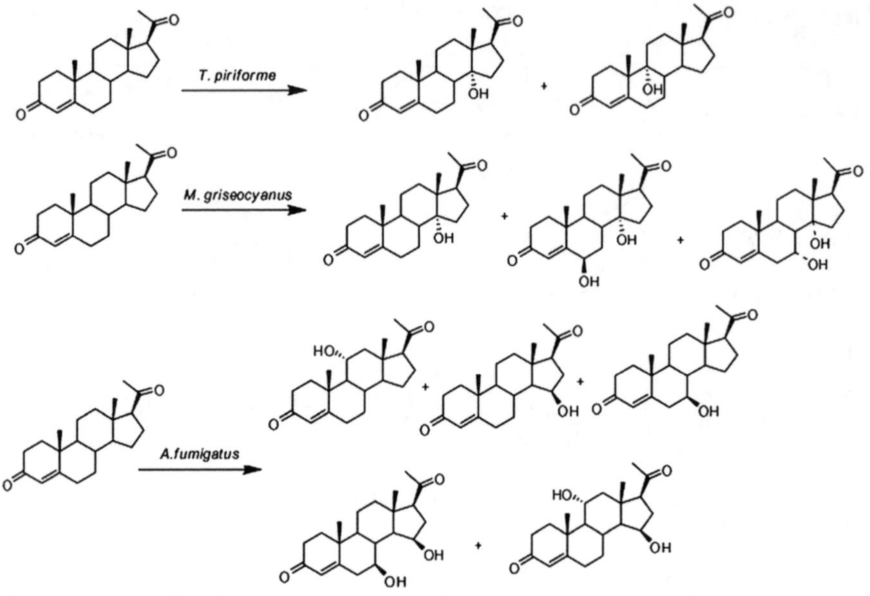

Figure 12.2 Examples of whole cell microbial steroid hydroxylations of progester-
one (adapted from ref. 9).

Many other similar biotransformations have been characterized, particularly selective hydroxylations (Figure 12.2), typically by membrane-bound enzymes. These biotransformations avoid complex multistep chemical syntheses, allowing functionality to be introduced at a late stage in synthesis and formed the backbone of pharmacologically active steroid production. The area is gradually yielding to metabolic engineering strategies and *de novo* synthesis of therapeutic steroids from carbohydrates by diversion of the ergosterol synthesis pathway in yeast is a distinct possibility.

A rare example of a commercially successful whole cell biotransformation to bulk products is the Mitsubishi Rayon process to produce acrylamide from acrylonitrile using the enzyme nitrile hydratase from *Rhodococcus rhodochrous*.[10] The success of this process lies partly in the poor conversion and selectivity of the chemical transformation, whereas this whole cell process exhibits 99.9% conversion and selectivity, thus reducing waste and yielding a higher purity product. Remarkably, it has been reported that acrylamide can be produced at a concentration 400 g L^{-1}, which clearly benefits the economics of downstream purification. Clearly, the lack of cofactor involvement means that this could have been operated as an isolated enzyme reaction, but the cost of enzyme recovery and immobilization is presumably outweighed by the efficiency of the whole cell process.

A nice example of both the benefits of biocatalysis over chemical catalysis and the continuing improvement of an industrial process comes from the Dorzolamide (Piramal Healthcare) process. A key step in the synthesis of this agent for the treatment of glaucoma is the production of reduction of a chiral ketosulfone to the correct diastereomer for further processing. Chemical reduction not only produced a mixture of both hydroxyl group enantiomers, but the presence of the chiral methyl group produced an unfavourable diastereoselectivity. A strain of *Neurospora crassa* was found to produce a ketosulfone reductase which produced the correct diastereomer and initially developed as a whole cell biotransformation process, fed by glucose to provide the necessary reducing equivalents to drive the reaction. Subsequently the gene encoding the enzyme has been cloned and expressed to a high level in *E. coli*, allowing an intensified whole cell process to be operated, again using glucose to maintain the recycling of reduced NADPH for the reaction. More recently this has been converted into a purified enzyme process in which catalytic amounts of NADPH are added and NADP$^+$ recycled by glucose dehydrogenase (Figure 12.3).

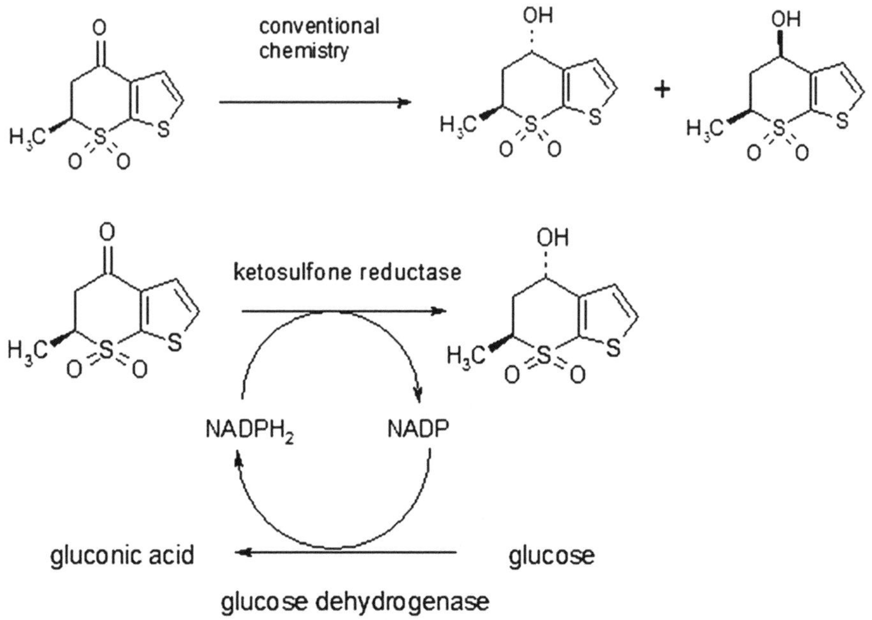

Figure 12.3 The Piramal process for ketosulfone reduction used in the production of Dorzolamide.

12.4.2 Free and Immobilized Enzyme Biocatalysis

The production of a pure or partially pure enzyme entails the cost of protein purification. Many commercial enzymes are sold on the basis of their predominant activity, but detailed analysis shows that they are often relatively impure. For an intracellular enzyme, purification involves the additional cost of cell recovery and disruption to release the enzymes from the cells, compared with extracellular enzymes, which can be recovered directly from the medium. It is therefore not surprising that many recombinant enzymes are produced in secreted form from organisms such as yeast and *Bacillus* spp. which are good secretors.

Many of the early industrial enzymes (lipases, proteases, glycosyl hydrolases) were naturally secreted, being involved in the breakdown of materials which are too bulky to transport into cells, including polymers such as cellulose, starch and proteins as well as oils and fats. Their natural degradative function means that they found application in large quantities for brewing, the food and drink industry, the textile industry, biological detergents and more recently in second-generation biofuel production. However, the ready availability of inexpensive enzymes such as lipases and proteases in both free and immobilized form, together with an extensive literature of their application, can often be a good reason to develop biotransformations around their activity.

12.4.2.1 Extracellular Hydrolytic Enzymes. Given that their natural function is to hydrolyse substrates, the simplest application of hydrolytic enzymes in biocatalysis is where the enzymes offer regio- or stereoselectivities which are difficult to achieve chemically without extensive modification. Thus lipases, esterases and proteases have frequently been used for chiral resolution of racemic mixtures such as kinetic resolution of *R,S* glycidyl butanoate by porcine pancreatic lipase to *R*(−)glycidyl butanaote and *R*(+)glycidol.[11] This process worked commercially because the enzyme was sufficiently specific (k_{cat}/K_m) for hydrolysis of the *S* enantiomer that both products could be obtained in high enough yield and enantiomeric purity for further processing and both had uses. A more recent example is in the lipase catalysed ring opening of lactides,[12] where both enantiomeric products can be used in the synthesis of co-polymers. However, it is rare for both pure enantiomeric products to have commercial use and, more typically a kinetic resolution will achieve, at best, a 50% yield of a single enantiomer, with the other product being wasted or recycled. To achieve such a high yield of a single enantiomeric product requires a very high specificity for one and not the other substrate enantiomer.

The ability of an enzyme to discriminate between enantiomers is measured by the ratio of the two specificity constants $(k_{cat}/K_m)_R/(k_{cat}/K_m)_S$ which is given the symbol E (the enantioselectivity ratio). Where the pure enantiomers of a substrate are available E is straightforward to measure. Where they are not, E may be estimated from progress curves using the relationship:

$$E = \frac{\ln[1 - c(1 + eep)]}{\ln[1 - c(1 - eep)]} = \frac{\ln[1 - c(1 - ees)]}{\ln[1 - c(1 + ees)]} \tag{12.5}$$

where c = extent of conversion (0–1), and eep and ees are the enantiomeric excesses of the product and substrate, respectively.[13] It can be seen from the progress curves in Figure 12.4 that to obtain both good yield and high enantiomeric excess of the residual substrate in a kinetic resolution, an E value of 50–100 is required, otherwise high eep or ees values can only be achieved at the expense of yield by running the reaction to less than or greater than 50% conversion, respectively.

Where only one of the enantiomers is useful, then recycling the unwanted substrate/product to form a racemic mixture is clearly of economic benefit. However, if this can be done in combination with the kinetic resolution, the resulting dynamic resolution is a very powerful approach.[14] This has been exploited with hydantoinases, where spontaneous chemical or enzyme catalysed racemization is exploited to continuously generate the racemic starting material, resulting in 100% conversion to single enantiomer D or L amino acids.

12.4.2.2 Exploiting Hydrolases in Synthesis. While some hydrolases function by promoting nucleophilic attack of a hydroxyl group directly on the substrate, others function in two steps forming a covalently bound intermediate while displacing one of the products. This covalent intermediate is subsequently hydrolysed. These enzymes typically involve either a nucleophilic serine (*e.g.* in serine proteases and lipases) or aspartate/glutamate (*e.g.* in 'retaining' glycosyl hydrolases and epoxide hydrolases).

In the natural hydrolytic function of the enzyme a hydroxyl group displaces the original nucleophile, regenerating the original enzyme and releasing the second part of the hydrolysed product. With the exception of epoxide hydrolases and some similar enzymes, where the hydroxyl group attacks the original nucleophilic carboxyl directly,

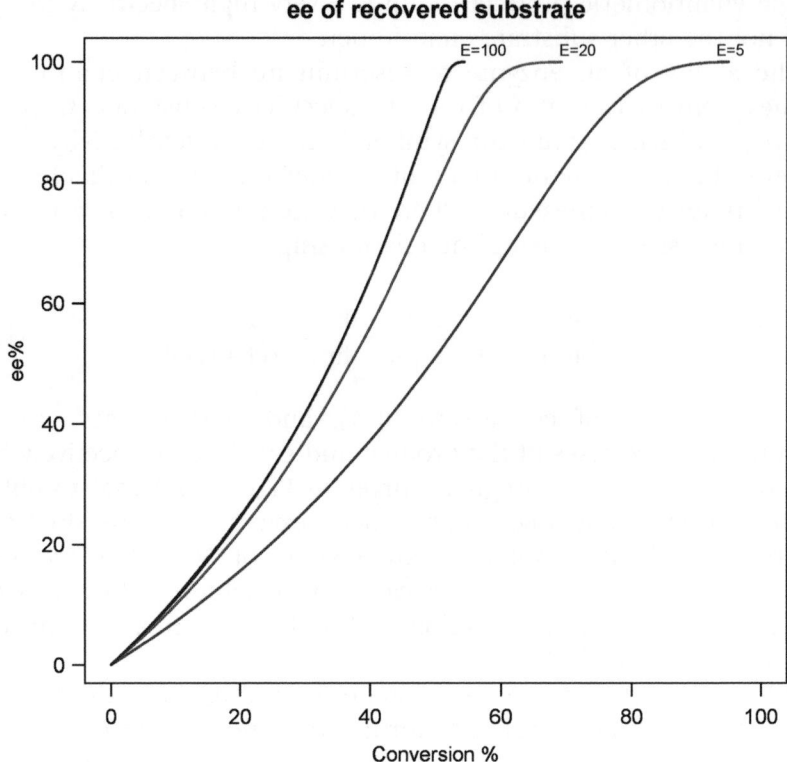

Figure 12.4   The effect of the enantioselectivity ratio (E) on the enzymatic kinetic resolution of a racemic substrate. Enantiomeric excess of residual substrate is plotted against the extent of conversion for enzymes/reactions with differing E values.

the second step involves the direct displacement of the original nucleophile, thus retaining the original stereochemistry of the substrate. In a low water environment it is possible to replace the hydroxyl group with an alternative nucleophile,[15] facilitating synthesis of new compounds (Figure 12.5). This strategy has been extensively exploited using lipases,[16] which naturally function at an oil–water interface and therefore function well in a low-water environment (a small amount of water is always required for enzyme function) such as organic media and ionic liquids. This 'thermodynamic' approach works well for lipases and proteases but is less effective with glycosyl hydrolases, which appear to require a minimum of 10% water to function,[17] making hydroxyl groups the most abundant nucleophile. However, here a set of elegant kinetic strategies have been developed involving substrate analogues and mutant enzymes. Instead of forming an

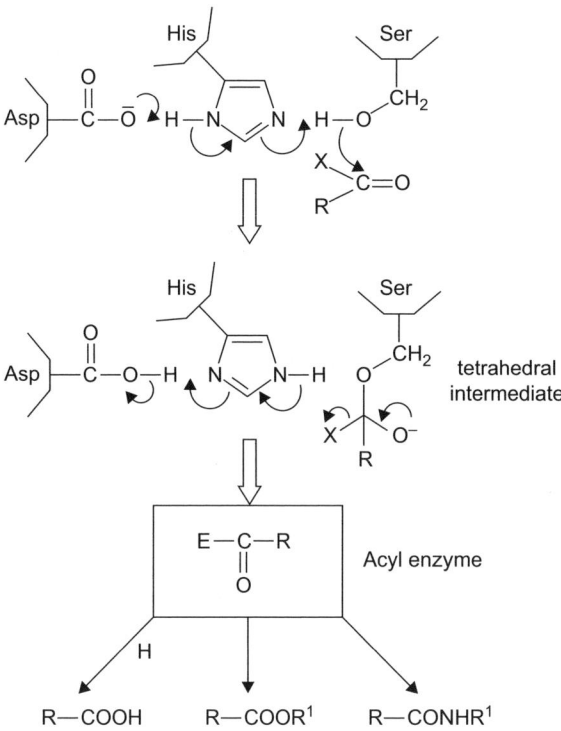

Figure 12.5 Formation of an acyl-enzyme intermediate in enzymes that employ a two-step hydrolysis mechanism using serine as the nucleophile (*e.g.* serine protease, lipase).

enzyme adduct, the catalytic nucleophile of retaining glycosyl hydrolases (*i.e.* those which go through a two-step process) is replaced forming 'glycosynthases'.[17,18] Fluorinated analogues of the substrate, but with the fluorine in the opposite anomeric configuration, are used to mimic the natural intermediate and can be attacked by a variety of nucleophiles (displacing the fluoride) without risk of re-hydrolysis by the enzyme. As well as synthesis of carbohydrates this approach has been used to synthesize glucosphingolipids and flavinoid glucosides.[17]

12.4.3 The Co-factor Problem

Many enzymes require non-protein components (co-factors) for their activity. Prosthetic groups remain bound to the enzyme (often loosely, so loss of prosthetic group can be a problem) throughout the catalytic cycle and are regenerated by the end of the cycle, whereas co-enzymes are effectively substrates which are used in the catalytic cycle and regenerated elsewhere in the cell. In applied biocatalysis the most commonly used coenzymes are the reduced nicotinamides NADH and NADPH, and the flavins FADH and FMNH, although ATP may also be encountered. Provision of a pure co-factor as a substrate is prohibitively expensive so they need to be provided at low concentrations and regenerated *in situ*. In some instances this is best done in whole cells where co-factor regeneration is a natural metabolic process, which can be driven by the provision of a cheap substrate such as glucose. However, to avoid side reactions or transport limitations, isolated enzyme systems are preferable. In these cases an effective system for co-factor regeneration needs to be included.[19] The nicotinamides and flavins are typically involved in redox reactions, particularly chiral reductions or oxygen activation for mono-oxygenases. Regeneration requires the provision of a sacrificial substrate (*e.g.* alcohol) and an enzyme capable of oxidizing it (*e.g.* alcohol dehydrogenase) and with the necessary co-factor specificity. In some cases (*e.g.* ketone reduction) it may be feasible to use the same enzyme for co-factor regeneration, although typically this would require high concentrations of the sacrificial alcohol donor to drive the reaction in the direction required and may involve complex separations at the end of the reaction. A more elegant solution is to use formate dehydrogenase as the oxidation of formate is effectively irreversible as the resulting CO_2 is lost as a gas. The most commonly used formate dehydrogenase is that from the yeast *Candida boidinii*,[20] which has been engineered to improve its stability and also engineered to change its co-factor specificity from NADH to NADPH[21] so that both forms are now available for use.

12.5   EXPLOITING THE POWER OF ENZYME EVOLUTION

Biocatalysis endeavours to exploit the potential of enzyme-based catalysis to deliver reactions with a defined stereochemistry and high catalytic rates. However, typically, the desired conversions use non-natural substrates which may bind poorly to the active site, give relatively low rates of conversion and give lower product selectivity than seen with the natural substrate. Two complementary approaches are available to help improve this situation. The first recognizes that there is a huge natural variation in enzyme specificity present in the environment and that better enzymes may exist. Historically, this meant extensive screening of cultures of organisms isolated from the environment, sometimes with very little selective pressure available to bias the cultures isolated towards the enzyme characteristics required. However, the development on metagenomics[22] and the reduction in cost of DNA sequencing has meant that this exercise is now much more targeted. Furthermore, the massive increase in genome sequence information means that a huge amount of comparative sequence information is already available in databases. So, using the sequence of your protein as a template it is simple to identify large numbers of closely or more distantly related functionally equivalent sequences which can provide the starting point for a screening exercise.

The second approach is the use of 'evolution in the laboratory'.[23] The basic principle of natural selection involves mutation, recombination and selection, and it is possible to mimic these to great effect in the laboratory. The polymerase chain reaction (PCR) enables amplification of DNA sequences based on a parent template and, under sub-optimal conditions, can be made increasingly error-prone. Where the sequences of homologous genes are quite similar it is possible to 'shuffle' the sequences together, mimicking the natural process of recombination, while methods have also been developed for recombination of less similar sequences and removing bias from natural recombination mechanisms. Although, in some instances, it has been possible to devise methods in which the activity of the improved enzyme can be linked to the improved 'fitness' of a host organism; typically 'selection' is replaced by screening, making use of liquid handling robotics.

Given the number of amino acids in a protein and the number of possible variations of each amino acid, purely random use of mutation, recombination and screening is not a practical way to approach subtle changes such as modifying the stereo-selectivity of an enzyme (but may be used for factors such as thermostability, which are difficult to predict, even with a detailed knowledge of sequence and structure of an enzyme). Factors affecting substrate recognition, turnover number and selectivity of product formation are typically associated with entry to, or interactions within, the active site of a protein and modelling the binding of substrates within the active site is increasingly effective in understanding enzyme-substrate reactivity. Hence predicting sites at which changes should alter these interactions if fairly accurate, even if predicting the optimal amino acid change is less simple, due to the connectivity of structures within the active site. Therefore, many 'directed evolution' approaches start from a knowledge of active site interactions and focus on changing defined residues by saturation mutagenesis (*i.e.* to all possible amino acids) or to specific families of amino acids.[24] Increasingly sophisticated structure-activity based algorithms are being employed in defining the optimal combinations of mutations leading to some remarkable examples of enzyme engineering and industrialization.

12.6   ENZYME IMMOBILIZATION

Enzymes have advantages over most non-biological catalysts such as high efficiency, specificity and selectivity, and the ability to function under mild conditions.[25] However, some drawbacks of native enzymes have limited their application. Enzyme function is dependent on operating conditions such as temperature and pH, so it is very easy for them to lose their activity. In addition, it is difficult to separate free enzymes from substrate and product for repeated use.[26] Enzyme immobilization has been shown to overcome these disadvantages. By definition, enzyme immobilization is the conversion of enzymes from a water-soluble, mobile state to a water-insoluble immobile state,[25] which is usually achieved by restricting enzymes to an inert and insoluble carrier. Through immobilization, the enzyme can be reused and it can also inherit mechanical and thermal stability from its carrier, making it less susceptible to changes in exterior environment. In addition, immobilization can simplify the producing technology. The immobilized enzyme can easily be separated from products, thus leading to simple separation and purification of products, enhancing the yield and qualification of products. Due to the aforementioned advantages, enzyme immobilization has gained more attention in recent years.[27,28] One drawback with enzyme immobilization is that you will always lose some of the activity in the process, as well as potentially reducing the available surface area.

12.6.1   Immobilization Methods

Numerous techniques have been developed for enzyme immobilization. These can essentially be divided into three categories: (1) binding to a support (carrier); (2) entrapment (encapsulation); and (3) cross-linking.[26] A classification of immobilization methods according to different physical and chemical principles[29] is shown in Figure 12.6.

12.6.1.1 Binding to a Support. Enzyme can be attached to a carrier by covalent or physical binding. Physical binding usually includes three interactions: hydrophobic, Van der Waals and ionic interaction. Although physical binding is simple and the support

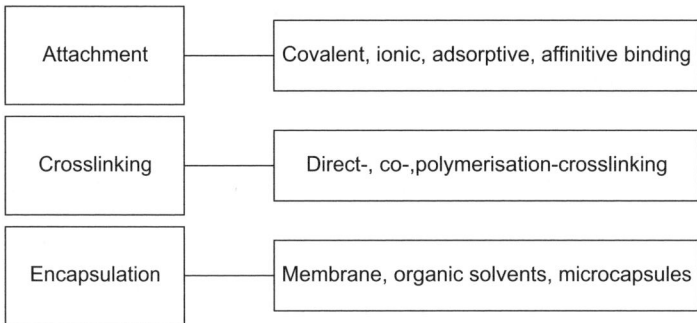

Figure 12.6 Classification of immobilization methods.

can be easily recycled, immobilization *via* such methods usually gives insufficient stability under operational conditions and leakage of enzymes often occurs under high reactant or product concentrations and high ionic strength. Therefore, physical binding is not suitable for industrial application. In contrast, covalent binding is much stronger, so the immobilized enzyme is more stable and does not leak markedly from the support surface. However, there are also some disadvantages: (1) covalent binding is usually irreversible and the support cannot be reused once the enzymes have been deactivated, thus increasing the cost; and (2) the chemical reactions involved in immobilization may deactivate the enzyme. Therefore, the reagents and conditions for chemical binding should be thoroughly selected and optimized.

12.6.1.2 Entrapment. The aim of entrapment is to restrict the enzymes to a polymer network such as an organic polymer or a silica sol-gel,[30,31] or a membrane device such as a hollow fibre.[32] Similar to physical binding, the physical restraints are generally too weak to prevent enzyme leakage entirely. Therefore, additional covalent binding is often necessary to stabilize the immobilized enzymes.

12.6.1.3 Cross-linking without Support. The use of a carrier unavoidably causes a dilution in enzyme activity due to the introduction of a large amount of non-catalytic groups, which result in lower space–time yields and productivities.[26] Recently, a new technique has been proposed to prepare macroparticles without carriers *via* cross-linking of enzyme aggregates or crystals using a bifunctional reagent.[33,34] The resulting enzyme aggregates or crystals have some advantages such as high volumetric productivity, high stability and lower production costs due to the absence of an additional carrier.

In summary, of the above three immobilization methods, covalent binding and cross-linking can produce strong bonds between enzyme and supports or enzymes, but are relatively expensive and can damage enzymes, reducing activity. Physical binding and entrapment are simple and efficient, but are too weak to prevent the enzyme from leakage, making them less applicable for continuous operation in industry. By combining physical binding and cross-linking, and through proper design and optimization of the cross linking step, a protocol that minimizes activity loss whilst retaining the long-term activity has been developed[35–38] making it suitable for long-term operation.

12.6.2   Enzyme Functional Groups for Immobilization

Among the 20 common amino acids, the alkyl side chains of the hydrophobic residues are chemically inert. Generally, only 10 side chains have useful chemical activity for immobilization, including: the guanidinyl group of arginine; the sulfhydryl group of cysteine; the β- and γ-carboxyl groups of glutamic and aspartic acid, respectively; the hydroxyl groups of serine and threonine; the thioether moiety of methionine; the ε-amino group of lysine; the phenolic hydroxyl group of tyrosine; and the indolyl group of tryptophan.[29] Hence, most covalent binding and cross-linking immobilized protocols are designed based on these active side chain groups.

Most immobilization methods choose the amine groups of enzymes as a reaction target, including ε-amine of lysine residues and the terminal amine groups, since these residues are present on the surface of most enzymes, and are quite reactive with supports without any pre-activation. The number of surface lysine residues usually exceeds that of terminal amine groups.[39] Exposed lysine residues have a *pKa* close to that of the free amino acid, around 10.5–10.7. Internal lysine residues have a different *pKa*, but usually do not participate in the immobilization reaction due to steric hindrance. However, the *pKa* of terminal amine groups is around 7–8.[40] Research has shown most of the reactive groups used for enzyme immobilization, such as glutaraldehyde and cyanogen bromide, can produce stable binding between enzyme and support under mild conditions.[41] But the high reactivity of these groups also makes them very unstable at alkaline pH, where reactivity of lysine residues is more appropriate for the reaction. Therefore, enzymes should ideally be immobilized on supports through the terminal amine groups at neural pH.[40]

12.7 ENZYME BIOREACTORS

12.7.1 Conventional Enzymatic Reactor

12.7.1.1 Batch Stirred Tank Reactor. The batch stirred tank reactor (BSTR) has traditionally been a popular enzyme reactor type due to its simple configuration, ease of operation and maintenance. During the reaction, the substrate and immobilized/free enzyme are stirred mechanically in a reaction vessel to maintain reactant and temperature homogeneity. At the end of the batch reaction the immobilized enzyme is separated from the reactant by filtration or centrifugation (regenerated if required) and reused.

Although they can usually achieve good substrate conversions, the productivity of BSTRs is low due to the time taken for emptying, cleaning and filling between batches. Consequently, a range of alternative enzyme reactor configurations have been proposed to overcome such disadvantages.

12.7.1.2 Packed Bed Reactor. Packed bed reactors (PBRs), also known as fixed bed reactors, are used for large-scale enzymatic reactions due to their high efficiency, low cost and ease of configuration. In such reactors, the immobilized enzyme on either particulate or fibre supports is packed inside a thermostat column or pipe, which renders a large surface area for reaction. Warmuth *et al.* first reported the hydrolysis of tributyrin by immobilized lipase on spheres in a packed bed reactor with a partial recycling operational mode.[42] The reactor volume was 10 mL and the feed rate was 2.5 g min^{-1}. Later, Carrara and co-workers successfully hydrolysed lactose with immobilized β-galactosidase on chitosan spherical particles in a PBR. The reactor was a column with 14 cm×1.2 cm and could be operated in a continuous mode.[43,44] Xu *et al.* produced inter-esterified lipids with immobilized lipase on beads in a continuous packed bed reactor at pilot scale. They found that a continuous PBR could effectively minimize side reactions in comparison with BSTRs. The immobilized lipase lost 40% of its activity after a 4-week run.[45] With a similar PBR configuration, refined olive pomace oil was converted to cocoa butter-like fat by acidolysis with fatty acids and 41.8% conversion was achieved under optimal conditions.[46]

Apart from enzymes immobilized on particles, PBRs have also been used enzymes immobilized on fibre supports. For example, β-galactosidase immobilized on cotton cloth was randomly packed in a stainless steel tube (8×93 mm) to form a PBR with batch operational mode and then successfully applied in lactose hydrolysis.[47] Then this reactor system was further scaled up to a continuous pilot-scale module with the cylinder dimensions increased to 140×1080 mm. As the residence time increased from (11.8 to 43.7) min, the conversion of lactose increased from 30.23% to 64.76%, indicating that the extent of hydrolysis could be adjusted by changing the flow rate.[48] Recently, Zhang *et al.* performed the enantioselective hydrolysis of racemic α-hydroxy-γ-butyrolactone with immobilized lactonase on cotton cloth in a PBR with a semi-continuous operational mode, where the reactor was a glass column with volume of 20 mL (1.5 cm $\times 12$ cm). An interesting finding was that the immobilized enzyme was much more stable in a PBR than in a stirred tank reactor (STR): the half-life in the PBR was at least six-fold higher than that in a STR.[49] However, the main drawbacks of PBRs are the associated large pressure drops (if the packing density is too high) as well as potential bypassing and channelling if the catalyst is inadequately packed. To reduce the pressure drop, large particles are required; however this decreases the amount of enzyme per volume in the reactor, thus decreasing overall reactor efficiency.[38]

12.7.1.3 Fluidized Bed Reactors. The fluidized bed reactor (FBR) is a loosened packed bed reactor (PBR). In FBR, when the fluid flows at high velocity, the bed is loosened and the particle–fluid mixture behaves like a fluid. Therefore, FBRs offer some of the advantages of STRs such as effective mixing of fluid and particles, and uniform temperature distribution across the reactor. However, FBRs have not been as widely as used in enzymatic reactions as PBRs, and only a few applications have been reported in literature.

One early report was the hydrolysis of ricebran oil by immobilized lipase from *Candida cylindracea* (now *C. rugosa*) on nylon-6 beads in a FBR with a semi-batch operational mode. The study showed that the FBR gave a smaller pressure drop than a PBR under the same operational conditions; for example, at flow rate of 0.7 m s^{-1}, the pressure in FBR was 20% smaller than that in PBR. Additionally, the FBR also showed almost double the conversion of ricebran oil compared to the PBR under optimal conditions.[50] Roy and Gupta studied the lactose hydrolysis in whole milk by β-galactosidase immobilized on cellulose beads in a FBR with continuous operational mode, with the reaction reaching a stable 60% conversion after 5 h.[51]

One advantage of FBRs over PBRs is that smaller particles can be used since the pressure drop is unaffected, although larger particles are usually required anyway due to the low density difference between fluid and particles used in immobilized enzyme FBR systems, and the high viscosity of the fluids usually used. This again decreases the amount of enzyme per volume in the reactor, decreasing overall reactor efficiency. As with PBRs, FBRs can also suffer from channelling and bypassing of the particles as well as significant particle attrition leading to enzyme loss and deactivation.[38]

In summary, conventional reactor configurations such as BSTRs, PBRs and FBRs, have been widely used in enzymatic reactions. However, each of them has drawbacks and a common problem for these reactors is mass transfer limitation at high conversion. Much research has focused on overcoming these problems from a process intensification perspective, as discussed below.

12.7.2 Process Intensification

Process intensification has been recognized as a promising development path for the chemical process industry, aiming to improve production efficiency, reduce cost, enhance safety and reduce environmental pollution.[52,53] It can be achieved in two ways: improved reactor design and optimization of operating parameters. In terms of the reactor design, published papers mainly focus on the reconfiguration of conventional reactors, providing incremental contributions to process intensification.[54] However, some innovative reactors have been proposed; for example, microengineering and microtechnology have been used to develop microreactors in enzymatic reactions with process intensification.[55,56] It has been shown that high heat and mass transfer rates in micro-fluidic systems resulted in higher reaction yields than conventional reactors under harsher conditions.[57] However, immaturity and high financial costs are the main disadvantages of this technology. One of the most established enzyme process intensification technologies is the enzyme membrane reactor.

12.7.2.1 Enzyme Membrane reactor. In an enzyme membrane reactor (EMR), the enzyme is immobilized on a membrane either as a flat sheet or hollow fibre, so that the membrane functions both in separation and as an immobilization support, providing process intensification through integration of the reaction and separation process. In the EMR, the product can be separated continuously from the reactant feed, which is very beneficial in cases where the chemical equilibrium significantly influences the reaction proceeding or the product inhibits the reaction. In addition, the EMR can also be operated with two liquid phases without emulsification problems. Given these advantages, the EMR has been applied in many fields such as the food, pharmaceutical and biomedical industries.[58] However, EMRs typically experience a reduction in reaction rate and yield during operation caused by loss of catalyst and mass transfer efficiency. Enzyme leakage may happen depending on the enzyme properties and pore size distribution. For enzyme immobilized on a membrane surface, reactants can accumulate and form a gel layer on the membrane surface, creating a diffusion barrier to the reactants and thus decreasing the mass transfer efficiency. In addition, fouling problems often appear caused by the deposition or adsorption of particles inside or on the surface of the membrane, limiting their industrial applications.[38,59]

12.7.2.2 Spinning Mesh Disc Reactor. The spinning disc reactor (SDR) is a novel type of PI- technology, recently applied to enzymatic reactions by Feng *et al.*,[35-38] which overcomes mass transfer problems. It consists of a rotating disc with a jet of liquid impinging onto the centre of the disc surface. The centrifugal force of the spinning disc forces this liquid to form a thin and highly sheared film on top of the rotating surface, leading to rapid mixing and short residence times. Research has shown that the heat and mass transfer in such devices can be significantly enhanced by the fluid dynamics within these films.[60-62] The SDR has been applied in a wide range of chemical reactions such as polymerization,[63] nanoparticle preparation,[64-66] photocatalysis,[67,68] and transesterification in biodiesel synthesis.[69]

The SDR concept has been adapted for immobilized enzymatic reactions as a novel rotating reactor system: the spinning mesh disc reactor (SMDR).[37,38] The SMDR also uses centrifugal forces to spread a thin film across a spinning horizontal disc; however this disc has a cloth (thickness 1.5 mm) with immobilized enzyme resting on top of it. The SMDR therefore produces a flow of thin film both on top of, as well as through the cloth, providing more effective surfaces for reaction. Feng and co-workers have shown that the cloth is critical to increasing the potential of immobilized enzymes since it accelerates reaction rates due to high mass transfer rates and rapid mixing on top and within the cloth, with the cloth helping protect the attached enzymes from excessive hydrodynamic forces, as well as providing an additional structure that can promote mixing and turbulence at the appropriate spinning speeds and feed flow rates.[36-38] Under comparable reaction conditions, tributyrin hydrolysis in the SMDR proceeded at a higher rate than in a batch stirred tank reactor (BSTR), giving a higher conversion over the entire hydrolysis process at all the investigated concentrations; for example, for 20 g L^{-1} tributyrin, the final conversion was 52.2% in the BSTR and 65.4% in the SCDR after 240 min. The SMDR has also shown to be a robust technology, where 80% of the original activity was maintained after 15 continuous runs. Studies to date have used a cloth; however, any material that can be used for enzymes can be used, opening up this technology for a wide range of support structures, enzymes and thus reactions. Overall, the SMDR is an innovative, superior and robust technology for enhancing enzyme reactions, taking enzyme reactors beyond the current state-of-the-art.

12.8 CONCLUSIONS

The dramatic advances in DNA sequencing and our ability to manipulate protein sequence, either specifically through site directed mutagenesis, or by combining knowledge-based and directed evolution approaches mean that 'doing chemistry' with enzymes and organisms is now a reality. There are now many commercial large-scale enzyme catalysed processes where incremental improvements in extending the lifetime and properties of already robust industrial catalysts are being made. At the other end of the scale, biocatalysis is making significant inroads in fine chemical synthesis. Where yields and productivity are perhaps less important they are already a significant part of the toolbox of academic chemistry. However, there are also an increasing number of examples of biocatalysis forming a key component of pharmaceutical biosynthesis. In this case, the enzymes have often been modified, sometimes extensively, to give them the necessary specificity and activity. This is often done under contract by specialist companies. One of the challenges for the future is to speed up the process of modifying enzymes for their desired purpose, so that this fits in with the typical timescale of new compound synthesis. This could lead to a change in mind set, where synthesis is designed with biocatalytic steps in mind, rather than as an afterthought.

REFERENCES

1. D. Voet, J. G. Voet and C. W. Pratt, *Principles of Biochemistry*, Wiley, Hoboken, NJ, 2013.
2. J. Porath, *Protein Expression Purif.*, 1992, **3**, 263.
3. T. Ishige, K. Honda and S. Shimizu, *Curr. Opin. Chem. Biol.*, 2005, **9**, 174.

4. J. S. Hwang and H. N. Chang, *Biotechnol. Bioeng.*, 1989, **34**, 380.
5. D. M. Lima, P. Fernandes, D. S. Nascimento, R. D. L. F. Ribeiro and S. A. de Assis, *Food Technol. Biotechnol.*, 2011, **49**, 424.
6. S. B. Mahato and S. Garai, *Steroids*, 1997, **62**, 332.
7. P. F. Cook and W. W. Cleland, *Enzyme Kinetics and Mechanism*, Garland Science Publishing, New York, 2007.
8. P. P. Taylor, D. P. Pantaleone, R. F. Senkpeil and I. G. Fotheringham, *Trends Biotechnol.*, 1998, **16**, 412.
9. H. N. Bhatti and R. A. Khera, *Steroids*, 2012, **77**, 1267.
10. Y. Asano, T. Yasuda, Y. Tani and H. Yamada, *Agric. Biol. Chem.*, 1982, **46**, 1183.
11. M. Kloosterman, V. H. M. Elferink, J. Vaniersel, J. H. Roskam, E. M. Meijer, L. A. Hulshof and R. A. Sheldon, *Trends Biotechnol.*, 1988, **6**, 251.
12. S. Matsumura, K. Mabuchi and K. Toshima, *Macromol. Rapid Commun.*, 1997, **18**, 477.
13. C. S. Chen, Y. Fujimoto, G. Girdaukas and C. J. Sih, *J. Am. Chem. Soc.*, 1982, **104**, 7294.
14. O. May, S. Verseck, A. Bommarius and K. Drauz, *Org. Process Res. Dev.*, 2002, **6**, 452.
15. E. Busto, V. Gotor-Fernandez and V. Gotor, *Chem. Soc. Rev.*, 2010, **39**, 4504.
16. K. E. Jaeger, B. W. Dijkstra and M. T. Reetz, *Annu. Rev. Microbiol.*, 1999, **53**, 315.
17. Z. Armstrong and S. G. Withers, *Biopolymers*, 2013, **99**, 666.
18. B. Cobucci-Ponzano and M. Moracci, *Nat. Prod. Rep.*, 2012, **29**, 697.
19. R. D. Woodyer, T. W. Johannes and H. Zhao, *Enzyme Technol.*, 2005, 85.
20. A. S. Bommarius, M. Schwarm, K. Stingl, M. Kottenhahn, K. Huthmacher and K. Drauz, *Tetrahedron-Asymmetr.*, 1995, **6**, 2851.
21. V. I. Tishkov and V. O. Popov, *Biomol. Eng.*, 2006, **23**, 89.
22. H. L. Steele, K. E. Jaeger, R. Daniel and W. R. Streit, *J. Mol. Microb. Biotech.*, 2009, **16**, 25.
23. G. A. Strohmeier, H. Pichler, O. May and M. Gruber-Khadjawi, *Chem. Re.v*, 2011, **111**, 4141.
24. M. T. Reetz, S. Prasad, J. D. Carballeira, Y. Gumulya and M. Bocola, *J. Am. Chem. Soc.*, 2010, **132**, 9144.
25. A. M. Klibanov, *Science*, 1983, **219**, 722.
26. R. A. Sheldon, *Adv. Synth. Catal.*, 2007, **349**, 1289.
27. Y. H. Wang, J. C. Zhang and J. M. Yin, *Desalination Water Treat.*, 2009, **1**, 157.

28. C. Mateo, J. M. Palomo, G. Fernandez-Lorente, J. M. Guisan and R. Fernandez-Lafuente, *Enzyme Microb. Technol.*, 2007, **40**, 1451.
29. W. Tischer and F. Wedekind, in *Biocatalysis – From Discovery to Application*, ed. W. D. Fessner, Springer, Berlin and London, 1999, vol. 200, pp. 95–126.
30. A. C. Pierre, *Biocatal. Biotrans.*, 2004, **22**, 145.
31. M. T. Reetz, A. Zonta and J. Simpelkamp, *Biotechnol. Bioeng.*, 1996, **49**, 527.
32. F. X. Malcata, H. S. Garcia, C. G. H. Jr. and C. H. Amundson, *Biotechnol. Bioeng.*, 1992, **39**, 647.
33. P. Gupta, K. Dutt, S. Misra, S. Raghuwanshi and R. K. Saxena, *Bioresour. Technol.*, 2009, **100**, 4074.
34. C. Mateo, J. M. Palomo, L. M. van Langen, F. van Rantwijk and R. A. Sheldon, *Biotechnol. Bioeng.*, 2004, **86**, 273.
35. X. Feng, D. A. Patterson, M. Balaban and E. A. C. Emanuelsson, *Colloids Surf. B: Biointerfaces*, 2013, **102**, 526.
36. X. Feng, D. A. Patterson, M. Balaban and E. A. C. Emanuelsson, *Chem. Eng. Res. Design*, 2013, **91**, 1684.
37. X. Feng, D. A. Patterson, M. Balaban and E. A. C. Emanuelsson, *Chem. Eng. J.*, 2014, **235**, 356.
38. X. Feng, D. A. Patterson, M. Balaban, G. Fauconnier and E. A. C. Emanuelsson, *Chem. Eng. J.*, 2013, **221**, 407.
39. J. An, *Enzyme Immobilization on Woolen Cloth*, PhD thesis, University of Auckland, 2010.
40. C. Mateo, O. Abian, M. B. Ernedo, E. Cuenca, M. Fuentes, G. Fernandez-Lorente, J. M. Palomo, V. Grazu, B. C. C. Pessela, C. Giacomini, G. Irazoqui, A. Villarino, K. Ovsejevi, F. Batista-Viera, R. Fernandez-Lafuente and J. M. Guisan, *Enzyme Microb. Technol.*, 2005, **37**, 456.
41. J. M. Guisan, *Immobilization of Enzymes and Cells. Methods in Biotechnology, vol. 22*, 2nd edn, Human Press, Totowa, NJ, 2006.
42. W. Warmuth, E. Wenzig and A. Mersmann, *Chem. Eng. Sci.*, 1994, **49**, 5087.
43. C. R. Carrara and A. C. Rubiolo, *Chem. Eng. J.*, 1997, **65**, 93.
44. C. R. Carrara, E. J. Mammarella and A. C. Rubiolo, *Chem. Eng. J.*, 2003, **92**, 123.
45. X. Xu, S. Balchen, C. E. Hoy and J. Adler-Nissen, *J. Am. Oil Chem. Soc.*, 1998, **75**, 1573.
46. O. N. Çiftçi, S. Fadıloğlu and F. Göğüş, *Bioresour. Technol.*, 2009, **100**, 324.
47. Q. Z. K. Zhou, X. D. Chen and X. M. Li, *Biotechnol. Bioeng.*, 2003, **81**, 127.

48. X. Li, Q. Z. K. Zhou and X. D. Chen, *Chem. Eng. Process.*, 2007, **46**, 497.
49. X. Zhang, J.-H. Xu, D.-H. Liu, J. Pan and B. Chen, *Biochem. Eng. J.*, 2010, **50**, 47.
50. P. Padmini, K. K. S. Iyengar and A. Baradarajan, *J. Chem. Technol. Biotechnol.*, 1995, **64**, 31.
51. I. Roy and M. N. Gupta, *Process Biochem.*, 2003, **39**, 325.
52. T. Van Gerven and A. Stankiewicz, *Ind. Eng. Chem. Res.*, 2009, **48**, 2465.
53. H. Z. Liu, X. F. Liang, L. R. Yang and J. Y. Chen, *Sci. China-Chem.*, 2010, **53**, 1470.
54. W. Zhang, *Chinese J. Chem. Eng.*, 2009, **17**, 688.
55. S. Kundu, A. S. Bhangale, W. E. Wallace, K. M. Flynn, C. M. Guttman, R. A. Gross and K. L. Beers, *J. Am. Chem. Soc.*, 2011, **133**, 6006.
56. S. Kataoka, Y. Takeuchi, A. Harada, M. Yamada and A. Endo, *Green Chem.*, 2010, **12**, 331.
57. J. C. Charpentier, *Chem. Eng. Technol.*, 2005, **28**, 255.
58. L. Giorno and E. Drioli, *Trends Biotechnol.*, 2000, **18**, 339.
59. D. M. F. Prazeres and J. M. S. Cabral, *Enzyme Microb. Technol.*, 1994, **16**, 738.
60. R. J. J. Jachuck and C. Ramshaw, *Heat Recovery Syst. CHP*, 1994, **14**, 475.
61. K. V. K. Boodhoo and R. J. Jachuck, *Green Chem.*, 2000, **2**, 235.
62. I. Boiarkina, S. Norris and D. A. Patterson, *Chem. Eng. J.*, 2013, **222**, 159.
63. K. V. K. Boodhoo and R. J. Jachuck, *Appl. Thermal Eng.*, 2000, **20**, 1127.
64. C. Y. Tai, Y. H. Wang and H. S. Liu, *AICHE J.*, 2008, **54**, 445.
65. H. S. Liu, Y. H. Wang, C. C. Li and C. Y. Tai, *Chem. Eng. J.*, 2012, **183**, 466.
66. H. S. Liu, K. A. Chen and C. Y. Tai, *Chem. Eng. J.*, 2012, **197**, 101.
67. I. Boiarkina, S. Pedron and D. A. Patterson, *Appl. Catal. B-Environ.*, 2011, **110**, 14.
68. I. Boiarkina, S. Norris and D. A. Patterson, *Chem. Eng. J.*, 2013, **225**, 752.
69. Z. Qiu, J. Petera and L. R. Weatherley, *Chem. Eng. J.*, 2012, **210**, 597.

CHAPTER 13

Bioelectrochemical Systems

UWE SCHRÖDER

Institute of Environmental and Sustainable Chemistry, Technische
Universität Braunschweig, Hagenring 30, 38106 Braunschweig, Germany
Email: uwe.schroeder@tu-braunschweig.de

13.1   INTRODUCTION

Microbial bioelectrochemistry is a research field dedicated to the
study and engineering of the interaction of living microbial cells and
electrodes, and to the development of microbial bioelectrochemical
systems (often abbreviated as BES).[1] The last decade has seen a
dynamic and rapid development of this research field with particular
emphasis on microbial fuel cells (MFCs),[2] microbial electrolysis cells
(MECs), microbial desalination cells (MDCs) and microbial solar
cells.[3] These bioelectrochemical systems can be subsumed under the
term microbial electrochemical technology (MET) or microbial
electrotechnology.

Figure 13.1 shows the number of papers that have been published
annually on the topic of microbial fuel cells over the past 33 years. The
impressive increase over the last decade is a synonym for the growing
research activity and the growing research community. This community has recently founded the *International Society for Microbial
Electrochemistry and Technology* (ISMES). It recruits from multiple
areas including environmental technology, civil engineering, electrochemistry and microbiology.

Chemical Processes for a Sustainable Future
Edited by Trevor M. Letcher, Janet L. Scott and Darrell A. Patterson
© The Royal Society of Chemistry 2015
Published by the Royal Society of Chemistry, www.rsc.org

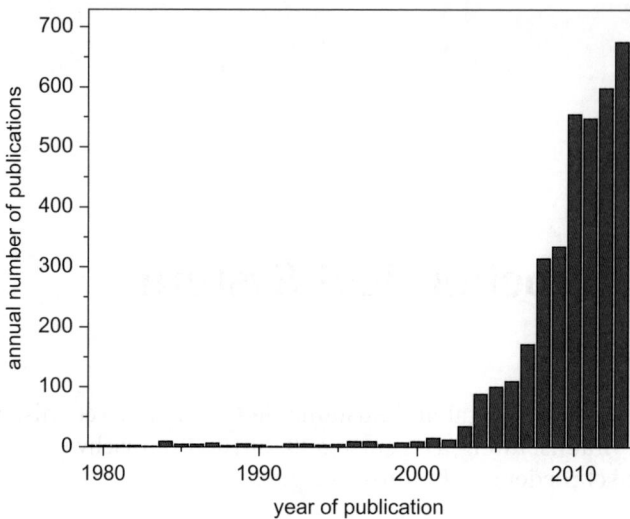

Figure 13.1   Number of publications per year from 1980 to 2013 containing the phrase 'microbial fuel cell' (Scopus, January 2014).

This chapter summarizes major developments in the field of microbial bioelectrochemical systems. It highlights major mechanisms in microbial electrochemistry, illustrates the development of new electrode concepts, which includes enhanced electrode performance, and highlights potential applications.

13.2   A MAJOR DRIVING FORCE

Energy and clean water represent major challenges for human kind. Both are strongly connected. Thus, increasing amounts of energy are currently consumed to provide clean water and to clean wastewater before it is released into our environment. As an example, in Germany, the treatment of municipal wastewater (10.1 km^3 wastewater per year) uses about 4.4 TWh electric energy—corresponding to the output of a large coal-fired power station.[4,5] In the United States, wastewater treatment uses about 1.5% of the total electricity consumption.[6] Furthermore, with an average of 3.3 kWh m^{-3} of stored chemical energy,[7] the energy content of municipal wastewater is considered to be nine times higher than the energy demand necessary for its treatment.[8]

These numbers illustrate that, despite the growing efforts to recover energy from wastewater (*e.g. via* anaerobic digestion), energy is afforded to abolish the valuable energy content in our wastewater. In this context, bioelectrochemical systems like microbial fuel cells have been proposed to combine wastewater treatment and energy recovery. In these bioelectrochemical cells, wastewater serves as the fuel for the bioelectrochemical energy conversion. Electric current is generated and organic matter, quantified by chemical oxygen demand (COD), is simultaneously removed. Besides the removal of COD, the removal of micropollutants such as antibiotics,[9] the recovery of nutrients like ammonia[10] and the recovery of metals[11] are under investigation. This direct combination of wastewater treatment and energy and resource recovery has become the major driving force in the development of microbial fuel cells and of deriving relevant technologies.[1,6,12–15]

13.3 MICROBE-ELECTRODE INTERACTIONS

Microbial electrochemistry is based on the functional electronic, interaction of (living) microorganisms and electrodes (Figure 13.2). For a long time, there has been the (mis-)conception that the transfer of electrons between bacterial cells and electrodes can be established only by means of artificial electron mediation.[16]

For this purpose, a large number of electron shuttling compounds (usually referred to as exogenous redox mediators) have been studied (Table 13.1). Although the use of artificial mediators helped to establish the concept of microbial fuel cells (as the archetype microbial BES), the mediators possess several crucial disadvantages. Thus, at higher concentrations most redox mediators are toxic to bacteria. Consequently, the concentration level (determining the limiting

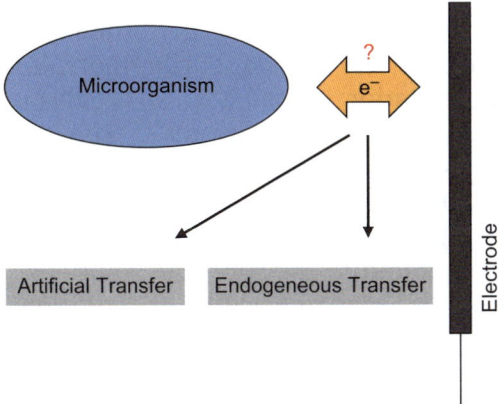

Figure 13.2 Microbial electrochemistry: cell–electrode interaction.

Table 13.1 Exemplary list of exogenous (artificial) redox mediators (adapted from ref. 16).

Substance Class	Redox Mediator	Redox Potential $E^{o\prime}$/V
Phenazines	Neutral Red	− 0.32
	Safranine	− 0.29
	Phenazine ethosulfate	0.06
Phenothiazines	New Methylene Blue	− 0.02
	Toluidine Blue O	0.03
	Thionine	0.06
	Phenothiazinone	0.13
Phenoxazines	Resorufin	− 0.05
	Gallocyanine	0.02
Quinones	2-Hydroxy-1,4-naphthoquinone	n.n.
	Anthraquinone-2,6-disulfonate	n.n.

current density) has to be kept in the low micromolar range and possibilities to increase the electrode performance *via* an increased mediator level are thus very limited. Furthermore, it has been shown that the current generation in mediated MFC substantially inhibited the growth of the used microorganisms.[17] The reason is that the mediator diverts ('steals') electrons from the metabolism of the bacteria and thus limits the energy gain for the microorganisms. The greatest disadvantage, however, is the necessity for a regular supply of the exogenous compound to the substrate solutions. This is technologically and economically unfeasible. In addition, it is environmentally questionable since the mediators are not consumed or degraded during the bioelectrochemical processes and would thus have to be removed from the effluent solution of a microbial fuel cell or separated from the product solution of a microbial electrosynthesis process. For microbial fuel cells, these disadvantages not only made any large-scale application unthinkable but lead to the general abandonment of artificial redox mediation.

As discovered by Kim and colleagues at the turn of the last century,[17,18] nature has developed different means for exocellular electron transfer. The basis are electrochemically active bacteria (often referred to as electricigens,[19] exoelectrogenic bacteria[20] or electrode respiring bacteria[21]) that can transfer electrons actively out of their cell membrane or into their cell membrane. Two fundamental types of electron transfer can be distinguished: direct electron transfer (DET) and mediated electron transfer (MET).[16] Depending on the electron transfer abilities of the involved bacterial species, the distance of electron transfer can reach up to hundreds of micrometres.

Direct electron transfer (DET) obviously requires the physical contact of bacteria or their organelles with an electrode. In the great majority of cases this means a permanent cell-electrode connection by means of the formation of a bacterial biofilm at the electrode. In spite of this, DET *via* temporary cell–electrode contact (*e.g.* of *Shewanella* species) has been reported.[22] Figure 13.3 depicts the two principal pathways of the direct electron transfer. DET based on trans-membrane electron transfer proteins is probably the most straightforward electron transfer path: the bacterium is adjacent to the electrode (or to another cell) and transfers the electrons *via* trans-membrane proteins such as cytochromes. This electron transfer mechanism has been described for numerous microorganisms, including *Geobacter* and *Shewanella* as prominent members.[19,20] Recently it has been demonstrated that some metal-reducing bacteria such as *Geobacter* and *Shewanella* strains can evolve electronically conducting molecular pili,[23–26] denominated as

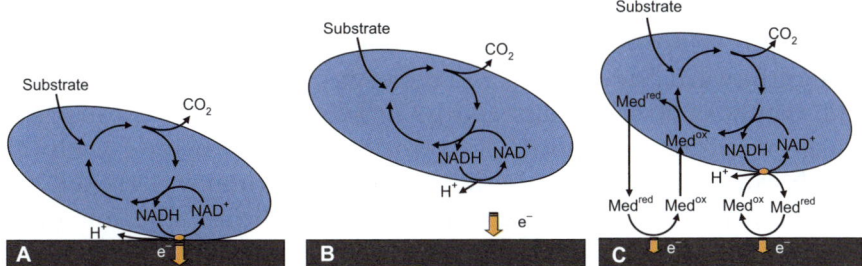

Figure 13.3 Natural, exocellular electron transfer mechanisms: (a) direct electron transfer (DET) *via* membrane bound redox proteins; (b) DET *via* electronically conductive pili; and (c) mediated electron transfer *via* endogenous (bacterial) redox mediators (adapted from ref. 16).

nanowires that allow the microorganism to reach and utilize more distant solid electron acceptors (and thus a BES anode) without physical whole cell contact [Figure 13.3(b)]. It is just necessary to attach the nanowires, which are linked to a redox active membrane bound protein that facilitates the trans-membrane electron transfer, to the electrode. Obviously, such nanowires may allow the development of thicker electroactive biofilms and thus higher anode performances.

Numerous organisms can produce low-molecular, electron shuttling compounds *via* secondary metabolic pathways.[27] Examples for such secondary metabolites, which have been shown to be involved in extracellular electron transfer processes, are bacterial phenazines such as pyocyanine and 2-amino-3-carboxy-1,4-naphthoquinone, ACNQ (see Table 13.2) as well as flavins.[28-30]

These so-called endogenous redox mediators are of great interest, as their synthesis makes the electron transfer independent of the presence of exogenous redox shuttles. The mediator serves as a reversible terminal electron acceptor for mediated electron transfer (MET), transferring electrons between the bacterial cells and the respective BES electrode, where it becomes re-oxidized or re-reduced and is again available for subsequent redox processes [Figure 13.3(c)].

13.3.1 Electroactive Biofilms

The core unit (the 'beating heart') of the majority of microbial fuel cells is the electroactive biofilm, *i.e.* a bacterial biofilm consisting of electrochemically active bacteria (Figure 13.4). Based on the above, natural electron transfer mechanisms, multiple layers of bacteria form electroactive biofilms of up to 100 μm thickness. The biofilm may be composed of either a single species (*e.g.* Geobacter sulfurreducens) or a bacterial community.

The procedure for the formation electroactive biofilms is quite straightforward (Figure 13.5) and the bacterial inoculum wastewater, soil, compost or sediments can be employed directly, without any bacterial extraction steps.[31-33] Together with a substrate solution, which can be either a synthetic medium or natural wastewater, the inoculum is placed in an electrochemical cell (*e.g.* the microbial fuel cell) in which the bacteria grow at the provided electrode. Cathodic biofilms grow at negatively polarized electrodes in the presence of electron acceptor substrates, and anodic biofilms grow at positive electrode potentials in the presence of electron donor substrates.

Table 13.2 Exemplary list of endogenous (bacterial) redox mediators (adapted from ref. 16).

Name	Structure	Redox potential $E^{o\prime}/V$
Phenazine-1-carboxamide		− 0.115
Pyocyanine (phenazine)		− 0.03
ACNQ(2-amino-3-carboxy-1,4-naphthoquinone)		− 0.071
Riboflavin		− 0.21

Primary biofilms can be used to inoculate further electrochemical cells; in most cases, the secondary biofilm outperforms the primary biofilm.[33]

The composition, the performance and the properties of the electroactive biofilm depend on a set of parameters. The greatest influence is the nature of the substrate (*e.g.* of the electron donor).

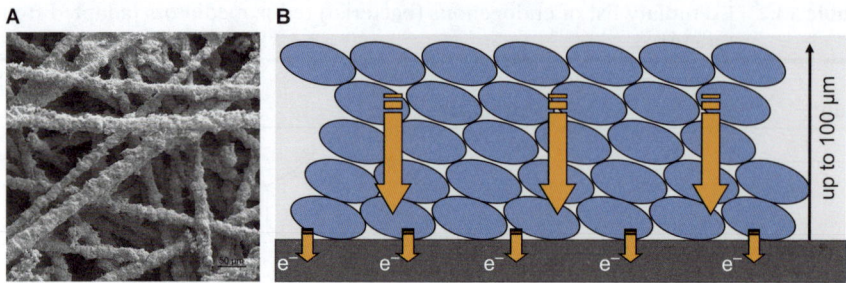

Figure 13.4   Scanning electron micrograph image (a) and schematic illustration (b) of an electroactive bacterial biofilm.

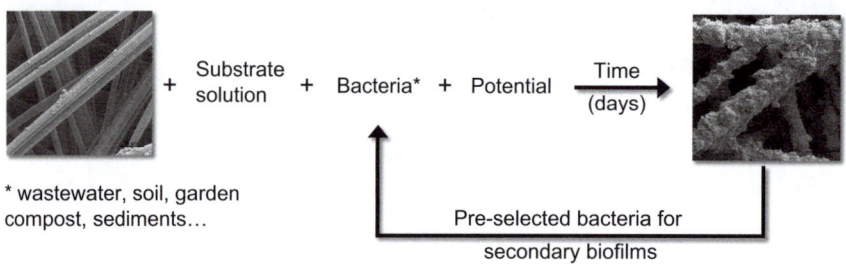

Figure 13.5   Schematic illustration of the procedure to grow an electroactive microbial biofilm.

Thus, the presence of single, simple substrates like acetate, which can be degraded by single species like *Geobacter*, lead to biofilms often dominated by one species.[34] In the case of substrate mixtures of complex substrates (*e.g.* complex carbohydrates), complex mixed communities are formed.

13.4   ELECTRODES FOR BIOELECTROCHEMICAL SYSTEMS

The development of electrodes for microbial electrochemical processes is highly dynamic. The current focus lies in the development of materials for biofilm electrodes. A major requirement for the electrode material is its biocompatibility. The material may range from graphite to metals like gold, platinum and stainless steel. For up-scaling reasons, graphite and stainless steel appear to be the most promising materials.[35]

For the improvement of the electrode performance, two major strategies are followed: (1) surface modification and (2) three-dimensional (3D) structuring. Surface modification has been achieved by various means including ammonia treatment,[36] polymer or cross-linker modification, and surface oxidation.[36,37] These measures aim to improve the biofilm–electrode interactions for increased electrochemical reversibility (rate of electron transfer) and to enhance the rate of colonization of new electrodes by bacteria.

The current density of high-performing anodic microbial biofilms at flat, macroscopically smooth electrode surfaces (*e.g.* smooth poly-crystalline graphite) reaches about 1 mA cm^{-2}, a value that is assumed to be determined by the limits of the bacterial metabolism and by of the kinetics of electron and proton transfer within the biofilms.[16,38] In order to increase the surface area available for the biofilm formation and thus to increase bacterial turnover and current density, three-dimensional electrodes are currently being intensively developed. Several 3D materials have been proposed, mostly based on carbon, but also some based on stainless steel.[35]

Examples for 3D electrodes are backed graphite granule electrodes,[40] carbon felt/paper electrodes,[41,42] carbon brush electrodes (Figure 13.6),[39] carbon fibre[42] and foams.[43] Some materials, like electrospun carbon fibre mats[44,45] have shown great performance (projected current densities of about 3 mA cm^{-2}) at a porosity of 98% and a minimum specific weight.[46]

Another interesting option is the use of natural templates such as plant stems for electrode preparation.[47] In this case, the natural, porous biopolymer scaffold has to be transformed into a carbonaceous form by thermal treatment at a temperature around (700–1000) °C in an inert gas atmosphere—the procedure is often referred to as carbonization.

A recent promising approach is the use of so-called layered corrugated carbon (Figure 13.7).[48] The material is prepared *via* the

Figure 13.6 Bottle brush electrode as a typical three-dimensional electrode material[39] (Reprinted with permission from ref. 39, Copyright 2007, American Chemical Society).

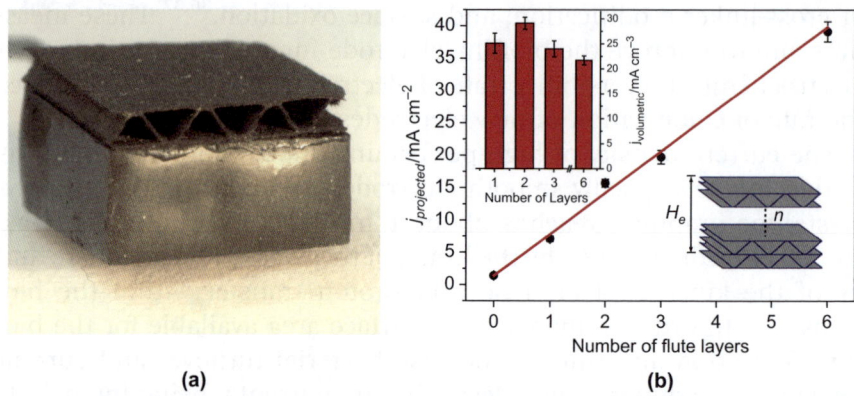

Figure 13.7   (a) Photographic picture of a layered corrugated carbon electrode. (b) Dependence of the steady state current density at layered corrugated carbon (main figure: projected, inset figure: volumetric current density) on the number of electrode layers (adapted from ref. 48).

carbonization of a material that everyone in industrial countries knows: corrugated fibreboard (we consume around $(30–70)$ kg a^{-1}). Corrugated fibreboard is a paper-based, lignocellulosic material that, in the case of single wall corrugated cardboard, consists of one fluted (corrugated) sheet sandwiched between two flat linerboards. By a simple carbonization step, this material is transformed into a carbon scaffold that retains the original cardboard structure, a structure that can be denoted as layered corrugated carbon. The material is stackable, *i.e.* it can be used in stacks of several layers, with a linearly increasing projected current density (and at almost constant volumetric current density (Figure 13.7). With this stacked electrode setup, a current density of 40 mA cm^{-2} has already been achieved.

As the result of ongoing efforts to enhance the performance of biofilm electrode by sophisticated biofilm growth procedures and improved electrode designs, the current density of microbial biofilm anodes has improved by four orders of magnitude, from a few microampere per cm^2 in the year 2000 to more than 20 mA per cm^2 in 2014 (Figure 13.8). It is unforeseeable as to what level this performance increase can still be driven.

13.5   TYPES OF BIOELECTROCHEMICAL SYSTEMS

13.5.1   Microbial Fuel Cells

Probably more than 90% of current BES research is dedicated to microbial fuel cells (MFCs), the archetype microbial bioelectrochemical

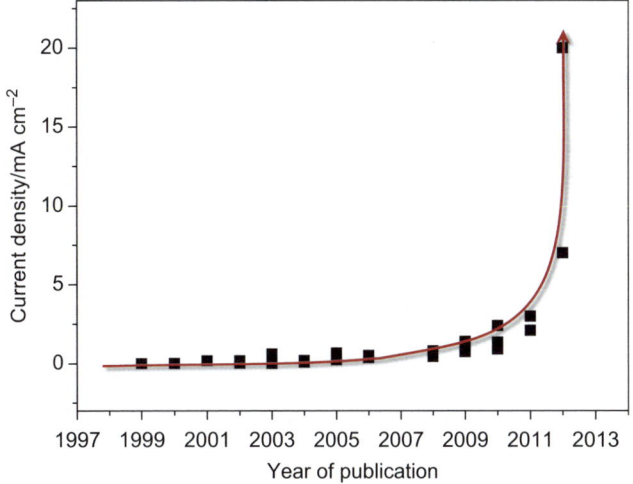

Figure 13.8 Development of the projected current density of microbial biofilm anodes over the past 15 years.

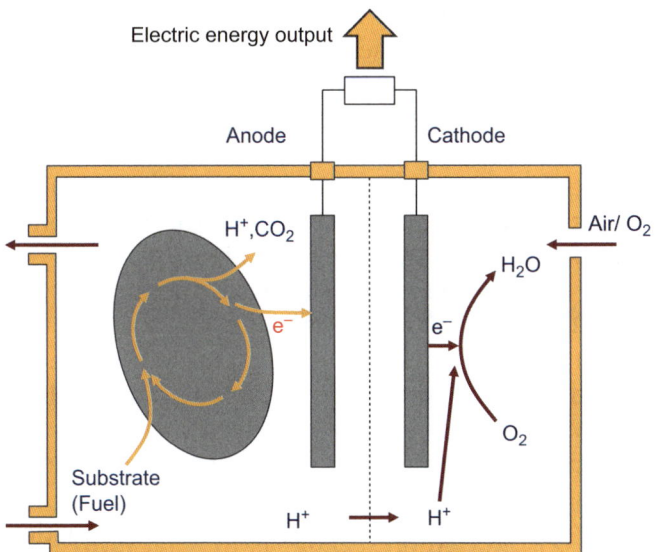

Figure 13.9 Schematic illustration of a microbial fuel cell.

system. Microbial fuel cells (Figure 13.9) are microbial bioelectrochemical systems that aim to generate electricity from the anodic oxidation of, for example, wastewater organics.[1] The cathode reaction is usually the oxygen reduction reaction. Oxygen possesses a high oxidation potential (the biological standard potential, $E^{o\prime}$, is 0.82 V), is non-toxic and its availability is virtually unlimited. In principle, both the anode and cathode reactions may be microbially catalysed. Yet, it is only at the anode that there is a need for microbial electrocatalysis (there is no chemical electrocatalyst available for the oxidation of organic matter at ambient conditions), whereas for oxygen reduction, a number of abiotic electrocatalyst options are available.[49–51]

Depending on the fields of application envisaged, different concepts and types of microbial fuel cells have been developed. These include, for instance, benthic fuel cells (for the powering of marine sensors),[52–56] microbial fuel cells for autonomous robots,[57–59] microbial solar cells[60–63] and microbial fuel cells for use as biosensors.[64–68] The major emphasis, however, is on the development of MFC technology for wastewater treatment.

13.5.2 Microbial Electrolysis Cells, Microbial Desalination Cells and beyond

For a long time BES research focused exclusively on microbial fuel cells. Recently several further bioelectrochemical systems and new applications have been proposed. A major step was the introduction of microbial electrolysis cells (MECs) that in contrast to electricity producing MFCs use electric current to accomplish endergonic cell reactions such as the synthesis and upgrading of chemical products. One of the initial examples was the electrolytic hydrogen generation, supported by the microbially catalysed anodic oxidation of organic substrates such as wastewater organics (Figure 13.10).[69–71]

Considering the biological standard potentials $E^{0\prime}_{acetate/CO2}$ (anode) $= -0.290$ V and $E^{0\prime}_{H2/H+}$ (cathode) $= -0.414$ V, the standard cell voltage for the cell drawn in Figure 13.10(a) is -0.124 V. Compared with conventional water electrolysis (with a standard cell voltage of -1.23 V), only one tenth of the electric energy is necessary to drive the hydrogen evolution reaction. In many cases the boundaries between MFCs and MECs are blurred, as in the case of the cathodic production of hydrogen peroxide,[72] where the cell reaction (anodic, microbially catalysed oxidation of, for example, acetate; and cathodic abiotic reduction of oxygen to H_2O_2) is exergonic. Thus, the production of hydrogen peroxide can be achieved in MFC mode, *i.e.* under the generation of excess electric energy. Yet, in order to accelerate the reaction to an economically feasible rate, the transition into MEC mode may become necessary.

Microbial desalination cells (MDCs) are hybrid cells, consisting of a microbial fuel cell (or a microbial electrolysis cell) into which an

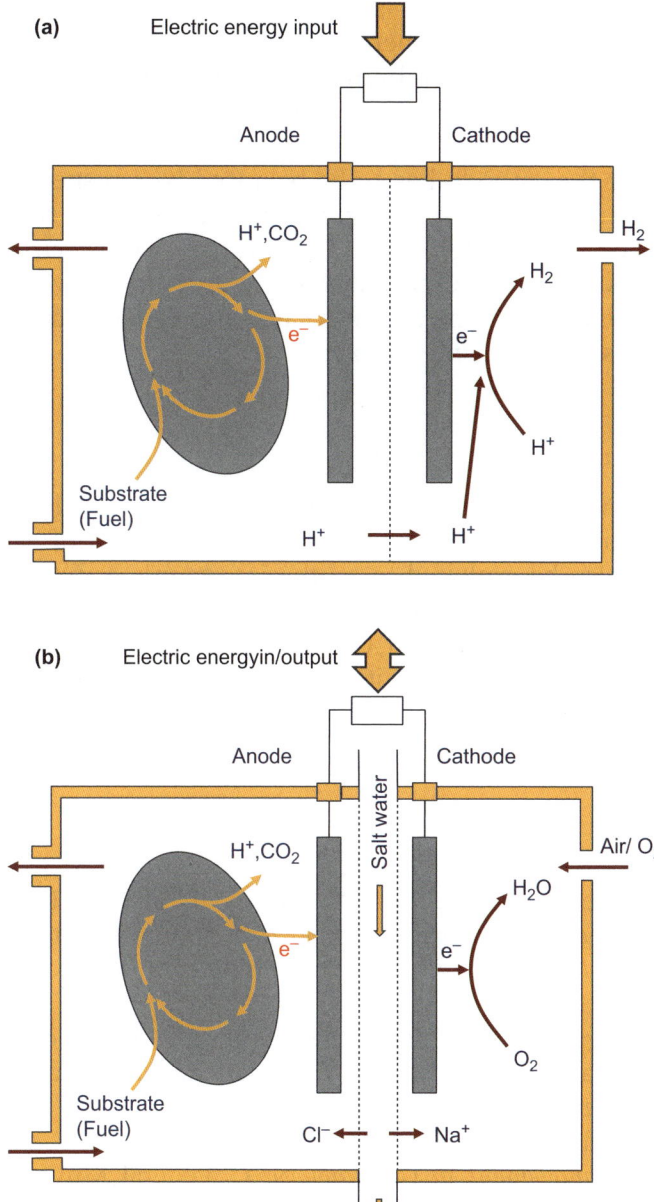

Figure 13.10 Schematic illustration of (a) a microbial electrolysis cell for the production of hydrogen and (b) microbial desalination cell.

electrodialysis compartment is integrated. This compartment consists of a cation exchange membrane facing cathode compartment and an anion exchange membrane facing the anode compartment of the MFC [Figure 13.10(b). During the operation of the MFC, the charge balancing ion flux (every electron liberated at the anode and consumed at the cathode has to be balanced by a respective charge balancing transfer of anions or cations) can be utilized to remove salts from an incoming saline solution.[73,74] This type of bioelectrochemical technology is of interest especially for arid areas in this world, where besides wastewater treatment, freshwater and surface water salinization becomes an increasing problem.

13.5.3 Microbial Electrosynthesis '

The use of microbial electrochemistry goes far beyond the above described degradation processes, used for the liberation of electric energy. It can also be exploited to synthesize organic compounds. This may be a simple reduction of butyrate to butanol or the complex synthesis of organic matter from CO_2.[75-78] Although microbial electrosynthesis is still in its infancy, it can be expected that this technology will greatly broaden the applicability of microbial bioelectrochemical systems.

REFERENCES

1. K. Rabaey, L. Angenent, U. Schröder and J. Keller (ed.), *Bioelectrochemical Systems: From Extracellular Electron Transfer to Biotechnological Application*, IWA Publishing, London and New York, 2010.
2. B. E. Logan, B. Hamelers, R. Rozendal, U. Schröder, J. Keller, S. Freguia, P. Aelterman, W. Verstraete and K. Rabaey, *Environ. Sci. Technol.*, 2006, **40**, 5181–5191.
3. U. Schröder, *J. Solid State Electrochem.*, 2011, 1–6.
4. *Waste Management in Germany*, N. C. Federal Ministry for the Environment, Building and Nuclear Safety, Bonn, Germany, 2011.
5. J. Görres, *Stuttgarter Berichte zur Siedlungswasserwirtschaft, Institut für Siedlungswasserbau*, Universität Stuttgart, Stuttgart, 2007, **191**, pp. 89–104.
6. B. E. Logan, *Microbial Fuel Cells*, John Wiley & Sons, New York, 2008.

7. T. H. Pham, K. Rabaey, P. Aelterman, P. Clauwaert, L. De Schamphelaire, N. Boon and W. Verstraete, *Eng. Life Sci.*, 2006, **6**, 285–292.
8. I. Shizas and D. M. Bagley, *J. Energy Eng.*, 2004, **130**, 45–53.
9. F. Harnisch, C. Gimkiewicz, B. Bogunovic, R. Kreuzig and U. Schröder, *Electrochem. Commun.*, 2013, **26**, 77–80.
10. K. Y. Cheng, A. H. Kaksonen and R. Cord-Ruwisch, *Bioresour. Technol.*, 2013, **143**, 25–31.
11. A. T. Heijne, F. Liu, R. V. D. Weijden, J. Weijma, C. J. N. Buisman and H. V. M. Hamelers, *Environ. Sci. Technol.*, 2010, **44**, 4376–4381.
12. L. T. Angenent, K. Karim, M. H. Al-Dahhan, B. A. Wrenn and R. Domíguez-Espinosa, *Trends Biotechnol.*, 2004, **22**, 477–485.
13. B. H. Kim, I. S. Chang, M. S. Hyun, H. S. Kim, H. S. Park, *World Pat.*, WO0104061, 2001.
14. K. Rabaey and W. Verstraete, *Trends Biotechnol.*, 2005, **23**, 291–298.
15. S. T. Oh, J. R. Kim, G. C. Premier, T. H. Lee, C. Kim and W. T. Sloan, *Biotechnol. Adv.*, 28, 871–881.
16. U. Schröder, *Phys. Chem. Chem. Phys.*, 2007, **9**, 2619–2629.
17. B. H. Kim, H. J. Kim, M. S. Hyun and D. H. Park, *J. Microbiol. Biotechnol.*, 1999, **9**, 127–131.
18. H. J. Kim, H. S. Park, M. S. Hyun, I. S. Chang, M. Kim and B. H. Kim, *Enzyme Microbial. Technol.*, 2002, **30**, 145–152.
19. D. R. Lovley, *Nat. Rev. Microbiol.*, 2006, **4**, 497–508.
20. B. E. Logan, *Nat. Rev. Microbiol.*, 2009, **7**, 375–381.
21. C. I. Torres, A. K. Marcus and B. E. Rittmann, *Biotechnol. Bioeng.*, 2008, **100**, 872–881.
22. H. W. Harrisa, M. Y. El-Naggarb, O. Bretschgerc, M. J. Wardd, M. F. Rominee, A. Y. Obraztsovad and K. H. Nealson, *Proc. Natl. Acad. Sci. U. S. A.*, 2010, **107**, 326–331.
23. Y. A. Gorby, S. Yanina, J. S. McLean, K. M. Rosso, D. Moyles, A. Dohnalkova, T. J. Beveridge, I. S. Chang, B. H. Kim, K. S. Kim, D. E. Culley, S. B. Reed, M. F. Romine, D. A. Saffarini, E. A. Hill, L. Shi, D. A. Elias, D. W. Kennedy, G. Pinchuk, K. Watanabe, S. i. Ishii, B. Logan, K. H. Nealson and J. K. Fredrickson, *Proc. Natl. Acad. Sci. U. S. A.*, 2006, **103**, 11358–11363.
24. G. Reguera, K. D. McCarthy, T. Mehta, J. S. Nicoll, M. T. Tuominen and D. R. Lovley, *Nature*, 2005, **435**, 1098–1101.
25. N. S. Malvankar and D. R. Lovley, *Curr. Opin. Biotechnol.*, 2014, **27**, 88–95.
26. N. S. Malvankar and D. R. Lovley, *ChemSusChem*, 2012, **5**, 1039–1046.

27. B. A. Haddock and C. W. Jones, *Bacteriol. Rev.*, 1977, **41**, 47–99.
28. T. H. Pham, N. Boon, P. Aelterman, P. Clauwaert, L. De Schamphelaire, L. Vanhaecke, K. De Maeyer, M. Höfte, W. Verstraete and K. Rabaey, *Appl. Microbiol. Biotechnol.*, 2008, **77**, 1119–1129.
29. E. Marsili, D. B. Baron, I. D. Shikhare, D. Coursolle, J. A. Gralnick and D. R. Bond, *Proc. Natl. Acad. Sci. U. S. A.*, 2008, **105**, 3968–3973.
30. S. Yamazaki, K. Kano, T. Ikeda, K. Isawa and T. Kaneko, *Biochim. Biophys. Acta*, 1999, **1428**, 241–250.
31. S. F. Ketep, A. Bergel, M. Bertrand, W. Achouak and E. Fourest, *Biochem. Eng. J.*, 2013, **73**, 12–16.
32. B. Cercado, N. Byrne, M. Bertrand, D. Pocaznoi, M. Rimboud, W. Achouak and A. Bergel, *Bioresour. Technol.*, 2013, **134**, 276–284.
33. Y. Liu, F. Harnisch, K. Fricke, R. Sietmann and U. Schröder, *Biosens. Bioelectron.*, 2008, **24**, 1012–1017.
34. F. Harnisch, C. Koch, S. A. Patil, T. Huebschmann, S. Mueller and U. Schroeder, *Energy Environ. Sci.*, 2011, **4**, 1265–1267.
35. D. Pocaznoi, A. Calmet, L. Etcheverry, B. Erable and A. Bergel, *Energy Environ. Sci.*, 2012, **5**, 9645–9652.
36. S. Cheng and B. E. Logan, *Electrochem. Commun.*, 2007, **9**, 492–496.
37. K. Scott, G. A. Rimbu, K. P. Katuri, K. K. Prasad and I. M. Head, *Process Saf. Environ. Protect.*, 2007, **85**, 481–488.
38. C. I. Torres, A. K. Marcus, H. S. Lee, R. Krajmalnik-Brown and B. E. Rittmann, *FEMS Microbiol. Rev.*, 2010, **34**, 3–17.
39. B. E. Logan, S. Cheng, V. Watson and G. Estadt, *Environ. Sci. Technol.*, 2007, **41**, 3341–3346.
40. D. Sell, P. Kraemer and G. Kreysa, *Appl. Microbiol. Biotechnol.*, 1989, **31**, 211–213.
41. S. Srikanth, E. Marsili, M. C. Flickinger and D. R. Bond, *Biotech. Bioeng.*, 2008, **99**, 1065–1073.
42. Y. Liu, F. Harnisch, U. Schröder, K. Fricke, V. Climent and J. M. Feliu, *Biosens. Bioelectron.*, 2010, **25**, 2167–2171.
43. Y. Zhao, K. Watanabe, R. Nakamura, S. Mori, H. Liu, K. Ishii and K. Hashimoto, *Chem. Eur. J.*, 2010, **16**, 4982–4985.
44. G. He, Y. Gu, S. He, U. Schröder, S. Chen and H. Hou, *Bioresour. Technol.*, 2011, **102**, 10763–10766.
45. S. Chen, G. He, A. A. Carmona-Martinez, S. Agarwal, A. Greiner, H. Hou and U. Schröder, *Electrochem. Commun.*, 2011, **13**, 1026–1029.
46. S. Chen, H. Hou, F. Harnisch, S. A. Patil, A. A. Carmona-Martinez, S. Agarwal, Y. Zhang, S. Sinha-Ray, A. L. Yarin,

A. Greiner and U. Schroeder, *Energy Environ. Sci.*, 2011, **4**, 1417–1421.

47. S. Chen, G. He, X. Hu, M. Xie, Q. Liu, Y. Zhou, S. Wang, H. Hou and U. Schröder, *ChemSusChem*, 2012, **5**, 1059–1063.

48. S. Chen, G. He, Q. Liu, F. Harnisch, Y. Zhou, Y. Chen, M. Hanif, S. Wang, X. Peng, H. Hou and U. Schröder, *Energy Environ. Sci.*, 2012, **5**, 9769–9772.

49. R. A. Rozendal, F. Harnisch, A. W. Jeremiasse and U. Schröder, in *Bio-electrochemical Systems: From Extracellular Electron Transfer to Biotechnological Application*, ed. K. Rabaey, L. Angenent, U. Schröder and J. Keller, IWA Publishing, London and New York, 2010.

50. B. Logan, D. Call, M. Merrill and S. Cheng, *US Pat.*, 20100119920A1, 2010.

51. F. Harnisch and U. Schröder, *Chem. Soc. Rev.*, 2010, **39**, 4433–4448.

52. M. E. Nielsen, C. E. Reimers and H. A. Stecher Iii, *Environ. Sci. Technol.*, 2007, **41**, 7895–7900.

53. M. E. Nielsen, C. E. Reimers, H. K. White, S. Sharma and P. R. Girguis, *Energy Environ. Sci.*, 2008, **1**, 584–593.

54. L. M. Tender, S. A. Gray, E. Groveman, D. A. Lowy, P. Kauffman, J. Melhado, R. C. Tyce, D. Flynn, R. Petrecca and J. Dobarro, *J. Power Sources*, 2008, **179**, 571–575.

55. Z. Lewandowski, H. Beyenal, A. Dewan, H. Gao and A. Meehan, Microbial fuel cells to power submersed electronic devices, in Abstracts of Papers of the American Chemical Society 238, American Chemical Society, Washington DC, 2009, vol. 238.

56. Y. Gong, S. E. Radachowsky, M. Wolf, M. E. Nielsen, P. R. Girguis and C. E. Reimers, *Environ. Sci. Technol.*, 2011, **45**, 5047–5053.

57. S. Wilkinson, *Autonomous Robots*, 2000, **9**, 99–111.

58. I. Ieropoulus, C. Melhuish and J. Greenman, Energetically autonomous robots, in *Proceedings of the 8th Intelligent Autonomous System Conference (IAS-8)*, Amsterdam, 2004, pp. 128–135.

59. I. Ieropoulus, C. Melhuish, J. Greenman and I. Horsfield, Artificial symbiosis: towards a robot-microbe partnership, presented at Towards Autonomous Robotic Systems 2005 (TAROS 2005), London, 2005.

60. M. Rosenbaum, U. Schröder and F. Scholz, *Environ. Sci. Technol.*, 2005, **39**, 6328–6333.

61. M. Rosenbaum, U. Schröder and F. Scholz, *Appl. Microbiol. Biot.*, 2005, **68**, 753–756.

62. M. Rosenbaum, Z. He and L. Angenent, *Curr. Opin. Biotech.*, 2010, **21**, 259–264.

63. M. Rosenbaum and U. Schröder, *Electroanalysis*, 2010, **22**, 844–855.

64. D. Odaci, S. Timur and A. Teleconcu, *Bioelectrochemistry*, 2009, **75**, 77–82.

65. M. Di Lorenzo, T. P. Curtis, I. M. Head and K. Scott, *Water Res.*, 2009, **43**, 3145–3154.

66. J. M. Tront, J. D. Fortner, M. Ploetze, J. B. Hughes and A. M. Puzrin, *Biosens. Bioelectro.*, 2008, **24**, 586–590.

67. G. Kunze and K. Riedel, in *Biosensors and Bioassays Based on Microorganisms*, ed. K. Riedel, G. Kunze and K. H. R. Baronian, Research Signpost, Trivandrum, India, 2006, pp. 105–129.

68. H. J. Kim, M. S. Hyun, I. S. Chang and B. H. Kim, *J. Microbiol. Biotechnol.*, 1999, **9**, 365–367.

69. B. E. Logan, D. Call, S. Cheng, H. V. M. Hamelers, T. H. J. A. Sleutels, A. W. Jeremiasse and R. A. Rozendal, *Environ. Sci. Technol.*, 2008, **42**, 8630.

70. D. Call and B. Logan, *Environ. Sci. Technol.*, 2008, **42**, 3401–3406.

71. R. A. Rozendal, H. V. M. Hamelers, G. J. W. Euverink, S. J. Metz and C. J. N. Buisman, *Int. J. Hydrogen Energy*, 2006, **31**, 1632–1640.

72. K. Rabaey, S. Bützer, S. Brown, J. Keller and R. A. Rozendal, *Environ. Sci. Technol.*, 2010, **44**, 4315–4321.

73. K. S. Jacobson and D. M. Drew, *Bioresour. Technol.*, 2011, **102**, 376–380.

74. X. Cao, X. Huang, P. Liang, K. Xiao, Y. Zhou, X. Zhang and B. E. Logan, *Environ. Sci. Technol.*, 2009, **43**, 7148–7152.

75. K. Rabaey and R. A. Rozendal, *Nat. Rev. Microbiol.*, 2010, **8**, 706–716.

76. K. P. Nevin, S. A. Hensley, A. E. Franks, Z. M. Summers, J. Ou, T. L. Woodard, O. L. Snoeyenbos-West and D. R. Lovley, *Appl. Environ. Microbiol.*, 2011, **77**, 2882–2886.

77. D. R. Lovley, *Environ. Microbiol. Rep.*, 2011, **3**, 27–35.

78. K. P. Nevin, T. L. Woodard, A. E. Franks, Z. M. Summers and D. R. Lovley, *mBio*, 2010, **1**.

CHAPTER 14

Fermentations and Sustainable Technologies: From Free Enzymes to Whole Cells, from Fine Chemicals to Bulk Commodities

PABLO DOMÍNGUEZ DE MARÍA

Sustainable Momentum, Ap. Correos 3517. 35004, Las Palmas de Gran Canaria, Gran Canaria, Canary Islands, Spain
Email: Dominguez@itmc.rwth-aachen.de;
dominguez@sustainable-momentum.net

14.1 INTRODUCTION: WHITE BIOTECHNOLOGY

The use of enzymes and whole cells as biocatalysts for chemical synthetic processes, broadly speaking the so-called 'white bio-technology', has gained increasing importance over the past two decades. The motivation can be found in the high selectivities (enantio-, regio- and chemo-) that natural catalysts often display, together with the mild and environmentally friendly reaction conditions needed for an efficient catalytic performance. In this respect, both economically feasible and more sustainable processes can be designed, and as a result a number of industrially implemented protocols are already available on a practical scale.[1-3]

A key aspect for the implementation of biocatalytic strategies at an industrial level has been the tremendous development of molecular

Chemical Processes for a Sustainable Future
Edited by Trevor M. Letcher, Janet L. Scott and Darrell A. Patterson
© The Royal Society of Chemistry 2015
Published by the Royal Society of Chemistry, www.rsc.org

biology techniques, which nowadays enable the identification, cloning, overexpression and improvement of a given enzyme (biocatalyst) towards a tailored application, *e.g.* to deliver a variant with a much higher 'on spec' substrate specificity, or enzymes active under required (more severe) reaction conditions. As is shown below with selected examples, such biological approaches make it possible to have variants with enzymatic activity for different industrial biotransformations with improvements of greater than 20 000 fold. Furthermore, the possibility of cloning an enzyme gene in heterologous microbial hosts, together with their overexpression through straightforward fermentative strategies, enables the accessibility of the desired biocatalysts at large and at a sustainable scale (*i.e.* based on *simple* fermentations to produce biomass and biocatalysts).

Once the enzymes have been identified, cloned and overexpressed, and genetically designed for a specific synthetic application, questions will emerge for discussion. The questions include: the type of reactor; the adequate form of the biocatalyst, either as free or immobilized isolated enzyme or used directly as (recombinant) whole cell, or fermentations with living cells; and the best set-up for a desired work-up. In this respect, the choice of aqueous or non-aqueous media for the reaction will be of utmost importance, as there are a number of enzymes and more recently, even whole cells (see below)[4] able to conduct efficiently organic reactions in neat substrates or in other virtually water-free media (*e.g.* organic solvents, supercritical fluids, ionic liquids). This obviously broaden the options that biotransformation may have for industrial purposes, as many substrates and products will definitely be insoluble in aqueous solutions (challenging the technical use of enzymatic processes). Once the development phase of a biocatalytic process is complete (enzyme design and process set-up) and all other aspects are holistically combined, a scaled-up economically efficient industrial biotransformation can be envisaged (Figure 14.1). As stated above, the pipeline involves an enzyme discovery phase, its further genetic design and finally aspects related to medium engineering and integration within a given industrial plant or production step.

Upon implementation of a biotechnological process, different benefits can be reached. First of all the above-mentioned high selectivity of enzymes, which results in the formation of highly added value products, which in some cases cannot be accessed by using more conventional chemo-catalysts. Interesting examples include the chiral induction delivered by enzymes, which are multi-chiral catalysts, and their regioselectivity among different functional groups.

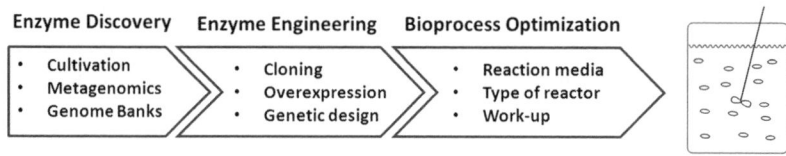

Figure 14.1   Conceptual approach for an industrial biotransformation, starting from enzyme identification (biocatalyst) and the genetic design, ending up with optimization of bioprocess parameters.

These represent important assets for the use of natural catalysts in industrial processes. In addition the mild reaction conditions applied to enzymes lead to much lower energy consumption during chemical (industrial) processes, providing potential for a less polluting process while at the same time contributing to energy and cost savings. Such mild reaction conditions can also lead to diminished by-product formation and less waste. And finally, the implementation of bio-catalytic processes may allow, in some cases, the labelling of green or natural products, with the possibility of fulfilling certain important regulations to further introduce such compounds into strategic markets.

As stated above, there is already an increasing number of biocata-lytic processes implemented at an industrial scale.[1] In the following sections, several examples encompassing the use of isolated enzymes (non-growing, resting) whole cells, or fermentative processes with living cells are briefly discussed. Rather than providing a com-prehensive overview of the current state-of-the-art, the emphasis is on selected examples addressing the integration of the enzymes with other aspects of a process and medium engineering, ending up with a robust and economically viable biotransformation.

14.2   USE OF ISOLATED ENZYMES IN CHEMICAL PROCESSES

The simplest way of implementing an industrial biocatalytic process is the direct use of isolated enzymes as catalysts. In this case, after the fermentation for biomass generation of recombinant strains over-expressing the desired enzyme, the produced cells are typically dis-rupted (*e.g. via* French press or sonication) to liberate the enzymes in the aqueous crude broth; in some cases enzymes are extracellular and therefore cell disruption is not needed. Subsequently, in some cases the crude fermentative broth can be directly used as a

biocatalyst for a given reaction as such or simply lyophilized to produce a solid powder. The concentration of the overexpressed enzyme
is high enough and no other secondary reactions are observed due to
other enzymes present in the broth, with the substrates or with the
products. In some other cases, however, some enzyme purification
process is needed, typically by using chromatography. Finally, whether such free enzymes must be immobilized or not will directly depend on the market price of the produced chemicals. As is shown
below, there are remarkable industrial cases in which the high market
price of the product allows the single use of the biocatalyst, being
subsequently discarded as a biodegradable waste. In other cases,
however, the introduction of immobilizing steps appears mandatory
to reach the desired economics.

There are no defined rules in the selection of isolated enzymes or
(non-growing, resting) whole cells for organic synthesis, and industrial examples of all kinds are known.[1–3] Furthermore, isolated enzymes are commonly reported in practical applications when the
desired enzymes do not require the aid of organic cofactors for their
performances, and/or are extracellular, and/or are particularly robust
towards non-conventional media or other severe industrially-sound
reaction conditions (*e.g.* high substrate loadings, presence of organic
co-solvents, relatively high temperatures). Furthermore, enzymes
using cofactors have also been successfully shown as free biocatalysts
in industrial processes (see below in this section). Another aspect to
be considered is the know-how and capabilities of a certain laboratory
or industrial plant to manipulate (living) biological material. While
the use of free enzymes allows the outsourcing of its production being
isolated enzymes, they can be used as a simple solid biocatalysts like
any other chemo-catalysts. The handling of recombinant whole cells
needs qualified personal and some regulatory and safety aspects for
the laboratory and/or the reactor plant, though in any case, these are
not extremely severe for biocatalysis.

Hydrolases (lipases, esterases, proteases, *etc.*) are remarkable types
of enzymes and can be used in industrial processes as free, isolated
enzymes. These cofactor-free enzymes are very robust and allow their
use in non-conventional media, thus opening up the lead for
innovative approaches. Thus, when hydrolases are used in aqueous
solutions, water is taken as the nucleophile of the reaction and
(selective) hydrolytic processes can be conducted. In fact, hydrolases
in nature are responsible for the hydrolysis of esters, amides, fats, *etc.*
in aqueous solutions. Conversely, when non-aqueous media are
involved, the addition of alcohols or amines as nucleophiles for

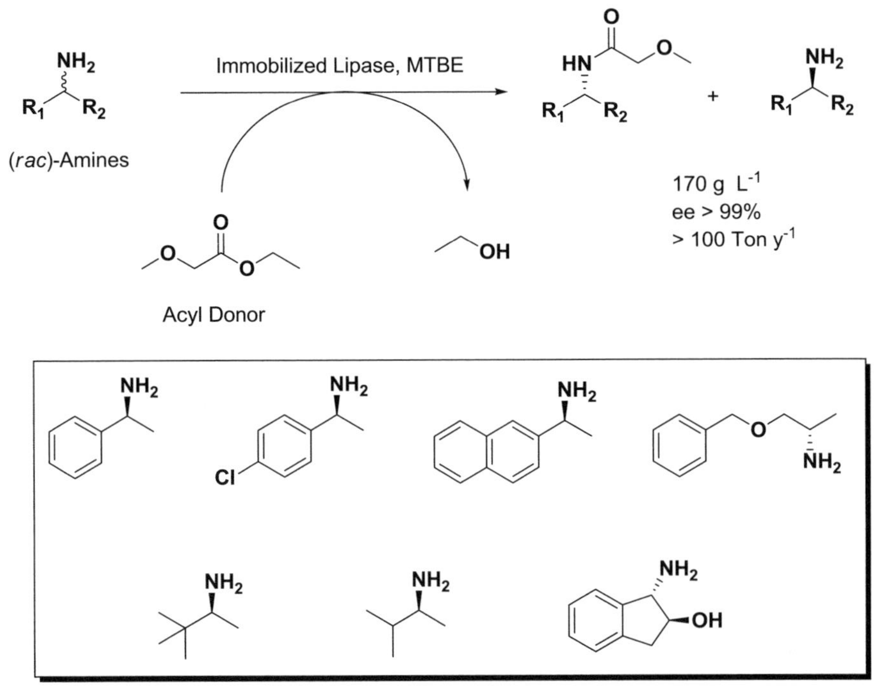

Figure 14.2 Process of amine synthesis at BASF catalysed by immobilized lipases and selected examples of the relevant building blocks produced.[3]

hydrolases enables the establishment of synthetic processes. In this respect, an outstanding case is the production of chiral amines industrially established at BASF, from which lipases and esters as acyl donors in methyl *tert*-butyl ether (MTBE) as solvent are used (Figure 14.2).[3] By means of an immobilized lipase from *Burkholderia plantarii*, a broad range of structurally different amines can be produced at high substrate loadings (typically higher than 150 g L^{-1}) with excellent enantioselectivities (>99%) and at a multi-tonne scale (>100 t a^{-1}, a here refers to annum and t to metric tonne). Remarkably, the same biocatalyst is able to accept a broad range of different racemic amines as substrates with a high stereo-bias.

Another interesting industrial example of the use of hydrolases is the production of enantiopure β-amino acids, which are highly valuable building blocks with a further broad range of applications in the pharmaceutical chemistry. In this area, Evonik Industries has developed a biocatalytic platform using different non-immobilized lipases in aqueous media to perform the enantioselective hydrolysis of racemic β-amino acid esters, again through kinetic resolution (analogous process to the one at BASF for amines, Figure 14.2). To

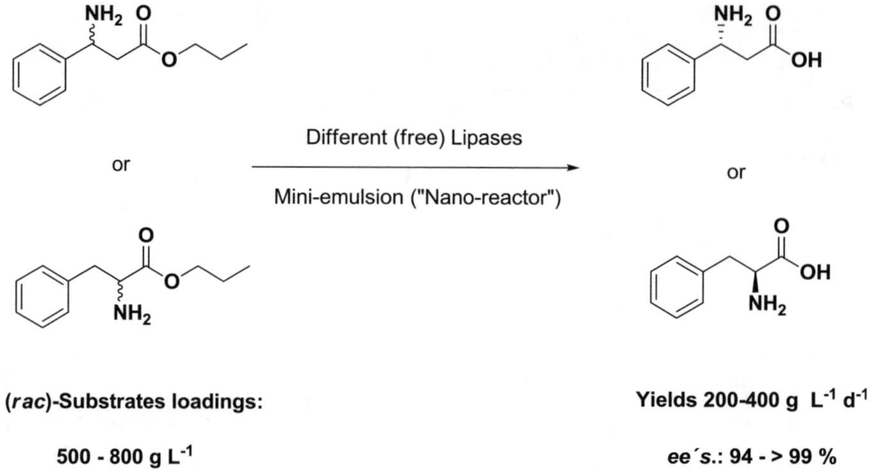

Figure 14.3   Enzymatic production of β-amino acids using mini-emulsions (or nano-reactors) developed by Evonik.[5]

further optimize the substrate loadings in aqueous media, a biphasic system (often using MTBE as the organic phase) was initially set up. This resulted in the addition of high substrate loadings, thus avoiding solubility problems or inhibitory effects created in the enzyme either by the substrates or the products (as most of the product or substrate remains in the organic phase). On the basis of this approach, a further innovation is the formation of mini-emulsions or the so-called 'nano' emulsions by adding a surfactant to the two phases. Thus, by dramatically diminishing the mass transfer limitations among phases, under these new conditions outstanding productivities of up to 400 g L^{-1} d^{-1} of enantiopure β-amino acids at >99.9% enantiomeric excess (ee) have been reported (Figure 14.3).[5]

Another important aspect in the above-reported process is its simplicity. Considering practical, industrial processes, the feasibility of scaling up is a given proof-of-concept and represents a crucial asset. Here, the synthesis of the starting materials is straightforward and is based on inexpensive and readily available substrates (benzaldehyde, malonic acid, ammonia, *etc.*). Free enzymes are not immobilized and its single use is affordable given the high market price of β-amino acids. Being a kinetic resolution, the highest theoretical yield is 50%. The remnant, ∼50% of the other enantiomeric form, can be either marketed as such or subsequently racemized for further use. These aspects, combined with medium engineering techniques (*e.g.* selection of biphasic systems and/or mini-emulsions, optimization of temperature, substrate loadings and way of dose), enable a straightforward and robust industrial biotransformation, with the enzymatic process easily integrated among other chemical steps.

Apart from cofactor-free and robust hydrolases, other enzymes can also be used as isolated biocatalysts for industrial biotransformations. For instance, transaminases have recently experienced a tremendous development with an outstanding focus on industrial applications for the synthesis of optically active amines using inexpensive and largely available (prochiral) ketones as substrates. As a remarkable example of this, a novel industrial biocatalytic process for the synthesis of sitagliptin has been established in a joint effort carried out by Merck and Codexis.[6] Several technical aspects of this concept must be considered from a process development viewpoint. First of all, wild-type transaminases were unable to accept the highly bulky ketone substrate needed for the sitagliptin synthesis and therefore several rounds of genetic design of the enzyme by means of directed evolution approaches, together with computational protein engineering programs, were needed to adapt the enzyme's active site to these challenging steric requirements. Furthermore, transaminases able to accept inexpensive isopropylamine as an amino donor, instead of more expensive amino acids, were chosen. This also allows the formation of acetone as a by-product, which can be easily stripped out to shift the thermodynamic equilibrium inherent in transamination reactions. The next challenge was the choice of reaction medium for such a reaction. By performing the biocatalytic process in aqueous conditions, as transaminases are less stable than hydrolases in non-aqueous media, the bulky substrates are not soluble and thus the need to add organic co-solvents was mandatory. In this case, 50% dimethyl sulfoxide (DMSO) (v/v) in the buffer was used, enabling a high enzyme stability combined with excellent substrate dissolution.[7] Once all the aspects were optimized and integrated, impressive final figures were achieved, reaching a >20 000 fold genetic improvement in enzymatic activity towards the desired bulky substrate (it must be mentioned that the starting wild-type transaminase was not active towards the substrate). The genetic improvement involved 27 amino acidic substitutions in the protein scaffold, ending up with a variant which accepted substrate loading of $\sim$200 g L^{-1} without inhibitory effects. The yield was >90% with an enantiomeric excess of >99.9% and was produced in a reaction time of less than 24 h (Figure 14.4).[6] Overall, this case represents not only a useful industrial biotransformation, but also a very valuable (academic) example on how biocatalysis can be envisaged, developed and finally implemented at practical level for commercialization.

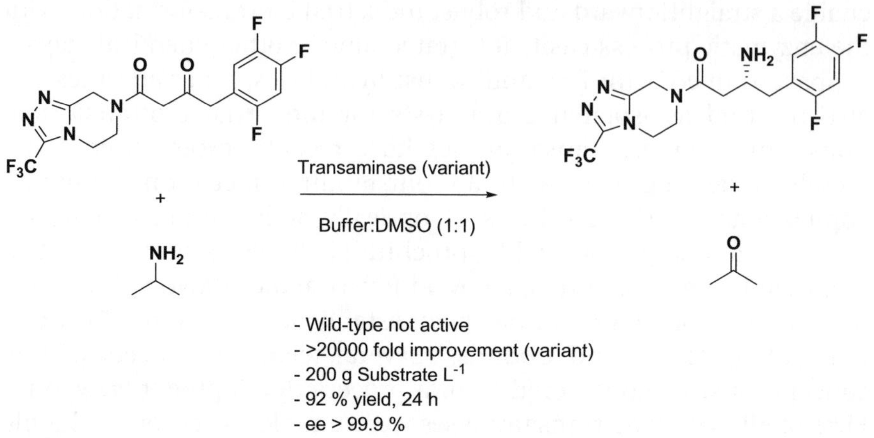

- Wild-type not active
- >20000 fold improvement (variant)
- 200 g Substrate L^{-1}
- 92 % yield, 24 h
- ee > 99.9 %

Figure 14.4   Transaminase-catalysed synthesis of sitagliptin, recently developed by Merck and Codexis.[6] The organic cofactor for transaminases (pyridoxal phosphate, PLP) was catalytically added in the reaction system as well.

Like transaminases and hydrolases, oxidoreductases represent an important group of enzymes with many industrial applications already successfully implemented (see also next section for whole cells).[8] Since these biocatalysts need organic cofactors for redox electron transport (*e.g.* NAD(P)$^+$, FAD$^+$), the involvement of a cofactor regeneration system is mandatory to assure an economic biotransformation (redox cofactors are very expensive for a stoichiometric addition in a reactor). For instance, alcohol dehydrogenase from *Lactobacillus kefir* was genetically improved through up to eight rounds of genetic directed evolution, creating a successful enzymatic variant able to catalyze the enantioselective reduction of tetrahydrothiophene-3-one to form a relevant building block for the synthesis of sulopenem (an antibacterial agent, Figure 14.5).[9] Remarkably for this example, the biocatalytic approach was feasible despite the nearly spatially symmetrical (β-CH$_2$ *vs.* β-S) ketone used as substrate. The production of such a molecule by means of chemocatalysts would be synthetically much more challenging. In this process, *L. kefir* alcohol dehydrogenase was coupled with glucose dehydrogenase to regenerate the cofactor upon stoichiometric addition of inexpensive glucose for electron provision and the process was conducted under buffered aqueous conditions. The optimized reaction was successfully scaled up to 100 kg of product (Figure 14.5).[9]

By starting with the same wild-type enzyme (*L. kefir* alcohol dehydrogenase), the industrial production of optically active alcohols through aldehyde enantioselective reduction has also been reported

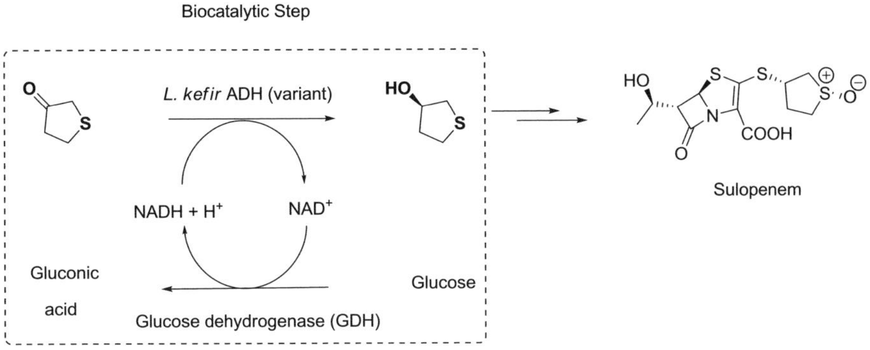

Biocatalytic Step

Figure 14.5 Biocatalytic oxidative step for the sulopenem synthesis.[9]

(in this case, the racemic group is adjacent to the aldehyde).[9,10] As it is typically observed when industrial biotransformations are considered, the activity and selectivity of the wild-type enzyme was not sufficient to warrant an economically attractive production of chiral alcohols. The problems related to such issues included acceptance of low substrate loadings, inhibitory effects, low catalytic activity and unacceptable long reaction times. To overcome these challenges, up to five rounds of genetic improvement through directed evolution were conducted towards specific substrates. Once an optimal variant had been designed an efficient biocatalytic strategy for the production of optically active (R)-2-methylpentanol was achieved *via* reductive aldehyde kinetic resolution. In this particular case, for cofactor regeneration purposes, the same enzyme also accepted isopropanol to yield acetone (which, as stated above, can easily be stripped out to shift the thermodynamic equilibrium to a desired production level). Under optimized reaction conditions, high yields (46%, for a kinetic resolution) with excellent substrate loadings ($\sim$220 g L^{-1}) were reached in a reaction time of less than 24 h and with outstanding enantiomeric excesses (>98%) (Figure 14.6).[10]

Apart from *L. kefir* alcohol dehydrogenase, another interesting example is *Rhodococcus erythropolis* alcohol dehydrogenase, which has been employed for the enantioselective synthesis of (S)-3,5-bistrifluoromethylphenylethanol (a building block for the pharmaceutical synthesis of NK-1 receptor antagonists). Instead of a biphasic system as used with hydrolases, in this case the substrate (ketone) was added directly to aqueous solutions to form a homogeneous emulsion upon vigorous stirring. To regenerate the cofactor, formate dehydrogenase (FDH) from *Candida boidinii* was initially used leading to productivities of $\sim$100–110 g L^{-1} d^{-1}. The cofactor regeneration

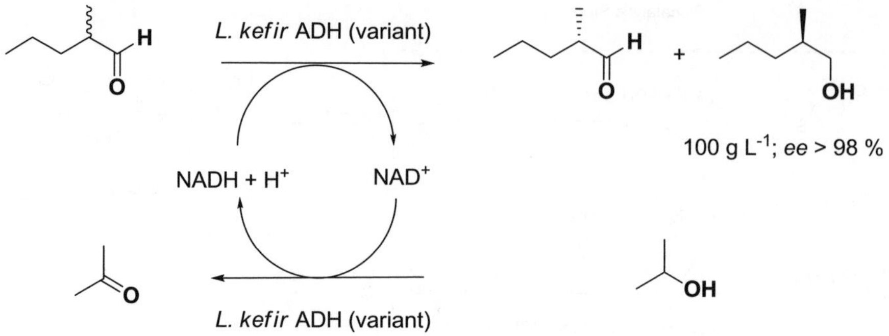

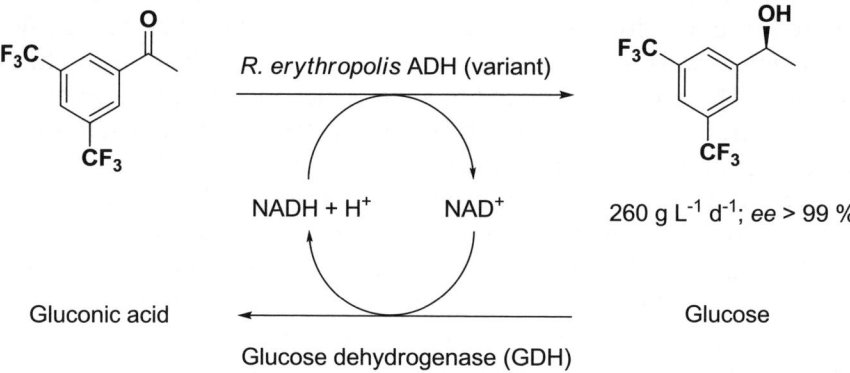

Figure 14.6 Production of (R)-2-methylpentanol by enantioselective kinetic resolution of the racemic aldehyde, catalyzed by mutants of *L. kefir* alcohol dehydrogenase.[10]

Figure 14.7 Enzyme-catalysed enantioselective ketone reduction to produce enantiopure (S)-3,5-bis-trifluoromethylphenylethanol by using free oxidoreductases and coupled enzymes for cofactor regeneration.[11]

reaction could be significantly optimized by changing FDH to glucose dehydrogenase (GDH), since that enzyme resulted in 10-fold activity increase over and above that found with FDH. Under these new optimized conditions, excellent productivity of ~ 260 g L^{-1} d^{-1} was reported with excellent enantioselectivities (>99%) (Figure 14.7).[11]

With regard to oxidoreductases being applied as free enzymes for industrial applications, another relevant case is the synthesis of montelukast, a leukotriene receptor antagonist used to control the symptoms of asthma and allergies.[12] In this particular process, the major challenge was the highly hydrophobic nature of the substrate (with logP in the range of ~ 7). Therefore, by working in aqueous solutions with oxidoreductases, the use of organic (co-)solvents was mandatory to provide efficient substrate dissolution in the reaction media. Again in this case, an alcohol dehydrogenase from Codexis (CDX-026, biological origin not disclosed) was genetically designed (with several rounds of directed evolution) towards the desired substrate and towards a stability to cope with high volumes of organic (co-)solvents in the aqueous solution. As a result of the genetic rounds, an enzyme variant was produced which resulted in an improvement in activity and stability of ∼3000-old compared with the wild-type. For the process set-up, optimal conditions were reached in a 1:5:3 (v/v/v) mixture of the toluene–isopropanol–buffer system, leading to a robust process able to compete with other chemical approaches, in terms of selectivity and productivity (Figure 14.8).[12]

Free oxidoreductases have also been employed in the synthesis of cholesterol lowering drugs (*e.g.* Lipitor®), in combination with other enzymes and chemo-catalysts. The optimized biocatalytic reduction displayed excellent yields (96%) and on spec enantiomeric excesses (>99.5%), resulting in an economically attractive industrial process. Again in this case, glucose dehydrogenase and stoichiometric (inexpensive) glucose were used for cofactor regeneration (Figure 14.9).[13]

Apart from the alcohol dehydrogenases discussed above, other oxidoreductases such as amino acid dehydrogenases—to catalyse the reductive amination of α-keto acids to deliver (non-natural) amino acids—have been extensively used at an industrial scale for decades. An outstanding example is the synthesis of L-tert-leucine in combination with FDH from *C. boidinii* as a coupled enzyme for cofactor regeneration (formate to carbon dioxide to shift the thermodynamic equilibrium). For this process, yields of >70% and excellent productivities (>600 g L^{-1} d^{-1}) have been reported.[14] Likewise, phenylalanine dehydrogenase from *Thermoactinomyces intermedius* has been used for the production of an intermediate for the pharmaceutical synthesis of saxagliptin (BMS-477118), a dipeptidyl peptidase IV inhibitor for the treatment of type 2 diabetes mellitus. Also in this case, FDH was used for cofactor regeneration, using ammonium formate both as electron and nitrogen donor for the amino acid synthesis (Figure 14.10).[15]

Lyases (for C–C bond forming reactions) represent another group of enzymes that have found industrial use as isolated enzymes. In this area, hydroxynitrile lyases are very versatile biocatalysts able to form optically active cyanohydrins, which are useful building blocks, starting from inexpensive aldehydes and ketones and HCN.[17] These enzymes are essential in plants, where apparently they act as defensive tools by releasing HCN upon insect attack. Their practical use at

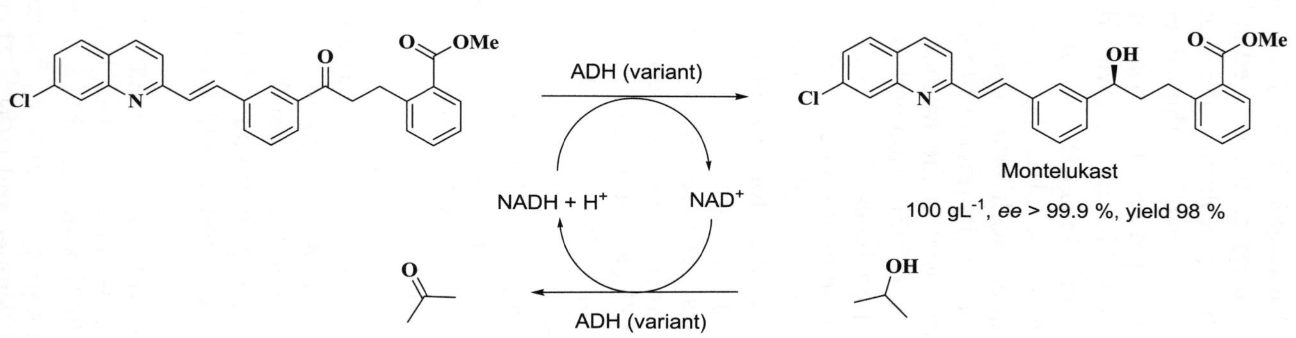

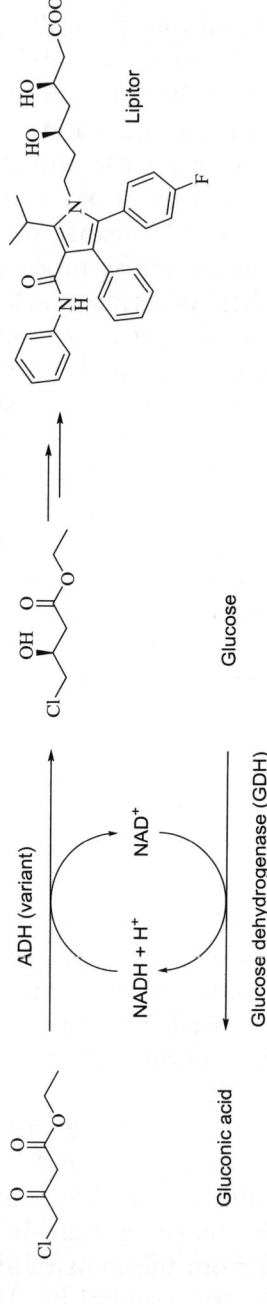

Figure 14.8 Biocatalytic process for the synthesis of montelukast, developed at Codexis.[12]

Figure 14.9 Biocatalytic redox step involved in the synthesis of the cholesterol lowering drug Lipitor® .[13]

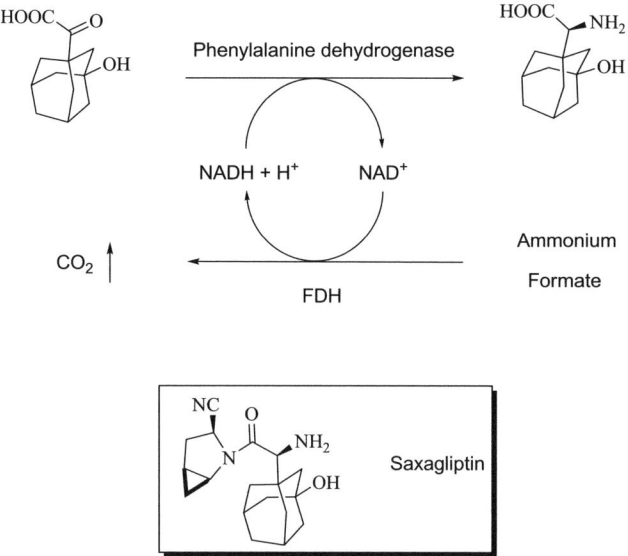

Figure 14.10 Synthesis of a precursor of saxagliptin involving a biocatalytic oxidative step by means of phenylalanine dehydrogenase.[15] Other uses of phenylalanine dehydrogenases can be found in the literature.[16]

industrial level has been due to the above-mentioned development in molecular biology, resulting in the cloning and overexpression of these plant enzymes in heterologous microbial hosts, together with their genetic design based on directed evolution strategies. At Dutch State Mines (DSM N.V.) several optically active cyanohydrins have been produced by this means with excellent productivities and stereoselectivities (Figure 14.11).[17] The market price of these building blocks is so high, and the overexpression of the (genetically tailored) enzymes proceeds in such a straightforward manner, that the single use of free enzymes is enough to support an economic biotransformation and as a result there is no need for biocatalyst immobilization.[17]

14.3 USE OF WHOLE CELLS AS BIOCATALYSTS: FROM FINE CHEMICALS TO BULK COMMODITIES

Apart from the above-discussed use of free, isolated enzymes (immobilized or not), in many cases the direct use of whole cells as biocatalysts appears a useful option. This is typically the case when the desired overexpressed enzymes are intracellular and the cells need to be disrupted if free enzymes are to be used. Furthermore, they need

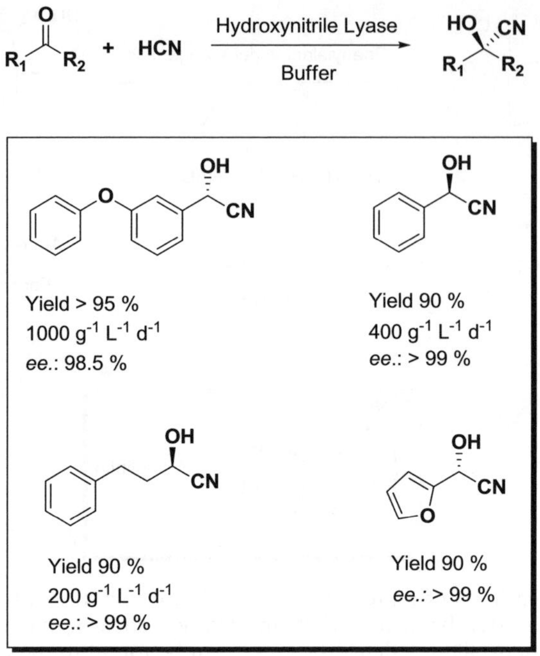

Figure 14.11   Biocatalysis with hydroxynitrile lyases at industrial scale.[17]

organic cofactors for the catalytic cycle, giving rise to the need to set up efficient cofactor regeneration strategies. Although industrial examples for all these free enzyme types do exist (see previous section), the application of whole cell biocatalysis is often preferred. The reasons may be economic (no need for cell disruption and enzyme purification), as well as simplicity. Once the fermentation to produce the whole cells has been completed, the whole cells, can be frozen before being used as solid or suspended biocatalysts.

A particularly interesting example of useful industrial whole cell biocatalysis is the oxidoreductase platform developed at Evonik.[18] The strategy consists of the cloning and overexpression of two enzymes in the same recombinant strain (the so-called *designer bug*). The two enzymes are the desired oxidoreductases with the required enantioselectivity for the intended final product, *R* or *S,* as well as the second enzyme responsible for the cofactor regeneration, which is typically a glucose dehydrogenase. By means of this approach and applying a biphasic media (often with MTBE) to enhance the substrate loadings, remarkable productivities with excellent enantioselectivities have been achieved (Figure 14.12).[18] Overall, the example emphasizes again the different aspects that must be optimized when envisaging an industrial

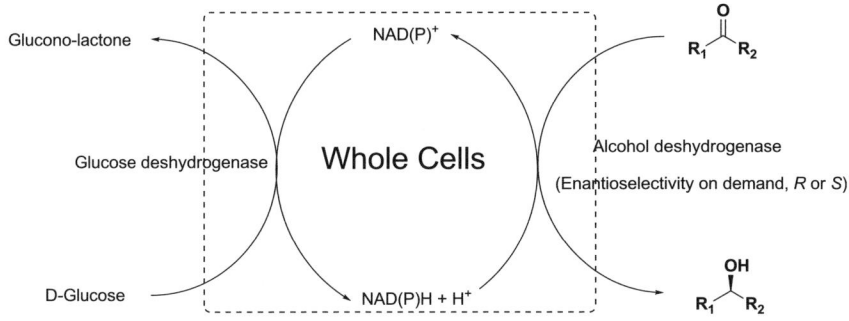

Figure 14.12 The designer bug concept developed at Evonik involving the cloning and overexpressing of two enzymes in the same recombinant whole-cell.[18]

biotransformation (Figure 14.1): from cloning and genetic design and improvement, to process set-up conditions.

To further improve the above-described concept, novel developments in medium engineering strategies have recently been reported.[4] Thus, instead of performing reactions in aqueous solutions, it has been observed that whole cells overexpressing oxidoreductases are able to efficiently catalyze enantioselective reductions by working in neat substrates (virtually water-free systems). Hence, the work-up procedures are simplified and involve simple filtration of the suspended cells and distillation of the volatile substrate products. Furthermore, this approach opens up the option of performing biocatalytic reactions with substrates or products that are water sensitive. As an example, the biocatalytic reduction of highly reactive unsaturated alkynes is schematically shown in Figure 14.13.[4]

Apart from oxidoreductases, several cofactor-dependent lyases have also been assessed in whole cell biocatalysis under industrially sound conditions. For instance, Evonik has shown that thiamine-diphosphate dependent lyases (ThDP-lyases) are useful biocatalysts able to perform the enantioselective C–C bond ligation of two aldehydes to afford

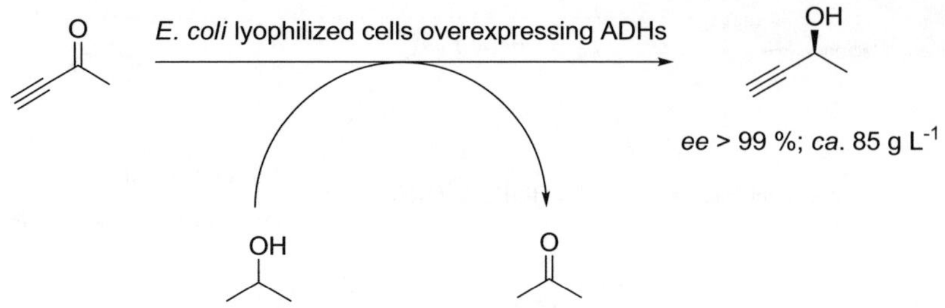

Figure 14.13   Conceptual water-free biocatalytic enantioselective reduction using *E. coli* whole cells overexpressing enzymes.[4]

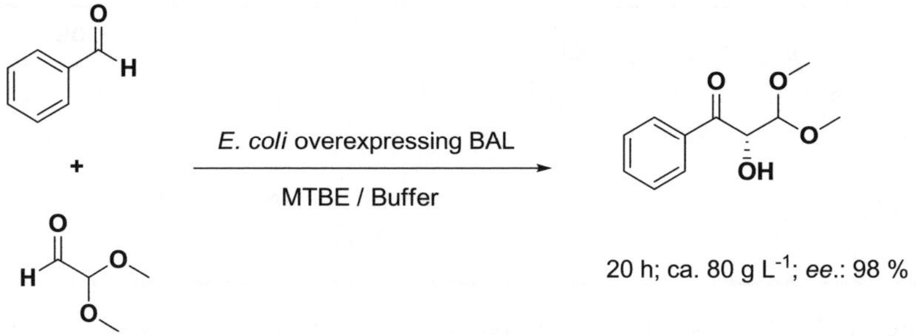

Figure 14.14   Model reaction of whole cells-based lyase reactions under industrially sound reaction conditions.[19]

useful optically active α-hydroxy-ketones. By means of whole cells overexpressing lyases (*e.g.* benzaldehyde lyase from *Pseudomonas fluorescens*, BAL) and biphasic systems involving MTBE and a buffer, high productivities and excellent enantioselectivities of useful building blocks have been produced (Figure 14.14).[19] Because whole cells were involved, no cofactor (thiamine diphosphate) was added to the reaction media thus reducing the costs of the biotransformations.

The examples presented in this chapter have been applied to chiral synthesis, and because of the use of free enzymes or whole cells and the fact that the market price of the final products was high, the processes were economical. Remarkably, biocatalysis can be also useful for the synthesis of low-value products and even bulk commodities. To reach a successful competitive biotransformation in those challenging areas, the use of inexpensive overexpressing whole cell systems, together with cell immobilization, is mandatory. As a rule-of-thumb, a production of more than 1 kg of product per g of dried cells must be produced from commodity chemicals with market prices typically about ~€1 kg^{-1}. The dried cells weight (or DCW), commonly used as unit in biocatalysis, is calculated by centrifuging wet cells, removing the supernatant and drying cells at 80 °C until a constant weight is achieved.

In this area, a paradigmatic and well-known case is the biocatalytic synthesis of acrylamide developed by Mitshubishi Rayon, using immobilized cells of *Rhodococcus rhodocrous* overexpressing a genetically designed nitrilase.[20] After many rounds of optimization, both at biocatalyst level (enzyme and strain development), as well as at technical scale (type of cell immobilization, reactor type, substrate concentration, temperature, *etc.*), impressive figures for yields, selectivities and conversions of >99.9% have been reported.[20] The process is conducted at low temperature (<10 °C) to overcome typical secondary reactions observed when highly reactive alkenes are used (Figure 14.15).[20] Here, the option for biocatalytic reactions to take place in mild conditions is a further asset to assure the product quality.

Another example has recently been reported by DuPont for the synthesis of glycolic acid.[21] The overall process involves several chemical steps, starting from a C1-based condensation of formaldehyde and HCN to produce glycolonitrile, which is subsequently hydrolyzed by means of *E. coli* cells overexpressing a nitrilase. A further ion exchange chromatographic step and the production of glycolic acid is complete (Figure 14.16).[21]

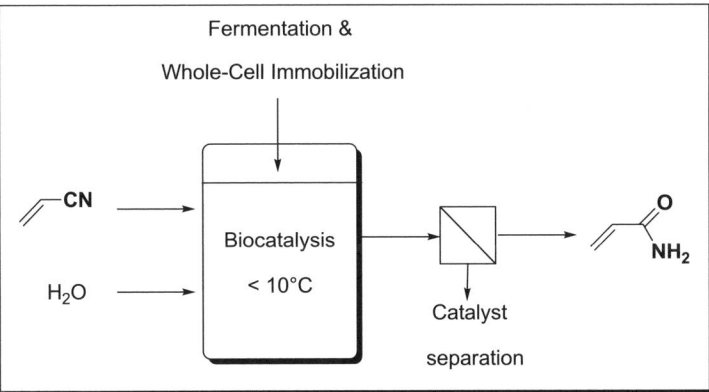

Figure 14.15 Acrylamide industrial synthesis performed with whole cells overexpressing nitrilases.[20]

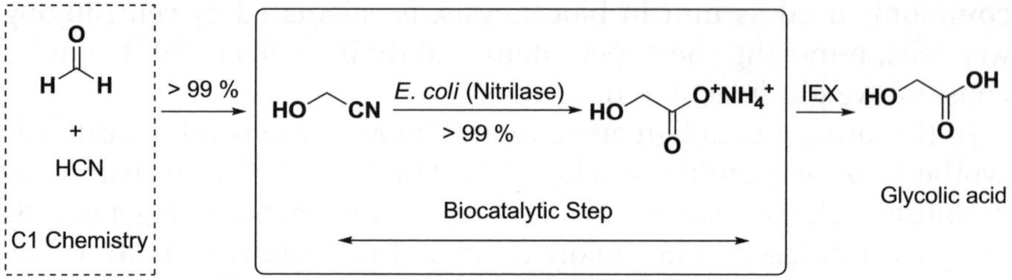

Figure 14.16 Chemo-enzymatic glycolic acid synthesis developed at *DuPont*.[21]

This latter biocatalytic process is an excellent example on how an industrial biotransformation is envisaged and built. First of all, a broad number of nitrilases were screened towards the intended substrate, glycolonitrile. It was found that one of them, a nitrilase from *Acidovorax facilis* 72W isolated and characterized from a soil sample, was actually catalysing the desired reaction. Yet, as is typical in such cases, it was observed that the wild-type enzyme was not sufficiently active to envisage an efficient and robust industrial process. Thus, several rounds of directed evolution were conducted to develop a proper, tailored variant. Later on, whole cells were immobilized on carragenane to allow their use over a period of months with high stability and efficiency under the real reaction conditions. Overall, the final process led to the production of >1 kg glycolic acid g dried cells^{-1}, assuring the economic viability for the production of a commodity chemical under environmentally friendly conditions.[21]

14.4 BEYOND: METABOLIC ENGINEERING

Following these latter options for the biotechnological production of low-added value commodity chemicals (as well as other more refined products), and apart from the use of immobilized resting non-growing cells (previous section), another option is the use of living cells (fermentations) using cheap fermentable sugars as the carbon source for the microorganisms. The optimization of the different biochemical pathways within living cells, the so-called metabolic engineering, enables the construction of highly efficient and sophisticated growing biocatalysts for many different purposes, ranging from the production of biofuels (*e.g.* ethanol, butanol) to the provision of monomers or platform chemicals for further use (*e.g.* itaconic acid, 1,3-propandiol).[22–24] In these fermentative cases, cells grow and produce the desired chemical through an optimized multi-enzymatic pathway. A certain enzyme or even a group of enzymes can be attenuated or overexpressed within the microorganism genome to foster the production of the desired compound and/ or to diminish the formation of unwanted by-products at the same time.

Metabolic engineering processes comprise typically two main parts: (1) the upstream part, whereby the (modified) microorganism biosynthesizes the target molecule in an optimized way; and (2) the downstream processing, whereby the compound is purified from the crude fermentative broth to on spec values. It is important to note that only the full integration and optimization of both approaches enables the set-up of an efficient and economic biosynthetic concept. Thus, strong cooperation among disciplines (biology, microbiology, chemistry and chemical engineering) is needed. Furthermore another crucial aspect is the (large-scale) provision of cheap substrates for cell growth and chemical production, especially when low-cost commodities are intended to be produced on a large scale. Here, glucose and related saccharides are often used, and the use of lignocellulose-based sugars for fermentations (second-generation approaches) will be considered in the next few years for large-scale applications that do not compete with the food industry. The challenge is, however, that the production of sugars from first-generation sources (*e.g.* starch) is technically more straightforward (and cheaper) than the (ligno)cellulose pre-treatment producing pentoses and hexoses from agricultural residues. Tremendous efforts in research and development are currently being made in these areas.[22–24]

An interesting example of fermentation is the biosynthesis of ethanol as biofuel, which has resulted in a massive worldwide industry and is currently an important research field and an area attracting much interest from economists.[22–24] In this fermentation process, the upstream process is normally carried out by genetically designed and optimized yeasts (*e.g.* *Saccharomyces cerevisiae*) through the glycolysis cycle, combined with the enzymatic pyruvate decarboxylation and subsequent enzymatic reduction of formed aldehyde to ethanol, which is secreted by the cells and accumulate in the fermentation broth. The downstream processing involves the distillation of ethanol from the aqueous solution (Figure 14.17).

In the quest for novel processes aligning biology pathways with engineering, Global Bioenergies very recently developed a fermentation process to produce alkenes, such as propylene or isobutylene,

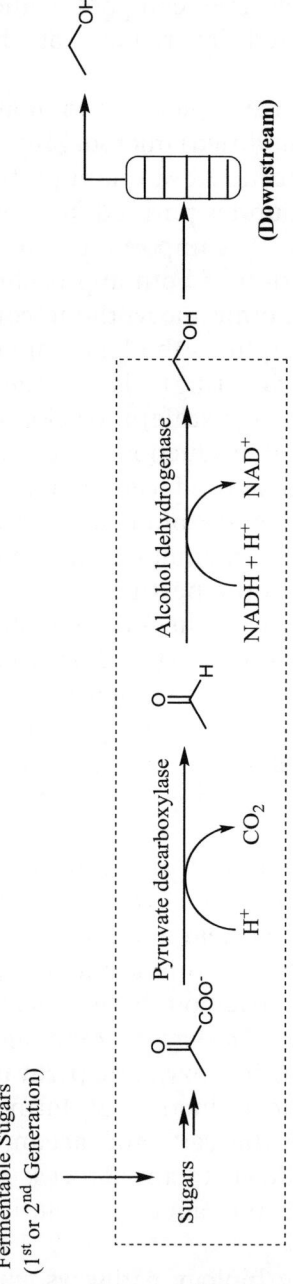

Whole-Cell - Upstream

Figure 14.17 Biosynthesis of ethanol.[22]

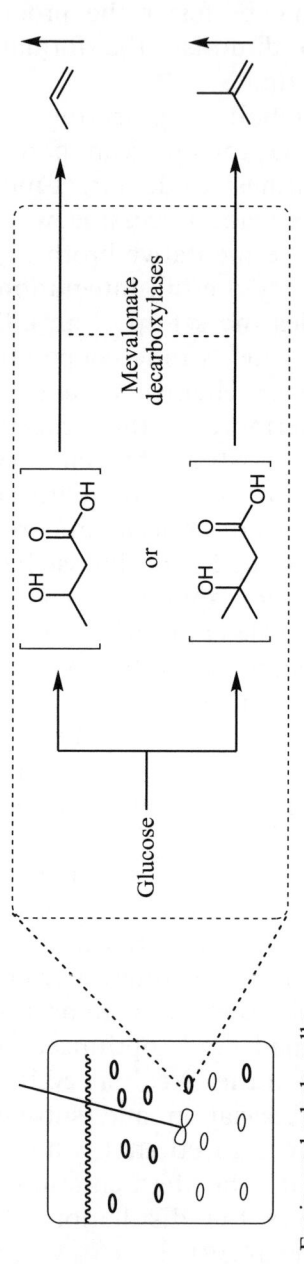

Engineered-whole-cells

Figure 14.18 Engineered *E. coli* cells to produce different alkenes from glucose.[23]

using glucose as the carbon source.[23,24] The introduction in the cell of overexpressed and genetically designed mevalonate decarboxylases leads to the *in situ* decarboxylation and dehydration of biologically formed β-hydroxy acids resulting in alkenes. Remarkably, the downstream processing of short-chain alkenes from the aqueous fermentation broth is straightforward and involves the stripping out of the gases from the system together with their collection, thus reducing costs and enhancing the economics of the process (Figure 14.18).[23]

14.5 CONCLUSIONS

This chapter briefly discusses the potent of white biotechnology for sustainable catalysis in the next few decades. First it introduces important aspects of the use of enzymes and cells in organic synthesis. This is related to the possibility of cloning and overexpressing enzymes, and improving them by means of genetic tools to provide variants with tailored activities and stabilities towards the desired reaction conditions or substrates. Secondly, the need to integrate such new enzymes (biocatalysts) within a process set-up is emphasized. Finally, several examples comprising the use of free enzymes, non-growing resting cells or fermentation approaches are presented. Rather than giving a comprehensive overview of all the applications, the selection of examples is intended to emphasize the requirements for process development: the need of free enzymes or whole cells, the set-up of biphasic or monophasic systems, or solvent-free approaches, convenience of immobilizing enzymes or cells, and finally, the need to combine upstream and downstream processing in fermentations for biorefineries. Being a complement of other chemo-catalytic strategies, white biotechnology is expected to play an important role in the future of sustainable (industrial) chemistry over the next few decades.

REFERENCES

1. U. T. Bornscheuer, G. Huisman, R. J. Kazlauskas, S. Lutz, J. Moore and K. Robins, *Nature*, 2012, **485**, 185–194.
2. J. M. Woodley, *Trends Biotech.*, 2008, **26**, 321–327.
3. R. M. Patel, *Coord. Chem. Rev.*, 2008, **252**, 659–701.
4. A. Jakoblinnert, R. Mladenov, A. Paul, F. Sibilla, U. Schwaneberg, M. B. Ansorge-Schumacher and P. Domínguez de María, *Chem. Commun.*, 2011, **47**, 12230–12232.

5. H. Gröger, O. May, H. Hüsken, S. Georgeon, K. H. Drauz and K. Landfester, *Angew. Chem., Int. Ed.*, 2006, **45**, 1645–1648.

6. C. K. Savile, J. M. Janey, E. C. Mundorff, J. C. Moore, S. Tam, W. R. Jarvis, J. C. Colbeck, A. Krebber, F. J. Fleitz, J. Brands, P. N. Devine, G. W. Huisman and G. J. Hughes, *Science*, 2010, **329**, 305–309.

7. More recently immobilized transaminase variants highly stable in isopropanol as co-solvent have been reported, with the aim of replacing DMSO for a more environmentally-friendly and downstream-useful solvent. See: M. D. Truppo, H. Strotman and G. Hughes, *ChemCatChem*, 2012, **4**, 1071–1074.

8. D. Gamenara, G. Seoane, P. Saénz Mendez and P. Domínguez de María, *Redox Biocatalysis: Fundamentals and Applications*, John Wiley & Sons, Hoboken, NJ, 2012.

9. J. Liang, E. Mundorff, R. Voladri, S. Jenne, L. Gilson, A. Conway, A. Krebber, J. Wong, G. Huisman, S. Truesdell and J. Lalonde, *Org. Proc. Res. Dev.*, 2010, **14**, 188–192.

10. O. W. Gooding, R. Voladri, A. Bautista, T. Hopkins, G. Huisman, S. Jenne, S. Ma, E. C. Mundorff and M. M. Savile, *Org. Proc. Res. Dev.*, 2010, **14**, 119–126.

11. D. Pollard, M. Truppo, J. Pollard, C. Y. Chen and J. Moore, *Tetrahedron: Asymmetry*, 2006, **17**, 554–559.

12. J. Liang, J. Lalonde, B. Borup, V. Mitchell, E. Mundorff, N. Trinh, D. A. Kochrekar, R. N. Cherat and G. G. Pai, *Org. Proc. Res. Dev.*, 2010, **14**, 193–198.

13. S. K. Ma, J. Gruber, C. Davis, L. Newman, D. Gray, A. Wang, J. Grate, G. W. Huisman and R. A. Sheldon, *Green Chem.*, 2010, **12**, 81–86.

14. S. A. de Wildeman, T. Sonke, H. E. Schoemaker and O. May, *Acc. Chem. Res.*, 2007, **40**, 1260–1266.

15. R. L. Hanson, S. L. Goldberg, D. B. Brzozowski, T. P. Tully, D. Cazzulino, W. L. Parker, O. K. Lyngberg, T. C. Vu, M. K. Wong and R. N. Patel, *Adv. Synth. Catal.*, 2007, **349**, 1369–1378.

16. R. L. Hanson, J. M. Howell, T. L. LaPorte, M. J. Donovan, D. L. Cazzulino, V. Zannella, M. A. Montana, V. B. Nanduri, S. R. Schwarz, R. F. Eiring, S. C. Durand, J. M. Wasylyk, W. L. Parker, M. S. Liu, F. J. Okuniewicz, B. C. Chen, J. C. Harris, K. J. Natalie, K. Ramig, S. Swaminathan, V. W. Rosso, S. K. Pack, B. T. Lotz, P. J. Bernot, A. Rusowicz, D. A. Lust, K. S. Tee, J. J. Venit, L. J. Szarka and R. N. Patel, *Enzyme Microb. Technol.*, 2000, **26**, 348–358.

17. T. Purkarthofer, W. Skranc, C. Schuster and H. Griengl, *Appl. Microbiol. Biotechnol.*, 2007, **76**, 309–320.

18. H. Gröger, F. Chamouleau, N. Orologas, C. Rollmann, K. H. Drauz, W. Hummel, A. Weckbecker and O. May, *Angew. Chem., Int. Ed.*, 2006, **45**, 5677–5681.
19. P. Domínguez de Maria, T. Stillger, M. Pohl, M. Kiesel, A. Liese, H. Gröger and H. Trauthwein, *Adv. Synth. Catal.*, 2008, **350**, 165–173.
20. H. Yamada and M. Kobayashi, *Biosci. Biotechnol. Biochem.*, 1996, **60**, 1391–1400.
21. A. Panova, L. Mersinger, Q. Liu, T. Foo, D. Roe, W. Spillan, A. Sigmund, A. Ben-Bassat, L. Wagner, D. ÓKeefe, S. Wu, K. Petrillo, M. Payne, S. Breske, F. Gallagher and R. DiCossimo, *Adv. Synth. Catal.*, 2007, **349**, 1462–1474.
22. B. Kamm, P. R. Gruber and M. Kamm (ed.), *Biorefineries—Industrial Processes and Products. Status Quo and Future Directions*, Wiley-VCH, Weinheim, Germany, 2010.
23. P. Domínguez de María, *ChemSusChem*, 2011, **4**, 327–329.
24. A. J. J. Straathof, *Chem. Rev.*, 2014, **114**, 1871–1908.

Sustainability of Biocatalytic Processes

DEEPIKA MALHOTRA,[a] JOYEETA MUKHERJEE[b] AND
MUNISHWAR N. GUPTA*[a]

[a] Department of Biochemical Engineering and Biotechnology, Indian
Institute of Technology Delhi, New Delhi-110016, India; [b] Department of
Chemistry, Indian Institute of Technology Delhi, New Delhi-110016, India
*Email: munishwar48@yahoo.co.uk

15.1 INTRODUCTION

In the beginning, there was biotechnology. Today, we have red, green, blue, black, grey and white biotechnology![1] While biotechnology in almost all its hues directly or indirectly concerns itself with biocatalysis; the theme of this chapter largely relates to what is known as white biotechnology.[2,3] Broadly, 'it involves the use of enzymes and organisms for the processing and production of chemicals, materials, and energy including biofuels'.[1] In an editorial overview of a thematic issue of a journal, Jaeger wrote:[4]

> 'White biotechnology rather than fire-and-sword-chemistry ... meets "green chemistry", "environmentally benign production processes" or "sustainable development", which have become important political prerequisites for worldwide industrial activities'.

Chemical Processes for a Sustainable Future
Edited by Trevor M. Letcher, Janet L. Scott and Darrell A. Patterson
© The Royal Society of Chemistry 2015
Published by the Royal Society of Chemistry, www.rsc.org

Those were the early days of white biotechnology. As is often pointed out, white biotechnology does not include bioremediation (end-of-pipe solutions); that approach is in the domain of grey biotechnology. An early report on white biotechnology by EuropaBio published in 2003 is still relevant today and needs to be known more widely.[5] In its foreword, the promises of white biotechnology are succinctly summarized as:[5]

'reductions in greenhouse gas emissions, energy and water usage are examples of the benefits brought about by cleaner, greener and simpler bioprocesses'.

The key findings stated in this report and which are pertinent to our present context are:

- use of biocatalysts makes industrial processes more sustainable;
- biorefinery replacing chemical refinery will mean that starting materials are obtained from renewable resources;

The report also noted that flavours and fragrances were already produced *via* biocatalysis.

In a 2004 paper, Jenck *et al.*[6] pointed out that chemical industries in the world had already set their own targets in enhancing sustainability of their chemical processes. Both the Europa Bio report[5] and Jenck *et al.*[6] emphasize that switching over to the use of biocatalysis in many cases would encourage innovation and improve profitability in the long run.

This chapter outlines the general parameters which are currently available for evaluating 'greenness' or 'sustainability' of a process, and the way biocatalysts are playing an increasing role in enhancing sustainability of industrial processes.

15.2 PARAMETERS FOR MEASURING SUSTAINABILITY

The first set of rules were the 12 principles of green chemistry which are by now fairly well known and discussed in various places.[7] The commonality between green chemistry and white biotechnology is that both aim at preventing pollution rather than waste remediation. The quantitative parameters mentioned in these rules were atom efficiency and E-factor.[7] The former encourages the chemists to devise reactions which do not lead to any product(s) other than the one which is wanted. E-factor quantifies waste produced compared with product production on a weight basis. Unfortunately, water used or produced as a waste stream is generally not included in E-factor estimates. That factor is now considered more important and we will come back to it shortly. Considering various industrial sectors, oil refining fares much better and the pharmaceutical sector is the worst one in terms of E-factor estimates.[7] Sheldon discussed the synthesis of pregabalin and the production of bioethanol in terms of E-factors of the processes.[8] Equally relevant (for the theme of this chapter) is his discussion on enzymatic *versus* chemical route for intermediates for atorvastatin and myristyl myristate (an emollient ester which is industrially produced). In both cases, biocatalytic routes yielded better product quality and were greener as estimated by many factors.

Some other parameters such as effective mass yield or mass intensity have also been proposed in place of the E-factor.[7] E-factor is, however, mostly widely accepted. Sheldon introduced the concept of environment quotient, EQ, which took into account that not all chemicals are equally harmful to the environment.[9] Thus, common salt (NaCl) can be considered relatively harmless and given a quotient Q of unity. In the real world, threat perception depends upon many factors including the context. So, no chemical can be assigned a permanent Q value! What EQ does help with is in making choices between various alternative processes, provided all other factors (*e.g.* scale of operation) are common.

Lancaster[10] lists parameters for 'pre-experimental route selection' which factors in energy requirement, process efficiency, *etc.* He also lists metrics such as waste treatment 'on site' and 'off site'. Lancaster[10] also provides a fairly good introduction to life cycle assessment (LCA), which aim to evaluate the greenness of the entire ('cradle to grave') process.

The water footprint (WFP) indicates direct and indirect use of water. A good discussion on WFP in the context of process industries is provided by Klemes.[11] Water use and wastewater discharge both have international standards set by the European Commission as well as the US Congress. While blue WFP refers to the consumption of surface and ground water, green refers to the use of rain water. Grey WFP refers to the volume of water required to ensure that diluted waste conforms to standards. Food processing industries seem to have been the early ones worrying about WFP.[11] WFP has to be a part of LCA studies, so any biocatalytic process ultimately will have to factor in WFP.

More recently, Clark *et al.*[12] also listed mass-based green chemistry metrics (Table 15.1). This work also provides a fairly comprehensive

Table 15.1 A summary of mass-based green chemistry matrices [adapted from ref. 12 with permission from Chemistry Central].

Metric	Calculation
Yield/%	Mass of desired product/Theoretical maximum mass of product
AE/%	Relative molecular mass (RMM) of desired product/RMM of reactants
RME/%	Mass of desired products/Total mass of reactants used
E-factor	Total mass waste/Mass of desired product
PMI	(Total mass required/Mass of desired product) = E-factor +1
gRME/%	(Mass of desired product/ Total mass required) = (1/PMI)

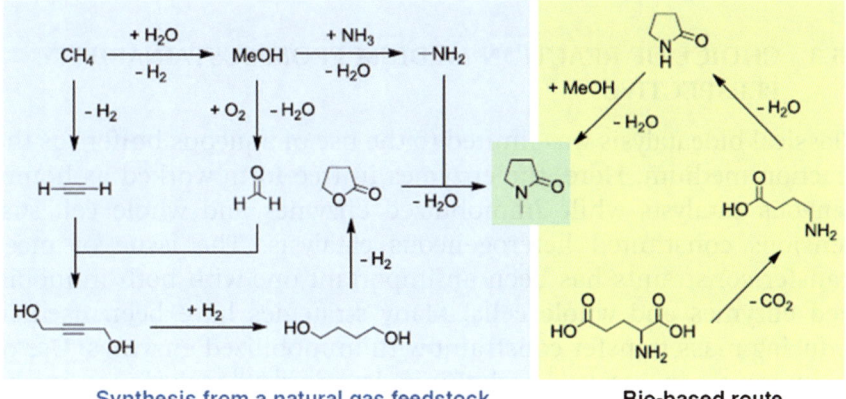

Synthesis from a natural gas feedstock   Bio-based route

Figure 15.1 Synthesis of natural gas derived and bio-based *N*-methyl pyrrolidone (equations not balanced) [reproduced from ref. 12 with permission from Chemistry Central].

discussion on the use of metrics to evaluate the production of *N*-methyl pyrrolidone *via* a bio-based route by replacing natural feedstock (Figure 15.1). This intermediate in turn was used for the synthesis of an ionic liquid, which in turn was used for catalysing Fischer esterification reactions.

With global warming as a major issue, 'the carbon footprint' of a process is another important parameter.[10] In reality, this is also a part of LCA but has become important in its own right. Any greenhouse gas emission can be converted to CO_2 equivalent.

Winterton[13] has discussed how chemical engineers and chemists have different terminologies. While a chemist will talk of reactions, schemes and molecules, a chemical engineer will think in terms of reactors and processes. The scale at which work is carried out in laboratories and industries also obviously differs. The approach of process intensification to enhance sustainability requires largely chemical engineering inputs.[14] Another good discussion on the need for close collaboration between chemists and engineers to design sustainable chemical processes is provided by Mulhollard *et al.*[15] Hence, biochemical engineers with their twin expertise on understanding biocatalysis and chemical engineers have a major role to play here.

Adoption of the green chemistry principles at industrial scale requires that engineering designs are in tune with the chemical strategy. That has resulted in 12 principles of engineering for sustainable development accepted by the Royal Academy of Engineering.[16]

15.3   CHOICE OF REACTION MEDIUM FROM SUSTAINABILITY PERSPECTIVE

Classical biocatalysis was limited to the use of aqueous buffers as the reaction medium. Here, the enzymes in free form worked as homogeneous catalysts while immobilized enzymes and whole cell suspensions constituted heterogeneous catalysis. The issue of mass transfer constraints has been an important one with both immobilized enzymes and whole cells. Many strategies have been used in reducing mass transfer constraint with immobilized enzymes. Use of non-porous supports[17,18] and affinity layering[19,20] techniques can be mentioned as two examples.

Currently, enzymes are also used in nearly anhydrous organic solvents,[21,22] aqueous-organic co-solvent mixtures,[23,24] two phase[25,26] or even multiphase systems,[27] reverse micelles,[28,29] supercritical fluids[30] and ionic liquids.[31–34] Use of organic solvents may give rise to volatile organic compounds (VOCs) which hurt the ozone layer. In fact, ease of evaporation/distillation during the downstream phase makes more volatile organic solvents a preferred choice in organic synthesis. In non-aqueous enzymology, the choice of organic solvents is dictated by the effect of the solvent's properties on the rate of the reaction. In general, more hydrophobic solvents give higher reaction rates as less hydrophobic solvents strip the enzyme of its essential water layer. While no satisfactory solvent-related parameter has been found over the entire range, log P has been considered to be the best possible parameter.[35–38] As Halling[39] writes, log P is definitely useful when the partition is the main mechanism. Solvent can influence the substrate specificity (by partitioning) and even enantioselectivity.[40] In nearly anhydrous organic solvents, the amount of water present is a critical parameter and needs to be optimized for obtaining maximum reaction rates. To quantify the amount of water, water activity (a_w) is the best parameter, at least for water immiscible solvents.[38] In nearly anhydrous media, even the free enzyme is insoluble and hence these systems are examples of heterogeneous catalysis. Linking the enzyme to polymers like polyethylene glycol (PEG) and some smart polymers makes them soluble in a few organic solvents.[41–43] Whether this is worth the effort in terms of increase in overall percentage conversion needs further investigation with more variety of systems. Surfactant–lipase complexes are also soluble in organic solvents in many cases.[44]

Aqueous-organic co-solvent mixtures generally consist of (5–20)% (v/v or w/w) water-miscible organic solvents. Any increase in percentage water beyond this generally leads to enzyme inactivation. Some analysis of 'good solvents' and 'bad solvents' in this context has been carried out.[45] The advantage of adding organic solvent in such systems is to increase the solubility of substrates which have limited solubility in water.

Extensive reviews on two-phase systems and multi-phase systems are available.[25–27,46,47] Solubilization of water-insoluble substrates, relieving product inhibition or keeping the biocatalyst in aqueous milieu constitute various useful features of such systems. Reverse micelles, with their limited water pool, again make it possible to use hydrolases for synthetic purposes.[48] Supercritical fluids, mostly supercritical carbon dioxide (sCO_2), have been fairly well studied as a green solvent for carrying out biocatalysis.[49,50] The cost of sCO_2 and the need to use closed reactors under controlled conditions are two major hurdles in larger adoption of sCO_2 as the reaction medium at industrial scale.

Room temperature ionic liquids have emerged as a very versatile and promising reaction media for biocatalysis.[31–34] These solvents with their high boiling points, release practically zero volatile organic solvents (VOCs). As the permutation and combination of different cations and anions can create an ionic liquid with a desirable property (*e.g.* polarity), these solvents have been called designer solvents. Some ionic liquids denature enzymes but many ionic liquids do not. There was a lag period in their use in biocatalysis but numerous reports are now available. Biotransformations involving carbohydrates have become increasingly important in the past decade or so. Very few organic solvents dissolve monosaccharides/disaccharides, *e.g.* pyridine, dimethyl formamide (DMF), acetonitrile and dimethyl sulfoxide (DMSO), in a limited fashion. Hence, ionic liquids which dissolve carbohydrates have become the reaction media of choice in such biotransformations.[33] Of special interest is the use of ionic liquids in the pre-treatment and fractionation of lignocellulosic biomass.[51]

Finally, 'solvent free' systems in which the substrate itself is a solvent offers various advantages: greenness since no additional solvent is required; high molarity of the substrates; and cost reduction as the solvent cost is saved.[52,53]

The possibility of altering reaction rates and substrate specificity by change of the reaction medium has made biocatalysis even more versatile. This has been sometime referred to as 'medium engineering'.[54–56] Meanwhile, the search for more options for green solvents continues. A few years back, Aldrich introduced cyclopentyl methyl ether (CPME) as a green organic solvent.[57] Ethyl lactate has been suggested as another green solvent.[58] It would be interesting to explore use of such solvents for biocatalysis. As Lancaster[10] points out:

'It is now widely accepted that no solvent is a *priori* green. Its greenness is a combination of its inherent properties, like acute toxicity, ozone depletion potential and persistency in the environment, its method of production, and the application it is being used in, including the method of recovery and reuse. A number of solvent selection tools, essentially taking a life cycle assessment (LCA) approach and taking into account the above are available, but are often complex to use and compare'.

Another interesting approach is 'solid-to-solid biocatalysis' wherein synthesis using biocatalyst has been shown to be feasible using highly concentrated substrates.[59,60]

15.4 BIOCATALYST ENGINEERING

Over the years, many techniques have been developed to improve the performance of the enzymes. A biocatalyst can only replace a chemical catalyst if it is more economical in the industrial context. Stability of an enzyme under the process conditions is hence a crucial factor. To make enzymes and whole cells more cost effective, reusability is an equally important consideration. Hence, immobilization of biocatalyst continues to be an active area of research. Immobilization quite often enhances biocatalyst stability to a significant level, but 'easier removal of the enzyme from the product' is more often the only advantage which is gained.[61] Some current reviews and books on immobilization are available.[44,62] In the context of sustainability, naturally occurring supports (which are also biodegradable) require a special mention. Alginates and κ-carrageenans have been often used for whole cell entrapment.[63–65] Less frequently, chitosans, alginates and even κ-carrageenans have been used as soluble supports for enzyme immobilization.[66] As each of these biopolymers can be made reversibly soluble–insoluble by simply altering the pH or by the addition or removal of cations, bioconjugates of such polymers with enzymes are also called 'smart biocatalysts'.[66,67]

It is not often realized that immobilization serves a somewhat different function when enzymes are used in low water media. Enzymes in such media perform much more poorly than chemical catalysts. For example, alkali-catalysed transesterification of oils and fats to form biodiesel takes at the most only couple of hours.[68] On the other hand, biodiesel synthesis by the lipase-catalysed route generally takes >24 h.[69,70] Hence, if we have to replace chemical catalysts with biocatalysts for industrial processes, the need for innovative efforts in biocatalyst engineering is much greater when catalysis requires low water media. Immobilization of enzymes, when it improves reaction rates in low water media, is often attributed to enzyme stabilization. In the absence of detailed investigations (which are seldom reported), it is difficult to rule out that possibility. However, in most of the commonly used immobilization approaches, it generally originates from an increase in catalytic surface. Enzyme molecules, rather than being present as lumps (in powder forms), are now spread over the support surface. As Halling points out, in many cases like celite, it may even be the case of simple dispersion rather than true immobilization.[38]

One key factor responsible for poor performance of enzyme powders (in low water media) is the reversible structural changes which protein molecules undergo when dried by freeze drying or spray drying.[71,72] In general, a decrease in α-helical context and an increase in β-sheet content have been reported. The use of cryo-protectant (*e.g.* many osmolytes) and lyo-protectant (which can replace bound water and form hydrogen bonds with protein molecules, *e.g.* sucrose) has been suggested to prevent such structural damages.[71] Use of a combination of a lyo-protectant and cryo-protectant during freeze drying is indeed found to result in lyophilized powders of enzymes which show higher initial rates in low water media.[73] Another strategy is to use alternative ways of 'drying' enzymes. Precipitation with organic solvents to obtain enzyme precipitated and rinsed with propanol (EPRP),[74,75] or enzymes precipitated and rinsed with organic solvents (EPROS),[76] propanol rinsed enzyme preparations (PREP)[77,78,32] represent the results of this approach and have often shown to be far superior than just lyophilized powders. Cross-linking such precipitates *in situ* results in cross-linked enzyme aggregates (CLEA).[79–81]

However, precipitation in the presence of a sugar or salt (which precipitates ahead of the enzyme to create microcrystalline cores) results in protein coated microcrystals (PCMC).[82–84] In some cases, cross-linked PCMC (CLPCMC) have been reported to perform even better.[85,86] It has been known for a long time that the side chains of amino acids (especially the side chains of active site residues) in the enzyme molecule being in the right ionic form (as these are at pH optimum) is critical to the activity of the enzymes in low water media too. Hence, incorporating solid state buffers or using solid state buffers as core in PCMC gives better catalytic rates.[87]

Finally, use of nanomaterials, with their advantage of high surface to volume ratio, as immobilization matrices is reported to result in high-performance biocatalysts in both water-rich and water-poor media.[88–91]

15.5 MICROWAVE ASSISTED BIOCATALYSIS

Microwave assistance has been extensively used to facilitate the synthesis of organic compounds and, lately, nanomaterials.[92,93] It is certainly an important component of green chemistry. Microwave assistance in enzyme catalysed reactions has also been reported.[94,95] However, it is definitely an underexploited approach. In our opinion, the main reasons for this are as follows.

The general impression is that it is not a scalable technology and hence cannot be useful at industrial level. One tends to overlook the fact that textile and food industries have been using large-scale microwave reactors for quite some time.[96] Nevertheless, a reactor suitable for large-scale biocatalytic processes is not readily (commercially) accessible and that is an important hurdle.

Many results related to microwave assisted biocatalytic processes are not readily accepted as many scientists do not believe in the 'non thermal effects' of microwaves.[97,98] When initial rates obtained for biotransformations with conventional heating and microwave heating are compared, the enhancement in the case of the latter mode is there but not really impressive (Table 15.2).[99]

'True power' microwave ovens are expensive. With domestic microwaves, it is not possible to control power and high power denatures biocatalysts.

The interplay between reaction medium and microwave radiations is not widely appreciated. As we pointed out recently:[99]

'It has not been often appreciated by enzymologists that heating characteristics of the solvent by microwave irradiations are

Table 15.2 Effect of microwave radiations on initial rates of (a) esterification of palmitic acid with *n*-propanol and (b) transesterification reaction of ethyl oleate with *n*-propanol catalyzed by Novozyme 435 compared with conventional heating in different organic solvents as a reaction medium [reproduced from ref. 99 with permission from the editor-in-chief, *Proceedings of the Indian National Science Academy*.]

Solvent	Initial rates /nmoles min^{-1} mg^{-1a}		Ratio of initial rates (B/A)
	Conventional heating (A)	Microwave heating (B)	
(a)			
n-Octane	1248	2132	1.7
n-Hexane	1248	2028	1.6
Toluene	1144	1782	1.6
CPME	936	1838	2.0
2-Pentanone	858	1638	1.9
n-propanol	1196	1404	1.2
Acetone	364	728	2.0
Dioxane	624	1352	2.2
DMF	1.1	2.6	2.4
Solvent free	1482	3978	2.7
(b)			
n-Octane	433	715	1.6
Toluene	182	433	2.4
CPME	411	1040	2.5
2-Pentanone	433	1040	2.4
Acetone	303	650	2.1
Dioxane	277	498	1.8

[a]Initial rates were based on conversions obtained by gas chromatography analysis.

dependent upon the dielectric properties of the organic solvents. The loss tangent, tan $\delta = \varepsilon'/\varepsilon'$ where ε' is the dielectric loss and ε' is the dielectric constant describing the polarizability of the molecules in the electric field. In general, non polar solvents have low tan δ, that is, these absorb the microwaves poorly. More polar solvents, with higher tan δ show more efficient absorption. So, the polar solvents have two advantages. They are better in converting microwave energy to heat energy. Simultaneously, as pointed out earlier, fewer microwave irradiations may be conducive to enzyme stability. Unfortunately, as is well known in non aqueous enzymology, these polar solvents also strip off the necessary water layer from the enzyme molecules and hence give poor initial rates'.

However, it is interesting to note that log P for the reaction media correlates well with the enhanced reaction rates observed for lipase catalysed esterification reaction in low water media (Figures 15.2

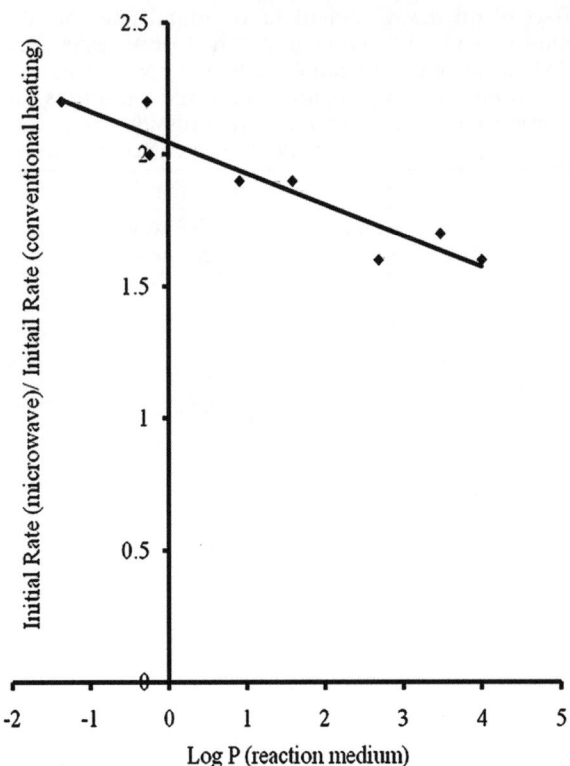

Figure 15.2 Correlation of log P with the enhancement of initial rates of esterification reaction of palmitic acid with *n*-propanol catalysed by Novozyme 435 by microwave radiations compared with conventional heating in low water media. Palmitic acid (0.975 mmoles) and *n*-propanol (4.87 mmoles) were taken in the ratio of 1 : 5 (mol/mol). For the reaction to be carried out with organic solvent as a reaction medium, 1 mL of the organic solvent was added to the above reaction mixture. The following solvents were varied: *n*-octane, *n*-hexane, toluene, cyclopentyl methyl ether, 2-pentanone, acetone, dioxane and DMF. 5% (w/w of palmitic acid) Novozyme 435 was added. These mixtures were incubated at 45 °C either in an orbital shaker at 200 rpm rotation or in a microwave reactor, magnetically stirred at medium speed. Aliquots were taken out periodically and were analysed by gas chromatography. Each reaction set was run in duplicate and the deviations within the pair of experimental data were 5% [reproduced from ref. 99 with permission from the editor-in-chief, *Proceedings of the Indian National Science Academy*].

and 15.3).[99] More extensive studies are needed before a generalization can be made.

Microwave assistance in few cases has been shown to reduce overall time and simultaneously give better percentage conversions in cases

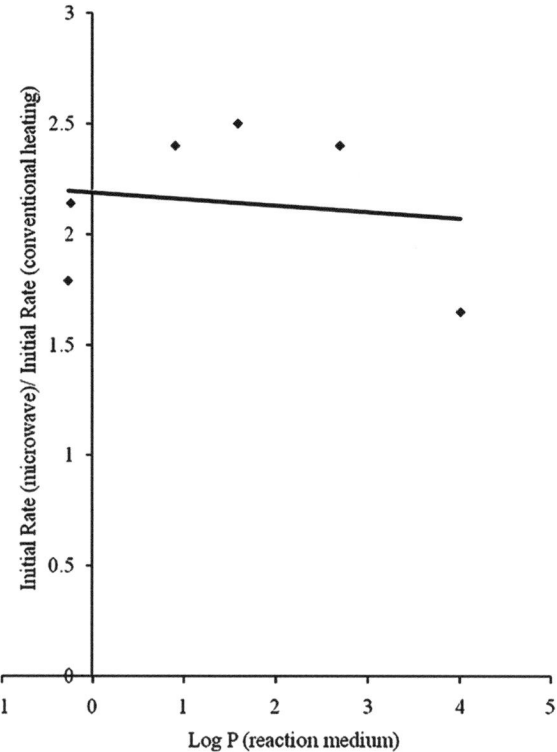

Figure 15.3 Correlation of log P with the enhancement of initial rates of transesterification reaction of ethyl oleate with *n*-propanol catalysed by Novozyme 435 by microwave radiation in low water media. Ethyl oleate (0.975 mmoles) and *n*-propanol (4.87 mmoles) were taken in the ratio of 1 : 5 (mol/mol). For the reaction to be carried out with organic solvent as a reaction medium, 1 mL of the organic solvent was added to the above reaction mixture. The following solvents were varied: *n*-octane, toluene, cyclopentyl methyl ether, 2-pentanone, acetone and dioxane. 5% (w/w of ethyl oleate) Novozyme 435 was added. These mixtures were incubated at 45 °C either in an orbital shaker at 200 rpm rotation or in a microwave reactor, magnetically stirred at medium speed. Aliquots were taken out periodically and were analysed by gas chromatography. Each reaction set was run in duplicate and the deviations within the pair of experimental data were 5% [reproduced from ref. 99 with permission from the editor-in-chief, *Proceedings of the Indian National Science Academy*].

of biodiesel formation,[100] the synthesis of important synthons such as β-ketoesters[101] and the kinetic resolution of secondary alcohols.[102] However, de Souza *et al.*[103] reported that microwave irradiation offered no advantage over conventional heating in kinetic resolution of ($\pm$)-1-phenylethanol. The possible reasons for these contradictory claims could be:

(a) We do not have enough data about stability of enzymes under microwave irradiations. Some limited studies are available.[104,105]

(b) Quite often, the conditions under which microwave treatment is carried out are not well defined. The studies carried out with domestic microwaves ignore the importance of power as a parameter.

(c) The interplay between microwave irradiation and other parameters is a fairly dark area. Even in true power microwave ovens, the actual irradiation time will depend upon the difference between ambient and 'set temp' (and power!).

There are few reports which suggest that microwave assistance can alter the kinetics and selectivity of biotransformations.[106–108]

The effect of microwave irradiation is less controversial when hydrolases have been used in aqueous buffers. In almost all the cases, microwave irradiation accelerates the hydrolysis compared with conventional modes of heating.[109,110] Mention may also be made of microwave pre-treatment of macromolecular substrates before enzyme catalysis. Here again, microwave irradiation offers clear advantage in obtaining better conversions.[111,112]

To sum up, we need more data generated under well-defined conditions before we can correctly assess the advantage of microwave irradiation in enhancing the overall sustainability of the biocatalytic processes. This is especially true for non-aqueous media. In water, the picture is somewhat clearer. Hopefully, we will see more applications of microwave assisted biocatalysis in the near future.

15.6 ULTRASONOENZYMOLOGY

Ultrasonication is very widely used for cell disintegration. Hence, traditional enzymologists generally view ultrasonication as a disruptive technique which damages biological structures. This bias has possibly hampered the growth of ultrasonoenzymology. Ultrasonication is now well established for facilitating organic synthesis[113] and obtaining nanomaterials.[114–116] The microcavitation produced due to ultrasonication creates high localized pressure and temperature, turbulence and shear stress, and leads to free radical formation. Some of these consequences need to be controlled (*e.g.* it is important that temperature is controlled) and most of these can be harnessed for assisting biocatalysis. Ultrasonicators that use irradiation with low power (there are some commercially available baths where irradiator power can be even varied) have been used to facilitate enzyme-catalysed reactions.

Sinisterra[117] wrote an early review, making the point that most of the interesting applications of this approach are in the area of non-aqueous enzymology; either pre-treatment of the enzyme powders (for better dispersion) or irradiating the system while the reaction is ongoing has been reported to enhance reaction rates and/or conversions. A later review is by Gupta and Roy.[118] Subsequent to this review, Liu *et al.*[119] carried out lipase-catalysed hydrolysis of soy oil. While they report that reaction rates were two-fold higher, a more interesting observation was that less water was required for hydrolysis in the solvent free system. As pointed out elsewhere, the water footprint of the process is going to become an increasingly important issue.

Sener *et al.*[120] studied milk lactose hydrolysis by lactase under ultrasonication. While sonication resulted in enhanced hydrolysis, enzyme inactivation up to 25% occurred. It will be good to see whether more stable lactases fare better so that the biocatalyst can be reused.

It is hoped that more comprehensive studies in many other cases, particularly with more stable enzymes will give more encouraging results.

15.7 EXPLOITING EXTREMOPHILES

15.7.1 Thermophiles and Alkalophiles as Sources of Enzymes

Many microorganisms survive or even thrive under very harsh conditions: extremely high temperature (*e.g.* hot springs), extremely cold conditions (*e.g.* Antarctica water) or highly saline conditions. These microorganisms are called thermophiles, psychrophiles and halophiles, respectively, and are collectively referred to as extremophiles. Also, included in this class are acidophiles and alkalophiles, which can grow under extreme pH conditions.[121,122] Thermophiles were one of the earliest to be investigated and have constituted the source of some very important enzymes. The polymerase chain reaction (PCR) technique would not be possible without the discovery of *Thermus aquaticus* (Taq) DNA polymerase. Extensive literature is available on the possible applications of enzymes from thermophiles.[123,124]

One of the important applications which is now industrially adapted is the use of amylases (from thermophiles) for starch degradation.[125] The use of lipases and proteases as additives for detergents for hot wash cycles in washing machines is another application with a huge market.[126] These enzymes simultaneously also have to tolerate the alkaline conditions during the wash and hence are obtained from alkalophiles. In such cases, mostly enzyme variants obtained by protein engineering are now used but earlier studies of enzymes from thermophile/alkalophiles have been very valuable in this regard. Phytase as an additive in feeds for animals and poultry is also required to be thermostable under pelletization conditions of high temperature and high moisture.[126,127] The pelletization of the feed is required so that the feed can be transported with minimum packing volumes.

15.7.2 Psychrophiles as a Source of Enzymes

Further advancement in making biocatalysis even more sustainable in many applications may come from switching over to enzymes from psychrophiles. Sometimes, microbiologists further classify psychrophiles as facultative psychrophiles (with an optimum temperature of 20 °C or higher) and obligate psychrophiles with $t_{optimum}$ of 20 °C or lower. However, there is no unanimity in the literature about these terms.[128] The catalytic activity of these enzymes at low temperature is achieved by enzymes being more flexible. More flexibility is achieved by multiple ways such as:[129]

'reduction in core hydrophobicity, decreased ionic and electrostatic interactions, increased charge of surface residues that promote increased solvent interaction, additional surface loops, substitution of proline residues by glycines in surface loops, a decreased arginine/lysine ratio, less inter-domain and subunit interactions and fewer aromatic interactions'.

Already many efforts based on protein engineering and directed evolution have been made to further improve their properties: higher catalytic activity and higher thermal stability.[130] There are two clear advantages of using enzymes from psychrophiles: (1) catalysis at a low temperature means saving the energy needed while using higher reaction temperature; and (2) in applications where hygiene and cleaning in place are crucial, the microbial growth in the reactor is minimized. Hence, it is not surprising that food biotechnologists have already shown keen interest in enzymes from psychrophiles.[129]

With emergence of metagenomics, the possibility of sourcing enzymes with diverse properties has taken a quantum leap. This should result in more enzymes replacing chemical catalysts in diverse applications.

15.8 APPLICATIONS OF BIOCATALYSIS IN CHEMICAL AND BIOCHEMICAL PROCESSES

15.8.1 Industrial Enzymology and Enzymes in Organic Synthesis

15.8.1.1 Why Enzymes are Underexploited?. In 1980s, Whitesides and co-workers discussed the potential of enzymes in organic synthesis in two articles which are still worth mentioning.[131,132] In the first one, the authors pointed out that often the nomenclature used by biochemists can deter organic chemists from exploiting the enzyme. Often, it does not reveal full information or possible applications. Today, the availability of databases like BRENDA and STRENDA[133–135] is useful to organic chemists but, unfortunately, these systems are not mentioned in the organic chemistry literature as much as these should be. Also, the phenomenon of catalytic promiscuity has provided another complication to the issue of enzyme commission nomenclature.[136] It is interesting to observe that 'disadvantages' mentioned in the second article[132] have almost become non-issues with the passage of time!

15.8.1.2 Industrial Enzymology. Industrial enzymology developed considerably in the 20th century. Applications of enzymes in baking, brewing, dairy, detergents, textiles, paper and pulp, and leather processing industries are well documented.[137] Some case histories on the early successes of biocatalysis in production of fine chemicals, *etc.* are also discussed in a number of papers.[138–141] Hence, we will limit ourselves to some less known and/or more recent examples.

15.8.1.2.1 **Aqueous Enzymatic Oil Extraction.** Extraction of oil from seeds for edible purposes is largely carried out by mechanical pressing or solvent extraction using hexane.[142] The latter generates a significant amount of VOCs. Hence, the search for a safer solvent has been attempted but no viable option has emerged.[143] Simultaneously, aqueous oil extraction (AOE) was attempted with finely grounded seed material by many workers over the past several decades.[144,145] As thermodynamically partitioning of oil in water is not favoured, this has remained only an academic quest. However, pre-treatment of seeds by enzymes[146] or the use of enzymes during AOE, a process called aqueous enzymatic oil extraction (AEOE), have been reported to enhance the yield of oil to a significant extent.[144,147,148] The oil in the seeds is generally present as oil bodies. These consist of proteinaceous and hemicellulosic structures encasing oil. Hence, enzymes like proteases, pectinases and cellulases have generally been successful in AEOE. One success story at industrial level is the extraction of olive oil by pre-treatment with pectinase preparation. However, as pointed out many years ago, the real economical reason in this case is that less water is required if pectinase preparations are used during the olive oil extraction.[149]

Another oil extraction process in which use of enzymes has shown some advantage is enzyme-assisted three-phase partitioning (EATPP). Three-phase partitioning (TPP) consists of mixing of finely grounded seed material with about an equal volume of water containing ammonium sulphate and *tert*-butanol. After a couple of hours, liberated oil is obtained in the upper *tert*-butanol phase.[150] Carrying out TPP after enzyme pre-treatment (EATPP) has been shown to enhance oil yield.[151–153] In both AEOE and EATPP, one additional advantage is that spent seed material is also partially hydrolysed and can constitute a better food or feed material.

Incidentally, both microwave assistance and ultrasonic irradiation can be used in synergistic fashion with AEOE.[154–157]

15.8.1.2.2 Use of Phospholipases for Degumming. This is another well-established application of enzymes in oil and fat industry. Edible oils after extraction require refining before marketing. Degumming is one of the early steps in oil refining. This is especially necessary for oils rich in phosphorous content (*e.g.* soybean, corn and sunflower).[158] One of the traditional approaches is to add a large quantity of water so that phospholipids (responsible for the gummy nature) are removed in an hydrated form. However, this does not remove non-hydrated phosphatides. One of the early commercial enzyme preparation for degumming consisted of phospholipase A_2 from porcine pancrease.[159] In recent years, other phospholipases such as phospholipase A_1 from *Fusarium oxysporum*[160] and phospholipase C from *Pichia pastoris*[161] have been introduced for industrial level degumming. Some patents also suggest combining various phospholipases.[162,163]

15.8.1.3 Synthesis of Chemicals, Biochemicals and Drug Intermediates. The overview by Schoemaker *et al.*[164] points out that some wrong perceptions (such as availability and stability) limit the wider applicability of enzymes in industrial sectors. They list examples of both non-chiral and chiral molecules which are currently manufactured with the help of biocatalysts. Of particular interest are the uses of nitrilase and aldolase in synthesizing different enantiomers and epimers. It is also increasingly becoming clearer that a combination of both chemical catalysts and biocatalysts (chemoenzymatic) may be, in many cases, the most viable strategy.[165]

Schmid *et al.*[166] reviewed catalytic processes currently in place at various industrial companies such as BASF, DSM and Lonza. Synthesis of the *S*-analog of piperazine-2-carboxylic acid as a precursor of the HIV drug Crixivan by Merck using amidase activity in the form of whole cells from DSM is also mentioned. An important idea is to synthesise steroids with building blocks obtained from soya polysterols. A useful 'rule of thumb is that for products valued at less than US$ 20 kg^{-1}, the intended production should exceed 1000 t a^{-1} [tonnes per year] and volumetric activities should be above 100 international units per litre to warrant further work'.[166]

Tao and Xu[167] reviewed the use of biocatalysis in synthesis of drugs like simvastatin, atorvastatin, pregablin, Paxil™ (for depression) and Keppra™ (for epilepsy). In all the cases, biocatalytic routes were greener as judged by several parameters. In the case of pregablin, the switch to a lipase based route reduced the E-factor from 86 to 17.[167] Another recent review by Nestl *et al.*[168] mentions a few more examples such as steps in the synthesis of sitagliptin (antidiabetic drug) and hydroxyltriazoles (β-adrenergic receptor blockers). This review illustrates that the use of directed evolution to obtain enzymes or whole cells is already impacting on the synthesis of pharmaceuticals. Wohlgemuth[169] discussed the various reasons why biocatalysis should be the key to creating sustainable processes in chemical industries. By exploiting the selective nature of enzyme catalysis, one can avoid the necessity of protection and deprotection chemistries. Regioselectivity similarly means that sugars and other polyalcohols can be made to undergo reactions at one particular hydroxyl groups. The review classifies synthetic targets in terms of types of reactions involved: redox reactions, amination reactions, C–C bond formation reactions, *etc.* Wohlgemuth[169] points out the synthesis of 12-amino lauric acid by enzymatic transcrystallization could be carried out without organic solvent, acid or alkali. Another advantage of biocatalysis is the reduction of the number of steps, for example, synthesis of atorvastatin using a keto-reductase. Ogawa and Shimizu[170] focused on use of microbial enzymes in large-scale synthesis in Japan. Production of D-amino acids, DOPA (a drug used for Parkinson's disease) and polyunsaturated fatty acids (PUFA), the optical resolution of racemic lactones and the microbial reduction of chiral alcohols represent large-scale processes based on catalytic steps using enzymes or whole cells.

Some more recent examples at the lab scale illustrate future directions in this area. Chemo-enzymatic approaches using glucose isomerise for conversion of glucose to 5-hydroxymethyl furfural in sea water has been described as an example of how biorefineries can exhibit untreated sea water.[171] Enzymatic reduction of CO_2 to methanol in water without the recycling of $NADP^+$ has also been possible,[172] vanadium chloroperoxidise has been used to convert amino acids to useful nitriles,[173] isobutyric acid (currently produced from petroleum feedstocks) has been produced from glucose by metabolic engineering of *E. coli* cells[174] and, finally, chemo-enzymatic synthesis of sustainable polyurethanes using lipases avoids the use of toxic di-isocynates.[175]

The above does not constitute a comprehensive list of what is happening or what is possible in this area. It just provides a fair mix of diverse kinds of possibilities and technologies operating at various scales and at various stages of maturation. It appears that in principle anything can be obtained from non-petroleum based feedstocks. We should definitely see more of industrial processes based upon bio-catalysis in future.

15.8.1.4   Catalytic Promiscuity.   One key driver in use of biocatalysis has been that biocatalysts show substrate specificity, regiospecificity and stereospecificity. That undoubtedly has been the basis of their applications in organic synthesis. What has not been realized is that selectivity (or specificity) consists of something like a magician's card trick. Some cards are arranged and one is supposed to pick one. That is the way we have been looking at the enzyme specificity. Various esters are chosen, a lipase may show preference to the chain length range of the acid and/or alcohol component. What happens if some non-esters are included? It is now well established that enzymes show catalytic promiscuity.[176–179] It could be what is known as accidental promiscuity (observed with enzymes as they occur) or it could be induced (say by directed evolution).[176,177] Lipases have been most widely studied vis-à-vis accidental promiscuity and are shown to catalyse many C–C bond formation reactions. While water-rich media have been largely used in such studies, medium engineering in this context is underexploited. Some unexpected products can sometimes be obtained when this approach is attempted.[24]

The main hurdle here is that reaction rates tend to be low. The amount of enzyme normally needed is very large. That is reminiscent of the situation when early results with enzymes in nearly anhydrous media were described. Hopefully, exploiting induced catalyst promiscuity in combination with more efficient biocatalyst designs and medium engineering will result in new industrial level biocatalytic processes in the coming years.

15.8.2 Biorefineries

Fermentation to produce fine chemicals has a long history. However, in several cases the biorefinery approach is being used to revisit the industrial strategy to make it more sustainable. As pointed out by Banner *et al.*,[180] while spent *Aspergillus niger* biomass (after extraction of citric acid) was earlier mostly used for composting or as part of animal feed, Cargill now markets Regenasure[™] glucosamine derived from the hydrolysis of chitin present in the fungal biomass. Chitosan (partially deacylated chitin) and β-glucan (another polysaccharide present in the fungal biomass) are two other obvious valuable products.[181–185] Another important issue flagged by Banner *et al.*[180] is that, in cases like itaconic acid where supply exceeds demand, innovative chemical strategies can create new markets. Itaconic acid, for example, can replace platform chemicals of petrochemical origin such as maleic anhydride and fumaric acid.

The concept of biorefineries is also rooted in belief that a renewable source (like an algae or plant) can be exploited in totality. Ideally, no waste is generated; the valorization approach takes care of that. Simultaneously, the idea is to switch over from petroleum based chemistry to chemistry based renewable resources.[3,186] The premise and promise of white biotechnology is to use the toolbox of biology to achieve this. There have been some early successes of developing new routes to production of chemicals and (even novel) materials. These are already well documented at many places.[2,3,187–190] Biocatalysts have played an important role in many of these cases.

15.8.3 Valorization of Waste Products

A few decades back, whey generated during cheese production was considered a waste material. It was produced in huge amounts and its discharge in to water affected the biochemical oxygen demand (BOD) value of water and consequently adversely affected marine life. The need to produce low lactose milk for lactose-intolerant people gave rise to one of the early commercial applications of immobilized enzyme at a fairly large scale. β-Galactosidase, also called lactase, was immobilized on different matrices by different companies.[191,192] In another obvious outcome, the enzyme was also used for hydrolysis of lactose in whey (almost all the lactose in milk is left in whey during cheese production).[191,193,194] The sugars produced could be also used for yeast fermentation to produce ethanol. Ethanol or bioethanol (as it is sometimes called) has over the years become an important biofuel. This was an early application of biocatalysis in the valorization of a waste product. As the technology develops, eventually the tag 'waste' is removed from the product. Such products eventually become valuable raw materials.

An interesting story is that of biodiesel.[195,196] It is mostly produced by chemical catalysts. It is, however, widely believed that enzymatic transesterification is the future for this industry.[70,197] The main issue is that of the cost of the biocatalyst. A more efficient biocatalyst design for low water enzymology is one obvious solution.[198,199] It is well established that downstream processing is easier with the enzyme-based route; the glycerol produced is cleaner.[69,70] Many countries focused on jatropha seeds as a source of oil (for producing biodiesel), but the general and current perception is that it has been a disappointing choice.[200] Starting with a waste fat or oil-rich product like kitchen waste seems to be an idea worth pursuing further.[52,201] The use of an efficient biocatalyst design for production of biodiesel (esters of fatty acids with methyl and ethyl alcohols) from spent coffee grounds has been recently described (Figure 15.4).[198] The intake of coffee as a beverage is globally on the upswing and the amount of waste generated is significant and is likely to grow in the coming years.

The amount of glycerol generated as a by-product during biodiesel production has been large enough to lead to a shutdown of industrial plants that were producing glycerol for numerous uses.[202] The recent overview shown in Figure 15.5 gives a good idea of some interesting applications of glycerol.[203] The early attempts in this direction have been mentioned in few places.[3,202] Another aspect, which is worth more attention, is the use of glycerol as a reaction medium. The general perception has been that its high viscosity means poor mass transfer. However, some interesting results in biocatalysis show that good percentage conversions are possible.[204,205]

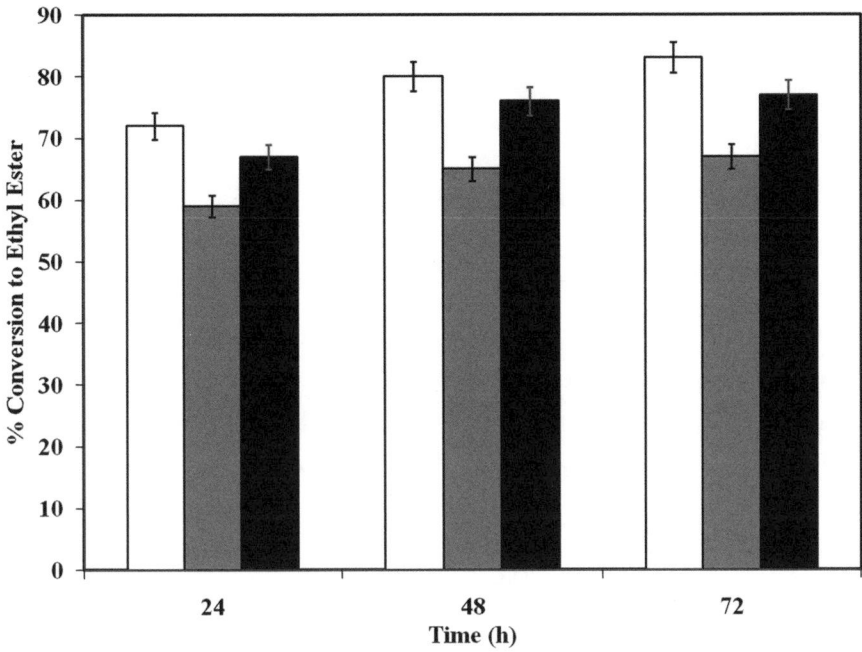

Figure 15.4 Biodiesel from clean coffee oil (coffee oil decolourized by refluxing with activated charcoal) catalysed by Combi-PCMC (white bars), a mixture of PCMCs (1 : 1) (grey bars) and CL-Combi-PCMC (black bars) of *Candida antarctica* lipase B (CALB) and palatase (*tert*-butanol was used for precipitation). Reactions were performed taking 0.5 g coffee oil. Combi-PCMCs and CL-Combi-PCMCs were made as described in the materials and methods section of ref. 199. For a mixture of PCMCs, the individual PCMCs were made and then they were mixed in the ratio 1 : 1. Ethanol was added in molar ratio 4 : 1 (ethanol : oil). Conversion rates to ethyl ester were measured by analysing aliquots withdrawn at different time intervals using gas chromatographic analysis. The experiments were performed in duplicate and the error between the two sets was within 3% [reproduced from ref. 198 with permission from Chemistry Central].

Luque and Clark[206] pointed out that:

'worldwide we throw away over one 1 billion tons of edible food waste each year and the total for food supply chain waste ("from farm to fork") will be several billion'.

It is also pointed out that this waste can be used to produce chemicals such as limonene, pectin, sugars, succinic acid, biodegradable polymers and biofuels. Either enzymes or microbes can catalyse these conversions, or chemo-enzymatic routes are available.[206] Some other possibilities in this regard have been discussed by Pleissner and Lin.[207]

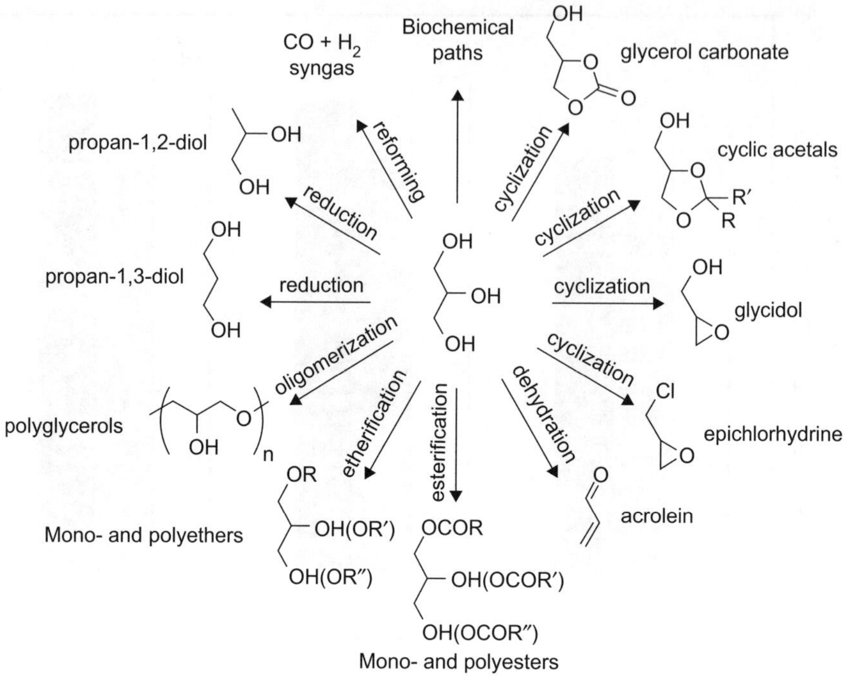

Figure 15.5 Most recent key continuous flow systems for glycerol valorization to valuable products using different types of catalysts and processes [reproduced from ref. 203 with permission from Chemistry Central].

15.9 CONCLUSIONS

It will be naïve to think that one can always unambiguously make a choice about what is a more sustainable option. In recent years, the debate on this with respect to both biodiesel and bioethanol is well documented.[208] So is the choice of jatropha oil as a starting material for biodiesel by some countries including India.[200] A less well-known older story is about the choice between paper and plastic as a material for bags and packaging materials.[209] It is not generally appreciated that treated paper can have a very low recycling value and its manufacturing can use up as much fossil fuel as a polymeric container. LCA studies are complicated and at times can make choices difficult. Implementation of environmental management systems (EMS) such as ISO 14001, the European Union's REACH (Registration, Evaluation and Authorization of CHemicals), regulation and the adoption of eco-labels are various initiatives noted by Lancaster[10] to help in making cleaner choices in terms of sustainable process/products.

Market forces can sometimes suddenly shift the strategic value of an approach. Making cocoa butter substitutes from cheap oils with lipase-catalysed reactions has been around for a while as a valuable biocatalytic approach.[210] Occasional swings in the relative market prices of cocoa butter and 'cheaper' oils have caught scientists in a despairing situation. One of us (M.N.G.) received the comment 'who says ionic liquids are green solvents' (on a manuscript submitted to RSC journal, *Green Chemistry*). Another leading journal had this to convey as a part of peer review of another manuscript: 'enzyme catalysed processes are not always green'. So, sometime greenness of a process lies in the eyes of the wise peer reviewer! Thus, a more robust and objective peer review would definitely help in catalysing the growth of this area.

What we need is the increasing use of metrics so that the discourse becomes more objective. Oxenbøll and Ernst[211] writing in 2008 provided an environmental assessment of the use of enzymes in the food industry with respect to a few specific examples. Increasingly, in the future we will have more definite quantitative data. By and large, it is believed that use of bicatalysis has huge potential in making chemical processes more sustainable. However, as Leresche and Meyer[212] pointed out, the issue of cost competitiveness in many cases (such as pharma industry) remains.

ACKNOWLEDGEMENTS

We thank the Department of Science and Technology, Government of India (Grant No. SR/SO/BB-68/2010) for financial support. J.M. thanks the Council of Scientific and Industrial Research, New Delhi, for the award of a Senior Research Fellowship.

REFERENCES

1. 'The Colours of Biotechnology', http://www.journals.elsevier.com/enzyme-and-microbial-technology/news/the-colours-of-biotechnology [accessed 17 February 2014].
2. M. N. Gupta and S. Raghava, *Chem. Cent. J.*, 2007, **1**, 17.
3. R. Ulber and D. Sell (ed.), *White Biotechnology*, Springer-Verlag, Berlin, 2007.
4. K. E. Jaeger, *Curr. Opin. Biotechnol.*, 2004, **15**, 269.
5. EuropaBio, *White Biotechnology – Gateway to a More Sustainable Future*, EuropaBio, Brussels, 2003, http://www.europabio.org/

white-biotechnology-gateway-more-sustainable-future [accessed 17 February 2014].

6. J. F. Jenck, F. Agterberg and M. J. Droescher, *Green Chem.*, 2004, **6**, 544.
7. I. Arends, R. Sheldon and U. Hanefeld, *Green Chemistry and Catalysis*, Wiley-VCH Verlag, Weinheim, Germany, 2007.
8. R. A. Sheldon, in *Biocatalysis for Green Chemistry and Chemical Process Development*, ed. J. Tao and R. J. Kazlauskas, Wiley-VCH, Weinheim, Germany, 2011, pp. 67–88.
9. R. A. Sheldon, *Chem. Ind. (London)*, 1992, 903.
10. M. Lancaster, *Green Chemistry: An Introductory Text*, RSC Publishing, Cambridge, 2nd edn, 2010.
11. J. J. Klemeš, *Curr. Opin. Chem. Eng.*, 2012, **1**, 238.
12. J. H Clark, T. J. Farmer, D. J. Macquarrie and J. Sherwood, *Sustainable Chem. Process.*, 2013, **1**, 23.
13. N. Winterton, *Curr. Opin. Chem. Eng.*, 2012, **1**, 225.
14. A. I. Stankiewicz and J. A. Moulijn, *Chem. Eng. Prog.*, 2000, 22.
15. K. L. Mulholland, R. W. Sylvester and J. A. Dyer, *Environ. Prog.*, 2000, **19**, 260.
16. J. Garcia-Serna, L. Perez-Barrigon and M. J. Cocero, *Chem. Eng. J.*, 2007, **133**, 7.
17. C. W. Wu, J. G. Lee and W. C. Lee, *Biotechnol. Appl. Biochem.*, 1998, **27**, 225.
18. G. Yalcin, B. Elmas, M. Tuncel and A. Tuncel, *J. Appl. Polym. Sci.*, 2006, **101**, 818.
19. M. Sardar and M. N. Gupta, *Enzyme Microb. Technol.*, 2005, **37**, 355.
20. S. Dalal and M. N. Gupta, *Chemosphere*, 2007, **67**, 741.
21. M. N. Gupta, in *Methods in Non-aqueous Enzymology*, ed. M. N. Gupta, Birkhauser Verlag, Basel, 2000, pp. 1–13.
22. M. Basri, K. Ampon, C. N. Razak and A. B. Salleh, in *Enzymes in Nonaqueous Solvents: Methods and Protocols*, ed. E. N. Vulfson, P. J. Halling and H. L. Holland, Humana Press, Totowa, NJ, 2001, pp. 65-76.
23. M. Kapoor, A. B. Majumder, J. Mukherjee and M. N. Gupta, *Biocatal. Biotransform.*, 2012, **30**, 399.
24. A. B. Majumder and M. N. Gupta, *Synth. Commun.*, 2014, **44**, 818.
25. A. N. Semenov, Y. L. Khmelnitsky, I. V. Berezin and K. Martinek, *Biocatalysis*, 1987, **1**, 3.
26. L. E. S. Brink, J. Tramper, K. C. A. M. Luyben and K. Van't Riet, *Enzyme Microb. Technol.*, 1988, **10**, 736.
27. P. J. Halling, *Biotechnol. Adv.*, 1987, **5**, 47.

28. K. Martinek, A. V. Levashov, N. Klyachko, Y. L. Khmelnitsky and I. V. Bcrezin, *Eur. J. Biochem.*, 1986, **155**, 453.
29. K. Martinek, I. V. Berezin, Y. L Khmelnitsky, N. L. Klyachko and A. V. Levashov, *Biocatalysis*, 1987, **1**, 9.
30. S. V. Kamat, E. J. Beckman and A. J. Russel, *Crit. Rev. Biotechnol.*, 1995, **15**, 41.
31. S. Park and R. J. Kazalauskas, *J. Org. Chem.*, 2001, **66**, 8395.
32. S. Shah and M. N. Gupta, *Bioorg. Med. Chem. Lett.*, 2007, **17**, 921.
33. H. Zhao, *J. Chem. Technol. Biotechnol.*, 2010, **85**, 891.
34. M. Naushad, Z. A. Alothman, A. B. Khan and M. Ali, *Int. J. Biol. Macromol.*, 2012, **51**, 555.
35. C. Laane, S. Boeren, V. Kees and C. Veeger, *Biotechnol. Bioeng.*, 1987, **30**, 81.
36. P. Adlercreutz and B. Mattiasson, *Biocatalysis*, 1987, **I**, 99.
37. M. N. Gupta, R. Batra, R. Tyagi and A. Sharma, *Biotechnol. Prog.*, 1997, **13**, 284.
38. P. Halling, in *Enzyme Catalysis in Organic Synthesis: A Comprehensive Handbook*, ed. K. Drauz and H. Waldman, Wiley-VCH, Weinheim, Germany, 2002, pp. 259–285.
39. P. Halling, *Enzyme Microb. Technol.*, 1994, **16**, 178.
40. G. Carrea, G. Ottolina and S. Riva, *Trends Biotechnol.*, 1995, **13**, 63.
41. G. Ljungar, P. Adlecreutz and B. Mattiasson, *Biocatalysis*, 1993, **7**, 279.
42. Y. Inada, M. Furukawa, H. Sasaki, Y. Kodera, M. Hiroto, H. Nishimura and A. Matsushima, *Trends Biotechnol.*, 1995, **13**, 86.
43. Y. Ito, N. Sugimura, O. H. Kown and Y. Imanishi, *Nat. Biotechnol.*, 1996, **31**, 2602.
44. P. Adlercreutz, *Chem. Soc. Rev.*, 2013, **42**, 6406.
45. Y. L. Khmelnitsky, V. V. Mozhaev, A. B. Belova, M. V. Sergeeva and K. Martinek, *Eur. J. Biochem.*, 1991, **198**, 31.
46. B. Mattiasson and O. Holst, *Extractive Bioconversions*, Marcel Dekker, New York, 1991.
47. S. Sharma, S. Teotia and M. N. Gupta, *Enzym. Microb. Technol.*, 2003, **32**, 337.
48. B. Orlich and R. Schomäcker, *Adv. Biochem. Eng. Biotechnol.*, 2002, **75**, 185.
49. P. Lozano, T. de Diego, D. Carrié, M. Vaultier and J. L. Iborra, *Chem. Commun.*, 2002, 692.
50. S. Cantone, U. Hanefeldb and A. Basso, *Green Chem.*, 2007, **9**, 954.
51. A. M. da Costa Lopes, K. G. João, A. R. C Morais, E. Bogel-Łukasik and R. Bogel-Łukasik, *Sustainable Chem. Process.*, 2013, **1**, 3.

52. A. F. Hsu, K. Jones, T. A. Foglia and W. N. Marmer, *Biotechnol. Appl. Biochem.*, 2002, **36**, 181.
53. S. Shah and M. N. Gupta, *Process Biochem.*, 2007, **42**, 409.
54. C. Laane, *Biocatalysis*, 1987, **1**, 17.
55. C. Laane, S. Boeren, V. Kees and C. Veeger, *Biotechnol. Bioeng.*, 1987, **30**, 81.
56. M. N. Gupta, *Eur. J. Biochem.*, 1992, **203**, 25.
57. Aldrich Chemistry, *Green Chemistry: Supporting the Advancement of Chemistry Through Sound Environmental, Social and Fiscal Responsibilities*, Sigma Aldrich, St Louis, MO, 2007.
58. J. Tao and R. J. Kazlauskas (ed.), *Biocatalysis for Green Chemistry and Chemical Process Development*, Wiley VCH, Weinheim, Germany, 2011.
59. R. V. Ulijn, A. E. M. Janssen, B. D. Moore and P. J. Halling, *Chem. Eur. J.*, 2001, **7**, 2089.
60. R. V. Ulijn and P. J. Halling, *Green. Chem.*, 2004, **6**, 488.
61. W. Tischer and V. Kasche, *Tibtech*, 1999, **17**, 326.
62. S. M. Minteer (ed.), *Methods in Molecular Biology. Enzyme Stabilization and Immobilization: Methods and Protocols*, Humana Press, Totowa, NJ, 2011.
63. S. Birnbaum, R. Pendleton, P. O. Larsson and K. Mosbach, *Biotechnol. Lett.*, 1981, **3**, 393.
64. H. Y. Wang and D. J. Hettwer, *Biotechnol. Bioeng.*, 1982, **24**, 1827.
65. F. Godia, C. Casas, B. Castellano and C. Sold, *Appl. Microbiol. Biotechnol.*, 1987, **26**, 342.
66. I. Roy and M. N. Gupta, *Chem. Biol.*, 2003, **10**, 1161.
67. I. Roy and M. N. Gupta, in *Methods in Biotechnology: Immobilization of Enzymes and Cells*, ed. J. M. Guisan, Humana Press, Totowa, NJ, 2nd edn, 2006, ch. 8, pp. 87–95.
68. S. Shah, S. Sharma and M. N. Gupta, *Ind. J. Biochem. Biophys.*, 2003, **40**, 392.
69. V. Singh, K. Solanki and M. N. Gupta, *Recent Pat. Biotechnol.*, 2008, **2**, 130.
70. M. S. Antczak, A. Kubiak, T. Antczak and S. Bielecki, *Renewable Energy*, 2009, **34**, 1185.
71. S. J. Prestrelski, N. Tedeschi, T. Arakawa and J. F. Carpenter, *Biophys. J.*, 1993, **65**, 661.
72. I. Roy and M. N. Gupta, *Biotechnol. Appl. Biochem.*, 2004, **39**, 165.
73. I. Roy, A. Sharma and M. N. Gupta, *Bioorg. Med. Chem. Lett.*, 2004, **14**, 887.
74. I. Roy and M. N. Gupta, *Bioorg. Med. Chem. Lett.*, 2004, **14**, 2191.

75. S. Shah and M. N. Gupta, *Biochim. Biophys. Acta*, 2007, **1770**, 94.
76. A. B. Majumder and M. N. Gupta, *Biocatal. Biotrasform.*, 2011, **29**, 238.
77. J. Partridge, P. J. Halling and B. D. Moore, *Chem. Commun.*, 1998, 7, 841.
78. A. B. Majumder, S. Shah and M. N. Gupta, *Chem. Cent. J.*, 2007, **1**, 10.
79. R. Schoevaart, M. W. Wolbers, M. Golubovic, M. Ottens, A. P. G. Kieboom, F. van Rantwijk, L. A. M. van der Wielen and R. A. Sheldon, *Biotechnol. Bioeng.*, 2004, **87**, 754.
80. S. Shah, A. Sharma and M. N. Gupta, *Anal. Biochem.*, 2006, **351**, 207.
81. M. N. Gupta and S. Raghava, in *Methods in Molecular Biology. Enzyme Stabilization: Methods and Protocols*, ed. S. D. Minteer, Humana Press, Totowa, NJ, 2010, pp. 133–145.
82. M. Kreiner, B. D. Moore and M. C. Parkar, *Chem. Commun.*, 2001, **12**, 1096.
83. S. Shah, A. Sharma, D. Varadani, B. Mehta and M. N. Gupta, *J. Nanosci. Nanotechnol.*, 2007, **7**, 2157.
84. V. Kumari, S. Shah and M. N. Gupta, *Energy Fuels*, 2007, **21**, 368.
85. S. Shah, A. Sharma and M. N. Gupta, *Biocatal. Biotransform.*, 2008, **26**, 266.
86. K. Solanki and M. N. Gupta, *Bioorg. Med. Chem. Lett.*, 2011, **21**, 2934.
87. M. Kreiner and M. C. Parker, *Biotechnol. Bioeng.*, 2004, **87**, 24.
88. S. Shah and M. N. Gupta, *Biochim. Biophys. Acta*, 2008, **1784**, 363.
89. M. N. Gupta, M. Kaloti, M. Kapoor and K. Solanki, *Artif. Cell Blood Sub. Biotechnol.*, 2010, **39**, 98.
90. K. Solanki and M. N. Gupta, *New J Chem.*, 2011, **35**, 2551.
91. J. Mukherjee and M. N. Gupta, *Chem. Cent. J.*, 2012, **6**, 133.
92. Mallikarjuna, N. Nadagouda and Rajender S. Varma, *Cryst. Growth Des.*, 2008, **8**, 291.
93. V. Polshettiwar and R. S. Varma (ed.), *Aqueous Microwave Assisted Chemistry Synthesis and Catalysis*, Royal Society of Chemistry, Cambridge, UK, 2010.
94. J. R. Carrillo-Munoz, D. Bouvet, E. Guibe-Jampel, A. Loupy and A. Petit, *J. Org. Chem.*, 1996, **61**, 7746.
95. I. Roy and M .N. Gupta, *Curr. Sci.*, 2003, **85**, 1685.
96. C. O. Kappe, D. Dallinger and S. S. Murphee, *Practical Microwave Synthesis for Organic Chemists: Strategies, Instruments and Protocols*, Wiley-VCH, Weinheim, Germany, 2009.
97. I. Roy and M. N. Gupta, *Tetrahedron*, 2003, **59**, 5431.

98. N. E. Leadbeater, L. M. Stencel and E. C. Wood, *Org. Biomol. Chem.*, 2007, **5**, 1052.

99. V. Singh, K. Solanki and M. N. Gupta, *Proc. Ind. Natl. Sci. Acad.*, 2012, **78**, 629.

100. D. Yu, L. Tian, D. Ma, H. Wu, Z. Wang, L. Wang and X. Fang, *Green Chem.*, 2010, **12**, 844.

101. G. D. Yadav and P. S. Lathi, *J. Mol. Cat. B: Enzym.*, 2004, **223**, 51.

102. D. Yu, P. Chen, L. Wang, Q. Gu, Y. Li, Z. Wang and S. Cao, *Process Biochem.*, 2007, **42**, 1312.

103. R. O. M. A. de Souza, O. A. C Antunes, W. kroutil and C. O. Kappe, *J. Org. Chem.*, 2009, **74**, 6157.

104. P. Jain, S. Jain and M. N. Gupta, *Anal. Bioanal. Chem.*, 2005, **381**, 1480.

105. B. R'ejasse, T. Bessonen, M. D. Legoy and S. Lamare, *Org. Biomol. Chem.*, 2006, **4**, 3703.

106. G. D. Yadav and I. V. Borkar, *Ind. Eng. Chem. Res.*, 2009, **48**, 7915.

107. Y. Fang, W. Huang and Y. M. Xia, *Process Biochem.*, 2008, **43**, 306.

108. W. Huang, Y. M. Xia, H. Gao, Y. J. Fang, Y. Wang and Y. Fang, *J. Mol. Cat. B: Enzym.*, 2005, **35**, 113.

109. W. Y. Chen and Y. C. Chen, *Anal Chem.*, 2007, **79**, 2394.

110. F. J. Izquierdo, I. Alli, V. Yaylayan and R. Gomez, *Int. Dairy J.*, 2007, **17**, 465.

111. B. N. Pramanik, U. A. Mirza, Y. H. Ing, Y .H. Liu, P. L. Bartner, P. C. Weber and A. K. Bose, *Protein Sci.*, 2002, **11**, 2676.

112. I. Roy, K. Mondal and M. N. Gupta, *Biotechnol Prog.*, 2003, **19**, 1648.

113. R. Miethchen, *Ultrasonics*, 1992, **30**, 173.

114. J. Bang and K. Suslick, *Adv. Mater.*, 2010, **22**, 1039.

115. J. P. Cheng, R. Ma, D. Shi, F. Liu and X. B. Zhang, *Ultrason. Sonochem.*, 2011, **18**, 1038.

116. D. Malhotra, J. Mukherjee and M. N. Gupta, *RSC Adv.*, 2013, **3**, 14322.

117. J. V. Sinisterra, *Ultrasonics*, 1992, **30**, 180.

118. M. N. Gupta and I. Roy, *Eur. J. Biochem.*, 2004, **271**, 2575.

119. Y. Liu, Q. Jin, L. Shan, Y. Liu, W. Shen and X. Wang, *Ultrason. Sonochem.*, 2008, **15**, 402.

120. N. Sener, D. K. Apar and B. Ozbek, *Process Biochem.*, 2006, **41**, 1493.

121. C. Schiraldi and M. de Rosa, *Trends Biotechnol.*, 2002, **20**, 515.

122. B. van den Burg, *Curr. Opin. Microbiol.*, 2003, **6**, 213.

123. T. de Miguel Bouzas, J. Barros-Velázquez and T. G. Villa, *Protein Pept. Lett.*, 2006, **13**, 645.

124. P. Turner, G. Mamo and E. N. Karlsson, *Microb. Cell Fact.*, 2007, **6**, 9.
125. I. S. Bentley and E. C. William, in *Industrial Enzymology*, ed. T. Godfrey and S. West, Macmillan Press, London, 2nd edn, 1996, pp. 339–357.
126. O. Kirk, T. V. Borchert and C. C. Fuglsang, *Curr. Opin. Biotechnol.*, 2002, **13**, 345.
127. G. Walsh and D. R. Headon, in *Protein Biotechnology*, John Wiley and Sons Ltd, Chichester, UK, 1994, ch. 9, pp. 337–364.
128. R. Y. Morita, *Bacteriol. Rev.*, 1975, **39**, 144.
129. R. Cavicchioli, K. S. Siddiqui, D. Andrews and K. R. Sowers, *Curr. Opin. Biotechnol.*, 2002, **13**, 253.
130. C. Gerday, M. Aittaleb, J. L. Arpigny, E. Baise, J. P. Chessa, G. Garsoux, I. Petrescu and G. Feller, *Biochim. Biophys. Acta.*, 1997, **1342**, 119.
131. G. M. Whitesides and C. H. Wong, *Aldrichim. Acta*, 1983, **16**, 27.
132. A. Aklyama, M. Bednarski, M. J. Kim, E. S. Simon, H. Waldmann and G. M. Whitesides, *Chemtech*, 1988, 627.
133. J. Barthelmes, C. Ebeling, A. Chang, I. S. Chomburg and D. Schomburg, presented at ESCEC, 19–23 March 2006, Rúdesheim am Rhein, Germany.
134. C. Sohngen, A. Chang and D. Schomburg, *BMC Bioinformatics*, 2011, **12**, 329.
135. D. Malhotra, J. Mukherjee and M. N. Gupta, *Proc. Ind. Natl. Sci. Acad.*, 2013, **79**, 291.
136. O. Khersonsky and D. S. Tawfik, *Annu. Rev. Biochem.*, 2010, **79**, 471.
137. T. Godfrey and S. West (ed.), *Industrial Enzymology*, Macmillan Press, London, 2nd edn, 1996.
138. S. West, in *Industrial Enzymology*, ed. T. Godfrey and S. West, Macmillan Press, London, 2nd edn, 1996, pp. 155–175.
139. P. S. J. Cheetham, in *Handbook of Enzyme Biotechnology*, ed. A. Wiseman, Ellis Horwood, Hertfordshire, 1995, pp. 419–552.
140. A. J. J. Straathof and P. Adlercreutz (ed.), *Applied Biocatalysis*, Harwood Academic Publishers, Amsterdam, 2nd edn, 2000.
141. G. Walsh and D. R. Headon, *Protein Biotechnology*, John Wiley and Sons, Chichester, UK, 1994.
142. Y. H. Hui (ed.), *Bailey's Industrial Oil and Fat Products: Edible Oil and Fat Products: Processing Technology*, John Wiley & Sons, New York, 5th edn, 1996, vol. 4.

143. E. W. Lucas, L. R. Watkins, S. S. Koseoglu, K. C. Ree, E. Hernandez, M. N. Riaz, W. H. Johnson Jr. and S. C. Doty, *Informatics*, 1997, **8**, 291.

144. A. Rosenthal, D. L. Pyle and K. Niranjan, *Enzyme Microb. Technol.*, 1996, **19**, 402.

145. K. A. Campbell and C. E. Glatz, *J. Agric. Food Chem.*, 2009, **57**, 10904.

146. H. Dominguez, M. J. Nunez and J. M. Lema, *Food Chem.*, 1994, **49**, 271.

147. A. Rosenthal, D. L. Pyle, K. Niranjan, S. Gilmour and L. Trinca, *Enzyme Microb. Technol.*, 2001, **28**, 499.

148. A. Sharma, S. K. Khare and M. N. Gupta, *J. Am. Oil. Chem. Soc.*, 2002, **79**, 215.

149. S. West, in *Industrial Enzymology*, ed. T. Godfrey and S. West, Macmillan Press, London, 2nd edn, 1996, pp. 293–300.

150. A. Sharma, S. K. Khare and M. N. Gupta, *Bioresour. Technol.*, 2002, **85**, 327.

151. S. Shah, A. Sharma and M. N. Gupta, *Ind. Crops Prod.*, 2004, **20**, 275.

152. R. Gaur, A. Sharma, S. K. Khare and M. N. Gupta, *Bioresour. Technol.*, 2007, **98**, 696.

153. N. N. Kurmudle, S. B. Bankar, I. B. Bajaj, M. V. Bule and R. S. Singhal, *Process Biochem.*, 2011, **46**, 423.

154. S. Shah, A. Sharma and M. N. Gupta, *Bioresour. Technol.*, 2005, **96**, 121.

155. A. Sharma and M. N. Gupta, *Ultrason. Sonochem.*, 2006, **13**, 529.

156. J. Li, Y. G. Zu, M. Luo, C. B. Gu, C. J. Zhao, T. Efferth and Y. J. Fu, *Food Chem.*, 2013, **138**, 2152.

157. J. Jiao, Z. G. Li, Q. Y. Gai, X. J. Li, F. Y. Wei, Y. J. Fu and W. Ma, *Food Chem.*, 2014, **147**, 17.

158. A. S. Hodgson, in *Bailey's Industrial Oil and Fat Products: Edible Oil and Fat Products: Processing Technology*, ed. Y. H. Hui, John Wiley & Sons, New York, 5th edn, 1996, vol. 4, pp. 172–187.

159. S. West, in *Industrial Enzymology*, ed. T. Godfrey and S. West, Macmillan Press, London, 2nd edn, 1996, pp. 295–299.

160. B. Yang, Y. H. Wang and J. G. Yang, *J. Amer. Oil Chem. Soc.*, 2006, **83**, 653.

161. Purifine PLC; http://www.verenium.com/ver_packet/purifine_productinfosheet.pdf [accessed September 2014].

162. E. Aalrust, W. Beyer, H. Ottofrickenstein, G. Penk, H. Plainer and R. Reiner, *US Pat.*, 5264367 A, 1993.

163. C. L.G. Dayton and F. Galhardo, *US Pat.*, 20080182322 A1, 2008.
164. H. E. Schoemaker, D. Mink and M. G. Wubbolts, *Science*, 2003, **299**, 1694.
165. D. Enders and K. E. Jaeger (ed.), *Asymmetric Synthesis with Chemical and Biological Methods*, Wiley-VCH, Weinheim, Germany, 2007.
166. A. Schmid, J. S. Dordick, B. Hauer, A. Kiener, M. Wubbolts and B. Witholt, *Nature*, 2001, **409**, 258.
167. J. Tao and J. H. Xu, *Curr. Opin. Chem. Biol.*, 2009, **13**, 43.
168. B. N. Nestl, B. A. Nebel and B. Hauer, *Curr. Opin. Chem. Biol.*, 2011, **15**, 187.
169. R. Wohlgemuth, *Curr. Opin. Biotechnol.*, 2010, **21**, 713.
170. J. Ogawa and S. Shimizu, *Curr. Opin. Biotechnol.*, 2002, **13**, 367.
171. P. M. Grande, C. Bergs and P. Dominguez de Maria, *ChemSusChem*, 2012, **5**, 1203.
172. A. Dibenedetto, P. Stufano, W. Macyk, T. Baran, C. Fragale, M. Costa and M. Aresta, *ChemSusChem*, 2012, **5**, 373.
173. A. But, J. LeNôtre, E. L. Scott, R. Wever and J. P. M. Sanders, *ChemSusChem*, 2012, **5**, 1199.
174. K. Zhang, A. P. Woodruff, M. Xiong, J. Zhou and Y. K. Dhande, *ChemSusChem*, 2011, **4**, 1068.
175. Y. Yanagishita, M. Kato, K. Toshima and S. Matsumura, *ChemSusChem*, 2008, **1**, 133.
176. K. Hult and P. Berglund, *Trends Biotechnol.*, 2007, **25**, 231.
177. E. Busto, V. Gotor-Fernandez and V. Gotor, *Chem. Soc. Rev.*, 2010, **39**, 4504.
178. M. N. Gupta, M. Kapoor, A. Majumder and V. Singh, *Curr. Sci.*, 2011, **100**, 1152.
179. M. N. Gupta and J. Mukherjee, *Curr. Sci.*, 2013, **104**, 1178.
180. T. Banner, A. Fosmer, H. Jessen, E. Marasco, B. Rush and J. Veldhouse, in *Biocatalysis for Green Chemistry and Chemical Process Development*, ed. J. Tao and R. Kaslauskas, Wiley and Sons, Hoboken, NJ, 2011, pp. 429–468.
181. M. N. V. Ravi Kumar, *React. Funct. Polym.*, 2000, **46**, 1.
182. M. Rinaudo, *Prog. Polym. Sci.*, 2006, **31**, 603.
183. C. Laroche and P. Michaud, *Recent Pat. Biotechnol.*, 2007, **1**, 59.
184. H. Vaghari, H. Jafarizadeh-Malmiri, A. Berenjian and N. Anarjan, *Sustainable Chem. Process.*, 2013, **1**, 16.
185. J. Chen, X. D. Zhang and Z. Jiang, *Anticancer Agents Med Chem.*, 2013, **13**, 725.

186. A. R. C. Morais and R. Bogel-Lukasik, *Sustainable Chem. Process.*, 2013, **1**, 18.
187. G. Frazzetto, *EMBO Rep.*, 2003, **9**, 835.
188. J. A. Glaser, *Clean Techn. Environ. Policy*, 2005, **7**, 233.
189. P. Lorenz and H. Zinke, *Trends Biotechnol.*, 2005, **23**, 570.
190. M. W. Bevan and M. C. R. Franssen, *Nat. Biotechnol.*, 2006, **24**, 765.
191. V. Gekas and M. Lopez-Leiva, *Process. Biochem.*, 1985, **20**, 1.
192. R. E. Huber, M. N. Gupta and S. K Khare, *Int. J. Biochem.*, 1994, **26**, 309.
193. T. Finocchairo, N. F. Olson and T. Richardson, in *Advances in Biochemical Engineering*, ed. A. Fiether, Springer-Verlag, New York, 1980, vol. 15, pp. 71–88.
194. N. Kosaric and Y. J. Asher, in *Advances in Biochemical Engineering. Agricultural Feedstock and Waste Treatment and Engineering*, ed. A. Fiether, Springer-Verlag, New York, 1985, vol. 32, pp. 25–80.
195. F. Ma and M. A. Hanna, *Bioresour. Technol.*, 1999, **70**, 1.
196. M. Balat and H. Balat, *Energy Convers. Manage.*, 2008, **49**, 2727.
197. T. Tan, J. Lu, K. Nie, L. Deng and F. Wang, *Biotechnol. Adv.*, 2010, **28**, 628.
198. A. Banerjee, V. Singh, K. Solanki, J. Mukherjee and M. N. Gupta, *Sustainable Chem. Process.*, 2013, **1**, 14.
199. M. N. Gupta, J. Mukherjee and D. Malhotra, *Universal Org. Chem.*, 2013, **1**, 1.
200. P. Kant and S. Wu, *Environ. Sci. Technol.*, 2011, **45**, 7114.
201. A. Demirbas, *Energy Convers. Manage.*, 2009, **50**, 923.
202. M. Pagliaro and M. Rossi, *Future of Glycerol. New Usages for a Versatile Raw Material*, Royal Society of Chemistry, Cambridge, UK., 2008.
203. C. Len and R. Luque, *Sustainable Chem. Process.*, 2014, **2**, 1.
204. A. Wolfson, C. Dlugy, D. Tavor, J. Blumenfeld and Y. Shotland, *Tetrahedron: Asymmetry*, 2006, **17**, 2043.
205. A. Wolfson, N. Haddad, C. Dlugy, D. Tavor and Y. Shotland, *Org. Commun.*, 2008, **1**, 9.
206. R. Luque and J. H. Clark, *Sustainable Chem. Process.*, 2013, **1**, 10.
207. D. Pleissner and C. S. K. Lin, *Sustainable Chem. Process.*, 2013, **1**, 21.
208. J. Tollefson, *Nature*, 2008, **45**, 36.

209. Committee on Polymer Science and Engineering, *Polymer Science and Engineering: The Shifting Research Frontiers*, National Academy Press, Washington, D. C., 1994, p. 25.
210. S. Sellapan and C. C. Akoh, in *Handbook of Industrial Biocatalysis*, ed. C. T. Hou, Taylor and Francis, Boca Raton, FL, 2005, ch. 9, pp. 9.1–9.38.
211. K. Oxenbøll and S. Ernst, *Food Sci. Technol.*, 2008, **22**, 35.
212. J. E. Leresche and H. P. Meyer, *Org. Process Res. Dev.*, 2006, **10**, 572.

EXTRACTIONS AND PREPARATIONS

CHAPTER 16
Biofuels from Microalgae

CHRISTOPHER J. CHUCK,* JONATHAN L. WAGNER AND
RHODRI W. JENKINS

Department of Chemical Engineering, University of Bath, Claverton Down,
Bath, BA2 7AY, UK
*Email: c.chuck@bath.ac.uk

16.1 INTRODUCTION

The transport sector is responsible for over 25% of all anthropogenic
greenhouse gas (GHG) emissions. To reduce the impact of this sector
without necessitating major infrastructure reforms, renewable liquid
transport fuels must be developed. First-generation biofuels, such as
biodiesel from crop oils and ethanol from sugar cane, compete with
food production while only satisfying a small fraction of fuel demand.
A promising alternative to these sources are fuels produced from
microalgae. Microalgae can grow far more quickly and efficiently than
crop biomass, and when coupled to waste streams such as water
treatment, offer a very credible source of biomass for advanced fuel
production.

While there are currently no large-scale production facilities producing algal fuels, the culturing of algae on this scale has been
thoroughly reviewed.[1] Microalgae can provide energy in a number of
ways. For example, dried algal biomass can be: co-fired with coal for
power generation; converted into gas and used for methane production; or used to produce hydrogen.[2] However, the biological

Chemical Processes for a Sustainable Future
Edited by Trevor M. Letcher, Janet L. Scott and Darrell A. Patterson

composition of algae allows for specific liquid fuel applications to be fulfilled and offers a potential source of replacement for diesel, gasoline or aviation kerosene.

16.2   LIPID DERIVED ALGAL FUELS

The vast majority of research on the production of fuels from algae has focused on developing algal biodiesel. Some algal species can accumulate lipid material in excess of 60% by dry weight. These glycerides can then be converted into fatty acid alkyl esters (biodiesel) by a base or acid catalysed transesterification.[3] Biodiesel has similar properties to mineral diesel and can be used in compression ignition (CI) engines with minimum modification. Biodiesel, sourced from terrestrial crops, is currently blended with all diesel used in the European Union up to a level of 7%, and in the USA of up to 5%. The vast majority of biodiesel is produced from rapeseed oil, soybean oil and palm oil, though these land-based crops have a much lower lipid yield per unit area than could potentially be achieved by algae. For example, if an alga containing 30 wt-% lipid was to be cultivated for fuel, it could potentially produce near $90\,000$ L ha^{-1} a^{-1}, in comparison to rapeseed (1190 L ha^{-1} a^{-1}) and soybean (446 L ha^{-1} a^{-1}).[4,5]

The properties of biodiesel, and therefore its suitability as a fuel, are determined by the fatty acid, methyl ester (FAME) profile. Generally long chain saturates have high viscosities and melting points, together with high cetane numbers. Mono-unsaturates have viscosities more similar to diesel, melting points below 0 °C, and lower cetane numbers, whereas polyunsaturated esters have extremely low viscosities, melting points significantly lower than conventional diesel and relatively low cetane numbers, and are extremely prone to oxidation.[6,7] While long chains are the most prevalent (C_{16} +), shorter chain esters tend to have lower viscosities, lower melting points and lower cetane numbers than their longer chain counterparts. However, the majority of fatty acids produced by terrestrial crops have carbon numbers ranging from C_{16} to C_{20}, with palmitic acid (16:0), stearic acid (18:0), oleic acid (18:1), linoleic acid (18:2), linolenic acid (18:3) and erucic acid (20:0) being the most common.

In comparison, microalgal lipids are far more variable and are dependent not only on the species of microalgae but also on the culture conditions. For example, some strains of *Chlorella vulgaris* can produce C_6 chains, while long chains of up to C_{24} are observed in the halophyte *Isochrysis galbana*.[8] Highly unsaturated esters are also commonly observed, for example, some strains of *Chlorella salina* produce C_{22} esters with up to six double bonds (Table 16.1).

There are many physical challenges to the adaptation of algal biodiesel. Due to the variable structure and unpredictable physical properties, there are technical and legislative hurdles in using higher blend levels without further large-scale testing. One of the key factors in limiting the current blend level is the reduced oxidative stability of biodiesel, especially of the polyunsaturated esters. Biodiesel oxidation leads to gum formation and blocked filters, and can cause substantial damage to an engine over a prolonged period of time.[15] In addition, biodiesel has a lower energy density than mineral diesel, as it contains approximately 10–12% oxygen. Although the oxidative stability can be increased by using more saturated esters, this will also have a major effect on the low temperature properties of the biodiesel, limiting the maximum blend levels, especially in winter.

16.2.1 Hydroprocessing of Algal Lipids

As an alternative to the production of biodiesel through transesterification, it is possible to hydroprocess the lipids to form alkanes which can be subsequently isomerized or cracked into suitable hydrocarbon fuels. Although these methods use higher temperatures than transesterification, the resulting fuels are more compatible with fossil fuels as they have a very similar composition and can therefore deliver higher performance and be used at higher blend levels.[16]

The majority of commercial processes are carried out in two separate reaction steps. In the first step, the triglycerides are deoxygenated under hydro-treating conditions, before they enter an isomerisation or cracking unit to improve the cold flow properties of the final fuel. The product properties can be fine-tuned by varying the conditions of the second stage so that the specifications for the target fuel are met.[17] Harsher operating conditions, which result in improved cold flow properties, must however be balanced against a reduction in yields, as a result of increased cracking leading to the formation of low molecular compounds unsuitable for fuel use. However, the hydro-treating catalysts are inhibited or poisoned by contaminants such as alkali metals and phosphorous compounds as well as impurities such as waxes, sterols, tocopherols and carotenoids which are often found in natural oils and fats, and additional pretreatment of the feedstock may be required. This is particularly problematic for lipids derived from microbial sources.

Table 16.1 Lipid profile of a range of oleaginous strains of microalgae, demonstrating the wide variation across different species.

	Fatty acid/(% of total)																		
	14:0	15:0	16:0	16:1	16:2	16:3	16:4	18:0	18:1	18:2	18:3	18:4	20:2	20:3	20:4	20:5	22:5	22:6	Ref.
Analipus japonicus	3.9	–	19.9	1.6	–	–	–	0.7	10.9	8.4	8.1	13.4	0.4	0.7	14.9	13.2	–	–	11
Chaetomorpha linum	9.7	–	23	1.3	–	–	13.8	–	9.8	28.6	1.5	0.6	1.3	–	1.2	3.2	–	–	11
Chlorella emersonii	0.6	0.1	18	1.4	2	4.8	0.5	4.3	38.5	6.9	23	–	–	–	–	–	–	–	12
Chlorella vulgaris	0.5	0.6	23.1	0.2	7.4	5.8	–	5.2	16.1	20.9	18	–	–	–	–	–	–	–	13
Coelastrella saipanensis	0.8	0.1	18.4	4.4	2.3	5.7	1	5.2	30	6.2	24.8	1	–	–	–	–	–	–	12
Cryptopleura violaceae	3.7	–	29.4	1.5	–	–	–	0.9	5.6	0.6	0.7	0.2	0.1	1.4	19.4	32.5	–	–	11
Dictyosphaerium pulchellum	2.5	–	11.4	1.9	4.3	0.6	–	0.6	1.4	7.5	30.6	2.3	–	–	–	–	–	–	10
Dunaliella primolecta	0.4	–	21.8	4.5	0.9	–	–	1.6	16.3	7	38.7	0.6	–	–	–	–	–	–	9
Emiliania huxleyi	18.8	–	10.3	–	–	–	–	10.8	42.2	–	–	8.7	–	–	–	–	–	9.2	9
Enteromorpha compressa	0.6	–	23.1	2.6	–	2.8	13.6	1.7	7.7	3.8	22.6	12.1	0.2	–	0.5	4.1	–	–	11
Gigartina harveyana	3	–	28.6	0.8	–	–	–	1.7	13.3	1.6	0.8	0.2	0.4	0.2	10.4	37.2	–	–	11
Hedophyllum sessile	5.8	–	20.1	1.2	–	–	–	2.5	40.7	6.5	3.3	3.6	0.4	0.5	9.7	0.1	–	–	11
Heterosigma akashiwo	6.6	–	40	12.7	4	–	–	–	–	4.5	6.7	5.2	–	–	3.5	14.8	–	–	9
Iridaea cordata	2.4	–	29.9	0.3	–	–	–	1.8	9.3	0.8	0.7	0.2	0.4	0.6	5.3	45.4	–	–	11
Laminaria dentigera	26	–	29.6	2.3	–	–	–	2.4	20	5.1	4.7	8.9	0.9	0.4	9.8	10.4	–	–	11
Mastigocladus laminosus	3.4	–	55.1	8.8	–	–	–	32.7	–	–	–	–	–	–	–	–	–	–	12
Microcoleus chthonoplastes	2.7	–	68.7	–	–	–	–	28.5	–	–	–	–	–	–	–	–	–	–	12
Nannochloris sp.	13.3	–	17.8	–	–	–	–	–	23.9	10.8	28.2	6.1	–	–	–	–	–	–	9
Nannochloropsis oculata	8.1	–	25.5	–	–	–	–	6.6	13.3	12.6	–	–	–	–	–	13.4	–	–	14
Nitzschia ovalis	3.2	–	18.8	28.2	–	4.4	–	0.3	0.7	0.2	0.4	0.1	–	–	2.6	23.7	0.4	4	10
Nostoc commune	–	–	25.3	24.1	–	–	–	–	–	12.5	38.1	–	–	–	–	–	–	–	9
Odonthalia floccosa	2.3	–	28.6	1.7	–	–	–	0.8	14.2	1.4	0.6	0.5	0.1	0.5	14.8	31.8	–	–	11
Parietochloris incisa	–	–	19.8	–	5.2	–	–	18.2	10.2	14.3	14.3	–	–	–	14	4.3	–	–	9
Pavlova lutheri	10.1	–	11.1	26.3	–	–	–	–	5.2	0.6	0.5	9.1	–	–	0.3	18	–	9.7	9
Phaeodactylum tricornutum	6.7	–	14.7	43.6	2	–	–	–	15.8	0.5	0.4	1.1	–	–	–	14.4	–	0.7	9
Plocamium violaceum	6.5	–	29.7	2.4	–	–	–	0.9	3.3	0.6	1	0.7	0.4	4.5	8.2	37.2	–	–	11
Prionitis linearis	3	–	26.7	0.6	–	–	–	0.9	9.2	0.7	0.8	0.2	1.3	1.5	23.4	27.8	–	–	11
Prionitis lanceolata	2.1	–	27.7	0.6	–	–	–	1	8.9	1	1.3	0.2	0.5	1.3	19.8	32.2	–	–	11
Synechocystis sp.	42.5	–	18.8	30.1	–	–	–	–	–	–	14.2	–	–	–	–	–	–	–	9
Thalassiosira weissflogii	8.8	–	36.6	40.5	–	–	–	–	14	–	–	–	–	–	–	–	–	–	9
Ulva lactuca	0.4	–	23.5	3	0.3	–	16.2	–	9.3	2.3	11.5	–	–	–	0.3	3.9	–	–	11

The fatty acids are given as the carbon number of the chain followed by the number of double bonds (for example oleic acid is given as 18:1).

16.2.2 Lipid Processing

Some of the largest costs in algal fuel technology (up to 40% of the fuel) involves the harvesting of the biomass, drying and lipid extraction.[18] There are many different strategies for the harvesting of algal biomass such as centrifugation, gravity sedimentation, filtration, flocculation, electrolysis and electrophoersis.[19] On harvesting, the algal lipids must be extracted. The algal cell structure is relatively robust and achieving a reasonable oil yield is far from trivial. Various pre-treatment methods have been investigated to aid the process including autoclaving, bead beating, microwave, sonication and using salt solutions to osmotically disrupt the cells.[19,20] While the exact individual efficiencies of the various pre-treatment methods are often species dependent, one study, focusing on *Botryococcus* sp. found that microwave disruption was the most efficient method whilst sonication was the least efficient.[21] Lipids themselves are soluble in most non-polar solvents, though the two most commonly used systems are hexane and chloroform–methanol. Industrial processes usually employ hexane, as it is more efficient, less toxic and has a higher selectivity for the neutral lipid fraction.[22,23]

16.2.3 Economics of Algal Lipid Production

Numerous estimations on the cost of production have been published over the last few decades. Varying estimates on size, variable cost of nutrients, the algal species used and technologies used for the growth, extraction and conversion of the lipids have led to wildly different estimates on the predicted cost of the fuels. Ribeiro and de Silva surveyed the techno-economic indicators of microalgae biofuel technologies, comparing and contrasting the differing production methods.[24] The lowest cost estimate for algal lipid was US\$0.43 L^{-1}, assuming an oil content of 50 wt-%, an oil yield of 30 g m^{-2} d^{-1} and culturing the algae in an open raceway pond bioreactor.[25] The cost of the algal oil was also reduced by the use of co-products from the same pond. The highest cost given was \$24.60 L^{-1} of algal oil in which the economics of a range of oil by weight (15, 25 and 50%), and different production methods (open bioreactor, photobioreactor and fermenter) were examined.[26]

For algal biodiesel to offer a competitive alternative, the lipids must cost less than \$0.74 L^{-1}.[24] A recent techno-economic analysis of autotrophic microalgae for hydrotreated fuel (HEFA) production estimated the cost of the algal lipid to be \$1.87 L^{-1} for microalgae grown in open ponds and \$3.98 L^{-1} for that grown in a photo-bioreactor. In this study, the costs were also calculated to include processing to HEFA. The total cost of the HEFA was then estimated to be \$2.16 L^{-1} for open pond production and \$4.52 L^{-1} for algae grown in a photobioreactor.[27] Though this is considerably more than is deemed competitive, it demonstrates that the major cost in the production of algae is not the chemical processing to a fuel, but the culturing, harvesting and extraction of the oil itself.

During culturing, a maximum biomass level can be achieved in a few days, though the lipid accumulation stage can take up to 2 weeks. Therefore, a possible method of reducing the culturing and extraction costs is to directly convert the biomass through thermochemical methods to produce liquid transportation fuels.

16.3   THERMOCHEMICAL CONVERSION OF WHOLE ALGAL BIOMASS

Among the main challenges related to the production of fuels from microalgae are the low growth rates, the high costs of drying and extraction of the lipids. Consequently, the thermochemical conversion of whole microalgae into bio-crudes, which can be further upgraded into chemicals and fuels, offers a highly credible route to the production of cost-effective transport fuels. Under these conditions, oil is formed not only from the lipid fraction, but from the entire biomass, including the protein and carbohydrate fractions. This would allow the use of non-lipid producing algae with much faster growth rates.

Thermochemical conversion can be conducted either in the presence (hydrothermal conversion) or absence of water (pyrolysis) and can be optimized for the production of solids (torrefaction), liquids (liquefaction) and gases (gasification). Whilst these methods have already been extensively studied and reviewed in the literature for lignocellulosic biomass,[28–32] the conversion of microalgae proceeds very differently, as a result of the large difference in biomass composition (Table 16.2). For example, the lower decomposition temperatures for lipids and proteins compared with carbohydrates[33,34] lead to a lower onset temperature for microalgae decomposition compared with terrestrial biomass.[35] In addition, whilst microalgal oils tend to contain less oxygen, they contain significantly higher concentrations of sulfur and nitrogen; these must be almost completely removed before the products can be applied to fuel use.

Table 16.2 Processing parameters for the production of bio-oil from algal biomass.

	Pyrolysis	Hydrothermal liquefaction
Temperature	~500 °C	280–370 °C
Pressure	Atmospheric	2.0–25.0 MPa
Feed quality	Dry feed with low moisture content (<10%)	Algae slurry with up to 95% water
Residence time	<2 s	5–6 min (batch)
Yield	Up to 60%	Up to 60%
Product quality	Highly viscous and acidic oil with high water content	Two-phase water/oil mixture

While liquefaction is the most relevant process for the production of liquid transport fuels, gasification for the production of syngas and subsequent Fischer–Tropsch synthesis has also been investigated as an alternative pathway to liquid fuels, although the high energy requirements for complete de-watering and gasification make this process economically unsuitable.[36]

16.3.1 Pyrolysis

The pyrolysis of biomass is generally conducted between 300 and 600 °C, at atmospheric pressure and in the absence of oxygen.[37] Maximum liquid yields of up to 55% have been reported at 500 °C for the conversion of a blue-green algal bloom, whilst lower temperatures favoured the formation of char and higher temperatures the formation of gases.[35] Depending on the residence time (RT) of the pyrolysis vapour, the process can be classed as slow pyrolysis (RT ~30 min), fast pyrolysis (RT ~10–20 s) and flash pyrolysis (RT ~1–2 s).[38] As the main products from slow pyrolysis are chars and char oils with yields of around 15–20%,[39] very short vapour residence times and high heating rates are required to minimize secondary condensation and polymerization reactions, and to obtain adequate liquid yields with suitable bio-oil properties.[40,41]

These rapid heating and cooling requirements for fast pyrolysis are a major processing challenge. Pyrolysis of biomass in fluidized beds provides high rates of heat transfer to the reacting particles and high mass-transfer coefficients,[41] but requires finely ground biomass with particles of around 2 mm and a low moisture content of <10%.[42] Alternative processes include ablative pyrolysis, where the biomass is pressed against a heated surface, and vacuum pyrolysis, where pyrolysis vapours are rapidly removed from slowly heated biomass.[37,42]

Another important factor influencing the yield of bio-oil is the composition of the microalgal biomass itself. Whilst liquid yields from autotrophically grown *Chlorella protothecoides* only reached 16.6% with char yields of 53.8%, the conversion of the same algae grown heterotrophically gave liquid yields of 57.2% with char yields of 11.2%.[43] These differences in yield can be correlated to a high lipid and low protein content (55% and 10%, respectively) in the heterotrophically grown algae and a low lipid and high protein content (15% and 53%, respectively) in the autotrophically grown algae. This is consistent with decomposition studies for proteins and lipids, which yielded much higher char residues from the protein fraction (37%) than from the lipid fraction (11%).[34] Similarly, a wide variation in liquid yields (ranging from 24% for *Dunaliella tertiolecta* to 43% for *Tetraselmis chuii*) was observed during the slow pyrolysis of six different species of microalgae at 500 °C, although a compositional breakdown was not provided.[44]

Depending on the condensation temperature of the reaction vapours, the crude bio-oil can contain a substantial quantity of water in the region of 20–25% of the oil weight.[42] In addition, more severe processing conditions in the presence of a catalyst resulted in reduced crude oil yields, albeit with a significantly reduced oxygen content.[45] This highlights the difficulty of comparing the results of different conversion studies.

16.3.2 Hydrothermal Liquefaction

An alternative to pyrolysis is hydrothermal liquefaction. This process does not require the extensive drying of the algal biomass prior to chemical conversion and the feed typically contains a water mass fraction of 80–95%.[46] The hydrothermal liquefaction of microalgae is carried out at conditions just below the critical point of water (280–370 °C, 2.0–25.0 MPa),[18,46,47] ensuring that the water remains in the liquid phase and thus minimising latent heat losses through vaporization. At these conditions water possesses a much lower dielectric constant and reduced hydrogen bonding, improving the solubility of many organic compounds and providing a high concentration of H^+ ions which may help to catalyse the reaction.[48] Indeed, the oils produced from hydrothermal liquefaction of microalgae have been reported to contain less oxygen and possess higher energy densities than the oils produced from fast pyrolysis.[49]

In many cases hydrothermal liquefaction experiments have been conducted in the presence of sodium carbonate or other water-soluble inorganic catalysts.[50–52] However, whilst these catalysts appear to selectively promote the decarboxylation of carbohydrates, they have been found to have a detrimental effect on oil formation from lipids and proteins.[52] A study on the conversion of albumin suggests that sodium carbonate may reduce the nitrogen distribution to the oil,[53] but this was not observed in other similar studies.[50] Liquefaction has also been studied in the presence of heterogeneous metal catalysts, which were found to increase liquefaction yields and have a significant influence on the colour and apparent viscosity of the product oils.[51] The highest liquid yields were obtained with Pd (57%), but Ni and Ru produced the oils with the lowest nitrogen content.[51] Pt, Ni and CoMo catalysts were also found to improve the degree of deoxygenation whilst reducing bio-crude yields.[54]

To date, the majority of microalgae liquefaction experiments have been conducted in batch, with residence times ranging from (5 to 60) minutes. Typical bio-oil yields range from (20 to 50)%, although maximum yields of 57% and 61.6% have been reported for *Botryococcus braunii*[55] and a planktonic algae collected from Hubing Lake, Hefei City, China,[56] respectively. Large variations in yields and oil composition have been observed for different algal species and under different growth conditions. A rough correlation between initial lipid composition and oil yields has been proposed which suggests that oil yields from lipids, proteins and carbohydrates fall within the ranges of 55–88%, 11–18% and 6–15% respectively.[52]

Recently it has been suggested that shorter reaction times, coupled with faster heating rates, may be sufficient to obtain comparable if not better bio-oil yields. By placing the reactors in a sand bath with a temperature significantly above the required setpoint temperature, very fast heating rates were obtained and liquid yields of up to 66% were achieved during the conversion of *Nannochloropsis* after only a one-minute reaction time.[57] Further analysis revealed that product yields pass through a maximum as the reaction severity, which is a function of reaction temperature and residence time, is increased. Continuous liquefaction studies have also shown that, while some practical operational issues of reactor blockage may occur at solid loadings above 5 wt-% in the feed,[58] it is possible to process microalgae slurries with a solid loading of up to 35% and obtain bio-crude yields of up to 63.6%.[59] A big advantage compared with lignocellulosic biomass or dry processing is that microalgae forms a slurry with water, which can be easily pumped into the reactor.

16.3.3 Bio-oil Upgrading

Whilst the bio-oils produced by the thermochemical upgrading of microalgae tend to display lower densities, a lower acid and a lower oxygen content than oils produced from the conversion of lignocellulosic biomass, the oils are also more viscous and contain much higher quantities of nitrogen and sulfur.[59,60] As the fuel standards require very low nitrogen and sulfur levels, substantial upgrading of the microalgal oils is necessary before they can enter the conventional fuel stream.

Chemical analysis of the resulting bio-oils is non-trivial, as they often contain several hundred compounds including phenols, alkanes, alkenes, fatty acids, ketones, aldehydes, nitriles, amides, and nitrogen-containing heterocycles such as indoles and pyridines.[35,61] Amino acid yields in turn tend to be low, as further decarboxylation and deamination reactions lead to the formation of carbonic acids, amines and ammonia.[46] Due to the complexity of the oils, as well as the high boiling point range, most reports have only partially characterized the oils produced to date.[46,60]

Despite an increased interest in the thermochemical conversion of microalgae, only a few studies have been published on the upgrading of the crude bio-oils (Table 16.3). Various publications focus on the upgrading of crude bio-oil produced by the liquefaction of *Nannochloropsis* sp.,[62,63] or *Chlorella* sp.[49] in the presence of supercritical water. One of the reasons for the addition of water is to eliminate a separation stage between the liquefaction and upgrading reaction, but it has also been suggested that water may react with the carbon oxides released in the reaction to generate hydrogen which further promotes the removal of heteroatoms.[64] To date, the best results were obtained over Pt/C at 530 °C and a reaction time of 6 h, with almost complete sulfur removal and a reduction in nitrogen content from (4.0 to 1.5)%.[62] Recently it was shown, however, that while the presence of water helps to reduce the formation of coke products, it has no beneficial effect on the denitrogenation performance, but rather results in an increased oxygen content in the reaction product.[49] Consequently, a similar reduction in nitrogen content from 5.32 to 1.6% was obtained during the upgrading of liquefaction oil from *Nannochloropsis* in the absence of water over HZSM-5 catalyst at 500 °C and 4 h reaction time.[65] These studies demonstrate that the reaction temperature is the most important variable affecting the denitrogenation performance, but unfortunately also results in a significant reduction in hydrogen content, which may raise the aromatic content of the reduction product above

Table 16.3 Current work in the chemical upgrading of algal derived bio-oils.

Feedstock (microalgae)	Biocrude quality/wt%			Catalyst	Reaction temperature	Pressure	Reaction mode	Reaction time	Product quality/wt%			Ref.	Comments
	N	S	O						N	S	O		
Bio-crude from HTL of *Nannochloropsis* at 350 °C for 1 h	5.32	0.56	8.35	HZSM-5 0-50 wt% loading	400-500 °C	4.35 MPa (hydrogen)	Batch	0.5-4 h	1.6-2.71	Bdl	0.39-2.81	65	
Bio-crude from HTL of *Nannochloropsis* at 340 °C for 4 h	4.80	0.48	8.07	Pt/C, Mo$_2$C, HZSM-5 5-20 wt%	430-530 °C	3.5 MPa (hydrogen)	Batch in super-critical water (water to oil mass ratio of 4:5)	2-6 h	1.50-3.61	Bdl	0.13-5.31	62	Temperature has biggest impact on oil properties
Bio-crude from HTL of *Nannochloropsis* at 320 °C for 4 h	4.89	0.68	6.52	Pt/C (25 wt%), HCl, NaOH	400 °C	3.4 MPa (hydrogen)	Batch in super-critical water (water to oil mass ratio of 1:1)	4 h	2.17-2.79	Bdl	4.31-4.71	63	
Bio-crude from HTL of *Chlorella* sp. at 350 °C for 1 h	7.3	Nd	7.8	Pt/γ-Al$_2$O$_3$ 0-40 wt% HCOOH 0-88 wt%	400 °C	6 MPa (hydrogen)	Batch in super-critical water 0-43 wt%	1 h	2.4-5.8	N/A	4.7-17.9	49	With formic acid, reacted with the crude algal oil resulting in increased yield
Bio-crude from HTL of *Desmodesmus sp.* at 200-375 °C for 5-60 min	0.2-6.3	Nd	10-39	HZSM-5 (SiO$_2$/Al$_2$O$_3$ = 280) (zeolite to sample mass ratio of 20:1)	600 °C		Pyrolysis probe (heating rate: 20 °C (°C s^{-1}), pyrolysis time: 10 s)		0.02-0.14	N/A	0.08-0.30	33	Elemental calculated from structure of compounds identified by GC
Bio-crudes from HTL of *Nannochloropsis* sp. at 344-362 °C, in continuous reactor	4.0-4.7	0.3-0.5	5.3-8.0	Co-promoted MoS$_2$ on fluorinated alumina support	125-170 °C (first quarter of reactor) 405 °C (main reactor)	13.6 MPa in hydrogen flow	Benchscale hydroprocessing system	LHSV of 0.14-0.20 h^{-1}	<0.05-0.16	<50× 10^{-6}	0.8-1.2	59	

Bdl = below detection limit; GC = gas chromatography; HTL = hydrothermal liquefaction; LHSV = liquid hourly space velocity; N/A = not applicable.

the desired levels.[62] The catalyst in turn helps to influence the physical properties of the reaction product, but has less influence on the denitrogenation performance itself.[62]

A potential reason for the poor denitrogenation performance in batch reactors is that not enough hydrogen was fed to the reaction or that the accumulation of ammonia limits the complete reduction of nitrogen compounds.[64] It is also possible that the high basicity of nitrogen compounds leads to deactivation of the active acidic catalyst sites.[66] In contrast, virtually complete removal of nitrogen was achieved during the hydro-treating of a crude bio-oil produced by the hydrothermal liquefaction of *Nannochloropsis* in a bench-scale hydroprocessing system employing a sulfided CoMo catalyst on a fluorinated alumina support.[59] The single-stage upgrading was undertaken at 405 °C with a velocity of 0.20 h^{-1} and a pressure of 14 MPa, in an excess of hydrogen. The nitrogen content of the resulting refined oil ranged from (0.07 to 0.25) wt-%. An even lower nitrogen content was achieved by pre-treating the oil at 125–170) °C in the first quarter of the reactor, prior to full conversion at 405 °C.

16.3.4 Pre-treatment and Co-product Extraction

To reduce the number of nitrogen compounds in the crude bio-oil and therefore reduce upgrading costs, a number of researchers have suggested microalgae pre-treatment to extract proteins or protein derivatives into the water phase.[52,67,68] Pre-treatment of a *Nannochloropsis occulata* slurry at temperatures above 200 °C produced an algal residue with a reduced nitrogen content and subsequent pyrolysis at 500 °C produced a much more uniform product compared with the untreated algae.[67] At the same time a carbon retention of only 40–50% was achieved at these conditions and therefore bio-oil yields are likely to be significantly lower. The microwave pretreatment of three algae species at temperatures ranging from (80 to 140) °C had no noticeable impact on the nitrogen content of the crude bio-oils produced by hydrothermal liquefaction at 300 °C.[68] However, this pre-treatment helped to recover up to 50% of phosphorus into the water phase, which may facilitate recycling. Pre-treatment may also be beneficial in reducing the ash content of the microalgae, which may have undesired catalytic properties during the hydrothermal liquefaction process,[52] or cause slagging and fouling problems during the thermochemical conversion.[69]

Finally, pre-treatment may be used to extract other desirable products from the algal biomass. For instance, pre-treating heterotrophically grown *Chlorella sorokiniana* at 160 °C for 20 min allowed the extraction of significant amounts of polysaccharides, without effecting yields from hydrothermal liquefaction.[36] On the contrary, significantly lower liquefaction temperatures were required to obtain the same oil yields, with a lower nitrogen content and reduced bio-char production.

An alternative strategy for producing low nitrogen oils is the stepwise liquefaction or pyrolysis of microalgae to exploit the differences in devolatilization temperatures for lipids, proteins and carbohydrates as determined by thermal degradation studies. Lipids were found to devolatize at temperatures ranging from (150 to 230) °C, proteins between temperatures of (230 to 400) °C and carbohydrates in the temperature range from (400 to 500) °C.[70] For example, the hydrothermal treatment of *Desmodesmus* sp. at temperatures below 250 °C produced oils with reduced nitrogen content, albeit with lower yields.[71] In contrast, at temperatures above 300 °C, the nitrogen content remained more or less constant whilst the highest yields were obtained at a temperature of 375 °C. The main challenge with employing this strategy lies in separating the different product fractions after each stage, especially if hydrothermal liquefaction is used. Nevertheless, further studies that investigate the relationship between liquefaction conditions, bio-oil yields as well as bio-oil composition to monitor the nitrogen distribution to the different reaction products will provide useful information.[71] In addition, the nature of nitrogen compounds is extremely important as presumably some of these compounds will be more reactive than others.

16.3.5 Nutrient Recycling

An essential aspect of sustainable biofuel production from microalgae is the recovery and recycling of nutrients.[72] To reduce costs substantially, especially on the scale required for fuel production, phosphorous and to a lesser extent nitrogen need to be retained and recycled.[73]

As a large proportion of the nitrogen and phosphorus present in the biomass partitions to the aqueous phase during hydrothermal liquefaction, recycling of the process water for further algal growth is a promising strategy. However, the process water also retains up to 40% of the carbon present in the biomass and organic molecules such as phenols are known growth inhibitors.[74] In addition, Ni leaching from the reactor may further inhibit growth.[74] In order to investigate the potential for process water recycling, a number of studies have been conducted to grow microalgae on recycled process water. It was found that high dilution factors of process water in deionized water of up to 1 : 100 are required for biomass to grow;[75] however, for some species mixotrophic growth, utilizing the carbon present in the recycled stream, resulted in higher growth rates than in the growth media.[74] In a different study, *Desmodesmus* sp. could be grown in the growth media with the addition of up to 5% of the recycled water phase, but were much less active in the absence of growth media.[76] These results suggest that, whilst the algae can utilize the recycled phosphorus and nitrogen, growth is inhibited by a lack of other essential nutrients such as S, Ca, Mg, Na, K and Cl, and not by the bio-toxicity of chemicals present in the recycled stream. Once again, further studies are required to better understand the individual factors influencing growth on recycled water.

16.4 FUTURE RESEARCH PERSPECTIVE

While microalgae have great potential as a source of renewable biomass, the process economics of producing algal biodiesel are not favourable. The largest factors in the cost of production are the slow growth rates of oleaginous microalgae, the de-watering and the lipid extraction. The cost of the culturing process could be significantly reduced by development of fuels produced by the hydrothermal liquefaction of the biomass. However, further research in a number of key areas is required to develop this process.

Firstly, a more in-depth understanding of the chemical mechanisms involved in the thermochemical conversion is essential. Subsequently, improving the process design to enable high pressure hydrotreatment and novel catalyst development will also play a key role. Even so, the high costs of this upgrading step are likely to significantly impact the process economics of the hydrothermal liquefaction route. Additional research is required to optimize this pathway to reduce the nitrogen content in the crude algal oils. This may be achieved by various pre-treatment steps to remove the microalgae protein fraction prior to liquefaction, or variation of the process conditions during liquefaction. Research is also needed to develop fast growing algal species with a relatively low protein content.

From a research perspective, it is extremely difficult to compare research published across different studies. It seems likely that only by analysing the system as a whole, including the culturing of the algae in the first place, that a viable process can be developed, though this is challenging as it requires complex infrastructure investment and building a large multidisciplinary team with complementary skills. Finally, more in-depth life cycle assessment is required to understand the environmental impacts and constraints in the separate stages and to guide further research into reducing the environmental impact of these key stages.

REFERENCES

1. M. Morweiser, O. Kruse, B. Hankamer and C. Posten, *Appl. Microbiol. Biotechnol.*, 2010, **87**, 1291–1301.
2. M. S. Ghayal and M. T. Pandya, *Energy Procedia*, 2013, **32**, 242–250.
3. G. Knothe, J. H. V. Gerpen and J. Krahl, *The Biodiesel Handbook*, AOCS Press, 2005.
4. A. Karmakar, S. Karmakar and S. Mukherjee, *Bioresour. Technol.*, 2010, **101**, 7201–7210.
5. A. L. Haag, *Nature*, 2007, **447**, 520–521.
6. G. Knothe, *Fuel Process. Technol.*, 2005, **86**, 1059–1070.
7. C. J. Chuck, C. D. Bannister, R. W. Jenkins, J. P. Lowe and M. G. Davidson, *Fuel*, 2012, **96**, 426–433.
8. S. K. Hoekman, A. Broch, C. Robbins, E. Ceniceros and M. Natarajan, *Renewable Sustainable Energy Rev.*, 2012, **16**, 143–169.
9. I. Lang, L. Hodac, T. Friedl and I. Feussner, *BMC Plant Biol.*, 2011, **11**, 124.
10. J. Pratoomyot and P. Srivilas, *Songklanakarin J. Sci. Technol.*, 2005, **27**, 1179–1187.
11. S. V. Khotimchenko, V. E. Vaskovsky and T. V. Titlyanova, *Bot. Mar.*, 2002, **45**, 17–22.
12. H. D. Smith-Badorf, C. J. Chuck, K. R. Mokebo, H. Macdonald, M. G. Davidson and R. J. Scott, *AMB Express*, 2013, **3**, 1–9.
13. M. J. Griffiths, R. P. van Hille and S. T. L. Harrison, *J. Appl. Phycol.*, 2012, **24**, 989–1001.
14. C.-H. Su, L.-J. Chien, J. Gomes, Y.-S. Lin, Y.-K. Yu, J.-S. Liou and R.-J. Syu, *J. Appl. Phycol.*, 2011, **23**, 903–908.
15. C. D. Bannister, C. J. Chuck, M. Bounds and J. G. Hawley, *P. I. Mech. Eng. D.-J. Aut.*, 2011, **225**, 99–114.
16. C. Wang, Z. Tian, L. Wang, R. Xu, Q. Liu, W. Qu, H. Ma and B. Wang, *ChemSusChem*, 2012, **5**, 1974–1983.
17. G. Knothe, *Prog. Energy Combust.*, 2010, **36**, 364–373.
18. I. Rawat, R. Ranjith Kumar, T. Mutanda and F. Bux, *Appl. Energy*, 2011, **88**, 3411–3424.

19. N. Pragya, K. K. Pandey and P. K. Sahoo, *Renewable Sustainable Energy Rev.*, 2013, **24**, 159–171.
20. A. Bahadar and M. B. Khan, *Renewable Sustainable Energy Rev.*, 2013, **27**, 128–148.
21. J.-Y. Lee, C. Yoo, S.-Y. Jun, C.-Y. Ahn and H.-M. Oh, *Bioresour. Technol.*, 2010, **101**, S75–S77.
22. R. Halim, B. Gladman, M. K. Danquah and P. A. Webley, *Bioresour. Technol.*, 2011, **102**, 178–185.
23. M. K. Lam and K. T. Lee, *Chem. Eng. J.*, 2012, **191**, 263–268.
24. L. A. Ribeiro and P. P. d. Silva, *Renewable Sustainable Energy Rev.*, 2013, **25**, 89–96.
25. J. R. Benemann and W. J. Oswald, *Systems and Economic Analysis of Microalgae Ponds for Conversion of CO$_2$ to Biomass*, Final report DOE/PC/93204–T5, Department of Energy, Pittsburgh Energy Technology Center, Pittsburgh, PA, 1996.
26. A. O. Alabi, M. Tampier, and E. Bibeau, *Microalgae Technologies and Processes for Biofuels/Bioenergy Production in British Columbia: Current Technology, Suitability and Barriers to Implementation*, British Columbia Innovation Council, Victoria, Canada, 2009.
27. R. Davis, A. Aden and P. T. Pienkos, *Appl. Energy*, 2011, **88**, 3524–3531.
28. K. Tekin and S. Karagöz, *Res. Chem. Intermediat.*, 2012, **39**, 485–498.
29. A. Kruse, A. Funke and M. M. Titirici, *Curr. Opin. Chem. Biol.*, 2013, **17**, 515–521.
30. W. N. R. W. Isahak, M. W. M. Hisham, M. A. Yarmo and T.-Y. Yun Hin, *Renewable Sustainable Energy Rev.*, 2012, **16**, 5910–5923.
31. A. V. Bridgwater, *Biomass Bioenergy*, 2012, **38**, 68–94.
32. J. Akhtar and N. Saidina Amin, *Renewable Sustainable Energy Rev.*, 2012, **16**, 5101–5109.
33. C. Torri, D. Fabbri, L. Garcia-Alba and D. W. F. Brilman, *J. Anal. Appl. Pyrol.*, 2013, **101**, 28–34.
34. K. Kebelmann, A. Hornung, U. Karsten and G. Griffiths, *Biomass Bioenergy*, 2013, **49**, 38–48.
35. Z. Hu, Y. Zheng, F. Yan, B. Xiao and S. Liu, *Energy*, 2013, **52**, 119–125.
36. C. Miao, M. Chakraborty and S. Chen, *Bioresour. Technol.*, 2012, **110**, 617–627.
37. R. H. Venderbosch and W. Prins, *Biofuel. Bioprod. Bioresour.*, 2010, **4**, 178–208.
38. A. Marcilla, L. Catalá, J. C. García-Quesada, F. J. Valdés and M. R. Hernández, *Renewable Sustainable Energy Rev.*, 2013, **27**, 11–19.

39. R. Maggi and B. Delmon, *Fuel*, 1994, **73**, 671–677.

40. X. Miao, Q. Wu and C. Yang, *J. Anal. Appl. Pyrol.*, 2004, **71**, 855–863.

41. J. Yanik, R. Stahl, N. Troeger and A. Sinag, *J. Anal. Appl. Pyrol.*, 2013, **103**, 134–141.

42. A. V. Bridgwater, D. Meier and D. Radlein, *Org. Geochem.*, 1999, **30**, 1479–1493.

43. X. Miao and Q. Wu, *J. Biotechnol.*, 2004, **110**, 85–93.

44. S. Grierson, V. Strezov, G. Ellem, R. McGregor and J. Herbertson, *J. Anal. Appl. Pyrol.*, 2009, **85**, 118–123.

45. P. Pan, C. Hu, W. Yang, Y. Li, L. Dong, L. Zhu, D. Tong, R. Qing and Y. Fan, *Bioresour. Technol.*, 2010, **101**, 4593–4599.

46. D. López Barreiro, W. Prins, F. Ronsse and W. Brilman, *Biomass Bioenergy*, 2013, **53**, 113–127.

47. I. Graça, J. M. Lopes, H. S. Cerqueira and M. F. Ribeiro, *Ind. Eng. Chem. Rev.*, 2013, **52**, 275–287.

48. P. E. Savage, *Chem. Rev.*, 1999, **99**, 603–622.

49. P. Duan, X. Bai, Y. Xu, A. Zhang, F. Wang, L. Zhang and J. Miao, *Fuel*, 2013, **109**, 225–233.

50. T. M. Yeh, J. G. Dickinson, A. Franck, S. Linic, L. T. Thompson and P. E. Savage, *J. Chem. Technol. Biotechnol.*, 2013, **88**, 13–24.

51. J. X. Han, J. Z. Duan, P. Chen, H. Lou, X. M. Zheng and H. P. Hong, *Green Chem.*, 2011, **13**, 2561–2568.

52. P. Biller and A. B. Ross, *Bioresour. Technol.*, 2011, **102**, 215–225.

53. Y. Dote, S. Inoue, T. Ogi and S.-Y. Yokoyama, *Biomass Bioenergy*, 1996, **11**, 491–498.

54. P. Biller, R. Riley and A. B. Ross, *Bioresour. Technol.*, 2011, **102**, 4841–4848.

55. Y. Dote, S. Sawayama, S. Inoue, T. Minowa and S.-Y. Yokoyama, *Fuel*, 1994, **73**, 1855–1857.

56. Y. Xu, H. Yu, X. Hu, X. Wei and Z. Cui, *Energy Source A*, 2014, **36**, 38–44.

57. J. L. Faeth, P. J. Valdez and P. E. Savage, *Energy Fuel.*, 2013, **27**, 1391–1398.

58. C. Jazrawi, P. Biller, A. B. Ross, A. Montoya, T. Maschmeyer and B. S. Haynes, *Algal Res.*, 2013, **2**, 268–277.

59. D. C. Elliott, T. R. Hart, A. J. Schmidt, G. G. Neuenschwander, L. J. Rotness, M. V. Olarte, A. H. Zacher, K. O. Albrecht, R. T. Hallen and J. E. Holladay, *Algal Res.*, 2013.

60. A. E. Harman-Ware, T. Morgan, M. Wilson, M. Crocker, J. Zhang, K. Liu, J. Stork and S. Debolt, *Renewable Energy*, 2013, **60**, 625–632.

61. U. Jena, K. C. Das and J. R. Kastner, *Bioresour. Technol.*, 2011, **102**, 6221–6229.
62. P. Duan and P. E. Savage, *Energy Environ. Sci.*, 2011, **4**, 1447.
63. P. Duan and P. E. Savage, *Bioresource Technol.*, 2011, **102**, 1899–1906.
64. P.-Q. Yuan, Z.-M. Cheng, X.-Y. Zhang and W.-K. Yuan, *Fuel*, 2006, **85**, 367–373.
65. Z. Li and P. E. Savage, *Algal Res.*, 2013, **2**, 154–163.
66. E. Furimsky and F. E. Massoth, *Catal. Rev.*, 2005, **47**, 297–489.
67. Z. Du, M. Mohr, X. Ma, Y. Cheng, X. Lin, Y. Liu, W. Zhou, P. Chen and R. Ruan, *Bioresour. Technol.*, 2012, **120**, 13–18.
68. P. Biller, C. Friedman and A. B. Ross, *Bioresour. Technol.*, 2013, **136**, 188–195.
69. A. B. Ross, J. M. Jones, M. L. Kubacki and T. Bridgeman, *Bioresour. Technol.*, 2008, **99**, 6494–6504.
70. P. Duan, X. Bai, Y. Xu, A. Zhang, F. Wang, L. Zhang and J. Miao, *Bioresour. Technol.*, 2013, **136**, 626–634.
71. L. Garcia Alba, C. Torri, C. Samorì, J. van der Spek, D. Fabbri, S. R. A. Kersten and D. W. F. Brilman, *Energy Fuels*, 2012, **26**, 642–657.
72. M. Nelson, L. Zhu, A. Thiel, Y. Wu, M. Guan, J. Minty, H. Y. Wang and X. N. Lin, *Bioresour. Technol.*, 2013, **136**, 522–528.
73. L. Lardon, A. Hélias, B. Sialve, J.-P. Steyer and O. Bernard, *Environ. Sci. Technol.*, 2009, **3**, 6475–6481.
74. P. Biller, A. B. Ross, S. C. Skill, A. Lea-Langton, B. Balasundaram, C. Hall, R. Riley and C. A. Llewellyn, *Algal Res.*, 2012, **1**, 70–76.
75. U. Jena, N. Vaidyanathan, S. Chinnasamy and K. C. Das, *Bioresour. Technol.*, 2011, **102**, 3380–3387.
76. L. Garcia Alba, C. Torri, D. Fabbri, S. R. A. Kersten and D. W. F. Brilman, *Chem. Eng. J.*, 2013, **228**, 214–223.

CHAPTER 17

Ocean Resources for the Production of Renewable Chemicals and Materials

FRANCESCA M. KERTON

Department of Chemistry, Memorial University of Newfoundland, St John's, NL, A1B 3X7, Canada
Email: fkerton@mun.ca

17.1 INTRODUCTION

As we move into a post-petroleum age, we need to learn how to obtain chemicals from nature in a sustainable fashion. This means that 'green' methods developed by chemists and chemical engineers will be employed on a frequent basis in order to transform biomass in environmentally friendly ways.[1,2] We need to be conscious of not creating new problems and pollution whilst attempting to reduce our dependence on petroleum. Significant advances have been achieved during the past two decades in terms of the complete utilization of biomass to yield fuels, chemicals and materials in a biorefinery.[3-5] At present, the majority of research in the development of biorefineries has focused on biomass originating from land with feedstocks including crops (such as maize and switch grass), and by-products of the forestry, paper and food industries.[6-8] There is some overlap

Chemical Processes for a Sustainable Future
Edited by Trevor M. Letcher, Janet L. Scott and Darrell A. Patterson
© The Royal Society of Chemistry 2015
Published by the Royal Society of Chemistry, www.rsc.org

between this chapter and the use of food waste as a source of chemicals,[9,10] as some of the opportunities highlighted herein focus on waste produced in the shellfish industries. In 2008, Arvanitoyannis and Kassaveti described environmental issues surrounding fish waste management and highlighted current important applications of the fishery by-products.[11] These applications included animal feed, biodiesel/biogas, chitosan, natural pigments, cosmetics (collagen), enzymes and fertilizers. Recently, the use of fish waste in the production of bio-oils and biofuels has been reviewed.[12]

In 2013, Kerton *et al.* highlighted opportunities surrounding the use of ocean- or marine-sourced biomass as a source of renewable fuels, chemicals and materials.[13] Ocean-sourced biomass includes such diverse feedstocks as micro- and macro-algae, waste from the shellfish industries (both crustacean and mollusc shells) and waste from finfish processing plants. At present, less than 5% of research into bio-sourced chemicals is focused on feedstocks obtained from our oceans and seas. Looking forward, sustainable industries will be tailored to their geographical region and the surrounding natural environment. In many areas of the world, including some with in-creasing populations *e.g.* Indonesia and the Philippines, implemen-tation of a biorefinery concept developed in the plains of North America will not be sustainable and alternatives will be needed. As oceans dominate the surface of our planet, many countries have extensive coastlines and this may lead to the development of new types of biorefineries that use marine plants or fishery/aquaculture by-products as their feedstocks.

It must be noted too that care should be taken not to over-exploit the materials produced in our oceans. In the past this has led to moratoriums on catching certain species of fish (*e.g.* cod in Newfoundland, Canada) and has also led to shortages in the production of certain hydrocolloids/biopolymers (see below) because of over-harvesting of seaweeds. One way around such shortages is to farm the fish, shellfish and algae using aquaculture methods. How-ever, such approaches are at present struggling with their own sustainability issues because of the rapid growth in this sector during the past decade. For example, there is a growing shortage of fish feed including valuable fish oils for salmon farming, but there is the potential for a symbiotic relationship whereby 'waste' of the fisheries is thought of as a valuable by-product and used in the production of fishmeal and other products. In short, introducing new targets for sustainability and waste reduction in the fishing and aquaculture industries may in fact help them to survive and thrive.

17.2 CHEMICALS AND MATERIALS FROM OCEANIC BIOMASS

Within this chapter, the aim is to highlight some of the unique possibilities available in the pool of biomass that can be sourced from the oceans, in contrast to materials obtained on land. However, the examples presented are not an exhaustive list of all the opportunities and chemicals available. It should also be noted that the utilization of micro- and macro-algae in fuel production is not included, as another chapter in this book focuses on that subject. In the next sections, the use of algae and waste crustacean shells as sources of non-cellulosic carbohydrates and other chemicals are presented, alongside opportunities surrounding the use of bio-sourced minerals from waste mollusc shells. The use of waste from finfish processing (*e.g.* effluent from salmon farms) is not covered but it should be noted that there are many prospects for adding-value to these waste streams too. Figure 17.1 provides an overview of some of the chemicals and materials available from ocean resources.

17.2.1 Non-Cellulosic Carbohydrates

A significant amount of research has focused on the development of technologies for converting cellulose into platform chemicals (chemical building blocks) and fuels.[2,7,14] This is because cellulose is the most abundant carbohydrate on Earth and present as the key structural material in all plants. However, oceanic plants and animals

Figure 17.1 Overview of chemicals and materials available from ocean-sourced biomass.

produce several other carbohydrates that can be used as interesting materials in their own rights or converted chemically or biochemically into other products.

17.2.1.1 Carrageenans, Agarose, Alginates and Chemicals from Algae. Algae form important parts of ocean ecosystems and, depending on their species, contain a wide variety of chemicals including many biologically active molecules.[15,16] Both wild and cultivated algae (seaweed) are harvested around the world and used as the source of three commercially important biopolymers (Figure 17.2). These are carrageenans, agarose and alginates, which are collectively known as phycocolloids. Carrageenans are linear polymers made up of galactose and anhydrogalactose monomers, which may or may not be sulfated. Different forms exist, primarily depending on the position of the sulfate group, with the kappa form shown in Figure 17.2 being the most widely used commercially. Agarose (agar) has a similar structure to carrageenans but

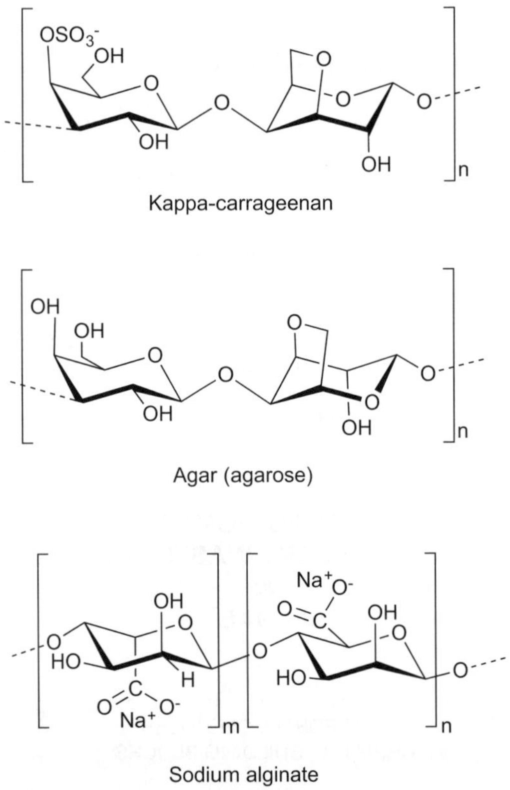

Figure 17.2 Valuable carbohydrate polymers obtained from seaweeds.

does not contain any sulfate groups. Alginates are copolymers of mannuronic and guluronic acids and are most commonly used as sodium and calcium salts. The exact quantities of each biopolymer present depend on the type of algae harvested. Carrageenans and agarose are obtained from red seaweeds and alginates from brown seaweeds. These polysaccharides are extracted from seaweed using water, with the pH of water adjusted to encourage dissolution. Carrageenans are soluble under basic conditions, alginic acid under acidic conditions (from which alginates are obtained by neutralization) and agarose is obtained *via* either an acidic or basic extraction route.[17] These biopolymers, especially carrageenans, have been used in food production as gelling agents for centuries. Their success in this area is due to the absence of a competitive man-made material. These three hydrocolloids alone had a market value of 1.02×10^9 (where 10^9 refers to billion) in 2009, with just over half of this market being the result of world carrageenan production.[18]

High and low grade biopolymers are produced from algae dependent on the resources and methods used. For example, high-grade alginates for speciality applications (*e.g.* printing, textiles) are produced in France, Norway and Scotland, whereas lower grade materials are produced in China and some South American countries. The popularity of natural fibres with the public has led to the development of new textiles based on sustainably harvested seaweeds such SeaCell™ produced by Smartfibre AG, Germany. However, these seaweed-sourced biopolymers are more commonly found in processed foods, dairy products, toothpastes, pet foods and other consumer products where gels are desired (*e.g.* cosmetics, air fresheners, drug capsules). It should also be noted that some bioactive carbohydrates are also accessible through extraction of algae using water and this may add further value to these feedstocks. For example, an extract containing a polysaccharide with rhamnose repeating units has the potential for treating type 2 diabetes and has been patented.[19] There are many websites dedicated to seaweeds (algae), their speciation, and existing and potential uses including The Seaweed Site (www. seaweed.ie), AlgaeBase (http://algaebase.org) and Suria-Link.com (http://surialink.seaplant.net). These are useful first ports of call to find out which strains of algae are native to your geographical region and what their chemical composition is.

Non-sulfated biopolymers from algae can be converted into small organic molecules *via* hydrolysis and dehydration processes in a similar fashion to cellulose. For example, agarose has been converted into 5-hydroxymethylfurfural (5-HMF) and ethyl levulinate using a solid-supported Brønsted acid catalyst in an ionic liquid,[20] and more recently, magnesium chloride has been used as catalyst to produce 5-HMF from agarose under aqueous conditions.[21] If algal bio-polymers contain sulfate groups, upon processing, care would have to be taken to avoid polluting either the atmosphere or aqueous streams with environmentally hazardous sulfur-containing by-products.

In addition to carbohydrates, algae can also be used as a source of lipids, fatty acids, vitamins and pigments.[22] Supercritical carbon dioxide, a 'green' solvent, has been employed in the extraction of these molecules. There is the potential to combine this method with the aqueous extraction of hydrocolloids and thereby maximize the productivity of a single feedstock. In the interests of sustainable development, it is important to consider the use of the whole plant. Of relevance in this regard is a recently published patent by FMC Corporation. The invention describes the isolation and purification of cellulose obtained after extraction of the majority of carrageenan present in the seaweed.[23] For example, if the cellulose-containing residue is treated *via* acid hydrolysis, microcrystalline cellulose is obtained. Using algae as a source of hydrocolloids, cellulose and other products (pigments, vitamins, lipids and trace elements) would greatly assist in maximizing the product pool and profitability of such processes.

Aside from water or supercritical fluid extraction of carbohydrates and other chemicals from seaweed, future processing options include pyrolysis and fermentation. Low temperature microwave-assisted pyrolysis has been shown to afford good yields of bio-oil and bio-char under mild processing conditions for three different types of macro-algae.[24] Fermentation of algae has typically afforded low molecular weight alcohols with potential applications as biofuels.[25] Fermentation of algae is an enormous and growing field, and as such the subject matter is beyond the scope of this chapter.

17.2.1.2 Chitin, Chitosana and Chemicals from Waste Crustacean Shells. As highlighted by Arvanitoyannis and Kassaveti,[11] chitosan (Figure 17.3) is one of the key materials currently produced in-dustrially from fishery waste streams. It is prepared *via* alkaline treatment of chitin, which causes deacetylation of the amide groups to give a polymer containing amine functionalities.[26] Chitin accounts for (20–25) wt% of crab and shrimp shells. The rest is mainly protein (40 wt%), calcium carbonate (30 wt%) and lipids (5–10) wt%. At present, methods used to isolate chitin from shells for

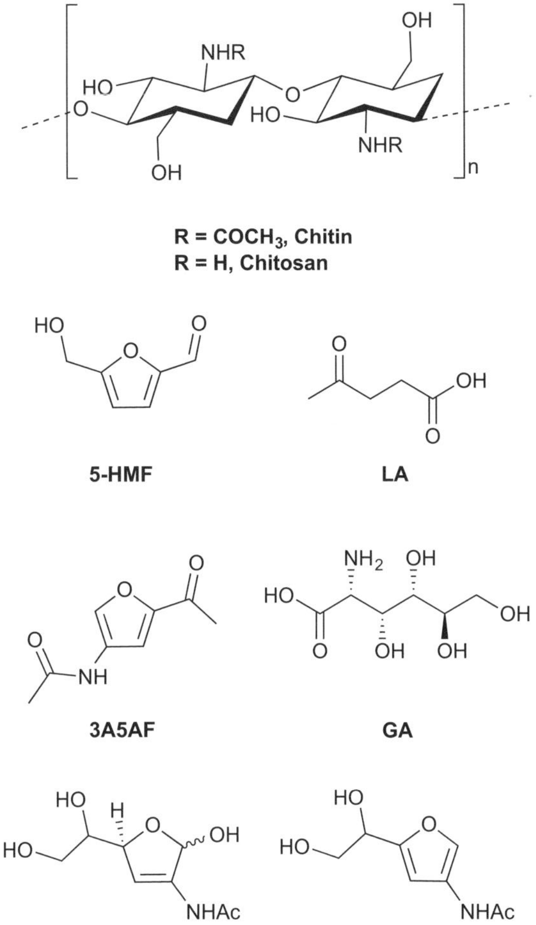

R = COCH$_3$, Chitin
R = H, Chitosan

5-HMF   **LA**

3A5AF   **GA**

Chromogen I and III

3A5AF, 3-acetamido-5-acetylfuran;
5-HMF, 5-hydroxymethylfurfural;
GA, glucosaminic acid; LA, levulinic acid.

Figure 17.3 Materials and chemicals accessible from waste crustacean shells.

subsequent chitosan production do not allow simultaneous co-product separation. The development of new approaches for the processing of shrimp waste with reduced environmental burdens (*e.g.* avoidance of strong acids and bases) was recently reviewed.[27] Microbial approaches including fermentation are particularly promising and may allow access to pigments, fatty acids and hydrolysate (oligopeptides and amino acids) if combined with the right separation processes. For example, in the fermentation of shrimp waste, Narayan and co-workers were able to isolate a chitin rich residue and then lyophilize the fermentation liquor to yield a powder rich in essential amino acids and carotenoid pigments.[28] The latter, they suggested, could be used in fishmeal.

Biochemical approaches have also been explored *via* trials in the conversion of chitin to chitosan. For example, chitin deacetylase (CDA) enzymes have been isolated from fungi and used in the deacetylation of shrimp chitin.[29] In this study, the chitin was pre-treated in several ways in an attempt to make it a more attractive substrate for the enzyme. The most effective pre-treatment was dissolution and a precipitation method that led to the formation of a de-crystallized chitin and the enzyme was able to produce chitosan with a 90% degree of deacetylation from this material. In a recent overview of this field, Zhao *et al.* identified three challenges that need to be addressed in order for such technologies to flourish.[30]

1. More fully characterized CDAs need to be available. In the long term, this will lead to enzymes with increased activity and lower costs.
2. Easy to use methods for the structural characterization of enzyme-deacetylated chitin need to be developed. This would allow researchers to understand the mode of activity and reaction mechanism for different CDAs.
3. An efficient method to break down the crystallinity of chitin needs to be identified. This would allow the enzymes easier access to the reactive groups within the biopolymer.

As biopolymers, chitin and chitosan have been used in many areas including catalysis,[31] medicine,[32] and the food industry.[33] Interestingly, in terms of developing sustainable processes, chitosan was recently shown to be an excellent flocculating agent in the dewatering of green algae for future use as a feedstock for biofuel production.[34] Not only was it superior in terms of life cycle assessment, it also afforded a superior technical performance compared with alum and other conventional flocculants.

For some applications, it would be useful to dissolve these biopolymers. Unlike the carbohydrates extracted from algae described above, chitin and chitosan are insoluble in water. Chitin is particularly recalcitrant and insoluble in nearly all solvents. It is soluble in hot concentrated solutions of CaX_2 (X = Cl, Br or I).[35] Chitosan, with its amino functional groups, is soluble in acidic organic or aqueous mixtures. A number of ionic liquids are able to disrupt the inter- and intra-molecular hydrogen bonding between polysaccharide chains and therefore are able to dissolve chitin and chitosan.[36,37] The anion of the ionic liquid has been shown to have a profound effect on the solubility of chitosan in these solvents.[38] Eight ionic liquids were studied for chitosan dissolution with the cation being 1-butyl-3-methylimidazolium and only the anion changing. A correlation between chitosan solubilising power and the hydrogen-bond accepting ability of the ionic liquid was observed with the acetate-containing ionic liquid being the most efficient at dissolving chitosan. Of particular note is the ability of some ionic liquids to dissolve not only chitin and chitosan but also the waste crustacean shell directly.[37] 1-Ethyl-3-methylimidazolium acetate was able to dissolve raw shrimp shells at concentrations of up to 0.46 g of chitin per 10 g of ionic liquid. Qin and co-workers also demonstrated that chitin fibres could be spun directly from these ionic liquid solutions and that the molecular weight of this processed chitin was higher than that obtained using traditional methods. Processes like this possess promise for the production of high quality materials from crustacean waste. More recently, carbon dioxide has been used to precipitate the chitin from this ionic liquid solution.[39] In this work, the authors proposed that the carbon dioxide reacts with the imidazolium liquid to yield a carboxylate zwitterion and acetic acid. The remaining acetate anions hydrogen-bonded preferentially with the acetic acid and this caused the biopolymer to precipitate. Luckily, the original ionic liquid could be regenerated in a simple and sustainable fashion by addition of water. In addition to playing a role in purifying crustacean waste, ionic liquids are starting to find applications in the preparation of composite and other more complex materials. For example, 1-butyl-3-methylimidazolium chloride has been used to dissolve mixtures of agarose, a biopolymer which can be obtained from algae and chitosan.[40] Addition of methanol as an anti-solvent allowed the composite biopolymer materials to precipitate and the ionic liquid could be recycled. These materials could also act as scaffolds for the formation of antimicrobial nanocomposites containing silver oxide nanoparticles distributed evenly throughout the biopolymer matrix. In another study, chitin microspheres were prepared by extrusion of a 1-butyl-3-methylimidazolium acetate solution of chitin into an aqueous or ethanol solution. The resulting microspheres then underwent a supercritical carbon dioxide drying process to yield an agglomerated but porous biopolymer material.[41] This was tested *in vitro* as a matrix for controlled drug release and as a scaffold for bone growth (apatite formation).

Chemocatalytic methods have been used recently in the conversion of chitin and chitosan into 5-HMF and levulinic acid (LA), which can also be produced from cellulose. In the first of these studies, microwave-assisted hydrolysis of chitosan suspensions in water using $SnCl_4 \cdot 5H_2O$ or SnO_2 nanoparticles in the presence of dilute HCl afforded up to 24 wt% LA or 10 wt% 5-HMF depending on concentrations and reaction temperatures.[42] Chitin processed in a similar way afforded up to 13 wt% LA. More recently, sulfuric acid afforded up to 22 wt% LA from chitin and 26 wt% LA from chitosan[43] and 10 mol% 5-HMF has been obtained from chitosan in concentrated aqueous $ZnCl_2$ reaction mixtures.[44] In such processes, it is known that the biopolymer first hydrolyses to form oligosaccharides and hexoses and that these subsequently dehydrate to afford 5-HMF, which further decomposes to yield LA.[42] The exact point in the mechanism where deamination of the polymer occurs is currently unknown. A limiting factor in such reactions is the heterogeneity of the reaction mixtures and a resulting slow rate of hydrolysis of the biopolymers. At the high reaction temperatures employed (120–200 °C), the biopolymer often chars and biodegrades in an uncontrolled fashion before hydrolysis occurs. Therefore, another option for conversion of chitin and chitosan into small organic molecules is to hydrolyse the biopolymers into their monomeric sugar units, namely, *N*-acetylglucosamine (NAG) and glucosamine ($GlcNH_2$). This reaction can be performed chemically using concentrated HCl or biochemically.[45] The resulting hexoses can then be transformed and higher yields of 5-HMF and LA are obtained.[42–44]

In addition to 5-HMF and LA, which can be obtained from cellulose, starch and other carbohydrates, *N*-containing chemicals have also been accessed using NAG or the parent biopolymers as feedstocks. 3-Acetamido-5-acetylfuran (3A5AF) can be isolated in up to 60 mol% yield *via* the dehydration of NAG in either organic solvents such as dimethyl acetamide (DMA) and DMF (dimethyl formamide) or ionic liquids (*e.g.* 1-butyl-3-methylimidazolium chloride). The key to obtaining high yields in these reactions is the presence of chloride ions and boric acid.[46,47] Similar trends have been observed for dehydration reactions of glucose and other carbohydrates.[48] Recently, 3A5AF has also been obtained directly from chitin in up to 7.5 mol% yield using *N*-methylpyrrolidinone (NMP) as the solvent in the presence of boric acid.[49] Chromogen I and III have been obtained in 23 mol% yield from NAG in a continuous process, with no catalyst, in high temperature water (120–220 °C and 25 MPa) at very short reaction times.[50] In all of the aforementioned reactions, the carbohydrates are dehydrated. In contrast, Ohmi *et al.* recently reported the catalytic oxidation of glucosamine hydrochloride and NAG to afford high yields of amino acid products.[51] Glucosaminic acid (GA), the product of glucosamine oxidation, has uses in the agricultural, medical, and food industries. Also, it should be noted that all products in this study are chiral and may be useful synthetic precursors for higher value products. Other products might also be accessible from these aminocarbohydrates and sugars, as in pyrolysis studies and the recent reactions of chitin in NMP, a number of chemicals have been identified albeit in small yields.[49,52,53] These include 3-acetamidofuran and other substituted furans, acetamidoacetaldehyde, acetamido-substituted pyrones, substituted pyrazines, pyridine and 4-(acetylamino)-1,3-benzenediol.

17.2.2 Minerals

The shells of crustaceans and molluscs contain significant amounts of calcium carbonate in order to afford their toughness (~30 wt% crustacean shells and ~97 wt% mollusc shells). Therefore, waste shells can be used as a renewable source of calcium carbonate or lime (calcium oxide). Due to the higher quantities of calcium carbonate present in mollusc shells compared with crustacean shells, it is not surprising that the majority of research on the production of bio-minerals has focused on the former feedstocks. However, the presence of calcium carbonate in crustacean shells cannot be ignored and is an important consideration for the development of sustainable processes for the production of chitin and chitosan.

17.2.2.1 Calcium Carbonate from Mollusc Shells. In some academic studies, the mollusc shells are prepared for use *via* washing in water and scrubbing with wire wool to remove organic (protein) residues. This would not be a viable method for large-scale production. In the region of Galicia, Spain, an industrial plant for the extraction of calcium carbonate from mussel shells has been constructed near to many shellfish processing facilities.[54] The industrial process involves removing water, salt, mud and residual meat from the shell to yield the mineral in high purity. The full process is described in detail in the paper by Barros *et al.*; however, a key step is the calcination process performed in a rotary kiln where temperatures are monitored to avoid going above 700 °C, which could lead to decarboxylation of the calcium carbonate.[54]

Biochemical methods for cleaning the shells also exist. A New Zealand patent describes the use of biocatalysts to remove meat from crushed mollusc shells[55] and claims that the dried excess meat is also available for future sale in addition to the powdered or crushed shells. The presence of protein residues would likely have a detrimental effect on some applications of waste mollusc shells. For example, ground mollusc shells (including clams, mussels, oysters and cockles) have been used in the production of a new cement product.[56] Portland cement was substituted in the range of (5 to 20) wt% with ground shells in the preparation of mortar and the resulting materials tested. The authors concluded that all mortars prepared had adequate strength, and due to slightly higher water content in shell-containing mortars, there was less shrinkage and this property might be useful in hot, dry climates. However, it should be noted that the shells were cleaned and dried before use, and the mortar's strength would likely have been affected if protein waste were still present as degradation would occur over time and lead to voids in the construction material. In another construction-related study focused on waste mussel shells, the new mortars prepared had enhanced mechanical properties because of the different morphologies of mussel shell calcium carbonate compared with quarried calcium carbonate aggregates.[57]

Materials from mussel shells have been used in a number of environmental remediation applications. Ground mussel shells and calcined ground mussel shells have been tested for mercury(II) adsorption and desorption from phosphate-free and phosphate-rich aqueous solutions under batch- and flow-conditions.[58] Of the materials tested, the calcined mussel shell was found to have the greatest potential for adsorption especially in the presence of phosphate salts in solution. Mussel shell ash, which is produced during the calcination of waste mussel shells, accounts for approximately 2 wt% of the waste shell and possesses a high adsorption efficiency for mercury(II) and arsenic(IV) in aqueous solution.[59] Bioreactors are an emerging technology for the treatment of acid mine drainage in coal mines. The effluent from such mines includes a range of metal cations (Fe, Al, Cu, Ni, Zn, Pb, Mn) and sulfate anions. In a published study, bioreactors using mussel shells as an alkaline amendment outperformed those using quarried calcium carbonate (limestone) to increase pH.[60] It would be interesting to see if mussel shells could be used in the treatment of other mining effluents.

Waste mussel shells have recently been used in the preparation of catalytic materials. For example, calcined shells, which are composed primarily of calcium oxide, have been used as transesterification catalysts for the production of biodiesel from soy oil and methanol.[61] Waste mussel shells have also been converted into hydroxyapatite, which has then been used as a photocatalyst for the treatment of wastewater.[62] In this study, it was shown that the biorenewable catalyst was competitive with commercially produced hydroxyapatite for photodegradation of the dye methylene blue.

17.3 CONCLUSIONS

Algae, as renewable feedstocks, provide access to unique biopolymers with valuable properties (*e.g.* biocompatible gelling agents), which have no competition from man-made materials. In addition to plants, fish, crustaceans and molluscs can also be used as renewable resources and, once meat has been removed, many other products might be accessible.

The oceans have provided man with food since the beginning of time but the modern food industries that they support (fishing and aquaculture) often produce waste and in many cases this is currently sent to landfill and does not have value. Some of the waste in these industries (*e.g.* shells from the processing of shrimp and crab) is already being used industrially (*e.g.* to produce chitosan). However, there are greater opportunities available and use of the whole plant or organism to afford foodstuffs alongside chemicals, fuels and materials will lead to greater sustainability. Clearly, care must be taken not to over-exploit or pollute the marine environment as has happened with industrial developments in the past. In pursuing whole use of marine resources, it will be important to employ the principles of green chemistry and engineering, and perform life cycle assessments when developing new, sustainable processes.

REFERENCES

1. J. H. Clark, F. E. I. Deswarte and T. J. Farmer, *Biofuels, Bioprod. Biorefin.*, 2009, **3**, 72–90.
2. R. A. Sheldon, *Green Chem.*, 2014, **16**, 950–963.
3. P. R. Stuart, Green Chemistry and Chemical Engineering: Integrated Biorefineries: *Design, Analysis, and Optimization*, CRC Press, Boca Raton, FL, 2012.
4. J.-L. Wertz and O. Bédué, *Lignocellulosic Biorefineries*, EPFL Press, Lausanne, Switzerland, 2013.
5. J. A. Melero, J. Iglesias and A. Garcia, *Energy Environ. Sci.*, 2012, **5**, 7393–7420.
6. J. J. Bozell and G. R. Petersen, *Green Chem.*, 2010, **12**, 539–554.
7. P. Gallezot, *Chem. Soc. Rev.*, 2012, **41**, 1538–1558.

8. M. Moshkelani, M. Marinova, M. Perrier and J. Paris, *Appl. Therm. Eng.*, 2013, **50**, 1427–1436.

9. C. S. K. Lin, L. A. Pfaltzgraff, L. Herrero-Davila, E. B. Mubofu, S. Abderrahim, J. H. Clark, A. A. Koutinas, N. Kopsahelis, K. Stamatelatou, F. Dickson, S. Thankappan, Z. Mohamed, R. Brocklesby and R. Luque, *Energy Environ. Sci.*, 2013, **6**, 426–464.

10. L. A. Pfaltzgraff, B. M. De, E. C. Cooper, V. Budarin and J. H. Clark, *Green Chem.*, 2013, **15**, 307–314.

11. I. S. Arvanitoyannis and A. Kassaveti, *Int. J. Food Sci. Technol.*, 2008, **43**, 726–745.

12. P. Jayasinghe and K. Hawboldt, *Renewable Sustainable Energy Rev.*, 2012, **16**, 798–821.

13. F. M. Kerton, Y. Liu, K. W. Omari and K. Hawboldt, *Green Chem.*, 2013, **15**, 860–871.

14. D. M. Alonso, J. Q. Bond and J. A. Dumesic, *Green Chem.*, 2010, **12**, 1493–1513.

15. A.-K. Kim, *Handbook of Marine Macroalgae: Biotechnology and Applied Phycology*, Wiley, Hoboken, NJ, 2011.

16. L. Brennan, A. Mostaert, C. Murphy and P. Owende, in *Biorefinery Co-Products : Phytochemicals, Primary Metabolites and Value-Added Biomass Processing*, ed. D. J. Carrier, S. Ramaswamy and C. Bergeron, Wiley, Hoboken, NJ, 2012, ch.10.

17. D. J. McHugh, *A Guide to the Seaweed Industry, Fisheries Technical Paper No.441*, Food and Agriculture Organization, Rome, 2003.

18. H. Bixler and H. Porse, *J. Appl. Phycol.*, 2011, **23**, 321–335.

19. B. A. Daniels, Seaweed extract composition for treatment of diabetes and diabetic complications, *World Pat.*, WO2004/103280, 2004.

20. B. Kim, J. Jeong, S. Shin, D. Lee, S. Kim, H.-J. Yoon and J. K. Cho, *ChemSusChem*, 2010, **3**, 1273–1275.

21. L. Yan, D. D. Laskar, S.-J. Lee and B. Yang, *RSC Adv.*, 2013, **3**, 24090–24098.

22. P. M. Foley, E. S. Beach and J. B. Zimmerman, *Green Chem.*, 2011, **13**, 1399–1405.

23. Z. Tan, M. Sestrick, W. Matakas and A. C. Venables, Process for making and using cellulose-containing seaweed residue and products made therefrom, *World Pat.*, WO2013/033598 A1, 2013.

24. V. L. Budarin, Y. Z. Zhao, M. J. Gronnow, P. S. Shuttleworth, S. W. Breeden, D. J. Macquarrie and J. H. Clark, , *Green Chem.*, 2011, **13**, 2330–2333.

25. R. P. John, G. S. Anisha, K. M. Nampoothiri and A. Pandey, *Bioresour. Technol.*, 2011, **102**, 186–193.

26. A. Domard, *Carbohydr. Polym.*, 2011, **84**, 696–703.
27. P. Kandra, M. M. Challa and H. K. P. Jyothi, *Appl. Microbiol. Biotechnol.*, 2012, **93**, 17–29.
28. B. Narayan, S. P. Velappan, S. P. Zituji, S. N. Manjabhatta and L. R. Gowda, *Waste Manage. Res.*, 2010, **28**, 64–70.
29. N. N. Win and W. F. Stevens, *Appl. Microbiol. Biotechnol.*, 2001, **57**, 334–341.
30. Y. Zhao, W.-T. Ju, G.-H. Jo, W.-J. Jung and R.-D. Park, in *Biotechnology of Biopolymers*, ed. M. Elnashar, InTech, Rijeka, Croatia, 2011, pp. 131–144, available from: http://www.intechopen.com/books/biotechnology-of-biopolymers/perspectives-of-chitin-deacetylase-research.
31. D. J. Macquarrie and J. J. E. Hardy, *Ind. Eng. Chem. Res.*, 2005, **44**, 8499–8520.
32. M. Kumar, R. A. A. Muzzarelli, C. Muzzarelli, H. Sashiwa and A. J. Domb, *Chem. Rev.*, 2004, **104**, 6017–6084.
33. F. Shahidi, J. K. V. Arachchi and Y.-J. Jeon, *Trends Food Sci. Technol.*, 1999, **10**, 37–51.
34. E. S. Beach, M. J. Eckelman, Z. Cui, L. Brentner and J. B. Zimmerman, *Bioresour. Technol.*, 2012, **121**, 445–449.
35. G. A. F. Roberts, *Chitin Chemistry*, Macmillan, Houndmills, Hampshire, UK, 1992.
36. H. Xie, S. Zhang and S. Li, *Green Chem.*, 2006, **8**, 630–633.
37. Y. Qin, X. Lu, N. Sun and R. D. Rogers, *Green Chem.*, 2010, **12**, 968–971.
38. Q. Chen, A. Xu, Z. Li, J. Wang and S. Zhang, *Green Chem.*, 2011, **13**, 3446–3452.
39. P. S. Barber, C. S. Griggs, G. Gurau, Z. Liu, S. Li, Z. X. Li, X. M. Lu, S. J. Zhang and R. D. Rogers, *Angew. Chem., Int. Ed.*, 2013, **52**, 12350–12353.
40. T. J. Trivedi, K. S. Rao and A. Kumar, *Green Chem.*, 2014, **16**, 320–330.
41. S. S. Silva, A. R. C. Duarte, J. F. Mano and R. L. Reis, *Green Chem.*, 2013, **15**, 3252–3258.
42. K. W. Omari, J. E. Besaw and F. M. Kerton, *Green Chem.*, 2012, **14**, 1480–1487.
43. A. Szabolcs, M. Molnar, G. Dibo and L. T. Mika, , *Green Chem.*, 2013, **15**, 439–445.
44. Y. Wang, C. M. Pedersen, T. Deng, Y. Qiao and X. Hou, *Bioresour. Technol.*, 2013, **143**, 384–390.
45. H. Sashiwa, S. Fujishima, N. Yamano, N. Kawasaki, A. Nakayama, E. Muraki and S.-I. Aiba, *Chem. Lett.*, 2001, 308–309.

46. M. W. Drover, K. W. Omari, J. N. Murphy and F. M. Kerton, *RSC Adv.*, 2012, **2**, 4642–4644.
47. K. W. Omari, L. Dodot and F. M. Kerton, *ChemSusChem*, 2012, **5**, 1767–1772.
48. T. Stahlberg, S. Rodriguez-Rodriguez, P. Fristrup and A. Riisager, *Chem. Eur. J.*, 2011, **17**, 1456–1464.
49. X. Chen, S. L. Chew, F. M. Kerton and N. Yan, *Green Chem.*, 2014, **16**, 2204–2212.
50. M. Osada, K. Kikuta, K. Yoshida, K. Totani, M. Ogata and T. Usui, *Green Chem.*, 2013, **15**, 2960–2966.
51. Y. Ohmi, S. Nishimura and K. Ebitani, *ChemSusChem*, 2013, **6**, 2259–2262.
52. J. Chen, M. Wang and C.-T. Ho, *J. Agric. Food Chem.*, 1998, **46**, 3207–3209.
53. R. A. Franich, S. J. Goodin and A. L. Wilkins, *J. Anal. Appl. Pyrolysis*, 1984, **7**, 91–100.
54. M. C. Barros, P. M. Bello, M. Bao and J. J. Torrado, *J. Clean. Prod.*, 2009, **17**, 400–407.
55. M. J. Frude, Method for removing excess meat from whole of crushed mollusk shellfish waste, *NZ Pat.*, NZ 555418A, 2008.
56. P. Lertwattanaruk, N. Makul and C. Siripattarapravat, *J. Environ. Manage.*, 2012, **111**, 133–141.
57. P. Ballester, I. Marmol, J. Morales and L. Sanchez, *Cement Concrete Res.*, 2007, **37**, 559–564.
58. S. Pena-Rodriguez, A. Bermudez-Couso, J. C. Novoa-Munoz, M. Arias-Estevez, M. J. Fernandez-Sanjurjo, E. Alvarez-Rodriguez and A. Nunez-Delgado, *J. Environ. Sci.*, 2013, **25**, 2476–2486.
59. N. Seco-Reigosa, S. Pena-Rodriguez, J. C. Novoa-Munoz, M. Arias-Estevez, M. J. Fernandez-Sanjurjo, E. Alvarez-Rodriguez and A. Nunez-Delgado, *Environ. Sci. Pollut. R.*, 2013, **20**, 2670–2678.
60. C. A. McCauley, A. D. O'Sullivan, M. W. Milke, P. A. Weber and D. A. Trumm, *Water Res.*, 2009, **43**, 961–970.
61. R. Rezaei, M. Mohadesi and G. R. Moradi, *Fuel*, 2013, **109**, 534–541.
62. J. H. Shariffuddin, M. I. Jones and D. A. Patterson, *Chem. Eng. Res. Des.*, 2013, **91**, 1693–1704.

PART C
SEPARATIONS AND PURIFICATIONS

CHAPTER 18

Overview of Separations, Purifications and Fractionations

DARRELL ALEC PATTERSON[a,b]

[a] Bath Process Intensification Laboratory, Department of Chemical Engineering, University of Bath, Claverton Down, Bath, UK, BA2 7AY;
[b] Centre for Sustainable Chemical Technologies, University of Bath, Claverton Down, Bath, UK, BA2 7AY
Email: d.patterson@bath.ac.uk; darrell.patterson@gmail.com

18.1 SEPARATIONS, PURIFICATIONS AND FRACTIONATIONS IN CHEMICAL PROCESSING

Separations processes are an often neglected unit operation when a new chemical, material and/or product are being developed. However, from an industrial process and processing perspective, it is crucial that they are considered as early on in the chemistry and chemical engineering development cycle as possible, since the ability to separate a wanted material from an unwanted one (*e.g.* the active pharmaceutical compound from a by-product that may be poisonous) may be the difference between a feasible industrial process for that product or not. Furthermore, the economic feasibility of a product and process may rely entirely on the efficacy of the separation process; in the past, separation processes have been found to account for 40–70% of industry capital and operating costs.[1] Consequently, the major separation processes and how these can make chemical processing more sustainable are considered in this volume.

Chemical Processes for a Sustainable Future
Edited by Trevor M. Letcher, Janet L. Scott and Darrell A. Patterson
© The Royal Society of Chemistry 2015
Published by the Royal Society of Chemistry, www.rsc.org

The main separation unit operations used in chemical processing include:

- distillation
- membrane separations
- extraction, including liquid–liquid extraction
- crystallisation
- stripping, including gas and steam stripping
- adsorption
- absorption
- chromatography
- ion exchange
- flotation
- leaching
- precipitation and cementation
- reactive separations
- hybrid systems combining two or more of the unit operations above

The following chapters cover membrane separations, liquid and gas extractions, and separations in mining and metal processing (including flotation, leaching, precipitation and cementation, solvent extraction and ion exchange unit operations). These separation unit operations cover most of the main ones that currently and in the future will have major impacts in making chemical processes more sustainable both as separate unit operations and integrated with other unit operations for process intensification (especially as a reaction–separation combination, such as in membrane reactors). Of the remaining unit operations that are not covered, this is not a reflection of their importance to sustainability, but rather due to both page limitations and coverage elsewhere. In particular, there four important separation unit operations not covered:

1: Distillation is the major separation unit used in chemical and process engineering (despite being energy intensive).[2] Distillation involves the separation of one or more components from a mixed solution utilising a difference in volatilities between the components. The conventional design and use of distillation columns has remained largely unchanged since the early 20th century. As a consequence, the basics of distillation, distillation column design and applications are well described elsewhere.[3–5] Distillation columns are been used to produce more sustainable separations and processing (compared with conventional distillation) and this has involved the integration of other unit operations with distillation including membrane distillations,[6,7] reactive distillations[8,9] and extractive distillation,[10] as well as novel internals and operating regimes.[2,11]

2 and 3: Absorption (the separation of one or more components from a mixed solution through uptake into the volume of another species) and adsorption (the separation of one or more components from a mixed solution through the adhesion of these components to a solid surface) are also very important unit operations that could be used where membrane separation and extractions are unsuitable. The sustainable applications of absorption and adsorption and their use in process intensification is covered well elsewhere.[12,13]

4: Crystallisation (the separation of one or more components from a mixed solution through a selective change to the solid phase of these components) is an extremely important separation and purification unit operation for the pharmaceutical and fine chemical industries in particular. These industries predominantly use batch (rather than continuous) processes, and one significant barrier preventing the conversion to continuous processes (and the advantages of doing so as outlined in Chapter 1.5.1) is the development of continuous crystallisation. This book does not cover crystallisation in detail, but does touch upon it in Chapter 24, in precipitation and cementation for the recovery mining and mineral streams. For pharmaceutical and fine chemical industry applications in particular, there is significant work in this area, covered well elsewhere.[14–18]

Even with these absences, overall this collection of chapters covering the chemistry and chemical engineering of separations provides a useful and broad coverage of the importance and contribution that good separation technologies give to the overall topic of sustainable chemical technologies. This is because improving the energy efficiency, the cost effectiveness and performance of separations is important for a wide range of industries, including biofuels and bio-derived molecules[19,20] (see also Chapters 19 and 20), water (and improving membrane separations in particular[21,22]), food and beverage (especially diary[23,24]), petrochemical,[25] pharmaceutical,[18,26] and fine chemical industries.[27,28] The techniques and technologies covered here therefore can have an impact in a wide range of industrial sectors—consequently please consider their use beyond the applications covered within these pages.

18.2 OVERVIEW OF SEPARATIONS IN THIS BOOK

There are six chapters covering separations as a sustainable chemical technology. Each chapter covers one topic in a complete way, but viewed together each complements and extends each other providing a more comprehensive overview of the basics of each technique and/or unit operation, looking at current and future sustainable chemical industry applications. In brief the content of each of these chapters is as follows.

Chapter 19 on 'Membrane Separations: From Purifications, Minimisation, Reuse and Recycling to Process Intensification' concentrates on molecular separations and purifications in the liquid phase by membranes. The basics of membrane operations is covered followed by the main applications of membranes for sustainable processing and process intensification. The chapter ends by looking at two major future applications of membranes that are currently challenging for other separation unit operations: (1) the use of membrane processes to concentrate dilute organics and biofuels from fermentation broths; and (2) membrane linked tandem chemo- and biocatalysed reactions using previously incompatible catalysts.

Chapter 20 concentrates on 'Ionic Liquids and their Application to a More Sustainable Chemistry'. This complements the coverage of ionic liquids for gas separations covered in Chapter 22. The basics of ionic liquids is covered, followed by the applications of ionic liquids in green chemistry, biomass processing (cellulose dissolution, fractionation of lignocellulosic biomass and extraction of valuable ingredients from biomass), catalysis, as well as separations of liquid and gas streams. Separations covered include the use of ionic liquids for separations for flue gas cleaning and removal of mercury.

Chapter 21 is an extensive overview and review of 'Liquid–Liquid Extraction' as used as a sustainable chemical technology. The chapter includes the use of supercritical and low vapour pressure (ionic liquid) as green solvents as well as traditional solvents. The applications discussed include the separation of similar boiling point petrochemicals, protein recovery, biomolecule recovery, recover of metals and terepene extraction from citrus oils. It has a major section on the recovery of organics (such as biofuels and biochemicals) from aqueous systems, which can be directly compared with the membrane operations used for the same application as covered in Chapter 19. An overview of contacting equipment is given, including batch, continuous, membrane and micro-channel contactors.

Chapter 22 looks at an alternative, potentially more sustainable technique and technology to the standard cryogenic distillation method for gas separations, extending Chapter 20 by covering the gas separations capability and applications of ionic liquids in more detail. In particular the chapter looks at solubility of ethane, ethylene, acetylene, propane, propylene and methyl acetylene in pure ionic liquids.

Chapter 23 provides more detail on the supercritical extractions briefly covered in Chapter 20 by looking at the extraction of functional compounds using supercritical carbon dioxide. This includes compounds that are of interest to a wide range of industries, including oleochemicals, terpenoids, alkaloids and metals.

Finally, Chapter 24 is a case study of producing more sustainable separation unit operations in the mining and metals processing industry. This chapter is the ideal end to this separations section since it covers a number of separation techniques and unit operations not considered in the previous chapters including flotation, leaching, precipitation and cementation, and ion exchange. The application of solvent extraction to mining and metal processing is also covered and this can be compared and contrasted with the extractions covered in Chapters 20, 21 and 23.

REFERENCES

1. J. L. Humphrey and G. E. Keller, *Separation Process Technology*, McGraw-Hill, New York and London, 1997.
2. Ö. Yildirim, A. A. Kiss and E. Y. Kenig, *Sep. Purif. Technol.*, 2011, **80**, 403.
3. J. H. Harker, J. R. Backhurst and J. F. Richardson, *Chemical Engineering*, Elsevier Science, Oxford, 5th edn, 2002.
4. P. C. Wankat, *Separation Process Engineering: Includes Mass Transfer Analysis*, Pearson Education, London, 2011.
5. A. A. Kiss, *Advanced Distillation Technologies: Design, Control and Applications*, Wiley, Chichester, UK, 2013.
6. E. Curcio and E. Drioli, *Sep. Purif. Rev.*, 2005, **34**, 35.
7. A. Alkhudhiri, N. Darwish and N. Hilal, *Desalination*, 2012, **287**, 2.
8. A. A. Kiss, in *Process Intensification for Green Chemistry: Engineering Solutions for Sustainable Chemical Processing*, ed. K. Boodhoo and A. Harvey, John Wiley & Sons, Chichester, UK, 2013, ch. 9, pp. 251–274.
9. G. J. Harmsen, *Chem. Eng. Process.*, 2007, **46**, 774.
10. Z. Lei, C. Li and B. Chen, *Sep. Purif. Rev.*, 2003, **32**, 121.
11. V. N. Maleta, A. A. Kiss, V. M. Taran and B. V. Maleta, *Chem. Eng. Process.*, 2011, **50**, 655.
12. J. Rouquerol, F. Rouquerol, P. Llewellyn, G. Maurin and K. S. W. Sing, *Adsorption by Powders and Porous Solids: Principles,*

Methodology and Applications, Elsevier Science, Amsterdam, 2nd edn, 2014.

13. A. A. Kiss, in *Process Intensification for Green Chemistry: Engineering Solutions for Sustainable Chemical Processing*, ed. K. Boodhoo and A. Harvey, John Wiley & Sons, Chichester, UK, 2013, ch. 11, pp. 289–306.

14. W. Beckmann, in *Crystallization: Basic Concepts and Industrial Applications*, ed. W. Beckmann, Wiley-VCH, Weinheim, Germany, 2013, pp. 173–185.

15. G. Hofmann and C. Melches, in *Crystallization: Basic Concepts and Industrial Applications*, ed. W. Beckmann, Wiley-VCH, Weinheim, Germany, 2013, pp. 203–233.

16. G. Hofmann and C. Melches, in *Crystallization: Basic Concepts and Industrial Applications*, ed. W. Beckmann, Wiley-VCH, Weinheim, Germany, 2013, pp. 305–324.

17. P. Moschou, H. J. M. de Croon Mart, J. van der Schaaf and C. Schouten Jaap, *Rev. Chem. Eng.*, **30**, 127.

18. J. Chen, B. Sarma, J. M. B. Evans and A. S. Myerson, *Cryst. Growth Des.*, 2011, **11**, 887.

19. I. M. Atadashi, M. K. Aroua and A. A. Aziz, *Renewable Energy*, 2011, **36**, 437.

20. H.-J. Huang, S. Ramaswamy, U. W. Tschirner and B. V. Ramarao, *Sep. Purif. Technol.*, 2008, **62**, 1.

21. W. Guo, H.-H. Ngo and J. Li, *Bioresour. Technol.*, 2012, **122**, 27–34.

22. K. P. Lee, T. C. Arnot and D. Mattia, *J. Membr. Sci.*, 2011, **370**, 1.

23. A. Y. Tamime (ed.), *Membrane Processing: Dairy and Beverage Applications*, Wiley-Blackwell, Chichester, UK, 2013.

24. B. W. Woonton, U. Kulozik, K. De Silva and G. W. Smithers, in *Advances in Dairy Ingredients*, ed. G. W. Smithers and M. A. Augustin, Wiley-Blackwell, Chichester, UK, 2013, pp. 137–159.

25. M. Takht Ravanchi, T. Kaghazchi and A. Kargari, *Desalination*, 2009, **235**, 199.

26. J. Vanneste, D. Ormerod, G. Theys, D. Van Gool, B. Van Camp, S. Darvishmanesh and B. Van der Bruggen, *J. Chem. Technol. Biotechnol.*, 2013, **88**, 98.

27. K. J. Carpenter and W. M. L. Wood, *Adv. Powder Technol.*, 2004, **15**, 657.

28. A. Bennett, *Filtr. Sep.*, 2013, **50**, 30.

Membrane Separations: from Purifications, Minimisation, Reuse and Recycling to Process Intensification

DARRELL ALEC PATTERSON,*[a,b] CHRISTOPHER JOHN DAVEY[a,b] AND ROSIAH ROHANI[c]

[a] Bath Process Intensification Laboratory, Department of Chemical Engineering, University of Bath, Claverton Down, Bath, UK, BA2 7AY; [b] Centre for Sustainable Chemical Technologies, University of Bath, Claverton Down, Bath, UK, BA2 7AY; [c] Department of Chemical and Process Engineering, Faculty of Engineering and Built Environment, Universiti Kebangsaan Malaysia, Bangi, 43600 Selangor, Malaysia
*Email: d.patterson@bath.ac.uk; darrell.patterson@gmail.com

19.1 INTRODUCTION

In recent years, a range of separations have been introduced into process industries based on one simple concept: a membrane. Membranes are defined as a selective barrier which separates a mixed fluid stream into two different streams through restricting the transport of the constituent chemical species either by size exclusion, or diffusion and partitioning by a selectivity of differing chemical properties. Membranes are now vital in many of the world's major industries including; dairy, food and beverage, chemical manufacture

Chemical Processes for a Sustainable Future
Edited by Trevor M. Letcher, Janet L. Scott and Darrell A. Patterson
© The Royal Society of Chemistry 2015
Published by the Royal Society of Chemistry, www.rsc.org

and wastewater treatment. The use of membranes has also revolutionised sustainable practices in some industrial sectors by allowing the recovery and reuse or sale of previously wasted materials.[1] This practice enables these industries to be more environmentally sustainable by reducing the amount of waste produced and more cost-effective, as less of these high-value materials are needed. Furthermore, the combination of membranes with other unit operations (such as reactors) produces new process intensification technologies, providing a means to selectively retain molecules in these unit operations, overcome equilibrium and as an *in situ* facile method for product separation, further enabling sustainability benefits. Membrane processes also require modest energy consumption compared with other more conventional separation techniques (such as distillation), making them an attractive alternative when considering cleaner more sustainable technologies. All of this makes membrane separations an attractive option when choosing between different separation unit operations. The advantages of the membrane separation processes compared with conventional separations such as distillation, absorption and extraction are highlighted in Table 19.1.

Membranes are used for the separation of both gaseous and liquid streams in chemical processing. To limit the scope of this chapter, it deals exclusively with liquid membrane separations and technologies. The separations described herein are therefore between particles and solutes in a solvent. There are, however, excellent reviews available on gaseous membrane separations (for example, see references 2 and 3).

There have also been many excellent books, book chapters and review papers on the basics of membrane separations and the application of membranes to sustainable chemical processes that complement the information in this chapter. Therefore for further reading on membrane separations and in particular their application to sustainable chemical processing, we refer readers to references 4–12.

Table 19.1 Advantages of membrane separation processes compared with conventional separations.

Conventional separations	Membrane processes
Complex systems	Process is based on molecular size/weight
May involve chemical separation	Require modest energy
May require addition of solvent and/or additives	Involve no additional waste products
May produce waste products	Have relatively low capital and running costs
Expensive for low concentration/ azeotropic conditions	Can be done directly without using additives
	Flexibility as the membrane is 'tailor-made' to meet individual requirements

This chapter does not intend to cover membrane separations from the same point of view as these excellent works. Instead, this chapter provides a basic overview of liquid separation membrane operations and investigates two topical case studies of developing and increasingly important applications of membranes that in the future will have significant impacts on increasing the sustainability of the chemical processes involved. These are:

- Case Study I: Concentration of dilute organics from fermentations
- Case Study II: Separation of incompatible reaction systems.

19.2   MEMBRANE SEPARATION BASICS

19.2.1   Membrane System Definitions

19.2.1.1   The Membrane Process.   Membrane separation processes involve the permeation of liquids or gases through a selective barrier under a specific driving force, *i.e.* pressure, concentration and potential. The type of driving force defines the type of membrane operation (see Section 19.2.2). A membrane selectively splits a feed into two streams (Figure 19.1):

1. The *retentate* or *concentrate*: the fraction that does not pass through the membrane (*i.e.* the portion retained). This consists of solutes that cannot be transported through the membrane for various

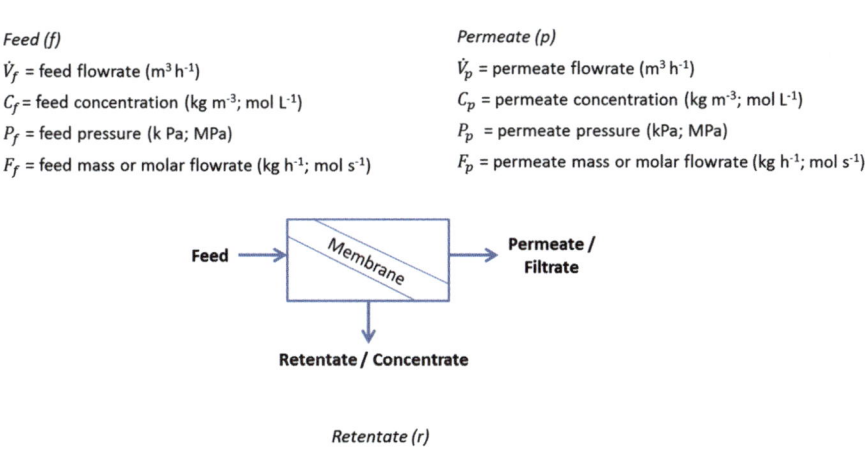

Feed (f)

$\dot{V}_f$ = feed flowrate (m³ h⁻¹)

C_f = feed concentration (kg m⁻³; mol L⁻¹)

P_f = feed pressure (k Pa; MPa)

F_f = feed mass or molar flowrate (kg h⁻¹; mol s⁻¹)

Permeate (p)

$\dot{V}_p$ = permeate flowrate (m³ h⁻¹)

C_p = permeate concentration (kg m⁻³; mol L⁻¹)

P_p = permeate pressure (kPa; MPa)

F_p = permeate mass or molar flowrate (kg h⁻¹; mol s⁻¹)

Feed → **Membrane** → **Permeate / Filtrate**

Retentate / Concentrate

Retentate (r)

$\dot{V}_r$ = Retentate flowrate (m³ h⁻¹)

C_r = Retentate concentration (kg m⁻³; mol L⁻¹)

P_r = Retentate pressure (kPa; MPa)

F_r = Retentate mass or molar flowrate (kg h⁻¹; mol s⁻¹)

Figure 19.1   Basic membrane operation and main stream parameters.

reasons (they are rejected by the membrane) and part of the solvent bound to these solutes as well as the solvent that did not have sufficient time and/or driving force to pass through the membrane.

2. The *permeate* or *filtrate*: the fraction that passes through the membrane (*i.e.* the portion permeated). This consists of the solvent and solutes that are able to pass through the membrane *via* one of the three transport mechanisms outlined in Section 19.2.4.

19.2.1.2   Important Characteristics of a Viable Membrane.   The ability of membranes to control the rate of permeation (*i.e.* flux and selectivity) of different chemical species is what enables their function for separations. For membranes to be useful for commercial separation processes, the membranes must exhibit the following characteristics:[13]

- high flux
- high selectivity
- mechanical stability
- complete tolerance to feed stream components (including fouling resistance, unless fouling is desired)
- tolerance to temperature variations

High flux and selectivity in fabricated membranes are desired for their use at large scale and maximum separation.[14] The membranes are also expected to have high mechanical stability and wide temperature variance to allow their use in a diverse array of applications.[12] Membranes, however, do have limited lifetimes and must be replaced periodically. The lifetime of a membrane depends strongly on the application and use conditions, but in a typical membrane application, it is expected that the membrane be replaced every 2–5 years. Fouling, which is a common limitation in membranes, must be minimised to ensure that the membrane possesses long-term stability without the necessity of frequent back flushing or replacement.[4,15] Otherwise, an increase in maintenance costs can be expected to run the system. Fouling is also of great concern to a majority of membrane applications because it is directly related to the decrease in product fluxes and the fractionating ability of the membranes.[16,17]

The rate at which the permeate flows through the membrane is usually represented by the volumetric or mass flux (J) through the area of the membrane perpendicular to the direction of flow, $J = F_p/A$, typically expressed in units of kg m^{-2} h^{-1} or L m^{-2} h^{-1} (or LMH), where h refers to hours:

$$J = \frac{1}{A}\frac{dV}{dt} = \frac{\dot{V}_p}{A} \tag{19.1}$$

Flux is a property of the membrane system and is a precise characteristic of the flow through the membrane area, but is only representative of a membrane at a given operating condition, since flux may vary non-linearly with the driving force applied to effect the separation (*e.g.* pressure). Therefore, when the driving force is a transmembrane pressure difference, often a more general flow characteristic of the system is used instead, called the 'permeance'. This normalises for transmembrane pressure, so is easier to measure and use (when measured and defined over a small set of pressures), but may not well represent the flow through the membrane for a wide range of operating pressures. It is a performance measurement, not a system property.

$$\text{Permeance} = \frac{J}{\Delta P} = \frac{1}{dP \cdot A}\frac{dV}{dt} = \frac{\dot{V}_p}{\Delta P \cdot A} \tag{19.2}$$

The selectivity of a membrane is the other important process property that needs to be characterised. The main parameter for describing the membrane selectivity in liquid separations is the rejection of the membrane. The rejection characterises how much solvent and other unwanted solutes can be separated from the solvent/solutes of interest. This therefore also indirectly defines the permeate quality produced by the membrane. Membrane rejection of a species i ($R_{j,i}$) is defined by Equation (19.3):

$$
\begin{aligned}
R_{j,i} &= \left(1 - \frac{\text{concentration of species } i \text{ in permeate}}{\text{concentration of species } i \text{ in feed}}\right) \times 100\% \\
&= \left(1 - \frac{C_{i,p}}{C_{i,f}}\right) \times 100\%
\end{aligned}
\tag{19.3}
$$

with concentrations in matching units.

The basic limits on rejection define the usefulness of the membrane or overall membrane module. So $R_j = 100\%$ for perfect rejection (all the solute is removed from the solvent) and $R_j = 0\%$ for no separation. Ultimately, the rejection is a limited measure of selectivity for

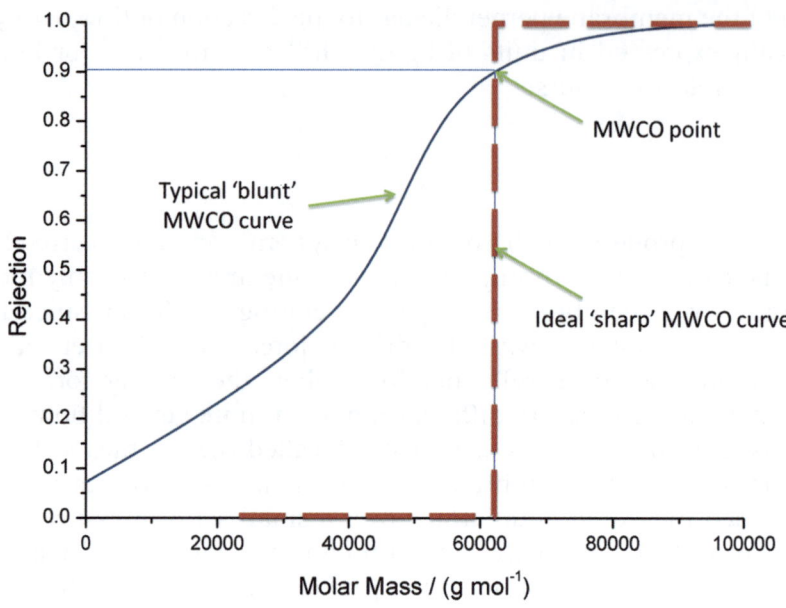

Figure 19.2 Ideal *versus* typical MWCO curves.

a particular species and like permeance is a performance measurement, not a system property that can be used to more generally characterise the separation capability of the membrane.

A more general approach to characterise the selectivity of a particular membrane (in a particular solvent) is the molar mass-cut off (MWCO) of the membrane. The MWCO is the molar mass (MW) for which 90% rejection of the solute is achieved by the membrane (Figure 19.2). In theory, this means that any molecule with a MW higher than the MWCO will have a high rejection. As shown in Figure 19.2, the ideal MWCO curve is a sharp curve that indicates a complete rejection of molecules at a particular MW. That implies that the membrane is highly selective as it gives a sharp rejection between close MW compounds. Unfortunately, this is not always the case, as MW is not the sole molecular property that controls membrane selectivity; shape, solubility, charge, concentration polarization and fouling also affect transport through a membrane. Therefore mass transport through membranes is not solely dependent on MW and most membranes do not give the ideal sharp separation. A blunt MWCO curve is therefore more typical: this means the membrane cannot give a sharp separation between molecules of similar molecular weights. Essentially, the MWCO curve gives an understanding of the difference in MW needed for a good separation. But since MW is not the only factor to affect separation, this may still not be a good predictor of the selectivity between two different solutes. Furthermore, the implication of this is that MWCO is not a universal classification of the separation that can be achieved by a membrane and can also sometimes be misleading (see ref. 18). Nevertheless, the MWCO curve is the best general characterisation of an individual membrane's selectivity available and therefore is used. For a more in-depth treatment of MWCO as well as a review of measurement techniques see ref. 1.

19.2.1.3 Modes of Membrane Filtration. There are two main modes of filtration utilised in membrane processes: dead-end (Figure 19.3) and cross-flow (Figure 19.4). In dead-end filtration, the feed flow is perpendicular to the membrane. As there is no retentate flow in this process, retentate builds up against the surface of the membrane until no more permeate can be collected. Therefore dead-end systems require regular backwashing, cleaning and/or replacement of membranes to overcome this fouling. Dead-end membranes, however, can better guarantee a complete rejection of molecules from the feed stream, since all flow must pass through the membrane. They are used in many industrial processes where guaranteed rejection of key species is needed, such as pre-treatment of desalination feedwater before reverse osmosis.

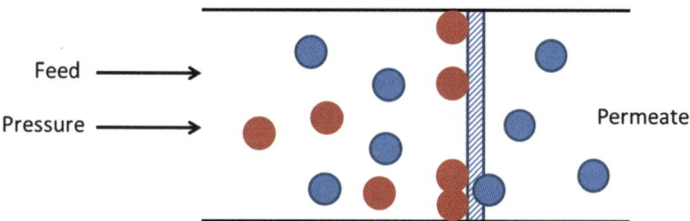

Figure 19.3 Schematic of a dead-end membrane filtration process.

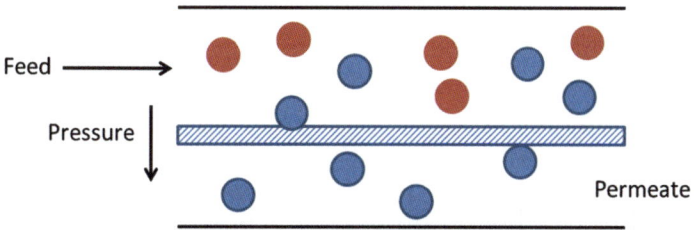

Figure 19.4 Schematic of a cross-flow membrane filtration process.

In cross-flow filtration, the membrane lies parallel to the flow of the feed and not all of the solvent passes through the membrane. In contrast to dead-end filtration, this allows a flow of the retentate stream across the surface of the membrane (hence 'cross-flow'), minimising concentration polarisation and fouling on the membrane. Higher feed/retentate flow rates can therefore be used to control fouling. Cross-flow systems are the most commonly used in industrial processes due to potential for fouling control.

19.2.2   The Different Membrane Processes

There are many different membrane operations available differentiated by the following factors:

1. *The type of process:* microfiltration (MF), ultrafiltration (UF), nanofiltration (NF), reverse osmosis (RO), forward osmosis (FO), dialysis/osmosis, electro-dialysis/electroosmosis, diafiltration, pervaporation and vapour permeation (see Section 19.2.3)
2. *The nature of the driving force:* the driving force uses differences in: pressure, concentration (or fugacity), vapour pressure, electromotive force (see Section 19.2.3)
3. *The separation mechanism:* the main three separation mechanisms are pore flow, solution diffusion and Donnan exclusion (see Section 19.2.4)
4. *Type of material the membrane is made of:* i.e. polymer, ceramic, biological, metallic; whether the membrane is symmetric or asymmetric; whether the membrane material is porous (and so functions *via* pore flow) or dense (and so functions *via* solution diffusion) or a mixture of the two, or if the membrane is charged (so functions *via* Donnan exclusion) or uncharged (see Baker[4] for further details).
5. *The type of membrane module used:* The membrane module is the complete membrane unit comprising: the membrane, a support structure (this will be a pressure support structure for pressurised membranes), feed inlet and outlet permeate and retentate ports and an overall support structure. Several types of membrane modules can be used, including:
   - spiral wound modules (incorporating layered flat sheet membranes),
   - plate and frame modules (for flat sheet membranes),
   - modules for tubular membranes, and
   - modules for hollow fibre membranes
     These are not detailed here, but further information can be found in references 4 and 19.

6. *The selectivity and size of separation achieved:* separation is typically measured in terms of molar mass-cut off (MWCO) as discussed previously (see Section 19.2.1.2)

19.2.3 Relationships between Driving Forces and Membrane Processes

The type of membrane process and driving force for transport across the membrane are closely related and are summarised together. The main driving forces and associated membrane processes are summarised in Table 19.2.

This chapter focuses on the two driving forces of pressure and vapour pressure and their associated membrane processes. Further details of other driving forces and their associated membrane processes can be found in references 4, 10 and 19.

19.2.3.1 Pressure Difference Driving Force. Here, the transmembrane pressure difference is the driving force for mass transfer across the membrane:

$$\Delta P_{tm} = \frac{P_f + P_r}{2} - P_p \qquad (19.4)$$

where ΔP_{tm} is the transmembrane pressure difference, P_i is the pressure of stream i, $i=f$, feed, r is retentate (concentrate) p is permeate.

The membrane separation processes operating under a pressure driving force are summarised hierarchically in Figure 19.5. The differences between the main processes are:

- Microfiltration is a pressure driven process that uses a porous membrane to separate molecules and/or particles between 0.1

Table 19.2 Main driving forces for separation across the membrane and the associated membrane processes.

Driving force: a difference in …	Membrane process
Pressure	Microfiltration, ultrafiltration, nanofiltration, reverse osmosis
Vapour pressure	Pervaporation, vapour permeation
Osmotic pressure	Forward osmosis
Electrical potential	Electrodialysis/electroosmosis
Concentration	Dialysis/osmosis

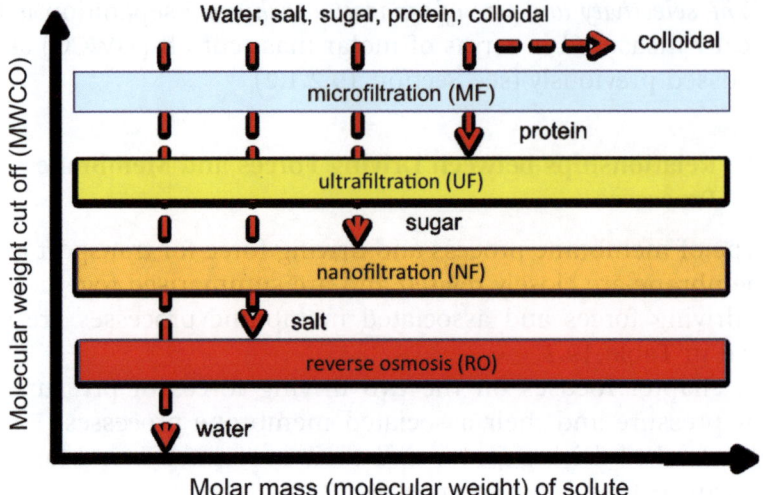

Figure 19.5 Overall summary of the differences between the various pressure-driven membrane processes (based on ref. 20).

and 10 μm in size. It can be used, for example, to retain solid particles such as yeast and bacteria.

- Ultrafiltration uses membranes with average pore sizes of between approximately 1 nm and 100 nm (or 10–1000 Å). Typically an UF membrane can be used for retaining proteins, cells, starch and enzymes and is applicable to molecules with molecular weights greater than 2000 g mol^{-1} up to the microfiltration range.

- Nanofiltration membranes typically have pore sizes in the 1–10 nm range and a MWCO range of 200–2000 g mol^{-1}. NF is considered attractive for various applications as it is commonly used to separate low MW organics/multivalent salts from monovalent salt water/solvents. These criteria are suitable for refinery processes and purification of fine chemicals to overcome the limitations from traditional separation techniques such as extraction and distillation.

- Reverse osmosis membranes are typically dense membranes without distinct pores. They are considered to be a barrier membrane, since they should only permeate the solvent leaving all solutes behind. The largest application is in desalination (see Section 19.3.1).

A typical dead-end and cross-flow pressure driven membrane process would be engineered with the unit operations and process flow shown in Figure 19.6.

(a)   **(b)**

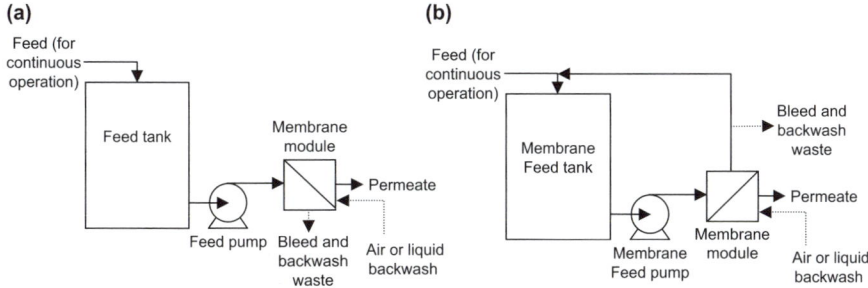

Figure 19.6   (a) Process schematic of a pressure driven dead-end membrane process.
(b) Process schematic of a pressure driven cross-flow membrane process.

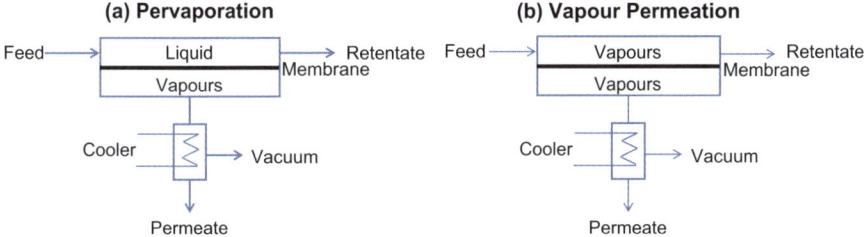

Figure 19.7   General schematics of the (a) pervaporation and (b) vapour permeation
processes.

19.2.3.2   Vapour Pressure Driving Force.   In pervaporation and vapour
permeation, a difference in partial vapour pressures across the membrane is the driving force for transmembrane transport. Figure 19.7
shows the general schematic for these processes. Combining evaporation and membrane permeation, pervaporation is unique among membrane separations as it involves a liquid–vapour phase change. A liquid
feed contacts one side of the membrane and permeate vapours are
withdrawn from the other. The difference in partial vapour pressures
can be induced by vacuum, temperature gradient or sweep gas.

Vapour permeation is analogous to pervaporation but differs in the
fact that no phase transition occurs. The feed, consisting of a mixture of
vapours or vapours and gases, contacts the membrane and the vapour
permeate is withdrawn from the other side. The difference in partial
vapour pressures is again induced by either a vacuum or sweep gas.

Membrane transport is generally described by the solution-
diffusion model[21,22] (see Section 19.2.4), occurring in three steps:

- selective sorption;
- selective diffusion through the membrane;
- desorption into a vapour phase.

Separation is therefore dictated by the chemical nature and physical structure of the membrane. The chemical interactions of the permeate with the membrane material achieves the selective sorption and diffusion required for separation.

Unlike distillation, the relative volatility of the permeate does not affect the separation of the process. For example, in a hydrophilic membrane, water will preferentially permeate over organic species, dehydrating the feed. In contrast, a hydrophobic membrane will enrich organic species over water in the permeate. This allows for the facile separation of azeotropic and close boiling mixtures. Being driven by a difference in vapour pressures, pervaporation can concentrate ethanol in aqueous solutions from 85 to 99 wt-%. For the same concentration using reverse osmosis, extremely high operating pressures would be required to overcome the osmotic pressure.

The main commercial application of pervaporation is thus for the dehydration of organic solvents such as ethanol and isopropanol, which form azeotropic mixtures with water.[23] High quality 'anhydrous' ethanol for the pharmaceutical industry can be produced by avoiding the use of any toxic third components used in typical azeotropic distillations. A further newer development is discussed in Section 19.4. Common industrial applications for vapour permeation include the recovery of recycling monomers and inert gases in the petrochemical industry, conditioning of natural gas to remove hydrocarbons and acid gases, and the recovery of liquefied petroleum gas (LPG).[18]

19.2.4 Solute and Solvent Transport and Selectivity Mechanisms

For many conventional membranes, three main solvent and solute transport mechanisms are thought to exist: pore-flow, solution diffusion and Donnan exclusion.

19.2.4.1 The Pore-flow Model. The pore-flow model is mainly used to describe the permeation through porous membranes *i.e.* UF and MF.[4,11] It describes the separation based on pressure-driven convective flow through tiny pores. Selectivity results from exclusion and is based on the incompatibility of molecule parameters such as size, shape and charge, with the same properties in the pores of the membrane. The basic equation covering the transport mechanism in the pore-flow model is Darcy's Law:[4]

$$J_i = K' c_i \frac{dp}{dx} \tag{19.5}$$

where J_i is the flux of component i, dp/dx is the pressure gradient existing in the porous medium, c_i is the concentration of component i in the medium and K' is a coefficient reflecting the nature of the medium.

The pore-flow model in NF membranes has also been proposed from the Hagen-Poiseuille model and the Jonsson and Boesen model.[15] These models have been mainly used to model solute flux in aqueous systems in UF and MF. However, these models only consider the solvent's viscosity as the solvent parameter. Other membrane characteristics such as porosity and pore size were not included.[15] So despite being used, the models may therefore not well describe the transport phenomena involved.

To overcome this, various groups have used the pore-flow model with modifications on these basic equations to describe the transport through NF membranes.[18,24-26] Since the pores in this membrane are a few times the size of water and dissolved solute molecules (less than 3 nm), the transport across NF membranes *via* convective flow, which applies for membranes with pores around 0.1 to 10 μm, is substantially hindered.[24] Therefore, Knudson diffusion and molecular sieving have been proposed to describe the transport properties in a porous membrane for nano range solutes.[24,26-29] Knudson diffusion is different from the molecular sieving, because the transport rate is inversely proportional to the square root of the solutes MW. Molecular sieving (also known as surface diffusion) involves diffusion of the absorbed species on the surface of the pores to diffuse through the membrane.[4]

19.2.4.2 The Solution-diffusion Model. The solution-diffusion model is typically used to describe transport through pore-less (dense) membranes, like those used in RO, gas separation, vapour permeation and pervaporation. It has also been used to describe the transport through polymeric NF membranes. It assumes that the permeating species dissolves in the membrane material and diffuses through the membrane down a concentration gradient. A separation is achieved between different solutes due to a difference in the amount of solute that dissolves into the membrane and the rate at which it diffuses through it.[4,21,22] The model is based on the solvent viscosity and its molar volume, the surface tension of the membrane material, and the sorption value describing the membrane–solvent interactions. Transport in the solution-diffusion model can be, in its most basic form, quantitatively described by Fick's first law:

$$J_i = -D_i \frac{dc_i}{dx} \tag{19.6}$$

where J_i is the flux of component i, D_i is the diffusion coefficient and dc_i/dx is the concentration gradient of component i across the media.

For further information see the excellent overviews and reviews of solution diffusion models and their applicability by Wijmans and Baker.[21,22]

19.2.4.3 Donnan Exclusion. The Donnan exclusion mechanism is used to explain the flux profile during filtration of charged solutes and is widely applied for charged or neutral membranes, in particular NF membranes. The membranes are made charged/conductive either by the addition of functional polymers (*e.g.* conducting polymers) or functional chemicals (*e.g.* acidic compounds). Donnan exclusion works on the exclusion of identically charged mobile ions to fixed groups called co-ions. The selectivity of the membranes results from the sorption of counter ions and the exclusion of co-ions of the membrane phase. These counter and co-ions can be cationic or anionic depending on the fixed charge groups.[30] An example of Donnan exclusion in NF is the Donnan steric pore model, proposed by Bowen *et al.*[27] The model showed successful prediction of NF performance in dye-salt diafiltration, which is a combination of both the Donnan exclusion and sieving mechanisms.[27] The ion transport was based on the extended Nernst–Plank equation, which was modified to include hindered transport, a combination of Donnan (electrical) and sieving (steric) mechanisms.[31]

19.2.4.4 Generalised Transport Models for Mixed Transport Systems. Overall, it is often a mixture of transport models that affect the transport through any one membrane and so specific models are not applied when trying to describe the flux through a particular membrane system. Instead, more general, empirical flux models are applied, with empirical overall mass transfer resistances (R_{ov}) or overall mass transfer coefficients ($K_{ov} = 1/R_{ov}$) used to describe the phenomena of the transport models outlined above. Simple versions that can be used for rough sizing of membrane area (which is the key sizing parameter for membranes) usually take the form of equations relating the driving force to the flux or flow rate through the membrane. For example, for a concentration (osmosis) driving force across the membrane, the relationship is as in Equations (19.7) and (19.8) and illustrated in Figure 19.8:

$$\dot{V}_p = J \cdot A = K_{ov} \cdot A \cdot (C_1 - C_2) = \frac{A \cdot (C_1 - C_2)}{R_{ov}} = \frac{A \cdot (C_1 - C_2)}{R_1 + R_m + R_2} \quad (19.7)$$

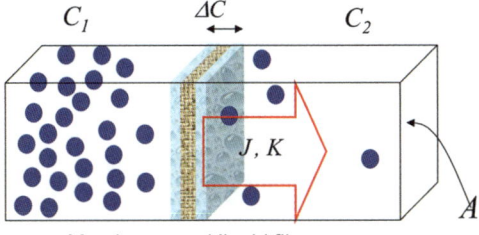

C_1 ΔC C_2

J, K

A

Membrane and liquid films

Figure 19.8 Schematic representation of the concentration differences, resistances and flows in a membrane system driven by a concentration difference.

$$\frac{1}{K_{ov}} = \frac{1}{k_1} + \frac{1}{k_m} + \frac{1}{k_2} \qquad (19.8)$$

The model indicates that there are at least three mass transfer resistances for transport of species across the membrane:

- $R_1 = 1/k_1$ and is mass transfer resistance due to the stagnant liquid film on the retentate side of the membrane;
- $R_m = 1/k_m$ is the mass transfer resistance due to the membrane (this can have other mathematical forms relating thickness, diffusivity and partitioning too);
- $R_2 = 1/k_2$ is the mass transfer resistance coefficient due to the stagnant film on the permeate side of the membrane.

Further resistances can be added to account for membrane fouling, concentration polarisation and other effects that inhibit flux.

The simplest and most universal model to use for transport when there is a pressure driving force for transport across a membrane is given in Equation (19.9) and summarised in Figure 19.9:

$$\dot{V}_{p,w} = J_W \cdot A = K \cdot A(\Delta p - \Delta \pi) \qquad (19.9)$$

where:

$\dot{V}_{p,w}$ is the permeate solvent volumetric transfer flow rate in units of volume/(time) *e.g.* L h^{-1}, m^3 s^{-1};

K is the overall mass transfer coefficient in units of volume/(time.area.pressure) at a particular temperature T, *e.g.* L d^{-1} m^{-2} kPa^{-1} @ 25 °C, where d refers to day;

A is the membrane cross-sectional area (length^2);

Δp is the average imposed transmembrane pressure gradient;

$\Delta \pi$ is the osmotic pressure difference between the feed and the permeate.

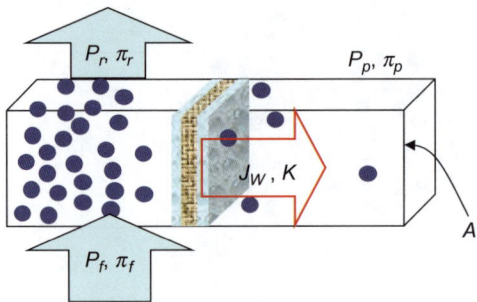

Figure 19.9 Schematic representation of the pressures and flows in the generalised pressure driven membrane system.

Osmotic pressure is the effective pressure that stops the unequal flow rates of water when the system is at equilibrium after osmosis. This pressure is equivalent to the pressure exerted on the membrane by the solute molecules. It is influenced by ionic concentration, temperature and the chemical species present.

These equations are typically used for rough design calculations to determine either the required membrane area for a particular production rate, or the particular production rate for a given membrane area and processing conditions. More detailed modelling and design uses the transport equations from Sections 19.2.4.1–19.2.4.3 (and the references therein).

19.3 OVERVIEW OF APPLICATIONS OF MEMBRANES IN CHEMICAL PROCESSING

Membranes are used in a wide range of chemical processes and the list below is by no means exhaustive. Examples of some major uses are described below.

19.3.1 Desalination by Reverse Osmosis

Osmosis is the movement of water (or other solvent) from a region of high water concentration (low solute concentration) to a region of lower water concentration (higher solute concentration) through a selectively permeable membrane. If a pressure greater than the osmotic pressure is applied to the solution with lower water concentration (higher solute concentration), osmosis is reversed, producing 'reverse osmosis'. This principle is used to produce clean water through desalination and has been applied in particular to the

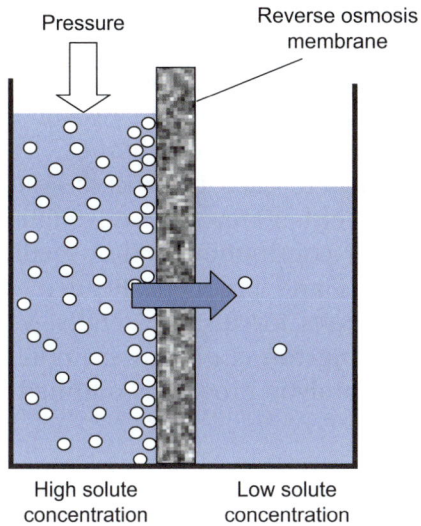

Figure 19.10 A basic schematic of the reverse osmosis process.

desalination of sea water at some of the largest scale applications of membrane technology worldwide. A basic schematic illustrating the principle of reverse osmosis with a semi permeable membrane is shown in Figure 19.10. Large RO desalination plants can produce desalinated water at 320 000 m^3 d^{-1} (Ashkelon, Israel).[32] As such, reverse osmosis is a major industrial application of membranes with a plethora of literature already available on the subject. For further information please refer to Greenlee *et al.*,[32] Lee *et al.*[33] and Baker.[4]

19.3.2 High Value Molecule Recovery

The advantage of some membrane processes is that they can selectively separate out molecules where other processes cannot. A key example of this is the use of membranes to selectively recover high value components from multi-component mixtures, especially when temperature sensitive components are present such as biomolecules, active pharmaceutical ingredients (APIs) and homogeneous transition metal based catalysts. This can be achieved by using any one of the membrane processes outline above. Examples include: the recovery of dilute biofuels from fermentation broths (see Section 19.4); whey protein recovery from milk; and the facile separation of products from unreacted substrate, reagents and by-products, in particular the separation, recovery and reuse of homogeneous catalysts.

The last example is being increasingly applied in non-aqueous systems in pharmaceutical and chemical applications which use organic solvent based reactions by using a new generation of membranes known as either 'organic solvent resistant' (OSN) or 'solvent resistant nanofiltration' (SRNF) membranes. These are pressure-driven membranes that are used in process systems such as those shown in Figure 19.6 to either recycle homogenous catalysts back to a reactor (either batch or continuous) and/or facilitate a facile separation between reactants and products. Readers are referred to the following excellent reviews for further details: Vandezande *et al.*,[11] Volkov *et al.*[34] and Livingston *et al.*[7] An example of the use of OSN membranes in a new catalytic process is outlined in Section 19.5.

19.3.3 Fractionations

Membranes can offer a lower energy, lower temperature and more scalable alternative to unit operations such as chromatography and distillation, which are used to fractionate and purify the individual constituents of multi-component mixtures. This is most readily applicable to streams that require the fractionation of different molar mass components. The technology that enables this is known as the membrane cascade technology.

Membranes are typically thought of as a unit operation that separates a stream into two separate streams. However, if a number of different membranes are put together in series and/or parallel, a fractionation cascade is produced (Figure 19.11). When the membranes all have different values of selectivity, this can be used to fractionate and purify a multi-component solute stream into its constituent components as long as the appropriate membranes are used, and crucially, are available. In the case of fractionation by molar mass, a range of membranes with MWCOs matched to those of the constituent

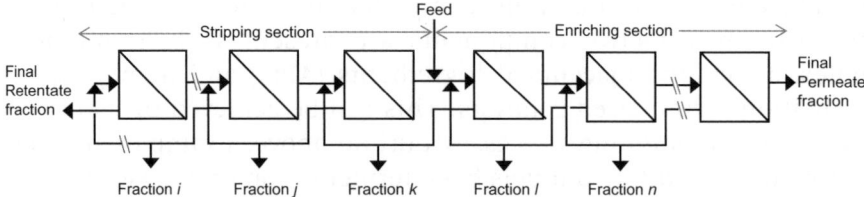

Figure 19.11 A counter-current multi-product membrane fractionation cascade (modified from ref. 35). For fractionation, each membrane module needs to have a different MWCO, matching the desired intermediate fractions.

molecular fractions are required. Membranes can be used in series and in parallel in order to purify a product or obtain greater yield. Parallel membranes may be used on the permeate streams to recover some of the lost product, increasing yield. These same ideas can be applied to when the product is in the permeate stream, *i.e.* low rejection where the retentate becomes the stream to which product is lost.

Fractionation cascades of membranes are a relatively new development and still only used at the laboratory scale level. They have only been considered for a limited set of separations, including gas separation,[35,36] continuous solvent exchange,[37] natural product extraction, olefin and paraffin separation,[38] a simple two-step cascade for rice bran purification[39] and for the purification of APIs.[40,41] However, as new membranes are developed and new configurations are tested, the known applications will no doubt expand significantly. Consequently, this is a membrane technology that could potentially have a significant impact in industries that are reliant on purifications and fractions (such as the chemical and pharmaceutical industry) in the near future.

19.3.4 Solvent Exchange by Diafiltration

Diafiltration has typically been used for the removal of salts and/or solvents from aqueous solutions containing molecules that are thermally liable. Therefore, it has been used primarily for solutions containing biomolecules, such as proteins, enzymes and peptides. In diafiltration, a dead-end membrane module and process configuration is used (*e.g.* Figure 19.3). A continuous feed of the solvent to be exchanged is added to the feed, diluting and with time eventually displacing the initial solvent in this closed system. With correct membrane selection, the biomolecule and any other molecules of interest can be retained, whilst the solvent and any unwanted molecules are removed. Typically UF membranes are used in biomolecule separation applications, as these will retain the biomolecules but allow the unwanted salts and/or solvents to permeate, thereby isolating the biomolecules in the new solvent.[42]

With the advent of solvent resistant NF membranes (Section 19.3.2), a whole new application of diafiltration has emerged, in particular for exchanging organic solvents. A key example is in the pharmaceutical industry, where the reaction is typically done in a different solvent to the isolation stages.[43] Distillation can be used, but this can be energy intensive (or indeed impossible) if the solvents have similar volatilities. Additionally, many pharmaceutical species are thermally liable and have molecular weights that could be retained by a NF membrane.[43] In these cases, diafiltration can be used to exchange the solvents in a low temperature membrane process, with the additional benefit of being able to isolate and purify the molecules of interest by the removal of any smaller molecules with the solvent being exchanged. This can be done in combination with a membrane fractionation cascade[44] (see Figure 19.11 and Section 19.3.3).

19.3.5   Upgrading of Liquid Streams

Membranes can be used as a lower energy process to upgrade liquid streams containing a contaminant or undesirable component. This is particularly relevant for beverage streams, which may contain temperature sensitive components (such as proteins, flavonoids, *etc.*) that may be denatured/degraded using the conventional heat-based treatments used in the food industry and/or avoid unnecessary waste by-products.

Two examples (and there are numerous others) of where membranes can and are used for liquid upgrading are:

- microfiltration membranes for the removal of bacterial spores from liquid food product streams (such as milk, in a 'cold pasteurisation' technique);[45,46]
- mixed matrix membranes (MMMs) for the continuous selective removal of the proteins that cause white wine to go cloudy ('continuous membrane wine fining').

MMMs immobilise the fining agent (bentonite or montmorillonite) in a polymer matrix, and through regeneration and reuse, alleviate the need to dispose of the fining agent after every batch of wine is fined.[47,48]

19.3.6   Wastewater Treatment

Since a membrane is a barrier, wastewaters can be very effectively cleaned using membrane processes. If the water is to be reused (in particular for drinking water in a complete reuse treatment system) nanofiltration and/or reverse osmosis (as in desalination above) can be used as selective barriers to remove practically all pollutants, whilst allowing the clean water to permeate. Examples of where this has or could be applied include:

- food and beverage wastewaters (*e.g.* dairy);[49]
- oily waste and wastewater streams;[50]

- polishing of evaporator condensate;[51,52]
- recovery and reuse of spent cleaning solutions, such as caustic for dairy process plant cleaning-in-place;[53]
- removal of emerging contaminants/recalcitrant organics such as endocrine disrupting compounds, pesticides and pharmaceuticals.[54,55]

19.3.7 Membrane Enhanced Reactors

As introduced in Section 19.1, the combination of membranes with other unit operations produces new process intensification technologies. Thereby providing means to: selectively retain molecules in these unit operations; overcome equilibrium; and provide facile *in situ* methods for product separation, further enabling sustainability benefits. This is particularly true for the combination of membranes with reactors.

The effectiveness and efficiency, in terms of yield and reaction rate, of the current state-of-the art in chemical and biochemical catalysts is currently limited by the constraints of conventional reactors. This is because both homogeneous and unsupported catalysts are challenging and uneconomical to use, since reuse in subsequent reactions is hindered due to difficulties in their removal from their associated products, especially when considered within continuous systems. This means that continuous reactors cannot be easily employed; therefore the associated benefits of smaller equipment, higher throughput and better economics cannot be obtained. Reactions are also sometimes product inhibited, with a need to continuously re-move the desired product in order to maintain high selectivity, conversion and yield. Equilibrium may need to be broken to obtain a high and selective yield.

A membrane coupled to the outlet stream of a reactor or built into the reactor as an intrinsic part of it (and both systems are known as a membrane reactor) can be used to overcome all of the afore-mentioned limitations. If the reactants undergo an order of magnitude change in size in the reaction, ionise or change polarity, and/or an appropriately sized and charged catalyst is used, then one or more membranes can be coupled to the reactor to selectively retain and/or recirculate both the catalyst and any insufficiently reacted molecules in the reactor. The membrane can also be used to effect a separation between reactants and products, enabling a facile purification directly from the reactor and also providing a solution to product inhibited reactions when needed. The membrane is chosen so that the

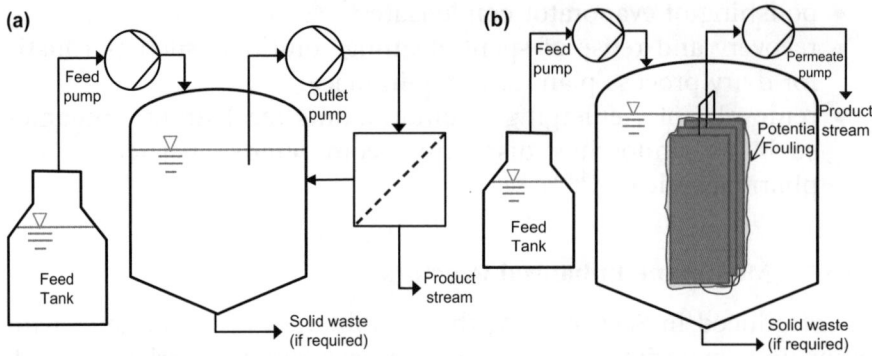

Figure 19.12   (a) Schematic of a membrane reactor with membrane module external to reactor. (b) Schematic of a membrane reactor with membrane module internal to reactor.

catalysts, reactants and insufficiently reacted molecules (intermediates) cannot permeate through the membrane, whilst the desired products are obtained (pure) in the permeate. A membrane reactor is therefore a perfect example of a process intensification technology, where reaction and separations are combined in a single unit operation. The membrane can be external or internal to the reactor (Figure 19.12).

For further information on membrane reactors, see the following excellent review papers and books: Vankelecom,[12] Westermann and Melin,[56] and Gallucci *et al.*[57] Furthermore, membranes used directly in a bioreactor for wastewater treatment (known as 'membrane bioreactors') are a significant application area in industry and are also covered well elsewhere.[8,58–60]

A further two key applications related to membrane reactors are outlined in further detail below.

19.4   CASE STUDY I: CONCENTRATION OF DILUTE ORGANICS FROM FERMENTATIONS

19.4.1   Fermentation and the Downstream Recovery of Organics

Even with the wide range of separation technologies available, the downstream recovery of dilute organics from fermentation broths is dominated by distillation technology. This involves the heating of large volumes of water, requiring considerable amounts of energy, to separate the desired products. These problems can be exacerbated further by the formation of azeotropic mixtures and the recovery of high boiling organics. For example, the recovery of 2,3-butanediol from a dilute fermentation broth can be attributed to more than half of the cost of its microbial production.[61] The two hydroxyl groups give it an affinity for water; this coupled with its high boiling point (180–184 °C) means it is recovered from the bottom of the distillation column. The use of energy efficient technologies could equate to large savings in the total downstream recovery costs in fermentative processes, particularly within smaller scale production facilities where the economies of scale and heat integration are lost.[62,63] Membrane technologies can provide an energy efficient alternative or aid to traditional thermal separations.

19.4.2 Nanofiltration and Reverse Osmosis Processes

As mentioned previously, nanofiltration and reverse osmosis membranes exhibit MWCOs of 150–1000 Da and <100 Da, respectively. Therefore these two pressure-driven processes in principle provide the correct rejections for concentration of dilute organic mixtures from fermentation broths. Able to be performed at room temperature, NF and RO can also be used with heat sensitive media (*e.g.* cells) present in the fermentation broth. With the majority of NF and RO membranes exhibiting selectivity towards water (hydrophilic), they are mainly used for water purification applications. For the purification of water from distillery condensates, low concentrations of low MW organics can be removed and the wastewater recycled.[64–66] Therefore it can also be envisaged that NF and RO membranes can be used for the concentration of dilute fermentation broths to achieve energy savings in downstream purification steps or for complete recovery of desired products.

Research has looked at the permeation of binary water alcohol mixtures through nanofiltration membranes.[67] Their use in the recovery of ethanol from fermentation media has also been proven.[68] Recovery of other fermentation products such as lactic acid, commonly used within the food, cosmetic, pharmaceutical and bulk chemical industries by NF and RO has also been studied. Models for the transport of lactic acid through NF and RO membranes have been developed by Timmer and co-workers and subsequently investigated for recovery from a 4 wt- broth.[69,70] The rejections for NF and RO membranes ranged from 50–70% and 80–99%, respectively. Fouling was also studied to gain an understanding of the long-term operation of a spiral wound NF membrane to demonstrate the ability of NF and RO processes for scale-up.

19.4.3 Pervaporative Separation

As discussed in Section 19.2.3, pervaporation has found commercial use for the dehydration of organic solvents. Considered one of the most effective and energy efficient processes for separating azeotropic mixtures, it is commonly used for the production of high purity ethanol for pharmaceutical applications. Pervaporation is also well-established for the recovery of ethanol from fermentation broths.[63,71] A typical set-up for the continuous pervaporative recovery of ethanol is shown in Figure 19.13. Microfiltration is often used to remove heat sensitive media from the broth and/or minimise potential fouling of the pervaporation membrane. It is possible, however, for pervaporation to be applied directly to an untreated fermentation broth. Ethanol recovery from the dilute broth is then performed using a hydrophobic membrane and the retained aqueous broth returned to the fermenter. A hydrophilic membrane is then used to dehydrate the concentrated ethanol feed up to the purity required.[71]

Pervaporation has also been studied for application within acetone, butanol, ethanol (ABE) fermentations. Butanol is an attractive alternative biofuel to ethanol as it is less volatile, flammable and hazardous, whilst also having a higher energy content.[72] Butanol, however, inhibits microbial growth and therefore its continuous extraction from a dilute fermentation broth (such as that shown for ethanol in Figure 19.13) can be used to increase the efficiency of its fermentative production. Pervaporation is effective for this due to its high efficiency in removing species of low concentrations from a feed. There have been many successful studies on the recovery of butanol by pervaporation, mainly focusing on the identification of suitable

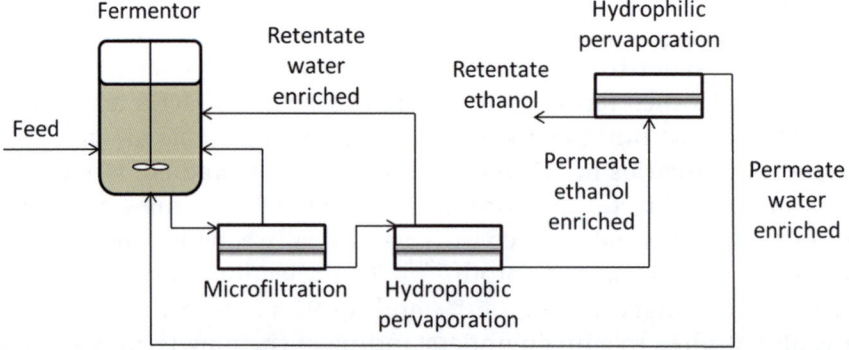

Figure 19.13 Schematic of pervaporation for recovery of ethanol from dilute fermentation broth.

membranes.[73] However, significant challenges need to be overcome to allow the implementation of pervaporation into the ABE process. Energy inputs and capital costs must be reduced and membranes demonstrating higher fluxes, selectivities and stabilities towards fermentation media identified.[73]

19.4.4 Membrane Materials for Fermentative Separations

Separation of the components of a liquid feed mainly relies on the vapour–liquid equilibrium in conventional distillation techniques. For membrane processes, however, the separation of liquid components depends on the different affinities towards the membrane material and diffusivity through it. The stability of the material towards the feed components is also of high importance for effective separation. This can be especially problematic when concentrating fermentation media due to the large amount and variety of species within the feed. Engineering of the membrane material can therefore be conducted to allow selective permeance of a desired species and limit potential fouling from the feed. For the concentration of organic solvents from fermentation broths, traditional membrane materials have been utilised such as polydimethylsiloxanes (PDMS) or polyimides (PI) for pervaporation and polyamides or cellulose acetate for nanofiltration/reverse osmosis. However, there has been much investigation into alternative materials.

Zeolitic membranes, formed of microporous silicalite or aluminosilicate, are ideal membrane materials due to their uniform, molecularly sized pores and high thermal, mechanical and chemical stability. High performances have been demonstrated for pervaporative separation of alcohol–water solutions.[74,75] However fabrication of these zeolite membranes generally incurs high cost and a trade-off between the high performances of zeolite membranes *versus* the cheaper lower performing polymeric membranes. The development of mixed matrix membranes (Figure 19.14) has alleviated some of the issues prevalent in the fabrication of zeolitic membranes whilst retaining some of their high performance. Inorganic fillers are incorporated within a polymeric matrix to take advantage of the high performance of inorganic materials and the ease of production of polymeric membranes. The inclusion of zeolites in these MMMs has been shown to increase the flux and separation factor of a polymer membrane for alcohol–water separations.[76,77]

Utilising this same principle, many other inorganic fillers have been incorporated into the matrix of polymer membranes to achieve

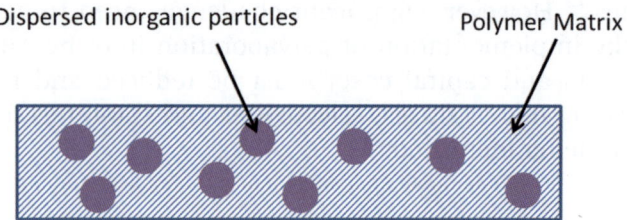

Dispersed inorganic particles   Polymer Matrix

Figure 19.14   Schematic of a mixed matrix membrane (MMM) showing the inorganic filler particles dispersed throughout a polymer matrix.

an increase in performance. Examples of metal-organic frameworks (MOFs),[78–80] carbon nanotubes,[81] hydrophobic silica[82] and single crystal adsorbents[83] have been studied for the separation of alcohol–water mixtures. Many of these MMMs have also been studied for the pervaporative separation of butanol from fermentation.[84]

Graphene has also been investigated as a potential membrane material. Graphene oxide sheets have been prepared as a membrane and have exhibited the sole unimpeded permeation of water.[85] This unimpeded permeation of water could exhibit extremely high fluxes for concentration of fermentation media as well as many other applications in desalination and water purification; however, problems of fouling and fabrication cost remain to be overcome.

The development of these new superior performing membranes and membrane materials shows that there is a bright future for the application of membranes in concentrating dilute organics from fermentation broths. Combined with the advantages of lower temperature (and therefore lower energy) dewatering (compared with the conventional high temperature routes) promise to make this a more economically viable and more environmentally sustainable process option in the near future.

19.5   CASE STUDY II: SEPARATION OF INCOMPATIBLE REACTION SYSTEMS

19.5.1   The Challenge of Incompatible Catalysts

There are many catalyst systems that are incompatible with each other, but which if linked in tandem, would produce an overall reaction with a higher selectivity and/or conversion or even a reaction that would otherwise not occur. The traditional approach to doing such tandem reactions is to run them in the same reactor—in so called 'one-pot' reactions. However, this approach means that tandem reactions are limited to a small set of compatible catalysts and re-action/reactor conditions that may not be optimal for achieving the desired reaction outcomes (*e.g.* high yield, reaction rate, selectivity).

Membranes are a potential solution to this problem. Work at Imperial College London (led by Professor Andrew Livingston) and the University of Warwick (led by Professor Paul Taylor) to develop a membrane enhanced reactor system has shown the way forward for combining catalyst systems that are incompatible.[86–91] Furthermore, this is an example of how membranes can not only enable process intensification, but also provide a new route to reactions and products not previously possible.

19.5.2 Dynamic Kinetic Resolution

Many essential biological molecules are inherently chiral and biological activity is highly dependent on enantiomeric purity.[92] Consequently, chiral molecules have applications in several industries such as pharmaceuticals, agrochemicals and food. However, for these applications, synthesis of a racemic compound is inefficient, as one enantiomeric form can have little or no activity. Furthermore, the presence of one (unwanted) enantiomer may have adverse side effects, *e.g.* (−)-(*S*)-thalidomide. The goal is therefore to synthesise enantiomerically pure compounds. These can be produced by asymmetric synthesis, but this is often difficult and, due to the use of expensive reagents, uneconomic. Alternatively, enantiomerically pure compounds can be generated by resolution of the racemic mixture, for example, a kinetic resolution or a chemical resolution using a chiral metal catalyst. When this is achieved by differential rates of reactions of the two enantiomers, the process is known as a kinetic resolution. However, since, by definition, a racemic mixture contains only 50% of the desired product, a maximum yield of only 50% can be obtained.[92–95]

The maximum possible yield of a kinetic resolution can be raised from 50 to 100% by using a tandem reaction combining the kinetic resolution and the racemisation process (Figure 19.15), converting the non-resolving isomer into the resolving isomer. A number of reviews have been written describing this process.[92–98] Dynamic kinetic resolution (DKR) is only possible when the chiral starting material racemises rapidly $(k_{rac} \gg k \gg k_{ent})$ and when racemisation of the product is very slow.

In general, the DKR process requires two reactive systems, *viz.* a racemisation system and a resolution system. These typically are an

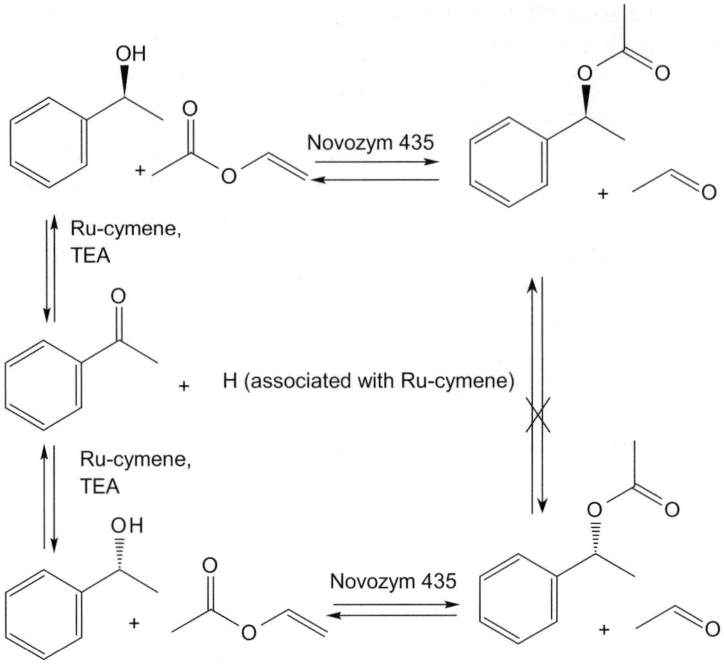

Figure 19.15 Example reaction scheme of a DKR showing 1-phenylethanol (a secondary alcohol) being transformed into a 1-phenylethylacetate (a secondary acetate).

organometallic catalyst in base and an enzyme catalyst. The inherent incompatibility between the enzyme (bio) and the other catalysts (chemo) severely limits the combination of bio and chemo catalysts that can be used and so the applicability of the process is limited to a small number of systems where the chemo and bio catalysts do not deactivate each other.[93] DKRs are therefore an extreme example of the incompatibility of catalysts in tandem reactions.

19.5.3   Membrane Enhanced Dynamic Kinetic Resolution

As introduced in Section 19.3.2, it has been shown that membranes may be applied to the field of organic synthesis, allowing facile separation of products from unreacted substrate, reagents and by-products. The teams at Imperial College and Warwick have extended this concept and have shown that membranes can be introduced to separate the two catalytic systems (Figure 19.16), but allow a complete conversion of the substrate which may pass freely across the membrane. They called this system 'membrane enhanced DKR'. This

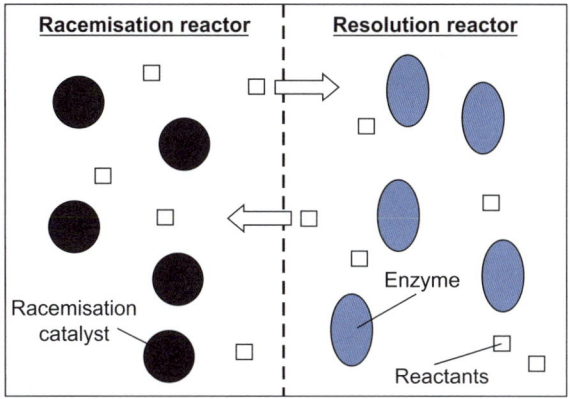

Figure 19.16 The principle of membrane enhanced DKR.[91]

membrane linked tandem reactor system removes the need for the two catalysts to have to tolerate each other under compatible (and potentially suboptimal) conditions and produces a means to also separate the products from the reactants and catalysts—something many reaction systems need even beyond DKRs.

The membrane enhanced DKR concept was engineered into a laboratory-scale continuous membrane enhanced reactor system that allowed both reactions (racemisation and resolution) to be performed in separate well-mixed pressure vessels with membranes at the outlets so that the catalyst reagents may be retained. The substrates and products are permeable to the membranes and so circulate between the two reactors. Two configurations that could be used in a continuous flow reactor system using this principle are shown in Figure 19.17.

Using this system, not only can incompatible catalysts be used, but also incompatible racemisation and resolution reactor conditions. This therefore opens up the possibility of using different reactor temperatures, pressures, *etc.* for a tandem reaction. This concept was demonstrated by Roengpithya *et al.*,[86] where the continuous dynamic kinetic resolution of 1-phenylethylamine was achieved using a membrane-enhanced two-vessel process. 1-Phenylethylamine was racemised in a higher temperature vessel (at 100 °C using the ruthenium-based Shvo complex). This was done at conditions incompatible with *Candida antarctica* lipase B (in immobilised form as Novozym 435) used for the kinetic resolution, which was reacted at 30 °C with ethyl acetate as the acyl donor. The kinetic resolution without the racemisation stage after 48 hours produced a 38%

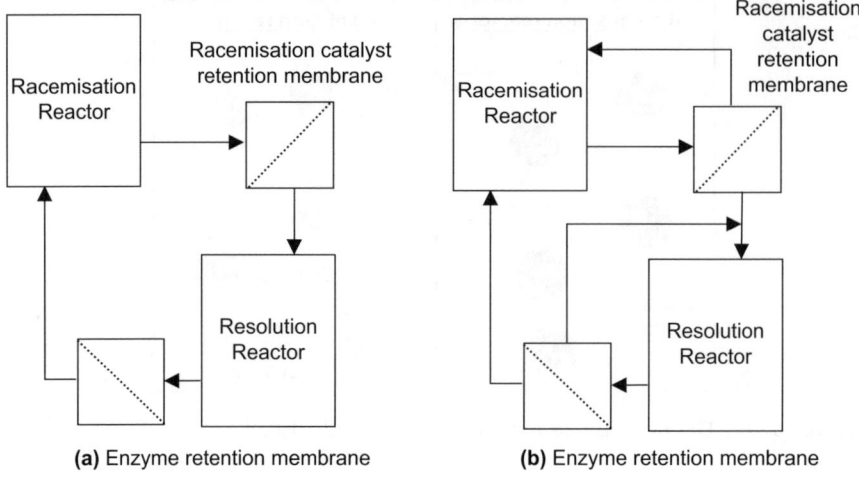

Figure 19.17 Schematics of two possible configurations for the continuous membrane enhanced DKR system: (a) using dead-end membrane filtration cells/modules; and (b) using cross-flow membrane cells/modules.

conversion with an enantiomeric excess of 99%. In contrast, the membrane enhanced DKR reaction (with the high temperature racemisation) after 72 hours produced a 91% conversion with an enantiomeric excess of 99%. However, the reaction was not; isolated yields were reported to be poor and the mass balance on this result was poor at only 55%, indicating that material was likely being lost from the system. Further work addressing this has not yet been published, but despite these limitations, it still can be said that this paper proved the principle that a tandem catalytic process with incompatible reaction conditions 'can be run in membrane separated chambers'.[86]

Overall, membrane-separated two (or more) vessel processes show great promise in overcoming the catalyst and reaction condition compatibilities that currently limit the scope and application of multiple catalyst 'tandem' reactions. Furthermore, this is an excellent example of how membranes can facilitate process intensification. Further work is needed to develop these systems for a wider set of reactions and reactors. This work also needs to be done as this system can be truly game changing, since it opens up the possibility of using new combinations of catalysts and reactions to produce tandem reactions and the products from these not possible with other technologies. In particular, the membrane enhanced tandem reactor gives us the ability of being able to use a wider range of chemo and bio catalysts together than ever before. The potential implications this will have to sustainable chemical processes therefore has only just begun.

19.6 CONCLUSIONS

Membrane separations have become a new essential unit operation for effecting selective, low temperature separations either where other unit operations are not suitable (or capable) of achieving a separation and/or as a lower energy, lower waste replacement for more conventional separations. With new developments in membrane materials (*e.g.* for the use of membranes in organic solvents and for the more selective transport of biofuels from dilute fermentation broths), membrane configurations (*e.g.* fractionation cascades and membrane enhanced tandem reactors) and membrane-based process intensification technologies (*e.g.* membrane reactors), the range of operating conditions and applications for sustainable processing is ever widening. Furthermore, the possibilities of new applications and products produced as a result of these new membranes and processes (such as membrane linked tandem chemo and bio catalysed reactions using previously incompatible catalysts) further shows that the impact of membranes in some process industries is only just being realised.

REFERENCES

1. R. Rohani, M. Hyland and D. Patterson, *J. Membr. Sci.*, 2011, **382**, 278.
2. P. Bernardo, E. Drioli and G. Golemme, *Ind. Eng. Chem. Res.*, 2009, **48**, 4638.
3. Y. Yampolskii, *Macromolecules*, 2012, **45**, 3298.
4. P. W. Baker, *Membrane Technology and Applications*, Wiley, Chichester, UK, 3rd edn, 2012.
5. E. Drioli, A. Brunetti, G. Di Profio and G. Barbieri, *Green Chem.*, 2012, **14**, 1561.
6. E. Drioli, A. I. Stankiewicz and F. Macedonio, *J. Membr. Sci.*, 2011, **380**, 1.
7. A. Livingston, L. Peeva, S. Han, D. Nair, S. S. Luthra, L. S. White and L. M. Freitas Dos Santos, *Ann. N. Y. Acad. Sci.*, 2003, **984**, 123.
8. R. Mazzei, E. Piacentini, E. Drioli and L. Giorno, in *Process Intensification for Green Chemistry: Engineering Solutions for Sustainable Chemical Processing*, ed. K. Boodhoo and A. Harvey, John Wiley & Sons, Chichester, UK, 2013, pp. 227–250.

9. R. Mazzei, E. Piacentini, E. Drioli and L. Giorno, in *Process Intensification for Green Chemistry: Engineering Solutions for Sustainable Chemical Processing*, ed. K. Boodhoo and A. Harvey, John Wiley & Sons, Chichester, UK, 2013, pp. 311–354.

10. J. Mulder, *Basic Principles of Membrane Technology*, Springer, Dordrecht, The Netherlands, 2nd edn, 1996.

11. P. Vandezande, L. E. M. Gevers and I. F. J. Vankelecom, *Chem. Soc.y Rev.*, 2008, **37**, 365.

12. I. F. J. Vankelecom, *Chem. Rev.*, 2002, **102**, 3779.

13. I. Pinnau and B. D. Freeman, in *Membrane Formation and Modification*, ed. I. Pinnau and B. D. Freeman, American Chemical Society, Washington DC, 1999, vol. 744 , pp. 1–22.

14. J. H. Hanemaaijer, T. Robbertsen, T. Van Den Boomgaard, C. Olieman, P. Both and D. G. Schmidt, *Desalination*, 1988, **68**, 93.

15. B. Van der Bruggen, M. Mänttäri and M. Nyström, *Sep. Purif. Technol.*, 2008, **63**, 251.

16. M. Meireles, P. Aimar and V. Sanchez, *J. Membr. Sci.*, 1991, **56**, 13.

17. A. I. Schäfer, A. G. Fane and T. D. Waite, *Desalination*, 2000, **131**, 215.

18. D. A. Patterson, L. Yen Lau, C. Roengpithya, E. J. Gibbins and A. G. Livingston, *Desalination*, 2008, **218**, 248.

19. G. Tchobanoglous, F. L. Burton, H. D. Stensel, Metcalf and Eddy, *Wastewater Engineerin.: Treatment and Reuse*, McGraw-Hill Education, Boston and London, 2003.

20. L. P. Raman, M. Cheryna and N. Rajagopalan, *Chem. Eng. Prog.*, 1994, **90**, 68.

21. J. G. Wijmans and R. W. Baker, in *Materials Science of Membranes for Gas and Vapor Separation*, ed. Y. Yampolskii, I. Pinnau and B. Freeman. John Wiley & Sons, Chichester, UK, 2006, ch. 5, pp. 159–189.

22. J. G. Wijmans and R. W. Baker, *J. Membr. Sci.*, 1995, **107**, 1.

23. A. Jonquières, R. Clément, P. Lochon, J. Néel, M. Dresch and B. Chrétien, *J. Membr. Sci.*, 2002, **206**, 87.

24. R. R. Sharma, R. Agrawal and S. Chellam, *J. Membr. Sci.*, 2003, **223**, 69.

25. Y. H. See-Toh, F. C. Ferreira and A. G. Livingston, *J. Membr. Sci.*, 2007, **299**, 236.

26. A. R. Hassan, N. a. Ali, N. Abdull and A. F. Ismail, *Desalination*, 2007, **206**, 107.

27. W. R. Bowen, A. W. Mohammad and N. Hilal, *J. Membr. Sci.*, 1997, **126**, 91.

28. W. R. Bowen and J. S. Welfoot, *Desalination*, 2002, **147**, 197.
29. A. Wahab Mohammad and M. Sobri Takriff, *Desalination*, 2003, **157**, 105.
30. M. M. Nasef and E.-S. A. Hegazy, *Prog. Polym. Sci.*, 2004, **29**, 499.
31. W. R. Bowen and J. S. Welfoot, *Chem. Eng. Sci.*, 2002, **57**, 1393.
32. L. F. Greenlee, D. F. Lawler, B. D. Freeman, B. Marrot and P. Moulin, *Water Res.*, 2009, **43**, 2317.
33. K. P. Lee, T. C. Arnot and D. Mattia, *J. Membr. Sci.*, 2011, **370**, 1.
34. A. V. Volkov, G. A. Korneeva and G. F. Tereshchenko, *Russ. Chem. Rev.*, 2008, **77**, 983.
35. R. Agrawal, *Ind. Eng. Chem. Res.*, 1996, **35**, 3607.
36. R. Pathare and R. Agrawal, *J. Membr. Sci.*, 2010, **364**, 263.
37. J. C.-T. Lin and A. G. Livingston, *Chem. Eng. Sci.*, 2007, **62**, 2728.
38. M. S. Avgidou, S. P. Kaldis and G. P. Sakellaropoulos, *J. Membr. Sci.*, 2004, **233**, 21.
39. I. Sereewatthanawut, I. I. R. Baptista, A. T. Boam, A. Hodgson and A. G. Livingston, *J. Food Eng.*, 2011, **102**, 16.
40. W. E. Siew, A. G. Livingston, C. Ates and A. Merschaert, *Chem. Eng. Sci.*, 2013, **90**, 299.
41. W. E. Siew, A. G. Livingston, C. Ates and A. Merschaert, *Sep. Purif. Technol.*, 2013, **102**, 1.
42. L. Schwartz, Diafiltration: a fast, efficient method for desalting, or buffer exchange of biological samples, Scientific & Technical Report PN 33289, Pall Life Sciences, Ann Arbor, MI, 2003, http://www.pall.com/pdfs/Laboratory/02.0629_Buffer_Exchange_STR.pdf [accessed 8 September 2014].
43. J. P. Sheth, Y. Qin, K. K. Sirkar and B. C. Baltzis, *J. Membr. Sci.*, 2003, **211**, 251.
44. I. Sereewatthanawut, F. W. Lim, Y. S. Bhole, D. Ormerod, A. Horvath, A. T. Boam and A. G. Livingston, *Org. Process Res. Dev.*, 2010, **14**, 600.
45. L. E. Head and M. R. Bird, *J. Food Process Eng.*, 2013, **36**, 113.
46. G. Gésan-Guiziou, in *Membrane Processing: Dairy and Beverage Applications*, ed. A. Y. Tamime, Wiley-Blackwell, Chichester, UK, 2013, pp. 128–142.
47. A. T. T. Tran, D. A. Patterson and B. J. James, *J. Food Eng.*, 2012, **112**, 38.
48. D. A. Patterson, M. Bowstead, A. Tran and B. J. James, *Procedia Eng.*, 2012, **44**, 131.
49. A. Chollangi and M. M. Hossain, *Chem. Eng. Process.*, 2007, **46**, 398.

50. M. Cheryan and N. Rajagopalan, *J. Membr. Sci.*, 1998, **151**, 13.

51. F. Lipnizki, *Desalination*, 2010, **250**, 1067.

52. K. Xie, H. J. Lin, B. Mahendran, D. M. Bagley, K. T. Leung, S. N. Liss and B. Q. Liao, *Environ. Technol.*, 2010, **31**, 511.

53. M. Dresch, G. Daufin and B. Chaufer, *Lait*, 1999, **79**, 245.

54. Y. Yoon, P. Westerhoff, S. A. Snyder, E. C. Wert and J. Yoon, *Desalination*, 2007, **202**, 16.

55. D. Dolar, K. Košutić, D. M. Pavlović and B. Kunst, *Desalination Water Treat.*, 2009, **6**, 197.

56. T. Westermann and T. Melin, *Chem. Eng. Process.*, 2009, **48**, 17.

57. F. Gallucci, A. Basile and F. I. Hai, in *Membranes for Membrane Reactors*, ed. A. Basile and F. Gallucci, John Wiley & Sons, Chichester, UK, 2011, pp. 1–61.

58. P. Le-Clech, V. Chen and T. A. G. Fane, *J. Membr. Sci.*, 2006, **284**, 17.

59. T. Stephenson, K. Brindle, S. Judd and B. Jefferson, *Membrane Bioreactors for Wastewater Treatment*, IWA Publishing, London, 2000.

60. B. Marrot, A. Barrios-Martinez, P. Moulin and N. Roche, *Environ. Progress*, 2004, **23**, 59.

61. Z. L. Xiu and A. P. Zeng, *Appl. Microbiol. Biotechnol.*, 2008, **78**, 917.

62. W. J. Koros, *AIChE J.*, 2004, **50**, 2326.

63. L. M. Vane, *J. Chem. Technol. Biotechnol.*, 2005, **80**, 603.

64. R. A. Diltz, T. V. Marolla, M. V. Henley and L. Li, *Bioresour. Technol.*, 2007, **98**, 686.

65. C. Sagne, C. Fargues, R. Lewandowski, M.-L. Lameloise, M. Gavach and M. Decloux, *Chem. Eng. Process.*, 2010, **49**, 331.

66. N. Aydogan, T. Gurkan and L. Yilmaz, *Sep. Sci. Technol.*, 2005, **39**, 1059.

67. J. Geens, K. Peeters, B. Van der Bruggen and C. Vandecasteele, *J. Membr. Sci.*, 2005, **255**, 255.

68. T. Brás, M. C. Fernandes, J. L. C. Santos and L. A. Neves, *Desalination Water Treat.*, 2013, **51**, 4333.

69. J. M. K. Timmer, H. C. van der Horst and T. Robbertsen, *J. Membr. Sci.*, 1993, **85**, 205.

70. J. M. K. Timmer, J. Kromkamp and T. Robbertsen, *J. Membr. Sci.*, 1994, **92**, 185.

71. C. Abels, F. Carstensen and M. Wessling, *J. Membr. Sci.*, 2013, **444**, 285.

72. N. Qureshi and T. C. Ezeji, *Biofuels Bioprod. Biorefin.*, 2008, **2**, 319.

73. V. García, J. Päkkilä, H. Ojamo, E. Muurinen and R. L. Keiski, *Renewable Sustainable Energy Rev.*, 2011, **15**, 964.

74. M. Weyd, H. Richter, P. Puhlfürß, I. Voigt, C. Hamel and A. Seidel-Morgenstern, *J. Membr. Sci.*, 2008, **307**, 239.

75. Q. Liu, R. D. Noble, J. L. Falconer and H. H. Funke, *J. Membr. Sci.*, 1996, **117**, 163.

76. S. Mosleh, T. Khosravi, O. Bakhtiari and T. Mohammadi, *Chem. Eng. Res. Des.*, 2012, **90**, 433.

77. D. Oh, S. Lee and Y. Lee, *Desalination Water Treat.*, 2013, **51**, 5362.

78. X.-L. Liu, Y.-S. Li, G.-Q. Zhu, Y.-J. Ban, L.-Y. Xu and W.-S. Yang, *Angew. Chem., Int. Edn*, 2011, **50**, 10636.

79. X. Liu, H. Jin, Y. Li, H. Bux, Z. Hu, Y. Ban and W. Yang, *J. Membr. Sci.*, 2013, **428**, 498.

80. S. Liu, G. Liu, X. Zhao and W. Jin, *J. Membr. Sci.*, 2013, **446**, 181.

81. A. F. Ismail, P. S. Goh, S. M. Sanip and M. Aziz, *Sep. Purif. Technol.*, 2009, **70**, 12.

82. S. Claes, P. Vandezande, S. Mullens, K. De Sitter, R. Peeters and M. K. Van Bael, *J. Membr. Sci.*, 2012, **389**, 265.

83. S. Takamizawa, C. Kachi-Terajima, M.-a. Kohbara, T. Akatsuka and T. Jin, *Chem. –Asian J.*, 2007, **2**, 837.

84. G. Liu, W. Wei and W. Jin, *ACS Sustainable Chem. Eng.*, 2013.

85. R. R. Nair, H. A. Wu, P. N. Jayaram, I. V. Grigorieva and A. K. Geim, *Science*, 2012, **335**, 442.

86. C. Roengpithya, D. A. Patterson, A. G. Livingston, P. C. Taylor, J. L. Irwin and M. R. Parrett, *Chem. Commun.*, 2007, 3462.

87. D. A. Patterson, C. Roengpithya, P. C. Taylor and A. G. Livingston, presented at Chemeca 2007, 35th Australialian Chemical Enginering Conference, Melbourne, Australia, 2007.

88. C. Roengpithya, D. A. Patterson, E. J. Gibbins, P. C. Taylor and A. G. Livingston, *Ind. Eng. Chem. Res.*, 2006, **45**, 7101.

89. C. Roengpithya, D. A. Patterson, P. C. Taylor and A. G. Livingston, *Desalination*, 2006, **199**, 195.

90. E. J. Gibbins, J. L. Irwin, A. G. Livingston, J. C. Muir, D. A. Patterson, C. Roengpithya and P. C. Taylor, *Synlett*, 2005, **2005**, 2993.

91. E. Gibbins, J. L. Irwin, A. G. Livingston, J. C. Muir, D. A. Patterson, C. Roengpithya and P. C. Taylor, *Abstr. Papers Am. Chem. Soc.*, 2005, **229**, U550.

92. B. A. Persson, A. L. E. Larsson, M. Le Ray and J.-E. Bäckvall, *J. Am. Chem. Soc.*, 1999, **121**, 1645.

93. M. T. El Gihani and J. M. J. Williams, *Curr. Opin. Chem. Biol.*, 1999, **3**, 11.

94. M.-J. Kim, J. Park and Y. K. Choi, in *Multi-Step Enzyme Catalysis: Biotransformations and Chemoenzymatic Synthesis*, ed. E. Garcia-Jenceda, Wiley-VCH, Weinheim, Germany, 2008, pp. 1–19.
95. O. Pàmies and J.-E. Bäckvall, *Chem. Rev.*, 2003, **103**, 3247.
96. M. Ahmed, T. Kelly and A. Ghanem, *Tetrahedron*, 2012, **68**, 6781.
97. F. F. Huerta, A. B. E. Minidis and J.-E. Backvall, *Chem. Soc. Rev.*, 2001, **30**, 321.
98. J. H. Lee, K. Han, M.-J. Kim and J. Park, *Eur. J. Org. Chem.*, 2010, **2010**, 999.

CHAPTER 20

Liquid–Liquid Extraction

STEPHEN TALLON* AND TERESA MORENO

Callaghan Innovation, 69 Gracefield Rd, Lower Hutt, New Zealand
*Email: Stephen.Tallon@CallaghanInnovation.govt.nz

20.1 INTRODUCTION

Liquid–liquid extraction (LLE) is a process for the separation of one or more components from a liquid solution by selectively adsorbing these components into a second liquid phase. It is a versatile processing operation, but one that is often overlooked and only considered as a less favourable alternative or workaround when distillation of the components is impractical or too costly. LLE adds additional complexity to a process compared with distillation, including the need for multiple unit operations and use of additional solvent phase(s), and these factors need to be evaluated carefully against distillation or other alternative processes before LLE is selected as the method of choice. There are, however, many opportunities where LLE could be favourable even when distillation is practical to carry out and established in industry practice, and it would be remiss only to consider LLE as a secondary option after eliminating other alternatives. Design according to green chemistry and engineering principles is an emerging practice that is driving the need to accurately access all processing methods including LLE, and published literature is increasingly beginning to reflect a better

Chemical Processes for a Sustainable Future
Edited by Trevor M. Letcher, Janet L. Scott and Darrell A. Patterson
© The Royal Society of Chemistry 2015
Published by the Royal Society of Chemistry, www.rsc.org

understanding of how to evaluate LLE opportunities for both new and established applications.

The schedule of green chemistry principles according to Anastas and Warner[1] lists 12 concepts. Of these, the concepts relating to minimisation of energy consumption (principle 6), minimisation of solvent use and use of benign solvents (principle 5), and overall design for human and environmental safety (principle 12) are particularly relevant to evaluation of LLE processes and are described further below. Other principles relating to the selection of chemical synthesis routes and product materials are considered to be higher level pre-defined criteria for evaluation of a LLE separation process and are not the focus of discussion in this chapter. However, it is appreciated that, in practice, there is likely to be an iterative optimisation between process chemistry and process engineering, including for example the opportunity for integrating LLE with other processing steps as in *in situ* fractionation of enzyme reaction products, or continuous stripping of products from fermentation systems.

- *Energy efficiency.* LLE simply replaces one solvent system with another and downstream separation of solutes from the new solvent system is still generally required, in addition to recovery of the extraction solvent and remediation of other secondary streams. These are all energy intensive steps, although there is potential for these to be lower than for processing the original solvent system if a solvent with high heat of vaporisation like water is replaced with one with lower energy demands, or if the solute is present in higher concentrations in the second solvent due to favourable selectivity. Different solvent systems can also have widely different separation behaviour and different opportunities for energy efficiency.
- *Solvent use and selection.* LLE is flexible and versatile because of the almost unlimited scope to choose solvents systems to match the target application. In practice many will be unsuitable due to toxicity and eco-toxicity considerations, but even suitable solvents represent an additional energy and material consumption cost associated with their manufacture and from losses or disposal of solvent as part of the process. It is important therefore to choose solvents and design processes that maximise solvent recovery. In some applications, such as in production of consumer products, selection of suitable solvents may be even more limited.
- *Human and environmental safety.* Three main considerations are incorporated under safety aspects: use of solvents that are

non-toxic if incorporated into the product, and do not generate or contribute to toxic artefacts in the product; use of solvents that can be produced sustainably with minimal input costs, and can be disposed of safely back to the environment without persistence or eco-toxicity; and use of chemicals and processes that do not present hazards to operators of processing equipment. Industrial hazards such as flammability or use of pressurised equipment are undesirable, but can generally be managed using established industrial practice appropriate to the hazard, and low levels of risk need to be weighed against other potential advantages. In some cases safety risks are unavoidable such as in the processing of fuels intended for combustion or the manufacture of cytotoxic drugs. In these cases risks should be minimised where possible through careful process design.

Many of the principles above correlate directly with design principles that minimise manufacturing cost including minimising energy consumption, and minimising costs associated with replenishing and disposing of waste solvent. Other green chemistry principles correlate with the value and quality of the product, including meeting consumer expectations for product safety and sustainability.

Some design decisions, for example, using additional energy to further remediate a process waste stream before disposal, may require a trade-off between several principles. Government initiatives intended to encourage green processing can also distort the economics of process optimisation and in some cases these can even incentivise processes that are non-optimal in terms of green chemistry principles.

20.1.1 Industrial Applications of LLE

LLE is often used when distillation cannot reasonably be carried out. This typically occurs when the components to be distilled have similar vapour pressure or form azeotropes; when the more volatile component has a high latent heat of vaporisation that is energy intensive to remove; or when a low volatility solute is present at low concentration, for example in evaporation of large volumes of water to recover low volumes of low volatility solutes. Distillation may also be undesirable where the components to be separated are heat sensitive and unstable under standard distillation temperatures.

LLE also has the potential to be favourable even when distillation or other separation methods are possible. In some cases the advantages of LLE may only emerge, or become significant, if it is fully integrated into the design of upstream and downstream processing operations that are optimised to favour the overall process. This highlights the importance of considering the full life cycle of the whole process when designing with green chemistry principles in mind.

Examples of LLE in current industrial practice span all sectors including nuclear, petrochemicals, through to biofuels, food and pharmaceuticals industries. Removal of aromatic hydrocarbons from kerosene using liquefied sulfur dioxide was carried out as early as 1909,[2] and is widely used in the petroleum industry today for separation of similar boiling point compounds such as aromatics from paraffins using solvents including sulfur dioxide, furfural, diethylene glycol and dimethylsulfoxide.

Examples in the chemicals and materials industry include separation of caproplactam, used predominantly in Nylon 6 production, from aqueous solutions containing ammonium sulphate using benzene as the solvent. This has been carried out since the 1960s by EniChem.[3] Phosphoric acid is predominantly produced by the 'wet phosphoric acid' process from rock phosphate, using liquid–liquid solvent extraction to concentrate and separate phosphoric acid from impurities.[4–8] LLE is also commonly used for recovery of acetic acid from dilute aqueous streams,[9–13] typically using phosphine oxide mixtures as the extraction solvent.[14] LLE is used widely in extraction and separation of metal ions.[15–29] For example, in the extraction of copper, dilute copper sulfate solutions are commonly extracted using a chelating organic solvent. The copper ions are then re-extracted back into aqueous solution under more acidic conditions in a second LLE process and copper is recovered from the new, more concentrated solution by methods such as electrolytic deposition. A wide range of proprietary solvent systems are available for this purpose such as those based on salicylaldoximes.

Applications involving natural organic materials include waste stream processing,[10,30] such as carboxylic acid recovery. Fractionation of oils and fats commonly utilises LLE for removal of undesirable components or concentration of desired ones[31–43] as the oils are easily processed in liquid form, but are often unsuited to distillation due to the high temperatures and high vacuum required and the temperature instability of the products. An example is de-oiling of lecithins using either a water extraction (degumming) process or LLE extraction using a nonpolar solvent such as hexane or supercritical carbon dioxide as the solvent.[44] Extraction of enzymes directly from aqueous systems commonly employs LLE[45,46] as an alternative to drying and solid phase extraction.

20.1.2 Current Trends in LLE

Green chemistry and processing principles are increasingly influencing, directly or indirectly, current research and development trends. This is occurring on several fronts. There is a natural cycle of maturing technology that gives an improved operational efficiency and reduced energy cost through process optimisation. Secondly, there is an increasing expectation from consumers and increasing local and global regulatory pressure for greener solvent systems. Thirdly, there is a trend towards increased recovery of secondary products from what previously were waste streams, or for improved remediation of waste streams before disposal.

Optimisation of LLE processing is being advanced through improved availability of phase equilibrium data for a wider range of solvent systems, and through improved and more integrated modelling of extraction and distillation steps. Advances are also being made in contacting equipment including, for example, new better performing materials for membranes used for the contacting interface, and wider availability of mixing and separation equipment for operation under pressurised solvent conditions.

A significant body of research has been carried out in developing new extraction solvent systems, including those relating to use of non-volatile and customisable ionic liquid (IL) solvents, use of non-miscible or switchable miscibility aqueous phase solvent systems, and use of near- or supercritical fluid solvents.

Extraction and fractionation of solutes from aqueous systems continues to be a growing area of research, driven in part by growth in use of microbiological fermentation systems as new sources of chemical production.

This chapter steps through the key aspects of LLE design for industrial separations, paying particular attention to green chemistry aspects that influence the selection of contacting equipment and solvent(s). Key research trends and opportunities in these areas are discussed. The chapter concludes with examples of processing aqueous organic systems to produce products for fuel, food or health applications, which exemplify many of the key design requirements relevant here, particularly for energy minimisation, benign solvent selection, and product safety. This chapter does not cover analytical applications of LLE, although there is a substantial body of literature on use of LLE to provide selectivity and resolution for many complex and useful assays of chemical composition.

20.2   LLE FUNDAMENTALS

All liquid–liquid extractions involve at least two operations: mixing of two fluid streams, during which mass transfer of species between the streams occurs; and separation of the two fluid phases. For a complete operation, many extraction processes will also involve one or more solvent regeneration steps to separate products from the extraction solvent, as indicated in Figure 20.1. Some of these operations may be carried out in the same physical equipment, for example, extraction and phase separation in a counter-current extraction column. Further complexity to that shown in Figure 20.1 is also possible. For example the process may include additional 'wash' solvent steps to enrich product in the extract phase, heat recovery stages, or chemical addition or reaction to aid mass transfer or selectivity of the process. In Figure 20.1 the feed stream to the process, A + B, is a fluid mixture of components which may be a simple solution of two chemical species or may be a complex solution or suspension of many compounds. In a simple system, the major component may be called the feed solvent, and minor dissolved components referred to as solutes.

The feed stream is mixed with an extraction solvent, chosen to have selective solubility for one or more of the feed stream components. After mixing and separation, two phases are produced. The solvent phase is often called the *extract* and the phase containing the non-extracted feed stream is called the *raffinate*, although this terminology is not universally used or applicable to all applications. Separation of the two phases is typically dependent on a density difference between the phases, and the extract may be either the lighter phase or the heavier phase. In some cases the principle aim of the process is recovery (extraction) of a solute from the feed stream. In other cases the

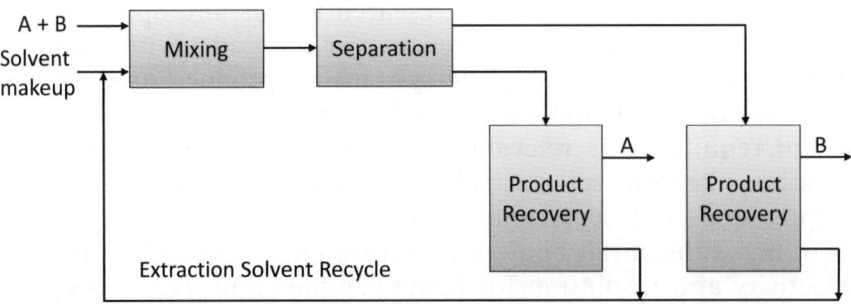

Figure 20.1   Components of a generic liquid–liquid extraction process.

purpose may be to produce a raffinate that has been cleaned or stripped of an undesirable solute, and the process may be referred to as stripping rather than extraction.

To characterise the extraction step, the equilibrium distribution of the solute between the extraction solvent and the feed solvent, K_{solute}, known as the distribution or partition coefficient, is defined as the mass fraction of solute in the extract, $x_{extract}$, divided by the mass fraction of solute in the raffinate, $x_{raffinate}$ (Equation (20.1)). Mole fractions, or mass ratios of solute to solvent, are also used to represent solute distribution. The distribution coefficient is a function of operating conditions as well as composition and does not have a single value for any given system, but is a useful relative indicator of the degree of separation that can be achieved and the number of separation stages that might be required to meet a given product specification.

$$K_{solute} = \frac{x_{extract}}{x_{raffinate}} \tag{20.1}$$

Where more than one solute exists, the relative distribution co-efficients, or selectivity, $\alpha_{a,b}$, for solutes a and b is an important design property, see Equation (20.2). Higher selectivity allows cleaner separation with fewer stages and lower solvent ratios.

$$\alpha_{a,b} = \frac{K_a}{K_b} \tag{20.2}$$

Variations of LLE processing include *fractional extraction* where the feed stream is mixed centrally with a counter-current flow of two immiscible solvents which each have a selective preference for different components. *Dissociation extraction* involves chemical reaction of some of the species to create or enhance selectivity, for example, protonation of amines to improve water solubility or conversely conversion of amine salts to free base to enable solvent extraction.

It is also possible for the extraction process to create a solid precipitated phase. This may occur when one or more dissolved solutes in the feed stream are not soluble in the extraction solvent, and the extraction solvent either strips sufficient feed solvent to cause supersaturation of the solutes or the dissolution of extraction solvent into the feed solvent reduces the solubility of the solute sufficiently to cause precipitation. This may be an unintended consequence of the separation process, as in the precipitation of minor contaminants, or in some cases the primary product may be in solid form. In the latter case the process may be referred to as anti-solvent precipitation or extraction crystallisation, but it involves many of the principles and considerations of LLE design. The handling of any solid phase formed requires special consideration including, for example, use of filtration or solids conveying equipment.

It is possible to use a wide range of solvents in liquid–liquid extraction including fluids under supercritical conditions and non-volatile fluids such as ionic liquid salts. These solvents have wide variations in physical as well as solvent properties and allow alternate processing options that may be advantageous for improved green chemistry processing. These are discussed further below.

Potential alternatives to liquid–liquid extraction include gas or steam stripping, and solid phase adsorption. Gas stripping, in which the solvent or stripping phase is in a gas state, is characterised generally by low solubility and low selectivity of solutes in the stripping phase and high gas volume throughput, but also simpler separation of solvent from raffinate phases and gas stripping can be well suited to removal of low levels of volatiles. Solid phase adsorbents, on the other hand, can provide strong interactions and very selective separations, potentially with very high loadings. Separation of the extract from a solid phase is usually as easy as filtration or decanting, but regeneration of the adsorbent may be more difficult and incur a greater energy and chemical use cost. Solid phase adsorption is likely to be favourable where high selectivity is required to achieve separation and when liquid extraction solvents are difficult to separate from other process streams. In many cases it is possible to incorporate or bond a liquid solvent onto a solid substrate, blurring the boundaries between these different separation methods. In some cases size or mobility based separations such as pervaporation may also be considered.

20.2.1 Solvent Selection

The choice of, and availability of, a suitable solvent is a key determinant for successful liquid–liquid extraction, particularly when considering green chemistry criteria. Choice of extraction solvent should be weighed against the following requirements.

1. *Ability to separate solutes.* This is often measured in terms of a distribution coefficient or equilibrium ratio of solute in the two liquid phases, as described above. If two or more solutes need to be separated in a single liquid–liquid extraction step, then the ratio of distribution coefficients of the solutes, or selectivity, is important.

2. *Ability to separate the extraction solvent from the feed solvent.* The extraction solvent must be at least partially immiscible with the feed solvent and have a different density for phase separation to occur. Greater immiscibility gives cleaner separation and lower costs for downstream solvent recovery, which needs to be weighed against poorer mass transfer between the phases normally associated with low miscibility. A higher density difference makes the phase separation process easier and more efficient. High interfacial tension can give mixed benefits: Coalescence will be more rapid, but separation of emulsions may be more challenging and a high interfacial tension may also inhibit mass transfer.

3. *Recoverability.* Ideally the extraction solvent has a low heat of vaporisation and a large difference in vapour pressure to the dissolved solute(s), so that distillation can be carried out with low energy input. The solvent should be thermally stable under extraction and recovery conditions, and non-reactive with the solutes. On final disposal the solvent should ideally be biodegradable and not eco-toxic.

4. *Physical properties.* Viscosity, vapour pressure, freezing point and flammability will all affect handling requirements for the process, capital cost for storage, energy consumption, and potentially safety considerations.

5. *Availability and cost.* The solvent needs to be available and cost-effective for the process to be viable. Cost should include environmental and human toxicity considerations, as well as the manufacturing process used to produce the solvent.

20.2.1.1 Solute Partitioning. Distribution coefficients and selectivity can be determined experimentally, or can be predicted using equations of state for systems that do not have good experimental data available or for systems that involve multiple components for which data is unavailable. This information needs to be combined with known or predicted miscibility (phase equilibria) data to enable selection of a shortlist of potential extraction solvents. The shortlist can then be weighed up against other criteria and carried through to a full process design evaluation. Non-random two liquid (NRTL) and universal quasi-chemical (UNIQUAC) semi-empirical equations have become acceptable for estimating properties of non-electrolyte systems where some experimental data are available. Purely predictive models, such as regular solution or UNIFAC (UNIQUAC Functional-group Activity Coefficient) models may also be used, at least for initial screening of solvents.

Table 20.1   Distribution coefficient of acetic acid at 20 °C in selected solvents.

Solvent	Distribution coefficient	Azeotrope/ wt% water	Boiling point/°C	Enthalpy of vaporisation/kJ kg^{-1}
n-Butanol	1.6	45	118	430
Ethyl acetate	0.8	8	77	395
MTBE	0.8	4	55	322
MIBK	0.5	24	116	488
Toluene	0.06	20	111	410
Hexane	0.01	9	81	350

Solvent design using group contribution methods, and methods that involve selection based on known interacting groups,[47] have also been used for pre-selection of solvent candidates. Tables of selected ternary systems suitable for LLE have also been established.[48]

As an example, the distribution coefficient and miscibility of potential solvents for extraction and concentration of acetic acid from aqueous solution are given in Table 20.1. Solvents with high distribution coefficient, such as propanol, will give rapid extraction of acetic acid and require low extraction solvent ratios. However, propanol also dissolves large quantities of water; the concentration of acetic acid in the extract will not be substantially enhanced and the energy cost of separating the product and the extraction solvent may be high. Solvents such as ethyl acetate and methyl *tert*-butyl ether (MTBE) are better suited as the carryover of water in the extract is lower and there is a wide difference in boiling point between the solvent and acetic acid (boiling point 118 °C). Both have relatively low toxicity, are not prone to formation of peroxides, and are used industrially for recovery of acetic acid from aqueous waste streams. Methyl isobutyl (MIBK) may be suitable. It has a moderate distribution coefficient but similar boiling point to acetic acid and high water solubility. The solvents in Table 20.1 with lower distribution coefficients are generally not suitable to use when better options exist.

It is important to note that separation factors are functions of operating conditions, including changes in composition during the extraction process, and that full thermodynamic and phase equilibria calculations are required to fully evaluate a solvent's potential. Separation factors by themselves are useful for screening, but can also be misleading[49] and result in missed opportunities. It is difficult to obtain or predict phase equilibria reliably for all solvent possibilities, but for some established LLE processes more detailed studies are being carried out, for example, the study by Magda *et al.*[8] gives a robust assessment of data for wet phosphoric acid extraction processes.

20.2.1.2 Solvent Miscibility. Design of a LLE process generally requires knowledge of the phase equilibria of the ternary system containing the extraction solvent, the feed solvent and the solute. Characteristic ternary diagrams for a type I partially miscible solvent system are shown in Figure 20.2. The feed solution (F), a mixture of the feed solvent and the solute, is plotted along the feed-solvent/solute axis. When mixed with the solvent, the mixture has a composition (M) in the two phase region, which under equilibrium conditions separates along the tie-lines to form extract and raffinate liquid phases. The composition of the two phases can be determined from the diagram: the extract phase is enriched in solute as well as containing some feed solvent, while the raffinate is depleted in feed solute.

Figure 20.2 shows diagrams for two hypothetical solvents, A, and B. Solvent B has a greater miscibility with the feed solvent, indicated by the smaller two phase region, and the distribution coefficient of the solute is lower, indicated by the shallower gradient of the tie-lines and the more dilute solute concentration in the extract. Solvent A would give higher solute concentrations with fewer separation steps, and lower solvent ratios, than solvent B. Subject to other considerations, solvent A would be the solvent of choice.

Ternary diagrams, as in Figure 20.2, can be used to graphically determine the separation that will be achieved. One approach is the Hunter and Nash graphical equilibrium stage method, which involves plotting out progressive equilibrium stages of separation. This gives a calculation for the number of theoretical equilibrium mixing stages that a contacting column needs to achieve a desired degree of separation, enabling the contactor to be designed. More commonly for industrial design the calculations will be done computationally using a fitted equation of state, enabling rapid assessment of different separation criteria and solvent to feed stream ratios. This also enables multiple solute systems to be more easily handled.

It is possible to adjust solvent miscibility during extraction, or subsequent to extraction, to aid recovery of the solute from the extraction solvent. This is useful in some cases and is used, for example, in reactive extractions including use of chelating or complexing agents, or through addition of salts or gelling agents.

Solvent properties are functions of temperature, pressure and pH, and this should be considered when screening solvents. Solvents that are gases at atmospheric pressure, for example, only have potentially useful solvent properties for LLE when liquefied or under super-critical fluid conditions. Because their standard properties are those

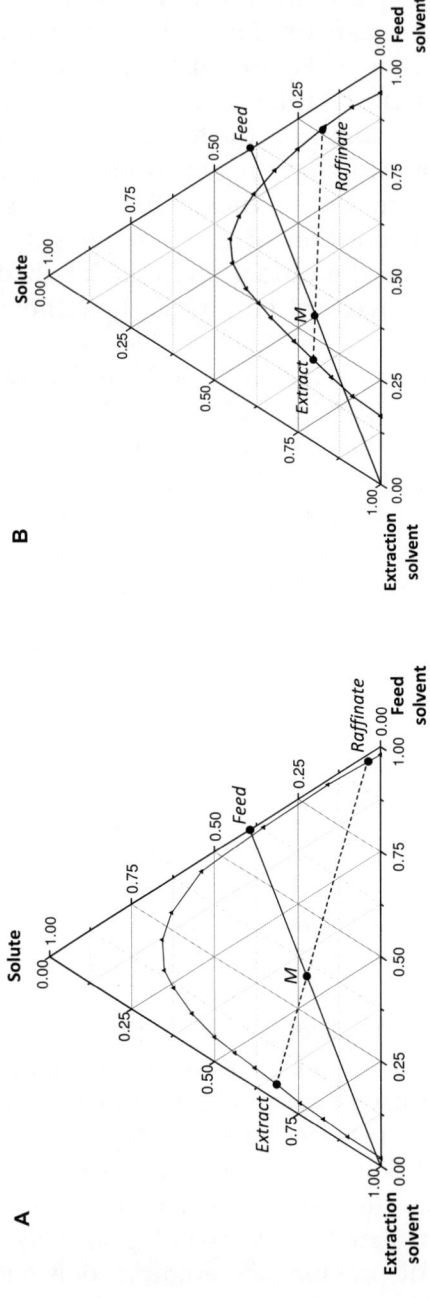

Figure 20.2 Ternary phase equilibrium diagrams for two hypothetical extraction solvents A,B, showing one equilibrium stage of LLE mixing and phase separation.

of gases, gas solvents may often be omitted from solvent lists used in screening processes for solvent selection. This is unfortunate as the inherent volatility and high vapour pressure of gases meets many green chemistry criteria for ease of separation and solvent reuse. Supercritical fluid solvents, along with contrasting solvents that have low volatility such as ionic liquids, and biphasic aqueous systems, are increasingly of interest as solvents for LLE applications. These solvents are discussed further below.

20.2.1.3 Supercritical Fluid Solvents. It is possible to use LLE extraction solvents that are under supercritical fluid conditions.[42,50–61] See also Chapter 4. The principles of liquid–liquid extraction remain the same, but with several key distinctions. Above the supercritical temperature or pressure there is no discrete phase change from gas to liquid conditions, and it is possible to control density-dependent solvent properties continuously over a wide range of values by varying operating pressure or temperature. It is possible, for example, to extract into a supercritical fluid solvent under high density, high solubility conditions, and then to progressively reduce pressure and solvent properties of the extract phase to fractionate different dissolved compounds with different affinity for the solvent.

Supercritical fluids have distinctive physical properties compared with liquid or gas states. This includes low viscosities even at high density, which enables higher mass and heat transfer rates, and zero surface tension. Many supercritical solvents used are gases under atmospheric conditions, leading to facile and potentially energy efficient separation and recovery of the solvent from less volatile product streams.

Of the negative aspects, supercritical fluid systems typically require operation under pressurised conditions to achieve useful solvent densities, and this adds capital cost and safety considerations to the process. For some applications this added cost is compensated for by other benefits, particularly when high reuse of solvents and low, or even zero, solvent residues in the product are design targets.

A commonly used supercritical fluid solvent is carbon dioxide. CO_2 is supercritical at temperatures above 32 °C and is often used in the range (40–60) °C for extraction of lipophilic organic compounds.[52,53,58,62] This temperature range is easy and efficient to achieve in industrial processing, and is also a useful temperature range for processing thermally labile organic compounds. Solvent removal can be carried out by reducing pressure, rather than by increasing temperature. At 40 °C and pressures above about 200 bar, CO_2 has a density greater than 800 kg m^{-3} and is an effective solvent for many low polarity substances, but has a viscosity of only 80×10^{-6} Pa s (*cf.* water at $\sim 600 \times 10^{-6}$ Pa s). If the pressure is reduced then the density reduces, and CO_2 becomes a poor solvent. At atmospheric pressure CO_2 is a gas and will separate cleanly from liquid or solid solutes.

An example of industrial LLE using supercritical CO_2 is de-oiling of lecithins. Oils extracted from soya or other sources are contacted in liquid form with supercritical CO_2, which extracts triglycerides and concentrates residual polar lipids (lecithins) in the raffinate. De-pressurisation of both extract and raffinate phases to (3.0–5.0) MPa is then typically carried out, allowing easy separation of the majority of the CO_2 (now in a gaseous state) and recycle of CO_2 back to the process. Further depressurisation to atmospheric pressure allows complete separation of CO_2 and recovery of the product(s) from the process.

Other known industrial operations include recovery of organics from wastewater and recovery of terpenes from citrus oils. Liquid–liquid contacting equipment for operation under supercritical fluid CO_2 conditions, including options for mechanical mixing or scraped surface internal components, is available as standard design equipment from several established equipment manufacturers.

Supercritical CO_2 is generally considered to be a green solvent due to its non-toxicity at low exposure levels and its prevalence in the natural environment to which most products are already exposed. In some cases CO_2 is not considered to be an organic solvent at all and CO_2 solvent extracted products may claim to have been made using a 'solvent free' process. Supercritical CO_2 complements water in its wide acceptability as a benign solvent, solubilising many non-polar compounds that do not dissolve in water. If ethanol is included as a relatively benign and widely accepted solvent for most applications, then binary or ternary mixtures of the three solvents can be used to form a basic 'green' solvent system that spans a wide polarity range and range of solutes that can be processed (see Figure 20.3).

CO_2 as a solvent also has relatively favourable lifecycle statistics. It is captured almost entirely as a by-product of existing petrochemical or biochemical processing, and it can be argued that the only energy and environmental cost of its use is that required for separating, compressing and transporting it. Use of CO_2 derived from sustainable bioprocessing, such as fermentations that produce bioethanol is possible if a sustainable source of CO_2 is required, but the volume of

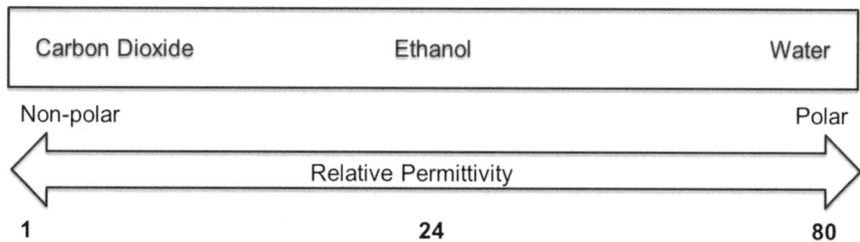

Figure 20.3 Spanning the solvent polarity range using benign solvents by using supercritical CO_2 for low polarity applications along with ethanol and water.

CO_2 that is ever likely to be used in solvent extraction is a minute proportion of un-utilised CO_2 handled currently by industry and it is unlikely that its use in LLE will ever require generation of new CO_2 that is not already being produced. See also Chapter 4.

Other compressed gas solvents can be used, including those operating below their critical temperature under liquid conditions (sometimes called sub-critical or near critical solvents). Light hydrocarbons (*e.g.* propane, butane), ethers, hydrofluorocarbons and sulfur hexafluoride[63] are examples. Ammonia is also an effective polar solvent but is difficult to make use of industrially.[64] These gases have a range of different solvent properties and can be liquefied at lower pressures than CO_2, which may confer some processing and capital cost advantages. Liquid solvents such as methanol or water can also be used under elevated temperature and pressure including supercritical fluid conditions, although the high temperatures, and increased reactivity, will only suit specific applications.

Safety considerations apply when using pressurised solvents. Hydrocarbon solvents can also be highly flammable.

20.2.1.4 Ionic Liquid Solvents. Ionic liquids are salts that exist in molten form at or close to ambient temperatures. They are typically composed of large organic anions and cations, and have distinctive properties that make them ideal solvents for some operations.[65] The range of potential applications is large and includes chromatography,[66–68] reactions and synthesis,[69,70] separations,[15,71] CO_2 capture,[72] dissociation and processing of cellulosic materials,[73,74] enzyme reaction media,[75] and use as lubricating fluids.[76] See also Chapter 22.

The largest body of literature relates to synthesis in ionic liquids, but many proposed extraction and separation examples exist[65] including the processing of proteins, spent nuclear fuel,[15,77] petro-chemicals[71] and biofuels,[78] sugars,[79] polymers such as chitan,[80] breaking azeotropes,[81] heavy metals[82] and separation of rare earths.[24] Examples of industrial use of ionic liquids for LLE exist such as BASF's BASIL™ process which uses N-alkylimidazole for scavenging acid re-action by-products in place of forming solid salts. In spite of these applications and the large body of research in the field, industrial applications outside of chemical synthesis are still slow to develop.[69]

Ionic liquids are touted as green solvents. Favourable properties include wide ranges of selectivity (including stereoselectivity), solu-bility in a broad range of other solvent systems, non-coordination with metals, and acceleration of reactions. Properties that are spe-cifically of importance to green chemistry relate to the ability of ionic liquids to displace less desirable solvents, and to be effectively and energy efficiently reused in the process. Selected operational prop-erties of ionic liquids include:

- No significant vapour pressure: This, in principle, simplifies separation and allows energy efficient and near-complete re-covery of the ionic liquid for reuse in the process.
- Stability under wide temperature ranges and exposure to other solvents: These properties also support high levels of recovery and recycling of the ionic liquid phase.
- Low temperature separations, allowing processing of thermally or chemically unstable compounds without production of arte-facts, for example in processing of organic products.

There has been some debate as to the true green credentials of ionic liquids. Some studies have highlighted the challenges in achieving high levels of ionic liquid recovery and reuse.[83] Accumulation of contaminants in the ionic liquid can potentially render it ineffective, and degradation of the ionic liquids themselves has been observed in some cases.[84] Many ionic liquids used in separations are not miscible with water, but are often hygroscopic and potentially unstable in the presence of water. Commonly used PF_6 based ionic liquids have been shown to be susceptible to degradation with exposure to even trace levels of water.[85] Meine et al.[86] have demonstrated temperature in-stability of 1-butyl-3-methylimidazoliumbromide at 200 °C.

Another important green chemistry principle is life cycle assess-ment of the solvent in terms of toxicity, biodegradability and manu-facturing process. Preparation of some ionic liquids has been demonstrated to be up to 100% atom efficient,[83] but allowing for waste solvent generated in manufacture, the overall environmental impact may still be high. Improvements in manufacturing process and use of sustainable feed stocks to produce ionic liquids provide the focus of current research, which may also reduce the cost of ionic liquids which is currently high compared with traditional organic solvents. A number of studies have examined the toxicity and eco-toxicity of ionic liquids themselves.[87] Non-toxicity is often attributed to the non-volatility of the ionic liquids, absence of solvent residues in product fractions, and complete recycling of the solvent so that it does not get introduced to the environment. These criteria are difficult, if not impossible, to meet in practice. Few ionic liquids currently available are considered readily biodegradable. Guidelines for designing greener ionic liquids including improved biodegradability are emerging,[83] although improved rates of biodegradability, for example using longer lipophilic alkyl chains, may coincide with increased eco-toxicity. Cheaper and more sustainable ionic liquids that have been investigated include choline chloride based ionic liquids.[88,89]

Compressed or supercritical gases including CO_2 are soluble at high levels in ionic liquids and applications in CO_2 capture have been studied. Conversely, because the ionic liquids are not soluble in CO_2, applications in LLE for recovery of solutes from the IL phase are also promising, and extraction of a range of aromatic and aliphatic solutes,[90] and metal chelates,[91] from ionic liquids has been investigated. Selective extraction of unsaturated fatty acids into ionic liquids and recovery using supercritical CO_2 has been described.[92] Use of other supercritical solvents with ionic liquids has also been widely studied, although CO_2 is favourable in many cases for low cost, high availability, low temperature operation and ease of separation from extracted products. Increased dissolution of CO_2 in ionic liquids can also operate as a phase separation switch in some applications, or give a melting point depression.

To aid recovery of ionic liquids there is a trend towards use of solid supported ionic liquids,[93] or dispersal of ionic liquids in organic solvent to take advantage of the tuneable ion exchange properties of the salts but with much lower viscosity and potentially easier recovery of adsorbed solutes.

20.2.1.5 *Aqueous–Aqueous Solvent Systems.*

Aqueous two-phase systems (ATPSs) are novel LLE solvent systems based on mixtures of either water-soluble polymers that are immiscible in each other, mixtures of polymers with salt solutions, or mixtures of micellar phases.[94–96] They have been widely investigated for the separation of proteins,[97–99] enzymes[100] and antibiotics,[101] and for recovery of metals[26] (*e.g.* copper extraction from low grade ore[20]). The use of two aqueous phases has a number of potential benefits, and some favourable characteristics for green chemistry including displacement of organic solvent use as well as improved product quality.

Some LLE bio-separations, for example separation of proteins, are unsuited to use of organic solvents as extractants because the solvent may denature the proteins. This can be avoided in ATPSs. Protein isolation also often involves multiple energy and solvent intensive steps including chromatographic separation. ATPSs may be suitable for pre-concentration before chromatography, or with multiple separation stages may produce a moderately purified product.[102]

The earliest report of ATPSs is credited to Dutch microbiologist Martinus Willem Beijerinck who observed formation of a two-phase agar–gelatin–water system. Modern ATPSs can be classified as mixtures of: non-ionic polymer pairs such as polyethylene glycol (PEG)–dextran and PEG–polyvinyl alcohol; non-ionic polymers and an electrolyte such as sodium sulfate dextran–PEG; two electrolytes such as sodium sulfate dextran–sodium carboxymethyl dextran; or a nonionic polymer with a low molar mass compound such as potassium phosphate or glucose. Many of these systems use compounds that are non-toxic and suitable for processing food products, and are safe for environmental disposal. In other industries an aqueous two-phase system may replace the use of less desirable, or hazardous, solvents such as sulfides.[20]

Polymer solutions and high concentration ionic solutions used as ATPSs typically have high viscosities, and phase separation can be slower and more difficult to achieve. To compound these challenges, the highest selectivity for target solutes is often achieved using higher polymer concentrations and higher molar mass solutions,[103,104] where viscosities are highest. Many research papers have studied selectivity of various solutes in different ATPS systems and the stability of compounds such as enzymes during extraction. Less work has focused on methods for recovery of products from the solvent phases, and recovery and reuse of the solvent. These are essential considerations for cost effective and green chemistry operation.

Micellar ATPS systems under consideration include, for example, Triton X-114–sodium phosphate buffer for extraction of pectinase from a fermentation broth.[105,106] In this case a phase split is induced by raising the temperature giving a micelle rich dense phase and a lighter extract phase containing a high concentration of enzyme.

Some undesirable compounds acting as enzyme inhibitors were also able to be selectively separated into the raffinate phase.

20.2.2 Contacting

Contacting of the two liquid phases is a fundamental part of a LLE process and is one that needs to be carefully designed to obtain optimal performance from the process. LLE depends on the two liquid phases being largely immiscible, which inherently presents a challenge to achieving good rates of mass transfer between the phases. The two phases need to be physically dispersed in each other to create as much interfacial surface area as possible, and for systems that rapidly coalesce, continued agitation may be required to maintain the high surface area for a sufficient time period for mass transfer to occur. The consequences of poor contacting include poor separation and overdesign of equipment, leading to low energy efficiency and excessive solvent use. This also has downstream ramifications in product recovery and solvent recycling steps of the process. On the other hand, effective contacting can give cost and environmental efficiencies, or in some cases, novel design of contactors may enable use of solvent or feed solutions that are difficult to handle but offer overall process improvement. Examples include highly viscous solvents, such as polymer solutions, or highly concentrated feed solutions.

Contacting equipment is categorised here into four types: single or multiple stage batch processing; continuous flow extractors; membrane contactors; and micro-channel contactors. Solid supported or bonded liquid phases can also be used for separations, employing the same chemistry and many similar operating principles, but the factors associated with contacting and separation of a solid phase are not described further here.

20.2.2.1 Batch Contactors. The most common batch contactor is a simple stirred tank or series of stirred tanks. Often called a mixer–settler process, the phase separation may be carried out in the same tank following batch mixing, or may be carried out in separate holding vessels. Semi-continuous operation and multiple extraction stages can be achieved by using a series of tanks in cross- or counter-current flow. This type of contactor suits smaller or intermittent scale operation such as: processing of some natural products with seasonal supply; extractions that require long contact times such as some metal separations;[2] and separations that involve formation of solid phase products. Mixing in a batch contactor may be achieved through use of impellers, through hydrodynamic forces using a recirculated liquid flow through a noz-zle or in-line static mixing element, or through centrifugal action.

Key design parameters in batch contactors are the intensity of mixing and the overall residence time. These are important considerations for products that are sensitive to either shear or long-term exposure to elevated temperature.

20.2.2.2   Continuous Contactors.   Many LLE operations are more effectively carried out in continuous contacting equipment, typically counter-current flow columns, and these are widely used in industry. The simplest operation is a spray column which involves simply feeding the lower density phase at the bottom of the column and the heavier phase at the top. Mixing is aided by atomisation of one the phases, and hydrodynamic and convective forces are used to maintain dispersion of the phases.

For more rapidly coalescing systems or where longer contact times are required, a packed column can be used. The packing creates local flow turbulence and gives continued mixing and dispersion of the phases. The height of column required to achieve an equilibrium stage of separation can be significantly reduced over that of a spray tower and packed columns are widely used. Packed columns are effective for systems that are not too viscous or prone to fouling of internal surfaces. The increased pressure drop and pumping requirements through a packed column are generally not significant in terms of overall energy consumption. The main issues associated with counter-current flow columns derive from the interdependence of the two fluid phases being contacted and include foaming, formation of emulsions, unloading and flooding.

In passive mixing systems such as those described above the mixing is determined by, and hence limited by, the density and physical properties of the liquid phases. For viscous or otherwise difficult to mix liquids, additional mixing energy can be applied by mechanical stirring, mechanical pulsing of plates or baffles, or for some systems, sonic or electrostatic forces can be used to induce droplet formation.[107] Mechanical stirring is typically applied with a central shaft and a series of rotating discs and baffles that create turbulent flow. It can enable very high stage efficiency and handling of viscous materials, but it adds mechanical complexity to the process and is generally only applied when passive mixing is unsuitable.

Column design typically involves determination of two key physical parameters, namely the height of the column needed to get the required degree of extraction, and the diameter of the column required to achieve the desired product throughput. Design of these physical parameters in turn depends on the number of effective equilibrium separation stages required, determined for example using Hunter–Nash graphical calculations, and on hydrodynamic behaviour which determines the height of column required to achieve each effective stage of separation. Methods for approaching these calculations are well described elsewhere.[108–110] Some solvent systems may be hard to model theoretically, for example, those that exhibit the Marangoni effect,[111,112] or where changing solute concentrations create large changes in hydrodynamic properties of the system.

Mixing of two liquid phases typically results in one of the phases being present as discrete droplets in a continuous matrix of the other liquid phase. Batch mixing generally results in the smaller volume phase forming the dispersed phase, but in column contactors, some control over which is the dispersed phase may be possible (*e.g.* by dispersing one of the phases at the column inlet) or by choosing materials of construction that favour wetting by one of the phases present.

Counter-current columns are well suited to extraction using supercritical fluid solvents. The low viscosity of the supercritical fluid phase enables efficient mixing stages, and smaller scale packing or taller columns can be used without excessive pressure drop. Continuous flow of liquid phases into and out of a pressurised process can be easily accommodated with available industrial equipment. Mechanically stirred systems create additional considerations, such as requirement for mechanical seals, but are available. External magnetic coupling of mechanical systems is also possible and is useful in some pressurised systems.

20.2.2.3 Membrane Contactors. Membrane contactors, also called liquid membranes or supported liquid membranes, are a hybrid technology. The process appears similar to conventional size-based membrane separation processes, but operation is fundamentally similar to that of liquid–liquid extraction and separation columns. A membrane contactor is operated with one fluid phase either side of the membrane, with mass transfer between the two phases driven by diffusion in the microporous internals of the membrane where the two phases are in contact. Very little pressure gradient, if any, is applied across the membrane, so no bulk mixing of the phases occurs and the mass transfer is controlled by the equilibrium kinetics of the two liquid phases. This allows high contact areas, rapid mass transfer, and potential for operating and cost efficiencies over column extraction for many systems. Further advantages of these types of contactor include high capacity without flooding or foaming issues independent of density difference between the phases. In their simplest form the membranes themselves are non-selective, but it is possible to use active membrane surfaces that favour wetting by one of the fluid phases present to enhance selectivity of the solvent system. Membrane contactors are generally unsuitable for highly viscous or gelling fluid phases, when solid phases may precipitate, or when high loadings of solute need to be recovered from the feed solution.

Membrane contactors are commonly used in tubular (hollow fibre) form, in which one fluid phase is passed through the internal core of the hollow fibres and the second phase on the outside of the fibres. Fibres can be assembled in bundles to make a contactor resembling a shell and tube configuration, and high contacting surface areas can be achieved in a small volume. Membrane contactors can be used in a flat sheet configuration, and moving membrane contactors also exist in which a continuous sheet or belt is drawn through one liquid phase to wet the membrane, which is then drawn through the second liquid phase in which mass transfer occurs. Membrane contactors can be used equally well for gas, liquid and supercritical fluid contacting systems and their use has been demonstrated in a wide range of applications in fermentation,[61] pharmaceuticals,[113] wastewater treatment,[114] chiral separations,[115] semiconductor manufacturing, carbonation of beverages, metal ion extraction,[116,117] protein extraction, volatile organic compound (VOC) removal from waste gas,[118] CO_2 capture[119] and osmotic distillation. It is a relatively new technology and much development is still occurring to improve the robustness and functionality of membranes available.[120,121]

When one or more of the fluid phases is a supercritical (or near-critical) fluid, the process is called porocritical extraction. This combines the porosity and high surface area of the membrane contactor with the high diffusivity of the supercritical or near-critical fluid state.[122] The porocritical extraction process is particularly suited for the extraction of VOCs present in low concentrations from aqueous streams.[61]

20.2.2.4 *Micro-channel Contacting.* Micro-channel contacting is a niche method of contacting fluid phases. Two immiscible fluids are introduced into a micro-channel bringing the two phases in contact with a high contact area between them. Because of the

Table 20.2 Operating ranges for different LLE contacting methods.

Condition	Mixer-settler	Centrifugal contactor	Static column	Agitated column	Membrane contactors	Micro-channel
Number of stages	Low	Low	Moderate	High	High	High
Flow rate	High	Low	Moderate	Moderate	High	Very low
Residence time	Very high	Very low	Moderate	Moderate	Low	Low
Interfacial tension	Moderate to high	Low to moderate	Low to moderate	Moderate to high	Low	Low to high
Viscosity	Low to high	Low to moderate	Low to moderate	Low to high	Low	Low
Density difference	Low to high	Low to moderate	Low to moderate	Low to high	Low to high	Zero to high
Plant size	High	Moderate	Low	Low	Low	Low

micro-channel flow, mixing is dominated by viscous and interfacial forces rather than gravitational forces. Micro-channel contacting has found good application in analytical and diagnostic applications, but challenges exist in scaling the process to preparative applications.[123] Solvents under supercritical conditions may be advantageous in some applications by reducing viscosity, surface tension, and pressure drop effects.

20.2.2.5 Operating Conditions. A summary of operating conditions suited to different contacting methods is given in Table 20.2.

Temperature and, in compressible fluid systems, pressure are important design parameters in liquid–liquid extraction affecting key performance variables such as phase equilibria and the mixing and separation properties of the solvents.

The pH is important in some metal and bio-extraction processes. For example in the extraction of penicillin, pH is controlled to improve the distribution coefficient and to minimise degradation of product. The kinetics of the process may be strongly influenced by pH, as in metal extractions. In other cases pH control is required to create desirable chemical changes, as in dissociation-based extraction of organic molecules (*e.g.* cresols) or to prevent undesirable reactions (*e.g.* the hydrolysis of ethyl acetate in the presence of mineral acids).

Residence time is an important parameter in reactive extraction processes (*e.g.* metals separations, formaldehyde extraction from aqueous streams) and in processes involving unstable molecules (*e.g.* antibiotics and vitamins).

Interfacial tension and miscibility affects the ability to disperse phases sufficiently to create good mass transfer. Mechanically agitated systems, or systems with surface wetting aids such as porous membranes contactors, are effective for systems with high interfacial tension. Use of a low viscosity solvent, such as a supercritical fluid, can also be effective.

In supercritical fluid CO_2 extraction systems there is a trend towards increasingly high pressure operation, which increases the solubility of many solutes allowing smaller footprint plants to be used, as well as broadening to some extent the range of soluble compounds.

20.2.3 Phase Separation

Liquid–liquid extraction generates two liquid solvent phases which must be first separated from each other, and then generally at least one of the phases must be further fractionated to recover extracted solutes from the solvent or to regenerate the solvent for reuse in the extraction process. Both of these operations can impact significantly on the overall energy and solvent waste footprint of a LLE process.

In some cases the phase separation is an integral part of the contacting equipment, as in the disengaging spaces or reflux sections of continuous flow extraction columns. In other cases phase separation may be carried out in separate unit operations.

Phase separation is a density driven process, and depends on a density difference between the phases as well as the viscosity and surface tension of the two phases. Poor phase separation lowers the overall efficiency of solute separation as well as resulting in increased levels of cross-mixing of the two solvents which adds to the energy requirements and cost of downstream solvent recovery. In some cases a very efficient phase separation may negate the need for further processing, for example, direct disposal of an aqueous raffinate stream that has been effectively stripped of contaminant solutes.

Poor phase separation can also disrupt process operation. This is sometimes manifested as foaming or flooding where liquid–liquid emulsion or gas–liquid foam fills the processing volume. A tendency for foaming or flooding creates throughput limits in both batch and counter-current column extraction. Highly viscous systems, for example concentrated or high molar mass polymer solutions, may also be slow to disengage. A practical multistage approach to phase separation for viscous aqueous two phase systems is described by Benavides and Rito-Palomares.[124]

Separation can be enhanced in two main ways: (1) by changing the composition or properties of the phases themselves (*e.g.* by changing the pressure or density if one of the phases is a compressible fluid) or by changing salt concentrations or pH; or (2) by changing the processing operation used to enhance separation such as through use of centrifugal forces, or through control of droplet coalescence with baffles or membranes. Passive enhancement methods such as use of membrane contactors, or simply allowing added residence time for separation, should be considered first where this is viable. If other alternatives are needed they should be evaluated against the added energy costs of processing such as the energy required for creating centrifugal separation forces, and the cost of added processing aids or solvents in the case of modifying the solvent properties. Solid-supported liquid phases can also be considered for difficult separations.

Some solvent systems may result in the precipitation of a solid phase. This adds requirements for additional phase separation, which may be achieved by density difference as in centrifugal separation or by size difference using filtering. Slurries of solids can often be transported through the process as a fluid using conventional or slightly modified pumping and valving systems. In other cases, dedicated solids handling equipment such as screw conveying may be required to achieve continuous processing. Batch processing is sometimes favourable for systems containing solids.

Phase separation of supercritical fluid solvent systems has some added considerations. Depressurisation of a liquid-supercritical fluid mixture allows easy separation of the low density supercritical fluid phase, but often high levels of the supercritical fluid may be dissolved in the liquid phase creating a highly expanded and relatively low density liquid phase that requires secondary depressurisation. Mixtures of two fluids also have different conditions under which the mixture is supercritical compared with those of the pure fluids. For example, the critical temperature of CO_2–ethanol mixtures increases monotonically with increasing concentration of ethanol. It is thus possible for a supercritical fluid solvent to enter a two-phase region when mixed with other solvents, often with only a very small density difference between them leading to challenges in phase separation or creation of flow and composition instabilities in a process.

20.2.4 Solvent and Product Recovery

The liquid–liquid extraction step is simply replacing one solvent system with another and recovery of the target solute(s) from the extract solution is still usually required. The feed solvent will contain at least some traces of extraction solvent that need to be recovered to achieve efficient recycling of the extraction solvent. Often the residual feed solvent, or raffinate, will also need further processing. It may need further processing to strip additional solutes before disposal, or to meet product specification if the raffinate is also a product.

The unit operations available for these secondary separations include all possible industrial separation processes. Often this will be distillation, but it could include concentration by evaporation, drying operations such as freeze or spray drying, membrane separations, crystallisation, or even a second LLE process such as in the re-extraction of metals into aqueous solution after solvent extraction from dilute aqueous streams.

20.2.4.1 General Considerations. The choice of secondary separation process influences the overall energy, cost and environmental suitability of the LLE process, but cannot be optimised in isolation. The ideal extraction solvent for selectivity towards target solutes, for example, may produce an extract phase that is difficult or energy intensive to recover, whereas another extraction solvent may still give satisfactory extraction but allow easier downstream recovery. The entire LLE process should be optimised before it is evaluated against alternative processing options.

The full range of processing operations cannot be usefully described here, but some general considerations for designing an integrated LLE process are highlighted. Where possible the following guidelines should be applied.

- Choose solvent systems where the solute, or the lower concentration component, is the most volatile species present. This allows separation by distillation or evaporation of smaller volumes of liquid.
- Choose solvent systems with large differences in volatility. Separation by evaporation can usually be carried out more energy efficiently than by distillation.
- Choose solvents with low heat of vaporisation, as the overall cost of solvent vaporisation is approximately proportional to this.
- Choose solvent systems with low miscibility. This may make contacting of the phases more challenging to achieve, but reduces the much more significant energy costs of separation for downstream processing.
- Design the liquid–liquid extraction process to give high solute concentration in the extract, rather than the minimum number

of contacting stages possible. This will make contacting equipment larger but produces lower volumes of solvent for further separation where most of the energy cost is incurred.

- Choose the lowest toxicity, lowest eco-toxicity, most sustainable and biodegradable solvents. The higher the environmental impact of the solvent, the greater is the requirement for high separation efficiency and high solvent recovery levels. High solvent recycle rates are more costly to achieve, and are also more prone to accumulation of trace contaminants to levels that may interfere with the separation process or product quality.
- Select the extraction solvent based on the assumption that it will contain impurities when it is recycled in the process due to incomplete separation from other process compounds. For example, for feed and extraction solvents that form azeotropes, it may be more energy and process efficient to use the azeotropic mixture as the extraction solvent as the azeotrope does not need to be broken during solvent regeneration. Experimental measurements of selectivity that only consider pure solvents may not identify the optimal extraction system.

20.2.4.2 Solvent Evaporation/Distillation. The secondary separation in some cases may be as simple as a single stage flash separation induced by a pressure or boiling point reduction. This is common in supercritical fluid extractions and other systems where there is a substantial difference in vapour pressure between the extraction solvent and other solvent or solute phases present. Other systems with wide difference in boiling point, including some systems with solid solutes, will suit evaporative separation. Similar boiling point systems typically require multiple effective stages of distillation.

Azeotrope forming systems add an additional load to separation by distillation, typically requiring addition of a third component such as benzene to break the ethanol–water azeotrope. There is trend towards other technologies such as molecular sieve dehydration for the final stages of recovery in these types of system.[125]

20.2.4.2.1 Liquid Solvents. Figure 20.4 shows a pressure enthalpy plot for a hypothetical extraction solvent. LLE using a liquid solvent at room temperature is indicated on the left side liquid phase region of the plot. In the example here, after separation from the raffinate, the extract solvent is vaporised to recover low volatility solutes and this is carried out in the gas phase region on the right hand high-enthalpy side of the diagram. Vaporisation is

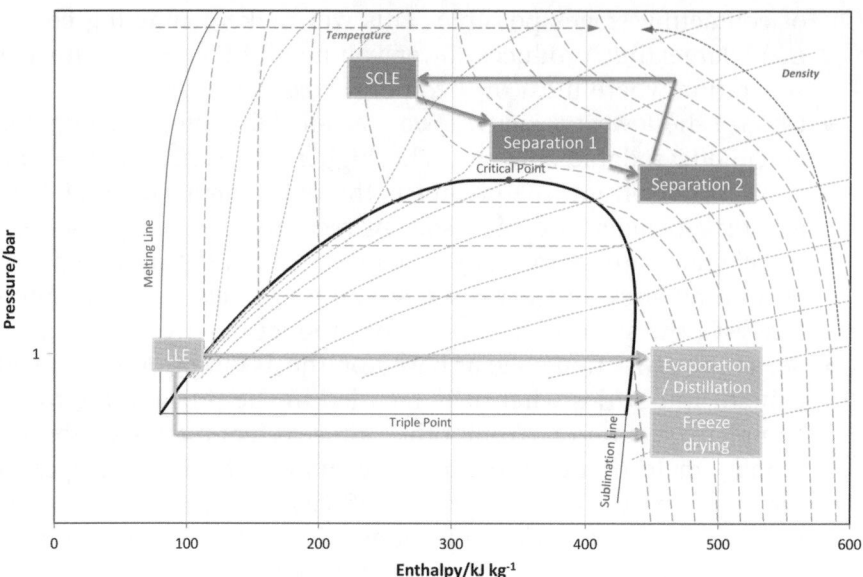

Figure 20.4 Extraction and distillation operations on a pressure enthalpy chart.

accomplished either by directly boiling the solvent at atmospheric pressure, at reduced temperature by evaporation or distillation under vacuum, or by sublimation if freeze drying is used. All cases require addition of heat, predominantly the enthalpy required for vaporisation. Latent heat of solutes or vaporisation of other components, as well as any interactions that occur as the solute concentration increases during solvent removal, all need to be considered. Generally, however, the heat of vaporisation of the solvent is the major energy sink.

20.2.4.2.2 Supercritical Fluid Solvents. Extraction using a supercritical fluid extraction solvent (SCLE) is also indicated in Figure 20.4, in the region above the critical point. In this type of extraction, pressure is generally elevated to a pressure above the critical point to a region where solvent density and solvation properties are just sufficient for the extraction process to occur at a reasonable rate. Pressures higher than this point need to be evaluated against increasing capital cost of pressure vessels and increased fluid compression costs. After phase separation the supercritical fluid extract phase is generally reduced in pressure to a region of sufficiently low solvent density that separation from dissolved solutes occur. Lower pressures than necessary are avoided as this introduces additional recompression costs to recover the solvent. As with liquid solvents, heat input is required to bring a supercritical fluid solvent from dense extraction conditions to low density (gas state) separation conditions. As indicated in Figure 20.4, the enthalpy change required is generally lower under supercritical fluid conditions than under liquid conditions. For example, between typical supercritical fluid extraction and separation conditions, CO_2 requires an enthalpy change ~ 150 kJ kg^{-1}. This compares with the enthalpy of vaporisation of typical liquid organic solvents (including liquid CO_2) which are in the range of 300–500 kJ kg^{-1}.

Depressurisation of a supercritical fluid can be carried out in several ways. Isenthalpic decompression through a valve gives a corresponding temperature drop, which may solidify or freeze other dissolved components, or may even cause phase separation of the solvent if it enters the two-phase region. After decompression, heat is applied to bring the solvent to a low density gas phase condition. Heat exchange and phase separation equipment that is used may resemble typical liquid phase evaporator or distillation columns, although the gas phase is still typically a dense gas phase with relatively low volumetric flow rates. The second method of depressurisation is to use multiple stages, shown in Figure 20.4, approaching isothermal decompression. The separation equipment used may share similarities in appearance with condensation/coalescence separators. Progressive depressurisation is of particular interest if solutes with different solubility and volatility are present, as fractionation of these different solutes can be achieved at each stage of decompression.

20.2.4.2.3 Ionic liquid Solvents. Ionic liquid extracts from a LLE process are generally non-volatile and high viscosity solutions, and solute recovery processes typically applied to solid ion-exchange adsorbents are often appropriate to use. If the solute is volatile then it can be removed by distillation or possibly gas stripping, but more generally the solute will be either non-volatile or strongly bound to the ionic liquid, and a secondary extraction step may be needed. Some ionic liquid solutions may have sufficient conductivity that electrodeposition of metals without need for re-extraction is possible.[88,89]

20.2.4.2.4 Heat Recovery. Solvent evaporation and distillation unit operations have a substantial heat load; however a large proportion of the heat flow can often be recovered so that the net input of energy to the process is less than the total heat of vaporisation. There are clear economic drivers that encourage use of heat recovery methods and heat recovery technology is very well developed in mature processing operations such as distillation. Economic drivers alone will encourage energy recovery to the point where incremental energy cost savings are matched by the incremental added cost of energy recovery, rather than to the absolute minimum energy consumption point where further heat recovery methods would consume as much energy as they recover. Green chemistry design considerations may push the optimal point of recovery further than when only direct energy costs are considered, although very large volume and highly efficient operations may already be close to the absolute energy optimum.

The net energy requirement for distillation or evaporation depends on a number of factors. In general the energy losses from the process will be in proportion to the total heat load, but other factors are also significant. These include:

- *Operating temperatures.* For a given solvent system, process temperatures substantially above room temperature result in higher rates of energy loss to the environment. Operating temperatures substantially below room temperature also result in increased energy loss to compensate for sensible heart absorbed from the environment. Temperatures further away from ambient are also generally more costly to provide heating and cooling services to. Moderate temperature heating sources, for example, can be produced with simple heated water loops, whereas higher temperatures may use steam boilers and pressurised heat exchange equipment, and lower temperatures would require a cooling tower or refrigeration system.
- *Residence time.* Long residence time operations and/or high reflux rates or number of stages result in larger plant surface area and increased rates of heat loss to the environment.
- *Solvent properties.* Solvents with high thermal conductivity and physical properties such as viscosity or compressibility that give lower heat transfer resistance or higher temperature gradients allow easier and more efficient heat recovery.

Evaporation equipment has many different layouts to cater for different fluid properties. Falling film evaporators give a good balance between low residence time and good heat transfer surface, and are suitable for many low viscosity systems and systems that do not scale or form a solid phase. Other flow arrangements suit more challenging fluid systems.

There are two main approaches to energy recovery in evaporators: (1) compression of the vapour phase exiting the evaporator so that it condenses at a higher temperature and can be used as the heat source for the evaporator; and (2) progressive reduction in operating pressure through multiple evaporation stages so that vapour from one stage provides heat for subsequent stages. Hybrid approaches also exist. Using heat recovery the net energy load can potentially be reduced to less than 10% of the total heat used for vaporisation. Distillation often makes use of vapour compression technology, or use of a secondary heat recovery loop, to recover heat from the condenser for use in the re-boiler.

The ability to recover heat, and to do this cost effectively, is affected by the choice of solvent system used in the LLE stage. Compression of solvents with a low vapour density is much less efficient than higher density vapour, and evaporation that requires low temperatures, for example for product stability, is generally less efficient. In some cases, for example, evaporation of fruit juice,[48] a secondary vapour phase can be used as an intermediary heat transfer fluid to improve energy efficiency. Common refrigerants such as Freon® or ammonia are used for this purpose as they have efficient compression cycles at temperatures close to ambient and the additional heat transfer losses due to use of a secondary fluid are more than offset through improved compression efficiency.

In the case of supercritical fluid solvent systems, for example CO_2 or propane systems, the solvent itself is often under ideal vapour recompression conditions where a large temperature rise is achieved with very small compression ratios. This can be seen in Figure 20.4 for re-condensing the supercritical fluid solvent back to extraction conditions. Condensation is taking place at a higher temperature than the vaporisation step(s) and the heat of condensation is easily and efficiently recovered by direct heat exchange between the two phases.

Further heat recovery can generally be applied by using hot outlet streams to provide sensible heat to incoming fluid streams. Integrated or multiple operation processing plants also have other options for optimising heat recovery across the entire plant.

20.3 RECOVERY OF ORGANICS FROM AQUEOUS SYSTEMS

Recovery of VOCs from aqueous systems is important to a number of large volume applications, including the recovery of products such as acetic acid from wastewater, and the production of biofuels and biochemicals from fermentation broths. Preparation of food ingredients and natural cosmetic products from plant, marine or microbiological systems also involves separation from high water content systems either before or after drying.

These aqueous separations exemplify many of the challenges of, and opportunities for, employing a liquid–liquid extraction approach to separation. Water has a large heat of vaporisation, leading to correspondingly high energy costs for processing by evaporative or distillation processes even when energy recovery systems are employed. Wastewater treatment and biofuel industries are ones that are already sensitive to environmental impact minimisation in addition to cost saving incentives, and biofuel production in particular needs to reduce the energy input per unit of fuel energy produced as much as possible.

Efficient recovery of organic compounds used in processing, or naturally present such as in wastewaters, is important where the ultimate fate of the processed water is to return it back to the environment, and it is important to understand toxicology and biodegradability ramifications of processing choices. Food and consumer product manufacturing is particularly conservative in its accepted use of processing solvents, with only a limited selection of solvents being acceptable and an even smaller selection considered to be favourable. For premium markets such as top end natural cosmetics, solvents, if used at all, must be non-toxic as well as not producing chemical artefacts through reaction or undue processing requirements such as high temperature.

In some cases LLE is already well established for aqueous extractions, such as the extraction of acetic acid using ethers or ethyl acetate. In other cases LLE is the subject of current evaluation as an alternative process in emerging processing industries such as bioethanol and biobutanol fermentation. This section explores some current trends and factors influencing use of LLE in selected aqueous extractions.

20.3.1   Extraction of Biofuels from Fermentation Broths

Production of ethanol, and 1-butanol (butanol hereafter), by fermentation is increasingly being used as a source of liquid biofuels. This includes fermentation of sugarcane and grains[126–129] to produce ethanol, with interest also in starch,[130] and lignocellulosic feedstocks. Production of butanol is commonly carried out using the acetone–butanol–ethanol (ABE) process by *Clostridium* fermentation. Butanol and propanediols are also produced as biochemical building blocks in addition to biofuels.

Distillation is still the most common method for recovery of the alcohols, generally after filtration or centrifugation of solid biomass from the fermentation broth,[131] and followed by pressure swing adsorption for full dehydration. Distillation is a well-developed and highly optimised processing technology. It is possible to recover ethanol from sufficiently concentrated fermentation systems using less than a third of the energy equivalent of the ethanol being produced,[125] and this has become an accepted industry starting point for viable operation. Competing separation methods have potential to reduce energy costs of separation, and would be attractive even if only small percentage point improvements in energy efficiency were achieved. It is challenging, however, to fully evaluate new technology that has not had the benefit of established industrial evaluation and fine tuning against the performance of a mature industry operation.[125]

Ethanol and butanol have generally favourable distillation curves and a high relative volatility in the gas phase at low concentrations, even though butanol has a higher boiling point than water. Less conveniently ethanol forms an azeotrope that requires a separate dehydration step to produce a dehydrated fuel grade ethanol. Butanol also forms an azeotrope; however, this forms above the solubility limit of butanol in water of about 8%. Initial distillate can be easily decanted, producing a butanol-rich supernatant that can be separately distilled to dryness.

A common approach to evaluating separation processes for fuel production is to express the energy consumption as energy input per unit mass of fuel produced (*e.g.* MJ kg^{-1} ethanol). A further distinction allows for different efficiencies from different input energy sources, for example, electricity produced by fuel combustion compared with heat directly produced by fuel combustion, by expressing energy input in terms of megajoules of liquid fuel energy equivalent required. For distillation the energy requirement increases asymptotically as the concentration of alcohol in the feed stream reduces. This is challenging to the viability of butanol production where toxicity of butanol to the bacterium typically limits concentrations to <20 g L^{-1},[132] compared with levels of 80–100 g L^{-1} that can be produced in ethanol fermentation. Mariano and Filho[133] calculate that butanol concentrations exceeding 36 g L^{-1} would be needed before butanol for fuel production is viable in the absence of an improved separation method. The high energy cost of distilling low concentration solutions opens up the opportunity for new separation technology to replace it, or hybrid separation methods where a first separation stage produces a more concentrated solution from the feed stream, followed by distillation of the concentrate.

Vane[125] evaluated a range of separation technologies for ethanol and butanol separation, including gas stripping, LLE and adsorption methods. As a first approximation, LLE is predicted to be favourable to distillation when using extraction solvents with a separation factor for ethanol exceeding approximately 10, although allowing for other process losses, a separation factor closer to 20 is estimated to be required. Solvents with separation factors in this range are available. Offeman and colleagues[49,134] reviewed and carried out measurements of separation factors for straight and branched chain alcohols and some esters all with separation factors in the range 10–20 depending on operating conditions. Boudreau and Hill[135] estimate the fatty acid nonanoic acid would have a separation factor of 21, enabling ethanol recovery from a 10 mass% solution using 2.8 MJ kg^{-1} ethanol. Feldman[136,137] proposes use of tetrahydroquinaline (THQ) with a calculated separation factor of only 5.3 and projected energy cost of (5–6) MJ kg^{-1} ethanol starting with a 4.5 mass% ethanol feed. These values are comparable with, or favourable to, the best distillation scenarios described by Vane,[125] and to distillation costs of (5–6) MJ kg^{-1} ethanol cited by Madson and Monceaux.[138]

LLE for the separation of butanol is expected to be reasonably favourable compared with the separation of ethanol. Due to the lower miscibility of butanol with water, separation factors into many organic solvents are typically several times higher for butanol than for ethanol, and separation factors of 30–50 are expected. Some solvent systems have been evaluated[78,133,139,140] and a patent for use of oleyl alcohol is held by Du Pont,[141] but few methods have been proved at pilot or commercial scale.

An attractive approach to butanol fermentation is to continuously harvest butanol from the fermentation system to keep accumulation of butanol in the system below levels toxic to the bacterium. This gives significant efficiencies in the fermentation process, but places challenges on the separation method and may add to the wastewater footprint if integrated water recovery is not applied. Distillation is not suited to this process due to high temperatures killing the bacteria and the challenges of processing slurries containing solid biomass, although periodic vacuum evaporation has been described by Qureshi *et al.*,[128] and Mariano *et al.*[142] as being able to recover butanol with energy consumption of 22 kJ kg^{-1} butanol. Gas stripping is effective[143,144] with reported values of selectivity between 5 and 10 using nitrogen.

LLE stripping is possible[145] if the solvent used is not toxic to the bacteria in the concentrations present in recycled water. Qureshi and Maddox[140] describe perstraction (using a membrane contactor) for continuous harvesting of biobutanol into oleyl alcohol producing a concentrate of 9.75g L^{-1}. Values of selectivity in excess of 250 have been demonstrated,[146] but may be subject to membrane stability and flux limitations.

The calculations of Vane[125] for the energy cost of LLE separation of alcohols are based on the assumption of a single separation factor value and that re-distillation of ethanol (and water) from the less volatile extraction solvent follows a similar energy demand to the distillation of ethanol from water. These assumptions mask some significant potential opportunities and it would be dangerous to use this basis alone as a screening approach for suitable technologies. As noted by Offeman *et al.*,[49] separation and loading factors are strong functions of solute concentration (for example, falling from 21 to 13 for 2-ethyl-1-hexanol, for ethanol concentrations in water increasing from dilute to 20 mass%), and also temperature and pressure conditions. Most significantly it is likely, for an optimised LLE process, that the regenerated extraction solvent will not be completely stripped from solute, particularly for azeotrope forming systems, and any calculations of extraction solvent loading and separating capacity should be based on this recycled composition and not those of pure solvents as is typically reported in published studies.

It is also misleading to assume typical distillation costs for extraction solvent regeneration. Distillation of alcohols from low volatility solvents such as longer chain alcohols, fatty acids, or even ionic liquids, will have significantly different energy demands and design considerations than for more volatile extraction solvents. Conversely, use of a very volatile extraction solvent such as a liquefied gas or compressed supercritical fluid requires different distillation considerations as described earlier in this chapter.

Compressed solvents for extraction of fermentations broths have been investigated, including use of CO_2,[57,58,61,62,147–150] propane,[147] ethane,[147,151] compressed nitrogen[151] and methyl ether.[61] Bothun *et al.*[147] and Moreno *et al.*[61] describe the use of hollow fibre membranes to enhance the separation. CO_2 dissolved in water produces acidic conditions toxic to many bacteria and is not ideally suited to continuous fermentation stripping where live bacteria are exposed to the solvent extraction conditions. Nitrogen and ethane have been shown to have no or limited detrimental effect on *Clostridium thermocellum* fermentation.[151] Propane[151] and other light organics may also be suitable, while maintaining values of selectivity for ethanol similar to that of CO_2 extraction. Gas solvents such as propane are able to be liquefied at much lower pressures than CO_2 offering further cost advantages.

Light hydrocarbons and ethers are very flammable gases, which raises process safety considerations, although these considerations are generally accommodated easily within standard industrial practice. In biofuel processing operations, the product itself is also by nature highly flammable. Extensive experience in the petrochemicals industry can be drawn on for safe handling of flammable materials. A demonstration plant for extraction of ethanol using propane based in Pennsylvania has been built by Thar, and promises to reduce energy cost to half of that required for distillation.[152]

Few authors give a convincing analysis of economic feasibility and energy consumption for compressed gas solvent extraction. Partitioning coefficients for ethanol between several compressed gases and water have been calculated by Bothun *et al.*[147] and others[153] and are reasonably low at around 0.05 for propane and ethane and rising to 0.25 for CO_2 at elevated pressure; however, separation factors are often high due to the low miscibility of these solvents with water. Longer chain alcohols (*e.g.* butanol) and acetone are more selectively extracted, and virtually complete extraction of butanol from simulated and actual ABE fermentation broths using CO_2 was demonstrated by Moreno *et al.*[61,154] Bothun *et al.*[147] demonstrated removal of >95% of acetone from 10 mass% solution using CO_2 with a solvent to feed molar ratio of 3. Assessment of the economics of extraction was carried out by McHugh and Krukonis[153] and was described as marginal by their calculations. The use of a co-solvent such as a triglyceride, fatty acid methyl ester[155] or 2-ethylhexanol[156] has been proposed to improve the energy efficiency and overall process economics. Extraction of propane 1,3-diol using CO_2 with addition of a surfactant modifier to enhance separation efficiency is described by Zhou *et al.*[157] and Yu *et al.*[158]

20.3.2 Extraction of Biochemicals from Fermentation Broths

Microbiological fermentation is a versatile method for production of a wide range of organic materials in addition to alcohols. Because they are generally cultivated systems, there is potential for the production to be sustainable and there is opportunity to target the process selectively towards individual compounds through selection or modification of the organism. Microbiological systems including bacteria, yeasts and fungi have been used to produce a range of products from pharmaceuticals, dietary and skin health ingredients, through to biochemical products and alternatives to animal sources of protein.

Recovery of products from fermentation broths is usually costly and energy intensive, and may form the larger part of the overall cost of manufacture. There are a number of steps that may be involved in product recovery including breakup of cell structures to release target compounds from the cell matrix, primary removal of solid biomass, extraction by solvent or solid phase adsorption, and final purification often by chromatography or crystallisation. Each of these steps can influence the effectiveness of following steps, so for design of a solvent extraction step that involves LLE it is useful to optimise up-stream processing steps as well as consider the effects of LLE design on any further downstream processing. These considerations may be as simple as selection or minimisation of antifoaming aids so as to prevent emulsification or reduction in selectivity of the solvent ex-traction step. It may also be possible to integrate a number of steps, for example, use of supercritical CO_2 as both a cell rupturing agent and an extraction solvent.

For high value biochemicals such as antibiotics, it is economically viable to manufacture these using energy and solvent intensive ex-traction, drying, evaporation and chromatography steps. There are, however, trends towards the use of solvents that are less toxic and more environmentally benign, towards improved energy efficiency, and towards recovery of more than one product from the fermen-tation, or at least effective recovery of as much water and nutrients as possible back to the process. LLE examples given below highlight these trends.

20.3.2.1 Penicillin.

Manufacture of penicillin has traditionally in-volved extraction of fermentation broth using *N*-butyl acetate, fol-lowed by re-extraction into a phosphate buffer. Use of *N*-butyl acetate is undesirable based on environmental considerations, as well as having non-optimal solvent properties for recovery and reuse.[159,160] Many solvent alternatives have been studied,[161,162] in-cluding aliphatic alcohols[163] as well as ketones, esters,[164] and fluorinated and sulfated hydrocarbons[165] targeting either a shift towards more benign solvents or improved efficiency.

Aqueous two-phase systems, including ionic liquids as the polymer forming phase, have also been studied.[159,166] Challenges still exist in achieving high extraction efficiency compared with conventional solvent extraction, maintaining product stability,[160] and in recovering penicillin from the ionic liquid.

Use of liquid-supported membranes has been considered for improving selectivity and aiding with solvent recycle.[167,168]

20.3.2.2 Polyhydroxyalkanoates. Polyhydroxyalkanoates (PHA) are biodegradable biopolymers, produced by bacteria for energy storage, with applications including packaging, medical devices and in chemical synthesis. Extraction methods typically involve extraction into acetone or a halogenated solvent, followed by anti-solvent precipitation using methanol or ethanol.[169] Environmental and solvent consumption concerns have led to alternatives being investigated, including use of 1,2-propylene carbonate as the primary extractant. Partitioning of PHA from lysed cells into an aqueous two-phase systems is described by Divyashree *et al.*[170] A selection of polymer–salt systems was evaluated with best results reported for PEG–phosphate phases with mass ratio partition coefficients up to 16 for pH of 8. They also describe use of a protease to digest the cells before extraction to provide functional benefits in extraction as well as replace chemical surfactants. The protease was extracted into the polymer phase along with PHA without loss of enzymatic activity. Isolation and recovery of product and enzyme from the PEG phase was not described. Separation and reuse of aqueous two-phase systems remains a challenge to industrial implementation.[124]

Khosravi-Darani *et al.*[171] describe use of supercritical CO_2 with 1% toluene co-solvent for pressure cycle disruption of cells, allowing improved LLE extraction of PHA from the lysed cells without need for lyophilisation.

20.3.2.3 Aromatics. There is a growing market for natural fragrances for flavouring applications as well as for products based on the antimicrobial properties of some essential oils.[42,172] Concomitant with these markets for natural products is a desire to produce these with minimal degradation from their natural state, and to produce these using benign and minimal quantities of solvent and an efficient extraction process. Drying is one of the steps that can be associated with both energy cost and degradation as well as loss of volatile aromatics. Extraction of aroma compounds from vinegars and wines is one area where LLE has been explored, using liquid or supercritical CO_2 to replace use of organic solvent. Extraction of wines,[173,174] vinegars[175] and brandy[176] has been successfully carried out using CO_2. Both batch contacting and continuous countercurrent extraction has been described. Processing economics and energy consumption for industrial application are not generally calculated, and solvent to feed solution mass ratios which directly affect overall process economics vary widely from 2 : 1 to in excess of 30 : 1. Downstream separation involves vaporisation of CO_2, a relatively energy efficient process if large ratios of CO_2 to feed material are not used, followed by further dehydration of the aroma fraction if required to remove co-extracted water.

20.3.2.4 Other Compounds. Lipophilic products including fatty acids, fat soluble vitamins and pigments, and complex lipids are of interest for a range of pharmaceutical, dietary and cosmetic applications. Lipids are often contained in cell wall structures, so that cell lysis is required before extraction. Because these compounds are generally hydrophobic and bound to the cell matrix, it is often possible to centrifuge or filter out much of the aqueous phase containing non-utilised fermentation media and other water-soluble compounds. The residual solid biomass can then be dried and extracted with a lipophilic solvent.[177]

It is also possible to carry out liquid solvent extraction of lipids from an aqueous system without needing to fully remove water and this may confer some environmental, energy reduction, and product quality benefits in some circumstances. Catchpole *et al.*[42,52,178] and Tallon and Catchpole[56] describe the use of liquefied dimethylether in counter-current flow extraction for recovery of oils from a range of undried substrates including virtually complete extraction of carotenoids and omega 3 fatty acids from algae, and fatty acids from fungi. The extraction can be operated using a low solvent to aqueous feed ratio to produce a lipid containing extract phase and an aqueous raffinate phase. It can also be operated at a higher solvent to water mass ratio, approximately 10 : 1 to 20 : 1, where water is fully extracted along with lipids into the solvent phase leaving dry precipitated insoluble material including proteins and carbohydrates. Water needs to be separated from extracted lipids in both cases. Billakanti *et al.*[179] describe extraction of the carotenoid fucoxanthin from a naturally harvested algae using liquefied dimethylether following an enzyme processing step. Dimethylether is a non-toxic solvent, efficiently and cheaply manufactured from methanol, and can be compressed and vaporised energy efficiently under conditions suitable for many organic extractions.

Lipophilic compounds can also be isolated as part of a multistep extraction using, for example, ethanol or water–ethanol mixtures to extract a broad range of compounds followed by LLE fractionation.

Brito Mariano *et al.*[38] describe a process for selective partitioning of free fatty acids from triglycerides extracted from macaúba pulp by water–ethanol. A low polarity solvent can also be used to give selective partitioning of oils and other low polarity compounds from more polar compounds such as phenolics which are recovered in the raffinate. Catchpole *et al.*[52] describe use of CO_2 to extract oil and phenolic fractions from a selection of herbs and propolis. A similar approach could be applied to fermentation biomass.

Hydrophilic compounds can potentially be extracted from aqueous systems using low polarity solvent systems with use of appropriate modifiers. For example, Cooper *et al.*[180] describe use of fluorinated dendrimers or siloxane based surfactants with supercritical CO_2 for removal of water soluble pigments from an aqueous solution.

20.4 CONCLUSIONS

LLE is a flexible processing technology for industrial separations and there are many existing and emerging applications. Viewed through a green chemistry lens, LLE has the potential in some industrial operations to replace energy-intensive distillation steps with low environmental impact and low energy use separations, or give better overall product quality or recovery of the product. LLE is also enabling some new separations that cannot be achieved by simple distillation.

Within LLE processing there is continued development towards greener operation, including: use of novel solvent systems that are either substantially more volatile (such as compressed gases) or less volatile (such as ionic liquids) so that solvent recovery is more efficiently carried out; use of less toxic solvents or solvents with lower life cycle impact, such as aqueous polymer or salt systems; use of solvents designed and optimised for selectivity; and use of solid supports or membrane structures to aid mixing, separation and potentially selectivity.

The design and evaluation of the benefits of LLE for a given process operation is challenging. This is due in part to the almost unlimited range and combination of solvents and processing methods and a corresponding shortage of comprehensive engineering and thermodynamic data on all the solvent systems concerned. There is also a need to design new and improved contacting equipment for increasingly challenging high viscosity or highly emulsifying systems. These challenges are being addressed by researchers providing a wider range of thermodynamic data and improved predictive models for a range of multicomponent solvent systems, and improved design of contacting equipment including membrane material compatibility. Most importantly, researchers are increasingly beginning to address full process cycle optimisation and evaluation of energy and solvent costs, and new LLE applications are likely to emerge in fields ranging from large-scale energy sensitive processes to niche high value compounds when these evaluations prove favourable to alternative methods of manufacture.

REFERENCES

1. P. T. Anastas and J. C. Warner, *Green Chemistry; Theory and Practice*, Oxford University Press, New York, 1998.
2. J. D. Thornton, in *Thermopedia*™, 2011, DOI: 10.1615/AtoZ.e.extraction_liquid-liquid.
3. V. Alessi, R. Penzo, M. J. Slater and R. Tessari, *Chem. Eng. Technol.*, 1997, **20**, 445–454.
4. M. A. Akl, A. M. Youssef, M. M. Ali and M. I. Amin, *Egypt. J. Chem.*, 2009, **52**, 525–540.
5. M. Feki, M. Fourati, M. M. Chaabouni and H. F. Ayedi, *Can. J. Chem. Eng.*, 1994, **72**, 939–944.
6. A. Hannachi, D. Habaili, C. Chtara and A. Ratel, *Sep. Purif. Technol.*, 2007, **55**, 212–216.
7. C. Ye and J. Li, *J. Chem. Technol. Biotechnol.*, 2013, **88**, 1715–1720.
8. A. A. Magda, M. Y. A. Fatah, M. A. Mohsen and I. A. Mostafa, *Period. Polytechn. Chem. Eng.*, 2010, **54**, 57–62.
9. T. Dobre, T. Ofiteru and A. Sturzoiu, *UPB Sci. Bull. B: Chem. Mater. Sci.*, 2006, **68**, 39–54.
10. S. P. R. Katikaneni and M. Cheryan, *Ind. Eng. Chem. Res.*, 2002, **41**, 2745–2752.
11. T. Pilhofer, Acetic acid recovery by liquid-liquid extraction or azeotropic distillation?, presented at International Solvent Extraction Conference, Munich, Germany, 1986.
12. D. H. Shah and K. K. Tiwari, *Inst. Chem. Eng. Symp. Ser.*, 1978, **54**, 257–266.
13. E. K. Watson, W. A. Rickelton, A. J. Robertson and T. J. Brown, Recovery of acetic acid, phenol and ethanol by solvent extraction with a liquid phosphine oxide, presented at International Solvent Extraction Conference, Munich, Germany, 1986.
14. H. G. Schutze, W. A. Quebedeaux and H. L. Lochte, *Ind. Eng. Chem.*, 1938, **10**, 675–677.
15. I. Billard, A. Ouadi and C. Gaillard, *Anal. Bioanal. Chem.*, 2011, **400**, 1555–1566.

16. P. J. Bailes, C. Hanson and M. A. Hughes, *Chem. Eng.*, 1976, **83**, 86–94.

17. A. N. Anthemidis and K. I. G. Ioannou, *Talanta*, 2009, **80**, 413–421.

18. A. W. Ashbrook, *Miner. Sci. Eng.*, 1973, **5**, 169–180.

19. T. W. Chapman, R. Caban and M. E. Tunison, *AIChE Symp. Ser.*, 1975, **71**, 128–135.

20. L. R. de Lemos, I. J. B. Santos, G. D. Rodrigues, L. H. M. da Silva and M. C. H. da Silva, *J. Hazard. Mater.*, 2012, **237–238**, 209–214.

21. R. K. Jyothi, J. Y. Lee, J. S. Kim and J. S. Sohn, *Solvent Extr. Res. Dev. Jpn.*, 2009, **16**, 13–22.

22. T. Kakoi, T. Toh, F. Kubota, M. Goto, S. Shinkai and F. Nakashio, *Anal. Sci.*, 1998, **14**, 501–506.

23. Y. Katayama, K. Nita, M. Ueda, H. Nakamura, M. Takagi and K. Ueno, *Anal. Chim. Acta*, 1985, **173**, 193–209.

24. Y. Liu, J. Chen and D. Li, *Sep. Sci. Technol.*, 2012, **47**, 223–232.

25. S. Memon, D. M. Roundhill and M. Yilmaz, *Collect. Czech. Chem. Commun.*, 2004, **69**, 1231–1250.

26. G. D. Rodrigues, M. d. C. H. da Silva, L. H. M. da Silva, F. J. Paggioli, L. A. Minim and J. S. d. Reis Coimbra, *Sep. Purif. Technol.*, 2008, **62**, 687–693.

27. K. Saito, Y. Masuda and E. Sekido, *Anal. Chim. Acta*, 1983, **151**, 447–455.

28. G. W. Stevens, J. M. Perera and F. Grieser, *Rev. Chem. Eng.*, 2001, **17**, 87–110.

29. A. E. Visser, R. P. Swatloski, S. T. Griffin, D. H. Hartman and R. D. Rogers, *Sep. Sci. Technol.*, 2001, **36**, 785–804.

30. J. M. Wardell and C. Judson King, *J. Chem. Eng. Data*, 1978, **23**, 144–148.

31. P. M. Florido, I. M. G. Andrade, M. C. Capellini, F. H. Carvalho, K. K. Aracava, C. C. Koshima, C. E. C. Rodrigues and C. B. Gonçalves, *J. Chem. Thermodyn.*, 2013.

32. D. Gonçalves, C. C. Koshima, K. T. Nakamoto, T. K. Umeda, K. K. Aracava, C. B. Gonçalves and C. E. D. C. Rodrigues, *J. Chem. Thermodyn.*, 2014, **69**, 66–72.

33. H. Guo, Z. Zhang, J. Qian and Y. Liu, *Ind. Crops Prod.*, 2013, **42**, 500–506.

34. Y. R. Lee, M. Tian, Y. Jin and K. H. Row, *Asian J. Chem.*, 2013, **25**, 7722–7726.

35. C. M. Oliveira, B. R. Garavazo and C. E. C. Rodrigues, *J. Food Eng.*, 2012, **110**, 418–427.

36. Y. Wei and H. Lei, Advanced upgrading of pyrolysis oil via liquid-liquid extraction, presented American Society of Agricultural and Biological Engineers Annual International Meeting, Kansas City, USA, 2013.

37. A. Arce, A. Marchiaro, J. M. Martínez-Ageitos and A. Soto, *Can. J. Chem. Eng.*, 2005, **83**, 366–370.

38. R. G. Brito Mariano, C. M. Da Silva, S. Couri, R. I. Nogueira and S. P. Freitas, *Defect Diffus. Forum*, 2010, **312–315**, 554–559.

39. C. Chiyoda, M. C. Capellini, I. M. Geremias, F. H. Carvalho, K. K. Aracava, R. S. Bueno, C. B. Gonçalves and C. E. C. Rodrigues, *J. Chem. Eng. Data*, 2011, **56**, 2362–2370.

40. C. M. Lakshmanan and G. S. Laddha, *J. Am. Oil Chem. Soc.*, 1960, **37**, 466–468.

41. R. H. McCormack and D. S. Bolley, *J. Am. Oil Chem. Soc.*, 1954, **31**, 408–412.

42. O. J. Catchpole, S. J. Tallon, W. E. Eltringham, J. B. Grey, K. A. Fenton, E. M. Vagi, M. V. Vyssotski, A. N. MacKenzie, J. Ryan and Y. Zhu, *J. Supercrit. Fluids*, 2009, **47**, 591–597.

43. J. Heidlas, *Agro Food Ind. Hi-Tech*, 1997, **8**, 9–11.

44. L. Teberikler, S. Koseoglu and A. Akgerman, *J. Food Sci.*, 2001, **66**, 850–853.

45. M. Dekker, K. Van't Triet, S. R. Weijers, J. W. A. Baltussen, C. Laane and B. H. Bijsterbosch, *Chem. Eng. J.*, 1986, **33**, B27–B33.

46. K. van't Riet and M. Dekker, Preliminary investigations on an enzyme liquid-liquid extraction process, presented at Third European Congress on Biotechnology, Munich, Germany, 1984.

47. L. A. Robbins, *Chem. Eng. Prog.*, 1980, **76**, 58–61.

48. R. H. Perry and D. Green, *Perry's Chemical Engineering Handbook*, McGraw Hill, London, 6th edn, 1984.

49. R. D. Offeman, S. K. Stephenson, G. H. Robertson and W. J. Orts, *Ind. Eng. Chem. Res.*, 2005, **44**, 6789–6796.

50. E. J. Beckman, *J. Supercrit. Fluids*, 2004, **28**, 121–191.

51. G. Brunner, *J. Supercrit. Fluids*, 2009, **47**, 373–381.

52. O. Catchpole, S. Tallon, P. Dyer, F. Montanes, T. Moreno, E. Vagi, W. Eltringham and J. Billakanti, *Am. J. Biochem. Biotechnol.*, 2012, **8**, 263–287.

53. O. J. Catchpole, N. E. Durling, J. B. Grey, W. Eltringham and S. J. Tallon, in *Current Trends of Supercritical Fluid Technology in Pharmaceutical, Nutraceutical and Food Processing Industries*, Bentham Science Publishers, Sharjah, UAE, 2009, pp. 71–79.

54. O. J. Catchpole, J. B. Grey, K. A. Mitchell and J. S. Lan, *J. Supercrit. Fluids*, 2004, **29**, 97–106.

55. O. J. Catchpole, S. J. Tallon, J. B. Grey, K. Fenton, K. Fletcher and A. J. Fletcher, *Chem. Eng. Technol.*, 2007, **30**, 501–510.
56. S. J. Tallon and O. Catchpole, Extraction of complex lipids from aqueous streams using near-critical dimethylether, presented at AIChE Annual Meeting, San Francisco, 2006.
57. M. Herrero, M. Castro-Puyana, J. A. Mendiola and E. Ibañez, *TrAC Trends Anal. Chem.*, 2013, **43**, 67–83.
58. M. Herrero, A. Cifuentes and E. Ibañez, *Food Chem.*, 2006, **98**, 136–148.
59. M. V. Palmer and S. S. T. Ting, *Food Chem.*, 1995, **52**, 345–352.
60. C. Pronyk and G. Mazza, *J. Food Eng.*, 2009, **95**, 215–226.
61. T. Moreno, S. J. Tallon and O. J. Catchpole, Supercritical CO_2 extraction of 1-butanol from aqueous solutions using a hollow fibre membrane contactor, presented at AIChE Annual Meeting, Pittsburgh, PA, 2012.
62. J. H. Clark, F. E. I. Deswarte and T. J. Farmer, *Biofuels Bioprod. Biorefin.*, 2009, **3**, 72–90.
63. M. Knez Hrnčič, M. Škerget and Ž. Knez, Sustainable processes using sub-and super critical fluids, presented at 20th International Congress of Chemical and Process Engineering, Prague, 2012.
64. M. D. Luque de Castro and M. M. Jiménez-Carmona, *TrAC Trends Anal. Chem.*, 2000, **19**, 223–228.
65. H. G. Joglekar, I. Rahman and B. D. Kulkarni, *Chem. Eng. Technol.*, 2007, **30**, 819–828.
66. V. Pino and A. M. Afonso, *Anal. Chim. Acta*, 2012, **714**, 20–37.
67. C. F. Poole and S. K. Poole, *J. Sep. Sci.*, 2011, **34**, 888–900.
68. Y. Wang, M. Tian, W. Bi and H. R. Kyung, *Int. J. Mol. Sci.*, 2009, **10**, 2591–2610.
69. R. D. Rogers and K. R. Seddon, *Science*, 2003, **302**, 792–793.
70. Suresh and J. S. Sandhu, *Green Chem. Lett. Rev.*, 2011, **4**, 289–310.
71. L. A. Aslanov and A. V. Anisimov, *Pet. Chem.*, 2004, **44**, 65–69.
72. J. L. Li and B. H. Chen, *Sep. Purif. Technol.*, 2005, **41**, 109–122.
73. S. Zhu, P. Yu, M. Lei, Y. Tong, Y. Lv, R. Zhang, J. Ji, Q. Chen and Y. Wu, *Energy Educ. Sci. Technol., Part A*, 2012, **30**, 95–106.
74. H. Tadesse and R. Luque, *Energy Environ. Sci.*, 2011, **4**, 3913–3929.
75. S. H. Ha and Y. M. Koo, *Korean J. Chem. Eng.*, 2011, **28**, 2095–2101.
76. M. D. Bermúdez, A. E. Jiménez, J. Sanes and F. J. Carrión, *Molecules*, 2009, **14**, 2888–2908.
77. X. Sun, H. Luo and S. Dai, *Chem. Rev.*, 2012, **112**, 2100–2128.

78. C. Z. Liu, F. Wang, A. R. Stiles and C. Guo, *Appl. Energy*, 2012, **92**, 406–414.
79. K. Tonova, *Sep. Purif. Technol.*, 2012, **89**, 57–65.
80. M. M. Jaworska, T. Kozlecki and A. Gorak, *J. Polym. Eng.*, 2012, **32**, 67–69.
81. A. B. Pereiro, J. M. M. Araújo, J. M. S. S. Esperança, I. M. Marrucho and L. P. N. Rebelo, *J. Chem. Thermodyn.*, 2012, **46**, 2–28.
82. A. Stojanovic and B. K. Keppler, *Sep. Sci. Technol.*, 2012, **47**, 189–203.
83. M. Cvjetko Bubalo, K. Radošević, I. Radojčić Redovniković, J. Halambek and V. Gaurina Srček, *Ecotoxicol. Environ. Saf.*, 2014, **99**, 1–12.
84. E. M. Siedlecka, M. Czerwicka, S. Stolte and P. Stepnowski, *Curr. Org. Chem.*, 2011, **15**, 1974–1991.
85. R. P. Swatloski, J. D. Holbrey and R. D. Rogers, *Green Chem.*, 2003, **5**, 361–363.
86. N. Meine, F. Benedito and R. Rinaldi, *Green Chem.*, 2010, **12**, 1711–1714.
87. D. Zhao, Y. Liao and Z. D. Zhang, *Clean -Soil, Air, Water*, 2007, **35**, 42–48.
88. A. P. Abbott, G. Frisch, J. Hartley and K. S. Ryder, *Green Chem.*, 2011, **13**, 471–481.
89. F. Endres, D. MacFarlane and A. Abbott, *Electrodeposition from Ionic Liquids*, Wiley-VCH, Weinheim, Germany, 2008.
90. M. Roth, *J. Chromatogr. A*, 2009, **1216**, 1861–1880.
91. S. Mekki, C. M. Wai, I. Billard, G. Moutiers, J. Burt, B. Yoon, J. S. Wang, C. Gaillard, A. Ouadi and P. Hesemann, *Chem. -Eur. J.*, 2006, **12**, 1760–1766.
92. W. Eltringham and O. Catchpole, Use of ionic liquids for extraction or fractionation of lipids, *New Zealand Pat.*, NZ2008/000183, 2008.
93. X. Sun, Y. Ji, J. Chen and J. Ma, *J. Rare Earths*, 2009, **27**, 932–936.
94. J. Persson, H. O. Johansson, I. Galaev, B. Mattiasson and F. Tjerneld, *Bioseparation*, 2000, **9**, 105–116.
95. I. Y. Galaev and B. Mattiason (ed.), *Smart Polymers for Bioseparation and Bioprocessing*, Yaylor & Francis, London, 2002.
96. P. A. Albertsson, *Partition of Cell Particles and Macromolecules*, Wiley, New York, 1986.
97. P. G. Mazzola, A. M. Lopes, F. A. Hasmann, A. F. Jozala, T. C. V. Penna, P. O. Magalhaes, C. O. Rangel-Yagui and A. Pessoa Jr, *J. Chem. Technol. Biotechnol.*, 2008, **83**, 143–157.

98. A. F. Jozala, A. M. Lopes, P. G. Mazzola, P. O. Magalhães, T. C. Vessoni Penna and A. Pessoa Jr, *Enzyme Microb. Technol.*, 2008, **42**, 107–112.

99. B. Bolognese, B. Nerli and G. Picó, *J Chromatogr. Part B*, 2005, **814**, 347–353.

100. X. Chen, G. Xu, X. Li, Z. Li and H. Ying, *Process Biochem.*, 2008, **43**, 765–768.

101. M. M. Bora, S. Borthakur, P. C. Rao and N. N. Dutta, *Sep. Purif. Technol.*, 2005, **45**, 153–156.

102. T. S. Porto, T. I. R. Monteiro, K. A. Moreira, J. L. Lima-Filho, M. P. C. Silva, A. L. F. Porto and M. G. Carneiro-Da-Cunha, *World J. Microbiol. Biotechnol.*, 2005, **21**, 655–659.

103. K. E. Nandini and N. K. Rastogi, *Food Bioprocess Technol.*, 2011, **4**, 295–303.

104. T. S. Porto, G. M. Medeiros e Silva, C. S. Porto, M. T. H. Cavalcanti, B. B. Neto, J. L. Lima-Filho, A. Converti, A. L. F. Porto and A. Pessoa Jr, *Chem. Eng. Process.*, 2008, **47**, 716–721.

105. P. M. Duque Jaramillo, H. A. Rocha Gomes, F. G. De Siqueira, M. Homem-De-Mello, E. X. F. Filho and P. O. Magalhães, *Sep. Purif. Technol.*, 2013, **120**, 452–457.

106. K. Selber, F. Tjerneld, A. Collén, T. Hyytiä, T. Nakari-Setälä, M. Bailey, R. Fagerström, J. Kan, J. Van Der Laan, M. Penttilä and M. R. Kula, *Process Biochem.*, 2004, **39**, 889–896.

107. L. R. Weatherly, *Science and Practice of Liquid–Liquid Extraction*, ed. J. D. Thornton, Clarendon Press, Oxford, 1992, ch. 5.

108. J. D. Thornton and H. R. C. Pratt, *Trans. IChemE*, 1953, **31**, 289.

109. H. R. C. Pratt and G. W. Stevens, *Science and Practice of Liquid–Liquid Extraction*, ed. J. D. Thornton, Clarendon Press, Oxford, 1992, ch. 8.

110. S. Mohanty, *Rev. Chem. Eng.*, 2000, **16**, 199–248.

111. L. E. Scriven and C. V. Sternling, *Nature*, 1960, **187**, 186–188.

112. J. Wang, C. Yang and Z. Mao, *Sci. China, Ser. B: Chem.*, 2008, **51**, 684–694.

113. H. Tabesh, G. Amoabediny, M. Madani, M. H. Gholami, A. Kashefi and K. Mottaghy, *J. Mazandaran Univ. Med. Sci.*, 2012, **22**, 111–125.

114. T. Y. Cath, S. Gormly, E. G. Beaudry, M. T. Flynn, V. D. Adams and A. E. Childress, *J. Membr. Sci.*, 2005, **257**, 85–98.

115. C. A. M. Afonso and J. G. Crespo, *Angew. Chem., Int. Ed.*, 2004, **43**, 5293–5295.

116. G. R. M. Breembroek, A. van Straalen, G. J. Witkamp and G. M. van Rosmalen, *J. Membr. Sci.*, 1998, **146**, 185–195.

117. W. Zhang, C. Cui, Z. Ren, Y. Dai and H. Meng, *Chem. Eng. J.*, 2010, **157**, 230–237.

118. G. Obuskovic, S. Majumdar and K. K. Sirkar, *J. Membr. Sci.*, 2003, **217**, 99–116.

119. P. Luis, T. Van Gerven and B. Van der Bruggen, *Prog. Energy Combust. Sci.*, 2012, **38**, 419–448.

120. A. K. Pabby and A. M. Sastre, *J. Membr. Sci.*, 2013, **430**, 263–303.

121. H. Jang, S. Kim, Y. Lee and K. H. Lee, *Appl. Chem. Eng.*, 2013, **24**, 456–470.

122. M. Sims, *Membr. Technol.*, 1998, **97**, 11–12.

123. N. Assmann, A. Ladosz and P. Rudolf von Rohr, *Chem. Eng. Technol.*, 2013, **36**, 921–936.

124. J. Benavides and M. Rito-Palomares, *J. Chem. Technol. Biotechnol.*, 2008, **83**, 133–142.

125. L. M. Vane, *Biofuels Bioprod. Biorefin.*, 2008, **2**, 553–588.

126. N. Qureshi and T. C. Ezeji, *Biofuels Bioprod. Biorefin.*, 2008, **2**, 319–330.

127. T. Ezeji, N. Qureshi and H. P. Blaschek, *Process Biochem.*, 2007, **42**, 34–39.

128. N. Qureshi, V. Singh, S. Liu, T. C. Ezeji, B. C. Saha and M. A. Cotta, *Bioresour. Technol.*, 2014, **154**, 222–228.

129. Y. Wang, X. Wang, T. Ezeji, H. Feng and H. Blaschek, Improvement of fermentation of dried distillers' grains and solubles (DDGS) hydrolysates to acetone butanol and ethanol (ABE) with hydrolysate-adapted *Clostridium beijerinckii* BA 101, presented at Bioenergy Engineering Conference, Bellevue, WA, 2009.

130. T. C. Ezeji, M. Groberg, N. Qureshi and H. P. Blaschek, *Appl. Biochem. Biotechnol. A*, 2003, **106**, 375–382.

131. A. A. Pinto, I. C. Bicalho, J. L. Mognon, C. R. Duarte and C. H. Ataíde, *Sep. Purif. Technol.*, 2013, **120**, 69–77.

132. E. M. Green, *Curr. Opin. Biotechnol.*, 2011, **22**, 337–343.

133. A. P. Mariano and R. M. Filho, *Bioenergy Res.*, 2012, **5**, 504–514.

134. R. D. Offeman, S. K. Stephenson, G. H. Robertson and W. J. Orts, *Ind. Eng. Chem. Res.*, 2005, **44**, 6797–6803.

135. T. M. Boudreau and G. A. Hill, *Process Biochem.*, 2006, **41**, 980–983.

136. J. Feldman, Process for recovering ethanol from dilute aqueous solutions employing liquid extraction media, *US Pat.*, US 4346241 A, 1982.

137. J. Feldman, Process for recovering oxygenated organic compounds from dilute aqueous solutions employing liquid extraction media, *US Pat.*, US 4447643 A, 1984.

138. P. W. Madson and D. A. Monceaux, in *The Alcohol Textbook*, ed. T. P. Lyons, D. R. Kelsall and J. E. Murtagh.Nottingham University Press, Nottingham, 3rd edn, 1999, ch. 17.

139. A. M. Dadgar and G. L. Foutch, *Biotechnol. Bioeng. Symp.*, 1985, **15**, 611–620.

140. N. Qureshi and I. S. Maddox, *Food Bioprod. Process.*, 2005, **83**, 43–52.

141. D. Pont, Improvements in or relating to the separation of alcohols, *GB Pat.*, GB437210A, 1935.

142. A. P. Mariano, R. M. Filho and T. C. Ezeji, *Renewable Energy*, 2012, **47**, 183–187.

143. T. C. Ezeji, P. M. Karcher, N. Qureshi and H. P. Blaschek, *Bioprocess Biosyst. Eng.*, 2005, **27**, 207–214.

144. T. C. Ezeji, N. Qureshi and H. P. Blaschek, *World J. Microbiol. Biotechnol.*, 2003, **19**, 595–603.

145. W. J. Groot, R. G. J. M. van der Lans and K. C. A. M. Luyben, *Process Biochem.*, 1992, **27**, 61–75.

146. A. Thongsukmak and K. K. Sirkar, *J. Membr. Sci.*, 2007, **302**, 45–58.

147. G. D. Bothun, B. L. Knutson, H. J. Strobel, S. E. Nokes, E. A. Brignole and S. Díaz, *J. Supercrit. Fluids*, 2003, **25**, 119–134.

148. A. Güvenç, Ü. Mehmetoglu and A. Çalimli, *J. Supercrit. Fluids*, 1998, **13**, 325–329.

149. A. Güvenç, Ü. Mehmetoğlu and A. Çalimli, *Turk. J. Chem.*, 1999, **23**, 285–291.

150. A. Güvenç, Ü. Mehmetoğlu and A. Çalimli, *Process Technol. Proc.*, 1996, **12**, 463–468.

151. B. L. Knutson, H. J. Strobel, S. E. Nokes, K. A. Dawson, J. A. Berberich and C. R. Jones, *J. Supercrit. Fluids*, 1999, **16**, 149–156.

152. R. Kotrba, in *Ethanol Producer Magazine*, 2009.

153. M. McHugh and V. Krukonis, *Supercritical Fluid Extraction*, Butterworth-Heinemann, Oxford, 1994.

154. T. Moreno, S. J. Tallon, J. Ryan and O. J. Catchpole, Extraction of 1-butanol from aqueous solutions using supercritical CO_2, presented at Chemeca Annual Conference, Wellington, New Zealand, 2012.

155. B. J. J. Waibel and V. J. Krukonis, Energy efficient separation of ethanol from aqueous solution, *US Pat.*, US8263814, 2009.

156. J. M. Moses, Liquid CO_2/cosolvent extraction, *US Pat.*, US4770780, 1984.

157. D. Zhou, W. Yu and J. Z. Yin, *J. Chem. Eng. Data*, 2012, **57**, 1787–1793.

158. W. Yu, D. Zhou, J. Z. Yin and J. J. Gao, *RSC Adv.*, 2013, **3**, 7585–7590.

159. M. Matsumoto, T. Ohtani and K. Kondo, *J. Membr. Sci.*, 2007, **289**, 92–96.
160. Z. Ren, W. Zhang, J. Li, S. Wang, J. Liu and Y. Lv, *J. Chem. Eng. Data*, 2010, **55**, 2687–2694.
161. S. C. Lee, *J. Ind. Eng. Chem.*, 2009, **15**, 403–409.
162. S. C. Lee, *Biotechnol. Prog.*, 2006, **22**, 731–736.
163. B. Wang, J. Chen, Q. F. Liu, Z. T. An and H. Z. Liu, *Guocheng Gongcheng Xuebao*, 2001, **1**, 157.
164. W. Y. Yang and I. M. Chu, *Biotechnol. Tech.*, 1990, **4**, 191–194.
165. W. Lei and Z. Li, *J. Chem. Technol. Biotechnol.*, 2004, **79**, 281–285.
166. Y. Jiang, H. Xia, J. Yu, C. Guo and H. Liu, *Chem. Eng. J.*, 2009, **147**, 22–26.
167. Z. Ren, W. Zhang, Y. Lv and J. Li, *Biotechnol. Prog.*, 2009, **25**, 468–475.
168. S. C. Lee, *J. Ind. Eng. Chem.*, 2008, **14**, 207–212.
169. B. Kunasundari and K. Sudesh, *Express Polymer Lett.*, 2011, **5**, 620–634.
170. M. S. Divyashree, T. R. Shamala and N. K. Rastogi, *Biotechnol. Bioprocess Eng.*, 2009, **14**, 482–489.
171. K. Khosravi-Darani, E. Vasheghani-Farahani, S. A. Shojaosadati and Y. Yamini, *Biotechnol. Prog.*, 2004, **20**, 1757–1765.
172. A. Martín, S. Varona, A. Navarrete and M. J. Cocero, *Open Chem. Eng. J.*, 2010, **4**, 31–41.
173. S. Macedo, S. Fernandes, J. A. Lopesio, H. de Sousa, P. J. Pereira, P. J. Carmelo, C. Menduiña, P. C. Simões and M. Nunes Da Ponte, *Food Bioprocess Technol.*, 2008, **1**, 74–81.
174. F. J. Señoŕns, A. Ruiz-Rodríguez, E. Ibañez, J. Tabera and G. Reg.lero, *J. Agric. Food Chem.*, 2001, **49**, 1895–1899.
175. Z. M. Lu, W. Xu, N. H. Yu, T. Zhou, G. Q. Li, J. S. Shi and Z. H. Xu, *Int. J. Food Sci. Technol.*, 2011, **46**, 1508–1514.
176. F. J. Señoráns, A. Ruiz-Rodríguez, E. Ibáñez, J. Tabera and G. Reg.lero, *J. Supercrit. Fluids*, 2003, **26**, 129–135.
177. R. M. Couto, P. C. Simões, A. Reis, T. L. Da Silva, V. H. Martins and Y. Sánchez-Vicente, *Eng. Life Sci.*, 2010, **10**, 158–164.
178. O. J. Catchpole, S. J. Tallon, J. B. Grey, K. Fletc.her and A. J. Fletc.her, *J. Supercrit. Fluids*, 2008, **45**, 314–321.
179. J. M. Billakanti, O. J. Catchpole, T. A. Fenton, K. A. Mitchell and A. D. Mackenzie, *Process Biochem.*, 2013, **48**, 1999–2008.
180. A. I. Cooper, J. D. Londono, G. Wignall, J. B. McClain, E. T. Samulski, J. S. Lin, A. Dobrynin, M. Rubinstein, A. L. C. Burke, J. M. J. Fréchet and J. M. DeSimone, *Nature*, 1997, **389**, 368–371.

CHAPTER 21

Ionic Liquids and their Application to a More Sustainable Chemistry

KATHARINA BICA

Institute of Applied Synthetic Chemistry, Vienna University of Technology, Getreidemarkt 9/163, A-1060, Vienna, Austria
Email: katharina.bica@tuwien.ac.at

21.1 INTRODUCTION

'Combined with green chemistry, a new paradigm in thinking about synthesis in general, ionic liquids provide an opportunity for science/engineering/business to work together from the beginning of the field's development.'[1]

This statement from the NATO Advanced Research Workshop in Crete 2000 perfectly summarizes hopes and expectations for ionic liquid applications towards a more sustainable chemistry. Compared with conventional organic solvents, the use of ionic liquids can provide a number of advantages determined by the unique set of properties inherent in liquid salts. As a result, research in this area has flourished since the mid-1990s. Close cooperation between academia and industry led to impressively diverse applications, resulting in several industrial processes, some in use and some soon to be implemented.[2] With this rapid development, it is nowadays almost impossible to review all ionic liquid technologies that hold the

Chemical Processes for a Sustainable Future
Edited by Trevor M. Letcher, Janet L. Scott and Darrell A. Patterson
© The Royal Society of Chemistry 2015
Published by the Royal Society of Chemistry, www.rsc.org

potential for a more sustainable chemistry; some of the most intriguing and exciting developments merging solvent, catalytic and separation properties of ionic liquids are highlighted in this chapter.

21.1.1   Ionic Liquids: Facts and Figures

As a very broad definition emerging from the 2000 NATO Advanced Research Workshop, ionic liquids are commonly accepted as being salts or mixtures of salts that have melting points below 100 °C.[3] The combination of large ions with high conformational flexibility results in small lattice enthalpies and large entropy changes that favour the liquid state and the typical depression of the melting points of these salts. Additionally, non-symmetrical cations (*e.g.* the typical 1-alkyl-3-methylimidazolium cations) can be used to avoid efficient packing in the solid state.[4-6] Interestingly the prolongation of the alkyl chain of 1-alkyl-3-methylimidazoliums first decreases and then increases the melting point of the corresponding ionic liquid. This indicates a rivalry between increasing entropic effects due to more possible configurations of the cations and an increase of van der Waals attraction between the cation with its surrounding cations and anions.[7] Apart from alkylimidazolium $[R_1R_2Im]^+$ derivatives, typical examples of ionic liquids include alkylpyridinium, $[RPy]^+$, tetra-alkylammonium $[NR_4]^+$ and tetraalkylphosphonium $[PR_4]^+$ cations with linear, branched or functionalized side chains as well as protic cations in combination with inorganic or organic anions (Figure 21.1). Novel ionic liquid cations and anions are constantly being developed and are reported in the literature.

Ionic liquids are characterized by the complex network and a charge-ordered structure of the composing ions resulting in some common characteristic features that differ from conventional solvents.[8] Due to their low volatility and low flammability as well as their high thermal stability and large electrochemical window combined with favourable solvating properties for a range of polar and non-polar compounds,[9] ionic liquids were thought to be ideal replacements for common organic solvents.[10,11] While many classical organic solvents will freeze or boil across such a large temperature range, ionic liquids maintain their volume and fluidity. Since it has been estimated that there may be up to 10^{18} ionic liquids or ionic liquid mixtures, it should be clearly stated that these characteristic properties may apply to many, but certainly not all, cation–anion combinations.[12] Still, the synthetic flexibility that is not available for

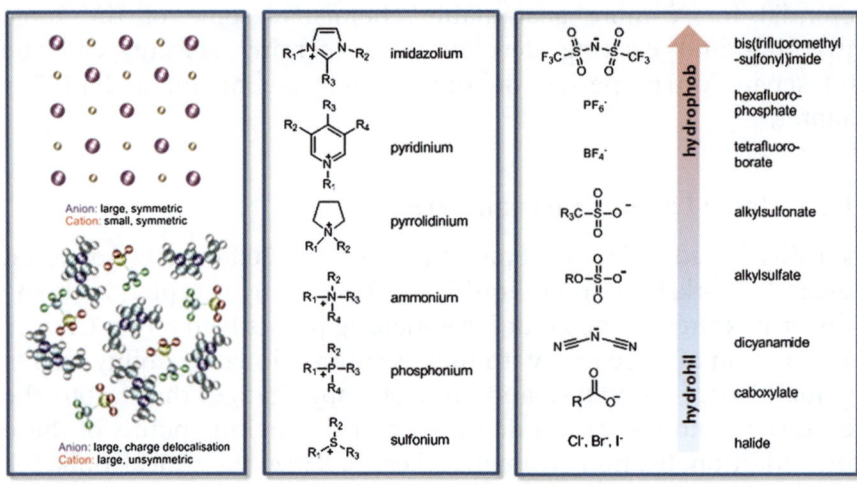

Figure 21.1 Left: schematic view of the structure of sodium chloride and 1-ethyl-3-methylimidazolium trifluoromethansulfonate rationalizing the decrease of melting points in ionic liquids from $T = 1074$ K to $T = 262$ K. Middle: typical ionic liquid cations. Right: typical ionic liquid anions and their effect on hydrophilicity/hydrophobicity of an ionic liquid.

conventional molecular solvents allows for simple modifications of ionic liquid characteristics by changing the structure of the component ions. Figure 21.2 lists typical properties and recent applications of ionic liquids.

The broad tuneability of ionic liquids permits the development of smart solvents, materials and electrolytes with tailored applications for a variety of applications. Excellent reviews have appeared covering the use of room temperature ionic liquids as reaction media for synthesis and catalysis emphasizing how ionic liquids can affect the reactivity of solute materials.[13,14] Solvent uses are not limited to synthesis and catalysis, but range from biomass dissolution,[15] protein solubilisation[16] and nanoparticles stabilization[17,18] to the large field of separations[19] where they find uses in solid–liquid, liquid–liquid and gas–liquid extraction processes and also as membrane materials in gas separations and as azeotrope breakers.[20–22]

Some of the most notable examples for the sustainable use of ionic liquids rely on their electrolyte properties. Ionic liquids and ionic liquid-derived polymeric materials[23] provide solutions for viable electrochemical challenges and applications where conventional media fails, as has been expertly reviewed by Armand *et al.*[24] Key properties of ionic liquids such as their conductivity and wide electrochemical window in combination with their negligible vapour

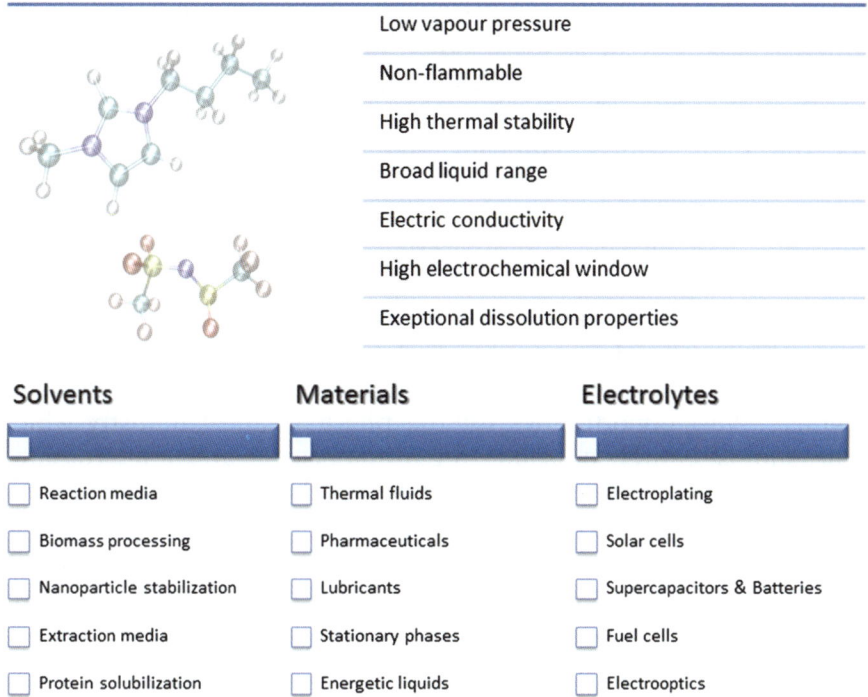

	Low vapour pressure
	Non-flammable
	High thermal stability
	Broad liquid range
	Electric conductivity
	High electrochemical window
	Exeptional dissolution properties

Solvents	Materials	Electrolytes
☐ Reaction media	☐ Thermal fluids	☐ Electroplating
☐ Biomass processing	☐ Pharmaceuticals	☐ Solar cells
☐ Nanoparticle stabilization	☐ Lubricants	☐ Supercapacitors & Batteries
☐ Extraction media	☐ Stationary phases	☐ Fuel cells
☐ Protein solubilization	☐ Energetic liquids	☐ Electrooptics

Figure 21.2   Typical properties and recent applications of ionic liquids.

pressure and their non-flammability make them ideal electrolytes; hence the flourishing research activity aimed at their use in electro-deposition,[25] supercapacitors,[26] lithium ion battery design[27,28] as well as in solar[29] and fuels cells.[30]

21.1.2   Ionic Liquids and Green Chemistry

While ionic liquids have been over enthusiastically classified as 'green solvents' in the early stages of ionic liquid research, merely based on their low vapour pressure, it soon became apparent that more caution rather than this *per se* definition would be appropriate.[31] After an unfortunate research period when far too many papers featuring ionic liquids were labelled 'green chemistry', ionic liquids and their applications have moved on towards a more sustainable future and are nowadays highly respected.[32]

Papers investigating (aqua-)toxicity and biodegradability appeared and clearly demonstrated that the biological compatibility of some ionic liquids may be an issue, which is as to be expected from such as large class of liquid materials.[33,34] In fact, ionic liquids cover the entire range from edible to extremely toxic compounds. While the antimicrobial activity of many ionic liquids has been the subject of heated discussions concerning their 'greenness', it has also opened novel opportunities for the development of new bioactive materials for the fight against clinically significant microbial pathogens, including methicillin-resistant *Staphylococcus aureus* (MRSA).[35] Other examples include ionic liquids as antiseptics and antifouling reagents, and also as combinatorial crop protecting agents with improved fungi toxicity against a range of potato tuber pathogens.

The potential of biological activity became apparent later on, when the groups of Rogers, MacFarlane and others deliberately designed ionic liquids composed of pharmaceutically active cations or anions.[36] Vast opportunities for the pharmaceutical industry which relies almost entirely on solid drug forms became apparent. To name but a few, pharmaceutically active ionic liquids do not only provide the possibility to circumvent polymorphism of a formerly solid drug, but can be used to modify solubility, lipophilicity or stability.[37–39] Novel delivery options arose from these modifications and improved membrane transport through lipophilic membranes has been reported by the proper design of ionic liquid forms of a pharmaceutically active compound.[40,41] Furthermore, novel dual formulations of two activities can be made by combining appropriate cation–anion pairs and synergistic, antagonistic or pharmacologically independent action may be observed.

21.2 IONIC LIQUIDS FOR BIOMASS PROCESSING

Biomass dissolution in molten salts dates back to 1934 when Graenacher partially dissolved cellulose using the molten salt *N*-ethylpyridinium chloride.[42] In the early 1980s, the dissolution of kerogen, a fossilized organic material present in sedimentary rocks which had been insoluble in all known solvents except for hydrofluoric acid with chloroaluminate ionic liquids, was reported.[43] Several years later, a ground-breaking paper on the dissolution of cellulose in imidazolium-based ionic liquids such as 1-butyl-3-methylimidazolium chloride ([C$_4$mim]Cl) was reported by Rogers and co-workers.[44] This platform strategy to dissolve and process cellulose for advanced new materials was awarded the 2005 Presidential Green Chemistry Challenge Award (Academic) and kick-started research in the area of biomass processing.[45] Figure 21.3 shows the extent of research interest in ionic liquids and biomass processing based on the results of a search on SciFinder™ in January 2014.

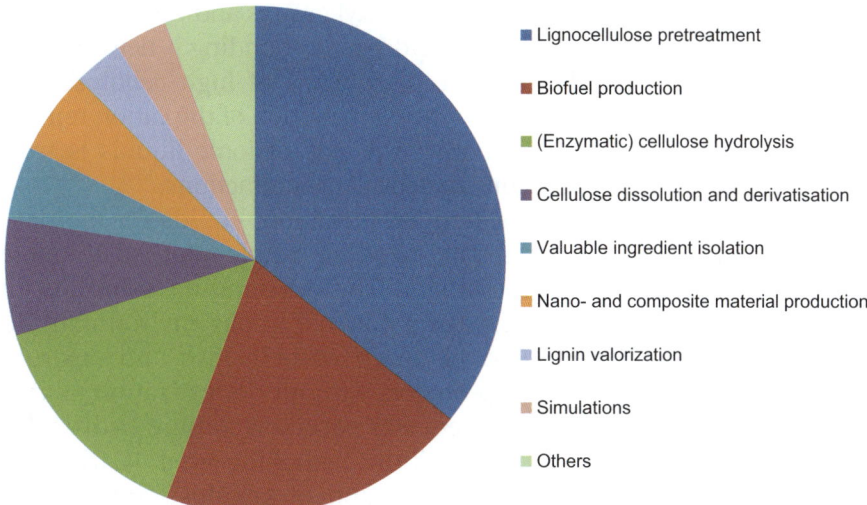

Figure 21.3   Research interest in ionic liquids and biomass, performed on Sci-Finder™, January 2014.

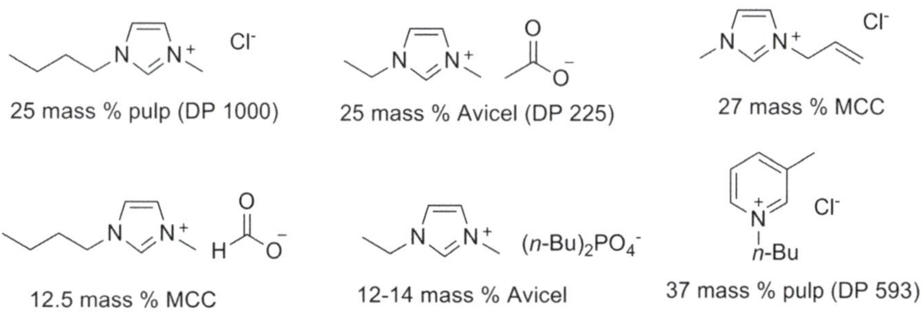

25 mass % pulp (DP 1000)   25 mass % Avicel (DP 225)   27 mass % MCC

12.5 mass % MCC   12-14 mass % Avicel   37 mass % pulp (DP 593)

Figure 21.4   Selected examples of the use of ionic liquids for cellulose dissolution. MCC, microcrystalline cellulose.
Adapted from ref. 63.

21.2.1   Cellulose Dissolution with Ionic Liquids

Prominent ionic liquids for cellulose dissolution are typically based on alkylmethylimidazolium or alkylpyridinium cations in combination with hydrogen-bond accepting anions (*e.g.* halides, carboxylates or dialkylphosphinates) and detailed reviews with elaborate list of cellulose solubility have been published elsewhere.[46–49] Figure 21.4 shows some selected examples of the use of ionic liquids for cellulose processing. As a tentative explanation and mechanism for the dissolution process, nuclear magnetic resonance (NMR) studies as well as molecular dynamics simulations showed that hydrogen bonding between the carbohydrate hydroxyl groups and the ionic liquid anion breaks intramolecular hydrogen bonding network of the biopolymer, thus allowing the dissolution of high cellulose.[50–52] However, some open questions on the contribution of the cation remain and doubt has been issued whether the role of ionic liquids is truly innocent and entirely based on non-covalent interactions.[53,54]

The initial interest in room temperature ionic liquids for biomass dissolutions originates from the need to implement sustainable processes for cellulose processing.[55] The outstanding solubility of cellulose, the most abundant natural polymer in our environment, in ionic liquids was expected to result in cost-effective and environmentally friendly dissolution protocols without derivatization as it is required in some commercial processes based on sodium hydroxide–carbon disulfide (Viscose) or sodium hydroxide–urea (CarbaCell) mixtures.[56] Additionally, after dissolution of cellulose and further recovery upon precipitation with water, the recovered cellulose had lost part of its crystallinity and thus was more accessible to depolymerisation under milder conditions, providing fermentable sugars for further use in (bio-)refineries.[57]

The processing of biomass is, however, not limited to the dissolution of cellulose and biopolymers as diverse as wood,[16] lignin, chitin,[17] silk fibroin,[18] wool keratin,[19] starch and zein protein,[20] cotton and bamboo,[21] chitosan[17] and cork biopolymers[22] have been reported to be at least partially soluble in various ionic liquids. Opportunities for biomass processing with ionic liquids are not limited to cellulose processing or biofuel production and regenerated cellulosic materials have been widely applied in materials chemistry. Biomass-derived materials including membranes nanoparticles, beads, films and highly porous materials have been designed based on cellulose or cellulose blends *via* ionic liquid technologies.[58,59] These materials have been applied in diverse areas and offer smart solutions for sensing technologies,[60] drug delivery[61] or (enzyme-)catalyst immobilization.[62]

21.2.2 Fractionation of Lignocellulosic Biomass

Lignocellulosic biomass in particular is receiving increasing attention as it is a widely abundant renewable feedstock and provides a great opportunity for valorisation. Recent reviews[63–65] have highlighted the potential of chemical transformation into biofuels, cheap fermentable sugars and platform chemicals with less significant by-product formation aided by ionic liquids, but also identified challenges in lignocellulose processing.

Differences in the chemical composition and structure, variable composition of the main components lignin, hemicellulose and cellulose and their different reactivity complicate the conversion towards target fuels and chemicals in high yield and quality. Fractionation of the three major components from lignocellulosic biomass represents a grand challenge in biorefinery, as it requires cleaving the primary bonds between the biopolymer components prior to their separation.[66] As for cellulose, a correlation between hydrogen bond basicity of the anion the dissolution power for wood chips has been observed, indicating that similar ionic liquids are also suitable for lignocellulose fractionation as well as for successive cellulose depolymerization to provide fermentable sugars of high purity and low cost.[67]

Several new trends have appeared adding to the overall energy and cost balance of lignocellulose processing with ionic liquids. Welton's group expanded the range of pure ionic liquids as solvents to include ionic liquid–water mixtures.[68] Although water typically reduces biomass solubility in ionic liquids, efficient lignin/cellulose fractionation was observed when significant quantities of water were present. A pretreatment step of ground lignocellulosic biomass from pine or willow with ionic liquid–water mixtures solubilised lignin and a significant portion of hemicellulose. A solid biomass fraction enriched in cellulose was obtained that was further subjected to enzymatic hydrolysis, while the ionic liquid layer contained most of the lignin and hemicellulose components. The enzymatic saccharification of Miscanthus pulp after pretreatment with mixtures of 1-butyl-3-methylimidazolium methyl sulfate and water resulted in high glucose yields, thus eliminating the need for anhydrous conditions. Figure 21.5 summarizes the approach to using an ionic liquid for lignocellulosic processing in a biorefinery.

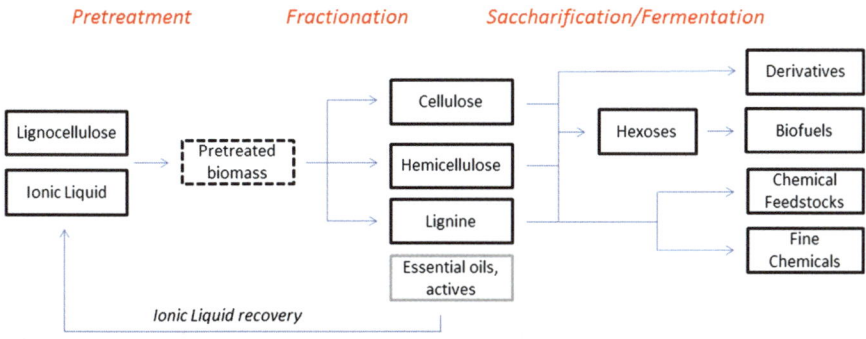

Figure 21.5 Ionic liquid strategies in a lignocellulose biorefinery.

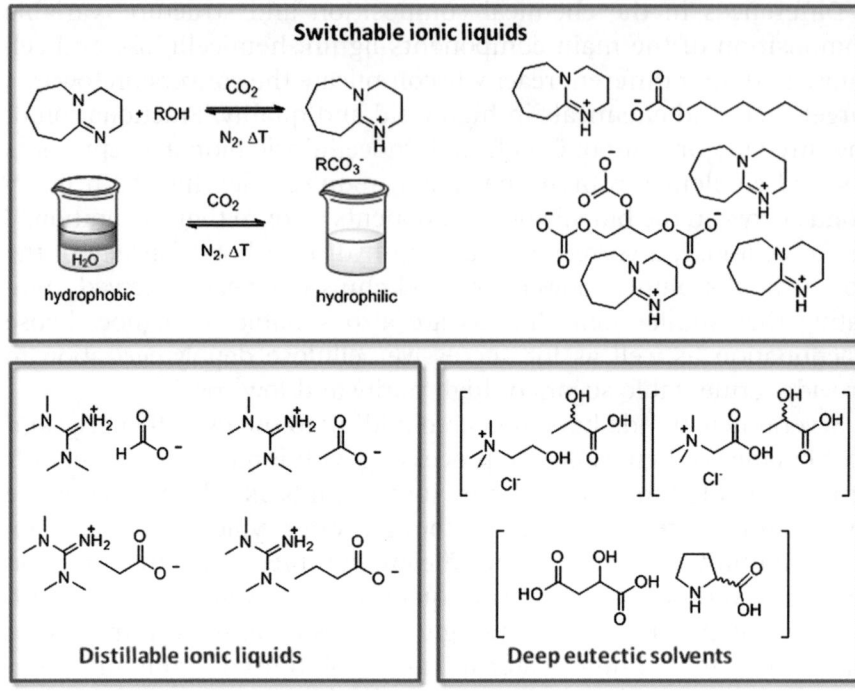

Figure 21.6   Emerging neoteric fluids for lignocellulose fractionation.

Likewise innovative and elegant trends in this area focus on the development of novel ionic media for (ligno-)cellulose processing other than conventional ionic liquids for future use in biorefineries and biomass valorisation (Figure 21.6).[69] Promising examples of distillable ionic liquids have been reported by King *et al.* that hold the potential to solve the problems associated with common imidazolium-based ionic liquids, particularly in the light of novel recycling strategies.[70] Additionally, switchable ionic liquids (Figure 21.6) are being developed based on alcohols, strong organic bases and CO_2.[71] A reversible switch between neutral and ionized form can be obtained by bubbling CO_2 or N_2, resulting in the transformation from low polarity to high polarity and thus producing a tuneable solubility for lignocellulosic biomass.[72]

Selective extraction of hemicelluloses from spruce and birch wood was obtained *via* pretreatment with undec-7-ene (DBU) glycerol carbonates, indicating that these switchable ionic liquids can be applied for the selective delignification and separation of hemicellulose but also to de-crystallize microcrystalline cellulose.[73] The concept of switchable ionic liquids could even be adopted for direct dissolution of microcrystalline cellulose in combination with CO_2 and dimethyl sulfoxide (DMSO) as a base *via* a reversible reaction with the –OH groups of cellulose.[74–76] In parallel, low-cost and bio-based deep eutectic solvents (DES) that are obtained *via* complexion of quaternary ammonium salts (*e.g.* choline chloride with hydrogen bond donors such as urea) offer novel opportunities that might overcome economic and environmental concerns that continue to exist with conventional solvents. Although the cellulose dissolution properties clearly range behind the values obtained for conventional imidazolium based ionic liquids, several deep eutectic solvents based on choline, betaine or amino acids could solubilize up to 15% lignin.[77,78]

21.2.3 Extraction of Valuable Ingredients from Biomass

While the growing interest in ionic liquids as solvents for biomass is mostly related to (ligno-)cellulose processing and to biofuel production, relatively little attention has been paid to the extraction of valuable ingredients from plant material (Figure 21.7). Yet, ionic liquids can be highly beneficial for this process. Apart from their extraordinary solvent properties compared with organic solvents, the ability of ionic liquids to swell or dissolve biomass can lead to a benign and higher yielding access to the valuable ingredient embedded in biopolymer matrices.[79]

A benchmark contribution in this emerging field was made by Lapkin *et al.* who compared emerging solvents including ionic liquids, supercritical carbon dioxide (scCO_2) and fluorinated solvents with conventional volatile organic solvents for the feasible extraction of important anti-malaria drug artemisinine from the herb *Artemisia annua.*[80] A careful evaluation of extraction efficiency, energy demand, costs and environmental factors revealed that ionic liquids are indeed very promising candidates for the extraction of pharmaceutically active small molecules from plant material. Over the past few years, several examples of active ingredient extraction with different ionic liquids have been reported, including the extraction of alkaloids, terpenes, flavones and flavonoids, essential oils and value-added substances of relevance for the pharmaceutical, fragrance or nutrition industry.

While the extraction of active ingredients with ionic liquids on an analytical scale is particularly well established and offers novel opportunities, *e.g.* for quality control, only a few papers deal with large-scale isolation of the valuable ingredients. Scale-up and isolation

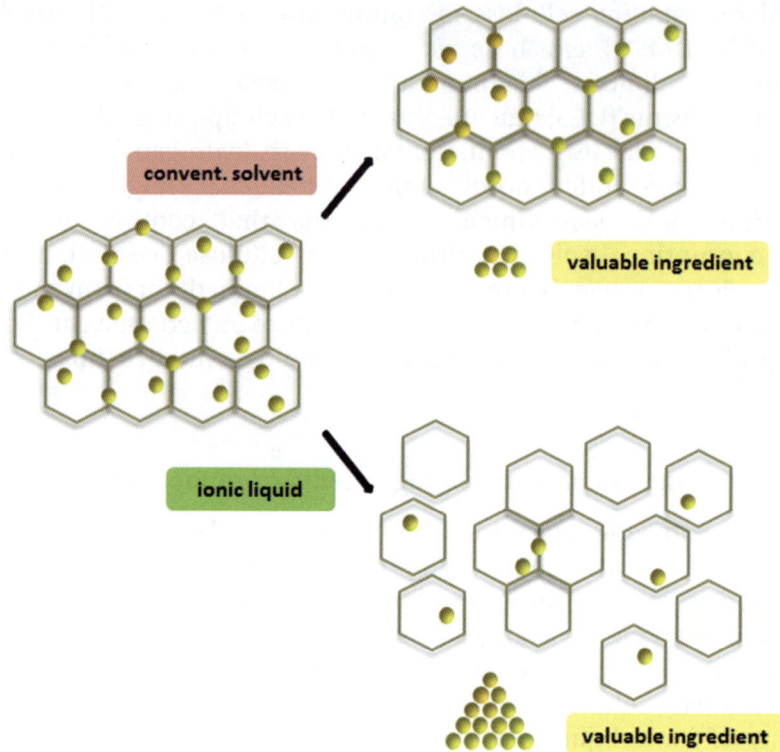

Figure 21.7   Ionic liquid benefits for the extraction of valuable ingredients from plant matter.
Reproduced with permission from Ref. 70.

faces the problem of separating the ionic liquid from the bioactive component, but also the challenge of recovery and recycling of the ionic liquid that might be mandatory for future applications on an industrial scale. Different strategies for the separation of the active ingredient and recovery of the ionic liquid have been developed that typically rely on extraction[81–85] or precipitation of the active ingredient with co-solvents.[86,87] Other technologies include ion exchange resins,[88] macroporous resins[89] or silica-confined ionic liquids for the adsorption and separation of the active ingredient.[90] Table 21.1 lists examples of the use of different isolation strategies.

In most cases, the ionic liquid is recovered as solution in water or in water–ethanol mixtures and has to be isolated *via* evaporation of the volatile co-solvents, a problematic process step that might have to be improved before further implementation on a large scale. A benchmark contribution to solve this problem was reported by MacFarlane and co-workers, who presented an elegant and efficient

Table 21.1 Isolation strategies for active ingredients after biomass dissolution in ionic liquids.

Active ingredient	Biomass	Isolation strategy	Ref.
Artemisinin	*Artemisia annua*	Precipitation	81
Betulin	Birch bark	Precipitation	82
Artemisinin	*Artemisia annua*	Extraction	83
Shikimic acid derivatives	*Illicium verum*	Extraction	84
Various lactones	*Ligusticum chuanxiong* Hort.	Extraction	85
Caffeine	Guaraná seeds	Extraction	86
Piperine	Black pepper	Extraction	87
Shikimic acid	*Ginkgo biloba* leaves	Anion exchange resin	88
Shikimic acid	*Illicium verum*	Anion exchange resin	69
Tannins	*Galla chinensis*	Sorption	89
Oxymatrine and matrine	*Sophora flavescens* Ait.	Sorption	90
Tannins	*Acacia catechu*, *Terminalia chebula*	Distillation of ionic medium	91

isolation strategy for tannins from the plant precursors, *Acacia catechu* and *Terminalia chebula*, using a distillable protic ionic liquid, *N*,*N*-dimethylammonium *N*,*N*-dimethylcarbamate (DIMCARB), as a novel extraction medium.[91]

Apart from active ingredients, plant matter may also contain a variety of fragrances and essential oils that are typically a complex mixture of individual fragrance components. The versatile features of ionic liquids do not only allow the dissolution of plant materials for efficient fragrance release, but enable the direct and mild distillation directly from the ionic liquids media without excessive steam production, solvent contamination or losses from water solubilisation.[92–94]

21.3   IONIC LIQUIDS AS INNOVATIVE FLUIDS IN CATALYSIS

21.3.1   Liquid Phase Processing

The original motive and most important driver for the interest in ionic liquids as solvents has been their intrinsic lack of vapour pressure and the opportunity to replace volatile organic solvents that are likely to become banned for industrial processes. Apart from their exceptional dissolution power, the tuneable properties of ionic liquids can be favourably exploited by rendering a homogeneous reaction mixture biphasic, thus combining the advantages of homogeneous and heterogeneous catalysis. Immiscible products can easily be 'decanted' enabling the recycling of a catalyst immobilized in the lower ionic liquid phase.[95]

However, it soon became apparent that advantages of ionic liquids are not limited to a mere replacement of volatile solvents, to the immobilization of catalysts in an ionic liquids layer or to facilitate the work-up of the product. In fact, some of the most notable aspects of sustainable ionic liquid applications in catalysis describe enhancements in reactivity or catalytic performance.[96] The critical aspects of ionic liquids can give rise to a distinct chemistry of their own compared with traditional solvents, and benefits extend beyond simple reductions in solvent losses. Nothing demonstrates this better than the famous BASIL™ (biphasic acid scavenging utilising ionic liquids) process launched by BASF that went on stream at the end of 2004 in Ludwigshafen, Germany.[97] Although not the first ionic liquid process run on an industrial scale,[2] the BASF's BASIL™ process had received the most publicity and was awarded several important prizes (*e.g.* the ECN Innovation Award in 2004), as it perfectly illustrates the benefits of replacing classical materials with ionic liquids in a formerly problematic process[98] (Figure 21.8)

BASF's BASIL™ process relies on the replacement of the previously used triethylamine with *N*-methylimidazole in the production of the photo-initiator intermediate, diethoxyphenylphosphine. This simple switch of the acid scavenger results in the formation of a liquid salt under the operating conditions, and as a consequence, two clear liquid layers are formed compared with the thick and non-stirrable slurry obtained with the common base triethylamine. Phase

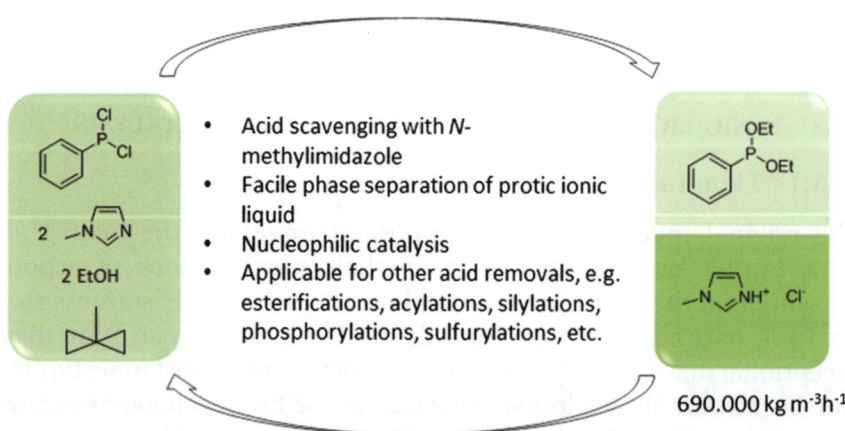

Figure 21.8 Acid scavenging with BASF's BASIL™ process.

separation results in the pure product in the upper layer without any need of additional solvents. Furthermore, despite the mere benefits of process improvement, *N*-methylimidazole functioned as nucleophilic catalyst. A dramatic increase in yield was observed allowing the use of a continuously operated thumb-sized jet reactor instead of the 20 m^3 batch vessels that had been used previously. Ultimately, the space-time yield was increased from 8 kg m^{-3} h^{-1} to 690 000 kg m^{-3} h^{-1}, while the 1-methylimidazol could easily be recycled *via* neutralization of the protic ionic liquid 1-H-3-methylimidazolium chloride with sodium hydroxide.

The concept of catalytically active ionic liquids can be further exploited by the introduction of catalytically active species such as Lewis acidic anions.[99] The pioneering work of Osteryoung, Wilkes and Hussey and co-workers on Lewis acidic ionic liquids was a major breakthrough for the development of catalytically active ionic liquids.[100,101] Although initially designed for electrochemical studies, Lewis acidic chloroaluminate ionic liquids can act as a combination of solvent and catalyst, and may provide an environmentally sustainable alternative for important acid-catalysed processes that are always associated with environmental problems caused by strong acids.[102] In general, Lewis acidic ionic liquids incorporating a metal in the anion can easily be prepared from imidazolium, ammonium or any other organic halide salt and the corresponding Lewis acid MX_n (eqn 21.1); examples including metal halide salts of Al(III), Fe(II), Fe(III), Cu(I), Cu(II), Ag(I), Au(III), Cd(I), Sn(II), Zn(II), Pd(II), Co(II), Ni (II) and Ti(IV) have been described.[103]

$$[\text{cation}]^+\text{X}^- + MX_n \rightarrow [\text{cation}][MX_{n+1}]^- \tag{21.1}$$

The actual anion form, and the acidity and physical properties in these types of ionic liquids, depend very much on the mole fraction of metal halide used. For chlorometallate ionic liquids, an equimolar amount of organic halide salt and $AlCl_3$ [corresponding to a mole fraction of $\chi(AlCl_3) = 0.5$] results in the formation of $AlCl_4^-$ as the sole anion, whereas multinuclear species such as $Al_2Cl_7^-$ and $Al_3Cl_{10}^-$ are formed when $\chi(AlCl_3)$ exceeds 0.5 (eqn 21.2, 21.3). Accordingly, these ionic liquids are typically classified as basic when $\chi(AlCl_3) < 0.5$, neutral at $\chi(AlCl_3) = 0.5$, or acidic when $\chi(AlCl_3) > 0.5$:[104]

$$2[AlCl_4]^- \rightarrow [Al_2Cl_7]^- + Cl^- \quad \text{if } \chi(AlCl_3) > 0.5 \tag{21.2}$$

and

$$2[Al_2Cl_7]^- \rightarrow [Al_3Cl_{10}]^- + [AlCl_4]^- \quad \text{if } \chi(AlCl_3) > 0.67 \tag{21.3}$$

In 2008, PetroChina announced a refinery process for high-quality gasoline production featuring ionic liquids deliberately designed to be catalytically active.[105,106] Acidic chloroaluminate(III) ionic liquids have been previously used as a homogeneous catalyst for isobutane alkylation in gasoline production, with the benefits of simple separation of gasoline from the ionic liquid layer and improvements in diffusion limitations that are present with solid-acid catalyst systems.[107–109] PetroChina's recent 'ionikylation process' provides an environmentally friendly and energy efficient isobutane alkylation process relying on a Lewis acidic ionic liquid composite catalyst with $AlCl_4CuCl^-$ as the presumably most active species. The conventional alkylation processes for the desirable trimethyl pentanes with high research octane number (RON < 100) was relying on either H_2SO_4 or anhydrous HF acid as catalysts, thereby raising serious environmental, health and safety concern. Careful design of a liquid composite catalyst featuring a chloroaluminate ionic liquid and CuCl, gave improved C_8 yield and trimethyl pentane selectivity with a relatively small percentage of undesirable side reactions such as isomerization and cracking, which might be explained by the high concentration of high concentration of $AlCl_4CuCl^-$ in the composite catalyst. While operating conditions for the alkylation unit were similar to those used with commercial H_2SO_4 and HF processes, the ionic liquid composite catalyst could run in a much simpler and cheaper design reactor and was found to be non-corrosive; carbon steel can be used for the hardware and reported corrosion rate was less than 0.001 mm a^{-1} where a refers to annum.

A successful demonstration in a continuous-flow pilot unit was carried out for 60 days in a retrofitted plant in an existing 65 000 t a^{-1} process (where t refers to metric tonne). This was part of PetroChina's recent ionikylation strategy and made the H_2SO_4 based alkylation unit the largest ionic liquid-based process running to date.[2]

21.3.2 Solid Supported Ionic Liquids: The SILP Concept

While ionic liquids can clearly simplify product separation and enable catalyst recovery, biphasic ionic liquid/organic liquid systems generally require a large amount of ionic liquid. Additionally, the high viscosity of some ionic liquids can induce mass transfer limitations, particularly if the chemical reaction is fast. These issues can be solved with the concept of solid-supported ionic liquid phases (SILP), a technology that allows the combining of the selectivity, specificity and synthetic availability offered by a homogeneous catalyst with the ease

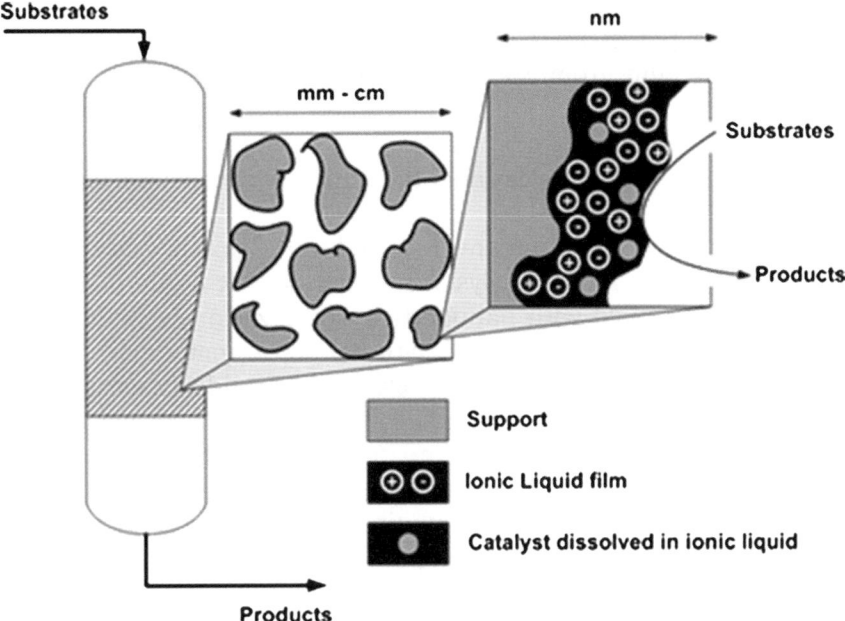

Figure 21.9 Schematic drawing of a supported ionic liquid-phase (SILP) catalyst. Reproduced with permission from ref. 110. Copyright 2005 Springer Verlag.

of processing and the robustness of heterogeneous catalysts[110–112] (Figure 21.9). A relatively small amount of ionic liquid is dispersed on a mesoporous support material with a high internal surface, creating a physisorbed thin layer thus allowing more efficient utilization of the ionic liquid. Silica-supported ionic liquid systems offer not only significantly simpler handling of ionic liquids compared with liquid-phase processing, but allow the immobilization and stabilization of transition metal catalysts in a thin film in the nanometre range.[113] The low volatility of ionic liquids in combination with the high solvation power for precious transition metal catalysts has led to extension of homogeneous catalysts being used for continuous-flow operated fixed bed processing in gas phase reactions, but also for slurry phase or supercritical CO_2 fixed bed applications.[114]

Important applications in the field of catalysis including gas and liquid phase hydroformylations in fixed bed reactors,[115–118] carbonylations of methanol and amines, olefin metathesis,[119] (asymmetric) hydrogenations,[120] hydroaminations,[121] and Pd-catalysed cross-coupling reactions[122] have been reported by the groups of Wasserscheid and Fehrmann as well as by others. While applications of SILP catalysis in gas phase processes is relatively straightforward, liquid phase reactions often suffer from catalyst or ionic liquid leaching. In this case, the ionic liquid can be covalently anchored to the (silica) support, so that the ionic liquid is rather considered as covalently anchored ligand, a concept that is also known under the term supported ionic liquid-like phases (SILLP).

However, recent developments are not limited to catalysis with well-defined transition metal catalysts. Supported ionic liquid phases can be further combined with nanoparticle stabilization in ionic liquids for the development of novel catalytically active materials.[123] Unlike conventional molten salt aggregates, ionic liquids possess pre-organised structures with an extended hydrogen bond network in the liquid state. This effect was found to be highly beneficial for the formation of small-scale metallic nanomaterials, as the presence of hydrophobic and hydrophilic domains allows for a controlled synthesis and a modification of nanoparticles.[124]

Likewise, it is also possible to immobilize biocatalysts *via* supported ionic liquid phases or *via* covalently supported ionic liquid-like phases to obtain efficient catalyst systems with good activity and robustness. A large number of enzymes and enzyme-catalysed reactions (*e.g.* esterification, kinetic resolution, reductions, oxidations) have been evaluated in various ionic liquid systems and most water-immiscible ionic liquids act as suitable reaction media for enzymatic catalysis with often improved activities and selectivities.[125] The coating or suspension of biocatalysts in an ionic liquid was found to be an efficient approach to protect enzymes against adverse effects of $scCO_2$. While the $scCO_2$ phase is highly soluble in ionic liquids, ionic liquids are hardly solubility in $scCO_2$, thus enabling the use of the supercritical fluid as the medium to deliver products for enzyme immobilised in the ionic liquid and to extract the obtained products from them.[126,127] Figure 21.10 shows the reactor design for a supported ionic liquid–$scCO_2$ biphasic system.

As a recent example, Lonzano *et al.* covalently anchored an imidazolium-based ionic liquid onto a polymer support and absorbed the enzyme *Candida antarctica* lipase B (CALB) in the ionic liquid phase. The obtained material served as a bioreactor for continuous transesterification of vinyl propionate and citronellol using $scCO_2$ as reaction medium.[128] The concept was further extended to include dynamic kinetic resolution: CALB immobilized on an acidic zeolite *via* an covalently bound hydrophilic ionic liquid allowed the resolving of *rac*-1-phenylethanol with excellent results in terms of yield (92%)

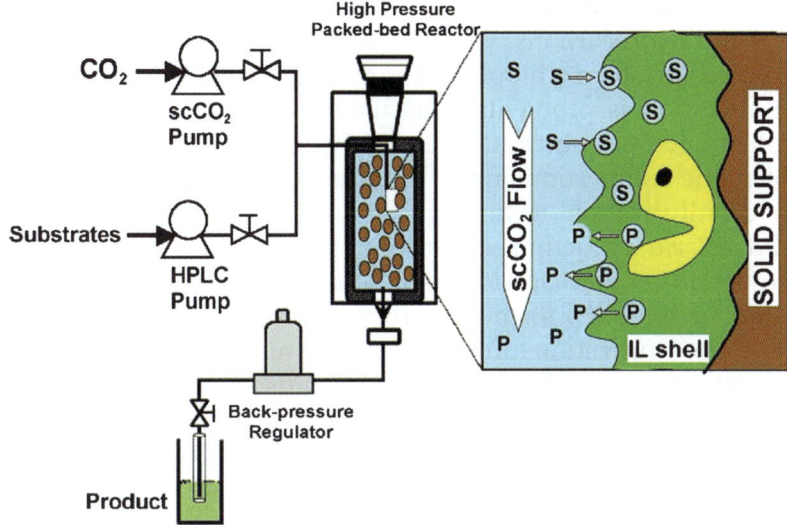

Figure 21.10   Reactor design for biocatalytic processes in supported ionic liquid-scCO₂ biphasic systems.
Reproduced from ref. 125.

and enantioselectivity (>99.9% enantiomeric excess) under flow conditions.[129]

Additionally, supported ionic liquid technologies have recently entered the world of pharmaceuticals, as biologically active compounds in ionic liquid form can immobilized onto mesoporous silica to form stable, easily handled solids with fast and complete release from the carrier material when placed into an aqueous environment.[130]

21.4   SEPARATION PROCESSES: FROM FLUE GAS CLEANING TO DE-MERCURIZATION

Some of the most exciting developments of ionic liquid applications deal with their use in separations, particularly when petrochemical processes are concerned. Supported ionic liquid phases offer opportunities in separation processes where conventional media fail (*e.g.* in gas purification) and enable new and economically viable processes to clean emission gasses. Since 2009, environmental regulations by the European Union require the removal of sulfur components that may generate sulfur oxide (SO_X), to a level below 10×10 parts per million (ppm).[131] Ultra-low sulfur fuels in oil refinery are typically prepared *via* catalytic hydro-desulfurization, where sulfur compounds such as

thiols, thioethers or disulfides are efficiently removed *via* conversion
to H_2S and hydrocarbons. When aromatic sulfur compounds such as
thiophene or benzothiophenes are concerned, catalytic hydro-
desulfurization is problematic as it requires harsh conditions and
high hydrogen pressures. Consequently, selective extraction of these
problematic sulfur compounds has been suggested as an alternative
deep desulfurization technology, and ionic liquids can be prime
candidates *via* extraction, adsorption, bioprocesses or oxidative desulfurization as they provide high solubility for sulfur components
but limited solubility for hydrocarbons.[132]

Deep desulfurization of oil refinery streams by extraction with
imidazolium based halogen-free ionic liquids showed that even sulfur
compounds such as dibenzothiophenes that are hard to remove by
hydrodesulfurization can be extracted with ionic liquids.[133] Oxidative
desulfurization involving chemical oxidation of divalent organic
sulfur compounds to the corresponding sulfones or sulfoxides can
further improve sulfur removal (Figure 21.11). The application of
Lewis acidic chloroferrate-based ionic liquids of the type [C$_n$mim]Cl/
FeCl$_3$ in combination with hydrogen peroxide result in a Fenton-like

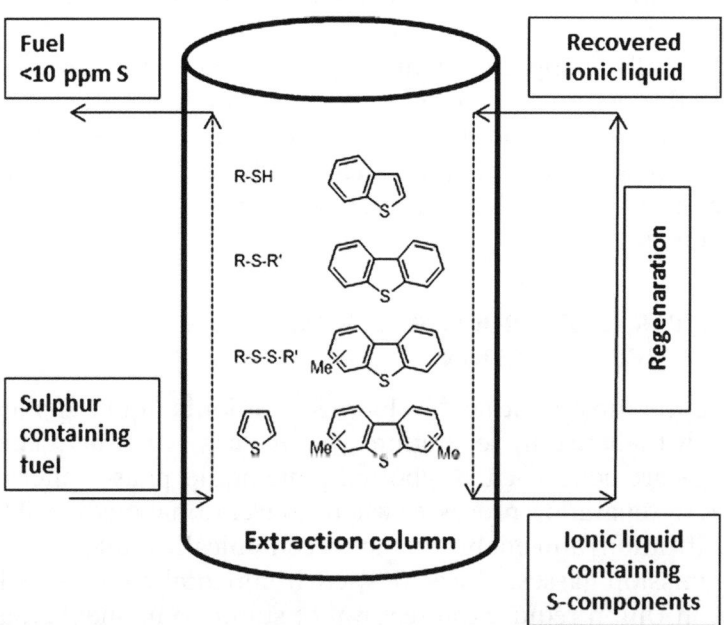

Figure 21.11   Concept of deep desulfurization of oil refinery streams by extraction
with ionic liquids.
Adapted from ref. 133.

oxidation process of sulfur components to sulfoxides, resulting in high desulfurization efficiency, highlighting again the multi-functional role of chlorometallate ionic liquids as catalyst, extractant and reaction media for combined catalysis and separation pro-cesses.[134] Similarly, a set of Keggin-type polyoxymetallate-based ionic liquids with phosphotungstate ($PW_{12}O_{40}^{3-}$), phosphomolybdate ($PMo_{12}O_{40}^{3-}$) or silicotungstate ($SiW_{12}O_{40}^{4-}$) has been successfully developed for this purpose, combining efficient catalytic properties with the extraction of sulfur components of gasoline.[135]

A comparative study on refinery desulfurization *via* liquid–liquid extraction and supported ionic liquid phase suspension processes has been published by Wasserscheid and co-workers, who found that chlorometallate ionic liquids of the type $[C_4mim]Cl/SnCl_2$ and $[C_4mim]Cl/SnCl_4$ exhibited good extraction performance for sulfur compounds.[136] Regeneration of the ionic liquid phase after each liquid–liquid extraction step was achieved either by distillation or re-extraction of the sulfur components. In comparison, supported ionic liquid phase (SILP) technologies exhibited a significantly higher desulfurization performance, reducing the sulfur content to less than 10 ppm in one stage, whereas multistage extractions were required in a liquid–liquid extraction process to lower the sulfur content to a similar level. Additionally, the supported ionic liquid materials were insusceptible to changes in viscosity; the higher viscosity of some chlorometallate ionic liquids affected the classical extraction experi-ments due to a slower mass transfer.

In a subsequent paper, the group expanded their work on con-tinuous gas cleaning processes and presented the desulfurization of a model gas stream in a fixed-bed sorption column, thus eliminating the problems associated with the ionic liquid under slurry phase conditions.[137] The selective removal of *n*-butyl mercaptan vapour from a hexane gas phase was investigated with 1-alkyl-3-methylimi-dazolium-based chlorometallate ionic liquid based on $SnCl_2$ and ZnCl with variable metal content loaded onto mesoporous alumina with a pore filling of 20%. A positive influence of longer-chain imidazolium cations, *i.e.* $[C_{12}mim]^+$, on the sulfur breakthrough times was ob-served, which recently was also reported for liquid–liquid extraction of sulfur components.[138] Best desulfurization performance and *n*-butyl mercapatan retention was obtained with the system $[C_{12}mim]Cl/SnCl_2$ with $\chi(SnCl_2) = 0.5$. Most importantly, the SILP material could be recovered *via* a simple change in temperature and pressure, and several sorption/desorption cycles could be run before the performance was compromised.

Emissions of acidic gases (*e.g.* NO_X, CO_X and SO_X) are a major concern in flue gases from fossil or biofuel based power plants or waste incineration units. Relatively energy intensive and resource demanding technologies *via* selective catalytic reduction (SCR) of NO_X with ammonia or urea, with gypsum formation after SO_2 wet scrubbing or organic amine-based absorbents in CO_2 scrubbers, are currently required to reduce the emission of air pollutants. Fehrmann's group at the Centre for Catalysis and Sustainable Chemistry at the Technical University Denmark investigated the selective catalytic NO_X removal utilizing ionic liquids as selective gas absorbers in a possible end-of-pipe installation.[139] While the mechanism of absorption varied, problematic flue gases like SO_2, NO, NO_2 and CO_2 could be reversibly and selectively absorbed in various ionic liquids, absorption/desorption processes could be tuned by temperature, gas pressure, carrier type and ionic liquid design. Remarkably, it was found that when the absorption of NO takes place in presence of atmospheric oxygen and small amounts of water, NO is not only absorbed in the ionic liquid but also converted very rapidly into nitric acid requiring no further oxidant than the oxygen present in the gas stream and no further addition of water.[140] This solid-supported ionic liquid strategy opens promising and attractive ways for NO scrubbing, as the captured NO could easily be released in a controlled manner as HNO_3 by increasing the temperature.

Recently, a fruitful cooperation between the Queen's University Ionic Liquid Laboratories Research Centre (QUILL) in Belfast and multinational oil and gas corporation PETRONAS resulted in one of the most exciting large-scale applications of ionic liquids that has been launched in record time. Varying levels of mercury present in hydrocarbon streams are a serious threat for gas processing plants and current commercially available mercury removal technologies suffer from robustness, capacity and effectiveness issues.[141] The design and careful development of a task-specific ionic liquid with an oxidizing moiety for mercury capture, in combination with immobilization strategies on a solid support, provided a novel mercury removal technology that could simultaneously oxidise elemental and organic mercury to mercury(II) species, extract inorganic mercury and capture all mercury species present in the gas. The solid-supported ionic liquids technology performed well under laboratory and actual plant conditions with a considerable improvement in performance compared with the commercial mercury removal techniques based on supported sulfides, halides or silver. Most importantly, the developed solid-supported ionic liquids can be retrofitted to existing fixed-bed mercury removal units without additional modifications to plant infrastructure. This novel solid-supported ionic liquid mercury removal unit triumphed at the 2013 global Institution of Chemical Engineers (IChemE) Awards, where QUILL, UK, in partnership with PETRONAS, Malaysia, received the Award for Outstanding Achievement in Chemical and Process Engineering, the Sustainable Technology Award as well as the Chemical Engineering Project of the Year Award.[142]

21.5 CONCLUSIONS

From tentative beginnings in the late 1990s, ionic liquid technologies have emerged into a significant field of research. More than 10 000 papers have appeared, as have numerous reviews and patents. Perspectives of ionic liquids as environmentally benign replacement for conventional organic solvent continue to exist, as the exposure risk and environmental pollution with ionic liquids is much lower than it is for a volatile solvent. More importantly, the unique properties of ionic liquids inherent in their structure will increasingly help to develop chemical processes that simply cannot be realized with molecular solvents and allow the development of feasible processes where conventional media fail.

Increasing attention will be spend on the design of inherently safer ionic liquids with low toxicology and improved biodegradation, but also on the development of cost-efficient ionic liquid-like materials. The boundaries carefully protecting the definition of ionic liquids are about to fall, thus enabling the development of novel neoteric solvents with improved properties, economics, availability and ecology that share many interesting properties with other conventional ionic liquids. Water and ionic liquids will overcome their long-standing hostility to combine the advantages of both solvents, thereby allowing novel approach to synthesis and catalysis in a manner that cannot be realized with a single solvent.[143,144] Temperature-sensitive lower-critical solution temperature (LCST) behaviour of ionic liquid-water mixtures as observed by, in particular by Fukumoto and Ohno, widens the potential of conventional ionic liquids.[145] Switchable ionic liquid–water functional fluids with reversible phase transitions *via* both temperature change and gas bubbling will be further investigated.

We expect that these recent development to provide novel aspects for multiple technological application and merge solvent, catalyst separation and electrolyte applications. In this sense, new ionic liquid technologies aimed at a more sustainable future, which improve energy efficiencies and reduce the emission of chemicals to the environment, can clearly be anticipated.

REFERENCES

1. R. D. Rogers, K. R. Seddon and S. Volkov (ed.), *Green Industrial Applications of Ionic Liquids*, NATO Science Series II: Mathematics, Physics and Chemistry, vol. 92, Kluwer, Dordrecht, 2002.
2. N. V. Plechkova and K. Seddon, *Chem. Soc. Rev.*, 2008, **37**, 123.
3. M. Freemantle, *An Introduction to Ionic Liquids*, RSC Publishing, Cambridge, UK, 2009.
4. I. Krossing, J. M. Slattery, C. Daguenet, P. J. Dyson, A. Oleinikova and H. Weingärtner, *J. Am. Chem. Soc.*, 2006, **128**, 13427.
5. H. Tokuda, K. Hayamizu, K. Ishii, M. Abu Bin Hasan Susan and M. Watanabe, *J. Phys. Chem. B*, 2005, **109**, 6103.
6. P. Wasserscheid, R. van Hal and A. Bösmann, *Green Chem.*, 2002, **4**, 400.
7. I. López-Martin, E. Burello, P. N. Davey, K. R. Seddon and G. Rothenberg, *Chem. Phys. Chem.*, 2008, **8**, 690.
8. M. G. Del Pópolo and G. A. Voth, *J. Phys. Chem. B*, 2004, **108**, 1744.
9. A. A. H. Pádua, M. F. Costa Gomes and J. N. A. Canongia Lopes, *Acc. Chem. Res.*, 2007, **40**, 1087.
10. K. R. Seddon, *J. Chem. Tech. Biotechnol.*, 1997, **68**, 351.
11. J. G. Huddleston, A. E. Visser, W. M. Reichert, H. D. Willauer, G. A. Broker and R. D. Rogers, *Green Chem.*, 2001, **3**, 156.
12. A. R. Katritzky, R. Jain, A. Lomaka, R. Petrukhin, M. Karelson, A. E. Visser and R. D. Rogers, *J. Chem. Inf. Comput. Sci,.*, 2002, **42**, 225.
13. J. P. Hallett and T. Welton, *Chem. Rev.*, 2011, **111**, 3508.
14. P. Wasserscheid and T. Welton (ed.), *Ionic Liquids in Synthesis*, Wiley-VCH, Weinheim, Germany, 2nd edn., 2007.
15. Y. Caoa, J. Wub, J. Zhang, H. Li, Y. Zhang and J. He, *Chem. Eng. J.*, 2009, **147**, 13.
16. K. Fujita, D. R. MᴀᴄFarlane and M. Forsyth, *Chem. Commun.*, 2005, 4804.
17. J. Dupont and J. D. Scholten, *Chem. Soc. Rev.*, 2010, **39**, 1780.
18. A. S. Pensado and A. A. H. Pádua, *Angew. Chem., Int. Ed.*, 2011, **50**, 8683.
19. G. W. Meindersma, A. B. De Haan, in *Ionic Liquids UnCOILed*, ed. N. V. Plechkova and K. R. Seddon, Wiley, Hoboken, NJ 2013, pp. 119–179.

20. M. F. Costa, P. Husson, in *Ionic Liquids: From Knowledge to Application*, ed. N. V. Plechkova, R. D. Rogers and K. R. Seddon, ACS Symposium Series, American Chemical Society, Washington DC, 2009, vol. 1030, pp. 223–237.

21. M. Malik, H. Ahmad, A. Mohd and F. Nabi, *Chem. Eng. J.*, 2011, **171**, 242.

22. A. B. Pereiro, J. M. M. Araujo, J. M. S. S. Esperanca, I. Marrucho and L. P. N. Rebelo, *J. Chem. Thermodyn.*, 2012, **46**, 2.

23. Y.-S. Ye, J. Rick and B.-J. Hwang, *J. Mater. Chem. A*, 2013, **1**, 2719.

24. M. Armand, F. Endres, D. R. MacFarlane, H. Ohno and B. Scrosati, *Nat. Mat.*, 2009, **8**, 621.

25. A. P. Abbott and K. J. McKenzie, *Phys. Chem. Chem. Phys.*, 2006, **8**, 4265.

26. A. Brandt, S. Pohlmann, A. Varzi, A. Balducci and S. Passerini, *MRS Bull.*, 2013, **38**, 554.

27. G. B. Appetecchi, M. Montanino, S. Passerini, in *Ionic Liquids: Science and Applications*, ed. A. E. Visser, N. J. Bridges and Robin D. Rogers, ACS Symposium Series, American Chemical Society, Washington DC, 2012, vol. 1117, pp. 67–128.

28. A. Lewandowski and A. Swiderska-Mocek, *J. Power Sources*, 2009, **194**, 601.

29. M. Gorlova and L. Kloo, *Dalton Trans.*, 2008, **20**, 2655.

30. H. Nakamotoa and M. Watanabe, *Chem. Commun.*, 2007, **24**, 2539.

31. T. Welton, *Green Chem.*, 2011, **13**, 225.

32. P. Wasserscheid and A. Stark, *Handbook of Green Chemistry – Green Solvents*, ed. P. T. Anastas, Wiley-VCH, Weinheim, Germany, 2013, vol. 6.

33. S. Stolte, M. Matzke, J. Arning, A. Boschen, W. R. Pitner, U. Welz-Biermann, B. Jastorff and J. Ranke, *Green Chem.*, 2007, **9**, 1170.

34. B. Jastorff, K. Moelter, P. Behrend, U. Bottin-Weber, J. Filser, A. Heimers, B. Ondruschka, J. Ranke, M. Schaefer, H. Schroeder, A. Stark, P. Stepnowski, F. Stock, R. Stoermann, S. Stolte, U. Welz-Biermann, S. Ziegert and J. Thoeming, *Green Chem.*, 2005, **7**, 362.

35. A. Busetti, D. E. Crawford, M. J. Earle, M. A. Gilea, B. F. Gilmore, S. P. Gorman, G. Laverty, A. F. Lowry, M. McLaughlin and K. R. Seddon, *Green Chem.*, 2010, **12**, 420.

36. J. Stoimenovski, D. R. MacFarlane, K. Bica and R. D. Rogers, *Pharm. Res.*, 2009, **27**, 521.

37. W. L. Hough, M. Smiglak, H. Rodríguez, R. P. Swatloski, S. K. Spear, D. T. Daly, J. Pernak, J. E. Grisel, R. D. Carliss,

M. D. Soutullo, J. H. Davis and R. D. Rogers, *New J. Chem.*, 2007, **31**, 1429.

38. K. Bica, C. Rijksen, M. Nieuwenhuyzen and R. D. Rogers, *Phys. Chem. Chem. Phys.*, 2010, **12**, 2011.

39. K. Bica and R. D. Rogers, *Chem. Commun.*, 2010, **46**, 1215.

40. J. Stoimenovski and D. R. MacFarlane, *Chem. Commun.*, 2011, **47**, 11429.

41. J. Shamshina, P. Barber and R. Rogers, *Expert Opin. Drug Discov.*, 2013, **10**, 1367.

42. C. Graenacher, Cellulose solution, *US Pat.*, 1943176, 1934.

43. R. C. Bugle, K. Wilson, G. Olsen, L. G. Wade Jr. and R. A. Osteryoung, *Nature*, 1978, **274**, 578.

44. R. P. Swatloski, S. K. Spear, J. D. Holbrey and R. D. Rogers, *J. Am. Chem. Soc.*, 2002, **124**, 4974.

45. S. K. Ritter, *Chem. Eng. News*, 2005, **83**, 40.

46. A. Pinkert, K. N. Marsh, S. Pang and M. P. Staiger, *Chem. Rev.*, 2009, **109**, 6712.

47. M. E. Zakrzewska, E. Bogel-Łukasik and R. Bogel-Łukasik, *Energy Fuels*, 2010, **24**, 737.

48. H. Wang, G. Gurau and R. D. Rogers, *Chem. Soc. Rev.*, 2012, **41**, 1519.

49. Y. Cao, J. Wu, J. Zhang, H. Li, Y. Zhang and J. He, *Chem. Eng. J.*, 2009, **147**, 13.

50. R. C. Remsing, R. P. Swatloski, R. D. Rogers and G. Moyna, *Chem. Commun.*, 2006, 1271.

51. T. G. A. Youngs, C. Hardacre and J. D. Holbrey, *J. Phys. Chem. B*, 2007, **111**, 13765.

52. J. Zhang, H. Zhang, J. Wu, J. Zhang, J. He and J. Xiang, *Phys. Chem. Chem. Phys.*, 2010, **12**, 1941.

53. T. Heinze, S. Dorn, M. Schoebitz, T. Liebert, S. Koehler and F. Meister, *Macromol. Symp.*, 2008, **262**, 8.

54. G. Ebner, S. Schiehser, A. Potthast and T. Rosenau, *Tetrahedron Lett.*, 2008, **49**, 7322.

55. H. Wang, G. Gurau and R. D. Rogers, *Chem. Soc. Rev.*, 2012, **41**, 1519.

56. T. F. Liebert, in *Cellulose Solvents: For Analysis, Shaping and Chemical Modification*, ed. T. F. Liebert, T. J. Heinze and K. J. Edgar, ACS Symposium Series, American Chemical Society, Washington, DC, 2010, vol. 1033, pp. 3–54.

57. A. P. Dadi, S. Varanasi and C. A. Schall, *Biotechnol. Bioeng.*, 2006, **95**, 904.

58. J. Han, C. Zhou, A. D. French, G. Han and Q. Wu, *Carbohyd. Polym.*, 2013, **94**, 773.

59. K. Prasad, S. Mine, Y. Kaneko and J. Kadokawa, *Polym. Bull.*, 2010, **64**, 341.
60. J. H. Poplin, R. P. Swatloski, J. D. Holbrey, S. K. Spear, A. Metlen, M. Grätzel, M. K. Nazeeruddin and R. D. Rogers, *Chem. Commun.*, 2007, 2025.
61. K. Prasad, Y. Kaneko and J. Kadokawa, *Macromol. Biosci.*, 2009, **9**, 376.
62. M. B. Turner, J. D. Holbrey, S. K. Spear and R. D. Rogers, *Biomacromolecules*, 2004, **5**, 1379.
63. N. Sun, H. Rodríguez, M. Rahman and R. D. Rogers, *Chem. Commun.*, 2011, **47**, 1405.
64. P. Mäki-Arvela, I. Anugwom, P. Virtanen, R. Sjöholm and J. P. Mikkola, *Ind. Crops Prod.*, 2010, **32**, 175.
65. A. Brandt, J. Gräsvik, J. P. Hallett and T. Welton, *Green Chem.*, 2013, **15**, 550.
66. A. M. Costa Lopes, K. G. João, A. R. C. Morais, E. Bogel-Łukasik and R. Bogel-Łukasik, *Sustainable Chem. Process.*, 2013, **1**, 3.
67. A. Brandt, J. P. Hallett, D. J. Leak, R. J. Murphy and T. Welton, *Green Chem.*, 2010, **12**, 672.
68. A. Brandt, M. J Ray, T. Q. To, D. J. Leak, R. J. Murphy and Tom Welton, *Green Chem.*, 2011, **13**, 2489.
69. P. Domínguez de María, *J. Chem. Technol. Biotechnol.*, 2014, **89**, 11.
70. A. W. T. King, J. Asikkala, I. Mutikainen, P. Järvi and I. Kilpeläinen, *Angew. Chem., Int. Ed.*, 2011, **50**, 6301.
71. P. G. Jessop, D. J. Heldebrant, X. Li, C. A. Eckert and C. L. Liotta, *Nature*, 2005, **436**, 1102.
72. P. Virtanen, I. Anugwom, J. P. Mikkola and P. Mäki-Arvela, Method for fractionating of lignocellulosic material and novel ionic liquids therefore, *World Pat.*, WO2012/059643, 2012.
73. I. Anugwom, P. Mäki-Arvela, P. Virtanen, S. Willför, R. Sjöholm and J. P. Mikkola, *Carbon Polym.*, 2012, **87**, 2005.
74. Q. Zhang, N. S. Oztekin, J. Barrault, K. De Oliveira Vigier and F. Jerome, *Chem. Sustainable Chem.*, 2013, **6**, 593.
75. I. Anugwom, P. Mäki-Arvela, P . Virtanen, S. Willför, P. Damlin, M. Hedenström and J. P. Mikkola, *Holzforschung*, 2012, **66**, 809.
76. Q. Zhang, N. S. Oztekin, J. Barrault, K. De Oliveira Vigier and F. Jerome, *Chem. Sustainable Chem.*, 2013, **6**, 593.
77. J. van Spronsen, G. J. Witkamp, F. Hollmann, Y. H. Choi and R. Verpoorte, Process for extracting materials from biological material, *World Pat.*, WO2011/155829, 2011.
78. M. Francisco, A. van den Bruinhorst and M. C. Kroon, *Green Chem.*, 2012, **14**, 2153.

79. R. Zirbs, K. Strassl, P. Gaertner, C. Schröder and K. Bica, *RSC Adv.*, 2013, **3**, 26010.
80. A. A. Lapkin, P. K. Plucinski and M. Cutler, *J. Nat. Prod.*, 2006, **69**, 1653.
81. A. K. Ressmann, P. Gaertner and K. Bica, *Green Chem.*, 2011, **13**, 1442.
82. B. Zhao, Y. Xia, J. Zeng, X. Wang, X. Yuan, Y. Wang, New method for extracting artemisinin from artemisia annua with ionic liquid as extraction solvent, *Chin. Pat.*, CN101597296A, 2009.
83. C. Yansheng, Z. Zhida, L. Changping, L. Qingshan, Y. Peifang and U. Welz-Biermann, *Green Chem.*, 2011, **13**, 666.
84. A. F. M. Claudio, A. M. Ferreira, M. G. Freire and J. A. P. Coutinho, *Green Chem.*, 2013, **15**, 2002.
85. A. K. Ressmann, R. Zirbs, M. Pressler, P. Gaertner and K. Bica, *Z. Naturforsch.*, 2013, **68b**, 1129.
86. K. Ressmann, K. Strassl, P. Gaertner, B. Zhao, L. Greiner and K. Bica, *Green Chem.*, 2012, **14**, 940.
87. Bioniqs Ltd, *Extraction of Artemisinin using Ionic Liquids*, Project Report 003-003/3, Bioniqs Ltd, York, UK, 2008.
88. T. Usuki, N. Yasuda, M. Yoshizawa-Fujita and M. Rikukawa, *Chem. Commun.*, 2011, **47**, 10560.
89. C. Lu, X. Luo, L. Lu, H. Li, X. Chen and Y. Ji, *J. Sep. Sci.*, 2013, **36**, 959.
90. W. Bi, M. Tian and K. H. Row, *J. Chromatogr. B*, 2012, **880**, 108.
91. S. A. Chowdhury, R. Vijayaraghavan and D. R. MacFarlane, *Green Chem.*, 2010, **12**, 1023.
92. Y. Zhai, S. Sun, Z. Wang, J. Cheng, Y. Sun, L. Wang, Y. Zhang, H. Zhang and A. Yu, *J. Sep. Sci.*, 2009, **32**, 3544.
93. K. Bica, P. Gaertner and R. D. Rogers, *Green Chem.*, 2011, **13**, 1997.
94. J. Jiao, Q.-Y. Gai, Y-J. Fu, Y.-G. Zu, M. Luo, W. Wang and C.-J. Zhao, *J. Food Eng.*, 2013, **117**, 447.
95. M. J. Muldoon, *Dalton Trans.*, 2010, **39**, 337.
96. J. H. Kim, J. W. Lee, U. S. Shin, J. Y. Lee, S.-G. Lee and C. E. Song, *Chem. Commun.*, 2007, **44**, 4683.
97. M. Maase, K. Massonne, K. Halbritter, R. Noe, M. Bartsch, W. Siegel, V. Stegmann, M. Flores, O. Huttenloch and M. Becker, Method for the separation of acids from chemical reaction mixtures by means of ionic fluids, *World Pat.*, WO 2003 062171, 2003.
98. J. Baker, *Eur. Chem. News*, 2004, 18.

99. J. Estager, J. D. Holbrey and M. Swadźba-Kwaśny, *Chem. Soc. Rev.*, 2014, **43**, 847.

100. H. L. Chum, V. R. Koch, L. L. Miller and R. A. Osteryoung, *J. Am. Chem. Soc.*, 1975, **97**, 3264.

101. J. S. Wilkes, J. A. Levisky, R. A. Wilson and C. L. Hussey, *Inorg. Chem.*, 1982, **21**, 1263.

102. T. Welton, *Chem. Rev.*, 1999, **99**, 2071.

103. J. Estager, A. A. Oliferenko, K. R. Seddon and M. Swadzba-Kwaśny, *Dalton Trans.*, 2010, **39**, 11375.

104. M. Currie, J. Estager, P. Licence, S. Men, P. Nockemann, K. R. Seddon, M. Swadzba-Kwaśny and C. Terrade, *Inorg. Chem.*, 2013, **52**, 1710.

105. Z. C. Liu, R. Zhang, C. M. Xu and R. G. Xia, *Oil Gas J.*, 2006, **104**, 52.

106. Z. Liu, C. Xu and C. Huang, Method for manufacturing alkylate oil with composite ionic liquid used as catalyst, *US Pat.*, 0133056, 2004.

107. Y. Chauvin, A. Hirschauer and H. Oliver, *J. Mol. Catal.*, 1994, **92**, 155.

108. K. Yoo, V. V. Namboodiri, R. S. Varma and P. G. Smirniotis, *J. Catal.*, 2004, **222**, 511.

109. C. P. Huang, Z. C. Liu, C. M. Xu, B. H. Chen and Y. F. Liu, *Appl. Catal., A: General*, 2004, **277**, 41.

110. A. Riisager, R. Fehrmann, M. Haumann and P. Wasserscheid, in *Regulated Systems for Multiphase Catalysis*, ed. W. Leitner and M. Hölscher, Topics in Organometallic Chemistry, Springer, Berlin and Heidelberg, 2008, vol. 23, pp. 149–161.

111. M. H. Valkenberg, C. de Castro and W. F. Hoelderich, *Green Chem.*, 2002, **4**, 88.

112. C. Van Doorslaer, J. Wahlen, P. Mertens, K. Binnemans and D. De Vos, *Dalton Trans.*, 2010, **39**, 8377.

113. A. Riisager, R. Fehrmann, M. Haumann and P. Wasserscheid, *Eur. J. Inorg. Chem.*, 2006, 695.

114. Supported Ionic Liquid Phase (SILP) Technology, www.silp-technology.com [accessed 20 February 2014].

115. A. Riisager, K. M. Eriksen, P. Wasserscheid and R. Fehrmann, *Catal. Lett.*, 2003, **90**, 149.

116. A. Riisager, P. Wasserscheid, R. van Hal and R. Fehrmann, *J. Catal.*, 2003, **219**, 252.

117. C. P. Mehnert, R. A. Cook, N. C. Dispenziere and M. Afeworki, *J. Am. Chem. Soc.*, 2002, **124**, 12932.

118. C. P. Mehnert, *Chem. Eur. J.*, 2005, **11**, 50.

119. H. Hagiwara, N. Okunaka, T. Hoshi and T. Suzuki, *Synletters*, 2008, 1813.

120. A. Wolfson, I. F. J. Vankelecom and P. A. Jacobs, *Tetrahedron Lett.*, 2003, **44**, 1195.

121. S. Breitenlechner, M. Fleck, T. E. Müller and A. Suppan, *J. Mol. Catal. A: Chem.*, 2004, **214**, 175.

122. H. Hagiwara, Y. Sugawara, K. Isobe, T. Hoshi and T. Suzuki, *Org. Lett.*, 2004, **6**, 2325.

123. M. Ruta, G. Laurenczy, P. J. Dyson and L. Kiwi-Minsker, *J. Phys. Chem. C*, 2008, **112**, 17814.

124. G. Machado, J. D. Scholten, T. de Vargas, S. R. Teixeira, L. H. Ronchi and J. Dupont, *Int. J. Nanotechnol.*, 2007, **4**, 541.

125. P. Lozano, *Green Chem.*, 2010, **12**, 555.

126. M. T. Reetz, W. Wiesenhofer, G. Francio and W. Leitner, *Chem. Commun.*, 2002, 992.

127. M. T. Reetz, W. Wiesenhofer, G. Francio and W. Leitner, *Adv. Synth. Catal.*, 2003, **345**, 1221.

128. P. Lozano, E. García-Verdugo, R. Piamtongkam, N. Karbass, T. DeDiego, M. I. Burguete, S. V. Luis and J. L. Iborra, *Adv. Synth. Catal.*, 2007, **349**, 1077.

129. P. Lozano, E. García-Verdugo, N. Karbass, K. Montague, T. De Diego, M. I. Burguete and Santiago V. Luis, *Green Chem.*, 2010, **12**, 1803.

130. K. Bica, H. Rodríguez, G. Gurau, O. A. Cojocaru, A. Riisager, R. Fehrmann and R. D. Rogers, *Chem. Commun.*, 2012, **48**, 5422.

131. EU Directive 98/70/EC of the European Parliament and of the Council of 13 October 1998 relating to the quality of petrol and diesel fuels and amending Council Directive 93/12/EC, *Off. J. Eur. Commun.*, L350, 28/12/98, 58, http://eur-lex.europa.eu/LexUriServ/LexUriServ.do?uri=CELEX:31998L0070:EN:HTML [accessed 20 February 2014].

132. P. S. Kulkarni and C. A. M. Afonso, *Green Chem.*, 2010, **12**, 1139.

133. J. Eßer, P. Wasserscheid and A. Jess, *Green Chem.*, 2004, **6**, 316.

134. W. Li, Y. Zhu, J. Wang, J. Zhang and Y. Yan, *Green Chem.*, 2009, **11**, 810.

135. W. Zhu, W. Huang, H. Li, M. Zhang, W. Jiang, G. Chen and C. Han, *Fuel Proc. Technol.*, 2011, **10**, 1842.

136. E. Kuhlmann, M. Haumann, A. Jess, A. Seeberger and P. Wasserscheid, *ChemSusChem*, 2009, **2**, 969.

137. F. Kohler, D. Roth, E. Kuhlmann, P. Wasserscheidt and M. Haumann, *Green Chem.*, 2010, **12**, 979.

138. N. H. Ko, J. S. Lee, E. S. Huh, H. Lee, K. D. Dung, H. S. Kim and M. Cheong, *Energy Fuels*, 2008, **22**, 1687.

139. S. Saravanamurugan, J. Due-Hansen, S. Kegnoes, T. Gretarsdottir, A. Riisager and R. Fehrmann, *Prepr. Symp. – Am. Chem. Soc., Div. Fuel Chem.*, 2010, **55**, 6.

140. A. Riisager, A. J. Kunov-Kruse, S. L. Mossin and R. Fehrmann, Absorption and oxidation of NO in ionic liquids, *PCT Int. Appl.*, WO 20 13079597A1, 2013.

141. M. D. Bigham, *SPE Prod. Eng.*, 1990, **5**, 120.

142. IChemE, Gas clean-up technology triumphs at global IChemE awards, press release 7 November 2013, www.icheme.org/media_centre/news/2013/gas-clean-up-technology-triumphs-at-global-icheme-awards.aspx#.UwXXv4VKabo [accessed 20 February 2014].

143. K. Bica, P. Gaertner, P. J. Gritsch, A. K. Ressmann, C. Schroeder and R. Zirbs, *Chem. Commun.*, 2012, **48**, 5013.

144. Y. Kohno and H. Ohno, *Chem. Commun.*, 2012, **48**, 7119.

145. K. Fukumoto and H. Ohno, *Angew. Chem., Int. Ed.*, 2007, **46**, 1852.

Gas Separations using Ionic Liquids

LEILA MOURA,[a,b] CATHERINE C. SANTINI[b] AND
MARGARIDA F. COSTA GOMES*[a]

[a] Institut de Chimie de Clermont-Ferrand, UMR 6296 CNRS and Université
Blaise Pascal, équipe Thermodynamique et Interactions Moléculaires,
BP 80026, 63171, Aubière, France; [b] LC2P2, UMR 5265 CNRS and
Université de Lyon 1, CPE LYON 43 Bd du 11 Novembre 1918,
69616 Villeurbanne, France
*Email: margarida.c.gomes@univ-bpclermont.fr

22.1 INTRODUCTION

Separation processes explore differences in the properties of the
species; these might be molecular (molar mass, dipole moment or
dielectric constant), thermodynamic (vapour pressure or solubility) or
transport (diffusivity) properties. Common separation techniques rely
on creating or adding a new phase, on introducing a barrier or a solid
agent, or on using a force field or a gradient.[1]

Distillation, crystallization, liquid–liquid extraction and absorption
are the most common examples of separation techniques based on
the creation or on the addition of a new phase that require an energy
transfer or the use of a mass transfer agent. The latter is usually less
energy demanding but technologically more challenging, as the mass
transfer agent needs to be effectively separated and then recycled with
reasonable purity. Membrane separation processes are based on the
use of liquid or solid (often polymeric) barriers[2] that have a different

Chemical Processes for a Sustainable Future
Edited by Trevor M. Letcher, Janet L. Scott and Darrell A. Patterson
© The Royal Society of Chemistry 2015
Published by the Royal Society of Chemistry, www.rsc.org

permeability for each of the components of a mixture. Membranes can be used in small, compact and clean units that require low energy to achieve separation but are usually difficult to scale up. Adsorption involves a solid agent, often a porous material with a large surface area, capable of selectively retaining at its surface (by physical or chemical adsorption) the constituents of a mixture. Because the active separating agents (*e.g.* alumina, ion exchanging resins, high surface area SiO_2, zeolites or molecular sieves) eventually become saturated and need to be regenerated, separation by adsorption is seldom conducted continuously. Centrifugation, electrolysis, electrodialysis and electrophoresis are examples of separation techniques that exploit the differences in the responses of the constituents of a feed to an external force or gradient. These techniques are used to separate both liquid and gaseous mixtures and are especially useful and versatile for separating biochemicals.

Cryogenic distillation is the most used separation process for mixtures of gases. It has been used in industry for ethane–ethylene and propane–propylene separation[1,3] since the 1960s; ethylene and propylene are used in large quantities (demand of 25×10^{12} t a^{-1} of ethylene where 't' refers to metric tonnes and 'a' to annum)[4] as precursors for the production of polymers, lubricants, rubbers, solvents and fuel components.[1,4,5] These separations require large distillation towers (120–180 trays), low temperatures ($-114\ °C$) and high pressures (1.5–3.0 MPa), resulting in high capital and energy demanding processes needing up to 126.61×10^{15} J a^{-1} (120×10^{12} Btu a^{-1}).[6,7] The implementation of alternative processes with improved economic and environmental performance would represent a major advance in the sector.

Ionic liquids have been suggested as separating agents for olefin–paraffin gas separation, as absorbents or as solvents for the chemical complexation of olefins with silver or copper salts. Ionic liquids are composed uniquely of ions and have a melting point below 100 °C. Many present unique properties such as negligible vapour pressure, high thermal, chemical and electrochemical stability, non-flammability and high ionic conductivity. Ionic liquid can be synthesized in a large variety of combination of cations and anions, leading to tuneable physical chemical properties. The cations and anions that constitute the ionic liquids proposed for gaseous olefin–paraffin separations are listed in Tables 22.1 and 22.2, respectively.

It is usually considered that unsaturated hydrocarbons present a higher solubility in ionic liquids than their saturated counterparts, as can be seen in Figure 22.1 where the Henry's law constants of

Table 22.1 Abbreviation, full designation and structure of some cations of ionic liquids. $C_{n/m}$ represent alkyl chains of variable size.

Abbreviation	Full designation	Structure
$C_nC_mIm^+$	1-C_n-3-C_mImidazolium	
$C_1(C_3H_5CH_2)Im^+$	1-(buten-3-yl)-3-methyl-Imidazolium	
$C_1(CH_2C_6H_5)Im^+$	1-benzyl-3-methyl-Imidazolium	
$(C_2SO_2C_2)C_1Im^+$	1-diethylsulfonyl-3-methylimidazolium	
$C_1C_3CNIm^+$	1-(3-cyanopropyl)-3-methylImidazolium	
$C_1C_1C_3CNIm^+$	1-(3-cyanopropyl)-2,3-dimethylImidazolium	
$(C_3CN)_2Im^+$	1,3-(3-cyanopropyl)-Imidazolium	
$C_1C_1COOC_5Im^+$	-1-(pentoxycarbonyl-methyl))-3-methylimidazolium	
$C_1C_1COOC_2OC_2Im^+$	1-(ethoxyethoxy-carbonylmethyl))-3-methylImidazolium	
$C_1C_1COOC_2O-C_2OC_4Im^+$	1-(2-(2-butoxyethoxy)-ethoxycarbonyl-methyl)-3-methyl-Imidazolium	

Table 22.1 (*Continued*)

Abbreviation	Full designation	Structure
C_nC_mPyrr$^+$	C_nC_mPyrrolidinium	
C_nPyr$^+$	C_nPyridinium	
P_{nmpq}^+	$C_nC_mC_pC_q$Phosphonium	
N_{nmpq}^+	$C_nC_mC_pC_q$Ammonium	
$N_{1132\text{-OH}}$	Propylcholinium	

different gases in one ionic liquid are presented. The exact reasons for the observed trends are still poorly known, but are essential for the development of new alkane/alkene/alkyne separation processes.[8] In this chapter, we present the current knowledge on the solubility of ethane, ethylene, acetylene, propane, propylene and methyl acetylene in pure ionic liquids. Whenever possible we have calculated the ionic liquid absorption capacity and ideal selectivity (solubility ratio) for ethane–ethylene, ethylene–acetylene, propane–propylene and propylene–methyl acetylene mixtures and so assess the potential of a particular ionic liquid as a separating agent.

The absorption capacity of gaseous hydrocarbons by pure ionic liquids is relatively low because, in most cases, it is only a physical process. The addition of silver(ı) and copper(ı) salts can significantly improve the amount of the unsaturated hydrocarbon absorbed by the ionic liquid phase as these metal cations are capable of selectively and reversibly reacting with the unsaturated gases. This concept has been explored for the separation of ethane from ethylene and propane from propylene.[7]

Table 22.2 Abbreviation, full designation and structure of some anions of ionic liquids. R_n represents alkyl chains of variable size.

Abbreviation	Full designation	Structure
NTf_2^-	Bis(trifluoromethylsulfonyl)imide	
$BETI^-$	Bis(perfuoroethylsulfonyl)imide	
DCA^-	Dicyanamide	
NO_3^-	Nitrate	
$C_nC_mPO_4^-$	C_nC_mPhosphate	
$C_nHPO_3^-$	C_nPhosphite	
PF_6^-	Hexafluorophosphate	
$C_nSO_4^-$	C_nSulfate	
$CF_3SO_3^-$	Trifluromethanesulfonate	
DBS^-	Dodecylbenzenesulfonate	
$C_nCO_2^-$	Carboxylate	
OAc^-	Acetate	
TFA^-	Trifluoroacetate	

Table 22.2 (*Continued*)

Abbreviation	Full designation	Structure
$AlCl_4^-$	Tetrachloroaluminate	
BF_4^-	Tetrafluoroborate	
$TMPP^-$	Bis(2,4,4-trimethylpentyl)-phosphinate	
FAP^-	Tris(pentafluoroethyl)-trifluorophosphate	

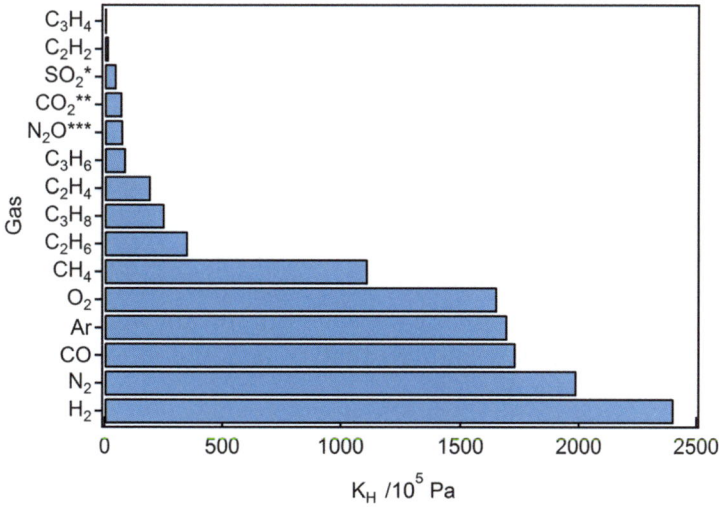

Figure 22.1 Henry's law constant, K_H, for several gases in ionic liquid $[C_1C_4Im][BF_4]$[9–13] at $T = 313$ K. * values at 293 K; **values at 314 K; *** values at 298 K.

22.2   GAS SOLUBILITY

The chemical potential of component i in a mixture, is defined as:[14,15]

$$\mu_i = \mu_i^{ref} + RT \ln(\gamma_i x_i) \tag{22.1}$$

where μ_i^{ref} is the chemical potential of component i in a chosen reference state and γ_i its activity coefficient.

The symmetric or the asymmetric convention are usually adopted to choose the appropriate combination of the reference state chemical potential and the activity coefficient, the choice depending on the physical state of the pure components at the thermodynamic conditions of the mixture.

The asymmetric convention is applied to solutions where the solute(s) or the solvent are not in the physical state of the solution at a given temperature and pressure, as in the case of the solutions of gases in ionic liquids considered in this chapter. For the solvent, the activity coefficient approaches unity when its mole fraction is approximately unity and for the solute its activity coefficient becomes unity when the solute is present at very low concentrations, $\gamma_i^H \to 1$ when $x_i \to 0$. In this case the solute is present in the limit of infinite dilution and the solution approaches ideal behaviour in the sense of Henry's law. The reference state adopted for the calculation of the chemical potential in this case is that of the pure solute at infinite dilution:[14,16]

$$\mu_i = \mu_i^{*H} + RT \ln(\gamma_i^H x_i) \qquad x_i \to 0 \Rightarrow \gamma_i^H \to 1 \tag{22.2}$$

where the activity coefficient in the asymmetric convention, γ_i^H, is a measure of the solute intermolecular interactions, the solvent and solute–solvent interactions being accounted for in the reference state.

The Gibbs energy of solution is defined as the difference in chemical potential when the solute is transferred, at constant pressure and temperature, from its pure state into the infinitely dilute solution. If the solute remains pure at equilibrium with the solution[†] and the solubility is low enough that $\gamma_i^H \approx 1$, the following, approximate relation can be used:

$$\Delta_{sol} G_i \approx RT \ln x_i \tag{22.3}$$

[†]This is the case for the solutions of gases in ionic liquids because the solvent has a negligible vapour pressure at moderate temperatures.

The Gibbs energy of solution expresses the difference between the solute–solute interactions in the pure species, which may be a condensed phase, and the solute–solvent interactions in an infinitely dilute solution. In order to isolate the role of the solute–solvent interactions in the process of dissolution, a thermodynamic transformation called solvation[17] can be defined as the difference in chemical potential when the solute is transferred from an ideal gas at standard pressure into the reference state at infinite dilution, Equation (22.3), leading to:

$$\Delta_{\text{solv}} G_i = RT \ln\left(\frac{K_{H,i}}{p^0}\right) \tag{22.4}$$

in which $K_{H,i}$ is the Henry's law constant defined as:

$$K_{H,i} \equiv \lim_{x_i \to 0} (f_i/x_i) \tag{22.5}$$

From the behaviour with temperature or with pressure of the Gibbs energy of solvation it is possible to calculate the other thermodynamic properties of solvation:[18]

$$\Delta_{\text{solv}} H_i = -T^2 \frac{\partial}{\partial T}\left(\frac{\Delta_{\text{solv}} G_i}{T}\right)_p$$

$$\Delta_{\text{solv}} S_i = -\left(\frac{\partial \Delta_{\text{solv}} G_i}{\partial T}\right)_p \tag{22.6}$$

$$\Delta_{\text{solv}} V_i = -\left(\frac{\partial \Delta_{\text{solv}} G_i}{\partial p}\right)_T$$

The thermodynamic properties of solvation provide insights into the molecular mechanisms determining the solution behaviour. The enthalpy of solvation reflects the energy of solute–solvent interactions and the entropic contribution is related to structural organization in the solution. These thermodynamic properties of solvation are approximately equal to the thermodynamic properties of solution in the case of gaseous solutes at low pressure; the differences becoming more important for liquid or solid solutes.[19] These solvation properties can be determined through experimentally accessible quantities namely from solubility measurements and the calculation of the Henry's law constants.

22.3   ETHANE AND PROPANE SOLUBILITY

The ethane and propane mole fraction solubilities, calculated for a 1 bar partial pressure of gas, in different ionic liquids at $T = 313$ K are depicted in Figure 22.2. For ethane, 27 different ionic liquids were studied (20 based on imidazolium cations) and the mole fraction gas solubility ranges from 1.50×10^{-3} in $[C_1C_2Im][DCA]$ to 52.6×10^{-3} in $[P_{(14)666}][TMPP]$. For propane, eight ionic liquids were studied (six based on imidazolium cations and two based on phosphonium

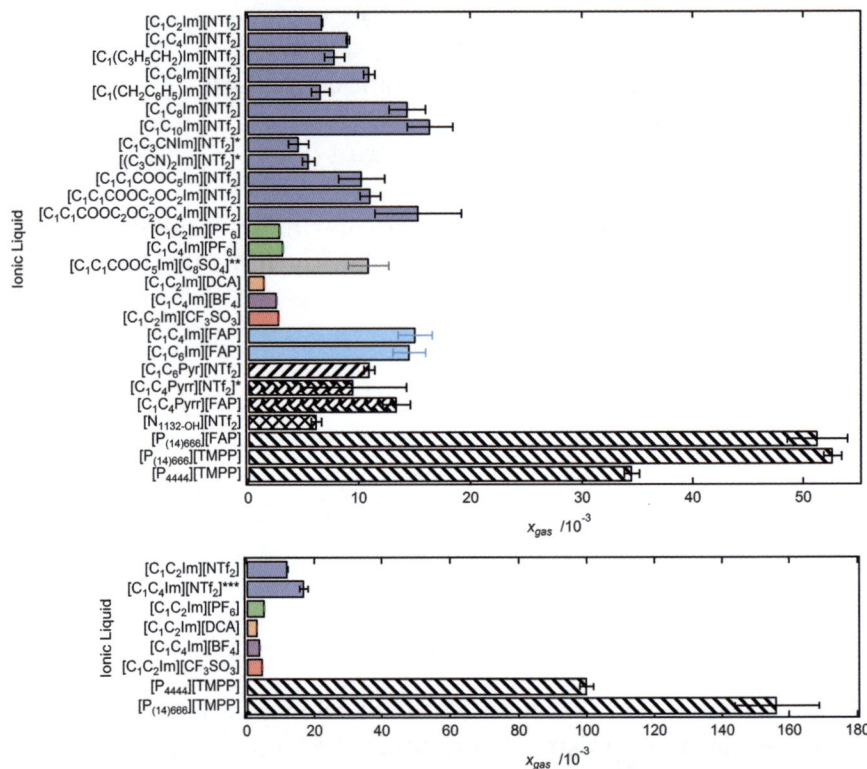

Figure 22.2   Mole fraction solubility for ethane (top) and propane (bottom) in several ionic liquids at $T = 313$ K. * values at 303 K; ** values at 323 K; *** values at 320 K. $[C_1C_2Im][NTf_2]$;[21,32] $[C_1C_4Im][NTf_2]$;[22,23] $[C_1(C_3H_5CH_2)Im][NTf_2]$;[24] $[C_1C_6Im][NTf_2]$;[25] $[C_1(CH_2C_6H_5)Im][NTf_2]$;[24] $[C_1C_8Im][NTf_2]$;[22] $[C_1C_{10}Im]$-$[NTf_2]$;[22] $[C_1C_3CNIm][NTf_2]$;[26] $[(C_3CN)_2Im][NTf_2]$;[26] $[C_1C_1COOC_5Im]$-$[NTf_2]$;[27] $[C_1C_1COOC_2OC_2Im][NTf_2]$;[27] $[C_1C_1COOC_2OC_2OC_4Im][NTf_2]$;[27] $[C_1C_2Im][PF_6]$;[12,28] $[C_1C_4Im][PF_6]$;[29] $[C_1C_1COOC_5Im][C_8SO_4]$;[27] $[C_1C_2Im][DCA]$;[29] $[C_1C_4Im][BF_4]$;[13,28] $[C_1C_2Im][CF_3SO_3]$;[29] $[C_1C_4Im][FAP]$;[20] $[C_1C_6Im][FAP]$;[20] $[C_1C_6Pyr][NTf_2]$;[30] $[C_1C_4Pyrr][NTf_2]$;[21] $[C_1C_4Pyrr][FAP]$;[31] $[N_{1132-OH}][NTf_2]$;[21] $[P_{(14)666}][FAP]$;[31] $[P_{(14)666}][TMPP]$;[32,33] $[P_{4444}][TMPP]$.[32]

cations) and the mole fraction gas solubility varied from 3.24×10^{-3} in $[C_1C_2Im][DCA]$ to 156×10^{-3} in $[P_{(14)666}][TMPP]$.

For both gases, the solubility is highest in phosphonium based ionic liquids followed by imidazolium bis(trifluoromethylsulfonyl)imide ionic liquids and imidazolium tris(pentafluoroethyl)trifluorophosphate ionic liquids; the latter having only been studied with ethane. Phosphonium based ionic liquids are capable of dissolving three times more ethane and eight times more propane than imidazolium based ones; these differences being less pronounced when the solubility is expressed in mass fraction, as can be observed in Figure 22.3.

For both imidazolium and phosphonium based ionic liquids, the gas solubility increases with increasing length of the alkyl side chains. For example, for imidazolium bis(trifluoromethylsulfonyl)imide ionic liquids, the mole fraction solubility of ethane increases, at 313 K, from 6.73×10^{-3} for $[C_1C_2Im][NTf_2]$ to 16.4×10^{-3} for $[C_1C_{10}Im][NTf_2]$ and the solubility of propane increases from 12.0×10^{-3} (at 313 K) for $[C_1C_2Im][NTf_2]$ to 16.9×10^{-3} (at 320 K) for $[C_1C_4Im][NTf_2]$. The increase in ethane solubility for the ionic liquids $[C_1C_nIm][NTf_2]$ can be explained by a more favourable entropy of solvation for the longer alkyl side chains of the cations. Using molecular simulation, it was observed that ethane is solvated in the non-polar domains of the ionic liquid, preferentially near the terminal carbon of the alkyl chain, and it becomes more mobile when the alkyl side chain is longer.[22] The same variation of ethane solubility is found for $[C_1C_nIm][PF_6]$ ionic

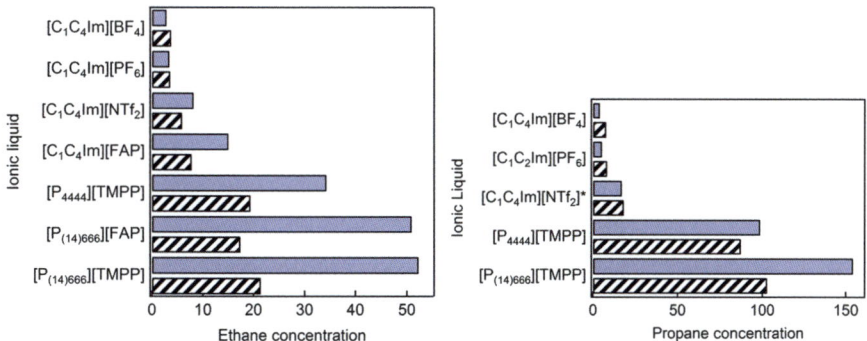

Figure 22.3 Mole fraction ($\times 10^3$, full bars) and mass fraction ($\times 10^4$, patterned bars) of ethane (on the left) or propane (on the right) in several ionic liquids at 1 bar partial pressure of gas and $T = 313$ K (* values at 320 K).

liquids but is much smaller for $[C_1C_nIm][FAP]$ ($n = 4$, 6) for which a more moderate increase of the ethane solubility is observed. This is attributed to a balancing effect between a more favourable enthalpy and less favourable entropy, resulting in similar Gibbs energies of solvation for $[C_1C_nIm][FAP]$ ($n = 4$, 6).[20]

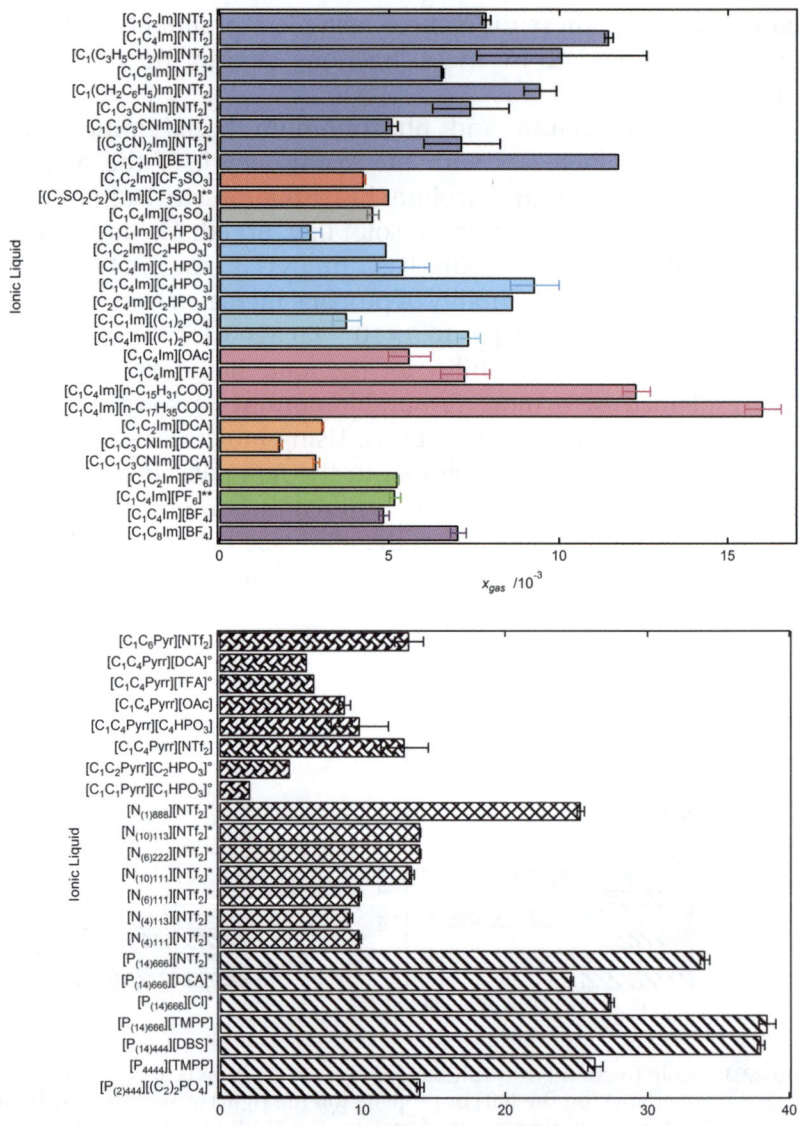

In the case of phosphonium based ionic liquids, a large increase in the solubility of ethane and propane with increasing size of the alkyl chains of the cation was observed. This was found to be independent of the anions studied.

The functionalization of the alkyl side chains in imidazolium based ionic liquids with cyano, ether or ester groups does not improve the solubility of ethane and propane for the ionic liquids studied so far. Furthermore, when ester or ether functions are included in the side chain of the imidazolium cations, the solubility increases more slowly with an increase in the chain length compared with non-functionalised alkyl chains.

Imidazolium-based ionic liquids containing large anions such as FAP^-, $C_8SO_4^-$ and NTf_2^- present higher ethane and propane solubility, followed by PF_6^-, $CF_3SO_3^-$, BF_4^- and DCA^-.

22.4 ETHYLENE AND PROPYLENE SOLUBILITY

The ethylene and propylene mole fraction solubility in different ionic liquids at 313 K is depicted in Figures 22.4 and 22.5, respectively. For ethylene, 53 different ionic liquids were studied (30 based on the imidazolium cation and 23 based on other cations) and the mole fraction gas solubility ranges from 1.78×10^{-3} in $[C_1C_3CNIm][DCA]$ to 38.5×10^{-3} in $[P_{(14)666}][TMPP]$. For propylene, 40 ionic liquids were studied (25 based on the imidazolium cation and 15 based on other

Figure 22.4 Mole fraction solubility for ethylene in imidazolium based ionic liquids (top) and in pyridinium, pyrrolidinium, ammonium and phosphonium based ionic liquids (bottom) at $T = 313$ K. * values at 303 K; ** values at 323 K; – uncertainty not indicated by the authors. $[C_1C_2Im][NTf_2]$;[32] $[C_1C_4Im][NTf_2]$;[24] $C_1(C_3H_5CH_2)Im][NTf_2]$;[24] $[C_1C_6Im][NTf_2]$;[34,35] $[C_1(CH_2C_6H_5)Im][NTf_2]$;[24] $[C_1C_3CNIm][NTf_2]$;[26] $[C_1C_1C_3CNIm][NTf_2]$;[36,37] $[(C_3CN)_2Im][NTf_2]$;[26] $[C_1C_4Im][BETI]$;[34] $[C_1C_2Im][CF_3SO_3]$;[12] $[(C_2SO_2C_2)-C_1Im][CF_3SO_3]$;[34,35] $[C_1C_4Im][C_1SO_4]$;[9] $[C_1C_1Im][C_1HPO_3]$;[9] $[C_1C_2Im]-[C_2HPO_3]$;[39] $[C_1C_4Im][C_1HPO_3]$;[10] $[C_1C_4Im][C_4HPO_3]$;[10] $[C_2C_4Im]-[C_2HPO_3]$;[39] $[C_1C_1Im][(C_1)_2PO_4]$;[9] $[C_1C_4Im][(C_1)_2PO_4]$;[9] $[C_1C_4Im][OAc]$;[9] $[C_1C_4Im][TFA]$;[9] $[C_1C_4Im][n-C_{15}H_{31}COO]$;[37] $[C_1C_4Im][n-C_{17}H_{35}COO]$;[37] $[C_1C_2Im][DCA]$;[12] $[C_1C_3CNIm][DCA]$;[36,37] $[C_1C_1C_3CNIm][DCA]$;[36,37] $[C_1C_2Im][PF_6]$;[12] $[C_1C_4Im][PF_6]$;[29,39] $[C_1C_4Im][BF_4]$;[9] $[C_1C_8Im][BF_4]$;[37] $[C_1C_6Pyr][NTf_2]$;[30] $[C_1C_4Pyrr][NTf_2]$;[40] $[C_1C_4Pyrr][DCA]$;[40] $[C_1C_4Pyrr][OAc]$;[9,40] $[C_1C_1Pyrr][TFA]$;[40] $[C_1C_1Pyrr][C_1HPO_3]$;[40] $[C_1C_2Pyrr]-[C_2HPO_3]$;[40] $[C_1C_4Pyrr][C_4HPO_3]$;[9] $[N_{(4)111}][NTf_2]$;[34,35,41] $[N_{(4)113}]-[NTf_2]$;[34,35,41] $[N_{(6)111}][NTf_2]$;[35,35,41] $[N_{(6)222}][NTf_2]$;[34,35,41] $[N_{(10)111}]-[NTf_2]$;[34,35,41] $[N_{(10)113}][NTf_2]$;[34,35,41] $[N_{(1)888}][NTf_2]$;[34,35,41] $[P_{(14)666}]-[NTf_2]$;[34,35,41] $[P_{(14)666}][DCA]$;[34,35,42] $[P_{(2)444}][(C_2)_2PO_4]$;[34,35,42] $[P_{(14)666}]-[Cl]$;[34,35,42,43] $[P_{(14)444}][DBS]$;[34,42] $[P_{4444}][TMPP]$;[32] $[P_{(14)666}][TMPP]$.[32,33]

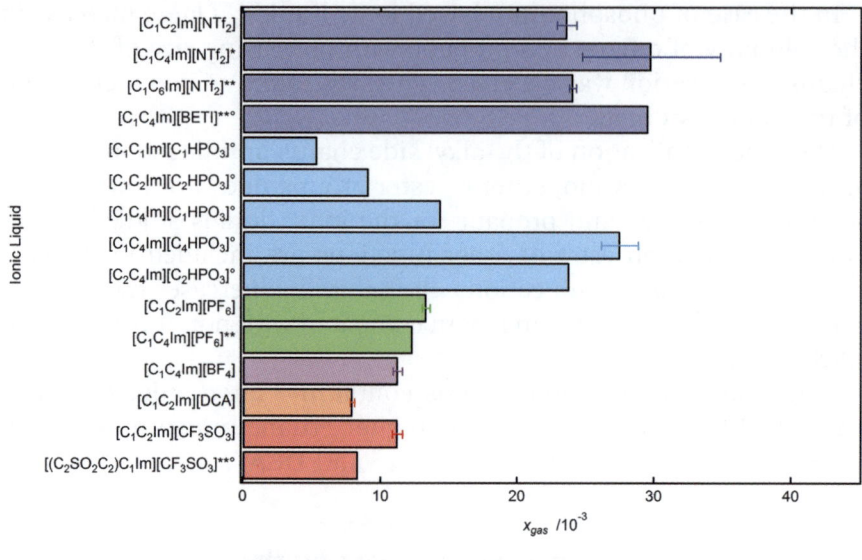

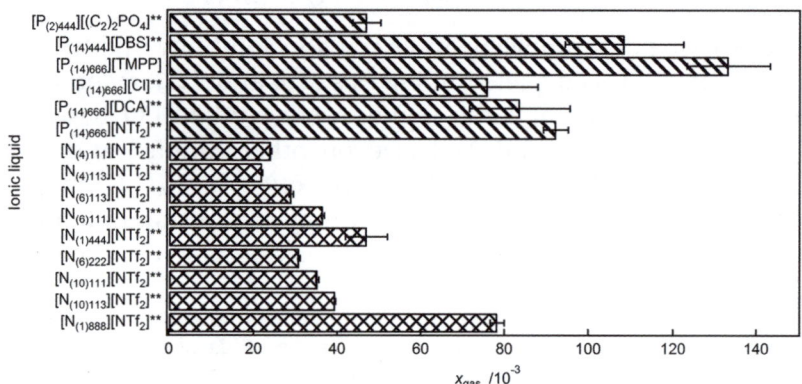

Figure 22.5 Mole fraction solubility for propylene in imidazolium based liquids (top) and in ammonium and phosphonium based ionic liquids (bottom) at $T = 313$ K. * values at 320 K; ** values at 303 K; – uncertainty not indicated by the authors. $[C_1C_2Im][NTf_2]$;[12] $[C_1C_4Im][NTf_2]$;[23] $[C_1C_6Im][NTf_2]$;[34,35] $[C_1C_4Im][BETI]$;[34] $[C_1C_1Im][C_1HPO_3]$;[38] $[C_1C_2Im][C_2HPO_3]$;[38] $[C_1C_4Im][C_1HPO_3]$;[38] $[C_1C_4Im][C_4HPO_3]$;[38] $[C_2C_4Im][C_2HPO_3]$;[38] $[C_1C_2Im][PF_6]$;[12] $[C_1C_4Im][PF_6]$;[34,35,43] $[C_1C_4Im][BF_4]$;[12] $[C_1C_2Im][DCA]$;[12] $[C_1C_2Im][CF_3SO_3]$;[12] $[(C_2SO_2C_2)C_1Im][CF_3SO_3]$;[34,35] $[N_{(4)111}][NTf_2]$;[34,35,41] $[N_{(4)113}][NTf_2]$;[34,35,41] $[N_{(6)111}][NTf_2]$;[34,35,41] $[N_{(6)113}][NTf_2]$;[34,35,41] $[N_{(1)444}][NTf_2]$;[42] $[N_{(6)222}][NTf_2]$;[34,35,41] $[N_{(10)111}][NTf_2]$;[34,35,41] $[N_{(10)113}][NTf_2]$;[34,35,41] $[N_{(1)888}][NTf_2]$;[34,35,41] $[P_{(14)666}][NTf_2]$;[34,35,42] $[P_{(2)444}][(C_2)_2PO_4]$;[34,35,42] $[P_{(14)666}][Cl]$;[34,35,42] $[P_{(14)666}][DCA]$;[34,35,42] $[P_{(14)444}][DBS]$;[34,42] $[P_{(14)666}][TMPP]$.[33]

cations) and the mole fraction gas solubility varies from 4.14×10^{-3} in $[C_1C_1Im][C_1SO_4]$ to 133×10^{-3} in $[P_{(14)666}][TMPP]$. Both ethylene and propylene solubility are less affected when changing from $[C_1C_2Im][DCA]$ to $[P_{(14)666}][TMPP]$ than are the solubilities of ethane or propane, respectively.

For both ethylene and propylene, the solubility is highest in phosphonium based ionic liquids followed by ammonium, pyridinium, pyrrolidinium and imidazolium ionic liquids. In the case of phosphonium based ionic liquids, several anions were studied and both ethylene and propylene solubility follow the order: $[P_{(14)666}][DCA] < [P_{(14)666}][Cl] < [P_{4444}][TMPP] < [P_{(14)444}][DBS] < [P_{(14)666}]$-$[TMPP]$. In the case of imidazolium and ammonium based ionic liquids, ethylene is more soluble in $[C_1C_4Im][NTf_2]$ than in $[N_{(4)111}][NTf_2]$ but significantly less soluble in $[N_{(1)888}][NTf_2]$, at $T = 303$ K.

As for ethane and propane, the solubility of ethylene and propylene increases with increasing length of the alkyl chain of the cation for imidazolium ionic liquids with NTf_2^-, BF_4^-, phosphate and phosphite anions. An outlier, ethylene in $[C_1C_6Im][NTf_2]$, was found to have a strangely low solubility. Ethylene solubility also increases with increasing length of the alkyl side chains in phosphonium and ammonium based ionic liquids but no precise contribution can be attributed to each supplementary $-CH_2-$ group. The ethylene solubility also increases with the increasing alkyl chain in the anion of imidazolium based ionic liquids as can be seen by comparing the gas solubility in $[C_1C_4Im][OAc]$ and in $[C_1C_4Im][n-C_{15}H_{31}COO]$ or in $[C_1C_4Im][n-C_{15}H_{31}COO]$ and in $[C_1C_4Im][n-C_{17}H_{35}COO]$. The same effect is observed when the alkyl chain in the anion is fluorinated; the solubility of ethylene being larger in $[C_1C_nIm][BETI]$ than in $[C_1C_nIm][NTf_2]$.

Functionalizing the alkyl side chains of an ionic liquid can have very different effects on the ethylene or propylene solubility. For example, ethylene solubility is lower in $[C_1C_2Im][CF_3SO_3]$ than in $[(C_2SO_2C_2)C_1Im][CF_3SO_3]$, but significantly higher than in $[C_1C_3CNIm][DCA]$. The presence of polar groups in the side chain of the cation of the ionic liquid can increase or decrease ethylene solubility, while evidence shows that it always seems to decrease the solubility of ethane. The absence of the imidazolium acidic proton in $[C_1C_1C_3CNIm][DCA]$ seems to significantly increase the solubility of ethylene as can be observed by comparing $[C_1C_3CNIm][DCA]$ and $[C_1C_1C_3CNIm][DCA]$.

22.5   ACETYLENE AND METHYL ACETYLENE SOLUBILITY

The acetylene and methyl acetylene mole fraction solubility in different ionic liquids at 313 K is depicted in Figure 22.6. For acetylene, 27 different ionic liquids were studied (20 based in the imidazolium cation and seven based in pyrrolidinium cations); the mole fraction gas solubility ranges from 42.4×10^{-3} in $[C_1C_2Im][NTf_2]$ to 204×10^{-3} in $[C_1C_4Im][(C_1)_2PO_4]$ and $[C_1C_4Pyrr][(C_4HPO_3]$. For methyl acetylene, 17 imidazolium based ionic liquids were studied and the mole

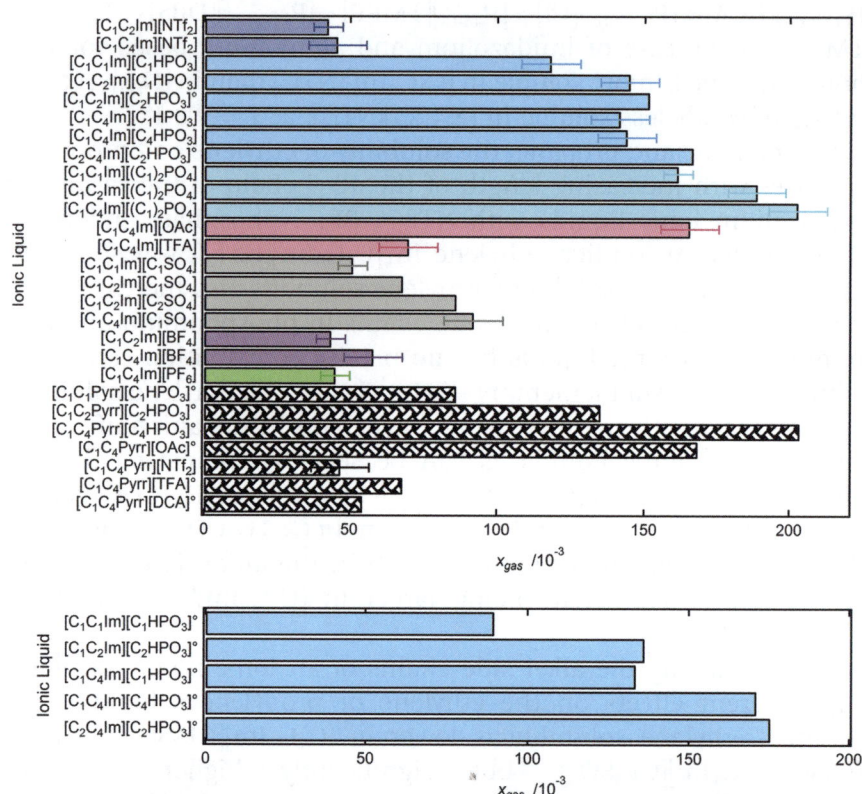

Figure 22.6   Mole fraction solubility for acetylene (top) and methyl acetylene (bottom) in different ionic liquids at 313 K. – Uncertainty not indicated by the authors. $[C_1C_2Im][NTf_2]$;[44] $[C_1C_4Im][NTf_2]$;[44] $[C_1C_1Im][C_1HPO_3]$;[38,44] $[C_1C_2Im][C_1HPO_3]$;[44] $[C_1C_2Im][C_2HPO_3]$;[38] $[C_1C_4Im][C_1HPO_3]$;[38,44] $[C_1C_4Im][C_4HPO_3]$;[9,38] $[C_2C_4Im][C_2HPO_3]$;[38] $[C_1C_1Im][(C_1)_2PO_4]$;[44] $[C_1C_2Im][(C_1)_2PO_4]$;[44] $[C_1C_4Im][(C_1)_2PO_4]$;[44] $[C_1C_4Im][OAc]$;[44] $[C_1C_4Im]$-$[TFA]$;[44] $[C_1C_1Im][C_1SO_4]$;[44] $[C_1C_1Im][C_1SO_4]$;[44] $[C_1C_1Im][C_1SO_4]$;[44] $[C_1C_1Im][C_1SO_4]$;[44] $[C_1C_2Im][BF_4]$;[44] $[C_1C_4Im][BF_4]$;[44] $[C_1C_4Im][PF_6]$;[44] $[C_1C_1Pyrr][C_1HPO_3]$;[40] $[C_1C_2Pyrr][C_2HPO_3]$;[40] $[C_1C_4Pyrr][C_4HPO_3]$;[40] $[C_1C_4Pyrr][OAc]$;[40] $[C_1C_4Pyrr][NTf_2]$;[44] $[C_1C_4Pyrr][TFA]$;[40] $[C_1C_4Pyrr]$-$[DCA]$.[40]

fraction gas solubility varies in this case from 58.5×10^{-3} in $[C_1C_1Im][C_1SO_4]$ to 156×10^{-3} in $[C_1C_4Im][(C_4)_2PO_4]$. Both gases, acetylene and methyl acetylene, are much more soluble in the ionic liquids studied than the four lighter gases reviewed above.

Contrary to what was found for ethane, ethylene, propane and propylene, the solubility of acetylene does not always increase when the side alkyl chain of the cation or the anion is increased. For example, almost no change in acetylene solubility is observed when comparing the solubility in $[C_1C_2Im][NTf_2]$ and $[C_1C_4Im][NTf_2]$, $[C_1C_2Im][C_1HPO_3]$, $[C_1C_2Im][C_2HPO_3]$, $[C_1C_2Im][C_1HPO_3]$ and $[C_1C_4Im][C_1HPO_3]$, $[C_1C_4Im][C_1HPO_3]$ and $[C_1C_4Im][C_4HPO_3]$, or $[C_1C_2Im][(C_1)_2PO_4]$ and $[C_1C_4Im][(C_1)_2PO_4]$. For pyrrolidinium based ionic liquids, a systematic increase in the solubility of acetylene with increasing length of the alkyl side chain of the cation or anion of the ionic liquid is observed.

For all the cations studied, the solubility of acetylene generally follows the order for the anions: $C_nC_mPO_4^- > OAc^- > C_nHPO_3^- > C_nSO_4^- > BF_4^- > TFA^- > PF_6^- = NTf_2^-$. Some exceptions are observed as, for example, no difference in acetylene solubility was found between $[C_1C_2Im][NTf_2]$ and $[C_1C_2Im][BF_4]$.

22.6 SELECTIVITY AND PERFORMANCE

The ideal selectivity of an ionic liquid for a particular gas in a mixture, α, can be defined as a ratio of Henry's law constants:

$$\alpha = \frac{K_H^1}{K_H^2} \tag{22.7}$$

where K_H^1 is the Henry's law constant of the lighter gas. The calculated ideal selectivity is plotted as a function of the absorption of the unsaturated gas for the pairs ethane–ethylene, ethylene–acetylene, propane–propylene and propylene–methyl acetylene in Figures 22.7 and 22.8, respectively.

In general, the highest ideal selectivities are found for the separation of unsaturated gases and the increase of the ionic liquid unsaturated gas absorption corresponds to a lower selectivity for this gas.

The ionic liquids, $[C_1C_1Im][C_1HPO_3]$ and $[C_1C_1Pyrr][C_1HPO_3]$, show the highest ethylene–acetylene ideal values of selectivity of 43 and 40, respectively. Ionic liquids containing NTf_2^-, DCA^-, TFA^-, BF_4^- or PF_6^- anions show both low acetylene absorption capacities and low ideal selectivity for the ethylene–acetylene separation.

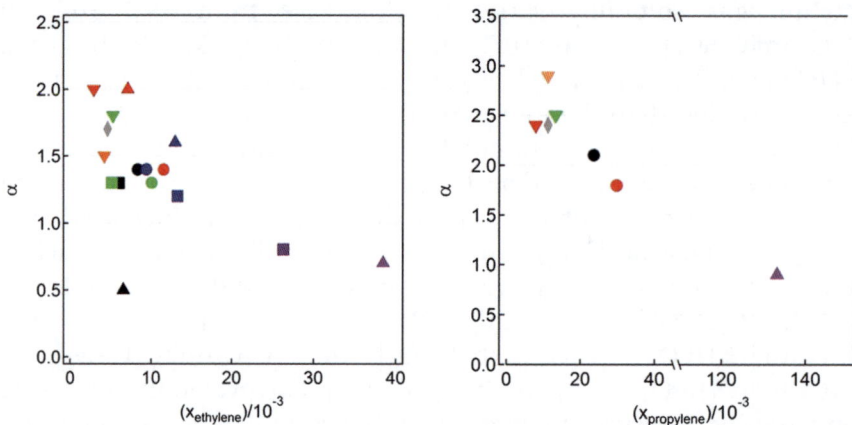

Figure 22.7   Ethane–ethylene ideal selectivity *versus* ethylene absorption (left-hand side) and propane/propylene ideal selectivity *versus* propylene absorption (right-hand side) in the ionic liquids: ●, [C₁C₂Im][NTf₂]; ●, [C₁C₄Im][NTf₂]; ●, [C₁(C₃H₅CH₂)Im][NTf₂]; ▲, [C₁C₆Im]-[NTf₂]*; ●, [C₁(CH₂C₆H₅)Im][NTf₂]; ■, [C₁C₃CNIm][NTf₂]*; ▲, [(C₃CN)₂Im][NTf₂]*; ▼, [C₁C₂Im][PF₆]; ■, [C₁C₄Im][PF₆]*; ▼, [C₁C₂Im][DCA]; ▼, [C₁C₄Im][BF₄]; ◆, [C₁C₂Im][CF₃SO₃]; ■, [P₄₄₄₄][TMPP]; ▲, [P(14)666][TMPP]; ▲, [C₁C₄Pyrr][NTf₂] and ■, [C₁C₆Pyr][NTf₂] at 313 K. * values at 303 K for the ethane–ethylene separation.

The maximum values for the ideal selectivities for the propylene–acetylene separation are half the values observed for ethylene–acetylene separation. As for ethylene–acetylene, the ideal selectivities for ionic liquids based on imidazolium cations having short alkyl chains and associated to phosphorous anions; the values for [C₁C₁Im][C₁HPO₃] and [C₁C₁Im][(C₁)₂PO₄] are 17 and 16, respectively.

With the lowest values for the ideal selectivity and absorption capacity, the ethane–ethylene separation seems to be the most difficult to perform with ionic liquids as far as these criteria are concerned. The ionic liquids [C₁C₂Im][DCA], [(C₃CN)₂Im][NTf₂] and [C₁C₂Im][PF₆] show the highest ideal selectivity for the ethane/ethylene separation; 2 for the first two and 1.8 for the third ionic liquid. The presence of a –CN group seems to have a positive effect in the ideal selectivity however this does lead to lower absorption capacities.

For all separations, an increase in the alkyl chain in the cation or in the anion leads to a decrease in the values of the ideal selectivity and to an increase of the absorption capacity of the ionic liquid. This increase seems to be less important in the case of ethane–ethylene separation.

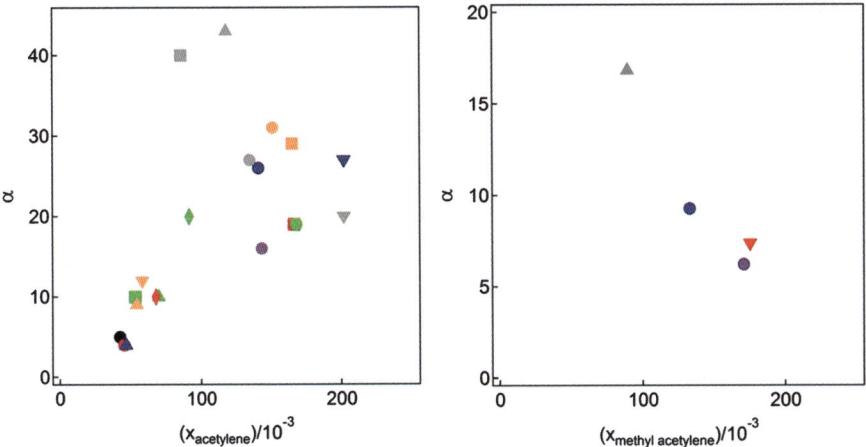

Figure 22.8 Ethylene–acetylene ideal selectivity *versus* acetylene absorption (left-hand side) and propylene–methyl acetylene ideal selectivity *versus* methyl acetylene absorption (right-hand side) in the ionic liquids. ●, $[C_1C_2Im][NTf_2]$; ●, $[C_1C_4Im][NTf_2]$; ■, $[C_1C_4Im][PF_6]$*; ◆, $[C_1C_2Im][BF_4]$ ▼, $[C_1C_4Im][BF_4]$; ▲, $[C_1C_4Im][TFA]$; ■, $[C_1C_4Im][OAc]$; ▲, $[C_1C_1Im][C_1HPO_3]$; ◆, $[C_1C_2Im]-[C_1HPO_3]$; ●, $[C_1C_2Im][C_2HPO_3]$; ●, $[C_1C_4Im][C_1HPO_3]$; ●, $[C_1C_4Im][C_4HPO_3]$; ▼, $[C_2C_4Im][C_2HPO_3]$; ▼, $[C_1C_1Im]-[(C_1)_2PO_4]$; ◆, $[C_1C_2Im][(C_1)_2PO_4]$; ▲, $[C_1C_2Im][(C_2)_2PO_4]$; ▼, $[C_1C_4Im][(C_1)_2PO_4]$; ■, $[C_1C_4Im][(C_4)_2PO_4]$; ◆, $[C_1C_1Im][C_1SO_4]$; ▼, $[C_1C_2Im][C_1SO_4]$; ●, $[C_1C_2Im][C_2SO_4]$; ◆, $[C_1C_4Im][C_1SO_4]$; ■, $[C_1C_1Pyrr][C_1HPO_3]$; ●, $[C_1C_2Pyrr][C_2HPO_3]$; ▼, $[C_1C_4Pyrr][C_4HPO_3]$; ●, $[C_1C_4Pyrr]-[OAc]$; ▲, $[C_1C_4Pyrr][NTf_2]$; ◆, $[C_1C_4Pyrr][TFA]$ and ▲, $[C_1C_4Pyrr][DCA]$ at 313 K. * values at 303 K for the ethylene–acetylene separation.

22.7 CONCLUSIONS

The solubility of ethane, ethylene, acetylene, propane, propylene and methyl acetylene in pure ionic liquids are presented and discussed. The potential of the use of different ionic liquids as separating agents for gaseous mixtures of saturated and unsaturated gases has been analysed on the basis of their ideal values of selectivity and absorption capacities.

The highest gas solubilities observed are for acetylene in $[C_1C_4Im][(C_1)_2PO_4]$ and $[C_1C_4Pyrr][C_4HPO_3]$, with a value of 0.20 mole fraction at $T=313$ K and 0.1 MPa. The maximum mole fraction solubilities reported for propane and propylene under the same conditions are 0.15 and 0.13 in $[P_{(14)666}][TMPP]$, respectively. For ethane and ethylene the maximum mole fraction solubilities are much lower in $[P_{(14)666}][TMPP]$ at 0.052 and 0.038, respectively.

When the cations forming the ionic liquids are considered, the solubility of ethane and propane follow the order: $P_{nmpq}^+ > C_nC_mIm^+ > C_nPyr^+ > C_nC_mPyrr^+ > N_{nmpq}^+$. For ethylene and propylene the solubility order is: $P_{nmpq}^+ > N_{nmpq}^+ > C_nPyr^+ > C_nC_mPyrr^+ > C_nC_mIm^+$. The solubility of acetylene has been reported only in imidazolium and in pyrrolidinium based ionic liquids and the solubility order is confirmed. For methyl acetylene, only imidazolium based ionic liquids have been studied.

Imidazolium based ionic liquids containing large anions such as large carboxylates, $BETI^-$, FAP^-, $C_8SO_4^-$, NTf_2^-, large phosphates and phosphites groups show higher ethane, ethylene, propane and propylene solubilities, followed by PF_6^-, $CF_3SO_3^-$, BF_4^- and DCA^- based ionic liquids. The order of solubility of acetylene and methyl acetylene relative to the anions in the ionic liquids is: $C_nC_mPO_4^- > OAc^- > C_nHPO_3^- > C_nSO_4^- > BF_4^- > TFA^- > PF_6^- = NTf_2^-$.

Ionic liquids present great potential for light hydrocarbon separation, especially in the case of ethylene–acetylene and propylene–methyl acetylene separations, with ideal selectivities of over 40 in the first case and over 15 in the second and corresponding mole fraction absorptions between 0.1 and 0.2.

REFERENCES

1. E. J. Henley, J. D. Seader and D. K. Roper, *Separation Process Principles*, John Wiley & Sons, Chichester, UK, 3rd edn, 2011.
2. M. S. Avgidou, S. P. Kaldis and G. P. Sakellaropoulos, *J. Membr. Sci.*, 2004, **233**, 21.
3. R. B. Eldrige, *Ind. Eng. Chem. Res.*, 1993, **32**, 2208.
4. R. W. Baker, *Ind. Eng. Chem. Res.*, 2002, **41**, 1393.
5. K. Keyvanloo, J. Towfighi, S. M. Sadrameli and A. Mohamadalizadeh, *J. Anal. Appl. Pyrolysis*, 2010, **87**, 224.
6. D. J. Safarik and R. B. Eldridge, *Ind. Eng. Chem. Res.*, 1998, **37**, 2571.
7. M. F. Asaro, Sorbents and processes for separation of olefins from paraffins, *US Pat.*, US2009 0143632A1, 2009.
8. Y. Hu, Z. Liu, C. Xu and X. Zhang, *Chem. Soc. Rev.*, 2011, **40**, 3802.
9. J. Palgunadi, H. S. Kim, J. M. Lee and S. Jung, *Chem. Eng. Process.*, 2010, **49**, 192.
10. M. Jin, Y. Hou, W. Wu, S. Ren, S. Tian, L. Xiao and Z. Lei, *J. Phys. Chem. B*, 2011, **115**, 6585.

11. M. B. Shiflett, A. M. S. Niehaus and A. Yokozeki, *J. Phys. Chem. B*, 2011, **115**, 3478.
12. D. Camper, C. Becker, C. Koval and R. Noble, *Ind. Eng. Chem. Res.*, 2005, **44**, 1928.
13. J. Jacquemin, M. F. Costa Gomes, P. Husson and V. Majer, *J. Chem. Thermodyn.*, 2006, **38**, 490.
14. K. Denbigh, *The Principles of Chemical Equilibrium*, Cambridge University Press, Cambridge, 4th edn, 1981.
15. J. M. Smith, H. C. Van Ness and M. M. Abbott, *Introduction to Chemical Engineering Thermodynamics*, McGraw Hill, New York, 5th edn, 1996.
16. J. M. Prausnitz, R. N. Lichtenthaler and E. Gomes de Azevedo, *Molecular Thermodynamics of Fluid-Phase Equilibria*, Prentice Hall, Hoboken, NJ, 3rd edn, 1999.
17. A. Ben-Naim, *Statistical Thermodynamics for Chemists and Biochemists*, Plenum Press, New York, 1992.
18. V. Majer, J. Sedlbauer and R. H. Wood, in *Aqueous Systems at Elevated Temperatures and Pressures: Physical Chemistry in Water, Steam and Hydrothermal Solutions*, ed. D. A. Palmer, R. Fernandez-Prini and A. H. Harvey, Elsevier, Amsterdam, 2004, ch. 4, pp. 99–147.
19. B. B. Benson and D. Krause Jr., *J. Solution Chem.*, 1989, **18**, 803.
20. D. Almantariotis, S. Stevanovic, O. Fandiño, A. S. Pensado, A. A. H. Padua, J.-Y. Coxam and M. F. Costa Gomes, *J. Phys. Chem. B*, 2012, **116**, 7728.
21. G. Hong, J. Jacquemin, M. Deetlefs, C. Hardacre, P. Husson and M. F. Costa Gomes, *Fluid Phase Equilibr.*, 2007, **257**, 27.
22. M. F. Costa Gomes, L. Pison, A. S. Pensado and A. A. H. Padua, *Faraday Discuss.*, 2012, **154**, 41.
23. B. Lee and S. L. Outcalt, *J. Chem. Eng. Data*, 2006, **51**, 892.
24. L. Moura, M. Mishra, V. Bernales, P. Fuentealba, A. A. H. Padua, C. C. Santini and M. F. Costa Gomes, *J. Phys. Chem. B*, 2013, **117**, 7416.
25. M. F. Costa Gomes, *J. Chem. Eng. Data*, 2007, **52**, 472.
26. H. Xing, X. Zhao, R. Li, Y. Yang, B. Su, Z. Bao, Y. Yang and Q. Ren, *ACS Sustainable Chem. Eng.*, 2013, **1**, 1357.
27. Y. Deng, S. Morrissey, N. Gathergood, A. Delort, P. Husson and M. F. Costa Gomes, *ChemSusChem*, 2010, **3**, 377.
28. D. Camper, C. Becker, C. Koval and R. Noble, *Ind. Eng. Chem. Res.*, 2005, **44**, 1928.

29. J. L. Anthony, E. J. Maginn and J. F. Brennecke, *J. Phys. Chem. B*, 2002, **106**, 7315.

30. J. L. Anderson, J. K. Dixon and J. F. Brennecke, *Acc. Chem. Res.*, 2007, **40**, 1208.

31. S. Stevanovic and M. F. Costa Gomes, *J. Chem. Thermodyn.*, 2013, **59**, 65.

32. X. Liu, W. Afzal and J. M. Prausnitz, *Ind. Eng. Chem. Res.*, 2013, **52**, 14975.

33. X. Liu, W. Afzal, G. Yu, M. He and J. M. Prausnitz, *J. Phys. Chem. B*, 2013, **117**, 10534.

34. P. K. Kilaru and P. Scovazzo, *Ind. Eng. Chem. Res.*, 2008, **47**, 910.

35. P. K. Kilaru, R. A. Condemarin and P. Scovazzo, *Ind. Eng. Chem. Res.*, 2008, **47**, 900.

36. Q. Zhang, Z. Li, J. Zhang, S. Zhang, L. Zhu, J. Yang, X. Zhang and Y. Deng, *J. Phys. Chem. B*, 2007, **111**, 2864.

37. J. Zhang, Q. Zhang, B. Qiao and Y. Deng, *J. Chem. Eng. Data*, 2007, **52**, 2277.

38. J. M. Lee, J. Palgunadi, J. H. Kim, S. Jung, Y. Choi, M. Cheong and H. S. Kim, *Phys. Chem. Chem. Phys.*, 2010, **12**, 1812.

39. J. L. Anthony, J. L. Anderson, E. J. Maginn and J. F. Brennecke, *J. Phys. Chem. B*, 2005, **109**, 6366.

40. S. Jung, J. Palgunadi, J. H. Kim, H. Lee, B. S. Ahn, M. Cheong and H. S. Kim, *J. Membr. Sci.*, 2010, **354**, 63.

41. R. Condemarin and P. Scovazzo, *Chem. Eng. J.*, 2009, **147**, 51.

42. L. Ferguson and P. Scovazzo, *Ind. Eng. Chem. Res.*, 2007, **46**, 1369.

43. D. Morgan, L. Ferguson and P. Scovazzo, *Ind. Eng. Chem. Res.*, 2005, **44**, 4815.

44. J. Palgunadi, S. Y. Hong, J. K. Lee, H. Lee, S. D. Lee, M. Cheong and H. S. Kim, *J. Phys. Chem. B.*, 2011, **115**, 1067.

The Application of Supercritical Carbon Dioxide Extraction of Functional Compounds

RAY MARRIOTT

Biocomposites Centre, Bangor University, Bangor, Gwynedd LL57 2UW, UK
Email: r.marriott@bangor.ac.uk

23.1 INTRODUCTION

The replacement of fossil fuel derived solvents with carbon dioxide (CO_2), either as a liquid or supercritical fluid, is perceived to be a greener and more sustainable alternative. Carbon dioxide is an ideal solvent for the extraction of functional molecules as excellent mass transfer rates result in rapid extraction and the solvent is easily removed, allowing products and residues to be recovered in a solvent-free state. Carbon dioxide is non-toxic, non-flammable, colourless, odourless and tasteless which makes it a good extraction solvent for natural products, especially when the extract or residual products are ultimately to be used in consumer products. Extraction of functional ingredients for use in food, beverages and personal care products has traditionally been carried out using organic solvents but changes in consumer preferences and legislation has reduced the number of permitted solvents and their maximum permitted residue levels. This has led to an increase in the use of liquid and supercritical carbon

Chemical Processes for a Sustainable Future
Edited by Trevor M. Letcher, Janet L. Scott and Darrell A. Patterson
© The Royal Society of Chemistry 2015
Published by the Royal Society of Chemistry, www.rsc.org

dioxide as an extraction solvent, being in some instances the only way to meet product specifications.

Carbon dioxide as an extraction solvent can be used in the liquid (sub-critical) or supercritical state. A supercritical fluid (SCF) is a substance above its critical temperature (T_c) and pressure (P_c), which for CO_2 is 304 K (31.8 °C) and 7.3 MPa (72.8 bar), making the supercritical phase easily accessible compared with other solvents.[1] Supercritical carbon dioxide has properties that are intermediate between gaseous and liquid carbon dioxide, exhibiting both high diffusivity and low viscosity with variable density and polarity achieved through changing pressure and temperature. Liquid and supercritical carbon dioxide are both weak, non-polar solvents most similar to short chain *n*-alkanes such as hexane. A comparison of physical properties with other commonly used extraction solvents is shown in Table 23.1 and it can be seen that for most parameters liquid and supercritical carbon dioxide compares most closely with *n*-hexane. The physical properties of liquid and supercritical carbon dioxide present both advantages and disadvantages for the extraction of functional molecules. The range of molecules that can be extracted is more limited than with traditional organic solvents because of its low polarity, but this can be mitigated by the use of more polar co-solvents.

To maintain this as a sustainable extraction process, ethanol and water or mixtures of both are most commonly used as co-solvents. However, the low polarity of the liquid and supercritical carbon

Table 23.1 Physical properties of carbon dioxide and common organic solvents.[2]

Solvent	Density/ g mL^{-1}	Viscosity/ Pa s	Flash point $t/$°C	Heat capacity/ kJ K^{-1} kg^{-1} at 25 °C	Richardson's polarity scale $E_T^{N(e)}$	Dielectric constant
Dichloromethane	1.326	4.06×10^{-4}	N/A	1.19	0.309	9.93
n-Hexane	0.655	2.95×10^{-4}	-23	2.27	0.009	2
Water	1	8.94×10^{-4}	N/A	4.18	1	80.4
Ethanol	0.789	10.74×10^{-4}	12	2.44	0.654	24.3
Ethyl acetate	0.894	4.31×10^{-4}	61	1.9	0.228	6.02
Supercritical CO_2	0.956^a	1.06×10^{-4c}	N/A	0.846^c	0.09 (var.)	1.1–1.5
Liquid CO_2	1.000^b	1.2×10^{-4e}	N/A	3.14^d	0.09 (var.)	1.5

^{a}40 °C, 40.0 MPa;
^{b}20 °C, 6.5 MPa;
^{c}40 °C;
^{d}10 °C;
^{e}25 °C. N/A, not applicable.

dioxide allows more selective extraction of less polar molecules, and when used to produce sequential extraction fractions at increasing pressure and temperature, can result in relatively pure extracts containing a narrow range of compounds.

Liquid and supercritical carbon dioxide extraction is not confined to small-scale extraction equipment and there are now over 150 commercial installations of various capacities worldwide using liquid and supercritical carbon dioxide to extract commodity products such as edible oils[3-5] as well as high-value products. Carbon dioxide extraction is an established technique for high-volume applications such as decaffeination of coffee and tea,[6,7] production of hop extracts for the brewing industry,[8,9] recovery of essential oils from herbs and spices[10,11] and nicotine extraction from tobacco.[12,13]

23.2 RENEWABLE AND SUSTAINABLE SOURCES OF CARBON DIOXIDE

Carbon dioxide is generated not only from anthropogenic activities but also as a by-product of a number of commercial processes including cement and fertiliser production, fermentation processes and, in some parts of the world, ground sources. These processes often produce relatively pure carbon dioxide and this reduces the downstream processing costs and hence represents the cheapest source of this extraction solvent. To be sustainable this should ideally be derived from renewable sources and the production of biofuels presents an opportunity to generate significant quantities of carbon dioxide; utilisation of this as a process solvent should be considered in preference to capture and storage in oceans or terrestrial sinks.[14] Carbon dioxide is an abundant by-product of bioethanol production and most plants already capture, purify and sell the carbon dioxide for a variety of applications as part of the overall economic mix. As the production of liquid biofuels such as ethanol increases (Figure 23.1), capturing and utilising the CO_2 generated as part of an integrated biorefinery will further enhance the benefit from using renewable feedstocks. In 2012, the bioethanol production volume reached 85 billion litres (85×10^9 L) which equates to approximately 64 million tonnes of high purity carbon dioxide. However, this equates to only 0.2% of annual global carbon dioxide emissions.

Carbon dioxide generated by fermentation is purified and dried in several stages, then compressed and cooled to produce liquid carbon dioxide which is then stored and transported in insulated bulk

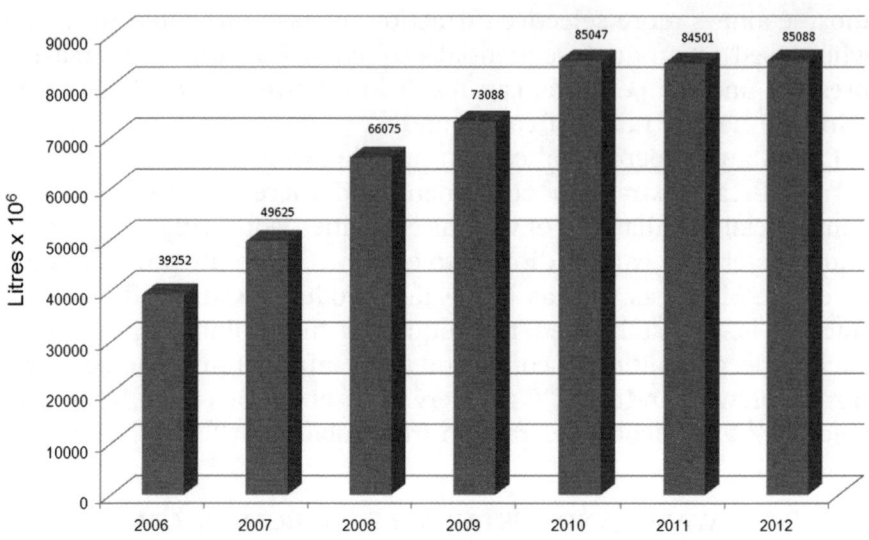

Figure 23.1 Annual production of fermentation ethanol 2006–2012.[15]

Table 23.2 Comparison of carbon dioxide and traditional bulk solvent costs (2013).

Solvent	Price
Hexane	£580–680 per tonne
Methanol	£185–220 per tonne
Ethanol	£680–720 per tonne
Acetone	£540–615 per tonne
Dichloromethane	£175–220 per tonne
CO_2 ex bioethanol plant (US)	$15–23 per tonne
CO_2 ex bioethanol plant (EU)	£15–25 per tonne
CO_2 ex bioethanol plant (EU)	£80–120 per tonne

containers held at approximately 2 MPa and −20 °C. The carbon dioxide captured as a by-product of bioethanol production is a very low cost solvent; Table 23.2 shows the bulk cost per tonne of carbon dioxide compared with traditional extraction solvents. Even with storage and distribution costs included it is considerably cheaper than dichloromethane and 20% of the cost of hexane, considered to be the most closely related solvent. At the present time, sufficient carbon dioxide is being produced from relatively pure process streams to be used as a low-cost extraction solvent, but if lower purity carbon dioxide sources needed to be used, then capture and purification costs would rise.

23.3 EXTRACTION USING LIQUID AND SUPERCRITICAL CARBON DIOXIDE

The advantages of using carbon dioxide as a benign and renewable solvent are to some extent counteracted by the capital investment required to utilise this solvent. The main cost is in the construction of equipment with a suitably high working pressure (typically 30–55 MPa), provision of compressors or pumps to deliver this pressure, and the heating and cooling capacity.

Using liquid carbon dioxide as the extraction solvent reduces costs somewhat as the operating pressure is lower. However, carbon dioxide needs to be delivered to high pressure pumps at not less than 5.0 MPa and reaches the critical point at 7.38 MPa, so operating pressures of (6.0–7.0) MPa are normal and a temperature range of (5–20) °C. Within this narrow band of processing conditions, the density of liquid carbon dioxide remains almost constant at 1.00 g cm^{-3} and exhibits properties similar to liquid hydrocarbon solvents.[16]

Liquid carbon dioxide is mostly used as an extraction solvent for botanical materials where the desired extract is composed of relatively small and non-polar molecules. It is therefore widely used to extract essential oils from herbs and spices including hops, which is the largest application.[17,18] In the 1980s, large-scale extraction plants were constructed in England, Australia and the United States,[19] predominantly for the extraction of hops using liquid carbon dioxide as an alternative to the methanol and hexane currently being used; a typical extraction scheme is shown in Figure 23.2.

One of the disadvantages of using liquid carbon dioxide as an extraction solvent is that the extract needs to be recovered by evaporation of the carbon dioxide and this necessitates the use of a recompression stage to recycle the carbon dioxide. This increases the energy required for the process, but in most extraction plants this is mitigated by using a heat pump which uses the heat of recompression to reduce the heat energy required in the evaporator stage.

The liquid carbon dioxide extraction plants built in the 1980s are still working today, but all plants built after that time were designed to operate with supercritical carbon dioxide so that a much greater range of temperature and pressure could be achieved and as a result a wider range of solvent polarity.

23.3.1 Supercritical Carbon Dioxide

The relatively low temperature and pressure required to reach the critical point together with the inert nature of the carbon dioxide

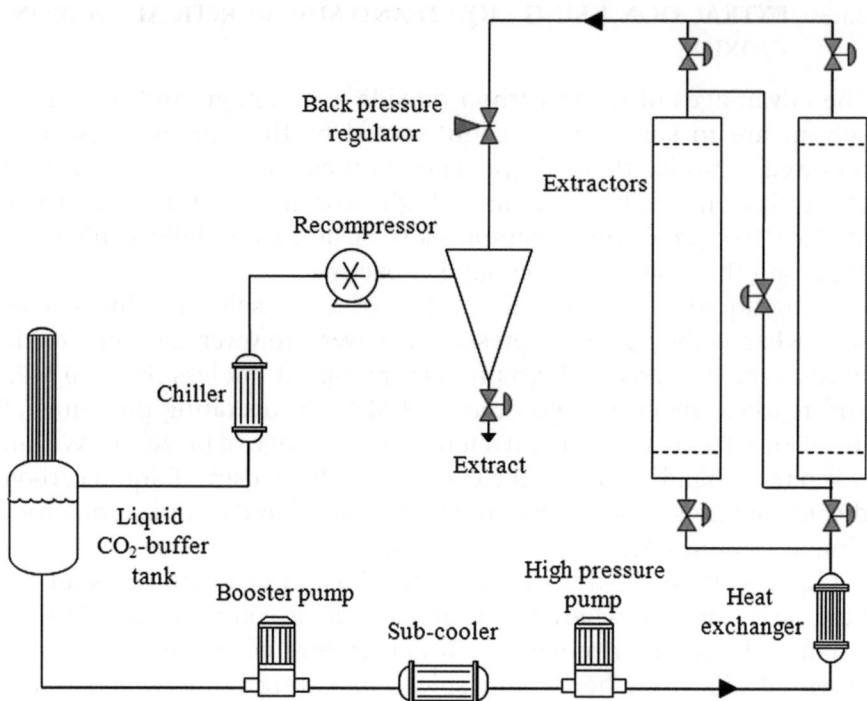

Figure 23.2   Typical extraction scheme using liquid (sub-critical) carbon dioxide.

molecule has led to a much wider variety of applications that use supercritical carbon dioxide compared with liquid carbon dioxide. Beyond the critical point, the density of the supercritical carbon dioxide rises rapidly and can be varied by changing the pressure and temperature. Solubility of molecules in supercritical carbon dioxide is closely related to solvent density. Many models calculate solubility— the most well-known is the Chrastil equation[20]—and in general solubility increases as pressure and temperature rise once past the crossover point. The solvent polarity also increases as density rises being close to *n*-pentane at low density and more similar to pyridine at high density.

The first large-scale applications were again the extraction of botanical material and the largest application is the decaffeination of coffee,[21] with one processor decaffeinating 60 000 t a^{-1} (60 000 metric tonnes per year). Large-scale extraction using supercritical carbon dioxide is carried out at pressures up to 55 MPa and differs from extraction using sub-critical carbon dioxide in that multiple separators are used at sequentially lower pressure to recover and

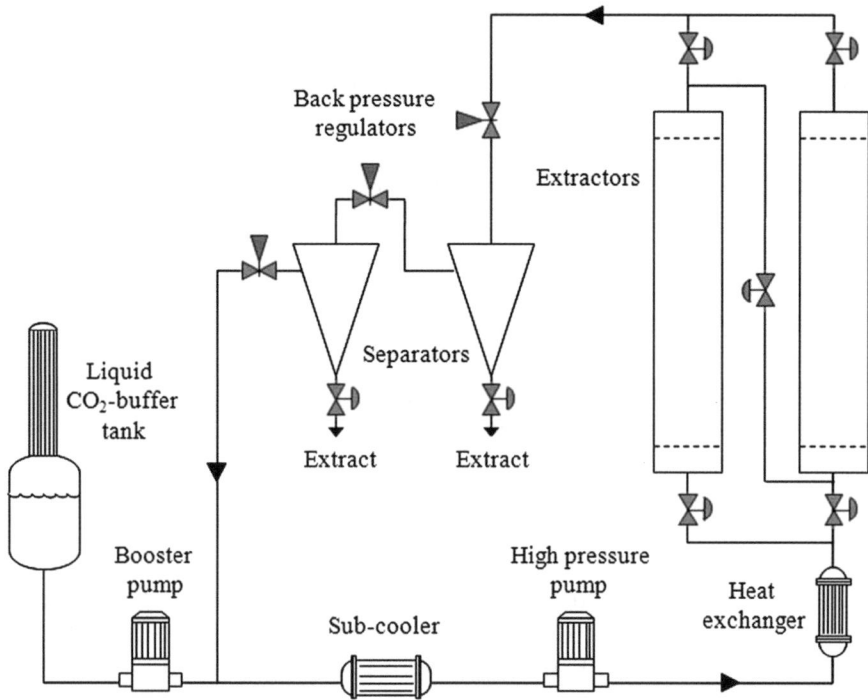

Figure 23.3 Typical extraction scheme using supercritical carbon dioxide.

partially fractionate the extract. The final separator is normally maintained above 5 MPa so that the carbon dioxide can be cooled and recycled into the pump without the need for a recompression stage as shown in Figure 23.3.

Since the solubility of the extracted molecules in the supercritical carbon dioxide is crucial to the rate of extraction and therefore the economics of the process, recent trends have been towards even higher pressure equipment.[22–24] Extraction of lipids such as seed oils have been carried out at pressures in excess of 150 MPa and the solubility of almond oil for example increases from 12.5 g L^{-1} at 30 MPa to 50 g L^{-1} at 150 MPa.[25] Small-scale laboratory extraction equipment capable of operating to 280 MPa already exists and commercial extraction at 110 MPa has been installed.[26] Greater solubility results in significantly shorter extraction times and therefore the energy used in the process, mostly for heating and cooling the supercritical carbon dioxide, is minimised.

Extraction of molecules that have very limited solubility in super-critical carbon dioxide can be further enhanced by the addition of entrainers. These are co-solvents added as a percentage of the supercritical carbon dioxide flow prior to the extractor to increase the solvating power of the extraction solvent. Ethanol is commonly used as a co-solvent and this can often be obtained alongside carbon dioxide within a bioethanol facility or more advanced biorefinery.

In addition to the temperature and pressure of supercritical carbon dioxide, extraction efficiency is also directly influenced by other process parameters.

1. Extraction time – equates to total mass of CO_2 per mass of extract
2. Extractor design
3. Raw material pre-treatment
   a. Particle size
   b. Particle porosity
   c. Bulk density

Extraction time is largely determined at the design stage of any equipment and the circulation pumps sized to meet a pre-determined batch extraction time. Extractor design is also optimised at the same time. Almost all carbon dioxide extraction plants are now individually designed to deliver optimum performance for their intended application with the lowest possible energy consumption and are built to accommodate the anticipated process volumes.

23.4   EXTRACTION OF OLEOCHEMICALS

Extraction using supercritical carbon dioxide ($scCO_2$) is ideally suited to hydrophobic molecules such as waxes, lipids and fatty acids. The potential to change from petrochemical solvents such as hexane to supercritical carbon dioxide is restricted not by technical barriers but by current capital equipment costs and the established use of depreciated solvent process plant. The greatest potential is in the extraction of vegetable oils, which in addition to being a large volume food ingredient, provide fuel (biodiesel), polymers, surfactants and many other smaller volume applications.

In the United States alone, most of the 8×10^6 t of oil obtained each year from seeds and grain commodities are extracted using hexane. Hexane had been recognised as a hazardous air pollutant by the US Environmental Protection Agency (US EPA) and it has been reported by the EPA Toxic Release Inventory (TRI) that more than 20 000 t of hexane are released to the atmosphere each year from the extraction of vegetable oils.[27] The US EPA also identified hexane as a potential carcinogen in the national emission standards for hazardous air pollutants.[28]

Although extraction with hexane has a lower unit cost than extraction with supercritical carbon dioxide, production costs increase as extra refining steps and energy input are needed to reduce solvent residues to meet legally enforced maximum residue levels and to recover the solvent. Supercritical carbon dioxide is often compared to hexane due to their similar polarity[29] and significant research has been carried out to demonstrate the extraction of lipids using supercritical carbon dioxide instead of hexane. Extraction yields and properties of supercritical carbon dioxide extracted oil (odour and colour) are comparable with or better than hexane.[30,31] Extraction of soybean, corn, wheat germ, sunflower seeds, safflower seeds and peanuts have been successfully demonstrated.[32]

Vegetable oils can also be obtained by cold pressing and the quality of cold pressed oil is considered to be higher than oil produced by solvent extraction.[33,34] Oil recovery from cold pressing is lower than hexane extraction and a second extraction of the pressed meal is often carried out using hexane as the solvent. This step can also be replaced by using supercritical carbon dioxide as the extraction solvent.

Waxes are produced in large volume for use in cosmetics, lubricants, polishes, surface coatings, inks and many other applications. Many of these are paraffin waxes of mineral origin but four in particular are derived from plant sources: beeswax, carnauba wax, candelilla wax and jojoba wax. In addition there are also other plant waxes that are commercially produced in smaller quantities such as citrus wax and apple wax; these are mostly used in cosmetic or personal care applications.[35]

Plant leaf and some stem surfaces are coated with a thin layer of waxy material that has many physiological functions. This layer is microcrystalline in structure and forms the outer boundary of the cuticle membrane; it is the interface between the plant and the atmosphere and provides protection from disease and insects and helps the plants resist drought. These plant waxes are complex mixtures of hydrocarbons, esters, ketones and alcohols,[36] and Table 23.3 illustrates some of the principal components.

The relative amount of each group and the chain lengths of the various molecular species vary greatly according to the botanical source. One of the most sustainable sources of plant waxes is straw and the leaf by-products of commercial crops. The composition of

Table 23.3 Major constituents of plant leaf waxes.[36]

Group	Generic formulae	No. of carbons	Odd or even
Hydrocarbons	$CH_3(CH_2)_nCH_3$	21–35	Odd
Wax esters	$CH_3(CH_2)_xCOO(CH_2)_yCH_3$	34–62	Even
Fatty acids	$CH_3(CH_2)_nCOOH$	16–32	Even
Primary alcohols	$CH_3(CH_2)_nCH_2OH$	22–32	Even
Aldehydes	$CH_3(CH_2)_nCHO$	22–32	Even
Ketones	$CH_3(CH_2)_xCO(CH_2)_yCH_3$	23–33	Odd
Secondary alcohols	$CH_3(CH_2)_xCHOH(CH_2)_yCH_3$	23–33	Odd
β-diketones	$CH_3(CH_2)_xCOCH_2CO(CH_2)_yCH_3$	27–33	Odd

straw waxes has been extensively studied and most work has been carried on the *Triticeae* and in particular wheat (*Triticum aestivum*), barley (*Hordeum vulgare*) and rye (*Secale cereale*). In most studies the level of total 'waxes' varied from (1 to 3)% depending on the variety, straw parts used (leaves, node or internode), crop year and method of extraction. The composition of 'waxes' from wheat straw has been extensively studied[37–39] and the wax fraction from wheat straw has also been extracted using supercritical carbon dioxide,[40] leading to a highly selective extraction of the lipid and wax fraction with minimal content of other compounds such as pigments. The composition of other straw waxes has also been reported including barley,[41] rye,[42] flax[43,44] and oil rape seed.[45,46]

Miscanthus, reed canary grass, hemp and giant reed (*Arundodonax*) are all being developed as potential energy crops. The lipophilic extractives of these have all been studied[47–52] and show a mixture of alkanes, fatty alcohols, fatty acids, aldehydes and sterols that are common to all four but vary in relative concentration according to species.

Using supercritical carbon dioxide as the extraction solvent, waxes can be produced with no organic solvent residues and furthermore can be certified as organic in many countries.[53] These waxes can be used as replacements for a wide range of existing products, such as cosmetics, as their physical properties can be varied by selecting the temperature and pressure of the supercritical carbon dioxide used in the extraction.[40] Importantly the microcrystalline structure of the waxes from straw is a key physical property for many cosmetic applications.

23.5 EXTRACTION OF TERPENOIDS

Terpenoids represent a diverse range of bio-active, secondary metabolites from monoterpenes found mostly in essential oils to tetraterpenoids having a C_{40} skeleton mostly represented by the carotenoids. Many of these molecules are non-polar hydrocarbons that are readily soluble in liquid or supercritical carbon dioxide, but others such as the sterols and xanthophylls contain oxygen in various functional groups and require either higher pressure supercritical carbon dioxide or the use of co-solvents such as ethanol to extract them.

Mono and sesquiterpenes are found mostly in essential oils. The use of liquid or supercritical carbon dioxide to extract these from a wide variety of aromatic plants was one of the earliest commercial uses[54] and is still widely used to produce solvent free extracts for food, cosmetic and personal care use. Of these the extraction of hops for use in the brewing industry is the largest application, with over 60% of the annual crop being processed using liquid or supercritical carbon dioxide. The extraction of hops not only concentrates the desirable essential oils and α-acids (humulones), but also delivers an extract that is stable for many years[55] compared with just a few months from dried hops. Compared with extraction using traditional solvents, the use of liquid carbon dioxide results in a lower overall yield though with a significantly higher concentration of essential oils. The α and β acids without the presence of solvent residues and tannins can cause chill haze in the final product (Table 23.4).

The greater range of pressure and temperature that are available when using supercritical carbon dioxide has been used to produce spice extracts from pepper and ginger, for example, which have variable levels of 'heat' and 'aroma' by using extraction at high pressure and sequentially reducing the pressure through a series of

Table 23.4 Comparison of solvent and liquid carbon dioxide extraction of hops.

	Extraction solvent		
	dichloromethane	ethanol	Liquid CO_2
α-acids	35–45%	30–40%	40–50%
β-acids	15–20%	10–15%	18–40%
Volatile oil	1–3%	1–2%	2–10%
Other soft resins	3–8%	3–8%	5–20%
Hard resins	2–5%	2–10%	None
Fats and Waxes	1–2%	Traces	0–5%
Tannins	Traces	1–5%	None
Chlorophyll	<1%	Traces	None
Inorganic salts	<1%	0.5–1%	Traces
Residual solvent	<1%	0.01–0.1%	None
Water	Traces	1–5%	1–5%
Typical yield	28%	38%	19%

separators to obtain fractions either rich in piperine or the essential oils in the case of pepper.[56] Using these techniques it was possible to obtain fractions containing over 60% piperine or less than 1%. Extensive reviews of essential oil applications using liquid or super-critical carbon dioxide have been published in recent years.[57–59]

Diterpenes represent a group of molecules that have proven anti-oxidant properties and extraction from coffee,[60] rosemary[61] and sage[62,63] have been particularly studied. Of these the phenolic diter-penes from rosemary have found the greatest application being used as oil soluble antioxidants in a wide range of food products and the use of supercritical carbon dioxide to extract these antioxidants has enabled higher purity extracts containing predominantly carnosol and carnosic acid (see Figure 23.4) to be produced without high levels of essential oils being present that contribute unwanted flavour.[64,65]

Sage contains similar phenolic diterpene antioxidant molecules but also high levels of manool and *trans*-ferruginol; yields were shown to be equivalent to solvent extraction.[62] Manool is an important precursor for the synthesis of spiroketals, which are used in fragrance formulations. The extraction of the diterpenes from coffee is signifi-cant as the cholesterol-raising effect of boiled coffee in humans has been linked to the diterpenes, cafestol and kahweol (see Figure 23.4).[66]

Sterols are examples of triterpenoids based on the cyclopenta[*a*]-phenanthrene carbon skeleton that are ubiquitous in plants. The most commonly occurring phytosterols are β-sitosterol, campesterol

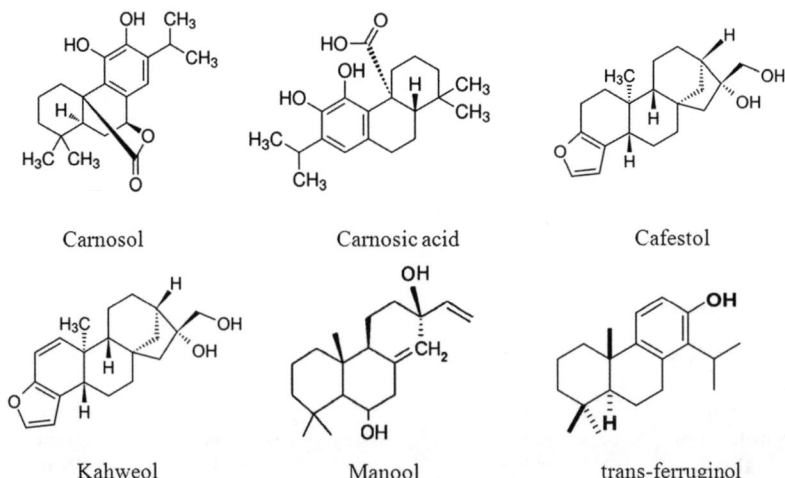

Figure 23.4   Diterpenes from rosemary, sage and coffee.

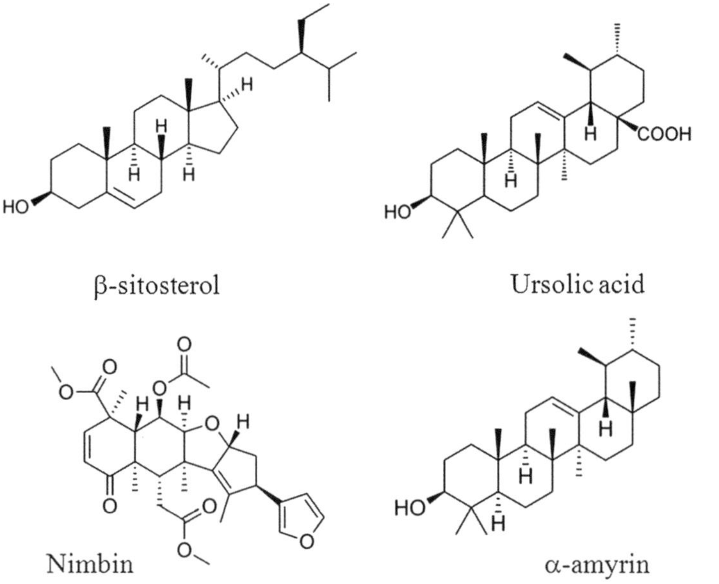

Figure 23.5 Pentacyclic and cyclopenta[*a*]phenanthrene triterpene structures.

and stigmasterol. In addition to being found in the unsaponifiable fraction of many vegetable oils,[67] these can be extracted from other sustainable non-food sources such as tree bark, leaves and wood,[68] and stems including sunflower[69] and cereal straw.[40] Sterols have been selectively extracted using supercritical carbon dioxide from a number of botanical sources[70–72] and are widely used either as free sterols or sterol esters in cholesterol lowering formulations and food products. Other groups of triterpenes such as the pentacyclic-triterpenes (see Figure 23.5) based on lupane, oleanane or ursane structures have also shown multi-target properties such as wound healing, anti-inflammatory and anti-bacterial combined with a low toxicity. These molecules are derived from a wide selection of plants[73] and have been successfully extracted using supercritical carbon dioxide without the use of co-solvents.[74]

The triterpenoids nimbin and salanin have also been extracted from neem seed,[75] and by varying the pressure and temperature, the yield and separation from azadirachtin could be optimised. Tocopherols and tocotrienols are polyterpenes often found alongside triterpenes and are potent antioxidants that have also been isolated and purified using supercritical carbon dioxide,[76,77] mostly from vegetable oil sources. Sources of natural antioxidants and the use of clean extraction techniques are becoming more important as other

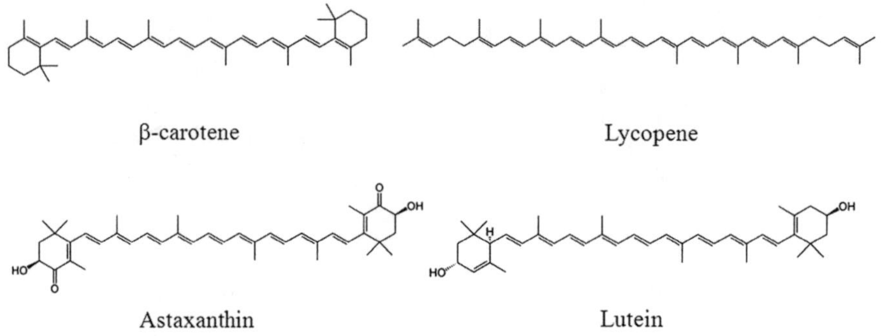

Figure 23.6   Carotenoids extracted using supercritical carbon dioxide.

synthetic antioxidants such as butylated hydroxyanisole (BHA) and butylated hydroxytoluene (BHT) are becoming restricted in their use.

The tetraterpenoids are mainly represented by the carotenoids, and higher pressure supercritical carbon dioxide or in combination with co-solvents is used to extract both the non-polar carotenes and the more polar xanthophylls such as astaxanthin and lutein (see Figure 23.6). Carotenoids are important because of their activity on cardiovascular disease and as precursors of vitamin A. Carotenoids are responsible for the colour of many foods, being found as pigments in a variety fruits, vegetables and seafood, with colours that vary from yellow to purple to red.

Extraction of the non-polar carotenoids β-carotene and lycopene can be achieved from both plant and algae sources without co-solvents[78,79] but pressures of 40–50 MPa are required. Extraction using ethanol or oils such as canola oil as co-solvents has also been demonstrated[80] and can be accomplished using lower pressures since the solvent polarity has been increased by the presence of the co-solvent. Polar groups on the terminal rings of the carotenoid structures significantly decrease the solubility of the xanthophylls in supercritical carbon dioxide. Extraction of lutein and lutein esters from carrot has been carried out using canola oil as co-solvent[81] and from marigolds using a variety of vegetable oils.[82] Astaxanthin is found in seafood, krill and microalgae such as *Haematocccus pluvialis* and is present in nature mostly as a diester, which enhances its solubility in supercritical carbon dioxide.[83] Commercial extraction is carried out mostly using co-solvents as the working pressure of most extraction equipment is insufficient to achieve an efficient extraction with supercritical carbon dioxide alone. However, recent develop-ments in plant design has resulted in medium size extraction plants working at pressures above 100 MPa,[22] and under these conditions, astaxanthin is soluble without the use of entrainers.

23.6   EXTRACTION OF ALKALOIDS

Alkaloids have important functional roles in medicines, food ingredients and personal care products. The purine alkaloids, caffeine, theobromine and theophylline (see Figure 23.7), are the most widely studied and also represent one of the largest applications of supercritical carbon dioxide extraction—the decaffeination of tea and coffee.

Historically, decaffeination was carried out using dichloromethane but this has now been completely replaced by the use of supercritical carbon dioxide or water. The process, first discovered in the 1970s,[84] is carried out on green coffee beans so that the flavour is not lost in the extraction process but up to 99% of the caffeine can be selectively removed. The extraction of caffeine from the green beans is not possible with dry carbon dioxide because the caffeine is bound into a complex with chlorogenic acid, so the process[85] involves pre-hydrating the green beans before extraction of the caffeine with supercritical carbon dioxide. The caffeine-rich stream is then passed through a counter-current scrubber unit which removes the caffeine and the supercritical carbon dioxide is recycled through the extractor as shown in Figure 23.8. The dilute aqueous caffeine stream is then concentrated using reverse osmosis or evaporation to recover the caffeine, which is then used in a wide range of products.

More recently, methods have been developed that can simultaneously extract caffeine and chlorogenic acid from green coffee beans[86] and the increased use of caffeine in non-beverage applications has resulted in methods for extracting caffeine from coffee husks.[87] The removal of theobromine from cocoa has potential application as a component in pharmaceutical products, as well as the

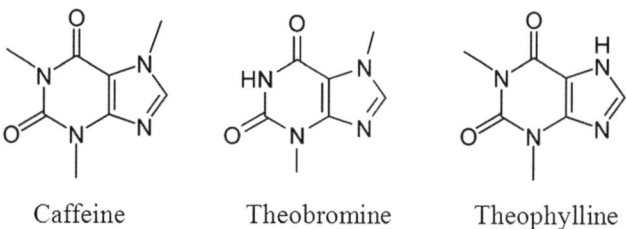

Caffeine   Theobromine   Theophylline

Figure 23.7   Purine alkaloids found in tea and coffee.

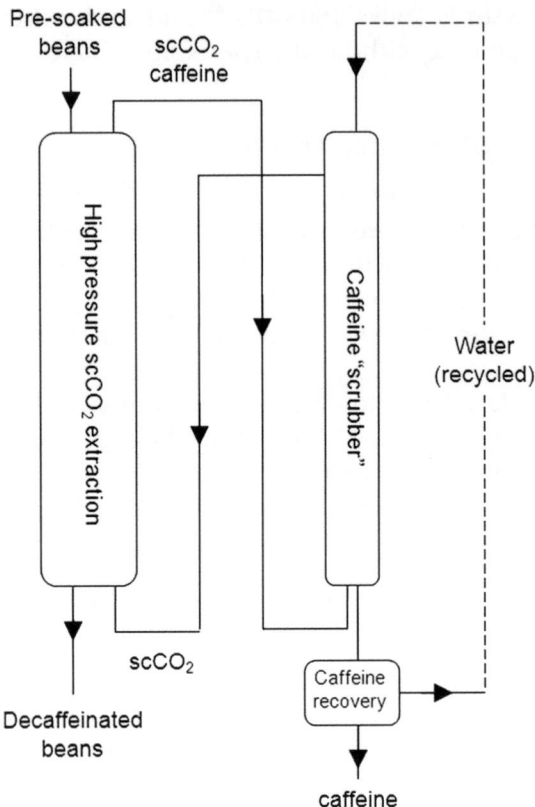

Pre-soaked beans

scCO₂ caffeine

High pressure scCO₂ extraction

Caffeine "scrubber"

Water (recycled)

scCO₂

Decaffeinated beans

Caffeine recovery

caffeine

Figure 23.8   Typical decaffeination scheme with caffeine recovery.

production of higher value methylxanthine-free products for human consumption. Extraction trials have shown that both caffeine and theobromine can be removed[88] without any loss of the more valuable polyphenols.

Extraction or partial removal of nicotine from tobacco has obvious health benefits and its use in alternative formulations for reducing tobacco dependence and its use as an insecticide creates a market for this molecule. Nicotine makes up (0.6–3)% of the dry weight of tobacco and its extraction with supercritical carbon dioxide was also first described in the 1970s.[89] Patents for nicotine reduction processes followed over the next 20 years.[90,91] Nicotine is also too polar to extract with supercritical carbon dioxide alone and hydration of the tobacco or the use of methanol or ethanol as a co-solvent is required.

Many other groups of alkaloids, mainly from plant sources, have also been investigated and these were reviewed by Kim and colleagues in 2001.[92] Some of these have commercial application as

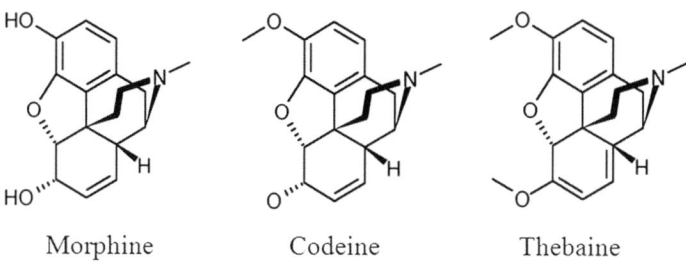

Morphine   Codeine   Thebaine

Figure 23.9   Morphine alkaloids found in poppy straw.

pharmaceutical molecules, the most notable being the morphine alkaloids, morphine, thebaine and codeine (see Figure 23.9), which are widely used analgesics and commercially extracted from poppy straw.

Early studies[93] showed that efficient extraction of these alkaloids needed the addition of a polar entrainer such as methanol, water or a combination of both, and the addition of basifying additives such as methylamine caused degradation of the alkaloids after extraction. Quinolizidine alkaloids from *Sophora flavescens* root have been described[94] with subsequent purification of individual alkaloids by high-speed counter-current chromatography and the extraction of colchicine and related alkaloids[95] from the seeds of *Colchicum autumnale* L. Supercritical fluid extraction of the terpenoid indole alkaloids vindoline, catharanthine and vinblastine from *Catharanthus roseus,*[96] which are used to treat a wide variety of neoplasms, has also been investigated and compared with other extraction techniques using experimental design methodology.

23.7   EXTRACTION OF METALS

Recovery of metals, particularly those that are becoming scarce, provides another area where the use of supercritical carbon dioxide extraction can be used to deliver a sustainable solution. Rare earth metals and some transition metals such as gold and indium in particular are used increasingly in electronic devices and the recovery of these is currently at a very low level.

Solvent extraction is already a well-established process within the hydrometallurgical industries for recovery of metals from aqueous solutions. This is accomplished by forming organic-soluble neutral complexes in the aqueous phase by reactions between the ionic species of interest and an appropriate organic compound so that the complex is soluble in a suitable solvent, normally a hydrocarbon.[97]

Three major types of complexing agents are used: cation exchangers, anion exchangers and solvating extractants. The fact that these are designed to transfer metal complexes to non-polar hydrocarbon solvents is itself an indication that this can be achieved using supercritical carbon dioxide. Supercritical carbon dioxide offers a number of advantages for metal extraction.

- Residual organic solvents can be eliminated.
- Low viscosity and high solute diffusivities may enhance rates of extraction.
- Supercritical carbon dioxide has a much lower surface tension than hydrocarbon solvents, which results in an increased dispersed phase surface area reducing the size of equipment required for a particular solvent-to-feed ratio.
- Solvency characteristics of supercritical carbon dioxide can be varied with small changes in temperature and pressure, and could be used to develop selective extraction processes.
- In some applications such as uranium extraction, the solvents are degraded by radiolysis and hydrolysis and using supercritical carbon dioxide may be beneficial due to its inert nature.

However, the solubility of the metal complexes is generally lower in supercritical carbon dioxide and this may limit the application to high value metals. Cation exchangers are mostly quite soluble in supercritical carbon dioxide and therefore have the greatest potential as commercial chelating agents. Initial work using cation exchangers was carried out with copper[98] and iron[99] solutions, but Lin and Wai[100] showed that trivalent lanthanides, lanthanum, europium and lutetium could all be recovered from aqueous solutions using thenoyltrifluoroacetone (TTFA) as the chelating agent with extraction efficiencies of up to 51%. Extraction of the actinides uranium and thorium has also been demonstrated using the same chelating agent[101] with 38% and 70% efficiency, respectively.

Trivalent lanthanides can also be extracted using solvating extractants; Dehgani *et al.*[102] showed that supercritical carbon dioxide modified with tributyl phosphate (TBP) could extract lanthanide nitrates. Lanthanum, cerium, praseodymium, neodymium, gadolinium, holmium, erbium and ytterbium were all extracted from an aqueous solution at $T = 313$ K and 15 MPa, with the amount extracted increasing with increased atomic number with a maximum at gadolinium. This trend had previously been observed when using hydrocarbon solvents. A later study[103] included samarium and dysprosium and established that extraction efficiencies increased from 43.6% for lanthanum to 91.8% for dysprosium before declining to 69.2% for lutetium. This work was carried out at 333 K and 30 MPa, and used a higher concentration of TBP in the supercritical carbon dioxide phase. Overall the level of extraction in the supercritical carbon dioxide based systems was found to be lower for light lanthanides and comparable for heavy lanthanides to the level of extraction in hexane at comparable fluid phase composition. The separation factors obtained with TBP–di(2-ethylhexyl)phosphoric acid (DEHPA) modifiers were generally higher in supercritical carbon dioxide than in hexane. Some of the reported metal concentrations in the fluid phase were unusually high and this may indicate that TBP and DEHPA are behaving as strong entrainers; this will assist the development of supercritical carbon dioxide based processes on an industrial scale.

Extraction of lanthanides and actinides from solid matrices is more problematic but the use of 2,2-dimethyl-6,6,7,7,8,8,8-heptafluoro-3,5-octanedione (HFOD) as a chelating agent[104] was shown to provide efficient extraction when methanol was used as a co-solvent in the extraction, and this was further enhanced if the solid matrix was wet. Extraction at 333 K and 15 MPa resulted in almost quantitative extraction efficiencies, and it was determined that the solubility of $La(FOD)_3$ and $Eu(FOD)_3$ complexes in pure 2,2-dimethyl-6,6,7,7,8,8,8-heptafluoro-3,5-octanedione were 5.5×10^{-2} mol L^{-1} and 7.9×10^{-2} mol L^{-1}, respectively. These values are considerably higher than those reported using other chelating agents.

Within the transition metals, gold and indium are of particular interest since these are widely used in consumer electronics, and recycling and recovery of these metals is at a very low level. Gold in its normal oxidation state, Au(I) and Au(III), forms the most stable complexes with ligands that have nitrogen or sulfur donor atoms. Gold also readily reacts with halogens and so solutions of gold chloride have generally been used to investigate extraction using supercritical carbon dioxide. Extraction of Au(III) has been demonstrated using bistriazolo-crowns in supercritical carbon dioxide with the addition of 5% methanol as co-solvent; when small quantities of water were present, up to 80% of the gold could be recovered.[105] Extraction of Au(III) using supercritical carbon dioxide without co-solvents has also been achieved using fluorinated calix[4]arenes[106] at a temperature of 333 K and pressures between 20 and 40 MPa with extraction efficiencies up to 75% under optimum conditions.

Indium is used extensively in the production of liquid crystal displays, semiconductors, infrared photodetectors and low temperature solders. More than half the indium used is for indium tin oxide (ITO) coatings, which are used in the manufacture of thin-film transistor liquid crystal displays (TFT-LCDs).[107] As well as the obvious need to recycle the indium, it is also discharged in etching wastewater and recovery from this source has been investigated not only to recover the metal but also to eliminate health risks.[108] Studies using industrial wastewater streams from the ITO process have shown that indium metal could be successfully extracted using supercritical carbon dioxide with fluorinated β-diketone chelating agents.[109] Extraction was investigated using three different chelating agents, acetyl acetone (ACAC), TTFA and HFOD. HFOD provided over 90% extraction efficiency under optimum conditions, which were found to be 353 K and 20.7 MPa. ACAC and TTFA exhibited significantly lower efficiency, suggesting that ACAC and TFFA have decreased abilities to form complexes with indium.

23.8 CONCLUSIONS

The use of sub-critical and supercritical carbon dioxide to extract and purify functional molecules for a wide range of applications has been the subject of extensive research over the past 40 years and many commercial processes have been developed that are now considered to be the best available extraction option. Carbon dioxide is itself a sustainable and low-cost solvent that is widely available and considerable advances have been made in the design of equipment to operate at pressures that can extract a diverse range of molecules. This review has shown that it is often the case that sub-critical and supercritical carbon dioxide can selectively extract the target molecules from complex raw materials often avoiding lengthy and costly purification steps. Many of these raw materials are themselves sustainable plant products or waste streams leading to integrated, sustainable and economic manufacturing with carbon dioxide providing the ideal process solvent.

REFERENCES

1. F. M. Kerton and R. Marriott, in *Alternative Solvents for Green Chemistry*, RSC Publishing, Cambridge, UK, 2nd edn, 2013, pp. 115–132.
2. J. H. Clark and S. J. Tavener, *Org. Process Res. Dev.*, 2007, **11**, 149–155.
3. J. P. Friedrich and G. R. List, *J. Agric. Food Chem.*, 1982, **30**, 192–193.

4. J. P. Friedrich, G. R. List and A. J. Heakin, *J. Am. Oil Chem. Soc.*, 1982, **59**, 288–292.
5. N. A. N. Norulaini, W. B. Setianto, I. S. M. Zaidul, A. H. Nawi, C. Y. M. Azizi and A. K. M. Omar, *Food Chem.*, 2009, **116**, 193–197.
6. K. Ramalakshmi and B. Raghavan, *Crit. Rev. Food. Sci.*, 1999, **39**, 441–456.
7. C. J. Chang, K.-L. Chiu, Y.-L. Chen and C.-Y. Chang, *Food Chem.*, 2000, **68**, 109–113.
8. Z. Zekovic, I. Pfaf-Sovljanski and O. Grujic, *J. Serb. Chem. Soc.*, 2007, **72**, 81–87.
9. R. Vollbrecht, *Chem. Ind.*, 1982, **19**, 397–399.
10. G. Leeke, F. Gaspar and R. Santos, *Ind. Eng. Chem. Res.*, 2002, **41**, 2033–2039.
11. D. Ehlers, E. Czech, K. W. Quirin and R. Weber, *Phytochem. Analysis*, 2006, **17**, 114–120.
12. J. Rincó, A. D. Lucas, M. A. García, A. García, A. Alvarez and A. Carnicer, *Sep. Sci. Technol.*, 1998, **33**, 411–423.
13. W. Huang, Z. S. Li, H. Niu, J. W. Wang and Y. Qin, *Biochem. Eng. J.*, 2008, **42**, 92–96.
14. K. Caldeira and M. Akai, *IPCC Special Report on Carbon Dioxide Capture and Storage*, ed. B. Metz, O. Davidson, H. de Coninck, M. Loos and L. Meyer, Cambridge University Press, New York, 2005, ch. 6, pp. 280–318.
15. F. O. Lichts, cited in the Renewable Fuels Association's 'Ethanol Industry Outlook' reports 2008–2013, www.ethanolrfa.org/pages/annual-industry-outlook [accessed 9 September 2014].
16. J. A. Hyatt, *J. Org. Chem.*, 1984, **49**, 5097–5101.
17. D. A. Moyler and H. B. Heath, *Dev. Food Sci.*, 1988, **18**, 41–45.
18. D. R. J. Laws, N. A. Bath, C. S. Ennis and A. G. Whelden, *US Pat.*, 4,218,491, 1980.
19. M. Gehrig, in *Proceedings of the 10th European Meeting on Supercritical Fluids: Reactions*, Materials and Natural Products Processing, Colmar, France, 2005, Plenary 3, www.isasf.net/fileadmin/files/Docs/Colmar/Summary.htm [accessed 9 September 2014].
20. J. Chrastil, *J. Phys. Chem.*, 1982, **86**, 3016–3021.
21. G. Brunner, *J. Food Eng.*, 2005, **67**, 21–33.
22. J. W. King, in *Proceedings of the 3rd Iberoamerican Conference on Supercritical Fluids* (PROSCIBA 2013), Cartagena, Colombia, 2013, vol. 5, pp. 1–14.
23. Jerry W. King, *Ann. Rev. Food Sci. Technol.*, 2014, **5**, 215–238.

24. C. Luetge, V. Steinhagen, M. Bork and Z. Knez. *Proceedings of the 9th International Symposium on Supercritical Fluids (ISSF-2009)*, Arcachon, France, 2009.

25. V. Steinhagen, C. Lütge, P. Kotnik and Ž. Knez, in *Proceedings of the 13th European Meeting on Supercritical Fluids (ISSF-2011)*, The Hague, 2011.

26. Nateco2, NATECO2 now offering sustainable industrial extraction at a maximum pressure up to 1.000 bar, Ingredients Insight, 31 July 2013, www.ingredients-insight.com/contractors/other-ingredients/nateco2-gmbh-co-kg/pressnateco2-now-offering-sustainable.html [accessed 26 February 2014].

27. J. M. DeSimone, *Science*, 2002, **297**, 799–803.

28. US Environmental Protection Agency, National Emission Standards for Hazardous Air Pollutants: Solvent Extraction for Vegetable Oil Production, *Fed. Reg.*, 2001, 66, 71.

29. Y. Ikushima, N. Saito, M. Arai and K. Arai, *Bull. Chem. Soc. Jpn.*, 1991, **64**, 2224–2229.

30. J. P. Friedrich and G. R. List, *J. Agric. Food Chem.*, 1982, **30**, 192–193.

31. A. P. Gandhi, K. C. Joshi, K. Jha, V. S. Parihar, D. C. Srivastav, P. Raghunadh, J. Kawalkar, S. K. Jain and R. N. Tripathi, *Int. J. Food Sci. Technol.*, 2003, **38**, 369–375.

32. M. McHugh and V. Krukonis, in *Biotechnology and Food Process Engineering*, ed. H. G. Schwartzberg and M. A. Rao, Basic Symposium Series, Institute of Food and Technology, Marcel Dekker, New York, 1990, pp. 203–212.

33. R. Usuki, T. Suzuli, Y. Endo and T. Kaneda, *J. Am. Oil Chem. Soc.*, 1984, **61**, 785.

34. C. Franzke, K. S. Grunert and H. Kroshel, *Nahrung*, 1972, **16**, 859.

35. E. Flemming and U. Hehner, *US Pat.* 5885561, 1999.

36. R. Marriott and E. Sin, in *The Role of Green Chemistry in Biomass Processing and Conversion*, ed. H. Xie and N. Gathergood, Wiley, Hoboken, NJ, 2013, pp. 181–204.

37. A. P. Tulloch and R. O. Weenink, *Can. J. Chem.*, 1969, **47**, 3119.

38. R. C. Sun and J. Tompkinson, *J. Wood. Sci.*, 2003, **49**, 1611–4663.

39. R. C. Sun and J. Tompkinson, *Cell. Chem. Technol.*, 2001, **35**, 471–485.

40. F. E. I. Deswarte, J. H. Clark, J. J. E. Hardy and P. M. Rose, *Green Chem.*, 2006, **8**, 39–42.

41. R. Sun and X.-F. Sun, 2001, 36, 3027–3048.

42. J. Xiufeng and J. Reinhard, *Phytochemistry*, 2008, **69**, 1197–1207.

43. W. H. Gibson, *Chem. Eng. Res. Des.*, 1931, **9a**, 30–35.

44. A. P. Tulloch and L. L. Hoffman, *J. Am. Oil Chem. Soc.*, 1977, **54**, 587–588.
45. F. Karaoosmanogalu, E. Tetick, B. Guerboy and A. Seanli, *Energy Sources*, 1999, **21**, 801–810.
46. W. W. Christie, Waxes: Structure, composition, occurrence and analysis, The ACOS Lipid Library, 2012, http://lipidlibrary.aocs. org/Lipids/waxes/index.htm [accessed 24 February 2014].
47. J. J. Villaverde, R. M. A. Domingues, C. S. R. Freire, A. J. D. Silvestre, C. Pascoal Neto, P. Ligero and A. Vega, *J. Agric. Food Chem.*, 2009, **57**, 3626–3631.
48. D. A. Boadi, S. A. Moshtaghi Nia, K. M. Wittenberg and W. P. McCaughey, *Can. J. Anim. Sci.*, 2002, **82**, 465–469.
49. J. J. Majnarich, *US Pat.* 3420935, 1969.
50. D. Coelho, G. Marques, A. Gutierrez, A. J. Silvestre and J. C. del Rio, *Ind. Crops Prod.*, 2007, **26**, 229–236.
51. A. Gutiérrez, I. M. Rodriaguez and J. C. del Rio, *J. Agric. Food Chem.*, 2006, **54**, 2138–2144.
52. G. Marques, J. Rencoret, A. Gutiérrez and J. C. del Río, *Open Agricult. J.*, 2010, **3**, 1–9.
53. Soil Association, *Soil Association Organic Standards,* clause 40.8.8, Version 16.4 (updated June 2011), Soil Association, Bristol, UK.
54. D. A. Moyler, in M. B. King and T. R. Bott (ed.), *Extraction of Natural Products Using Near-critical Solvents*, Blackie, Glasgow, UK, 1993, 140–183.
55. F. R. Sharpe and D. Crabb, *J. Inst. Brew.*, 1980, **86**, 60–64.
56. U. Nguyen, M. Anstee and D. A. Evans, presented at 5th International Symposium on Supercritical Fluids, Nice, France, 1998.
57. E. Reverchon, *J. Supercrit. Fluids*, 1997, **10**, 1–37.
58. T. Fornari, G. Vicente, E. Vázquez, N. R. García-Risco and G. Reglero, *J. Chromatogr. A*, 2012, **1250**, 34–48.
59. J. L. Martinez (ed.), *Supercritical Fluid Extraction of Nutraceuticals and Bioactive Compounds*, CRC Press, Boca Raton, FL, 2007.
60. J. Araújo and D. Sandi, *Food Chem.*, 2007, **101**, 1087–1094.
61. N. Babovic, S. Djilas, M. Jadranin, V. Vajs, J. Ivanovic, S. Petrovic and I. Zizovic, *Innovative Food Sci. Emerging Technol.*, 2010, **11**, 98–107.
62. S. Glisic, J. Ivanovic, M. Ristic and D. Skala, *J. Supercrit. Fluids*, 2010, **52**, 62–70.
63. Z. Djarmati, R. M. Jankov, E. Schwirtlich, B. Djulinac and A. Djordjevic, *J. Am. Oil Chem. Soc.*, 1991, **68**, 731–734.

64. M. Herrero, M. Plaza, A. Cifuentes and E. Ibáñez, *J. Chromatogr. A*, 2010, **1217**, 2512–2520.

65. G. Vicente, M. R. García-Risco, T. Fornari and G. Reglero, *Chem. Eng. Technol.*, 2012, **35**, 176–182.

66. M. A. Al Kanhal, *Int. J. Food Sci. Nutr.*, 1997, **48**, 135–139.

67. M. S. Kuk and M. K. Dowd, *J. Am. Oil Chem. Soc.*, 1998, **75**, 623–628.

68. N. Nasri, B. Fady and S. Triki, *J. Agric. Food Chem.*, 2007, **55**, 2251–2255.

69. E. Sin, PhD thesis, University of York, 2012, pp. 195–211.

70. M. Sajfrtová, I. Ličková, M. Wimmerová, H. Sovová and Z. Wimmer, *Int. J. Mol. Sci.*, 2010, **11**, 1842–1850.

71. C. J. Chang, Y. F. Chang, H. Z. Lee, J. Q. Lin and P. W. Yang, *Ind. Eng. Chem. Res.*, 2000, **39**, 4521–4525.

72. M. Eisenmenger and N. T. Dunford, *J. Am. Oil Chem. Soc.*, 2008, **85**, 55–61.

73. S. Jäger, H. Trojan, T. Kopp, M. N. Laszczyk and A. Scheffler, *Molecules*, 2009, **14**, 2016–2031.

74. J. Zhao, PhD thesis, University of York, 2011, pp. 47–109.

75. D. Mongkholkhajornsilp, S. Douglas, P. L. Douglas, A. Elkamel, W. Teppaitoon and S. Pongamphai, *J. Food Eng.*, 2005, **71**, 331–340.

76. E. Vági, B. Simándi, K. P. Vásárhelyiné, H. Daood, Á. Kéry, F. Doleschall and B. Nagy, *J. Supercrit. Fluids*, 2007, **40**, 218–226.

77. Y. Ge, H. Yan, B. Hui, Y. Ni, S. Wang and T. Cai, *J. Agric. Food. Chem.*, 2002, **50**, 685–689.

78. L. Jaime, J. A. Mendiola, E. Ibáñez, P. J. Martin-Álvarez, A. Cifuentes, G. Reglero and F. J. Señoráns, *J. Agric. Food. Chem.*, 2007, **55**, 10585–10590.

79. H. G. Daood, V. Illés, M. H. Gnayfeed, B. Mészáros, G. Horváth and P. A. Biacs, *J. Supercrit. Fluids*, 2002, **23**, 143–152.

80. J. A. Mendiola, *et al.*, *J. Supercrit. Fluids*, 2008, **43**, 484–489.

81. M. Sun and F. Temelli, *J. Supercrit. Fluids*, 2006, **37**, 397–408.

82. Q. Ma, X. Xu, Y. Gao, Q. Wang and J. Zhao, *Int. J. Food Sci. Technol.*, 2008, **43**, 1763–1769.

83. S. Machmudah, A. Shotipruk, M. Goto, M. Sasaki and T. Hirose, *Ind. Eng. Chem. Res.*, 2006, **45**, 3652–3657.

84. K. Zosel, US Pat. 4,247,570, 1981.

85. E. Lack and H. Seidlitz, in *Extraction of Natural Products Using Near-critical Solvents*, ed. M. B. King and T. R. Bott, Blackie, Glasgow, UK, 1993, pp. 101–139.

86. S. Machmudah, K. Kitada, M. Sasaki, M. Goto, J. Munemasa and M. Yamagata, *Ind. Eng. Chem. Res.*, 2011, **50**, 2227–2235.

87. J. Tello, M. Viguera and L. Calvo, *J. Supercrit. Fluids*, 2011, **59**, 53–60.
88. S. M. Rahoma, D. A. S. Marleny and M. Paulo, *Ind. Eng. Chem. Res.*, 2002, **41**, 6751–6758.
89. P. Hubert and O. G. Vitzthum, *Angew. Chem., Int. Ed. Engl.*, 1978, **17**, 710–715.
90. H. J. Gahrs, *US Pat.* 4,561,452, 1985.
91. H. J. Grubbs and R. Prasad, *US Pat.*, 5,497,792, 1996.
92. J. Kim, Y. H. Choi and K. P. Yoo, in *Alkaloids: Chemical and Biological Perspectives*, ed. S. W. Pelletier, Pergamon Press, Amsterdam, 2001, vol. 15, pp. 415–432.
93. J. L. Janicot, M. Caude, R. Rosset and J. L. Veuthey, *J. Chromatogr. A*, 1990, **505**, 247–256.
94. J. Y. Ling, G. Y. Zhang, Z. J. Cui and C. K. Zhang, *J. Chromatogr. A*, 2007, **1145**, 123–127.
95. E. Ellington, J. Bastida, F. Viladomat and C. Codina, *Phytochem. Anal.*, 2003, **14**, 164–169.
96. A. Verma, K. Hartonen and M. L. Riekkola, *Phytochem. Anal.*, 2008, **19**, 52–63.
97. C. Erkey, *J. Supercrit. Fluids*, 2000, **17**, 259–287.
98. J. M. Tingey, C. R. Yonker and R. D. Smith, *J. Phys.Chem.*, 1989, **93**, 2140–2143.
99. K. E. Laintz, C. M. Wai, C. R. Yonker and R. D. Smith, *Anal. Chem.*, 1992, **64**, 2875–2878.
100. Y. Lin and C. M. Wai, *Anal. Chem.*, 1994, **66**, 1971–1975.
101. Y. Lin, C. M. Wai, F. M. Jean and R. D. Brauer, *Environ. Sci. Technol.*, 1994, **28**, 1190–1193.
102. F. Dehghani, T. Wells, N. J. Cotton and N. R. Foster, *J. Supercrit. Fluids*, 1996, **9**, 263–272.
103. K. E. Laintz and E. Tachikawa, *Anal. Chem.*, 1994, **66**, 2190–2193.
104. Y. Lin, R. D. Brauer, K. E. Laintz and C. M. Wai, *Anal. Chem.*, 1993, **65**, 2549–2551.
105. S. Wang, S. Elshanl and C. M. Wai, *Anal Chem.*, 1995, **67**, 919–923.
106. J. D. Glennon, S. J. Harris, A. Walker, C. C. McSweeney and M. O'Connell, *Gold Bulletin*, 1999, **32**, 52–58.
107. A. M. Alfantatazi and R. R. Moskalyk, *Miner. Eng.*, 2003, **16**, 687–694.
108. W.-L. Chou and K.-C. Yang, *J. Hazard. Mater.*, 2008, **154**, 498–505.
109. H.-M. Liu, C.-C. Wu, Y.-H. Lin and C.-K. Chiang, *J. Hazard. Mater.*, 2009, **172**, 744–748.

CHAPTER 24

Sustainable Mining, Metals Processing and Recovery

JUSTIN SALMINEN,*[a] SAMI VIROLAINEN,[b] PÄIVI KINNUNEN[a] AND OLLI SALMI[a]

[a] VTT Technical Research Centre of Finland, Espoo, Finland;
[b] Lappeenranta University of Technology, Lappeenranta, Finland
*Email: justin.salminen@vtt.fi

24.1 INTRODUCTION

The global mineral market, driven primarily by the growing standard of living and urbanization of industrial and development countries, is expected to lead to a mineral resource gap in future years unless new mineral sources and more efficient technologies are developed. In addition to more efficient extraction processes, the looming resource gap requires more careful use of mineral raw materials over a wide range of products. To exacerbate the situation, the concentrations of metal found in primary deposits are now lower than that found before and furthermore the deposits are more complex than before. Also secondary sources, such as end-of-life products, mineral residue streams and urban landfills could offer competitive processes, starting with acceptable concentrations for a number metals which are very important in modern society together with base metals. These trends lead to a need to develop new technologies for economically sustainable extraction from low grade sources and proper

Chemical Processes for a Sustainable Future
Edited by Trevor M. Letcher, Janet L. Scott and Darrell A. Patterson
© The Royal Society of Chemistry 2015
Published by the Royal Society of Chemistry, www.rsc.org

management of the socio-environmental impacts related to general mineral extraction processes.

Within the European Union (EU), critical raw material (CRM) usage is expected to increase due to green technologies, electric vehicles, special bulk applications for materials such as steel, and the increased need for materials which can withstand extreme conditions.[1] To tackle this challenge, the European Commission is focusing its activities in the raw material sector on plans that include the raw materials initiative[2,3] and a strategy implementation plan (SIP) on raw materials by European Innovation Partnership (EIP). The SIP, for instance, outlines 24 specific actions for primary and secondary raw material and substitution activities during 2014 and 2015, some of which will be funded through the EU research programme Horizon 2020.

There is a clear need for tools and methods that allow for more rigorous and sustainable mining, processing and material design. Although a whole new value chain ought to be developed, the technological readiness level for a flexible utilization of low concentrate materials as a raw material is not high. Low energy efficiency and high economical costs are the main obstacles as to why low-concentrate raw materials, especially mining wastes, are not refined in larger quantities.

The general trend for the future is to increase expertise and knowledge in using low-concentrate mining waste from safe landfill and in particular to develop methods for recycling high-value functional materials. Substitution of critical metals may also play an important role in reducing material dependencies, but it should not lead to reduced product performance or increased environmental impacts. Thus there is a clear need for tools and methods that allow for more rigorous and sustainable material design.

Mining waste and tailings as well as metallurgical waste and industrial waste streams, combined with anthropogenic waste streams, are potential sources of raw materials. Material preparation in these areas is challenging because of the complex mixed chemistries, impurities and mineral structures, and these have a direct effect on the economics, energy consumption, chemical consumption and resultant wastewater issues.

This chapter looks at the hydrometallurgical separation methods needed in order to recover metals from secondary raw materials. The two main methods discussed are solvent extraction and ion exchange; (selective) leaching is also discussed. The purpose is also to enhance our knowledge of the secondary raw materials that are suitable as valuable metal sources from economic, environmental and social points of view.

New innovative ways to utilize secondary raw material sources requires skills, equipment, and knowledge of hydrometallurgy and materials sciences. A series of well thought out hydrometallurgical steps can lead to safe landfill and at the same time the extraction of valuable raw materials. The hydrometallurgical steps involving inter-relationships between primary raw materials, secondary materials (including anthropogenic wastes), by-products and even substitute materials can be designed in such a way that the raw material composition can be used directly without any addition to functional material preparation. The whole value chain needs to be clarified and technology gaps filled.

24.2 SUSTAINABLE MINING

24.2.1 Mining Wastes

The amount of mining waste worldwide, which involves flotation underflow, leaching residues, side rock and precipitates, is huge and is expected to increase. Average copper ore grades, for instance, have dropped to below 1.0% during recent decades and this has a direct link to increasing amounts of mining waste.[4]

Primary production and mining are the main producers of industrial waste. In Finland, 54.85 million tonnes (Mt) of waste are generated annually by the mining industry; this amounts to 53% of the total industrial waste of 104.34 Mt.[5] This mining waste amount is increasing annually and can result in significant and sometimes irreversible degradation to ecosystems.[6]

The problem with waste rock is that the content of metals are typically uneconomically low for processing but in the long term can generate environmentally harmful metal rich solution, *e.g.* due to acid formation through sulfide oxidation.[7] Major mine tailings sites, mineralogy development and major trace metals in Europe have recently been identified in an EU funded project called ProMine.[8]

Waste mineral (rock) typical contains base metals, iron, aluminium, silicon, *etc.* but also traces of valuable metals together with hazardous substances such as arsenic, chromium, lead and other metals which can cause environmental problems. One of these serious environmental problems is acid mine drainage (AMD).[9,10] The problem exists in operating mines as well as in closed mines.

The correct mining closure handling and maintenance procedure is crucial for the environment.

It is important to plan the safe closure of a mine which includes taking into account the post-mining life cycle, reducing adverse environmental impacts, and improving public knowledge and societal impacts. The goal is to give guidance on selecting the correct technologies and approaches crucial for successful mine closure. This is also useful for authorities to develop correct permitting procedures. Within this work, a wiki-based, modular and updatable web resource has been created.[10]

The mine waste and waste facilities are the most prominent sources of pollution after mine closure. In order to minimise the environmental impacts, it is important to predict the long-term behaviour of waste using proper characterisation methods and to select suitable solutions for closure of waste areas. Another important aspect is post-closure water treatment, where reliable but at the same time cost-effective methods with low maintenance are needed. Both active and passive water treatment solutions should be evaluated for their suitability, and selected passive methods studied in order to produce more information about their long-time performance.

To predict the long-term behaviour of disposed material and to minimize the amount of waste itself through efficient utilization, a proper characterization of the waste material is crucial. In addition, the methods available to successfully close a waste facility and to remediate the impacted areas are dependent on the geochemical processes derived from the properties of disposed material.

The target is to develop technologies that utilize nature's resources in environmentally, socially and economically sustainable ways. The main focuses are on sustainable utilization of raw materials and tailings, and understanding of the fate of wastes and waters generated by mining operations such as acid mine drainage. It involves mine waste management, the stability of solid waste in mine sites, and generally finding economically sustainable processes for low-grade ores.

There is a looming conflict between the growing amounts of waste, which includes mining waste and hazardous waste, and the simultaneous societal need to improve resource efficiency which includes recycling, improved recoveries of metals in a cost-effective manner and reduced dependency of foreign imports. The demand for improved treatment methods in the mining and extraction industry is increasing due to better public awareness. Unfortunately mining processes have not yet been developed to the level where there is no

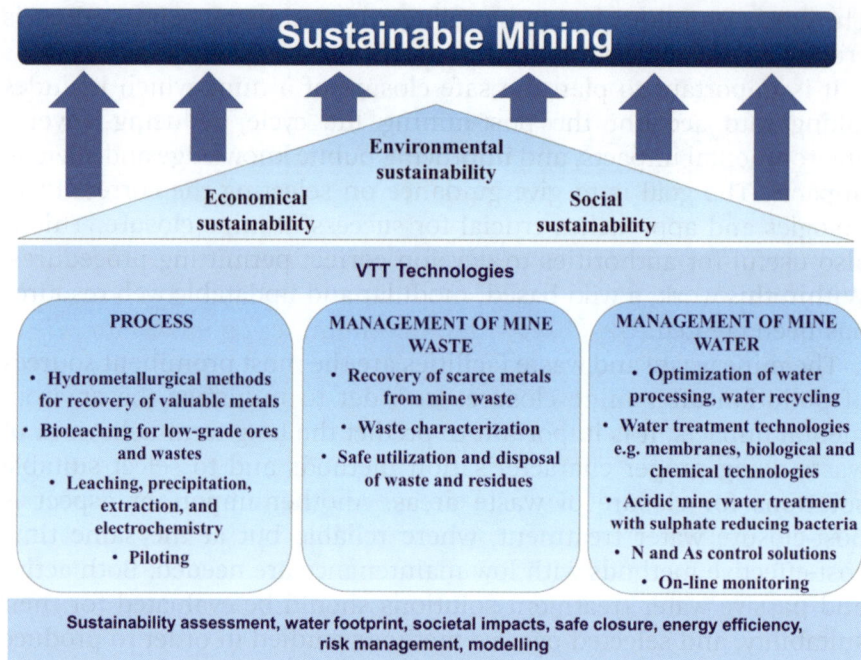

Figure 24.1   Focus areas in research on sustainable mining concept.

environmental impact. This is a global problem that must be dealt with. There is a growing demand and political will for closed loop production and towards zero waste society. This creates the need to improve methods and processes for utilization of waste, side streams and tailings in the whole chain. The mining and refining industry has a need to response to social interest in less hazardous waste and a simultaneous need for economical use of raw materials. An example of governmental concern covering technological and societal aspects of mining, is the action plan implemented by the Finnish government on sustainable mining,[11] whose brief includes a social licence for operating existing mines as well as opening new ones. Figure 24.1 summarizes key points of sustainable mining and how multi-disciplinary research is required to cover the whole life cycle of mining, process and operation.

One example of hydrometallurgical process development work is the improved recovery of rare earth metals from different sources. The unique properties of rare earth elements (REEs) make them essential for emerging sophisticated technologies in diverse areas related to environmental, energy efficiency and health issues. Unfortunately there are concerns worldwide about the future security and supply of these heavy rare earth elements. The use of primary and secondary sources and also wastes as raw material resources to improve REE supplies are being investigated by many stakeholders including an EU advisory group, the European Rare Earth Competence Network (ERECON),[12] and the Canadian Rare Earth Elements Network (CREEN).[13] Another instrument for addressing important raw material matters and knowhow and enabling new businesses through the whole values chain inside the EU is the European Institute of Technology Knowledge Innovation Communities on Raw Matters (EIT KIC RM).[14]

24.2.2 Water in Mining Operations

Water has always been a critical element for mining projects. Water is used in the mining industry for extraction and processing, grinding, screening, dust scrubbing, washing, dust suppression, pump gland seal and reagent mixing.[15,16] Mining operations are generally the highest consumers of water in any mining region.

Many processes were designed at a time when water was readily available and environmental regulations more relaxed. However, the current interest of the mining industry is to take care of the water issues in a sustainable way, so that the issues do not jeopardise the social licence to operate.[17] Reuse, recycling, reduced consumption, safe closure, and desalination of mine water has the potential to increase the public acceptance of mining.[16,18,19] If water resources become unavailable, it can lead to work stoppages or even mine shut downs. Environmentally beneficial water recycling and reuse projects can also lead to a significant economic return.

The requirements of process water are highly dependent on specific industrial applications. Every mine site is specific in terms of ore type, gangue material, plant situation, water quality, *etc.* Water quality may have significant impact on the performance of the separate unit processes and the whole process chain. Therefore, water recycling and reuse needs to be assessed individually on a case-by-case basis. Water of lower quality also has a value. It is often possible to replace fresh water with poorer quality water in the minerals industry and in some cases the effluent of one unit operation can be considered as feed or as makeup solution for another unit operation.[20] Novel cost-efficient water treatment technologies can be developed to obtain appropriate water quality for recycling. In addition, processes can be further developed to enable the use of lower quality water.

Water has always been a critical element for any mining project. Approximately 1% of the total water consumption in USA and Canada, and approximately 2% in Australia, is due to the mining industry.[21–23] Mining operations can have a serious impact on the water usage. For example, in Australia 95% of the water used in mines is sourced locally from surface and groundwater and this does have a significant local consequences.[23]

Typical average water consumption values in mineral processing are shown in Tables 24.1 and 24.2. Mines and concentrators consume typically (0.4–0.9) m^3 of water for every tonne of ore. In addition, smelting with refining and leaching with solvent extraction and electrowinning (SX/EW) consume (5–13) and (3–30) m^3 of water for every tonne of target metal, respectively. Flotation is often used for concentrating the ore and is one of the biggest water consuming unit processes. Gold, platinum, diamonds, nickel and copper are associated with the highest water consumption due to their presence in low concentration in the ores. Mining operations often take place in

Table 24.1 Water consumption data derived from literature representing typical average values.[23]

Metal	Process	Stage	Water consumption
Copper	Smelting/converting and electrorefining	Mine and concentrator	0.37 m^3 t^{-1} ore
		Smelting	7.8 m^3 t^{-1} Cu
		Refining	0.6 m^3 t^{-1} Cu
	Heap acid leaching and SX/EW	Mining and heap leaching	23.0 m^3 t^{-1} Cu
		SX/EW	6.4 m^3 t^{-1} Cu
Nickel	Flash furnace smelting and Sherritt–Gordon refining	Mine and concentrator	0.93 m^3 t^{-1} ore
		Smelting	0.81 m^3 t^{-1} conc
		Refining	7.16 m^3 t^{-1} matte
	Pressure acid leaching and SX/EW	Total all stages	3.4 m^3 t^{-1} ore
Lead	Blast furnace	Mine and concentrator	0.64 m^3 t^{-1} ore
		Smelting	4.85 m^3 t^{-1} Pb
		Refining	0.47 m^3 t^{-1} Pb
	Imperial smelting process	Mine and concentrator	0.64 m^3 t^{-1} ore
		Smelting	12.73 m^3 t^{-1} Pb
		Refining	0.47 m^3 t^{-1} Pb
Zinc	Imperial smelting process	Mine and concentrator	0.64 m^3 t^{-1} ore
		Smelting	12.73 m^3 t^{-1} Zn
		Refining	0.54 m^3 t^{-1} Zn
	Electrolytic process	Mine and concentrator	0.64 m^3 t^{-1} ore
		Electrolytic refining	12.33 m^3 t^{-1} Zn

Table 24.2 Major water uses in a 50 000 tonnes per day low-grade copper deposit mine model.[18]

Major water users	Flow/m^3 d^{-1}	Flow/% of total
Flotation process water (30% solids by mass)	115 646	84.4
SAG^a ill cooling water	4100	3.0
Ball mill cooling water	4100	3.0
Compressor cooling water	4100	3.0
Road-dust suppression	3520	2.6
Froth wash water	2880	2.1
Pump GSW	1440	1.1
Reagent dilution water	720	0.5
Primary crusher dump pocket – dust suppression	358	0.3
Coarse ore stockpile – dust suppression	121	0.1
Mine/mill/office staff domestic water	58.1	0

^{a}Semi-autogenous grinding.

water-scarce regions. Ore grades are declining globally resulting in higher water consumption per each unit of production.[15] On the other hand water reuse technologies can be improved.

Many water uses in the mines are insensitive to water quality and only the water volume is crucial. However, water quality is a crucial in mineral processing operations such as flotation and agglomeration. Despite this, research efforts have been rather limited in understanding and controlling the influence of water quality in mineral processing.[24] Recycled water can contain elements that are deleterious to process performance. Discharge regulations and attempts to reduce the use of fresh water have resulted in the use of more expensive technologies such as reverse osmosis and ion-exchange based technologies.[17] Water savings can also be created by reducing water losses and by utilizing dry ore upgrading technologies.[25]

24.2.3 Effluent Discharge Regulations

International effluent discharge standards applicable to metal mining operations are summarized in Table 24.3. Most stringent international limits are found: in South Africa for copper, zinc, iron and total ammonia; in Chile for lead and selenium; in the UAS for nickel; and in Canada for total suspended solids (TSS). New stricter limits have been proposed for several elements in Canada.

International effluent discharge regulations related to sulfate are summarized in Table 24.4. It is noteworthy that the concentrations of

Table 24.3 Effluent discharge standards applicable to metal mining operations.

Parameter	Unit	Finland[a] (single measurement in brackets)	World[b]	Canada MMER[c] (proposed new limit in brackets)	US[d] (maximum for 1 day)	US[d] (30 days average)	Most stringent international limit[3]
Arsenic	mg dm^{-3}	<0.5 to <1.0 (1.7) or no limit	0.1	0.5 (0.1)	1.0	0.5	0.1 (multiple)
Cadmium	mg dm^{-3}		0.1		0.1	0.05	
Chromium (hexavalent)	mg dm^{-3}	–	0.1				
Copper	mg dm^{-3}	<0.2 to <0.5 (1.0) or no limit	0.5	0.3 (0.05)	0.30	0.15	0.02 (South Africa)
Cyanide	mg dm^{-3}	0.4	1.00 0.1 free 0.5 WAD	1.0 (0.5)			0.5 (multiple)
Lead	mg dm^{-3}	–	0.2	0.2 (0.05)	0.6	0.3	0.05 (Chile)
Mercury	mg dm^{-3}	–	0.01		0.002	0.001	0.1 (US)
Nickel	mg dm^{-3}	<0.5 to <2.5 or no limit	0.5	0.5 (0.25)	0.2	0.1	0.3 (South Africa)
Zinc	mg dm^{-3}	<1.0 to <3.0 or no limit	2	0.5 (0.25)	1.0–1.5	0.5–0.75	
TSS	mg dm^{-3}	<10 to <25	50	15.0 (15.00)	30–50	20–30	15.00 (Canada)
Oil and grease	mg dm^{-3}		10				
Aluminium	mg dm^{-3}	<3.0 or no limit		none (tbd)			1.0 (multiple)
Iron	mg dm^{-3}	–	3.5	none (tbd)	2.0	1.0	0.3 (South Africa)
Selenium	mg dm^{-3}			none (tbd)			0.01 (Chile)
Total metals	mg dm^{-3}		10				
Total ammonia as nitrogen	mg dm^{-3}	–		none (tbd)		100 NH$_3$	1.0 (South Africa)
Chemical oxygen demand (COD)	mg dm^{-3}		150		200	100–500	
Phosphorus	mg dm^{-3}	–		no limit (no limit)			
Chloride	mg dm^{-3}			no limit (no limit)			
pH		5.5–10	6–9	6.0–9.5 (no change)	6–9.1	6–9.1	

[a]Range of limits from environmental permits of metal mines in Finland; the limits vary for each mine.
[b]World Bank 1998.
[c]Environment Canada 2012.
[d]US Code of Federal Regulations Title 40 Part 440; discharge limits vary according to type of ore mined, mining technique, and applicable treatment technology. MMER = Metal Mining Effluent Regulations; tbd = to be determined; WAD = Weak Acid Dissociable Cyanide.

Table 24.4 Effluent discharge regulations in the world related to sulfates.[25]

Country	Legal instrument	$SO_4^{2-}/$ mg dm^{-3}	TDS/ mg dm^{-3}	Conductivity/ mS m^{-1}
Chile	DS90	1000–2000		
	DS46			
	NCh1333	250–500		
Peru	DS-002-2008-MIN	300 (Cat 3)	500 (Cat 4)	
Ecuador	NCADE Lib VI, 1	1000	3000	
Brazil	Resolution 357	250 (Class 1)	500	
South Africa	Water Act 1998	–	–	70–150
Australia	ANZECC 2000/	1000	2400	

substances in effluents are generally much lower than the permitted environmental limits.

24.2.4 Water Reuse and Recycling

The recirculation of process water and changes in production can have considerable effects on the water quality. All additives and chemicals in contact with process water affect the water quality when a significant part of the process water is recirculated.[26] Appropriate points to recycle water should be properly identified. For example, the introduction of high alkalinity water into an acid circuit is inefficient.[24]

There are several challenges associated with water recycling in mineral processing unit operations. For example, the concentrations of dissolved and colloidal species in recycled water can change when different dams or sections of a dam are used and water evaporation alters seasonal water balances.[27] Since metallurgical processes are sensitive to variations in water quality, the quality fluctuations should be minimized.[25]

Water quality also has an influence on waste residues, which is considered as an important factor in industry. In the studies of Whittington *et al.* (2003) the use of saline process water in the pressure acid leaching of nickel laterite ore resulted in the increased formation of jarosite precipitates.[28]

24.2.5 Flotation

Flotation solutions typically contain (25–35)% by mass of solid material. If water is not recycled or reused, flotation would require (1.9–3.0) m^3 of water for every tonne of ore processed.[18] Flotation is one of the biggest water-consuming unit processes in the mining and mineral processing chain. For example, total water consumption in the Minera Esperanza mine in Chile is approximately 630 dm^3 s^{-1} and the major part of the water (approximately 600 dm^3 s^{-1}) is used in the concentrator plant.[29] There is an increasing need to recycle water from tailings dams and ponds (external recycled water), thickener overflows, dewatered and filter products (internal recycled water).[27,30,31] Recycling of water within the flotation process can be advantageous because: (1) the need to receive new water into the system is lowered; (2) the amount of discharged water is decreased; and (3) reagent consumption is reduced due to retention of some reagents.[25] However, process water usually contains several chemical compounds, which has raised concerns about the use of recycled process water in flotation.

Process water chemistry can have a critical impact on flotation efficiency. Ca^{2+} and SO_4^{2-} ions are typical components in the process water from the flotation of sulfide minerals originating from the ore and flotation reagents.[32] Minor species include reduced sulfur compounds such as SO_3^{2-}, $S_2O_3^{2-}$, $S_2O_5^{2-}$ and $S_4O_6^{2-}$, cations of ferrous and non-ferrous metals, anions of different compounds often of polymeric nature, frothing molecules, residual chemical reagents and products of their degradation.[33] Tables 24.5 and 24.6 show possible effects of the most important ions of concern and typical flotation reagents in flotation, respectively. Organic compounds released from residual reagents in water and dissolved inorganic species may alter water quality and flotation kinetics. In internal water recycling, suspended solids levels tend to be high having a negative impact on flotation.[31] Some constituents have more effect on flotation than others. For example, sulfate ions tend to have a more negative effect on flotation than do chloride ions.[25] Bacteria-laden water, such as

Table 24.5 Effect of different components on selective flotation.[20,32,35]

Component	Effect
Cu^{2+}, Fe^{2+}, Pb^{2+}	Undesired activation/flotation of sulfide minerals
S^{2-}	Undesired depression of sulfide minerals
Alkaline earth ions (Mg^{2+}, Ca^{2+} *etc.*)	Activation of the non-sulfide gangue, variation in the slurry pH and pulp potential, lower grade of sulfide concentrate obtained, lower recovery of oxides
Primary amines	High adsorption at fine particles causes slime flotation
Residual xanthates and their oxidation products	Selective absorption on most sulfides

Table 24.6 Typical flotation reagents and their consumptions.[34]

Use	Reagent	Purpose	Typical consumption/ kg t^{-1} ore
Copper flotation	AP3418A	Cu collector	0.040
	Methyl isobutyl carbonyl (MIBC)	frother	0.035
Copper–molybdenum concentrator	Lime	pH modifier	0.295
	Dowfroth 250	frother	0.007
	Pine oil	frother	0.007
	Potassium amyl xanthate (PAX)	Cu–Mo collector	0.016
	Fuel oil	Mo collector	–
	Sodium hydrogen sulfide	Cu depressant	6.63 (kg t^{-1} concentrate)
Copper–nickel flotation	Potassium isobutyl xanthate	Cu–Ni collector	0.11
	Dow Froth 250 °C	frother	0.06
	$CuSO_4$ (copper sulfate)	Ni activator	0.04
	NaCN (sodium cyanide)	Ni depressant	0.003
	Lime	pH modifier	1.6
	Percol	flocculant	0.002
	H_2SO_4 (sulfuric acid)	pH modifier	0.25
Copper–zinc flotation	Sodium isopropyl xanthate (SIX)	Cu–Zn collector	0.05
	MIBC	frother	0.03
	Na_2SO_3 (sodium sulfite)	Zn depressant	0.72
	$CuSO_4$	Zn activator	0.93
	Lime	pH modifier	2.00
	Percol	flocculant	0.008
Copper–lead–zinc flotation	SIX	Cu–Pb–Zn collector	0.170
	PAX	Cu–Pb–Zn collector	0.042
	Soda ash	Cu–Pb pH modifier	2.5
	$CuSO_4$	Zn activator	0.75
	Lime	Zn pH modifier	1.0
	Aerofloat 241	Pb collector	0.020
	Aero 3894	Cu collector	0.005
	Percol 351	flocculant	0.005
	SO_2 (sulfur dioxide)	pyrite depressant	0.9
Lead–zinc flotation	SIX	Pb–Zn collector	0.97
	$ZnSO_4$ (zinc sulfate)	Zn depressant	0.075
	NaCN	Zn depressant	0.040
	$CuSO_4$	Zn activator	0.30
	MIBC	frother	0.014
	Percol	flocculant	0.012

found untreated effluent, can have an effect on the flotation efficiency through microbiological activities. Precipitates such as calcium sulfate, magnesium carbonate, iron hydroxide and silica are the main contributions to the mineral surface contamination in many recycle streams.[31] With proper control of the components in process water, it is possible to recycle the water in a flotation circuit.[33]

The maintenance of consistent circuit water quality and elimination of short-term variation is essential in obtaining better flotation process performance. Uncontrolled variation in the concentrations, pH and Eh can result in: losses in recovery; a reduction in grade and separation efficiency; surface coatings on value minerals; competitive adsorption of activators and reagents with other species; loss of reagents due to precipitation; overcoating; and inadvertent reactions. Non-selective coatings on value minerals and gangue, ineffective separation of gangue minerals and ineffective activation can cause a reduction in grade.[27] In most of the cases where water quality has been found to affect flotation processes, however, the precise reason for the effect has remained unknown.[31]

Tailings water treatment and recycling back to the flotation process is a common practice in mineral flotation plants. Recycling of process water tends to be more appropriate for the plants producing only one concentrate. When more than one concentrate is produced, recycled water has to be matched to the particular concentrate or be treated, so that deleterious reagents are not present.[25] Effects of specific process water on flotation need to be evaluated case-by-case.

Dissolved ions and flotation chemical additives can accumulate in recycled water and may have either a positive or negative effect on the process. Zinc ions in the recycle water are beneficial for selective separation of galena from sphalerite in the lead–zinc flotation separation. Calcium and sulfate ions in the process water can cause scaling problems and affect mineral flotation separation. Process water with the recycle water from the tailings treatment can be saturated or supersaturated with gypsum. Gypsum in recycle water can affect the interactions between flotation reagents and sphalerite, and decreased flotation recovery. However, calcium and thiosulfate ions have been shown to have a positive effect on flotation in a study made by Kirjavainen et al. (2002)[37] by improving the flotation of sulfides after grinding in a steel mill. Calcium has been shown to activate nickel and copper sulfides, and thiosulfate has been shown to reduce the effect of hydrophilic compounds on sulfide particles, thus simultaneously improving flotation.[37] Other studies on the impact of impurities into flotation performance can be obtained from the literature.[31–34,36]

Reports from Anglo Platinum plants since 1999 indicate that waters from thickener overflows and return dams have little negative effect on flotation and they can be used without further treatment. When thickeners do not operate optimally, the suspended solid loadings increase and this has an impact on the reagent consumption of flotation processes. With return dam water, the problems are most likely organic in nature, as a result of organic material degradation.[25]

Other issues related to waste water include salinity issues. This is especially true when the only available water is seawater. For example, Minera Esperanza mine in Chile uses untreated seawater for flotation. Desalination and washing is, however, often necessary to prevent corrosion and provide fresh water feed.[29,38–40]

Examples of the reuse attempts of the use of process water in sulfide ore concentrators are given in Table 24.7.

Process water chemistry depends on ore type, gangue material, plant situation, *etc.* Different mines and processing plants may experience different problems related to the same water chemistry. Examples of water chemistry of process or waste waters from different parts of the process chain in the mines have been studied.[25,27,33,41–43]

24.3 SUSTAINABLE PROCESSING

24.3.1 Recovery of Metals from Solutions

Hydrometallurgy is defined as a field of metal processing that is connected to water[44] and hydrometallurgical separation methods are those in which one or more of the phases involved in the process is aqueous. It has been said that modern hydrometallurgy began in the 19th century[44] with the cyanide process for gold recovery and the Bayer process for aluminium recovery from bauxite. Hydrometallurgical solutions are typically generated by dissolving metals present in the raw materials in acids or bases. Metals can be and are often separated in the dissolution step when one or more of the metals is not soluble in the solvent used. The reverse process, precipitation (including cementation), is also widely used as a separation method. Alongside these, hydrometallurgical separation is often performed by solvent extraction or by means of solid ion exchangers. These methods were first used for the separation of rare earth elements and actinides, but soon their application spread to all kinds of metals.[45] Other hydrometallurgical separation methods include

Table 24.7   Reuse attempts of process water in sulfide ore concentrators.[25,33]

Site	Outcome
Kristineberg, Sweden	– significant problems during pumping of the froth due to insufficient destruction of frothers and collectors in the tailings pond
Kristineberg, Stekenjokk	– no disturbance in grinding – no disturbance to pyrite flotation – when SO_4^{2-} >100 mg L^{-1}, selectivity against zinc in copper and zinc flotation decreased
Laisvall, Sweden	– deterioration of lead flotation and frothing due to precipitation of calcium carbonate in pipes and spray nozzles (carbonate from mine water)
Zinkgruvan, Sweden	– reduction in lead recovery – increase of zinc in lead concentrate – reuse of water stopped
Falun, Sweden	– reuse limited to 25% due to blockage of pipes by gypsum precipitate – saturation of calcium sulfate solution in the tailings pond water
Benambra, Australia	– lower recovery, grade and flotation kinetics with Melbourne process water
Kidd Creek, Canada	– process water 500×10^{-6} thiosalt and 300×10^{-6} calcium – improvement of flotation by enhancement of depression of pyrite and recovery of copper due to presence of thiosalt and calcium ions
Fankou, China	– rest collector in process water beneficial to lead flotation
Mufulira, Zambia	– low pH underground mine water reduced copper recovery and increased copper loss to tailings – deleterious effect removed by addition of quicklime in the primary mill
Rustenburg, South Africa	– tests to assess viability of using treated sewage effluent from the Rustenburg municipality – overall platinum recovery and grade at similar level as with normal process water – slightly lower kinetic rate

separation using ionic liquids, membrane separation, and adsorption onto activated carbon or other natural materials. Once the metals have been separated from each other, pure metal can be produced by many methods including precipitation, cementation or electrolytically. Further purification is often done using electro-chemical methods.

Bioleaching was developed in the 1950s and biohydrometallurgy is today a widely studied area of hydrometallurgy.[46] It is often used in the treatment of mine effluent,[47–50] and for the recovery of metal from secondary raw materials.[51,52]

Since the mid-20th century, hydrometallurgy has become an ever increasingly viable option for replacing pyrometallurgy. The main

Table 24.8 Comparison of hydro- and pyrometallurgy.

Feature	Hydrometallurgy	Pyrometallurgy
Sulfur dioxide generation	+	−
Gaseous emissions to atmosphere	+	−
Aqueous emissions to environment	−	+
Energy consumption	+	−
Treatment of low grade and complex raw materials	+	−
Treatment of high-grade ores	−	+
Capital costs	+	−
Operation costs	−	+
Emission control	+	−
Scale of operation	−	+
Selectivity/product purity	+	−

reasons are: (1) the grade of primary ores for metal refining and also secondary raw materials has decreased over the years making hydrometallurgical methods more suitable; (2) concentrations of metals in secondary raw materials are not only small but are often highly variable, with complex chemical structures which make them suitable for the more adjustable hydrometallurgical recovery methods; and (3) hydrometallurgical methods have features and applications that can have significant positive effects on the environmental aspects of metallurgical plants.[50] Hydro- and pyrometallurgical methods are compared in Table 24.8, based on some typical considerations in extractive metallurgy.

24.3.2 Leaching

Leaching and precipitation were the first methods used in hydrometallurgy.[44] They are both based on solubility properties. In the leaching process, minerals are dissolved in suitable solvents while in precipitation processes, conditions are arranged for the metal or metals to form insoluble compounds. Typical leaching reagents include water, acids (H_2SO_4, HCl, *etc.*), bases (NaOH, NH_4OH), salt solutions and combinations of these. Processes may be enhanced by the addition of oxidizing (air, O_2, O_3, H_2O_2, Cl_2, HClO, NaClO) or reducing (Fe^{2+}, SO_2) agents.

The production of zinc is an important hydrometallurgical process and almost all the world's zinc production is derived from the treatment of sulfide concentrates in which sphalerite, (Zn,Fe)S, is the dominant zinc mineral. The concentrate, together with the slurry from the conversion process and acid from the electrolysis, is fed to reactors where leaching takes place by injecting oxygen into the slurry. The availability of dissolved oxygen for reactions has an important effect on the kinetics of the whole process as the oxidation of Fe^{2+} into Fe^{3+} controls the overall rate of leaching during the early stage of the process.[53,54] The overall leaching reaction rate is controlled by gas diffusion in the aqueous phase, the kinetics of the oxidation reaction, and the leaching reaction rate.[55] The oxidation of Fe^{2+} in acidic aqueous sulfate solutions with dissolved molecular oxygen is commonly employed in many hydrometallurgical processes such as leaching. One of most important challenges in hydrometallurgy is indeed technologies to precipitate or remove dissolved iron in a controlled way as, for example, iron sulfate, jarosite, goethite or hematite.

Equilibrium and kinetics of leaching are often favoured by more aggressive conditions, and thus elevated temperature and/or pressure is commonly used. Several methods to perform the leaching process exist including heap leaching,[56,57] dump leaching (heap without crushing), reactor leaching, vat leaching, autoclave leaching (high pressure and temperature) and *in situ* leaching (leachate pumped in the ore deposit).

Microorganisms can have an effect on the structure of the rocks. This feature is used in heap and dump leaching of, especially, sulfide ores of gold and copper, in a process called bioleaching.[47,58,59] The Talvivaara mine in Sotkamo, Finland, exploits bioleaching, producing nickel, zinc, copper and cobalt in Arctic conditions.[60] In bioleaching, bacteria oxidize Fe^{2+} to Fe^{3+}, which then oxidizes the minerals.[61] Bioleaching is expected to reduce investment and operating costs compared with conventional leaching processes. It is also often seen as a more environmentally sensitive process, since the use of strong chemicals, energy and aggressive conditions are reduced. Other methods that have been studied for improving the performance of conventional leaching are, for example, microwave[62,63] and ultrasound[64,65] treatments, but no industrial applications for these have been reported. Ultrasound treatment in particular has been mentioned in many cases related to the leaching of waste materials and/or precious metals,[63] but economic considerations have prevented, and are likely to continue to prevent, industrial application in the near future.

The appropriate leaching method depends on the ore and metal to be extracted. Different methods for copper leaching from primary sulfides can be divided into predominantly sulfate and chloride processes. Sulfate leaching processes can be further divided into atmospheric or super-atmospheric pressure, chemical or biological processes. Some sulfate based copper processes are listed in Table 24.9.[66]

Table 24.9 Sulfate-based copper hydrometallurgy processes.[66]

Process	$t/°C$	Pressure/ 0.101 MPa	Regrind D80/μm	Special conditions
Activox process	90–110	10–12	5–10	Fine grinding combined with high oxygen overpressure overcomes chalcopyrite passivation
Albion process	85	1	5–10	Atmospheric ferric leaching of very finely ground concentrate
Anglo American – University of British Columbia Process	150	10–12	10–15	Modest regrind combined with surfactants for chalcopyrite leaching
Bactech/Mintek low temperature bioleach	35	1	5–10	Low temperature bioleach (35–50 °C) requires very fine grind to overcome chalcopyrite passivation.
BIOCOP™	65–80	1	37	High temperature bioleach (65–80 °C) uses thermophilic bacteria
CESL copper process	140–150	10–12	37	Chloride catalysed leach of chalcopyrite producing basic copper sulfate precipitate in the autoclave
Dynatec process	150	10–12	37	Chalcopyrite is leached using low grade coal as an additive
Mt Gordon process	90	8	100	Pressure oxidation of chalcopyrite/pyrite ore or bulk concentrate in an iron sulfate rich electrolyte
PLATSOL process	220–230	30–40	15	Total pressure oxidation in the presence of 10–20 $g \cdot dm^{-3}$ NaCl. Precious metals leached at the same time as base metals
Sepon Copper process	80 (Cu)	Atm	100	Atmospheric ferric leach for copper from chalcocite. Pressure oxidation of pyrite concentrate to make acid and ferric sulfate for copper leach.
	220–230 (FeS_2)	30–40	50	
Total pressure oxidation process	200–230	30–40	37	Extreme temperature and pressure conditions designed to rapidly destroy chalcopyrite and other sulfides

Gold can be recovered by several methods. For over 100 years, cyanide has been the leach reagent of choice in gold mining. Substitutes for cyanide have been proposed and thiourea, thiocyanate and

thiosulfate have been regarded as the most realistic ones.[67,68] For re-
fractory ores various methods exist for the oxidation of the gold containing mineral matrix and for more efficient use of cyanide.[69]

Cyanide losses are minimized in gold leaching by operating the
leaching systems at a pH >10. Several compounds have an effect on
gold leaching. Chloride ions in feed water will form gold chloride
complexes in the autoclave, which reduces to gold on carbon. This
gold is not dissolved in leaching operations and is lost to the tailings.
Also metal concentrations in cyanide leaching have effect on leaching.
Studies by Kurama and Catalsarik[70] showed that the efficiency of
cyanide solution as a solvent decreases when the recycled solution
contains high concentrations of zinc cyanide. Lead ions (0.77×10^{-6})
and ferrocyanide ions $(200 \times 10^{-6}$ Fe$)$ have a significant positive effect
on gold dissolution in direct leaching experiments. Copper cyanide
$(50 \times 10^{-6}$ Cu$)$ was shown to have a slightly negative effect on the gold
dissolution. Aghamirian and Yen[71] and Bahrami et al.[72] used activated carbon for one hour to remove the dissolved metals from the
process water prior to reusing the process water in the gold leaching
step. The recoveries of Au and Ag were (93 and 72)%, respectively,
when leaching with fresh water, and (91 and 66)%, respectively, when
leaching with process water. Sulfide ions in solution result in the
passivation of the dissolution of gold in cyanide leaching. Lead can
promote the removal of sulfide by precipitation as PbS. However, too
much lead can also result in a passivation phenomenon in gold
cyanide leaching. A general belief is that the presence of sulfur species in the cyanide leaching solution results in high consumption of
cyanide and oxygen. The formation of copper cyanide species can
result in low free cyanide species and reduce the gold leaching rate.[73]

In bioleaching, the tolerance of the microorganisms to different
concentrations is crucial for a successful operation. However, the
microorganisms can be adapted to increasing concentrations of
metals. The difference in tolerance between un-adapted and adapted
microorganisms can be significant. Bio-heap leaching is based on recycling of the leaching liquor through the heap during several months
and can generally tolerate relatively high concentrations of different
compounds of recycled solutions. Many researchers have studied the
tolerance of mesophilic iron oxidizers to different ions.[74–85]

24.3.3 Precipitation and Cementation

Precipitation is an essential unit process in the hydrometallurgical
industry. Solubility defined by the solubility product or more precisely

the Gibbs energy minimum at a given temperature and pressure.[86,87] Typically solubility in leaching and precipitation is kinetically constrained and in practice the operation time is less than the time it would take to reach equilibrium.[86–91] Advanced methods to take kinetic constrains into account by means of chemical affinity and constrained Gibbs energy methods have been successfully applied to real processes.[86,87,91]

Typically precipitation is carried out by pH alteration, redox potential change or introducing a common ion into the solution to form an insoluble salt with the desired species. Commonly used precipitation agents include hydroxides [CaO or Ca(OH)$_2$, Mg(OH)$_2$, NaOH, NH$_4$OH];[88] sulfides, thiosulfides and sulfur dioxide (FeS, CaS, Na$_2$S, NaHS, NH$_4$S, H$_2$S, Na$_2$S$_2$O$_3$, SO$_2$);[89] and carbonates and carbon dioxide (CaCO$_3$, MgCO$_3$, Na$_2$CO$_3$, CO$_2$).[90,91] Precipitation with all of these alternatives is pH dependent and the desired precipitation process can be in many cases controlled by pH.[88,90,92] Table 24.10 shows performance characteristics of metal removal and recovery techniques.[93]

Cementation is an electrochemical process, in which a metal in the aqueous solution is reduced to its elemental form by some more electropositive metal in solid form. The general reaction equation for cementation is thus:

$$M_1^{x+} + \frac{x}{y}M_2^0 \rightleftharpoons M_1^0 + \frac{x}{y}M_2^{y+} \tag{24.1}$$

Zinc powder is often used as a cementation reagent because it is more electropositive than many metals. Other cementation reagents include aluminium, copper and iron. In industrial hydrometallurgy, cementation was used until the 1970s as a mainstream unit operation, and is also widely used today, in the recovery of gold and silver from cyanide solutions, known as the Merrill-Crowe process.[69]

Currently new Au leachates, for example thiosulfate, are being investigated to replace hazardous cyanide, and large scale operation with ion exchange as separation method has been launched in Barrick's Goldstrike mine in Nevada.[94] Cementation is also seen as a possible recovery method from such solutions, since carbon adsorption – currently a mainstream industrial unit process – is not suitable for thiosulfate solutions.[95,96] Copper is cemented – often with scrap iron – from various solutions, but industrial use has diminished recently.[97] Zinc sulfate solutions are often purified in industrial processes before electrowinning by cementation with zinc powder and some activators.[98] Figure 24.2 shows cementation process of silver from cyanide solution.[99]

Table 24.10 Performance characteristics of metal removal/recovery techniques.[93]

Technique	pH change	Metal selectivity	Influence of suspended solids	Tolerance of organic compounds	Working level for appropriate metal/ mg dm^{-3}	End product
Hydroxide precipitation	Limited tolerance	Non-selective	Tolerant	Tolerant, complexing agents may have adverse effect	>10	Gelatinous sludge
Sulfide precipitation	Limited tolerance	Limited selective, pH-dependant	Tolerant	Tolerant	>10	Dense sludge
Electrochemical operations	Tolerant	Moderate	Can be engineered to tolerate	Can be accommodated	>10	Metal deposit
Ion exchange	Limited tolerance	Chelate-resins can be selective	Fouled	Can be poisoned	<100	Concentrated liquid
Membrane processes	Limited tolerance	Moderate	Fouled	Intolerant	>10	Concentrated liquid
Adsorption	Limited tolerance	Moderate	Fouled	Can be poisoned	<10	Concentrated liquid or metal-containing adsorbent
Liquid–liquid extraction	Limited tolerance	Moderate	Tolerant	Tolerant	>10	Concentrated liquid
Foam flotation	Limited tolerance	Non-selective	Tolerant	Tolerant	>10	Dense sludge

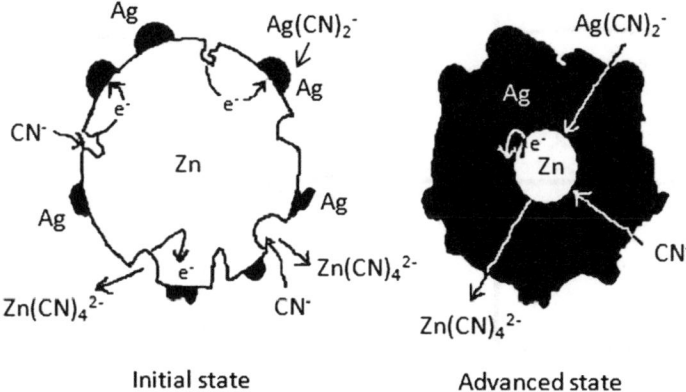

Initial state   Advanced state

Figure 24.2 Cementation of Ag from cyanide solution.[99]

24.3.4 Solvent Extraction

Industrial utilization of solvent extraction in hydrometallurgy began with the Manhattan project in the 1940s.[44] To the present day, solvent extraction has been applied to separate almost every element in the periodic table.[100] It is also widely applied in the separation of organic compounds.[101,102] In solvent extraction, metals are extracted with a suitable reagent to a water-insoluble organic phase, which is typically, for example, a kerosene-based hydrocarbon solvent. While mixing, one of the phases is dispersed within the other one creating large surface area, over which the extractable species moves to the other phase (Figure 24.3).[103]

The following three fundamental quantities are used to describe the efficacy of the solvent extraction process. The distribution ratio, D_i, is defined with analytical concentrations of the component i. Fraction extracted, E_i, describes the extent of extraction and may also be presented as a percentage. The separation factor, α_{ij}, describes selectivity between two components, and is by definition written so that it is larger than unity:[104]

$$D_i = \frac{\bar{c}_i}{c_i}$$
$$E_i = \frac{D_i}{D_i + \dfrac{V}{\bar{V}}},$$
$$\alpha_{i,j} = \frac{D_i}{D_j}$$

(24.2)

where bars above the symbols refer to the organic phase.

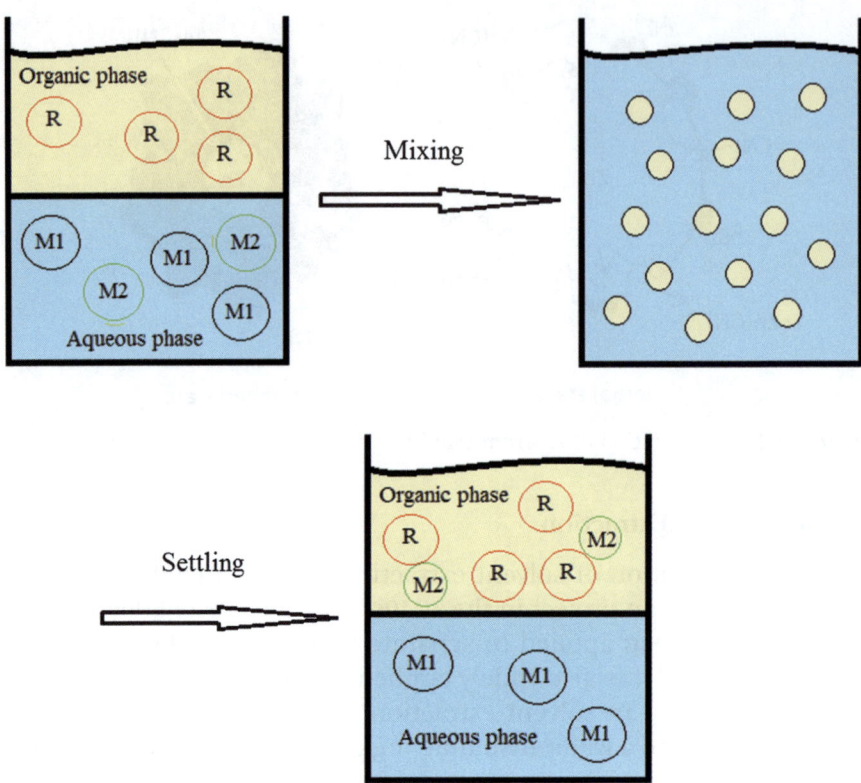

Figure 24.3   Principle of solvent extraction. R = reagent, M1 and M2 are different metals.[103]

In the authors' opinion, the separation factor describes well the relations of metals' theoretical affinities to the phases. Consider, for example, two metals *i* and *j* with equal total concentrations in a system, with fractions extracted of 99.9% and 90%, respectively. For this kind of equilibrium to take place, there is a need to have a much higher affinity to the target (organic) phase and thus the value of the separation factor (equal to 1000) describes the ratio of affinities well. However, if we consider practical application, the selectivity is not good and the purity of *i* in the organic phase is only 53%. Thus, in the authors' opinion, purities describe the selectivity better. Or then the ratio of fractions extracted, $\beta_{i,j}$, may be calculated and used as a quantity of selectivity:

$$\beta_{i,j} = \frac{E_i}{E_j} \qquad\qquad (24.3)$$

According to Ritcey,[105] solvent extraction reagents are divided into three categories as shown by Equations (24.4) to (24.6) based on their extracting mechanism. The extraction mechanisms and methods were studied by Virolainen[103] and Virolainen *et al.*[106,107]

1. Extractants involving compound formation:

$$M^{n+} + n\overline{HA} \rightleftharpoons \overline{MA_n} + nH^+ \qquad (24.4)$$

where M^{n+} refers to the metal cation with charge n', $\overline{HA}$ is the extractant and $\overline{MA_n}$ is the metal complex of extractant and the bars refer to the organic phase.

2. Extractants involving ion association:

$$\begin{array}{c} \overline{n(R_3NH^+ \cdot X^-)} + MA^{-n} \rightleftharpoons \overline{(R_3NH^+)_n \cdot MA^{-n}} + nX^- \\ \overline{n(R_4N^+ \cdot X^-)} + MA^{-n} \rightleftharpoons \overline{R_4N^+ \cdot MA^{-n}} + nX^- \end{array} \qquad (24.5)$$

where R_3N is the tertiary amine, R_4N^+ is the quaternary amine, and MA^{n-} is the anionic complex of metal M with charge n^-.

3. Extractants involving solvation. The mechanism cannot be generalized for a simple reaction equation, but, as an example, reaction equations for In extraction from HCl by TBP are:[108]

$$\begin{array}{c} In^{3+} + 3H^+ + 3Cl^- + \overline{nTBP} \rightleftharpoons \overline{InCl_3 \cdot nTBP} + 3H^+ \\ In^{3+} + 4H^+ + 4Cl^- + \overline{nTBP} \rightleftharpoons \overline{H(InCl)_4 \cdot nTBP} + 3H^+ \end{array} \qquad (24.6)$$

The first group consists of acidic extractants that extract metals either by chelating or a cation exchange mechanism, and includes organophosphoric acid derivatives or carboxylic acids or hydroxyoximes. The second group consists of amine extractants, which are either strong (quaternary amines) or weak (secondary and tertiary amines) anion exchangers. The mechanism in Equation (24.5) is given for a univalent counter ion. The third group includes ethers, esters, ketones, aldehydes and alcohols, together with organophosphoric acid derivatives with oxygen or sulfur as an electron donor atom. The mechanism involves the solvation of neutral inorganic molecules or complexes.[105]

Examples of the reagents, with their practical applications in hydrometallurgy, are given in Table 24.11.

Although nowadays there is a wide range of equipment available for solvent extraction,[101,105] the most common form is still a simple mixer-settler (Figure 24.4 (a)), in which the actual reaction takes place in a continuous mixing unit. After the mixer is a much larger settler

Table 24.11 Some industrially important and commercially available solvent extraction reagents.[105,109–114]

Group	Commercial name	Chemical name (active compound)	Structure	Manufacturer	Applications
Compound forming	D2EHPA	Di-(2-ethylhexyl) phosphoric acid	R = 2-ethylhexyl	Several	Zn, U, Mo, In, REE
	Cyanex® 272 Ionquest® 290	Bis-(2,4,4-trimethylpentyl) phosphinic acid	R = 2,4,4-trimethylpentyl	Cytec Rhodia	Co/Ni separation, REE, V
	LIX® 84-I	2-Hydroxy-5-non-ylacetophenone oxime	R = C₉H₁₉	BASF	Cu

Ion associating	Versatic™ 10	Neodecanoic acid	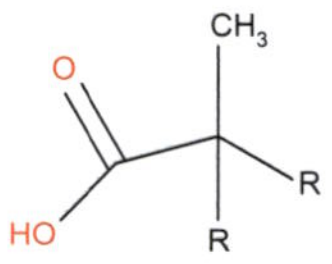	Resolution Performance Products Ltd.	Cu, Co, Ni, Fe
	e.g. Alamine® 336	Tri-*n*-octylamine	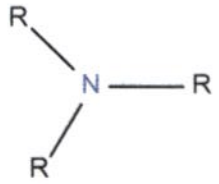 R = n-octyl	BASF	U, V, W, Mo
	Aliquat® 336	Quaternary ammonium salt	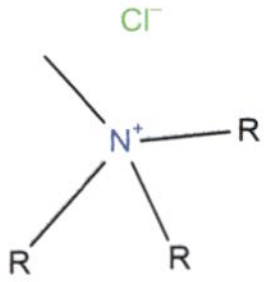 R = octyl	BASF	U, V, W, Mo
Solvating	TBP Cyanex® 921	Tri-*n*-butylphosphate Trioctylphosphineoxide	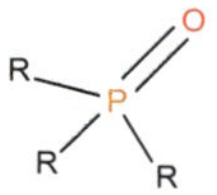 R = octyl	Several Cytec	U, V, Mo, Nb/Ta separation, Re, REE U, Nb/Ta

R = alkyl group

Cl^-

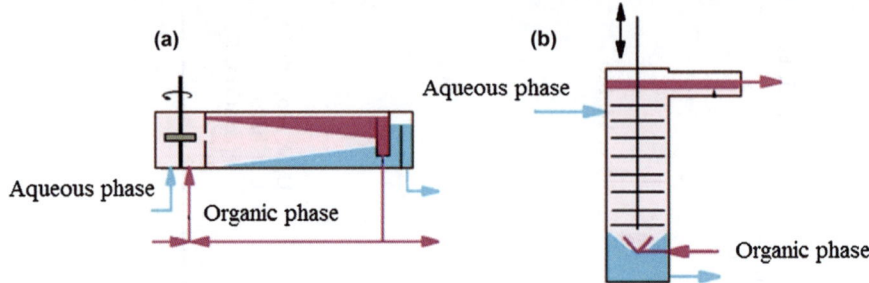

Figure 24.4   Schematics of (a) a mixer-settler unit and (b) a column extractor.[115] See respectively photos of these at www.kccl.co.ug/photos.htm[116] and www.alchemet.com/projects/index1.htm.[117]

unit, in which the phases are separated from each other. Another industrially important type of solvent extraction equipment is the column extractor (Figure 24.4 (b)), of which there are several variations.[105]

Applicability of solvent extraction in utilizing waste from electric and electronic equipment (WEEE) as a raw material for minor metals is demonstrated by the example of indium recovery from liquid crystal display (LCD) screens. This topic has been discussed in detail in the literature.[103,107] The electrode material in the surface of the glass of LCD panels is indium tin oxide (ITO = 90% In_2O_3 + 10% SnO_2). In and Sn can be leached from the glass by common mineral acids (H_2SO_4, HCl, HNO_3) in spite of the kinetics being slow. From the leachate, separation of In and Sn is possible by solvent extraction using di-2-ethylhexylphosphoric acid (D2EHPA), tributylphosphate (TBP) and a mixture of these two solutions. With TPB and the mixture, Sn can be selectively removed using an HCl solution, but the concentration of In in the solution is very low and some form of concentration is would be needed. Thus it was suggested that both of the metals should be extracted in the D2EHPA solution, which was known to have a high capacity for In. After that the In could be selectively stripped using HCl. This has resulted in an effective concentration of the Indium from $0.044\,g\,dm^{-3}$ to $6.5\,g\,dm^{-3}$. Figure 24.5 shows a process for recovery of indium from LCD screens by solvent extraction.

24.3.5   Ion Exchange

Ion exchange research began in the 1800s and early 1900s, but, as with solvent extraction, practical applications only began with the

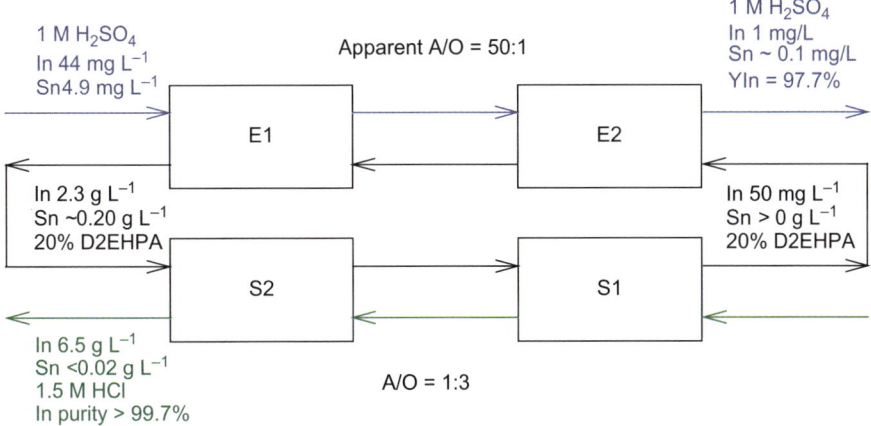

Figure 24.5   Process for separating and concentrating indium by solvent extraction from H_2SO_4 leachate of glass from an LCD panel.[103]

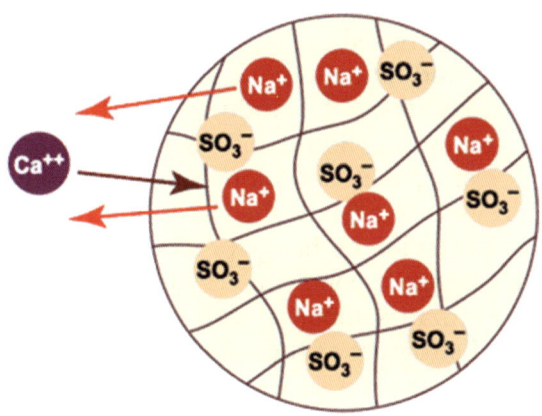

Figure 24.6   Principle of ion exchange.[119]

Manhattan project in the 1940s.[44,118] In ion exchange, an ion which is associated with the fixed charge in the ion exchange material, is re-placed by another ion (Figure 24.6). The process is stoichiometric. The process in which an adsorbent, which can also be an ion ex-changer, replaces some species from solution and is not, by defin-ition, ion exchange, but sorption.[118] Despite describing ion exchange as a separation method, usually it covers both mechanisms and in-volves only solid separation materials. And because the ion exchange process often involves sorption of electrolytes, *etc.*, and vice versa, sorption processes may be accompanied by ion exchange.[119]

As with solvent extraction, the distribution ratio, D_i, in ion exchange is defined in terms of analytical concentrations of the component i and the separation factor, $\alpha_{i,j}$, as a ratio of the distribution ratios. The selectivity coefficient, $k_{i,j}$, is corrected for the valences of the metals:[120]

$$D_i = \frac{q_i}{c_i}$$

$$\alpha_{i,j} = \frac{D_i}{D_j} \tag{24.7}$$

$$k_{i,j} = \frac{q_i^{|z_j|} c_j^{|z_i|}}{q_j^{|z_i|} c_j^{|z_i|}}$$

where q_i is the concentration of component i in the resin phase.

Ion exchangers are based on many different materials which include minerals, synthetic inorganic materials, ion exchange resins, ion exchange coals (activated carbon), liquid ion exchangers (solvent extraction reagents and ionic liquids), ion exchange membranes and cellulose ion exchangers.[118,121] Industrially, the most important group of these, excluding liquid ion exchangers, is ion exchange resins.[121] In the resins, as in solid ion exchangers in general, the reactive functional groups are attached to a matrix, which does not usually take part in the ion exchange process. The matrix is usually a polystyrene divinylbenzene copolymer (PS-DVB) or acrylic compound. Depending on the matrix material and fraction of the crosslinker (DVB), the physical structure of the resin may either be macroporous or a gel. In macroporous resin, the matrix has pores in which the solution can enter, while the gel type resin can be considered as a homogenous gel.[118,121] The resins are hydrophobic in nature, with the exception of their functional groups. Based on their functional groups, ion exchange resins can be divided into four groups. General reaction equations written for univalent ions (except for chelating resins) are also given for each group in the following list:

1. **Strong cation exchangers**

$$A^+ + \overline{B^+ R^-} \rightleftharpoons B^+ + \overline{A^+ R^-} \tag{24.8}$$

2. **Weak cation exchangers**

$$A^+ + \overline{H^+ R^-} \rightleftharpoons H^+ + \overline{A^+ R^-} \tag{24.9}$$

3. Strong anion exchangers

$$A^- + \overline{B^- NR_4^+} \rightleftharpoons B^- + \overline{A^- NR_4^+} \tag{24.10}$$

4. Weak anion exchangers

$$A^- + \overline{B^- NR_3 H^+} \rightleftharpoons B^- + \overline{A^- NR_3 H^+} \tag{24.11}$$

5. Chelating resins

$$A^{2+} + \overline{R(OH)_2} \rightleftharpoons 2H^+ + \overline{A^{2+} R(O)_2^{2-}} \tag{24.12}$$

In Equation (24.9) the proton is marked as another cation to highlight the fact that weak cation exchangers typically have good affinity, and thus they function better at a high pH than in very acidic solutions. Anion exchangers are not exclusively amines, but they are by far the most common ones in practice. Chelating resins act as multidentate ligands for polyvalent cations and often have good selectivity towards a certain metal because of the favourable structure of the cyclic complex that the functional groups form with the metal.

Chelating resins are often weak cation exchangers. Examples of some practically important commercial resins are given in Table 24.12.[103] It should be noted that the applications given are those provided by the manufacturer and that all these resins have potential in hydrometallurgical separations.

In practical applications, ion exchange is usually carried out in columns (Figure 24.7(a) and (b)). A good illustration of the principle of ion exchange columns is given in ref. 132. After the resin has been saturated with the desired metal(s) taken from the feed solution, the adsorbed metals are eluted. Finally the resin is converted to the desired ion form, if needed, by acid, base or salt solution. Between these process steps the resin bed is usually washed with water or a salt solution because the feed, elution and regeneration solutions should preferably not be mixed.

Ion exchange can be also performed so that there are continuous raffinate and eluent flows coming out from the process.[118,121] In this multicolumn, the so-called simulated moving bed (SMB) set-up, the locations of the incoming and outgoing solutions are changed periodically, either by valves or by moving the columns (Figure 24.7(c)). This set-up was used for the purification of a Ag–NaCl solution in a study by Virolainen *et al.*[131] A good illustration of the SMB set-up is also presented in ref. 133. SMB operation offers efficient utilization of the resin and eluent consumption is small.[134] In the

Table 24.12 Some industrially important and commercially available ion exchange resins.[122–128] For example, Amberlite™ IRA-743 was used in ref. 129 and Lewatit® TP-260 in references 130 and 131.

Group	Commercial name	Functional group	Structure	Manufacturer	Applications
Strong cation exchangers	Dowex™ 50	Sulfonic acid		Dow	Food processing
Weak cation exchangers	Purolite® C-104	Carboxylic acid		Purolite	Water treatment, transition metals
Strong anion exchangers	Amberlite™ IRA410	Quaternary ammonium		Rohm and Haas	Water treatment
Weak anion exchangers	WPGM® (WP-1®)	Polyamine		Purity Systems Inc.	Cu, and variety of other metals

Chelating resins				
Amberlite™ IRA-67	Tertiary amine		Rohm and Haas	Water treatment
Lewatit® TP-260	Aminomethylphos-phonic acid		Lanxess	Cu/Co, Zn/Co, U, Ti
Amberlite™ IRA-743	N-methylglucamine		Rohm and Haas	B, Ge

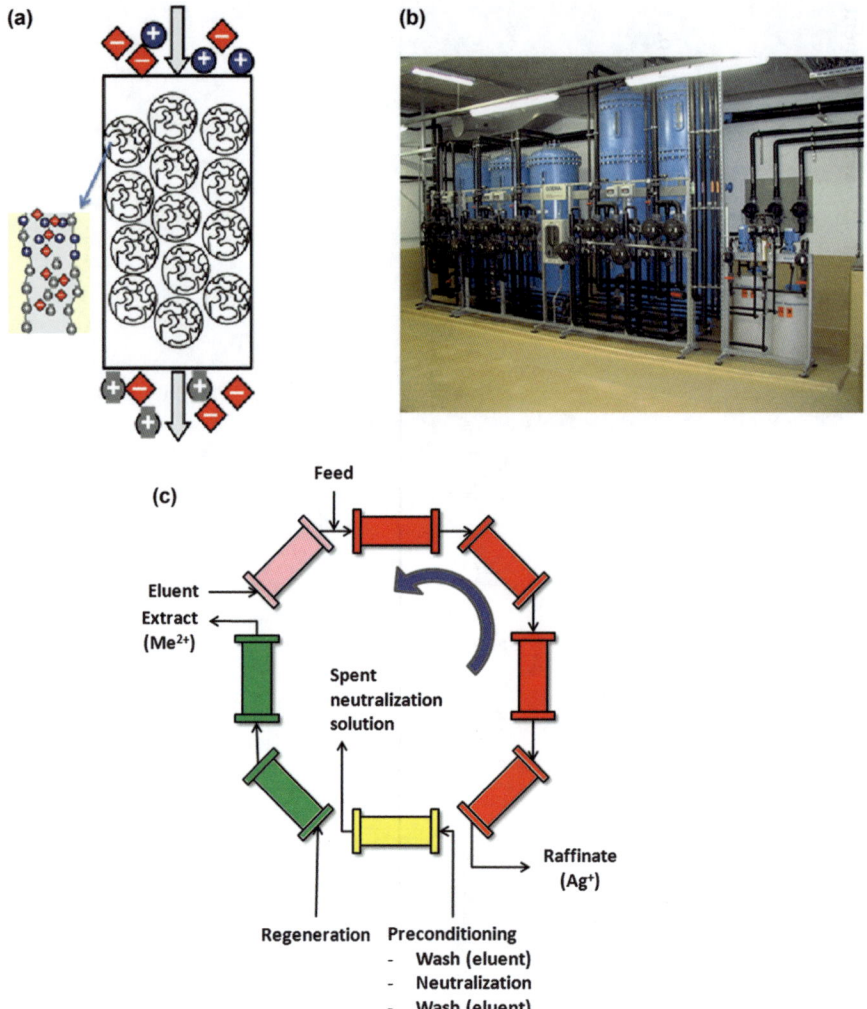

Figure 24.7   Ion exchange equipment: (a) schematic of ion exchange column; (b) real-life ion exchange columns;[136] and (c) example of SMB set-up.[103]

hydrometallurgical industry, one increasingly prevalent set-up in ion exchange is the so-called resin in pulp (RIP) process, in which the resin is mixed with the leaching slurry, thus reducing the number of process steps.[135]

Applicability of ion exchange to the recovery of minor metals from concentrated industrial base metal solutions is illustrated with the example of germanium (Ge) recovery. This topic has been discussed in detail by Virolainen[103] and Virolainen *et al.*[129] This example also shows the typical challenge in these types of separation tasks; the

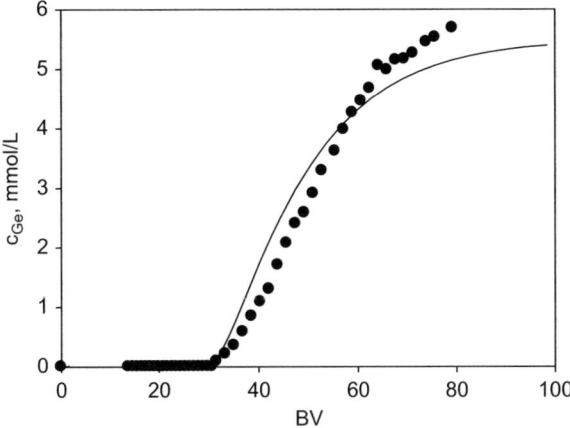

Figure 24.8 Recovery of Ge from CoSO$_4$ solution with *N*-methylglucamine functional resin IRA-743. Feed pH was 3.0. Circles represent the experimental data and the line is a modelling result.[129]

concentration of the target metal is low while the concentration(s) of
other metals are high. To effect such separations special separation
materials and/or techniques are needed, which highlights the need
for a good understanding of the field.

Recovery of Ge using *N*-methylglucamine functional IRA-743 resin
was observed to be highly dependent on the composition of the feed
solution, especially pH, but also the quality and quantity of other
metals. Based on the experimental data, a mechanism for this
chemically complicated ion exchange case was suggested and dynamic column separation was modelled. At the best, over 30 bed
volumes of 8.6 g dm^{-3} CoSO$_4$ solution were treated to recover all the
Ge, which had initial concentration of 0.400 g dm^{-3}. Figure 24.8
shows the recovery of germanium from base metal sulfate solutions
by the *N*-methylglucamine functional resin.

24.3.6 Recycling of Metals from Secondary Raw Materials

The term 'secondary raw material' is not formally defined but it
is widely used to describe a waste that has become a possible raw
material.[137–140,145] However, there is an official reverse term known as
the end-of-waste (EoW) or EoW criteria. End-of-waste status is defined
within the EU Waste Framework Directive 2008/98/EC;[139] EoW criteria
specify when a material is no longer after going through a recovery
operation based on certain legal conditions. After this, some material

formerly defined as waste is defined as a product or secondary raw material.[140] The EoW criteria are defined separately for different wastes. The purpose of the Waste Framework Directive is to guide society towards recycling, so that different wastes should be seen more as possible raw materials than a useless burden.

Today, hydrometallurgical separation methods are an essential part of extractive metallurgy and are utilized in various metal refining plants throughout the world. As rich ores for metals refining become scarce, the metallurgical industry has been turning to lower grade ores, solid and liquid side and waste streams in metal refining plants, and other forms of waste. For all these potential raw materials, hydrometallurgical separation methods are very suitable. They are known to be flexible, highly selective and environmentally friendly for the treatment of raw materials with varying amounts and compositions. Their energy consumption is also lower than more conventional pyrometallurgical methods. It is often stated that hydrometallurgical methods have major potential for resolving the future challenges in producing essential metals through environmentally and economically sustainable practices.

In principle, any object containing metals may be suitable as a raw material. In practice, the price of the obtainable metal together with the quantity and quality of the material define the viability of recovery. Some of the most important secondary raw materials for metals include: side streams and wastes (both solid and liquid) from the mining industry; WEEE including LCD panels and fluorescent lamps, batteries; and used catalysts. The valuable metals in these include: minor metals (Ge, In, Mo, Re); noble and precious metals (Ag, Au, Pd, Pt); valuable base metals (Co, Cu, Ni); and rare earth elements (REE). According to a United Nations Environment Program (UNEP) report,[141] current end-of-life (EoL) recycling rates of these metals are: $Ge < 1\%$, $In < 1\%$, $Mo = 30\%$, $Re > 50\%$, $Ag = (30–50)\%$, $Au = (15–20)\%$, $Pd = (60–70)\%$, $Pt = (60–70)\%$, $Co = 68\%$, $Cu = (43–53)\%$, $Ni = (57–63)\%$ and $REE < 1\%$. This clearly shows the potential of secondary raw materials as a source for many different metals.

As the world's population increases and people use more high-tech products, the usage of various natural resources in the coming decades requires more and more attention. In particular, the depletion of oil resources has been discussed widely. With metal resources, the situation is generally not as acute but, ultimately, they are not inexhaustible. It has been stated in a UNEP report concerning metal recycling[142] that the global demand of metals could increase up to nine-fold when developing countries reach the same standard of living as developed countries. The European Commission produced a list of 14 critical raw materials in 2010,[143] highlighting uncertainty in the balance of demand and supply for certain elements. As a consequence the Commission is studying a possibility of stockpiling these raw materials,[144] which include antimony (Sb), beryllium (Be), cobalt (Co), fluorspar (CaF_2), gallium (Ga), germanium (Ge), graphite (C), indium (In), magnesium (Mg), niobium (Nb), platinum group metals (PGMs), rare earth elements (REE), tantalum (Ta) and tungsten (W).

At the same time, the use of metals is increasing. The primary ore deposits that are currently being mined or will be mined have seen a drop in ore grade. Thus new primary metal deposits may require a larger investment, especially in the technological area. Therefore it makes sense to increase the recovery of metals from secondary raw materials, which may be of a much higher grade than primary ores and moreover are possibly more readily obtainable. But despite being seen as a promising metal resource, there are many practical difficulties in utilizing secondary raw materials, concerning mainly economics, logistics, technology and the complexity of raw materials. Today, the recovery rates for metals in secondary raw materials vary from a few per cent up to 75%, depending on the metal, the raw material and the processing route used.[142]

Currently a widely used term 'urban mining' simply means the utilization of anthropogenic wastes as metal resources. Jesse Stallone, an expert in the field of urban mining, gives a broader definition:[145] 'The process of reclaiming compounds and elements from products, building and waste'. The phrase 'urban mine' is an Australian trademark.[142] In addition to economic viability, metal recycling is also environmentally friendly, as many of the metals are categorized as hazardous waste and cannot be landfilled. The energy consumption, and therefore greenhouse gas emissions, of metals produced by recycling are also often lower than in primary production.[142]

Due to environmental risks, legislation concerning metal-containing waste, especially in developed countries, is being tightened; this is also a driving force for metal recycling. An approach called extended producer responsibility (EPR), introduced by the Organisation for Economic Co-operation and Development (OECD) in 2001[146] has become an important tool for controlling the end-of-life processing of many waste streams such as vehicles and WEEE. EPR implies that manufacturers are responsible for their products being disposed of appropriately when they become waste. EPR is especially followed in EU countries, Switzerland and Japan. In the United States and Canada, it is applied in only some states and provinces.[147]

To force environmentally sustainable manufacturing and EoL processing of electrical and electronic equipment (EEE), the EU launched two directives in January 2003. The first prevents the use of certain hazardous substances in EEE[148] and the other, which was recast in 2012, deals with EPR.[149,150] In Switzerland and Japan there are also several laws and directives concerning EPR, and in India and Thailand, for example, there are proposals to change the legislation towards EPR in WEEE management.[151]

24.4 CONCLUSIONS

Raw material issues have climbed up the political priority list and there is now more attention from various stakeholders such as local people, the tourism industry and environmental groups. These ongoing programmes highlight the need to increase resource efficiency, recycling rates, improve recoveries of metals in a cost-effective manner, and reduce dependency on foreign imports of certain metals and products.

Stricter effluent discharge limits have been proposed for metal mines in many countries. In Finland, the sulfate concentration of their effluents has been in the public focus. In some Finnish mines the challenge has been the excess of water after heavy rains and melting of snow. Globally, in many arid areas the challenge is related to water scarcity. In both these cases, the tendency and need in the mines is to reuse and recycle more water in order to gain public and regulatory acceptance. Sometimes lower quality water can be recycled directly back to the process, but often there is a need for different water treatment processes. New water treatment technologies such as forward osmosis may have potential applications in the mining industry.

Understanding the effects on process performance is crucial when deciding on water recycling. Extraction, leaching, flotation and hydroblasting represent over 80% of total water demand in mining. The whole process chain has to be taken into consideration when planning water recycling in minerals processing. Recycling of water in the mining industry is always case-specific, but good examples of successful operations in the world exist. Environmentally beneficial water recycling and reuse projects can also show a significant economic benefit.

Identified challenges related to mine water recycling in Finnish mines include pH, suspended solids and elevated metal, sulfate, chloride and nitrogen concentrations—all of which can disturb process performance. Calcium and sulfate can precipitate as gypsum, and block pipes and devices. Recycling is possible in places, where recycling does not form gypsum or other problems due to metal sulfate concentrations. Chloride can in some processes have a detrimental effect on autoclaves and reduce gold recoveries. Recycling of chloride containing tailings pond water into flotation solutions is possible, but often there is a limit to the chloride content in smelters.

Utilizing more and more secondary raw materials, such as liquid and solid side streams from the mining industry and different anthropogenic wastes like WEEE is an essential feature for sustainability in metallurgical industry. Using these secondary raw materials for metals' extraction is driven by legislative, political, environmental and economic motives.

Hydrometallurgical separation methods are increasingly being used in extractive metallurgy. This is because raw materials are becoming more low grade and complex; hydrometallurgical methods are more suitable for these than the more conventional pyrometallurgical methods. Hydrometallurgical separation methods also have many environmentally good features compared with pyrometallurgical methods such as lower gaseous emissions and energy consumption. Thus hydrometallurgy is likely to have a significant role in a sustainable metallurgical industry. Solvent extraction and ion exchange are popular tools in hydrometallurgical processes because of their effectiveness in metals' separation and purification. With these methods the target metal(s) may also be concentrated during the separation step. Solvent extraction and ion exchange have also been shown to be good methods for processing secondary raw materials because, among previously mentioned features, they are known to be flexible and can readily be modified to cater for complex and varying raw materials.

REFERENCES

1. CRM Innonet, Critical Raw Materials Innovation Network, www. criticalrawmaterials.eu/partners/ [accessed 5 February 2014].
2. (a) European Commission, *The Raw Materials Initiative – Meeting our Critical Needs for Growth and Jobs in Europe*, COM(2008) 699, European Commission, Brussels, 2008; (b) European Commission, *Tackling the Challenges in Commodity Markets and on Raw Materials*, COM(2011) 25, European Commission, Brussels, 2011.
3. European Innovation Platform on Raw Materials, http://ec.europa. eu/enterprise/policies/raw-materials/innovation-partnership/ [accessed 5 February 2014].
4. M. Vieira, M. Goedkoop, P. Storm and M. Huijbregts, *Env. Sci. Tech.*, 2012, **46**, 12772.

5. Eurostat statistics, http://epp.eurostat.ec.europa.eu/ [accessed 5 February 2014].
6. O. Salmi, *Science, Sulphur and Sustainability: Environmental Strategies of Mining in the Russian Kola Peninsula*, Doctoral thesis, HUT Dissertations 137, Espoo, Finland, 2008.
7. T. Lecher and J. Scott (ed.), *Materials for a Sustainable Future*, RSC Publishing, Cambridge, UK, 2012.
8. The ProMine project, http://promine.gtk.fi/ [accessed 5 February 2014].
9. J. Salminen, L. Räsänen, P. Blomberg and P. Koukkari, Thermodynamic models and experiments for mine water treatment, presented at 3rd International Conference By-Product Metals in Non-Ferrous Metals Industry, Wrocław, Poland, 15–17 May 2013.
10. P. M. Heikkinen, P. Noras and R. Salminen (ed.), *Mining Closure Handbook*, Geological Survey of Finland, Espoo, Finland, 2008.
11. Ministry of Employment and the Economy, *Making Finland a Leader in Sustainable Extractive Industry – Action Plan*, Ministry of Employment and the Economy, Helsinki, 2013.
12. European Rare Earth Competence Network (ERECON), http://ec.europa.eu/enterprise/policies/raw-materials/erecon/index_en.htm [accessed 5 February 2014].
13. Canadian Rare Earth Elements Network (CREEN), www.cim.org/en/RareEarth/Home/AboutUs.aspx [accessed 5 February 2014].
14. European Institute of Technology, Knowledge Innovation Communities Raw Matters, http://eit.europa.eu/kics/ [accessed 5 February 2014].
15. M. Miranda and A. Sauer, *Mine the Gap: Connecting Water Risks and Disclosure in the Mining Sector*, WRI Working Paper, World Resources Institute, Washington, DC, 2010.
16. A. J. Gunson, B. Klein, M. Veiga and S. Dunbar, *J. Clean. Prod.*, 2012, **21**, 71.
17. Mining Journal, Mining, People and the Environment, www.mpe-magazine.com/reports/reduce,-reuse,-recycle [accessed 6 March 2014].
18. R. F. Rubio, *Mine Water Environ.*, 2012, **31**, 69.
19. *Mine Closure Handbook, Environmental Techniques for the Extractive Industries*, ed. P. M. Heikkinen, P. Noras, R. Salminen, Espoo, Finland, 2008, http://arkisto.gtk.fi/ej/ej74.pdf [accessed September 2014].
20. CSIRO Land and Water – For the Smart Water Fund, *Guidance for the Use of Recycled Water by Industry*, Institute for Sustainability and Innovation, Victoria University, 2008.

21. US Geological Survey, Mining water use, http://ga.water.usgs. gov/edu/wumi.html [accessed 5 July 2013].
22. Environment Canada, http://www.ec.gc.ca/eau-water/default. asp?lang = En&n = 8456A142-1 [accessed September 2014].
23. T. E. Norgate and R. Lovel, *Water Use in Metal Production: A Life Cycle Perspective*, DMR-2505 report, CSIRO Minerals, Clayton, South Victoria, Australia, 2004.
24. H. Fleming and R. Radakovich, Water management in mining operations. Hatch water, presented at Mining Magazine Congress, Johannesburg, South Africa, 24–25 October 2011.
25. K. A Slatter, N. D. Plint, M. Cole, V. Dilsook, D. de Vaux, N. Palm and B. Oostendorp, Water management in Anglo Platinum process operations: Effects of water quality on process operations, in Abstracts of the International Mine Water Conference, Pretoria, South Africa, 19–23 October 2009, pp. 46–55, https:// www.imwa.info/docs/imwa_2009/IMWA2009_PlintSlatter.pdf [accessed 10 September 2014].
26. M. Westerstrand and B. Öhlander, *Mine Water Environ.*, 2011, **30**, 252.
27. G. Levay, R. Smart and W. M. Skinner, *J. South Afr. Inst. Min. Met.*, 2001, **101**, 69.
28. B. I. Whittington, R. G. McDonald, J. A. Johnson and D. M. Muir, *Hydrometallurgy*, 2003, **70**, 31.
29. International Council on Mining & Metals, *Water Management in Mining: A Selection of Case Studies,* Report ICMM 2012, International Council on Mining & Metals, London, 2012.
30. GE Power & Water, *Handbook of Industrial Water Treatment,* www. gewater.com/handbook/index.jsp [accessed 4 July 2013].
31. E. Muzenda, *Eng. Tech.*, 2010, **45**, 237.
32. F. K. Ikumapayi, *Recycling Process Water in Complex Sulphide Ore Flotation. Mineral Processing*, Doctoral thesis, Luleå University of Technology, Luleå, Sweden, 2013.
33. F. Ikumapayi, M. Mäkitalo, B. Johansson and K. H. Rao, *Miner. Eng.*, 2012, **39**, 77.
34. S. R. Rao and J. A. Finch, *Miner. Eng.*, 1989, **2**, 65.
35. Lakefield Research Limited & SENES Consultants Limited, *Applicable Technologies for the Management of Mining Effluents in the Northwest Territories*, QS-Y201-000-EE-A1, Department of Indian and Northern Affairs Canada, 2002.
36. M. Deng, Q. Liu and Z. Xu, *Miner. Eng.*, 2013, **49**, 165.
37. V. Kirjavainen, N. Schreithofer and K. Heiskanen, *Miner. Eng.*, 2002, **15**, 1.

38. P. A. Moreno, H. Aral, J. Cuevas, A. Monardes, M. Adaro, T. Norgate and W. Bruckard, *Miner. Eng.*, 2011, **24**, 852.

39. M. S. Manono, K. C. Corin and J. G. Wiese, *Miner. Eng.*, 2013, **40**, 42.

40. D. E. Ng'andu, *J. South Afr. Inst. Min. Met.*, 2001, **10**, 367.

41. N. P. Haran, E. R. Boyapati, C. Boontanjai and C. Swaminathan, *Dev. Chem. Eng. Min. Process.*, 2008, **4**, 197.

42. H. B. Dharmappa, M. Sivakumar, and R. Singh, in *Wastewater Recycle, Reuse and Reclamation*, ed. S. Vigneswaran, EOLSS, Paris, 2009, vol. 1, pp. 337–371.

43. J. Xu, R. Liu, W. Sun, Y. Hu and J. Dai, *J. Environ. Sci. Eng.*, 2012, **A1**, 279.

44. F. Habashi, *Textbook of Hydrometallurgy, Métallurgie Extractive*, Quebec, Canada, 1999.

45. M. B. Mooiman, K. C. Sole and D. J. Kinneberg, *Hydrometallurgy*, 2005, **79**, 80.

46. J. A. Brierley and C. L. Brierley, *Hydrometallurgy*, 2001, **59**, 233.

47. C. S. Gahan, H. Srichandan, D.-J. Kim and A. Akcil, *Res. J. Recent Sci.*, 2012, **1**, 85.

48. D. S. Holmes, *Hydrometallurgy*, 2008, **92**, 69.

49. D. B. Johnson, *Hydrometallurgy*, 2006, **83**, 153.

50. B. R. Conard, *Hydrometallurgy*, 1992, **30**, 1.

51. G. Wu, H. Kang, X. Zhang, H. Shao, L. Chu and C. Ruan, *J. Hazard. Mater.*, 2010, **174**, 1.

52. J.-C. Lee and B. D. Pandey, *Waste Manage.*, 2012, **32**, 3.

53. I. Palencia Perez and J. E. Dutrizac, *Hydrometallurgy*, 1996, **26**, 211.

54. J. Salminen, P. Kobylin, and T. Kaskiala, in *Developments and Applications in Solubility*, ed. T. Lecher, RSC Publishing, Cambridge, UK, 2007, pp. 261–272.

55. T. Kaskiala and J. Salminen, *Ind. Eng. Chem. Res.*, 2003, **42**, 1827.

56. Melrose Project, www.korabresources.com.au/melrose.htm [accessed 13 February 2014].

57. CST Mining Group Ltd, View of the top of the heap-leach pad, http://www.cstmining.com/image/view-top-heap-leach-pad [accessed 13 February 2014].

58. C. L. Brierley, *Hydrometallurgy*, 2010, **104**, 324.

59. V. Z. Arens and S. A. Chernyak, *Metallurgist*, 2008, **52**, 3.

60. M. Riekkola-Vanhanen, *Nova Biotech.*, 2010, **10**, 7.

61. T. Rohwerder, T. Gehrke, K. Kinzler and W. Sand, *Appl. Microbiol. Biotechnol.*, 2003, **63**, 239.

62. M. Al-Harahsheh and S. W. Kingman, *Hydrometallurgy*, 2004, **73**, 189.
63. S. W. Kingman and N. A. Rowson, *Miner. Eng.*, 1998, **11**, 1081.
64. J. L. Luque-Garcia and de C. Luque, *Trends Anal. Chem.*, 2003, **22**, 41.
65. K. L. Narayana, K. M. Swamy, R. Sarveswara and J. S. Murty, *Miner. Process. Extr. Metall. Rev.*, 1997, **16**, 239.
66. D. Dreisinger, *Hydrometallurgy*, 2006, **83**, 10.
67. G. Hilson and A. J. Monhemius, *J. Cleaner Prod.*, 2006, **14**, 1158.
68. M. G. Aylmore and D. M. Muir, *Miner. Eng.*, 2001, **14**, 135.
69. J. O. Marsden and C. I. House, *The Chemistry of Gold Extraction*, Society for Mining, Metallurgy, and Exploration, Littleton, CO, 2nd edn, 2006.
70. H. Kurama and T. Catalsarik, *Desalination*, 2000, **129**, 1.
71. M. M. Aghamirian and W. T. Yen, *Miner. Eng.*, 2005, **18**, 89.
72. A. Bahrami, M. R. Hosseini and K. Razmi, *Mine Water Environ.*, 2007, **26**, 191.
73. X. Dai and M. I. Jeffrey, *Hydrometallurgy*, 2006, **82**, 118.
74. P. Kinnunen, *High-rate Ferric Sulphate Generation and Chalcopyrite Concentrate Leaching by Acidophilic Microorganisms*, Doctoral thesis, Tampere University of Technology, Tampere, Finland, 2004.
75. A. W. Breed and G. S. Hansford, *Biochem. Eng. J.*, 1999, **3**, 193.
76. J. Romero, C. Yañez, M. Vásquez, E. R. Moore and R. T. Espejo, *Res. Microbiol.*, 2003, **154**, 353.
77. G. C. De, D. J. Oliver and B. M. Pesic, *Hydrometallurgy*, 1997, **44**, 53.
78. E. N. Lawson, C. J. Nicholas, and H. Pellat, In *Biohydrometallurgical Processing*, ed. T. Vargas, C. A. Jerez, J. V. Wiertz and H. Toledo, University of Chile, 1995, pp. 165–174.
79. N. S. Vardanyan and V. P. Akopyan, *Microbiology*, 2003, **72**, 438.
80. A. Das, J. M. Modak and K. A. Natarajan, *Miner. Eng.*, 1997, **10**, 743.
81. N. V. Fomchenko, O. V. Slavkina and V. V. Biryukov, *Microbiology*, 2003, **39**, 82.
82. I. A. Chisholm, L. G. Leduc and G. D. Ferroni, *Antonie van Leeuwenhoek*, 1998, **73**, 245.
83. L. G. Leduc, G. D. Ferroni and J. T. Trevors, *World J. Microbiol. Biotechnol.*, 1997, **13**, 453.
84. G. Rossi, *Biohydrometallurgy*, McGraw-Hill, Hamburg, Germany, 1990.
85. M. I. Sampson and C. V. Phillips, *Miner. Eng.*, 2001, **14**, 317.
86. J. Salminen, *Chemical Thermodynamics of Aqueous Electrolyte Systems for Industrial and Environmental Applications*, Doctoral thesis, Helsinki University of Technology, Finland, 2004.

87. P. Koukkari, R. Pajarre, H. Pakarinen and J. Salminen, *Ind. Eng. Chem. Res.*, 2001, **40**, 5014.
88. J. F. Blais, Z. Djedidi, C. Ben, R. D. Tyagi and G. Mercier, *Pract. Period. Hazard., Toxic, Radioact. Waste Manage.*, 2008, **12**, 135.
89. A. E. Lewis, *Hydrometallurgy*, 2010, **104**, 222.
90. P. Hetherington, *Process intensification : a study of calcium carbonate precipitation methods on a spinning disc reactor*, Thesis, Newcastle University, UK, 2006.
91. J. Salminen and J. M. Prausnitz, in *Developments and Applications in Solubility*, ed. T. Letcher, RSC Publishing, Cambridge, UK, 2007, pp. 337–349.
92. BioMineWiki, Sulfide *versus* hydroxide precipitation, http://wiki. biomine.skelleftea.se/wiki/index.php/Sulfide_*versus*_hydroxide_ precipitation [accessed 13 February 2014].
93. A. Kaksonen, *The Performance, Kinetics and Microbiology of Sulfidogenic Fluidized-Bed Reactors Treating Acidic Metal- and Sulphate-Containing Wastewater*, Doctoral thesis, Tampere University of Technology, Tampere, Finland, Publication 489, 2004.
94. P. Braul, *CIM Magazine*, 2013, **8**, 42.
95. H. Arima, T. Fujita and W.-T. Yen, *Mater. Trans.*, 2002, **43**, 485.
96. E. Guerra and D. Dreisinger, *Hydrometallurgy*, 1999, **51**, 155.
97. F. Habashi, *CIM Magazine*, 2006, **1**, 99.
98. B. S. Boyanov, V. V. Konareva and N. K. Kolev, *Hydrometallurgy*, 2004, **73**, 163.
99. G. Viramontes Gamboa, M. Medina Noyola and A. López Valdivieso, *J. Colloids Interface Sci.*, 2005, **282**, 408.
100. G. M. Ritcey, *Tsinghua Sci. Technol.*, 2006, **11**, 137.
101. T. C. Lo, M. H. Baird and C. Hanson, *Handbook of Solvent Extraction*, Krieger, Malabar, FL, 1991.
102. V. S. Kislik, *Solvent Extraction – Classical and Novel Approaches*, Elsevier, Oxford, UK, 2012.
103. S. Virolainen, *Hydrometallurgical Recovery of Valuable Metals From Secondary Raw Materials*, Doctoral Thesis, Lappeenranta University of Technology, Lappeenranta, Finland, 2013.
104. N. M. Rice, H. M. Irving and M. A. Leonard, *Pure Appl. Chem.*, 1993, **65**, 2373.
105. G. M. Ritcey, *Solvent Extraction: Principles and Applications to Process Metallurgy*, G.M. Ritcey & Associates, Ottawa, Canada, 2nd edn, 2006, vol. 1.
106. S. Virolainen, D. Ibana and E. Paatero, *Hydrometallurgy*, 2011, **107**, 56.

107. S. Virolainen, R. Salmimies, M. Hasan, A. Häkkinen and T. Sainio, *Hydrometallurgy*, 2013, **140**, 181.

108. M. Golinski, Extraction of tin and indium with tributyl phosphate from hydrochloric acid solutions, in *Proceedings of International Solvent Extraction Conference ISEC '71,* London, 1971, pp. 603–615.

109. Cyanex, Cyanex 272, 921 and 272, http://www.chemindustry. com/chemicals/0896429.html, http://www.cytec.com/businesses/ in-process-separation/mining-chemicals/industries-applications/ nickel-cobalt [accessed September 2014].

110. LIX 84-I Solvent Extraction Reagent, Technical Information Sheet TI/EVH 0129e, BASF, Ludwigshafen, Germany, 2013, www. mining-solutions.basf.com/ev/internet/mining-solutions/en/ function/conversions:/publish/content/mining-solutions/ download-center/technical-data-sheets/pdf/LIX_84-I_TI_EVH_ 0129.pdf [accessed 6 February 2014].

111. Momentive, Versatic Acid 10 technical data sheet, 2012, www. momentive.com/Products/TechnicalDataSheet.aspx?id = 6018, [accessed 6 February 2014].

112. Ionquest 290, Rhodia, http://www.flomin.com/es/SNF.nsf/ Ionquest290_Flomin.pdf [accessed September 2014].

113. Aliquat® 336, Product Specification, Sigma-Aldrich, www. sigmaaldrich.com/catalog/product/aldrich/205613? lang = fi&region = FI [accessed 6 February 2014].

114. Products & Technologies for Hydrometallurgy, BASF, http://www. mining-solutions.basf.com/ev/internet/mining-solutions/en_GB/ products-technologies/hydrometallurgy [accessed September 2014].

115. Halwachs, Precious metals refining by solvent extraction, www. halwachs.de/solvent-extraction.htm [accessed 6 February 2014].

116. Kasese Colbalt Company Limited (KCCL), Photo gallery, www. kccl.co.ug/photos.htm [accessed 13 February 2014].

117. Alchemet.com, Solvent extraction of uranium using columns commissioned with assistance from Alchemet, www.alchemet. com/projects/index1.htm, [accessed 13 February 2014].

118. F. Helfferich, *Ion Exchange*, Dover Publications, New York, 1995.

119. Ion Exchange Basics, http://dardel.info/IX/IX_Intro.html [accessed 13 February 2014].

120. Recommendations on ion-exchange nomenclature, *Pure Appl. Chem.*, 1972, 29, 619.

121. K. Dorfner, *Ion Exchangers*, Walter de Gruyter, Berlin, 1991.

122. Lewatit TP-260 Product information, Lanxess, 2010.
123. Rohm and Hass Company, IRA-743 Product data sheet, 2008. www.dow.com/assets/attachments/business/ier/ier_for_ industrial_water_treatment/amberlite_ira743/tds/amberlite_ ira743.pdf [accessed 6 February 2014].
124. Purity Systems Inc., WP-1 Product data sheet, 2004.
125. Dow Water Solutions, Dowex Fine Mesh Spherical Ion Exchange Resins, http://msdssearch.dow.com/PublishedLiteratureDOWCOM/ dh_006f/0901b8038006f232.pdf?filepath = liquidseps/pdfs/noreg/ 177-01509.pdf&fromPage = GetDoc   [accessed   7   February 2014].
126. Purolite C-104 - Technical data, http://www.reskem.com/pdf/ purolite-c104.pdf, [accessed 7 February 2014].
127. IRA-410 - Product data sheet, 2000. http://www.wallner-wasser.at/de/ downloads/Amberlite_IRA_410_Cl.pdf, [accessed 7 February 2014].
128. Dowex M4195 - Product information, Dow, 2008, http:// msdssearch.dow.com/PublishedLiteratureDOWCOM/dh_006f/ 0901b8038006f241.pdf?filepath = liquidseps/pdfs/noreg/177- 01817.pdf&fromPage = GetDoc [accessed 7 February 2014].
129. S. Virolainen, J. Heinonen and E. Paatero, *Sep. Pur. Tech.*, 2013, **104**, 193.
130. S. Virolainen, I. Suppula and T. Sainio, *Reactive Funct. Polym.*, 2013, **73**, 647.
131. S. Virolainen, I. Suppula and T. Sainio, *Hydrometallurgy*, 2014, **142**, 84–93.
132. How Morton System Saver Water Softeners Work, www. systemsaver.com/morton-website/softeners/principle-of-ion- exchange.htm [accessed 13 February 2014].
133. ISEP System, http://pdf.directindustry.com/pdf/calgon-carbon- corporation/isep-system/60944-92300.html [accessed 13 February 2014].
134. Kirk-Othmer, *Encyclopedia of Chemical Technology*, Wiley, New York, 4th edn, 1991.
135. J. van Deventer, *Solvent Extr. Ion Exch.*, 2011, **29**, 695.
136. Ovivo, Ion exchange, www.ovivowater.us/en/IonExchange [accessed 13 February 2014].
137. Eurometaux-European Association of Metals, *Recovered Metals, Metal Compounds, Alloys, By-Products and End-Of Waste Materials under REACH: Issues and Options*, Eurometaux-European Association of Metals, Brussels, 2008.
138. F. Costantino, G. di Gravio, G. Sciarra, and M. Tronci, in *Proceedings of International Conference on Sustainable Energy and*

Environmental Engineering, ICSEEE 2012, Guangzhou; China, 2013, vol. 295–298, pp. 1714–1719.

139. European Union, Directive 2008/98/EC of the European Parliament and of the Council of 19 November 2008 on Waste and Repealing Certain Directives, *Off. J. Eur. Union*, 2008, **L312**, 3.

140. Waste Framework Directive. End-of-waste criteria, http://ec.europa.eu/environment/waste/framework/end_of_waste.htm [accessed 7 February 2014].

141. T. E. Graedel, J. Allwood, J.-P. Birat, B. K. Reck, S. F. Sibley, G. Sonnemann, M. Buchert and C. Hagelüken, *Recycling Rates of Metals: A Status Report*, A Report of the Working Group on Global Metal Flows to the International Resource Panel, United Nations Environment Programme, Paris, 2011.

142. M. Reuter, C. Hudson, A. Van Schaik, K. Heiskanen, C. Meskers, and C. Hagelüken, *Metal Recycling: Opportunities, Limits, Infrastructure*, United Nations Environment Programme, Paris, 2013.

143. European Commission, *Critical Raw Materials for the EU*, Report of the Ad-hoc Working Group on Defining Critical Raw Materials, European Commission, Brussels, 2010.

144. Risk and Policy Analysts Limited, *Stockpiling of Non-energy Raw Materials*, Final report prepared for European Commission DG Enterprise and Industry, Brussels, 2012.

145. What is Urban Mining, http://urbanmining.org/ [accessed 7 February 2014].

146. OECD, *Extended Producer Responsibility: A Guidance Manual for Governments*, OECD, Paris, 2001.

147. A. J. Spicer and M. R. Johnson, *J. Cleaner Prod.*, 2004, **12**, 37–45.

148. European Union, Directive 2002/95/EC of the European Parliament and of the Council of 27 January 2003 on the Restriction of the Use of certain Hazardous Substances in Electrical and Electronic Equipment (RoHS), *Off. J. Eur. Union*, 2003, **L37**, 19.

149. European Union, Directive 2002/96/EC of the European Parliament and of the Council of 27 January 2003 on Waste Electrical and Electronic Equipment (WEEE), *Off. J. Eur. Union*, 2003, **L37**, 24.

150. European Union, Directive 2012/19/EU of the European Parliament and of the council of 4 July 2012 on waste electrical and electronic equipment (WEEE), *Off. J. Eur. Union*, 2012, **L197**, 38.

151. P. Kiddee, R. Naidu and M. H. Wong, *Waste Manage.*, 2013, **33**, 1237.

PART D
PROCESS INTEGRATION

Process Integration: An Overview

RAFIQUL GANI

CAPEC-PROCESS, Department of Chemical & Biochemical Engineering, Technical University of Denmark, DK-2800 Kgs. Lyngby, Denmark Email: rag@kt.dtu.dk

25.1 PROCESS INTEGRATION AND SUSTAINABILITY

The current emphasis on sustainable production has prompted much improvement within chemical and biochemical processing plants to minimize their raw material and energy usage as well as waste generation without compromising the economic value of the enterprise. While computer tools are available to assist in the sustainability assessment of processes, their applications are constrained to a specific domain of the synthesis–design–control problem such as alternative generation, economic and environmental impact analysis, process optimization or process control. Computer aided methodologies for sustainable process design are being developed and applied to find more sustainable alternatives to existing processes as well as to new processes. This has also resulted in a redefinition of the process synthesis–design problem as highlighted in Figure 25.1.

It is no longer enough simply to find a sequence of operations that can convert a set of raw materials to the desired products, given the identities of the raw materials and the desired specifications on the products one would like to produce. The process, operating in batch or continuous mode, in addition to converting the raw material (feed

Chemical Processes for a Sustainable Future
Edited by Trevor M. Letcher, Janet L. Scott and Darrell A. Patterson

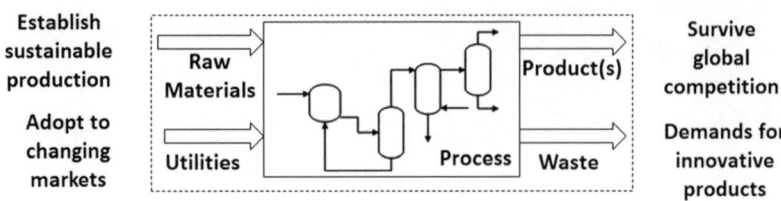

Figure 25.1 Sustainable process design: processes need to be sustainable (economically feasible, reduced waste, utility efficient, environmentally acceptable), safe, operable, *etc.*

streams) to the desired product (outlet streams) through a series of configured operations performed by appropriate equipment, must also minimize cost, waste, environmental impact subject to process operability, safety hazards, supply chain (of raw materials and/or products) and many more factors.

The synthesis issues usually deal with the questions of configuration of operations, while the design issues usually deal with the questions of performance of the operations. The synthesis–design problems may seek to improve an existing process, which is usually referred to as the retrofit problem. Or the synthesis–design problems may seek to find a completely new process. Because of the concerns about economics, energy, water, waste, safety and the environment, the designed process must also be sustainable, safe, flexible and reliable. Issues such as integration of heat and mass, process intensification and sustainability have therefore become important topics in process synthesis and design.

The generic mathematical definition of the process design problem, which also includes process integration and intensification, is given by:

$$F_{obj} = \{C_p\, f_p(x, y) + C_w\, f_w(x, y) + C_s\, f_s(x, y) + \cdots \cdots\} \qquad (25.1)$$

subject to:

$$D_1 = h_1(x, z, p) \qquad (25.2)$$

$$D_2 = h_2(x, z, p) \qquad (25.3)$$

$$0 \geq g_1(x, z, p, \theta) \qquad (25.4)$$

$$0 \geq g_2(x, z, p, \theta) \qquad (25.5)$$

$$By + Cx \leq d \qquad (25.6)$$

In the above generic problem definition, the performance index, Equation (25.1), needs to be minimized or maximized, and may include multiple terms (for example, process cost, penalty for waste, penalty for environmental impact). The process model, Equation (25.2), satisfies the conservation of mass and/or energy as a function of the process variables (x), design variables (z) and model parameters (p). The process performance model, Equation (25.3), predicts the behaviour of the process during its application. The process structure equation, Equation (25.6), generates the feasible flow sheet structures as a function of x and the decision (integer) variables (y) and, finally, the process constraints, Equations (25.4) and (25.5) define operational constraints (θ). Considering that the models may be multi-scalar, non-linear and the variables involved may be integers as well as real, the generic problem defined above could represent a variety of problems from a simple linear programming (LP) type to a complex multi-dimensional problem of the mixed integer non-linear programming (MINLP) type.

Several variations of the above problem have also been formulated and solved in process optimization,[1–4] heat-mass exchange networks[5–8] and in product–process design.[9–10] Chapters 26 and 27 highlight two variations of the generic problem. In Chapter 26, process control for sustainable processes with respect to exergy is highlighted, while in Chapter 27, sustainable process synthesis–design–intensification is highlighted.

Chapter 26 introduces the concept of sustainable process control as an integrated approach to process design, process control and sustainability that may be applied during the early process design stages. Sustainability as a concept covers multiple dimensions such as energy, ecological, political and cultural dimensions. In this chapter, only energy-related sustainability issues are considered. Several tools are also introduced to fulfil sustainable process control by integrating process control, process design and sustainability.

Chapter 27 proposes that, by considering process intensification as part of process synthesis, sustainable alternatives of existing as well as new processes can be determined that correspond to lower values of a set of targeted performance criteria. A systematic computer-aided multi-level framework for process synthesis, design and intensification that leads to more sustainable process designs is presented. The framework, operating at multiple levels, consists of a collection of methods and tools and evaluates alternatives through a set of targeted performance criteria in terms of economics, sustainability indices and life cycle factors. At the unit operations level, process flow sheets are synthesized by the sequencing of unit operations. At the task level, a task-based process synthesis method is used where tasks representing unit operations are identified, analyzed and recombined through a means-ends analysis. At the phenomena level, a phenomena-based process synthesis method is used, where the phenomena associated with each task are identified, manipulated and combined to generate new and/or existing unit operations. These unit operations are configured into flow sheet alternatives that are more sustainable and which correspond to even lower values of the set of targeted performance criteria compared with the base case design. The framework as well as associated methods and tools are presented, and their application highlighted through a case study involving the production of methyl acetate.

REFERENCES

1. K. P. Papalexandri and E. N. Pistikopoulos, *Comput. Chem. Eng.*, 1995, **19**, 71.
2. M. Hostrup, R. Gani, Z. Kravanja, A. Sorsa and I. E. Grossmann, *Comput. Chem. Eng.*, 2001, **25**, 73.
3. I. E. Grossmann, *AIChE J.*, 2005, **51**, 1846.
4. A. Quaglia, B. Sarup, G. Sin and R. Gani, *Comput. Chem. Eng.*, 2013, **59**, 47.
5. M. M. El-Halwagi and V. Manousiouthakis, *AIChE J.*, 1989, **35**, 1233.
6. M. J. Bagajewicz, *Comput. Chem. Eng.*, 2000, **24**, 2113.
7. K. C. Furman and N. V. Sahinidis, *Ind. Eng. Chem. Res.*, 2002, **41**, 2335.
8. R. Smith, *Chemical Process Design and Integration*, John Wiley and Sons, New York, 2005.
9. A. T. Karunanithi, L. E. K. Achenie and R. Gani, *Ind. Eng. Chem. Res.*, 2005, **44**, 4785.
10. R. Gani, *Comput. Chem. Eng.*, 2004, **28**, 2441.

CHAPTER 26

Process Control for Sustainable Processes with respect to Exergy

M. T. MUNIR, W. YU AND B. R. YOUNG*

Industrial Information & Control Centre (I²C²), The University of Auckland, New Zealand
*Email: b.young@auckland.ac.nz

26.1 INTRODUCTION

This chapter introduces to the process control community the concept that it is vital to consider sustainable process control in an integrated approach during early process design stages for a sustainable process. Process sustainability has earned an important place along with production, process design, process control, efficiency, quality and safety in manufacturing priorities. This chapter summarizes recent research activities related to sustainable process control.

An integrated approach to process design, process control and sustainability is becoming more popular and has many advantages over the classical approach in which process control and sustainability are considered separately during late stages of the process design. Applying an integrated approach means being able to change the flow sheet structure, employ different types of equipment and use alternative design parameters more easily and more cost-effectively. This integrated approach simultaneously allows changes to be made in the configuration of the process to permit easier and sustainable

Chemical Processes for a Sustainable Future
Edited by Trevor M. Letcher, Janet L. Scott and Darrell A. Patterson
© The Royal Society of Chemistry 2015
Published by the Royal Society of Chemistry, www.rsc.org

control and evaluation of the many control structure candidates in early process design stages. This new integrated approach may result in better process design, an easily controllable and sustainable process with low additional capital expenditure and few lost sales opportunities.[1,2]

Sustainability is a multi-faceted concept which has multiple dimensions. *e.g.* energy, ecological, political and cultural. In this chapter, only energy-related sustainability is considered.

26.2 BACKGROUND

26.2.1 Process Design

Process design practice is a major activity in a chemical processing plant. It requires a deep understanding and knowledge about all the subjects of chemical engineering including design, process control and thermodynamics. The process design philosophy is creative and iterative in nature. It starts with a conceptual understanding of the process followed by process synthesis and collection of data. Optimization and changes are made at the end of the classical design procedure. At the end of every stage of the process, the design is evaluated in terms of economics.[3,4]

Process design has a close connection with process control because the fundamental design of a process determines its inherent controllability. The inherent controllability of a process design determines its disturbance rejection (robustness) and the ease with which a process moves between operating conditions. Thus process control must be considered during early process design stages (process synthesis and design activities).[1,2]

Integrating concepts of process sustainability and process design are well explored in the literature. For example, thermodynamic properties such as exergy have been used to develop a systematic approach to optimal and sustainable process design with low exergy destruction.[5–7]

26.2.2 Process Control

Process control is a complex and multi-faceted task having four main stages: (1) control system structure design; (2) controller algorithm/configuration selection; (3) controller tuning; and (4) control hardware/infrastructure/selection and implementation. Control loop configuration/control pair selection occurs during the control system structure stage. The selection of the controller type (*e.g.* a proportional, integral and derivative type) pertains to the controller algorithm/configuration stage. The final stage of process control involves dealing with controller tuning. The selection of control hardware such as sensors, control valves and final control elements is the last stage of process control.[1,8,9]

We now return to considering the determination of the control system structure. The selection of control system structure is based on the type of control used for the process. A centralized multi-input multi-output (MIMO) controller or a set of decentralised multiple single-input single-output (SISO) controllers is used to control a process. A decentralized type of control system is more common than the centralized type of control system in the process industry as it has more industry-relevant advantages: it is easy to understand; uses uncomplicated hardware; and employs simple working algorithms.[8,10]

For a decentralized type of control system, the control system structure decides the best control scheme. Control scheme selection pertains to the pairing of manipulated variables (MV) and controlled variables (CV). For a process, control scheme selection is a straightforward task, provided that no interactions are present between the various control loops in the multi-loop control schemes of that process. However, this is rarely the case in process control design practice. Good control scheme selection is essential because the incorrect pairing of MV and CV will result in poor performance.

There are several classical techniques and methods for designing/ selecting decentralized control schemes, such as the relative gain array (RGA), the Niederlinski index (NI), singular value decomposition (SVD), the condition number (CN) and Morari's resiliency index (MRI).[9,11]

26.2.3 Sustainability

The sustainability of a process typically refers to its maintenance of well-being using renewable resources. As noted in Section 26.1, only energy-related sustainability is considered in this chapter. Energy sustainability is also a measure of progress in 'green' chemical engineering growth. The concept of energy sustainability can be traced back to the 1960s as the concept of environmental efficiency, or as a business when linked to sustainable development. Energy sustainability has a role in expressing how efficient economic activity is with regard to natural resources. Energy sustainability focuses on methods of resource use, obtaining economic and environmental progress through efficient use of resources and lower emissions.[12-14]

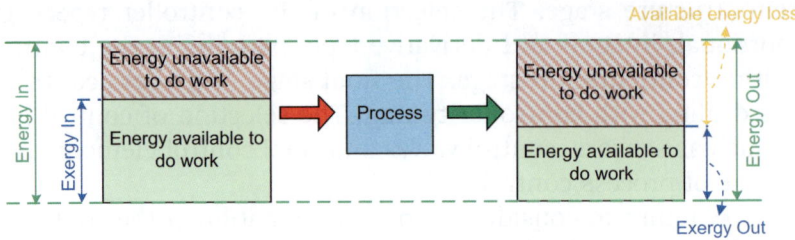

Figure 26.1   Energy *versus* exergy.

Thermodynamic properties such as exergy have been used as a measure of energy sustainability. There are many exergy-based concepts and methods such as efficiency concepts and exergy flow diagrams which can be used to describe, analyse and optimize systems in terms of energy sustainability.[6] A brief history of exergy use is given in ref. 15.

26.2.3.1 Energy versus *Exergy Analysis.*   The total energy flowing into and out of a process is equal, according to the first law of thermodynamics (*i.e.* law of conservation of energy). The total energy of a material stream or system has available and unavailable parts as shown in Figure 26.1. The thermodynamic property exergy is the component of energy which is available to do useful work and also accounts for energy quality (*i.e.* available energy). Unlike the first law of thermodynamics, the total exergy flow into and out of a process is always unequal due to entropy generation or exergy destruction during the spontaneous process (*i.e.* second law of thermodynamics) as also shown in Figure 26.1.

26.2.3.2 Exergy and its Calculation.   Formally exergy was defined by Kotas[16] as:

'The maximum amount of work/useful energy drawn from a process/material stream as it comes from its original state (process condition) to the ultimate dead state (reference state) during which it interacts only with the environment'.

At the ultimate dead state, the process or material stream is in thermodynamic (thermal, mechanical and chemical) equilibrium with the environment.

For an exergy analysis, the exergy value of every material stream coming into and going out of the considered process is required.

There are different methods for exergy calculation available in the literature.[17–22] In this chapter the exergy calculator developed by Munir *et al.*[21] is used for the exergy calculation of material streams. We do this for three major reasons: (1) it is based on a recent commercial process simulator with the most up-to-date thermodynamic data; (2) calculations using thermodynamic data are relatively easy to perform in a commercial process simulator; and (3) it divides the total exergy of the material stream into three main components (physical exergy, chemical exergy and exergy change due to mixing) which is helpful to identify the major component responsible for the total exergy destruction of a material stream. Readers are referred to textbooks and research articles for further details on exergy and its analysis.[6,23–28]

26.2.3.3 Energy versus Exergy Efficiency. In thermodynamics, efficiency is one of the most frequently used terms to indicate how well available energy is converted into useful work. Generally it is defined as the ratio of desired output to required input. In literature, there are a variety of other specific efficiency relations for different engineering systems or operations.[29]

Energy efficiency (*i.e.* ratio of output energy to input energy) is primarily based on the first law of thermodynamics. It helps to understand the system's behaviour to some extent and provides the relative magnitude of energy losses, but does not include the usability of the converted energy. In other words it only considers the quantity of energy but not the quality of the converted energy.

In contrast to energy efficiency, exergy efficiency (*i.e.* ratio of output exergy to input exergy) is based on combing the first and second laws of thermodynamics. It includes the quality of the converted energy, and thereby determines the true magnitudes of energy losses, their causes and their locations. Exergy analysis of a process and/or the individual units making up the overall process is performed using mainly their exergetic efficiencies. Recently exergy analysis has been used as a useful tool in process design, control, optimization and energy improvements.[22,30–35] There are several ways of defining exergy efficiency, depending on the type of the process information available or the desired comparison for the process under consideration.[6,12]

To analyse a process, both efficiencies (*i.e.* energy and exergy) are important as they consider the quality and quantity of the energy to achieve a given objective. But exergy analysis appears to be a more powerful tool than energy analysis because it can determine magnitudes of energy losses, their causes and locations simultaneously.

However, exergy calculations are much more complex than energy calculations and most commercial process simulators do not have a built-in ability to calculate exergy of a material stream.

26.2.3.4 Exergy and Irreversibility. The concept of irreversibility is not new in thermodynamics and it has been applied to different processes. It is the difference between reversible and useful work. Almost all real and complex natural processes have irreversiblities that cause exergy destruction leading to an increased entropy production. This makes exergy destruction a direct measure of the process irreversibility. In other words, a reduction in irreversibility of a given process increases the exergy or available energy to perform useful work and hence the exergetic efficiency of the process.[23]

Every spontaneous process causes exergy destruction due to irreversiblities, leading to increased entropy production. Exergy analysis has the ability to detect and evaluate the causes of the thermodynamic imperfections of the considered process. This ability of exergy analysis enables it to identify the inefficient parts of a process.[13]

26.2.4 Integration of Process Design, Control and Sustainability

Process control is usually delayed to the late stages of the process design as it requires a great deal of process information. With this approach to process design, changes in design are difficult and costly during start-up and operational periods. Sometimes it can lead to the processes that are not controllable or difficult to control because small design changes can have a large impact on the process control. To avoid these problems, an integrated approach to process design and control in early process design stages is recommended. It allows changes in process configuration of a process for easier control and evaluation of many control structure candidates simultaneously. This integrated approach results in better process design and control in a cost effective manner.[1,2,36]

The Venn diagram in Figure 26.2 shows the integration of process design, control and sustainability regions where region 'a' shows the integration of process control and sustainability, region 'b' shows the integration of process control and process design, region 'c' shows the integration of process design and sustainability, and region 'd' shows the integration of process design, control and sustainability. The integration of process control and sustainability (region 'a' in Figure 26.2) combines thermodynamic properties such as exergy and process control into a single domain to design sustainable and controllable control structures also called sustainable process control.

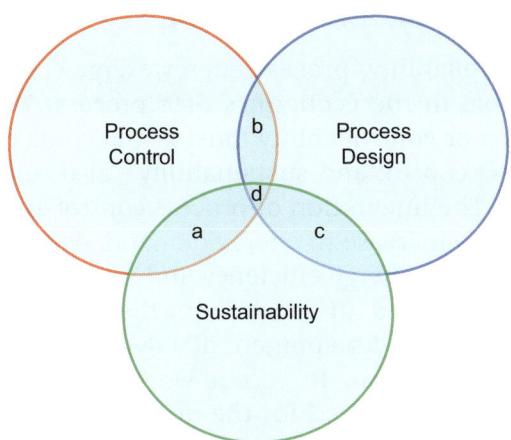

Region: a = Process control and sustainability   Region: b = Process control and process design

Region: c = Process design and sustainability   Region: d = Process design, control and sustainability

Figure 26.2   Integration of process design, control and sustainability.

The integration of process design and control (region 'b' in Figure 26.2) combines steady state design and dynamic control into a single optimization step. Process design has a strong connection with process control or controllability. Process design also has a substantial effect on the ability of a controller to respond to disturbances and parametric uncertainty. For example, in the presence of big disturbances, a control valve can quickly saturate with a small range of control action.[37]

The integration of process design and sustainability (region 'c' in Figure 26.2) combines thermodynamic properties such as exergy and process design to improve the energy efficiency of process design. Total exergy input/output analysis is used for the exergy analysis of a process by comparing the amount of exergy coming in and going out of the process.[38,39] Exergy has also been used to develop a systematic approach to optimal process design with low exergy destruction. Objective functions in terms of energy and exergy have also been constructed.[40] Sorin *et al.*[5] explained that process design with reduced energy consumption is a function of three aspects of process design: energy efficiency; efficient use of raw materials; and waste reduction. Process design alternatives can also be evaluated quickly by using the exegetic efficiency tools.[6]

Region 'd' in Figure 26.2 shows the integration of all three components: process design, sustainability and process control which are discussed in detail in the following section.

26.3 SUSTAINABLE PROCESS CONTROL

Along with controllability, process energy/exergy consumption plays an important role in the economics of a process. Thereby, control scheme selection or controllability must also include energy usage to integrate process control and sustainability—also called sustainable process control. The integration of process control and sustainability is necessary due to increase in energy demand, cost and government regulations regarding energy efficiency and ecological impacts.

Luyben et al.[1] provided an appendix in their book which acts as a basic framework for the development of a dynamic exergy balance for process control evaluation. In recent work by the authors of this chapter, exergy has been used for the development of control structures because exergy plays a role in evaluating the controllability of a process.[21,22,41,42] It is also important to know the exergy destruction rate changes due to changes in the dynamic state of the process. A summary of sustainable process control tools, recently developed by the authors, is given in Sections 26.3.1 to 26.3.3.

As exergy destruction is directly proportional to entropy production, actual/spontaneous processes normally occur in the direction of decreasing exergy or increasing entropy.[43] Some research has also been carried out on the theoretical development and application of process control linked to entropy production,[44–47] and readers may find more detailed information within these references.

26.3.1 Relative Exergy Array (REA)

The relative gain array (RGA) is a controllability index used to compare different control structure candidates during early process design stages. But the RGA only accounts for the controllability of a MIMO system. It does not combine sustainability of process with its controllability. The relative exergy array (REA) is a tool that combines sustainability and controllability. It is derived by placing exergy gain in place of simple process gain in the RGA.

The REA for an $n \times n$ MIMO system is defined analogously to RGA with relative exergy, γ_{ij}, defined as:

$$\gamma_{ij} = \frac{\left(\dfrac{\Delta B(y_i)}{\Delta B(u_j)} \right)_{\text{all loops open}}}{\left(\dfrac{\Delta B(y_i)}{\Delta B(u_j)} \right)_{\text{all loops closed (in perfect control) except } u_j \text{ loop}}} \qquad (26.1)$$

The relative exergy, γ_{ij}, in Equation (26.1) is the ratio of gain change in the steady state exergy of the material stream y_i with respect to that of the material/energy stream u_j when all loops are open to the gain change in the steady state exergy of the material stream y_i with respect to that of the material/energy stream u_j when all other loops are closed and in 'perfect control'. The numerator of Equation (26.1) shows the open loop exergy gain change in a given control loop, $(u_j - y_i)$. The denominator of Equation (26.1) shows the closed loop exergy gain change when all other control loops are closed except the $(u_j - y_i)$ loop. If there is no interaction from other loops then the value of γ_{ij} is close to 1. This case $\gamma_{ij} \sim 1$ indicates that the thermodynamic efficiency of the control loop considered is less affected by inter- actions from other loops. This would be a good pairing candidate from an exergy point of view as its thermodynamic efficiency is not affected by interactions from other control loops.

The REA is a controllability index like RGA, which can provide engineers with an idea of the effects of loop interactions of a control structure on the sustainability of the process during the early stages of process control design. Detailed information about REA development can be found in references 7, 21 and 22.

26.3.2 Exergy Eco-efficiency Factor (EEF)

The REA only measures the sustainability within the scope of single, decentralized control loops. To overcome this deficiency, the exergy eco-efficiency factor (EEF) was defined and employed for the sus- tainability analysis of the whole process. The EEF is based on an understanding of the total exergy of each material stream in and out of the process, and was developed for the sustainability analysis of a whole unit operation, process or plant. EEF helps to select the best control configuration for sustainability of the whole process.

The exergetic efficiency of a process is equivalent to eco-efficiency. For a general process as shown in Figure 26.3, exergetic efficiency is defined as the ratio of the exergy going out, to the exergy coming in,[13] namely:

$$\eta = B_{out}/B_{in} \tag{26.2}$$

where η is exergetic efficiency, B_{out} is the total exergy going out of a process and B_{in} is the total exergy coming in to a process.

However, the exergetic efficiency for the whole process shown in Figure 26.3 does not provide any information about how the control

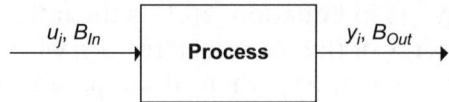

Figure 26.3   Exergy flows for a general process.

loop configuration affects this exergetic efficiency. The EEF was developed to estimate the effect of control loop configuration on exergetic efficiency.[48] The EEF for a control pair (u_j, y_i) is defined as:

$$\tau_{ij}(\Delta B_{Out} - \Delta B_{In}) \frac{\Delta u_j}{\Delta y_i} \qquad (26.3)$$

where Δu_j denotes a step change of the MV, (u_j), Δy_i denotes a response in the CV, (y_i), caused by a step change of the MV, (u_j), and ΔB_{out} and ΔB_{in} represent the exergy differences caused by the MV step change for exergy out and exergy in, respectively. For example, for a 2×2 system, if τ_{21} is less than τ_{22}, it means that for the same amount of CV change, Δy_2, using MV u_1, will cause less exergy loss/change than using MV u_2. The final interpretation is that the control pairing (u_1, y_2) is more eco-efficient than the pairing (u_2, y_2).

Several well-performing decentralized MIMO control scheme candidates using techniques such as RGA, NI and CN are normally selected. EEF can be used to select best control scheme among already selected control schemes based on controllability in the sense of sustainability. An array of EEF values for all possible control structures are called relative exergy eco-efficiency factor (REEFA). The details of EEF development are found in references 41 and 48.

26.3.3   Relative Exergy Destroyed Array (REDA)

The EEF determines the most sustainable control configuration by considering one control configuration at a time. Its results also require dynamic simulation validation because they are based on steady state information. Dynamic simulation validation of EEF increases the computation load on practicing engineers. As a result, the relative exergy destroyed array (REDA) was proposed to easily measure sustainability of all possible control schemes at a time.[42]

The EEF in Equation (26.3) can be re-arranged with basic definitions of steady state gain and destroyed exergy during a step input as:

$$\tau_{ij} = \Delta B_D / k_{ij} \qquad (26.4)$$

where k_{ij} is the steady state gain for a control pair (u_j, y_i), and $\Delta B_D = \Delta B_{Out} - \Delta B_{In} =$ exergy destroyed in a process during a step input.

The relative EEF (REEF) is defined as:

$$v_{ij} = \frac{\left(\tau_{ij}\right)_{u_k = const,\, k \neq j}}{\left(\tau_{ij}\right)_{y_k = const,\, k \neq i}} \tag{26.5}$$

Relative exergy destroyed (ζ_{ij}) is defined as:

$$\zeta_{ij} = \frac{\left(k_{ij}\right)_{u_k = const,\, k \neq j}}{\left(k_{ij}\right)_{y_k = const,\, k \neq i}} \times \frac{\left(\tau_{ij}\right)_{u_k = const,\, k \neq j}}{\left(\tau_{ij}\right)_{y_k = const,\, k \neq i}} = \lambda_{ij} \times v_{ij} \tag{26.6}$$

where k_{ij} is the steady state gain for a control pair (u_j, y_i), τ_{ij} is the EEF for a control pair (u_j, y_i), λ_{ij} is the relative gain for a control pair (u_j, y_i), and v_{ij} is the REEF for a control pair (u_j, y_i). After we put the relative exergy destroyed in Equation (26.6) into a matrix form, we obtain the relative exergy destroyed array (REDA). Readers can find more detailed information about REDA in ref. 42.

The REDA is used to compare the sustainability of a process under different possible control schemes. The REDA tool is simple and easy to use to quickly find an optimal and sustainable control design during early design stages. REDA is simpler than EEF as it helps to develop a rank for selecting a control configuration that is based on controllability and sustainability. It can also be used as a screening tool to preliminarily select sustainable control schemes. Dynamic simulation validation should be used for the validation of REDA results.

26.4 CASE STUDY

A monochlorobenzene (MCB) separation process distillation column studied by Seider and co-workers[49] was selected to illustrate the use of REA, EEF and REDA. A schematic of this distillation column is shown in Figure 26.4.

The feed of the distillation column contains a mixture of MCB and benzene. The distillate product (D) contains most of the benzene and the bottom product (B) contains most of the MCB. The symbols used for the manipulated variables (MVs) and control variables (CVs) are listed in Table 26.1. The top and bottom product compositions, benzene (x_D) and MCB (x_B), respectively, are the controlled variables.

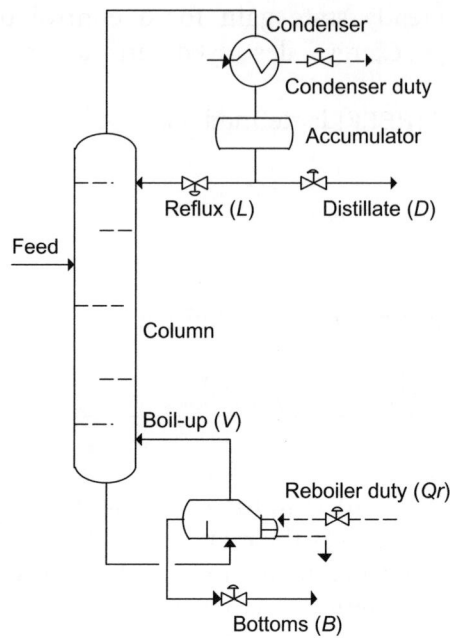

Figure 26.4 Distillation column schematic for case study.

Table 26.1 Control loop pairing nomenclature.

	Symbol	Description		Symbol	Description
MV	L	Reflux rate	MV	D	Distillate rate
	V	Boil-up rate	CV	$\begin{bmatrix} x_D \\ x_B \end{bmatrix}$	Distillate composition
	B	Bottom rate			Bottom composition

Three basic control configurations are defined, *LV, LB* and *DV*, for two-point composition control of this distillation column. In each control configuration, *L* (reflux rate) or *D* (distillate rate) is used to control the composition of the top product and *B* (bottom rate) or *V* (boil-up rate) is used to control the composition of the bottom product also shown in Figure 26.4. A list of the MVs and CVs for each of the control configurations is also presented in Table 26.1. Further details of control configurations and RGA, NI, CN, REA, EEF and REDA calculations from simulation experiments are found in references 9, 22 and 42.

Table 26.2 shows the controllability (RGA, NI and CN) and sustainability (REA, REEFA and REDA) tool results. According to the RGA results, the leading diagonal elements of the *LB* and *DV* control

Table 26.2 The controllability and sustainability tool results for the case study.

Configuration	Controllability tools			Sustainability tools			Results
	RGA	NI	CN	REEFA	REA	REDA	
LV	$\begin{bmatrix} 0.11 & 0.89 \\ 0.89 & 0.11 \end{bmatrix}$	1.2	31	$\begin{bmatrix} -0.12 & 1.12 \\ 1.12 & -0.12 \end{bmatrix}$	$\begin{bmatrix} -0.23 & 1.23 \\ 1.23 & -0.23 \end{bmatrix}$	$\begin{bmatrix} 0.98 & 0.02 \\ 0.02 & 0.98 \end{bmatrix}$	Uncontrollable Unsustainable
LB	$\begin{bmatrix} 0.97 & 0.03 \\ 0.03 & 0.97 \end{bmatrix}$	1.0	16	$\begin{bmatrix} -0.13 & 1.13 \\ 1.13 & -0.13 \end{bmatrix}$	$\begin{bmatrix} 0.88 & 0.12 \\ 0.12 & 0.88 \end{bmatrix}$	$\begin{bmatrix} -0.09 & 1.09 \\ 1.09 & -0.09 \end{bmatrix}$	Controllable Sustainable
DV	$\begin{bmatrix} 0.84 & 0.16 \\ 0.16 & 0.84 \end{bmatrix}$	1.0	24	$\begin{bmatrix} 0.01 & 1.01 \\ 1.01 & 0.01 \end{bmatrix}$	$\begin{bmatrix} 0.11 & 0.89 \\ 0.89 & 0.11 \end{bmatrix}$	$\begin{bmatrix} 0.15 & 0.85 \\ 0.85 & 0.15 \end{bmatrix}$	Controllable Unsustainable

configurations are positive and close to 1. The NI and CN results of the *LB* and *DV* control configurations are also positive and less than 50, respectively. These results from the controllability tools helped to further select *LB* and *DV* control configurations for further study of their sustainability. The off-diagonal elements of any control configuration for RGA close to 1 were not selected because pairing off-diagonal elements introduces a significant amount of dead time in the process. But for REEF and REDA, the off-diagonal elements close to 1 are preferred. The diagonal elements close to 1 are preferred for REA.

The results of only the *LB* control configuration show that the leading diagonal elements of REEFA and REDA are far from 1 and the leading diagonal elements of REA are close to 1. In other words, the *LB* configuration is the most sustainable control configuration. The *LB* control configuration is therefore a controllable and sustainable control configuration. Pairing of a control configuration which is not controllable but sustainable is normally not preferred. But in this case study, controllability results match with sustainability results by selecting the *LB* control configuration.

26.5 SUMMARY

Control structure selection/control scheme selection is an important task in process control. The control system structure selection process decides the best control scheme (easily controllable) for which there are a number of techniques such as RGA, NI, and CN. In the wake of energy crises, control system structure selection must not only focus on controllability and control quality but should also include sustainability and environmental impacts. For integrating controllability and sustainability, the potential of the thermodynamic property exergy is used. Exergy combines control scheme selection and sustainability into a single domain. There are many tools recently developed by authors such as REA, EEF and REDA to integrate control scheme selection and sustainability. These tools help in deciding controllable and sustainable control schemes for a process.

26.6 CONCLUSIONS

This work has introduced the necessity to consider sustainable process control in an integrated approach during early process design stages for a sustainable process. It has also emphasized that controllability and sustainability of a process should be considered simultaneously during early process design stages. Controllability tools such as RGA can only integrate process design and control. For sustainable process control, sustainability tools (REA, REEFA and REDA) were proposed which integrate process design, control and sustainability. The combined use of controllability and sustainability tools provides the measure of controllability and sustainability of the process under a certain design, though the final decision of control scheme requires dynamic simulation validation.

The controllability (RGA, NI and CN) and sustainability (REA, REEFA and REDA) tools were applied to an illustrative case study for the determination of its controllability and sustainability. A suitable control configuration was able to be identified as a controllable and sustainable control configuration using these tools for this case study. These results can be used during early process design stages to minimize the number of possible control configurations to a reasonable number for their detailed study.

REFERENCES

1. W. L. Luyben, B. D. Tyreus and M. L. Luyben, *Plantwide Process Control*, McGraw-Hill, New York, 1998.
2. P. Seferlis and M. C. Georgiadis (ed.), *The Integration of Process Design and Control, Computer Aided Chemical Engineering series*, Elsevier, Amsterdam, 2004, vol. 17.
3. G. D. Ulrich and P. T. Vasudevan, *Chemical Engineering Process Design and Economics: A Practical Guide*, Process Publishing, Durham, NH, 2004.
4. R. K. Sinnot, *Chemical Engineering Design*, Elsevier Butterworth-Heinemann, New York, 2005.
5. M. Sorin, A. Hammache and O. Diallo, *Energy*, 2000, **25**, 105–129.
6. G. Wall, *J. Power and Energy*, 2003, **217**, 125–136.
7. J. M. Montelongo-Luna, W. Y. Svrcek and B. R. Young, *Can. J. Chem. Eng.*, 2010, **89**, 545–549.
8. T. E. Marlin, *Process Control: Design Process and Control System for Dynamic Performance*, McGraw Hill, New York, 2nd edn, 2000.
9. W. Y. Svrcek, D. P. Mahoney and B. R. Young, *A Real-Time Approach to Process Control*, John Wiley & Sons, Chichester, UK, 3rd edn, 2014.
10. M. Morari and Z. Evanghelos, *Robust Process Control*, Prentice Hall, Englewood Cliffs, NJ, 1989.

11. D. E. Seborg, T. F. Edgar, D. A. Mellichamp and F. J. Doyle, *Process Dynamics and Control*, John Wiley & Sons, New York, 3rd edn, 2010.

12. I. Dincer and M. A. Rosen, *Exergy: Energy, Environment and Sustainable Development*, Elsevier, Amsterdam, 2007.

13. J. Szargut, D. R. Morris and F. R. Steward, *Exergy Analysis of Thermal, Chemical, and Metallurgical Processes*, Hemisphere, New York, 1988.

14. A. Valero and A. Valero, *Energy*, 2011, **36**, 1848–1854.

15. S. Enrico and W. Goran, *Int. J. Thermodyn.*, 2007, **10**, 1–26.

16. T. J. Kotas, *The Exergy Method of Thermal Plant Analysis*, Butterworth-Heinemann Ltd, London, 1985.

17. A. P. Hinderink, F. P. J. M. Kerkhof, A. B. K. Lie, J. De Swaan Arons and H. J. Van Der Kooi, *Chem. Eng. Sci.*, 1996, **51**, 4693–4700.

18. F. Abdollahi-Demneh, M. A. Moosavian, M. R. Omidkhah and H. Bahmanyar, *Energy*, 2011, **36**, 5320–5327.

19. A. Ghannadzadeh, R. Thery-Hetreux, O. Baudouin, P. Baudet, P. Floquet and X. Joulia, in *21st European Symposium on Computer Aided Process Engineering*, ed. E. Pistikopoulos, M. C. Georgiadis and A. C. Kokossis, Computer Aided Chemical Engineering series, Elsevier, 2011, vol. 29, pp. 1758–1762.

20. A. Ghannadzadeh, R. Thery-Hetreux, O. Baudouin, P. Baudet, P. Floquet and X. Joulia, *Energy*, 2012, **44**, 38–59.

21. M. T. Munir, W. Yu and B. R. Young, in *Plantwide Control: Recent Developments and Applications*, ed. G. P. Rangaiah and V. Kariwala, John Wiley & Sons, Chichester, UK, 2012, ch. 20, pp. 441–457.

22. M. T. Munir, W. Yu and B. R. Young, *Chem. Eng. Res. Des.*, 2012, **90**, 110–118.

23. A. Bejan, *J. Appl. Phys.*, 1996, **79**, 1191–1218.

24. Y. A. Gögüs, Ü. ÇamdalI and M. S. Kavsaoglu, *Energy*, 2002, **27**, 625–646.

25. A. Niksiar and A. Rahimi, *Energy*, 2009, **34**, 14–21.

26. R. Rivero and M. Garfias, *Energy*, 2006, **31**, 3310–3326.

27. I. H. Aljundi, *Appl. Therm. Eng.*, 2009, **29**, 324–328.

28. S. J. S. Borelli and S. de Oliveira Junior, *Energy*, 2008, **33**, 153–162.

29. M. Kanoglu, I. Dincer and M. A. Rosen, *Energy Policy*, 2007, **35**, 3967–3978.

30. N. A. Madlool, R. Saidur, N. A. Rahim, M. R. Islam and M. S. Hossian, *Renewable Sustainable Energy Rev.*, 2012, **16**, 921–932.

31. M. M. Joybari, M. S. Hatamipour, A. Rahimi and F. G. Modarres, *Int. J. Refrig.*, 2013, **36**, 1233–1242.
32. Y. Liu, Y. Zhao and X. Feng, *Appl. Therm. Eng.*, 2008, **28**, 675–690.
33. A. Aspelund, D. O. Berstad and T. Gundersen, *Appl. Therm. Eng.*, 2007, **27**, 2633–2649.
34. D. C. Sue and C. C. Chuang, *Energy*, 2004, **29**, 1183–1205.
35. K. Takeshita and M. Ishida, *Energy*, 2006, **31**, 3097–3107.
36. J. L. A. Koolen, *Design of Simple and Robust Process Plants*, Wiley-VCH, Weinheim, Germany, 2001.
37. L. A. Ricardez-Sandoval, H. M. Budman and P. L. Douglas, *Ann. Rev. Control*, 2009, **33**, 158–171.
38. J. M. Montelongo-Luna, W. Y. Svrcek and B. R. Young, *Asia-Pac. J. Chem. Eng.*, 2007, **2**, 431–437.
39. T. Muangnoi, W. Asvapoositkul and S. Wongwises, *Appl. Therm. Eng.*, 2007, **27**, 910–917.
40. L. T. Fan and J. H. Shieh, *Energy*, 1980, **5**, 955–966.
41. M. T. Munir, W. Yu and B. R. Young, *ISA Trans.*, 2013, **52**, 162–169.
42. M. T. Munir, W. Yu and B. R. Young, *Can. J. Chem. Eng.*, 2013, **91**, 1686–1694.
43. I. Dincer and Y. A. Cengel, *J. Entropy*, 2001, **3**, 116–149.
44. A. A. Alonso, B. E. Ydstie and J. R. Banga, *J. Process Control*, 2002, **12**, 507–517.
45. B. E. Ydstie, *Comput. Chem. Eng.*, 2002, **26**, 1037–1048.
46. M. Ruszkowski, V. Garcia-Osorio and B. E. Ydstie, *AIChE J.*, 2005, **51**, 3147–3166.
47. L. T. Antelo, I. Otero-Muras, J. R. Banga and A. A. Alonso, *Comput. Chem. Eng.*, 2007, **31**, 677–691.
48. M. T. Munir, W. Yu and B. R. Young, *ISA Trans.*, 2012, **51**, 827–833.
49. W. D. Seider, J. D. Seader and D. R. Lewin, *Product and Process Design Principles: Synthesis, Analysis, and Evaluation*, John Wiley, New York, 2nd edn, 2004.

CHAPTER 27

Systematic Computer Aided Framework for Process Synthesis, Design and Intensification

RAFIQUL GANI* AND DEENESH K. BABI

CAPEC Research Centre, Department of Chemical and Biochemical Engineering, Technical University of Denmark, Søltofts Plads, Building 229, DK-2800, Kgs. Lyngby, Denmark
*Email: rag@kt.dtu.dk

27.1 INTRODUCTION

The objective of process synthesis is to find the best processing route (process flow sheet) among numerous alternatives for converting given raw materials to specific desired products subject to predefined performance criteria. Sustainable process synthesis–design refers to the design of alternatives that correspond to lower values of a set of targeted performance criteria related to economical, operational and environmental factors. Process intensification (PI) can be defined as the improvement of a process through: (1) the integration of operations; (2) the integration of functions (tasks); (3) the integration of phenomena; and/or (4) the targeted enhancement of phenomena of a given operation.[1]

In the chemical and biochemical industry, there is a continuous need for improvements related to the use of sustainable

Chemical Processes for a Sustainable Future
Edited by Trevor M. Letcher, Janet L. Scott and Darrell A. Patterson
© The Royal Society of Chemistry 2015
Published by the Royal Society of Chemistry, www.rsc.org

technologies/processes,[32] the efficient use of raw materials, and the environmental and life cycle issues. To evaluate the feasibility of these improvements different sets of criteria can be employed such as in energy consumption, waste generation, number of pieces of equipment, and capital/operational cost. These improvements can be achieved through the use of different synthesis methods, which operate at the unit operations (unit-ops) level, task level and phenomena level. At the unit-ops level the synthesis methods can be classified into heuristic (or knowledge-based) methods;[2-5] mathematical optimization methods[6-9] and hybrid methods.[10-13] Heuristic methods and mathematical optimization methods mainly perform process synthesis at the unit-ops level and tasks level. The search space includes well known/established unit-ops. By incorporating PI into process synthesis, the potential use of intensified/hybrid unit-ops is realized through the application of the hybrid methods.[1,14] Therefore the search space becomes wider and includes established plus intensified/hybrid unit-ops. However, to go beyond the current search space of unit-ops (established plus intensified) and to find truly innovative and predictive solutions, process synthesis must be performed at lower levels of aggregation (*i.e.* the phenomena level) where in principle new unit-ops can be designed. This concept is shown in Figure 27.1.

From Figure 27.1 it can be seen that generating flow sheet designs through a phenomena-based approach includes the entire search space of unit-ops, *i.e.* well known, hybrid and innovative (new). An example of an innovative unit-op generated through the phenomena-based approach is shown in Figure 27.2.[13] This new unit-op is used for the production of isopropyl acetate. Compared with the base case design and one of the intensified designs,[15] it uses a tenth and a fifth of the energy requirement, respectively. Also compared with the amount of catalyst used in the intensified design, the innovative unit-op reduces the usage by 40%.

The concept and reasoning behind the phenomena-based synthesis method is analogous to computer-aided molecular design (CAMD) because in principle these two methods are multi-level, *i.e.* they both operate at different levels of aggregation. Consider as an analogy, the design of chemical-based products where a set of target property attributes of the product are matched by considering a set of molecules available in a database, and/or by considering functional groups (of atoms) that can be combined to form existing as well as new molecules that match the specified targets, and/or by considering a set of atoms that can be combined to generate even more existing as well as new molecules that match the specified targets.

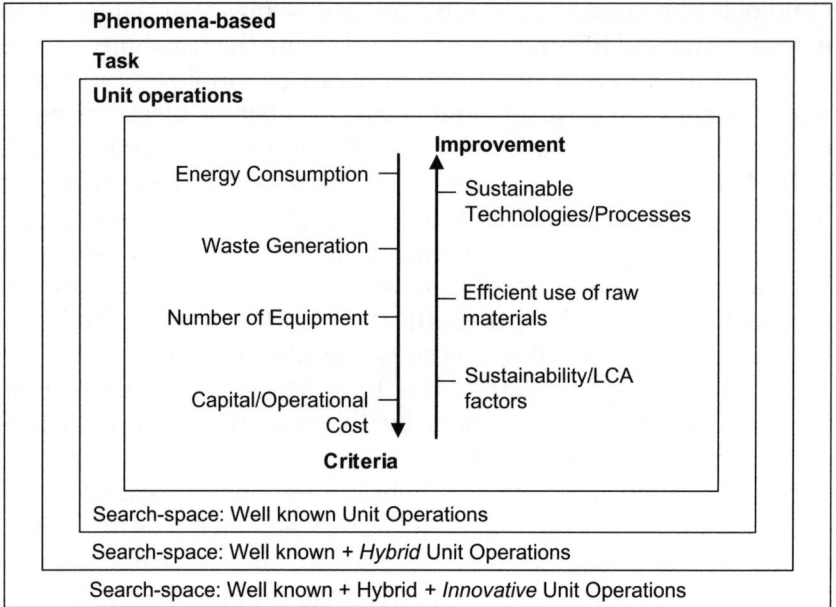

Figure 27.1 Role of phenomena-based process intensification for achieving sustainable process synthesis–design. LCA = life cycle assessment.

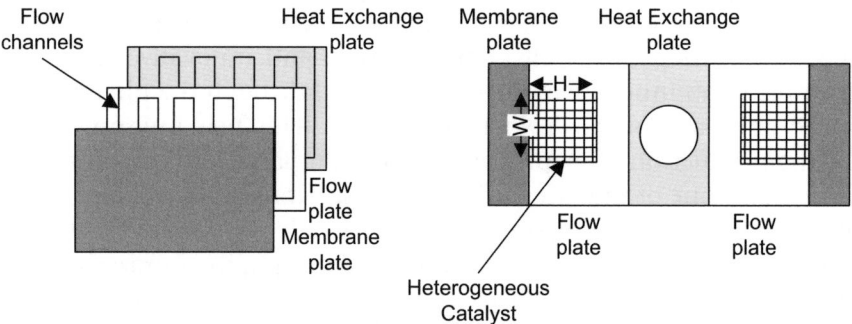

Figure 27.2 Detailed design of the rectangular plate-frame-flow reactor-pervapourator.[13]

In the same manner, now consider phenomena building blocks (PBBs) as atoms, which are combined to form functional groups, *i.e.* simultaneous phenomena building blocks (SPBs). These SPBs are combined to form operations in the same way that groups are joined together to form molecules. Therefore by combining building blocks at lower scales, new molecules as well as new unit-ops/process flow sheets are obtained. This concept is illustrated in Figure 27.3. First,

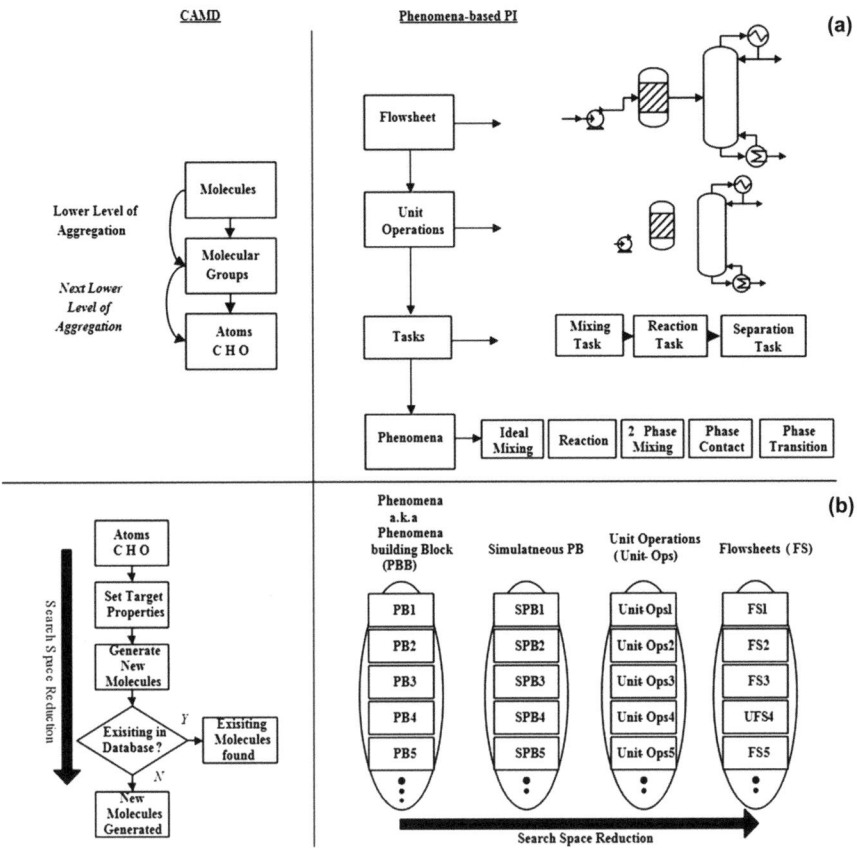

Figure 27.3 Concept of phenomena and CAMD for generating flow sheet alternatives:[16] (a) flow sheet to phenomena; and (b) phenomena to flow sheet alternatives.

similar to CAMD, the flow sheet is decomposed into unit-ops, tasks and phenomena (*i.e.* phenomena building blocks) (Figure 27.3(a)). Other phenomena that are beneficial are added to these phenomena, which are then screened and recombined to form unit-ops which constitute the final flow sheet alternatives (Figure 27.3(b)).

In this chapter process synthesis is performed through a systematic computer-aided multi-level framework for process synthesis, design and intensification. The developed framework allows process synthesis-intensification at the unit-ops level,[11] the task level[10] and the phenomena level.[13] First, a base case design is found. Second, this design is analysed in terms of economics (to identify cost related problems), sustainability (to identify waste and consumption related problems) and life cycle assessment (to identify environmental re-lated problems) to identify the process 'hot spots'. Hot spots are identified process limitations or bottlenecks associated with per-forming specific tasks; removal of hot spots leads to improvement of the overall process performance. These hot spots are then used to define a set of targeted performance criteria for process improvement and/or new design. Third, process synthesis is performed at the unit-ops level,[11] at the task level[10] and at the phenomena level[13,17] to determine alternatives that match the targets. This application of multi-level synthesis at different levels of aggregation is possible because a flow sheet can be represented in terms of unit-ops, tasks[10] and phenomena[13,17] (see Figure 27.3). Fourth, a more sustainable design corresponds to an alternative that matches the targeted performance criteria. Therefore this is also an optimal solution since the objective function is usually defined in terms of the performance criteria. The best intensified flow sheet which is also a more sustainable option is found and includes within the design novel and/or hybrid unit operations.

The objective of this chapter is to present and discuss the process synthesis–intensification framework. First, the concepts of phenomena-based process synthesis are presented illustrating how phenom-ena building blocks can be combined for generating flow sheet al-ternatives. An overview of the workflow of the framework and an explanation is given. Second, the process synthesis-intensification problem is defined. The mathematical description of the synthesis–intensification problem is presented and the proposed solution technique employed in solving this problem. Third, the computer-aided framework is presented and explained together with the dif-ferent methods and tools implemented within the framework. Fourth, an overview of the methods and tools involved in the framework is provided. Fifth, a step-by-step application of the framework is pre-sented through a case study. Finally the chapter ends with conclu-sions and suggestions for future work.

27.2   CONCEPTS FOR PHENOMENA-BASED SYNTHESIS

In this section the concept of phenomena based-synthesis is de-scribed and compared with computer-aided molecular design (CAMD). These two methods are multi-level, *i.e.* they both operate at different levels of aggregation. The concept of phenomena building blocks (PBBs) is introduced and the different types of PBBs that have been found in chemical process are stated. The concept is shown of how unit operations (unit-ops)/flow sheet designs inclusive of intensified/hybrid equipment can be generated through the combination of PBBs.

27.2.1 Phenomena Building Blocks

A set of building blocks to represent process flow sheets is needed, just like atoms, groups, bonds, *etc.* are needed to represent molecules. At the first level (*i.e.* the largest scale), unit-ops are used. At the next lower scale, tasks are used. At the next lowest scale, PBBs are used. Most chemical processes can be represented by different combinations of the following nine PBBs:[13]

1. Mixing (M)
2. Two-phase mixing (2phM)
3. Heating (H)
4. Cooling (C)
5. Reaction (R)
6. Phase contact (PC)
7. Phase transition (PT)
8. Phase separation (PS)
9. Dividing (D).

The inlet/outlet stream conditions of the PBBs are liquid (L), vapour and liquid (VL), liquid–liquid (L-L) vapour (V), vapour–liquid–liquid (VLL), solid (S) and solid–liquid (SL).

27.2.2 From Flow Sheet to Phenomena Building Blocks

To identify the PBBs in a flow sheet, the different unit-ops in the flow sheet are first identified. Next, the unit-ops are replaced by tasks by identifying and classifying each unit-op into the tasks they are supposed to perform. For example, a reactor unit-op is identified with a reaction task because the same compounds found in the inlet of the unit-op may also be found in the outlet together with additional compounds. Also a distillation unit-op is identified with a separation task because a minimum of two outlets are associated with the unit-op and the same compounds found in the inlet are also found in the outlets.

The PBBs in each task are then identified based on the type of unit-op. For example the PBBs involved in the reaction task for a reactor with an exothermic reaction are M (mixing), C (cooling) and R

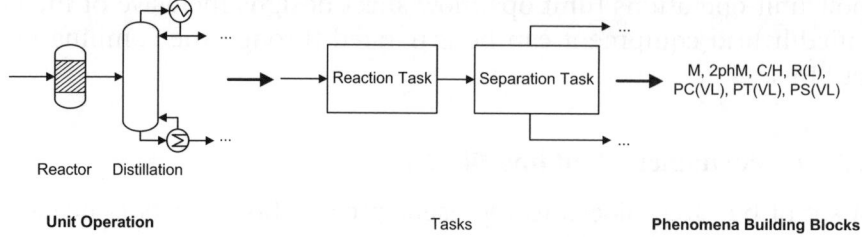

Figure 27.4 Task based flow sheet to phenomena based flow sheet.

(reaction). When all the PBBs involved in the different tasks of the flow sheets have been identified, a list of PBBs for the flow sheet is generated. This is illustrated in Figure 27.4.

27.2.3 Combination of PBBs to form SPBs

One or more PBBs are combined to form simultaneous phenomena building blocks (SPBs) which fulfil the objectives of any task subject to combination rules for ignoring/accepting certain combinations. For example:

1. Energy transfer PBBs, H and C, cannot be combined to form a SPB containing H=C because this combination is not thermo-dynamically feasible (simultaneous heating and cooling).
2. Reaction and energy transfer PBBs can be combined to form, for example, a R(L)=C SPB, which performs the tasks of isothermal single liquid phase reaction combined with cooling.
3. Reaction and separation PBBs can be combined to form inte-grated reactive-separation SPBs for example, M=2phM=R=PC(VL)=PT(VL)=PS(VL), which performs the tasks of a liquid phase reaction combined with simultaneous separation util-izing two phases, vapour and liquid. This is beneficial for equilibrium based reactions where simultaneous reaction–separation increases raw material conversion. An explanation of the SPB is as follows:
   a. A liquid phase reaction is present: R PBB needed.
   b. Ideal mixing of the reactants is assumed: M PBB needed.
   c. The reaction is happening in a two-phase system and there-fore mixing of the two phases is assumed: 2phM needed.
   d. The system is a two-phase system, vapour–liquid, and therefore, there is contact between the two phase: PC PBB needed.

e. Two-phases are present and in contact with each other, and therefore transition of one phase to the other is assumed: PT PBB needed.

f. Two phases are present, in contact and transition from one to the other, and therefore removal/separation of one of the phases is assumed: PS PBB needed.

It has been shown that the nine PBBs in Section 27.2.1 are combined to form different combination of SPBs. In the same way multiple SPBs can be combined to form different combinations of operations. Operations are combinations of SPBs, which are then translated into existing or new unit-Ops. Configuration unit-ops constitute the final flow sheet alternatives. Two examples of representing two unit-ops from a combination of SPBs are illustrated in Figure 27.5. A distillation column is obtained by combining three SPBs (Figure 27.5(a)) and a reactive distillation column having

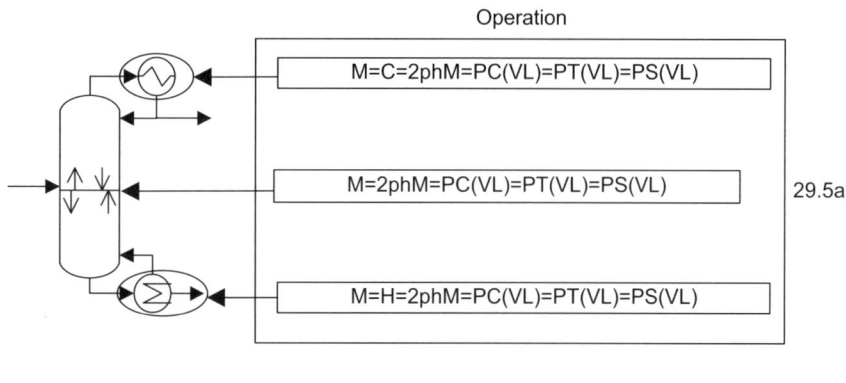

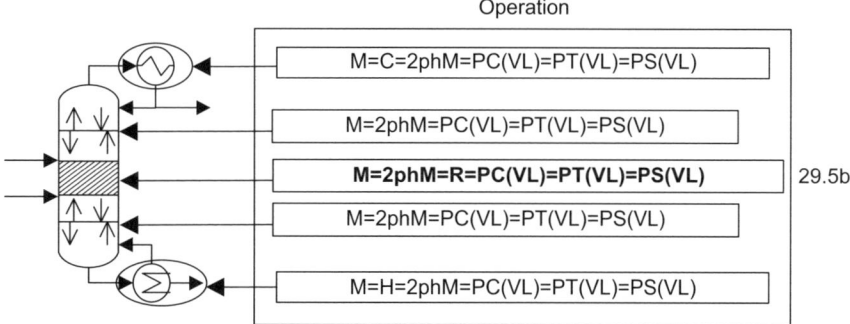

Figure 27.5 Combination of SPBs to form operations and translated into unit operations to fulfil particular tasks: (a) distillation; and (b) reactive distillation.

reactive and non-reactive sections is obtained by combining five SBPs (Figure 27.5(b)).

27.3   PROCESS SYNTHESIS–INTENSIFICATION: PROBLEM DEFINITION AND MATHEMATICAL SOLUTION APPROACH

In this section the process synthesis–intensification problem is defined together with the mathematical formulation. The proposed solution approach for solving the complex synthesis-intensification problem is discussed and illustrated through an example.

27.3.1   Process Synthesis–Intensification Problem Definition

The process synthesis–intensification problem is defined as: find feasible intensified flow sheet alternatives that either leads to a reduction in number of unit-ops and/or the use of intensified/hybrid unit-ops, subject to process specifications and targets related to a set of performance criteria.

27.3.2   Mathematical Description of the Synthesis–Intensification Problem

The process synthesis–intensification problem is expressed mathematically by Equation (27.1) to (27.8):[13,16]

$$\min/\max F_{obj} = F_{obj}(x, y, d, z, \Theta) \tag{27.1}$$

s.t.

$$g(x, z, \Theta) = 0 \tag{27.2}$$

$$f(x, y, d, z, \Theta) = 0 \tag{27.3}$$

$$b^l \le B_1 x + B_2 y \le b^u \tag{27.4}$$

$$h^l \le h(x, y) \le h^u \tag{27.5}$$

$$v^l \le v(x, y) \le v^u \tag{27.6}$$

$$w^l \le w(x, y) \le w^u \tag{27.7}$$

$$y_j = 0/1, \; j = 1, 2, \ldots n_y, \; x \ge 0 \tag{27.8}$$

where: x is a vector of design (optimization) variables, for example, composition, y is a vector of decision integer variables, d is a vector of equipment parameters, z is a vector of thermodynamic variables, and Θ is a vector of process/product specifications, respectively. Equation (27.1) represents the objective function, which can either be either minimized or maximized. Equations (27.2) and (27.3) represent the steady state process model, which can be linear or non-linear, depending on the derivation of the process model. Equations (27.4) and (27.5) represent the physical constraints and design specifications, respectively. Unique to this definition of the synthesis problem is Equation (27.6), which represents a set of performance criteria that the intensified process must satisfy[13] because process synthesis–intensification is performed together. Equations (27.7) and (27.8) represent the operational, structural and logical constraints and decision variables y (binary or integer), respectively.

The process synthesis–intensification problem given by Equations (27.1) to (27.8) represents a mixed integer non-linear programming (MINLP) problem, which can be difficult to solve because of the size of the problem where the optimization variables contain real and integer or binary (0-1) variables and the inherent non-linearity of the objective function and/or at least one of the constraint equations. Furthermore, generating flow sheet alternatives by first combining PBBs is a complex problem because of the large number of possible combinations. Therefore, an efficient and systematic solution approach is necessary.[13]

27.3.3 Solution Technique

To manage the complexity of the intensification–synthesis MINLP problem an efficient and systematic solution approach is used where the problem is decomposed into a set of sub-problems that are solved according to a pre-defined calculation order.[18] This solution approach is called the decomposition-based solution strategy (DBSS). It has the following properties and advantages:[19]

1. Properties:
   a. The problem is decomposed into manageable sub-problems.
   b. Each sub-problem, except for the final, requires only the solution of a subset of constraints from the original problem formulation.
   c. The final sub-problem is the solution of the objective function with the remaining constraints.
2. Advantages:
   a. The problem is broken down into manageable sub-problems,
   b. The solution of the broken down problem is equivalent to solution of the original MINLP problem (this is illustrated by the example in Section 27.3.4).

c. The MINLP problem is broken down into manageable
   sub-problems from which the final MINLP is solvable;
   the solution approach is more flexible in managing
   complexity.
d. Solution of each sub-problem acts as a screening tool within
   the search space and therefore, infeasible solutions are removed while each sub-problem is solved. Hence as stated
   previously, the final MINLP problem is solvable because the
   problem becomes smaller.
e. Since the solution of the sub-problems satisfies the constraints defined in the original problem, the finding of a
   global optimal solution from solution of the reduced MINLP
   can be guaranteed as long as a global optimization algorithm
   is used to solve it.

27.3.4   Conceptual Example

The application of the DBSS is highlighted through the following
MINLP problem:[19] Consider the minimization of the objective function in Equation (27.9) subject to the constraints defined in Equations
(27.10) to (27.17).

$$\min z = 2x_1 + 3x_2 + 1.5y_1 + 2y_2 - 0.5y_3 \tag{27.9}$$

s.t.

$$x_1^2 + y_1 = 1.25 \tag{27.10}$$

$$x_2^{1.5} + 1.5y_2 = 3 \tag{27.11}$$

$$x_1 + y_1 \leq 1.6 \tag{27.12}$$

$$1.333x_2 + y_2 \leq 3 \tag{27.13}$$

$$-y_1 - y_2 + y_3 \leq 0 \tag{27.14}$$

$$y_1y_2 = 1 \tag{27.15}$$

$$x_1, x_2 \geq 0 \tag{27.16}$$

$$y_1, y_2, y_3 = \{0,1\} \tag{27.17}$$

27.3.4.1   Sub-problem 1.   First the enumeration of the binary variables (y_1, y_2, y_3) is performed as given in Table 27.1.

27.3.4.2 Sub-problem 2. The linear constraint defined in Equation (27.14) is used to screen the feasible solutions from Table 27.1 to identify those that satisfy Equation (27.14). The feasible enumerated binary variable sets are given in Table 27.2.

27.3.4.3 Sub-problem 3. The non-linear constraint defined in Equation (27.15) is used to screen the feasible solutions from Table 27.2 to identify those that satisfy Equations (27.14) to (27.15). The feasible enumerated binary variable sets are given in Table 27.3.

27.3.4.4 Sub-problem 4. The problem is now reduced to the non-linear programming (NLP) problem and can be solved for x_1 and x_2

Table 27.1 Enumeration binary variables.

Enumeration No.	y_1	y_2	y_3
1	0	0	0
2	1	1	1
3	1	1	0
4	1	0	0
5	1	0	1
6	0	1	1
7	0	1	0
8	0	0	1

Table 27.2 Enumerated binary variables that satisfy the linear constraint of Equation (27.14).

Enumeration No.	y_1	y_2	y_3
1	0	0	0
2	1	1	1
3	1	1	0
4	1	0	0
5	1	0	1
6	0	1	1
7	0	1	0

Table 27.3 Enumerated binary variable sets that satisfy Equations (27.14) to (27.15).

Enumeration No.	y_1	y_2	y_3
2	1	1	1
3	1	1	0

Table 27.4 NLP problem for each set of the feasible solutions (see Table 27.3).

$\{y_1, y_2, y_3\} = \{1, 1, 1\}$	$\{y_1, y_2, y_3\} = \{1, 1, 0\}$
$\min z = 2x_1 + 3x_2 + 1.5(1) + 2(1) - 0.5(1)$	$\min z = 2x_1 + 3x_2 + 1.5(1) + 2(1) - 0.5(0)$
s.t.	s.t.
$x_1^2 + (1) = 1.25$	$x_1^2 + (1) = 1.25$
$x_2^{1.5} + 1.5(1) = 3$	$x_2^{1.5} + 1.5(1) = 3$
$x_1 + (1) \leq 1.6$	$x_1 + (1) \leq 1.6$
$1.333x_2 + (1) \leq 3$	$1.333x_2 + (1) \leq 3$
$x_1, x_2 \geq 0$	$x_1, x_2 \geq 0$

Table 27.5 Optimization results of the MINLP and NLP problems.

MINLP		$\{y_1, y_2, y_3\} = \{1, 1, 1\}$		$\{y_1, y_2, y_3\} = \{1, 1, 0\}$	
Variable	Value	Variable	Value	Variable	Value
y_1	1	y_1	1	y_1	1
y_2	1	y_2	1	y_2	1
y_3	1	y_3	1	y_3	0
x_1	0.5	x_1	0.5	x_1	0.5
x_2	1.31	x_2	1.31	x_2	1.31
F_{obj}	7.93	F_{obj}	7.93	F_{obj}	8.43

to determine the optimal solution. Table 27.4 gives the reduced NLP problem and Table 27.5 compares the solution of the reduced NLP problem using the DBSS with the solution of the problem as an MINLP. The DBSS provides the same result as the MINLP approach for the objective function (F_{obj}).

27.4 COMPUTER-AIDED FRAMEWORK

In this section the framework is presented together with an explanation of each step of the framework. The methods used in the framework and tools employed in the framework are also presented.

27.4.1 The Synthesis-Intensification Framework

The detailed workflow for the computer-aided multi-level multi-scale framework for performing process synthesis-intensification is illustrated in Figure 27.6. The framework consists of methods and different computer-aided tools for performing each step of the framework. The developed framework consists of eight steps (S1 to S8) and four integrated task-phenomena-based synthesis (IT-PBS) steps. Steps 1–8 operate at the integrated unit-ops-task outer levels,

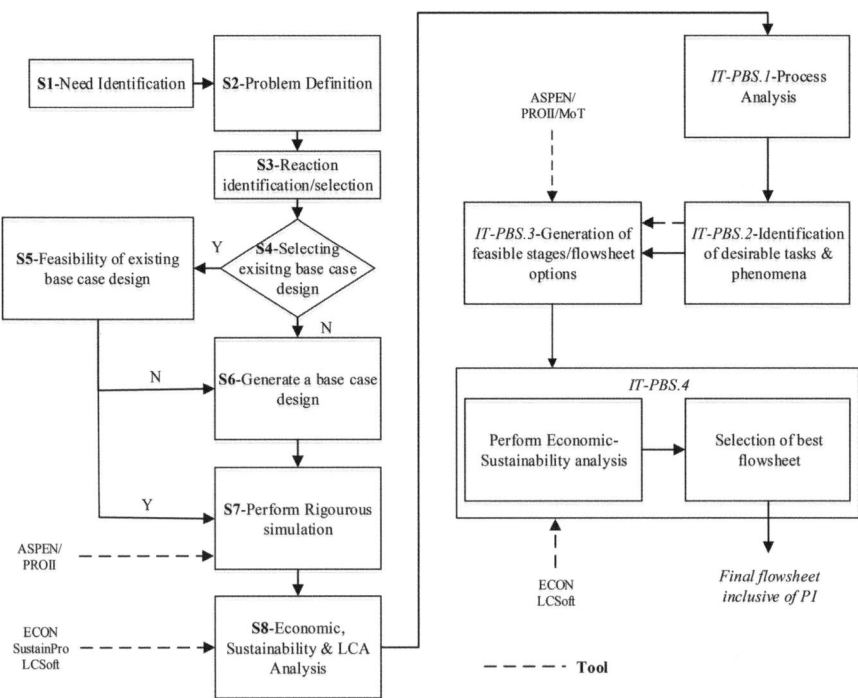

Figure 27.6 The Process Synthesis–Intensification Framework. S = Step; IT-PBS = Integrated Task-Phenomena-based Synthesis. Tools: ICAS/PROII/ASPEN = compound analysis/rigorous simulation; Sustain = pro-sustainability analysis; MoT = model based analysis; ECON = economic analysis; LCSoft = life cycle analysis.

while, IT-PBS 1 to IT-PBS 4 operate at the integrated task-phenomena inner level. Compared with the method of Lutze *et al.*,[13] the framework uses comprehensive analyses for identifying process hot spots and an integrated task-phenomena based approach for generating flow sheet alternatives. The objectives of each step, as well as the input data needed and the output data obtained for each of the steps, are explained in Sections 27.4.1.1 to 27.4.1.12. The associated methods and tools used in each step are presented in Section 27.4.2.

27.4.1.1 Step S1: Need Identification. The objectives of this step are to obtain information about the main uses of the product and to define the annual production rate of the product. The input data are the results of a literature survey using keywords related to uses of the product and its current total annual production. The output data are the main uses, and a defined desired annual production rate of the product. The main uses of the product provide general information about the importance of the product while the desired annual production rate of the product provides a design target that all flow sheet alternatives must satisfy.

27.4.1.2   Step S2: Problem Definition.   The objectives of this step are to define the problem statement and to define the mathematical problem statement. The input data are the decision whether the application is a new or retrofit design problem. This provides the definition of a suitable objective (cost) function which is to be minimized/maximized subject to performance criteria and constraints. The performance criteria are related to the improvement in sustainability/LCA factors for all generated feasible flow sheet alternatives.[17] The constraints are of the form, structural, logical and operational constraints. The output data are the defined problem statement and the objective function for evaluating flow sheet alternatives.

   Tools needed/used: PI knowledge base

27.4.1.3   Step S3: Reaction Identification/selection.   The objectives of this step are to select the reaction pathway, raw materials state and catalyst, and to determine the reaction type (*i.e.* exothermic or endothermic). The input data are the results of a literature survey using keywords related to feasible reaction pathways for producing a desired product using different raw materials as well as the equilibrium information and reaction kinetic data, if available. The heat of reaction is an important criterion which can be obtained from the heat of formation of each component. The latter are obtained from a database.[20] If a compound does not exist within the database, then the heat of formation is estimated by a property prediction tool, ProPred[21] which uses a group contribution approach. The output data are a set of suitable reaction pathways, the type, state and purity of the raw materials and the type of catalyst (if any) used in the reaction.

   Tools needed/used: ICAS (integrated computer-aided system) database (compound property), ProPred (pure component property prediction tool).

27.4.1.4   Step S4: Selecting Existing Base Case Design.   The objective of this step is to select a base case design based on either a literature survey or the information on processes stored in a PI knowledge-base. The input data are the reaction path and raw materials from step S3. If a base case design is available then step S5 is performed otherwise step S6 is entered.

   Tools needed/used: PI knowledge base tool

27.4.1.5 Step S5 (if the output of Step S4 is 'Yes'). The objective of this step is to verify the feasibility of the pre-selected design selected if step S4 to be used as a base case. The input data provide information about the selected design in step S4, *i.e.* the process type (batch or continuous), the overall input/outputs of the process (raw materials, products, purges, *etc.*), the separation structure (the unit-ops involved in recovery/separation of compounds) and the energy structure (the use of heaters, cooler, heat exchangers, *etc.*). The verification of the selected design is performed by applying the process synthesis method of Douglas.[5] If the pre-selected design cannot be verified, then step S6 is entered.

27.4.1.6 Step 6 (if the output of Step S4 is 'No'): Generate a Base Case Design. The objective of this step is to generate a feasible base case design for the production of the desired product. The input data provide the annual production target of the product (output data from step S1), the synthesis problem objective function and constraints (step S2) and reaction information (output data from step S3). The base case process design is generated using two methods[10,11] which operate at the unit-ops and task scales, respectively. First, task-based synthesis[10] is performed where tasks are identified and verified using analyses of pure component and mixture properties. These analyses are based on thermodynamic insights.[11] Next, using the results from the thermodynamic insights, the feasible unit-ops which fulfil a particular task are identified.[11] The set of feasible flow sheet alternatives are then rigorously simulated and the objective function is calculated. The base case design for the process is then selected based on the alternative which gives the best value of the objective function. The output data is a design which is used as the base case design for subsequent steps.

Tools needed/used: ICAS database (compound property), ProPred (pure component property prediction), CAPSS (computer-aided process synthesis system) (binary ratio matrix calculation), ASPEN/PROII (rigorous simulation).

27.4.1.7 Step S7: Perform rigorous simulation. The objective of this step is to simulate the base case design rigorously for obtaining the necessary output data for the economic, LCA and sustainability analyses. The input data are the annual production target of the product (output data from step S1), reaction information (output data from step S3), selected/generated base case design (output data from step S5 or step S6) and selection of a thermodynamic model for describing the phase equilibria of the system. The output data are detailed mass and energy balance results.

Tools needed/used: ASPEN/PROII (rigorous simulation).

27.4.1.8   Step S8: Economic, Sustainability and LCA Analysis. The objective of this step is to perform three analyses for setting the targets for sustainable design by first identifying the process hot spots. These analyses are economic, LCA (carbon footprint is also calculated) and sustainability analyses. The input data are detailed mass and energy balance data of the base case design, obtained from rigorous simulation (output data from step S7). The output data are the process hot spots of the base case design. The process hot-spots are targeted for process improvement with or without PI.

Tools needed/used: ECON (economic analysis), SustainPro (sustainability analysis), LCSoft (LCA)

27.4.1.9   IT-PBS 1: Process Analysis. The objectives of this step are to analyse the base case design in terms of tasks and PBBs, and to identify the PBBs involved in the process hot spots. The input data are the base case design at unit-ops scale (output data from step S5 or step S6), the process hot spots (output data from step S5 or step S7), the rules for translating the base case design at the unit-ops scale to the task scale (that is if S6 has not been performed) and finally the phenomena scale, the physical properties of the compounds and the system thermodynamic model (input to step S7). The output data are the base case flow sheet represented in terms of tasks, the identification of the PBBs involved in each task from which the base case flow sheet is represented in terms of phenomena and the tasks–PBBs involved in each hot spot. Therefore, the base case flow sheet at unit-ops scale is translated first into a task-based form and then into a phenomena-based form. From the hot spots identified in step S8, each hot spot is linked to a task and then to the corresponding PBB. For example, an equilibrium reaction hot spot is associated with a reaction task, which is associated with a reaction PBB. Ratios of properties of all binary pairs of compounds (called the binary ratio matrix) present in the system help to identify the task-specific PBBs.

Tools needed/used: ICAS database (compound property), ProPred (pure component property prediction), CAPSS (binary ratio matrix calculation), ASPEN/PROII (rigorous simulation)

27.4.1.10   IT-PBS 2: Identification of Desirable Tasks and PBBs. The objective of this step is to identify desirable tasks and the corresponding PBBs for overcoming the process hot spots. The input data are the binary ratio matrix (output from IT-PBS 1) and the process hot-spots (output from step S8). The output data are the identified tasks, the PBBs associated with these tasks for overcoming the process hot spots and selection of the total set of PBBs. First the tasks are identified followed by the PBBs which fulfil these tasks. Two types of tasks must be identified: the essential task and the desirable task. The essential task must be performed. The objective of a desirable task, which is optional, is to enhance or improve the primary task. For example, to produce a new compound from the interaction of two other compounds a reaction task (essential task) must be performed and, assuming the reaction is an equilibrium reaction, then to increase its conversion, a separation task (desirable task) for simultaneous removal of a by-product can be performed. Using the tasks (essential and desirable), the PBBs are identified. These identified PBBs related to the hot spots plus those identified from IT-PBS 1 (found in the base case design) comprise the total set of PBBs. The PBBs are then screened using the performance criteria and constraints defined in step S2.

27.4.1.11   IT-PBS 3: Generation of Feasible Flow Sheet Alternatives.
The objective of this step is to generate feasible flow sheet alternatives using an integrated, task–phenomena based approach. The input data are the selected PBBs (output from step IT-PBS 2), combination rules for combining PBBs to form SPBs, combination rules for combining SPBs to form operations, and rules for translating the operations into unit-ops. All possible combinations of PBBs are calculated using Equation (27.18) and screened using combination rules for obtaining a feasible set of SPBs (see Section 27.2.3). In Equation (27.18), $NSPB$ is the total number of calculated SPBs, $NPBB$ is the total number of PBBs, $n_{PBB,max}$ is the maximum number of PBBs that can be combined to form an SPB. This is calculated using Equation (27.19). In Equation (27.19) $NPBB_E$ and $NPBB_M$ are the total number of energy and mixing PBBs, respectively, and D is the dividing PBB.

$$NSPB_{\max} = \sum_{k=1}^{n_{PBB,\max}} \frac{(NPBB-1)!}{(NPBB-k-1)!k!} \qquad (27.18)$$

$$n_{PBB,\max} = NPBB - (NPBB_E - 1) - (NPBB_M - 2) - D \qquad (27.19)$$

Using means-ends analysis[10] and thermodynamic insights,[11] SPBs are used in an integrated manner to generate feasible flow sheet alternatives, *i.e.* these two methods are used to identify and verify the feasibility of tasks for performing specific functions. These tasks are then represented in terms of SPBs because, in principle, multiple SPBs can be identified for performing a specific task. Using connectivity rules, multiple SPBs are added to generate operations which achieve each task. These operations are then translated into unit-ops using the PI knowledge base (see Figure 27.5(a) and (b)). If a certain combination of SPBs generates an operation that cannot be found in the knowledge base, then a novel unit-op is proposed.

The main steps involved in the generation of the flow sheet alternatives are:

Step IT-PBS 3.1: Generate task-based superstructure
> S3.1.1 Identify all possible tasks that can performed for achieving reaction and recovery of the product(s) and un-reacted raw material(s).
> S3.1.2 Order tasks sequentially starting with the first task that must be performed (*e.g.* reaction) followed by tasks that allow the recovery of product(s) and raw material(s).

Step IT-PBS 3.2: Generate and select SPBs for generating task-based flow sheets
> S3.2.1 From the pure component properties and mixture properties, select the feasible path from the task-based superstructure for performing the desired tasks.
> S3.2.2 Different tasks selected in S3.2.1 can be performed in different ways. Therefore multiple SPBs are selected for performing different tasks. All SPBs containing reaction PBBs and all SPBs not containing reaction PBBs are used to achieve a reaction task and the remaining tasks (*e.g.* separation) in the task-based flow sheet.

Step IT-PBS 3.3: Generate unit-operation based flow sheet alternatives
> S3.3.1 For the SPBs from S3.2.2, add the remaining SPBs to generate operations.
> S3.3.2 Translate operations into Unit-Ops, which constitutes the final flow sheet alternative.

Step IT-PBS 3.4: Integrate and verify task-integration
> S3.4.1 Integrate and verify; through the use of integrated SPBs; neighbouring tasks in the task-based superstructure. Add these integrated tasks to the first generated task-based superstructure.
> S3.4.2 Repeat steps IT-PBS 3.2 to IT-PBS 3.4.1 until tasks cannot be integrated.

Step IT-PBS 3.5: Detailed analysis and screening of the flow sheet alternatives
  S3.5.1 For each flow sheet alternative analyse if the selected unit-op can be used for achieving the desired task. The analysis can be model based or with the use of graphical tools.
  S3.5.2 Screen the generated flow sheet alternatives using the defined operational and structural constraints.

The generated flow sheet alternatives are then screened using performance criteria and the constraints defined in step S2 (logical, structural and operational). For flow sheet alternatives containing hybrid/intensified unit-ops (*e.g.* membrane-based unit-ops), model-based analysis and membrane selection are performed to evaluate the feasibility of using these types of unit-ops. These models are either retrieved from a model library/database or obtained using a literature survey and then stored in the model library. The output feasible flow sheet alternatives are then rigorously simulated which is the output of this step.

Tools needed/used: ICAS-MoT (model-based analysis), ASPEN/PROII (rigorous simulation)

27.4.1.12 IT-PBS 4:-Perform Economic Sustainability Analysis. The objective of this step is to calculate the objective function and perform an economic, LCA and sustainability analysis to compare the selected flow sheet alternatives from IT-PBS 3 with the base case. From the results, the best designs among the feasible alternatives obtained from IT-PBS 3 are selected. The input data are the rigorous simulation results from IT-PBS 3. The output data are the calculation of the objective function, economic LCA and sustainability analysis. The flow sheet alternative(s) which produces the best value of the objective function while satisfying/improving the performance criteria are (is) selected as the best design(s).

Tools needed/used: ECON (economic analysis), LCSoft (life cycle assessment), SustainPro (sustainability analysis)

27.4.2 Methods and Tools used in the Synthesis–Intensification Framework

In this section the different methods and tools employed in the framework to obtain the relevant data to perform the action in each step of the workflow (see Figure 27.6) are presented. The methods operate at different scales: the unit-ops scale, task scale and phenomena scale. The computer-aided tools used are model-based.

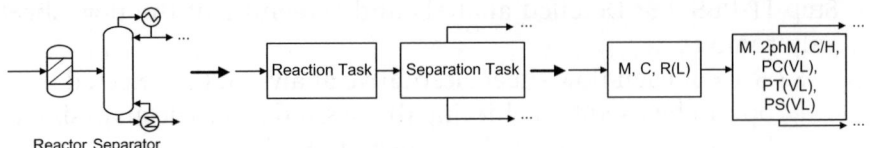

Figure 27.7   Task based flow sheet to phenomena based flow sheet.

27.4.2.1   Methods.   Five important methods are highlighted in this section. The methods are related to: (1) the means-ends analysis also used for flow sheet generation (step S6 and step IT-PBS 3); (2) the method of thermodynamic insights used for flow sheet generation (step S6 and step IT-PBS 3) and pure component property analysis (step IT-PBS 1); (3) the translation of the unit-ops flow sheet to the task and phenomena based flow sheets; (4) flow sheet generation using an integrated task-phenomena-based approach; and (5) the sustainability analysis. Finally, Table 27.7 lists the tools, methods and sub-methods available within the framework.

27.4.2.2   Means-ends Analysis.   The method[10] is an operator-based state transformation method, often used in automated goal-oriented problem solving. This is based on reducing or eliminating property differences between an initial state (*e.g.* raw materials) and goal state (*e.g.* products). Starting with the initial state, property differences are identified. Next transformation operators (tasks) are applied to reduce and eliminate these differences in order to produce intermediate states with fewer and fewer differences until the goal state (objective) is achieved.

27.4.2.3   Method of Thermodynamic Insights.   The method[11] consist of two levels where pure component properties and mixture properties analyses are used for identifying feasible separation tasks: unit-ops for performing these tasks; and arranging these unit-ops into flow sheet alternatives. The degree of complexity increases from one level to the next.

   In level 1, the pure component properties and mixture properties analysis is performed and the binary ratio matrix is calculated. The binary ratio is based on the ratio of pure component properties (*e.g.* boiling point). Using this information, separation techniques involving different unit-ops are identified and screened. Different unit-ops are directly related to specific pure component properties (*e.g.* distillation is related to boiling point, vapour pressure and heat of vaporization). Therefore based on the binary ratio value of these properties, a unit-op can be identified for separation of compounds. Then the separation tasks for different binary pairs are identified, because tasks are defined in terms of separation between two key compounds.

In level 2, first of all the redundant separation tasks identified in level 1 are removed. Second, mass separating agents are selected for those tasks that require them. Third, the tasks are screened using model-based calculations and graphical techniques. Fourth, the tasks that are linked to different unit-ops are combined to generate flow sheet alternatives.

27.4.2.4 Translation Unit-Ops Flow Sheet to Task and Phenomena Based Flow Sheet. For generating the task based flow sheet, first, each unit-op is identified and classified, for example:

- a reactor is identified if the inlet and outlet compositions of the unit-op are different and if some or all of the compounds in the inlet undergo a change of state to produce new compounds;
- a separator (distillation, liquid-liquid separation, *etc.*) is identified if the inlet and outlet compositions of a unit operation are different and has a minimum of two outlet streams.

Second, each unit-op is then translated into a task that it fulfils, for example:

- a reactor at the task level is referred to as a reaction task;
- a separator translated at the task level is referred to as separation task.

When all the tasks are identified, the unit-ops in the flow sheet are replaced by their respective tasks. Third, the PBBs in each task are identified based on the type of unit-op, for example:

- Continuously stirred tank reactor (CSTR) with an exothermic reaction: the involved PBBs in the reaction task are M, C, R.
- Distillation column (separator): the involved PBBs in the separation task are M, 2phM, C, H, PC(VL), PT(VL), PS(VL).

When all the PBBs are identified the tasks in the task-based flow sheet are replaced by their PBBs. An example is shown in Figure 27.7.

27.4.2.5 Flow Sheet Refinement (Generation) Using an Integrated Task Phenomena Based Approach. The means-ends analysis,[10] thermodynamic insights[11] and SPBs are used to generate the flow

sheet alternatives through the refinement of the selected task-based
flow sheet in Step IT-PBS 3. Starting with the raw materials (initial
state), tasks are identified followed by SPBs that fulfil these tasks.
Multiple SPBs are then combined using combination rules for generating operations which are translated into unit-ops. Flow sheets
are generated until the number of unit-ops in an alternative cannot
be further reduced.

The concept for generating flow sheet alternatives using multi-
level synthesis methods, *i.e.* the thermodynamic insights (unit-ops
based),[11] the means-ends analysis (task-based),[10] and PBBs (IT-PBS)
from IT-PBS 3, is explained using a conceptual example. Consider
the reaction between compounds A and B to produce compound C
that is $A + B \rightarrow C$. The order of boiling points from lowest to highest is
$C < B < A$, with B and C having close boiling points. The SPBs in
Table 27.6 are found to be feasible.

Using the means-ends analysis and thermodynamic insights, the
following tasks are identified for reaction and separation:

1. Reaction task: change of the state of the raw materials
2. Separation task 1: separation of A from B and C
3. Separation task 2: separation of C from B

Therefore the feasible SPBs for each task are:

1. Reaction Task: SPB.1
2. Separation Task 1: SPB.2
3. Separation Task 2: SPB.2 or SPB.5 because B and C have close
   boiling points a separation using a PT(VL) PBB might use more
   energy compared with a hybrid separation which uses a
   PT(PVL) PBB

The task based flow sheet can then be represented in terms of all
the necessary SPBs. Figure 27.8 illustrates the task based flow sheet

Table 27.6 Feasible SPBs for the flow sheet generation example.

SPB	Connected PBB	In	Out
SPB.1	M=R=C	1..n(L)	1(L)
SPB.2	M=2phM=PC(VL)=PT(VL)=PS(VL)	1..n(L,VL)	1(V/L)
SPB.3	M=H=2phM=PC(VL)=PT(VL)=PS(VL)	1..n(L,VL)	2(V;L)
SPB.4	M=C=2phM=PC(VL)=PT(VL)=PS(VL)	1..n(VL)	2(V;L)
SPB.5	M=C=2phM=PC(VL)=PT(PVL)=PS(VL)	1..n(L,VL)	2(V;L)

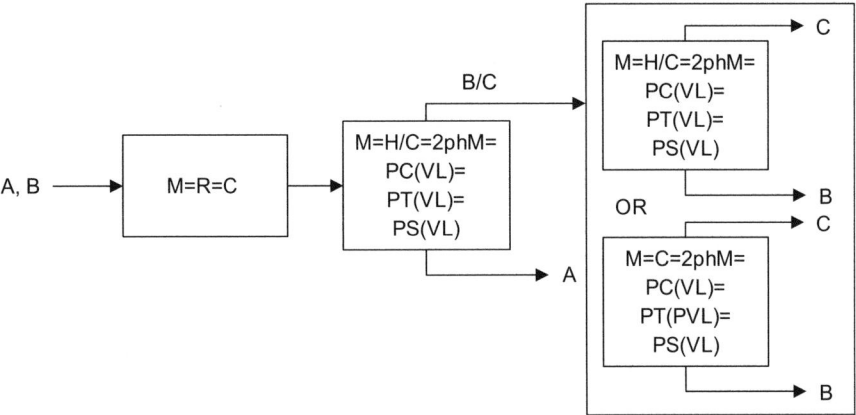

Figure 27.8 Tasks represented in terms of their main SPBs for the flow sheet generation example.

and the SPBs involved. The remaining SPBs needed to complete the operation are then added and these operations are translated into unit-ops. This is done using the following rules:

1. For a reaction SPB, ... = R= ..., this SPB becomes an operation and is translated into a CSTR reactor with a cooling jacket at the unit-ops level.
2. For the VL separation, SPB M=2phM=PC(VL)=PT(VL)= PS(VL), two other SPBs are added, M=C=2phM=PC(VL)= PT(VL)=PS(VL) (SPB.4) and M=H=2phM=PC(VL)=PT(VL)= PS(VL) (SPB.3). This is translated into a distillation column (see Figure 27.9(a)) at the unit-ops level.
3. For a PVL separation SPB, M=C=2phM=PC(VL)=PT(PVL)= PS(VL), this SPB becomes an operation and is translated into a pervaporation membrane at the unit-ops level (see Figure 27.9(b)).

The two generated flow sheet alternatives are shown in Figure 27.9.

27.4.2.6 Sustainability Analysis. The sustainability analysis uses an indicator-based methodology where a set of calculated closed and open path indicators are used to identify the structural bottlenecks within any process flow sheet. In brief the methodology determines and ranks a set of mass and energy indicators from steady-state process data which are then used to identify process bottlenecks. Having performed the analysis, one can then identify from the sustainability indicators the process hot spots because the indicators are ordered

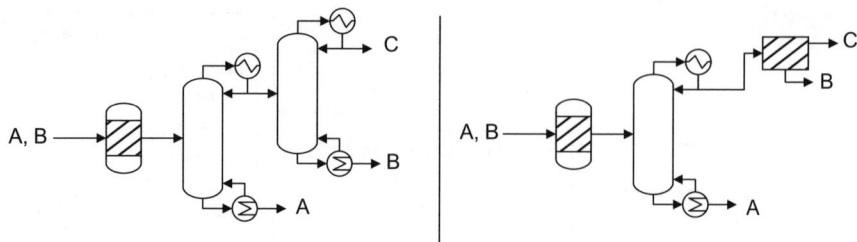

Figure 27.9   Flow sheet alternatives for the flow sheet example alternative.

in such a way, in terms of open paths and closed paths, as to show which parts of the process has the highest potential for improvement.[22] An open path (OP) is associated with the path a chemical takes in entering and leaving the process, while a closed path (CP) is associated with the recycle loop of a chemical within the process (this means that the chemical is unable to exit from the process in this path). A brief explanation of the sustainability indicators and what should be done to improve the process are presented:

1. *Material value added (MVA).* This indicator gives the value added between the entrance and the exit of a given compound, *i.e.* the value generated between the start and the end point of the path. Negative values of this indicator show that the component has lost its value in this open path and therefore point to potential for improvements.
2. *Energy and waste cost (EWC).* This indicator is applied to both open and closed paths. It takes into account the energy costs (EC) and the costs related with the compound treatment (WC). The value of EWC represents the maximum theoretical amount of energy that can be saved in each path within the process. High values of this indicator show high consumption of energy and waste costs, and therefore these paths should be considered in order to reduce the indicator value,
3. *Total value added (TVA).* This indicator describes the economic influence of a compound in a given path and is the difference between MVA and EWC. Negative values of this indicator show high potential for improvements in terms of decrease in the variable costs.

27.4.2.7 Tools.   The framework employs different computer-aided tools. Table 27.7 gives a selected list of the tools used for various calculations related to solving the synthesis-intensification problem.

Table 27.7 Summary of Tools used in the Synthesis-Intensification Framework.

Objective	Method	Tool name/ Tool type	Features
Phase diagram generation	Property model based	ICAS utility[a]/ Analysis	Group contribution based property models used for calculations of VLE, LLE, SLE, distillation boundary, residue curve, *etc.*
Solvent selection	CAMD; database search	ProCAMD[a]/ Selection	Searches for solvents for various types of solvent-based separation processes
Economic analysis	Model/ heuristic based	ECON[a]/ Analysis	Cost calculation based on the model from Peters, Timmerhaus and West[23]
Sustainability analysis	Model based	SustainPro[a]/ Analysis	Indicator based method using mass and energy balance data from rigorous simulation[22]
Life cycle assessment	Model based	LCSoft[a]	Indicator based method using cradle to the gate concept for LCA analysis[24]
Pure component properties analysis	Model based	CAPSS[a]/ Analysis	Calculated from the pure component properties from the system[11]
Pure component property prediction	Model based	ProPred[a]	Group contribution based property models[21]
Distillation (with or without reaction)	Driving force based; Equilibrium based	PDS[a]/Design, Analysis	Based on generated phase diagrams and driving force diagrams design of distillation columns
Modelling	Equation oriented problem solution	MoT[a]/ Analysis	Process and property models can be quickly generated and solved without spending time on programming
Process simulation	Model based calculations	Aspen Plus, PROII/ Analysis	Models for distillation and reactive distillation are available for use in analysis of hybrid distillation schemes (HDS)

[a]Part of ICAS.[20]

27.5 CASE STUDY

Each step of the framework is highlighted through this case study[17] involved with the production of methyl acetate (MeOAc) from an equilibrium-limited reaction between methanol (MeOH) and acetic acid (HOAc) with water (H_2O) as a by-product. This case study has been selected to highlight the application of the framework for generating new as well as existing flow sheet alternatives.

27.5.1   Step S1: Need Identification

MeOAc is an important chemical and is used as a solvent for glues, paints and nail polish removers. Therefore, the specified product rate of MeOAc is set to be 17 009 kg h^{-1} (122×10^6 t a^{-1}).[25] Here h, t and a refer to hour, metric tonne and annum, respectively.

27.5.2   Step S2: Problem and F_{obj} Definition

Problem statement: The identification of a process flow sheet alternative for the production of MeOAc, achieving an optimal conversion of HOAc and yielding a maximum highest profit, *i.e.* maximize Equation (27.20) subject to a set of constraints:

$$Max \ F_{obj} = \left(\sum m_j C_{Prod,j} - \sum m_j C_{RM,j} - \sum E_j C_{Ut,j} \right) / kg \ Prod \qquad (27.20)$$

The constraints are as follows:

- logical ($\underline{\theta}_1$), structural ($\underline{\theta}_2$), operational ($\underline{\theta}_3$) and performance criteria ($\underline{\varphi}$);
- reaction occurs in the first unit operation (defined by an element of $\underline{\theta}_1$);
- reactor effluent is connected to a separation sequence (defined by an element of $\underline{\theta}_1$);
- MeOAc purity is set at 99 mol-% (defined by an element of $\underline{\theta}_1$);
- H$_2$O purity is set at 99 mol-% (defined by an element of $\underline{\theta}_1$);
- phenomena are only connected using phenomena combination rules (defined by an element of $\underline{\theta}_2$);
- the use of solvents should be avoided (defined by an element of $\underline{\theta}_2$);
- recycle unreacted raw materials (defined by an element of $\underline{\theta}_2$);
- raw materials are in their pure state (defined by an element of $\underline{\theta}_3$);
- conversion of HOAc > equilibrium conversion (defined by an element of $\underline{\theta}_3$);
- production target of MeOAc is set at 17 009 kg h^{-1} (defined by an element of $\underline{\theta}_3$);
- number of unit-ops must be less (defined by an element of $\underline{\varphi}$);
- sustainability and LCA factors must be the same or better (defined by an element of $\underline{\varphi}$).

27.5.3 Step S3: Reaction Identification/Selection

The raw materials for the reaction are MeOH and HOAc. The reversible reaction that occurs in the liquid phase is catalysed by Amberlite 15^{32} and is exothermic having a calculated heat of reaction of $\Delta H = -5.42$ kJ mol^{-1}. The reaction is given in Equation (27.21).

$$MeOH + HOAc \leftrightarrow MeOAc + H_2O \qquad (27.21)$$

The raw materials for the reaction are assumed to be at their pure state. The equilibrium conversion is calculated to be 84 using the adsorption-based model.[26]

27.5.4 Step S4: Design Available?

From a literature survey a base case design is available[27] (Figure 27.11). It consists of 10 unit operations, one reactor, six distillation columns, one liquid–liquid extractor and decanter.

27.5.5 Step S5: Feasibility of Existing Base Case Design

The pre-selected design is then verified using the process synthesis method.[5] It is found that the design satisfies the requirements of the method, allowing it to be selected as a feasible design. Therefore it is used as a feasible design and step S7 is performed next.

27.5.6 Step 7: Perform Rigorous Simulation

The base case design was rigorously simulated and an overview of the results is presented in Table 27.8. The base case design is rigorously simulated using Aspen Plus and the UNIQUAC model for liquid activity coefficients. The reactor and separation columns are simulated using equilibrium based models. MeOH is fed in excess in order to achieve close to the equilibrium conversion.[26] An overview of the results is given in Table 27.8.

Table 27.8 Highlighted results from the base case design simulation.

	Value
Feed mole ratio (MeOH : HOAc)	2 : 1
MeOAc product/kg h^{-1}	17 009
Energy usage/MJ h^{-1}	372 198
Utility cost/\$ a^{-1}	12 343 384

27.5.7   Step 8: Economic, Sustainability and LCA Analysis

An economic, LCA and sustainability analysis are performed on the base case design. The economic and LCA analyses are shown in Figure 27.10. From Figure 27.10(a) it can be seen that the unit-ops which have the highest carbon footprint also have the highest utility costs (Figure 27.10(b)). This indicates that a non-tradeoff solution exists for this synthesis problem. The carbon footprint is calculated for each type of utility used for providing cooling and heating requirements. It is calculated using life cycle inventory data and global warming characterization factors.[24] The unit-ops that have a high utility cost relative to other units are identified as process hot spots in terms of operating cost.

The two most critical paths, represented in terms of streams, in the base case from the sustainability analysis[22] are given in Table 27.9 and their paths are illustrated in Figure 27.11.

As given in Table 27.9, MeOH in OP5 has a negative MVA value, which means that MeOH is losing its value as it exits the process through this path. Note that the achieved MeOAc purity in T2 is 99 mol%, where the remaining 1 mol-% is MeOH. For DMSO in

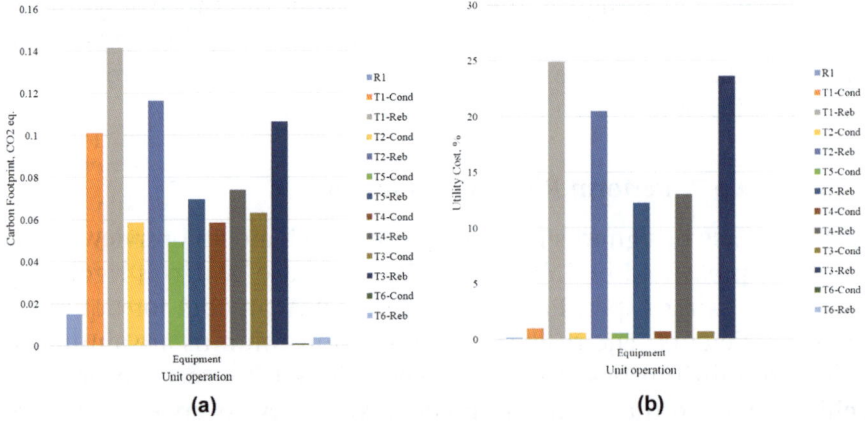

Figure 27.10   Carbon footprint (a) and utility cost (b) comparison with the base case design. Cond = condenser; Reb = reboiler.

Table 27.9   List of OP and CP with the highest potentials for improvement.

Path	Compound	Flow rate/kg h^{-1}	MVA/10^3\$ a^{-1}	TVA/10^3\$ a^{-1}	EWC/10^3\$ a^{-1}
OP5	MeOH	166	− 477	–	–
CP30	DMSO	78133	–	–	4440

DMSO = dimethyl sulfoxide.

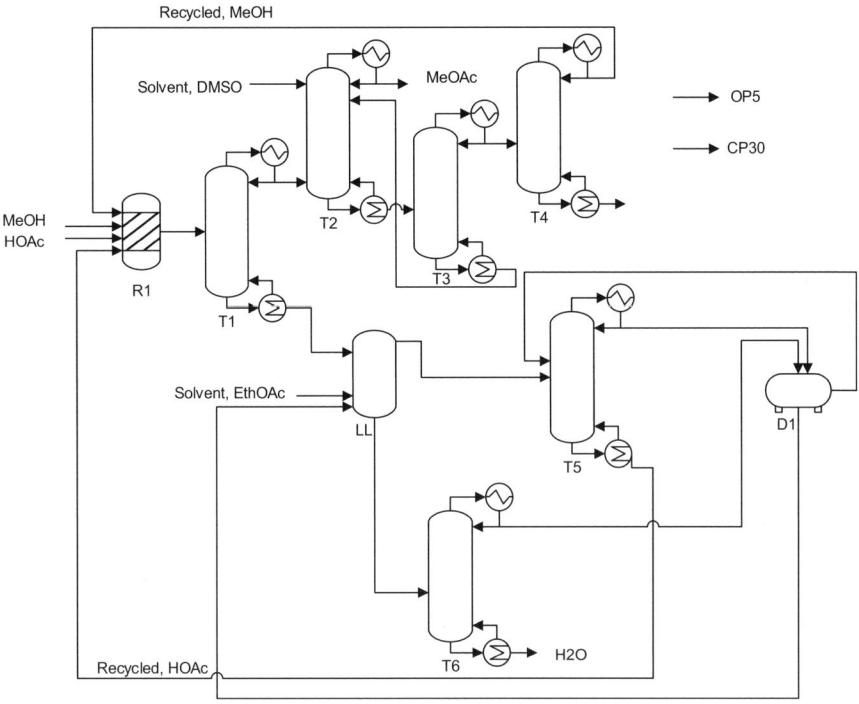

Figure 27.11   Paths in the base case design for the highlighted OP and CP from the sustainability analysis.

CP30 a high EWC value is found. This means that within this path too much material is being recycled resulting in high loads of energy and waste of utilities. Therefore, the unit-ops belonging to this CP are potential hot spots as they are using high amounts of energy and/or utilities. This can also be seen from Figure 27.10; for example, column T3-Reb, where DMSO is recovered, has a high carbon footprint and accounts for 24 of the utility costs. A list of the unit-ops according to their identification as a major hot spot in the three analyses (economic, LCA and sustainability) is given in Table 27.10. The columns where MeOAc is recovered from a mixture with MeOH and H_2O, T1, T2 and T3 have the highest carbon footprints. They are followed by T4 and T5 where MeOH and EthOAc are recovered, respectively.

Therefore the process hot spots are as follows:

- high energy consumption for separation between HOAc and MeOH (column T1), recovery of MeOAc from MeOH (columns T2 and T3) and recovery of the solvent for recovering HOAc from H_2O (column T5);

Table 27.10 Unit-ops identified as hot spots from the economic, LCA and sustainability analysis.

Unit	Label	Economic	LCA	Sustainability
Reactor	R1			+
Column 1	T1^{++}	++	+	+
Column 2	T2^{++}	++	+	+
Column 3	T3^{++}	++	+	+
Column 4	T4^{+}	++	+	+
Column 5	T5^{+}	++	+	+

Note: '++' and '+' are used to differentiate between the unit-ops having a higher impact compared with the others from the three performed analyses.

- limiting equilibrium (reversible) reaction, *i.e.* the raw materials are not fully converted;
- raw material loss of MeOH where MeOAc is recovered from MeOH using extractive distillation (column T2).

27.5.8 IT-PBS 1: Process Analysis

The base case is represented in terms of tasks and then the involved phenomena. The task based flow sheet and phenomena based flow sheet are shown in Figure 27.12 and Figure 27.13, respectively. The following PBBs are identified in the base case design:

- Reactor-1 phase: M, C, R
- Distillation (normal, extractive, azeotropic): M, 2phM, C/H, PC(VL), PT(VL), PS(VL)
- Liquid–liquid separation: M, 2phM, PC(LL), PT (LL), PS(LL)
- Decanter: M, PC(LL), PS(LL)

The operating window of each PBB is given in Table 27.11.

There are four compounds and therefore six binary pairs. The pure component properties are analysed and the binary ratio matrix is calculated and given in Table 27.12.

An analysis of the mixture properties of the binary pairs of compounds is performed and three minimum boiling binary azeotropes are found: HOAc/H$_2$O, MeOH/MeOAc and MeOAc/H$_2$O. The azeotropes are then further analysed for pressure dependence because this information may be useful for the flow sheet generation stage using the integrated task-phenomena based approach in IT-PBS 3. The result of the pressure dependence is illustrated in Figure 27.14. It can be seen that that all the azeotropes are pressure dependent with

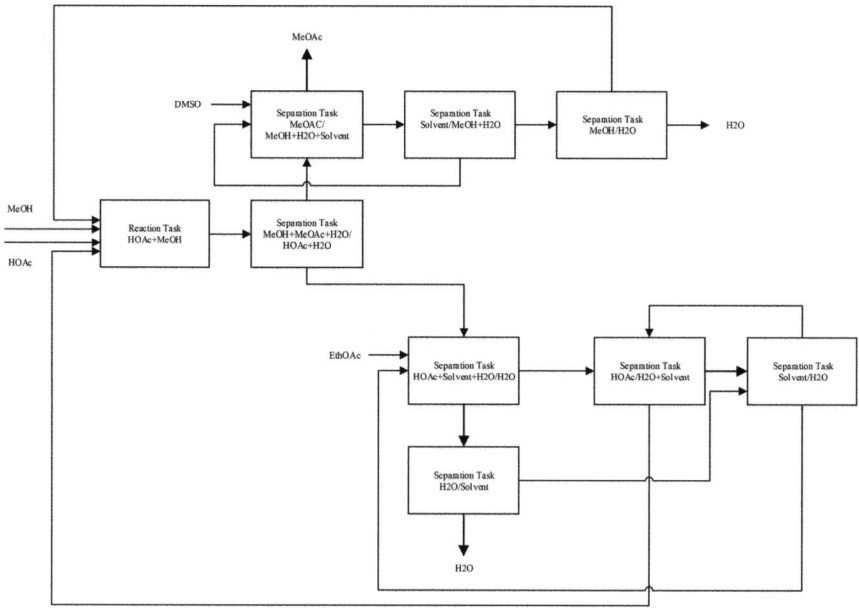

Figure 27.12 Task based flow sheet for the base case design.

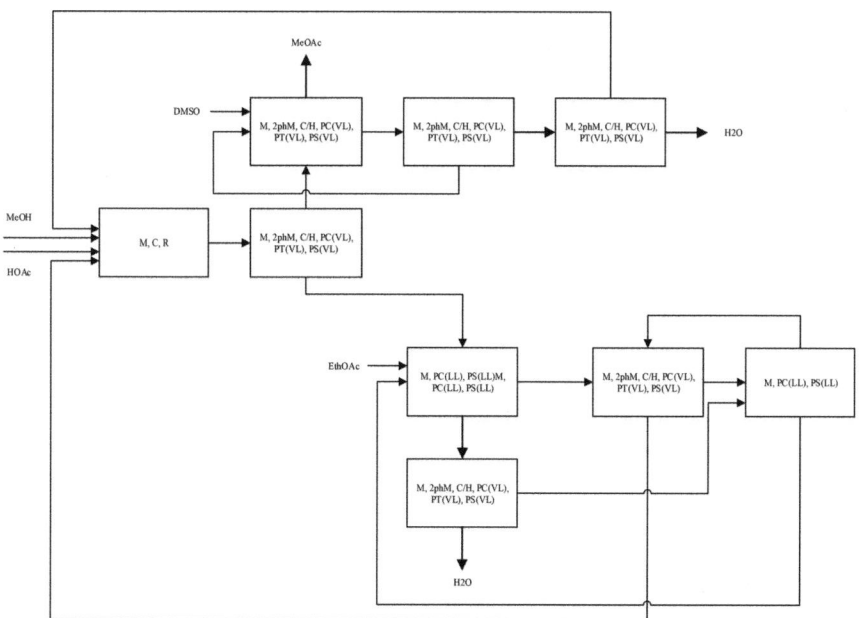

Figure 27.13 Phenomena based flow sheet for the base case design.

Table 27.11 Operating window of the identified phenomena in the base case design.

Unit-op	Phenomena	Operating window
Reactor	R	$T_{low} = 175.15$ K (lowest melter)
		$T_{high} = 378.15$ K (maximum T for reactor operation)
	M_V	$T_{low} = 330.05$ K (lowest boiler)
		$T_{high} = 391.05$ K (highest boiler)
		All concentrations are below the dew point line.
Distillation (all types)	M_{Id}	$T_{low} = 175.55$ K (lowest melter)
		$T_{high} = 391.05$ K (highest boiler)
	M_V	$T_{low} = 326.41$ K (lowest boiling azeotrope)
	PC(VL)	V-L present
	PT(VL)	$T_{low} = 326.41$ K (lowest boiling azeotrope)
		$T_{high} = 391.05$ K (highest boiler)
	PS(VL)	V-L present
		All concentrations are below the dew point line and above the bubble point line.
Liquid–liquid separation (LLS)	M_{Id}	$T_{low} = 175.55$ K (lowest melter)
		$T_{high} = 391.05$ K (highest boiler)
	PC(LL)	L-L present
	PS(LL)	L-L present
Decanter	M_{Id}	$T_{low} = 175.55$ K (lowest melter)
		$T_{high} = 391.05$ K (highest boiler)
	PC(LL)	L-L present
	PS(LL)	L-L present

Table 27.12 Binary ratio matrix for a selected set of properties.

r_{ij}	MW	T_m	T_b	RG	SolPar	VdW	V_m	VP
MeOH/HOAc	1.87	1.65	1.16	1.68	1.56	1.53	1.42	8.1
MeOH/MeOAc	2.31	1	1.02	1.93	1.53	1.96	1.97	1.7
MeOH/H$_2$O	1.78	1.56	1.1	2.52	1.62	1.76	2.25	5.31
HOAc/MeOAc	1.23	1.65	1.18	1.15	1.02	1.28	1.39	13.76
HOAc/H$_2$O	3.33	1.06	1.05	4.24	2.52	2.69	3.19	1.52
MeOAc/H$_2$O	4.11	1.56	1.13	4.87	2.47	3.44	4.42	9.02

MW = molar mass (molecular weight); T_b = normal boiling point; RG = radius of gyration; T_m = normal melting point; V_m = molar volume; SolPar = solubility parameter; VdW = Van der Waal volume; VP = vapour pressure.

the MeOAc/MeOH and HOAc/H$_2$O being the most and least dependent respectively.

27.5.9 IT-PBS 2: Identification of Desirable Tasks and PBBs

The desirable task for overcoming the limiting equilibrium reaction is a separation task of one of the products during the reaction. For example, based on the binary ratio of the solubility parameter and

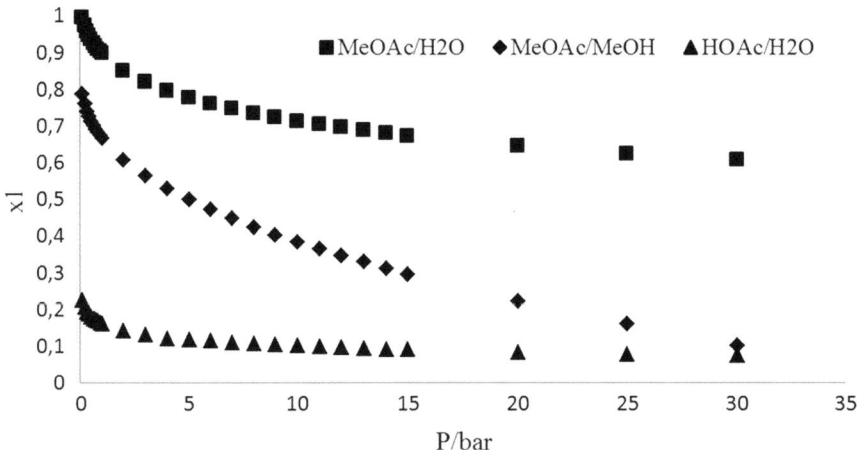

Figure 27.14 Pressure dependence analysis of the three minimum boiling azeotropes.

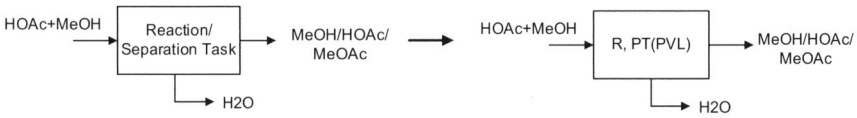

Figure 27.15 Identified desirable task for overcoming process hot spot limiting equilibrium and the involved phenomena.

Van der Waals volume, it is feasible to remove H_2O using a separation task related to permeability/affinity. This is because it is a compound that cannot be avoided in the reaction that is it is a by-product and it is present in two of the three identified minimum boiling azeotropes. This is shown in Figure 27.15.

The same concept is applied for the separation of the three identified azeotropes. For example, all the azeotropes can be separated using separation tasks related to permeability/affinity separation. Table 27.13 shows the connection between desirable tasks and PBBs.

The final number of identified PBBs is 15. These are R, M (assuming four types: ideal liquid, flow, rectangular, ideal vapour), 2phM, C, H, PC(VL), PT(VL), PS(VL), PT(PVL), PT(VV), PS(VV) and D. Note the total set of PBBs is obtained by adding the identified PBBs in this step to those identified in the base case (see step IT-PBS 1). The total set of PBBs is screened using the structural constraints ($\underline{\theta_2}$) and performance criteria (φ) defined in step S2, *i.e.* the use of solvents is to be avoided ($\underline{\theta_2}$) because of the potential effect on sustainability and LCA factors (φ).

Table 27.13 Translation of desirable tasks to PBBs.

Task	Feed condition	Separation type	PBB	Important pure component properties	Mixture properties analyses	Solvent required?
S-Task	L	V-separation	PT(PVL)	Molar volume, solubility parameter, dipolement	Identification of azeotropes	No
S-Task	V	V-separation	PT(VV)	Molar volume, solubility parameter, dipolement	Identification of azeotropes	No
S-Task	V/L/VL	VL-separation	PT(VL)	Boiling point, vapour pressure, heat of vaporization	-	No

S = separation; V = vapour; L = liquid.

27.5.10 IT-PBS 3: Generation of Feasible Flow Sheet Alternatives

The maximum number of PBBs that can be combined into a SPB, $n_{PBB,max}$ is calculated to be 11 using Equation (27.19). The number of SPBs is calculated using Equation (27.22).

$$NSPB_{max} = \sum_{k=1}^{n_{PBB,max}=11} \frac{(15-1)!}{(15-k-1)!k!} = 16278 \qquad (27.22)$$

The number of SPBs is then reduced by applying connectivity rules (see Section 27.2.3) and a total of 64 SPBs are found to be feasible. A selected list of the SPBs considering ideal mixing is given in Table 27.14.

The generation of a flow sheet alternative is exemplified below for one generated task based flow sheet. This task based flow sheet represents a large set of phenomena based flow sheets because, in principle, different SPBs can perform the same task. Therefore a large number of flow sheet options are generated.

27.5.10.1 Generation of Flow Sheet Alternatives. A task-based superstructure that covers all possible combinations of tasks representing the reaction between HOAc and MeOH to produce MeOAc and H_2O, and the recovery of MeOAc and un-reacted raw materials

Table 27.14 Partial list of feasible SPBs.

SPB	Connected PBB	In	Out	Task they may fulfil
SPB.1	M	1..n(L)	1(L)	Mix.
SPB.2	M=R	1..n(L)	1(L)	Mix. + React.
SPB.3	M=H	1..n(L)	1(L)	Mix. + Heat.
SPB.4	M=C	1..n(L)	1(L)	Mix. + Cool.
SPB.5	M=R=H	1..n(L)	1(L)	React. + Heat.
SPB.6	M=R=C	1..n(L)	1(L)	React. + Cool.
SPB.7	M=R=C=2phM=PT(VL)=PC(VL)	1..n(L,VL)	2(V/L)	React. + Sep.
SPB.8	M=R=H=2phMPT(VL)=PC(VL)	1..n(L,VL)	2(V/L)	React. + Sep.
SPB.9	M=R=C=2phM=PT(VL)=PC(VL)=PS(VL)	1..n(L,VL)	2(V;L)	React. + Sep.
SPB.10	M=R=H=2phM=PT(VL)=PC(VL)=PS(VL)	1..n(L,VL)	2(V;L)	React. + Sep.
SPB.11	M=R=C=2phM=PT(VL)=PC(PVL)=PS(VL)	1..n(L,VL)	2(V;L)	React. + Sep.
SPB.12	M=R=H=2phM=PT(VL)=PC(PVL)=PS(VL)	1..n(L,VL)	2(V;L)	React. + Sep.
SPB.13	M=R=2phM=PT(VL)=PC(VL)=PS(VL)	1..n(L,VL)	2(V;L)	React. + Sep.
SPB.14	M=2phM=PC(VL)=PT(VL)	1..n(L,VL)	2(V;L)	Cool. + Sep.
SPB.15	M=2phM=PC(VL)=PT(VL)	1..n(L,VL)	2(V;L)	Heat. + Sep.
SPB.16	M=C=2phM=PT(VL)=PC(VL)=PS(VL)	1..n(L,VL)	2(V;L)	Cool. + Sep.
SPB.17	M=H=2phM=PT(VL)=PC(VL)=PS(VL)	1..n(VL)	2(V;L)	Heat. + Sep.
SPB.18	M=C=2phM=PT(VL)=PC(PVL)=PS(VL)	1..n(VL)	2(V;L)	Cool. + Sep.
SPB.19	M=H=2phM=PT(VL)=PC(PVL)=PS(VL)	1..n(L,VL)	2(V;V)	Heat. + Sep.
SPB.20	M=C=2phM=PT(VL)=PC(VV)=PS(VV)	1..n(L,VL,V)	2(V;V)	Cool. + Sep.
SPB.21	M=H=2phM=PT(VL)=PC(VV)=PS(VV)	1..n(L,VL,V)	2(V;V)	Heat. + Sep.
SPB...	...	...	...	...
SPB.64	D	1(L;VL,V)	1..n(L;V; VL)	Stream Div.

Mix. = mixing; Cool. = cooling; Heat. = heating; React. = reaction; Sep. = separation; Div. = dividing.

(HOAc and MeOH) is shown in Figure 27.16. Raw materials HOAc (A) and MeOH (B) produce the products MeOAc (C) and H_2O (D), *i.e.* a change of state of raw materials must occur. In principle all SPBs that contain PBBs having PC, PT or PS or a combination of the three can fulfil a separation task. A processing path within the superstructure is obtained as follows. Consider S-Task 1, the

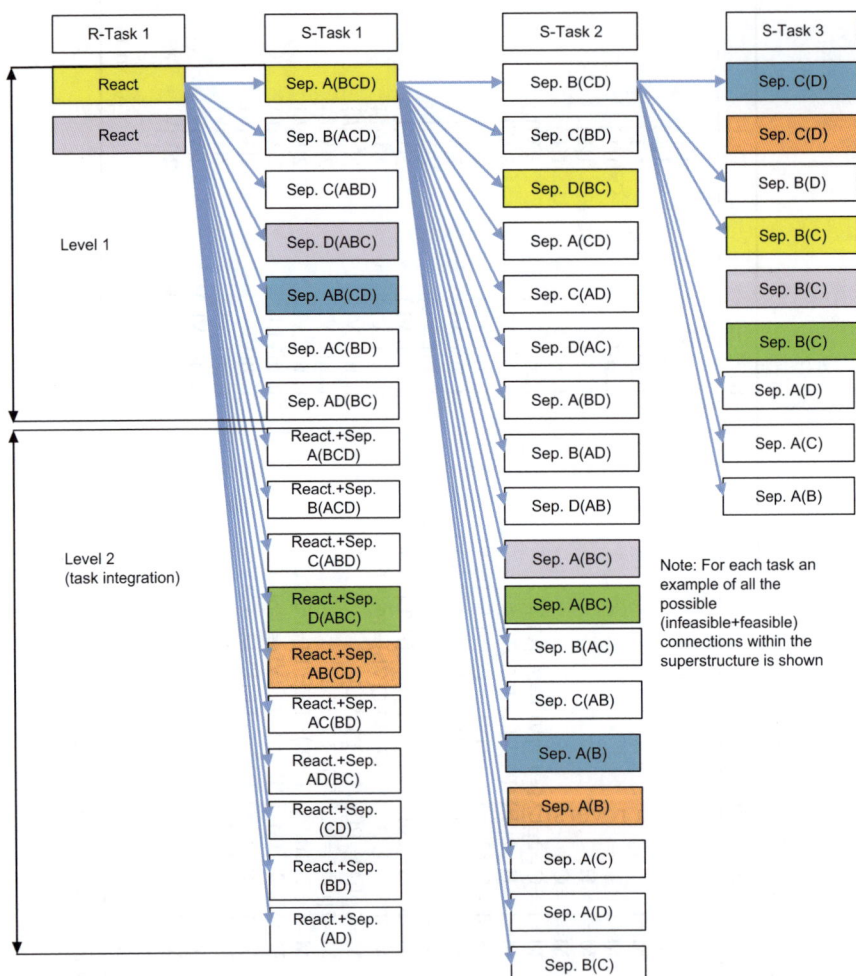

Figure 27.16 Task-based superstructure including task integration. The super-structure is highlighted in blue, flow sheet alternatives 1 and 2 are highlighted in yellow, flow sheet alternatives 3 and 4 are highlighted in green, and flow sheet alternatives 6 to 9 are highlighted in orange. React. = reaction; Sep. = separation.

separation of AB from CD (highlighted blue in Figure 27.16), followed by S-Task 2 that separates A from B, *i.e.* A(B), and S-Task 3 that separates C from D that is C(D). Using this task-based superstructure, the flow sheet alternatives found in Section 27.3.10.2 to Section 27.3.10.7 are generated.

27.5.10.2 Generation of Flow Sheet Alternative 1 and Flow Sheet Alternative 2 from the Task-based Superstructure.
Flow sheet alternative 1 is generated from the task-based superstructure. To produce the products a change of state of the raw materials, MeOH and HOAc, must occur. Therefore to achieve this, a reaction task is identified. There are no azeotropes between HOAc (A) and MeOH (B) and between HOAc and MeOAc (C); therefore a separation task is identified to remove A from BCD. Since two minimum boiling azeotropes exist between MeOH/MeOAc (B/C) and MeOAc/H_2O (C/D), the removal of D is selected, followed by separation of C from B. This alternative 1 is highlighted in yellow in Figure 27.16. This task-based flow sheet (alternative 1) is then further developed in terms of SPBs that can perform the tasks.

Similarly, flow sheet alternative 2 is also generated from the task-based superstructure. The removal of H_2O is selected because H_2O is present in two of the three identified azeotropes (see IT.PBS 1). HOAc is then separated followed by the separation of MeOAc from MeOH. This generated alternative is highlighted in purple in Figure 27.16.

27.5.10.3 Flow Sheet Alternative 1 and Flow Sheet Alternative 2 Refinement: Use of a Membrane.
For flow sheet alternative 1 and flow sheet alternative 2, a reactor feed (as in the base case) of a 2 : 1 mole ratio of MeOH to HOAc is used. The detailed task-based flow sheet for flow sheet alternative 1 is shown in Figure 27.17. It is generated as follows.

The reaction is a reversible reaction and therefore the reactor outlet contains both products and reactants. A recycle task is identified for recycling the unreacted raw materials. From the mixture properties analysis, three minimum boiling azeotropes that are also pressure dependent are present. From the binary ratio matrix, the r_{ij} values for the boiling points of the MeOH–HOAc and HOAc–/MeOAc binary pairs are 1.16 and 1.18, respectively (note that these binary pairs do not form azeotropes). Therefore a VL separation task using an SPB containing a PT(VL) PBB is assigned to separate MeOH and MeOAc from HOAc. However, HOAc and H_2O also forms an azeotrope and

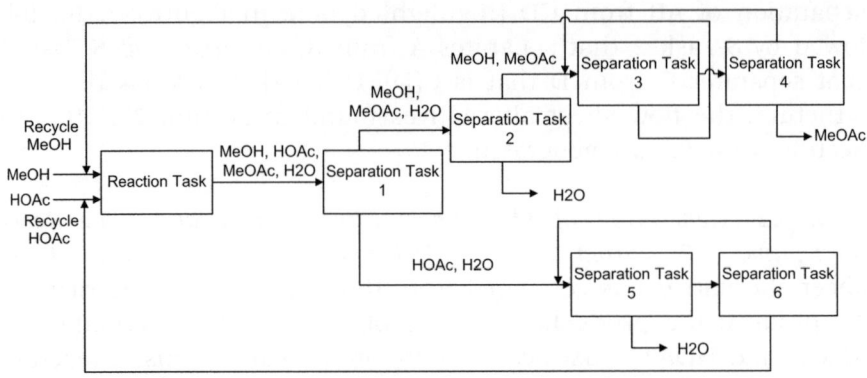

Figure 27.17   Final task based flow sheet for flow sheet alternative 1.

therefore H_2O will be found in the two output streams of this separation task.

The stream from separation task 1, which contains MeOH, MeOAc and H_2O, forms two azeotropes, MeOH–MeOAc and MeOAc–H_2O. From the binary ratio matrix, the r_{ij} values for the Van der Waals volume of the MeOH–MeOAc and MeOAc–H_2O binary pairs are 1.96 and 3.44, respectively. Therefore a separation task using an SPB containing a PT(VV) PBB is assigned to break the MeOAc–H_2O azeotrope.

The stream from separation task 2, which contains MeOH and MeOAc, forms one azeotrope, MeOH–MeOAc. From the mixture properties analysis (see Figure 27.14), this azeotrope is pressure dependent. Therefore two separation tasks using an SPB containing a PT(VL) PBB is assigned to break the MeOH–MeOAc azeotrope. The same concept is applied for the HOAc–H_2O azeotrope. At this point the unreacted raw materials are recovered and therefore the recycles are closed.

The SPBs associated with each separation task are then combined to form operations which are translated into unit-ops. The SPBs associated with each task, for flow sheet alternative 1 are:

- Reaction task
  - M=R=C (SPB.6)
- Separation task 1, 3, 4, 5, 6
  - M=2phM=PC(VL)=PT(VL)=PS(VL) (SPB.13)
- Separation task 2
  - M=C=2phM=PC(VL)=PT(VV)=PS(PVV) (SPB.16)

Using the SPBs the operations are then completed and translated into unit-ops. The unit-ops associated with each operation are:

- Reaction task
  - For the SPB M=R=C, this is complete and is translated into a reactor with an integrated cooling jacket. This reactor could be a flow reactor or a CSTR.
- Separation Task 1, 3, 4, 5, 6
  - For the SPB M=2phM=PC(VL)=PT(VL)=PS(VL), this is completed by adding two more SPBs, M=H=2phM=PC(VL)= PT(VL)=PS(VL) (SPB.17) and M=C=2phM=PC(VL)=PT(VL)= PS(VL) (SPB.16) and this is translated into a distillation column (see Figure 27.5(a)).
- For the SPB M=C=2phM=PC(VL)=PT(VV)=PS(VV), this is complete and is translated into a vapour permeation membrane.

The flow sheet alternative 1 at the unit-ops scale is shown in Figure 27.18(a).

Flow sheet alternative 2 is similar to flow sheet alternative 1 with the exception that H_2O is separated after the reactor. This simplifies the proceeding separation tasks as no azeotropes exist between the pairs HOAc–MeOH and HOAc–MeOAc. Flow sheet alternative 2 is generated through the removal of H_2O at the outlet of the reactor. The corresponding PBB for achieving this is PT(PVL). The final flow sheet is shown in Figure 27.18(b).

27.5.10.4 Generation of Flow Sheet Alternative 3 and Flow Sheet Alternative 4 from the Task-based Superstructure. From flow sheet alternative 2, H_2O was removed after the reactor because it is present in two of the minimum boiling azeotropes. Therefore an integrated task representing this separation is selected. The final task-based flow sheet is highlighted in green in Figure 27.16. The selected task-based flow sheet is then further refined in terms of the means to achieve each task and the associated SPBs for each task.

27.5.10.5 Flow sheet Alternative 3 and Flow Sheet Alternative 4 Refinement: Use of a Membrane Reactor. For flow sheet alternative 3 the reactor feed of a 2 : 1 mole ratio of MeOH to HOAc is used, as in flow sheet alternative 1 and flow sheet alternative 2. From the set of feasible the SPBs the question is considered, can reaction and separation be combined? From the list of feasible SPBs in Table 27.14, this is possible using, for example, the SPB M=R=C=2phM=PC(VL)=PT(PVL)=PS(VL) (SPB 11). This SPB is

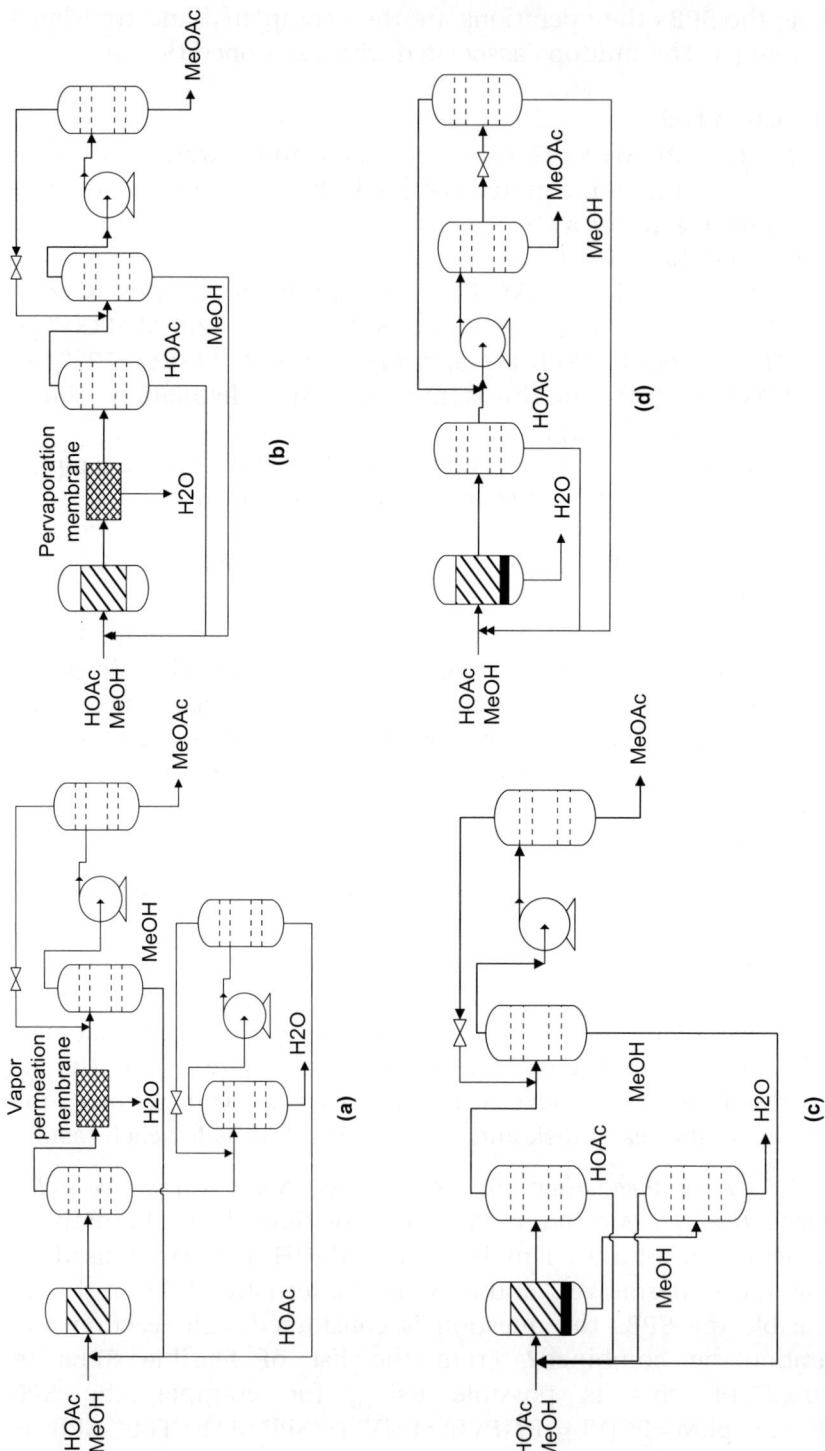

Figure 27.18 Flow sheet for alternative 1 (a), alternative 2 (b), alternative 3 (c) and alternative 4 (d).

complete and is translated into a membrane reactor at the unit-ops scale, *i.e.* the reaction and the first separation involving a perva-poration membrane assisted operation is now combined. The new flow sheet alternative (alternative 3), is shown in Figure 27.18(c).

The feed condition for the flow sheet alternatives generated to this point uses MeOH in excess. This is also the condition for the base case design and was used by Agreda *et al.*[27] to reach close to the equilibrium conversion. With a membrane assisted reactor, however, the equilibrium is further improved through the *in situ* removal of H_2O and the use of a 1 : 1 feed is possible. This gives rise to another flow sheet alternative (alternative 4). Flow sheet alternative 4 is shown in Figure 27.18(d).

27.5.10.6 Detailed Analysis of Flow Sheet Alternatives 1 and 2. For alternatives 1 and 2, the calculated equilibrium conversion values indicate that specified operational constraint (θ_3) is not satisfied, *i.e.* the conversion of HOAc is greater than equilibrium conversion (the equilibrium conversion is 71.4%). Therefore, these alternatives are not feasible with respect to the specified targets for improvement.

27.5.10.7 Detailed Analysis of Flow Sheet Alternatives 3 and 4. To test the feasibility of the membrane reactor, a model-based reactor analysis is performed using a semi-batch model.[28] The membrane chosen is a poly(vinyl alcohol) (PVA) membrane produced by Sulzer Chemtech, PERVAP 2201.[29,30] The semi-batch model was analysed and solved using ICAS-MoT[20] in order to study the feasibility of achieving a HOAc conversion greater than the equilibrium con-version. From the model-based membrane reactor analysis it was found that a conversion of HOAc of 92% can be achieved. The cal-culation details explained as follows.

It is recommended[27] that a semi-batch model is sufficient to study the pervaporation membrane reactor. For analysing the semi-batch model, a non-dimensionalized mass balance of the reactor is used:[29]

$$\frac{d\bar{N}_i}{dt} = v_i Da \left(\frac{a'_{HOAc} a'_{HOAc} - a'_{MeOAc} a'_{H_2O} / K_a}{a'_{HOAc} + a'_{HOAc} + a'_{MeOAc} + a'_{H_2O}} \right) - \frac{Da \delta a_i}{\beta_i} \quad (27.23)$$

The dimensionless parameters in Equation (27.23) are as follows:

1. Damkohler number, Da, is the dimensionless ratio of a char-acteristic liquid residence time to the reaction time. High values indicate that the forward reaction is fast and low values indicate that the reaction is kinetically controlled.

2. The rate ratio,[29] δ is the dimensionless ratio of the permeation rate to the reaction rate.
3. The membrane selectivity, β, is the dimensionless ratio of the membrane selectivity. High values mean that the membrane is not selective to a specific component.

The reaction rate using an adsorption-based model[26] is expressed in Equation (27.24), where K_i is the adsorption equilibrium constant and M_i is the molar mass of the compound.

$$r = m_{cat} \left(\frac{a'_{HOAc} a'_{HOAc} - a'_{MeOAc} a'_{H_2O} / K_a}{\left(a'_{HOAc} + a'_{HOAc} + a'_{MeOAc} + a'_{H_2O} \right)^2} \right)$$

$$a'_i = \frac{K_i a_i}{M_i}$$

(27.24)

The activities of the components are defined using Equation (27.25) where x_i is the mole fraction of the compound in the liquid phase and γ_i is the activity coefficient.

$$a_i = x_i \gamma_i$$

(27.25)

The liquid activity coefficients are calculated using the UNIQUAC thermodynamic model and are expressed as a function of temperature, binary parameters and molar fractions.

The ideal membrane should be selective to H_2O only; however, as reported by Assabumrungrat et al.,[29] the membrane is also selective to MeOH and weakly selective to MeOAc. Therefore a main parameter in Equation (27.23) is the membrane selectivity, β. Therefore, through the model-based reactor analysis, one can study the following process.

1. Set and verify that a set conversion of HOAc can be achieved.
2. Propose a recommended minimum value of β, i.e. a design criterion of the membrane. This can be useful because it provides a measure for which the membrane can be improved.
3. The effect of the dimensionless parameters of the HOAc conversion.

Equation (27.23) is solved assuming the membrane is only permeable to H_2O (β = 1) and MeOH (β = to be calculated). The conversion of HOAc is set at 92%, i.e. the minimum conversion to satisfy the operational constraint for HOAc conversion. The minimum value of β for MeOH is calculated using the following method.

1. Obtain the liquid hourly space velocity (LHSV) for the catalyst. This gives the value final time, t, to solve Equation (27.23) because it provides an estimate of the residence time for the reactor.
2. Investigate the effect of Da and δ on β.
3. Select the value of Da and δ and corresponding value of β for achieving the required conversion of HOAc.
4. Investigate the effect of the Da number and δ for the selected value of β. This shows the effect of each parameter on the conversion of the raw material considered.

Therefore Equation (27.23) is solved using the above method and the results are:

1. LHSV $= 5h^{-1}$ [28]
2. The effect of Da and δ on β is shown in Figure 27.19 and Figure 27.20, respectively.
   a. Increasing the Da number decreases the membrane selectivity (βMeOH) for MeOH. This happens because the rate of reaction increases, and therefore to ensure that MeOH is available for reacting with HOAc, the membrane should be less selective to MeOH. The reverse is also true, *i.e.* for low values of the Da number, βMeOH increases because the rate

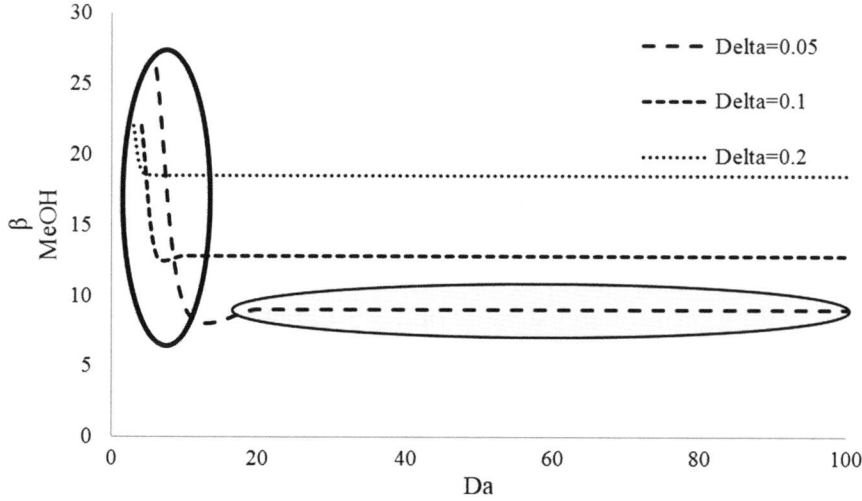

Figure 27.19 Effect of Da number on membrane selectivity (βMeOH) for different values of rate ratio (δ) for HOAc conversion $= 0.92$.

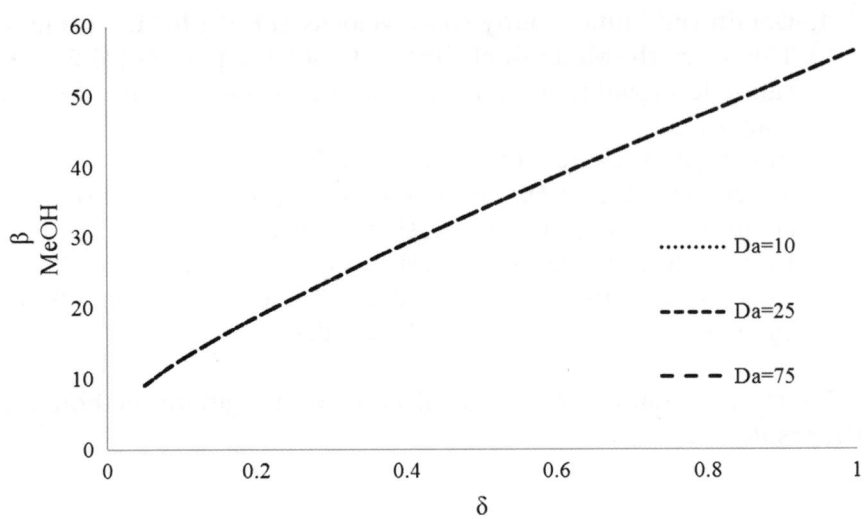

Figure 27.20 Effect of rate ratio (δ) on membrane selectivity (βMeOH) for different values of Da number for HOAc conversion $= 0.92$.

of reaction decreases and therefore the membrane should be least selective to MeOH. For increasing values of the rate ratio (δ), the value of βMeOH increases for the same Da number because the membrane must be least selective to MeOH in order to prevent MeOH loss through the membrane. Therefore one does not want to be in the operating window that is circled but the operating window highlighted grey in Figure 27.19.

b. Increasing the value of the rate ratio (δ) increases the membrane selectivity (βMeOH) for MeOH. This happens because MeOH will be lost through the membrane and not enough MeOH will be present to react with HOAc. Also the Da number has no effect on the rate ratio (δ) because the Da number governs the rate at which the reaction happens and not the behaviour of the membrane (Figure 27.20).

3. The selected values of Da and δ are 20 and 0.05, respectively, with a recommended value of $\beta = 9.15$ (calculated).

4. The effect of the Da number and δ for the selected (calculated) value of β on the HOAc conversion is shown in Figure 27.21 and Figure 27.22, respectively.

a. For a set value of δ, increasing the Da number increases the HOAc conversion. This happens because increasing the Da number increases the reaction residence time. For increasing values of δ and the Da number, it is possible to go above the

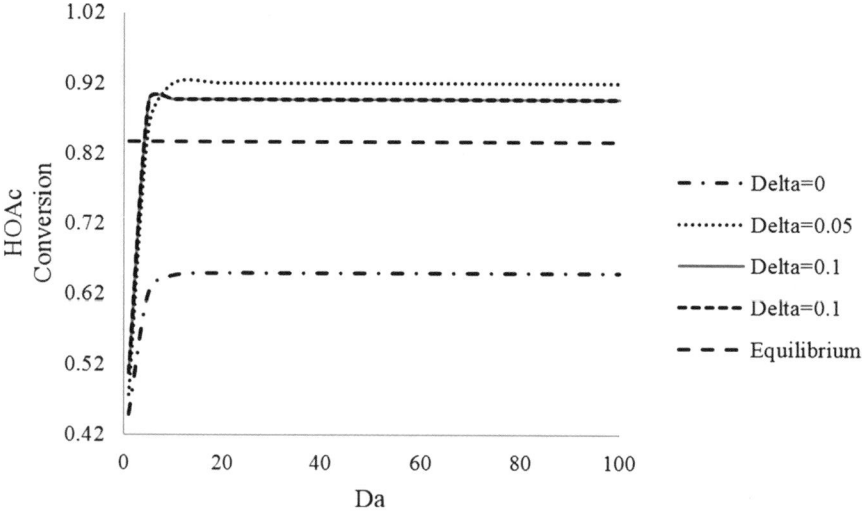

Figure 27.21 Effect of Da number on HOAc conversion for different values of rate ratio (δ).

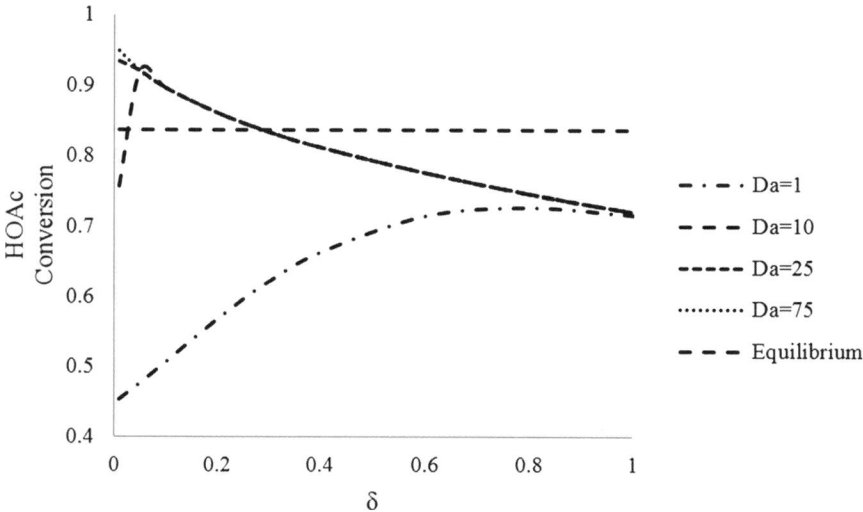

Figure 27.22 Effect of rate ratio (δ) on HOAc conversion for different values of Da number.

equilibrium conversion due to the increase rate of removal of H_2O. However, because the membrane is also permeable to MeOH, MeOH is also being removed from the reactor and therefore the steady state conversion is achieved (Figure 27.21).

b. For a set value of the Da number, increasing δ increases the HOAc conversion. This is because increasing δ increases the rate of removal of H_2O (but also MeOH) thereby shifting the equilibrium. For low values of the Da number, one cannot go beyond the equilibrium conversion because the reaction is kinetically controlled (Figure 27.22).

Therefore if the same feed condition is used as in flow sheet alternative 1 (MeOH : HOAc is 2 : 1), then MeOH also permeates through the membrane of the membrane reactor. Van Baelen *et al.*[30] studied the effect of the permeation of MeOH and H_2O through PERVAP 2201 and this is shown in Figure 27.23. Therefore, the permeation of MeOH must be considered when generating flow sheet alternative 3 (see Figure 27.18(c)). Considering the possibility of MeOH and H_2O permeating through the membrane based on the feed condition, the wt-% of the H_2O in the feed to the membrane is 32 and the wt-% of H_2O in the permeate is calculated to be 75 (see Figure 27.23). For flow sheet alternative 4 (MeOH : HOAc is 1 : 1), the wt-% of the H_2O in the feed to the membrane is 87 and the wt-% of H_2O in the permeate is calculated to be 99.9 (see Figure 27.23). Therefore both flow sheet alternative 3 and flow sheet alternative 4 are feasible; however, for flow sheet alternative 3, an extra column is required to recover the permeated MeOH through the membrane (Figure 27.18(c)).

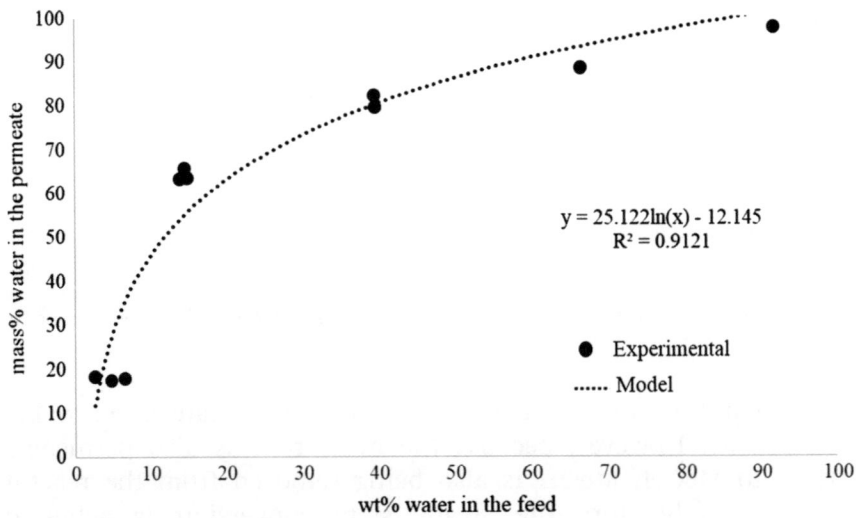

Figure 27.23   Mass-% of H_2O in the permeate for PERVAP 2201.

27.5.10.8 Flow Sheet Alternatives including Reactive Distillation.

Flow sheet alternatives 6 to 9, consisting of a reaction distillation column, are also generated through the application of IT-PBS 3; these are presented and discussed by Babi *et al.*[17] The final task-based flow sheet in Figure 27.17 is highlighted orange. In the same manner reaction and separation can be combined using the SPB M=R=2phM=PC(VL)=PT(VL)=PS(VL) (SPB 13) representing a reactive VLE. The unit-ops associated with this SPB is a reactive distillation column (RD). For the reactive distillation column, different configurations (single or double feed) and column structure can be investigated. That is, four flow sheet alternatives are possible: a single feed reactive distillation column fully packed or partially packed with catalyst, giving flow sheet alternatives 6 and 7. A double feed reactive distillation column fully packed or partially packed with catalyst, gives flow sheet alternatives 8 and 9. This is illustrated in Figure 27.24.

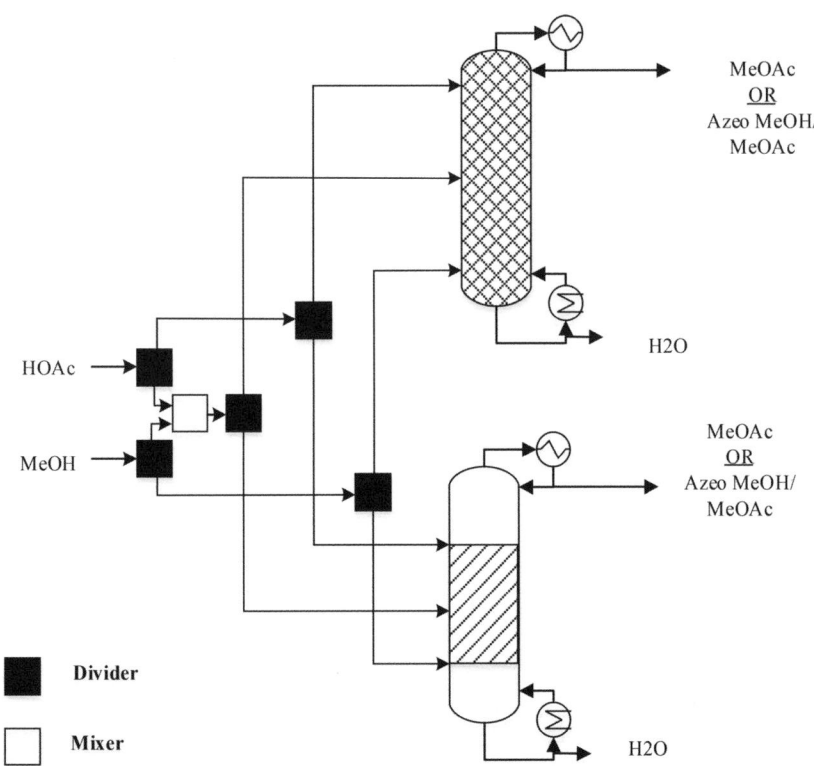

Figure 27.24 Superstructure for generated RD flow sheet alternatives.

For investigating the feasibility of the RD alternatives, the RD flow sheet alternatives are rigorously simulated and the elemental method of Pérez-Cisneros and co-workers[31] is used to investigate the column profiles. It was found that the double feed RD column a column with both reactive and non-reactive sections is preferred to a column with only reactive sections.[17,31]

27.5.10.9 Screening of the Flow Sheet Alternatives. The final nine flow sheet alternatives considered were screened in terms of a selected set of performance criteria defined in step S2. This is shown in Table 27.15. From the nine alternatives generated, four were found to be feasible for further investigation in IT-PBS 4. That is, the feasible flow sheet alternatives selected for further analysis in IT-PBS 4 are flow sheet alternative 3 (see Figure 27.18c)), flow sheet alternative 4 (see Figure 27.18d)), flow sheet alternative 5 (not shown) and flow sheet alternative 9 (Figure 27.24, double feed RD column with both reactive and non-reactive stages).

27.5.10.10 IT-PBS 4: Perform Economic Sustainability Analysis. The results of the economic–LCA–sustainability analysis for the four feasible alternatives are given in Table 27.16. The prices used for different utilities are: cooling water at $0.35 GJ^{-1}, low-pressure steam at $7.78 GJ^{-1} and electricity at $16.80 GJ^{-1}.[23]

The total membrane area calculated for each flow sheet alternative containing a membrane reactor is as follows:

- Flow sheet alternative 3 (MR4COL): 6304 m^2 (feed ratio $2:1$ MeOH:HOAc)
- Flow sheet alternative 4 (MR3COL): 3033 m^2 (feed ratio $1:1$ MeOH:HOAc)
- Flow sheet alternative 5 (MR2COL): 3033 m^2 (feed ratio $1:1$ MeOH:HOAc)

Table 27.15 Screening of flow sheet alternatives using logical and operational constraints.

Alternatives	Constraint	Remaining alternatives
1, 2, 3, 4, 5, 6, 7, 8, 9	Operational: conversion of HOAc > equilibrium conversion (θ_3)	3, 4, 5, 6, 7, 8, 9
3, 4, 5, 6, 7, 8, 9	Logical: the MeOAc purity is set at 99 mol-% (θ_1)	3, 4, 5, 8, 9
3, 4, 5, 8, 9	Logical: H$_2$O purity is set at 99 mol-% ($\underline{\theta}_1$)	3, 4, 5, 9

Table 27.16 F_{obj}, economic and LCA analysis results for feasible flow sheet alternatives: flow sheet alternative 3 (MR4COL), flow sheet alternative 4 (MR3COL), flow sheet alternative 5 (MR2COL) and flow sheet alternative 9 (RD).

	Measures of Comparison (units)	Base Case	MR4COL	MR3COL	MR2COL	RD
Inlet Information	Feed mole Ratio (MeOH:HOAc)	2 to 1	2 to 1	2 to 1	1 to 1	1 to 1
	Mole Feed to reactor/(kmol h^{-1})	560:280	499.13:249.57	499.13:249.57	249.57:249.57	230.15:230.15
	Mole Feed Make-up/(kmol h^{-1})	234.81:2293.66	229.64:229.6	229.62:229.64	229.72:229.61	–
	HOAc conversion	0.82	0.92	0.92	0.92	0.98
	Input-MeOH/(kg h^{-1})	7524	7358	7357	7361	7379
	Input-HOAc/(kg h^{-1})	13792	13788	13790	13789	13829
	Input-Total/(kg h^{-1})	21315	21146	21148	21150	21208
Outlet Information	MeOAc product/(kg h^{-1})	17009	17008	17008	17009	17009
	Operating Days	300	300	300	300	300
	Product purity/(MeOAc)/%	99.02	99.99	99.99	99.97	99.91
	By-product purity/(H$_2$O)/%	99.9	99.50	1.00	1.00	98.67
	Raw Material loss/(MeOH)/(kg h^{-1})	166.78	0.00	0.54	3.75	14.77
	Raw Material loss/(HOAc)/(kg h^{-1})	3.72	0.00	2.19	0.87	40.79
	Raw Material loss/(Total)/(kg h^{-1})	170.49	0.00	2.72	4.62	55.56
Results Summary	Energy usage/(MJ h^{-1})	372198	349862	325215	61295	38210
	Utility Cost/($/a^{-1})	12343384	11162783	9866031	1479538	959945
	Raw Material Cost/($ a^{-1})	107653098	106834774	106842343	106850799	107144908
	Operating Cost/($ a^{-1})	119996482	117997557	116708374	108330337	108104853

Table 27.16 (*Continued*)

Measures of Comparison (units)	Base Case	MR4COL	MR3COL	MR2COL	RD
Primary Performance Metrics					
Raw Material usage/[kg Raw Material (kg MeOAc Product)$^{-1}$]	1.25	1.24	1.24	1.24	1.25
Energy usage per kg of product/[MJ (kg of MeOAc product)$^{-1}$]	21.88	20.57	19.12	3.60	2.25
Raw Material Cost per kg of product/[\$ (kg of MeOAc product)$^{-1}$]	0.88	0.87	0.87	0.87	0.87
Utility Cost per kg of product/[\$(kg of MeOAc product)$^{-1}$]	0.10	0.09	0.08	0.01	0.01
Fobj-Profit/(\$/kg of MeOAc product)-max	2.06	2.08	2.09	2.16	2.16
Total Carbon Footprint [kg CO2(kg of MeOAc product)$^{-1}$]	0.647	0.558	0.517	0.088	0.055
LCA Results					
HTPI(1/LD 50)	9.19E-04	1.36E-03	2.93E-04	2.92E-04	7.52E-03
HTPE/(1TWA)	1.31E-05	1.16E-05	1.16E-05	1.16E-05	1.16E-05
GWP(CO$_2$ eq.)	2.060	1.783	1.741	1.313	1.284
HTC/(kg benzen eq)	5.90E-05	5.09E-05	5.09E-05	5.02E-05	5.03E-05
HTNC/(kg toluen eq.)	1.34E-02	2.23E-02	1.66E-03	1.20E-03	1.41E-01

HTPI = human toxicity potential by ingestion; HTPE = human toxicity potential by exposure; GWP = global warming potential; HTC = human toxicity (carcinogenic impacts); HTNC = human toxicity (non-carcinogenic impacts); TWA = time-weighted average.

The objective function (Equation (27.20)) value for each of the alternatives is given in Table 27.16.[17] Flow sheet alternative 5 and flow sheet alternative 9 show the best values of the objective function. These two alternatives also have the lowest carbon footprint (see LCA results in Table 27.16). From Table 27.16, flow sheet alternative 5 is a feasible alternative compared with the well-known reactive distillation process for MeOAc production.[27]

Figure 27.25 shows a radar plot where the boundary represents the base case while the feasible flow sheet alternatives (more sustainable solutions) lie inside this boundary (within the radar). For each variable plotted in this figure, the ratio (multiplied by 100) of the value of a specific variable for an alternative and the corresponding value for the base case have been used. In this way, it is possible to see that for all the considered criteria, the alternatives are better or the same (*i.e.* lie at the boundary).

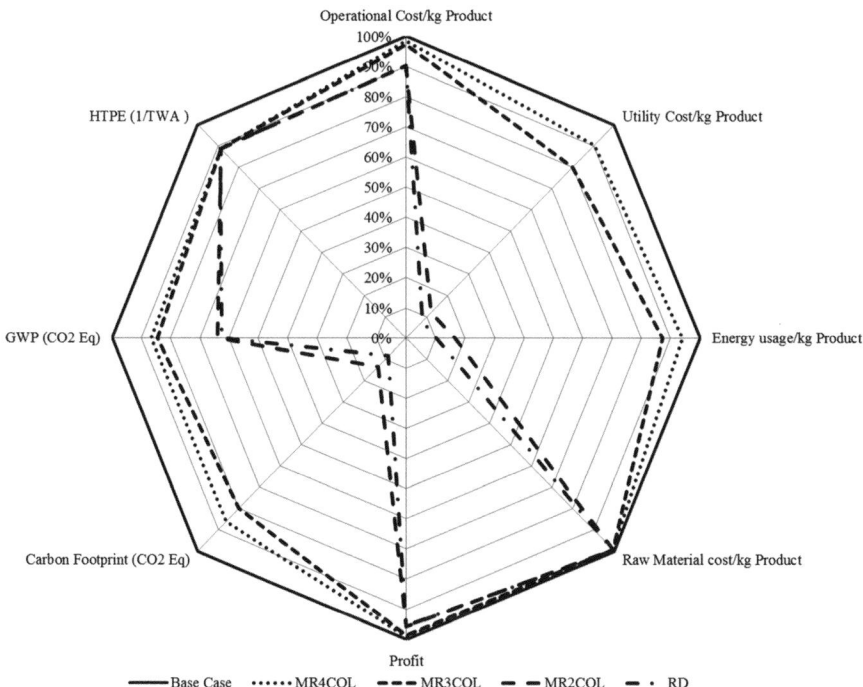

Figure 27.25 Economic and LCA improvements relative to the base case design. HTPI = human toxicity potential by ingestion; HTPE = human toxicity potential by exposure; GWP = global warming potential; HTC = human toxicity (carcinogenic impacts); HTNC = human toxicity (non-carcinogenic impacts); TWA = time-weighted average.

27.6 CONCLUSIONS AND FUTURE PERSPECTIVES

A computer-aided multi-level framework for process synthesis–intensification has been developed. The complex synthesis–intensification problem is solved systematically using a decomposition approach. It has been shown through application of the framework that intensified process alternatives can be generated by performing process synthesis and intensification together and operating at all three scales: unit-ops, tasks and phenomena.

Sustainable design has been evaluated through three analyses: economic, LCA and sustainability metrics. These analyses have proved to be useful because they provide a measure for which designs can be quantitatively evaluated. The application of the framework has been highlighted through a case study and it is shown that flow sheets which include membrane-based designs offer major improvements to existing processes in terms of economic, sustainability metrics as well as LCA factors.

REFERENCES

1. P. Lutze, J. M. Woodley and R. Gani, *Chem. Eng. Process.*, 2010, **49**, 547.
2. J. Siirola and D. F. Rudd, *Ind. Eng. Chem. Fundam.*, 1971, **10**, 353.
3. R. W. Thompson and C. J. King, *AIChE J.*, 1972b, **18**, 941.
4. M. D. Lu and R. L. Motard, *Comput. Chem. Eng.*, 1985, **9**, 431.
5. J. M. Douglas, *AIChE J.*, 1985, **31**, 353.
6. G. Stephanopoulos and A. W. Westerberg, *Chem. Eng. Sci.*, 1976, **31**, 195.
7. I. E. Grossmann and J. Santibanez, *Comput. Chem. Eng.*, 1980, **4**, 205.
8. K. P. Papalexandri and E. Pistikopoulos, *Comput. Chem. Eng.*, 1995, **19**(Suppl.), S71.
9. A. Quaglia, B. Sarup, G. Sin and R. Gani, *Comput. Chem. Eng.*, 2013, **59**, 47.
10. J. Siirola, *Comput. Chem. Eng.*, 1996, **20**(Suppl.), S1637.
11. C. Jaksland, R. Gani and K. M. Lien, *Chem. Eng. Sci.*, 1995, **50**, 511.
12. L. d'Anterroches and G. Rafiqul, *Comput.-Aided Chem. Eng.*, 2004, **18**, 379.
13. P. Lutze, D. K. Babi, J. M. Woodley and R. Gani, *Ind. Chem. Eng. Res.*, 2013, **52**, 7127.

14. D. K. Babi, J. M. Woodley and R. Gani, presented at 8th International Conference on Foundations of Computer-Aided Process Design, 2014.
15. I. K. Lai, S. B. Hung, W. J. Hung, C. C. Yu, M. J. Lee and H. P. Huang, *Chem. Eng. Sci.*, 2007, **62**, 878.
16. S. S. Mansouri, M. I. Ismail, D. K. Babi, L. Simasatitkul, J. K. Huusom and R. Gani, *Processes*, 2013, **1**, 167.
17. D. K. Babi, P. Lutze, J. M. Woodley and R. Gani, *Chem. Eng. Process.*, 2014, DOI: 10.1016/j.cep.2014.07.001.
18. P. Harper and R. Gani, *Comput. Chem. Eng.*, 2000, **24**, 677.
19. A. T. Karunanithi, L. E. K. Achenie and R. Gani, *Ind. Chem. Eng. Res.*, 2005, **44**, 4785.
20. R. Gani, G. Hytoft, C. Jaksland and A. K. Jensen, *Comput. Chem. Eng.*, 1997, **21**, 1135.
21. J. Marrero and R. Gani, *Fluid Phase Equilibr.*, 2001, **183-184**, 183.
22. A. Carvalho, H. M. Matos and R. Gani, *Comput. Chem. Eng.*, 2013, **50**, 8.
23. P. S. Peters, K. D. Timmerhaus and R. E. West, *Plant Design and Economics*, McGraw Hill, New York, 5th edn, 2003.
24. S. Kalakul, P. Malakul, K. Siemanond and R. Gani, *J. Cleaner Prod.*, 2014, **71**, 98.
25. R. S. Huss, F. Chen, M. F. Malone and M. F. Doherty, *Comput. Chem. Eng.*, 2003, **27**, 1855.
26. T. Pöpken, L. Götze and J. Gmehling, *Ind. Chem. Eng. Res.*, 2000, **39**, 2601.
27. V. H. Agreda, L. R. Partin and W. H. Heise, *Chem. Eng. Prog.*, 1990, **86**, 40.
28. T. Inoue, T. Nagase, Y. Hasegawa, Y. Kiyozumi, K. Sato, M. Nishioka, S. Hamakawa and F. Mizukami, *Ind. Chem. Eng. Res.*, 2007, **46**, 3743.
29. S. Assabumrungrat, J. Phongpatthanapanich, P. Praserthdam, T. Tagawa and S. Goto, *Chem. Eng. J.*, 2003, **95**, 57.
30. D. Van Baelen, B. Van der Bruggen, K. Van den Dungen, J. Degreve and C. Vandecasteele, *Chem. Eng. Sci.*, 2005, **60**, 1583.
31. O. S. Daza, E. Bek-Pedersen, E. S. Pérez-Cisneros and R. Gani, *AIChE J.*, 2003, **49**, 2822.
32. Rohm and Haas, Amberlite 15 catalyst, Industrial Grade Strongly Acidic Catalyst, Datasheet.

Subject Index

References to figures are given in *italic* type. References to tables are given in **bold** type.